# Bacterial Viruses

## Exploitation for Biocontrol and Therapeutics

**Edited by**

**Aidan Coffey**

Department of Biological Sciences & APC Microbiome Ireland
Cork Institute of Technology
Cork
Ireland

and

**Colin Buttimer**

APC Microbiome Ireland
University College Cork
Cork
Ireland

**Caister Academic Press** **www.caister.com**

ISBN: 978-1-913652-51-7 (paperback)
ISBN: 978-1-913652-52-4 (ebook)

DOI: https://doi.org/10.21775/9781913652517

# Dedication

This publication is dedicated to:

Elke Arendt
Charles Daly
Gerald Fitzgerald
Paul Ross
Colm Murphy

# Contents

## Theme 3: Bacteriophage Derived Agents



## Theme 4: Technology and Intellectual Property Considerations

# Preface

The potential of bacterial viruses (bacteriophages or phages) as antibacterial agents was recognised over a century ago. However, the success of antibiotic therapy from the 1940s onwards caused a general decline in the applications of phages as therapeutics. Nonetheless, recent decades have seen a disquieting proliferation of pathogenic bacteria resistant to multiple antibiotics commonly used in human and veterinary medicine, in turn leading towards a crisis in public health. The renewed exploration of phages and their molecular components for biocontrol and therapeutics offers a complementary approach for eliminating problematic bacteria across a range of sectors where bacterial infections and contaminations are a persistent problem.

Written by internationally-recognised scientists involved in the exploitation of bacterial viruses in diverse areas from around the world, this book provides comprehensive coverage of the current research in several phage applications for biocontrol of undesirable bacteria in human and veterinary medicine, horticulture, aquaculture and food. Chapters from the famous centres of phage therapy in Wrocław and Tbilisi detail some of the pioneering historical contributions to the topic. The book also examines the impact of phages on the human microbiome as well as the progress of research in phage engineering, phage enzymes, phage delivery systems, biodetection as well as intellectual property aspects.

Aidan Coffey and Colin Buttimer

# Editor Biographies

**Aidan Coffey** is a Principal Investigator and Professor at the Centre for Advanced Therapeutic Engineering at Cork Institute of Technology (Cork, Ireland) and an Adjunct Professor with the Microbiology Department at University College Cork (UCC). He is the author of over 200 research papers, and he has also served in scientific working groups at the European Food Safety Authority. His journey on bacterial virus research began with his PhD on bacteriophage resistance in lactic acid bacteria under the direction of Charles Daly and Gerald Fitzgerald at UCC, graduating in 1991. Following the completion of a Marie Curie postdoctoral fellowship at Wageningen University in the Netherlands where he worked with pathogenic bacteria at the laboratory of Tjakko Abee, he took up a Senior Research Officer post at the Irish Agriculture and Food Development Authority (Teagasc) with the group of Paul Ross. In 1998, he began isolating and characterising bacteriophages active against animal and human infectious bacteria, continuing the topic after transferring from Teagasc to his current post in 2002.

**Colin Buttimer** completed his PhD in 2018 in the research group of Aidan Coffey at Cork Institute of Technology. There, he worked on bacteriophages of Gram-negative phytopathogens of the genera *Pectobacterium* and *Erwinia*, collaborating and publishing with bacteriophage scientists at KU Leuven and ETH Zurich. His early career research made a significant contribution to the field by leading to the establishment of three novel phage genera. He currently works on phages of the human gut microbiome and their potential application for human phage therapy at the APC Microbiome Ireland, University College Cork in the research group of Colin Hill.

# Current books of interest

# Chapter 1

# An Overview of Current Phage Therapy: Challenges for Implementation

**Naomi Hoyle[1,3*], Randolph Fish[2,3], Nata Nakaidze[3], Satya Ambrose[4,5], Elizabeth Martin Kutter[3,6]**

[1]Eliava Phage Therapy Center, Eliava Foundation, 3 Gotua St., Tbilisi, Georgia 0160.
[2]Medical Staff, Grays Harbor Medical Center Wound Clinic, Aberdeen, WA, USA
[3]Phagebiotics Research Foundation 4610 Green Cove Ct. Olympia, WA, USA
[4]Sunnyside Collaborative Care, Happy Valley, OR, USA
[5]Oregon College of Oriental Medicine, National University of Natural Medicine, Chitari Foundation, Portland, OR, USA
[6]The Evergreen State College, 2700 Evergreen Parkway, Olympia, WA, USA

*Corresponding author: naomisulinger82@gmail.com

DOI: https://doi.org/10.21775/9781913652517.01

**Abstract**

Bacteriophages (phages) provide an increasingly recognised important option for the treatment of antibiotic resistant infections. This chapter will discuss key points in the history of phage therapy but will primarily focus on current advancements in phage therapeutics. We discuss compassionate use of phage therapy in the United States, Europe and Australia for severe infections caused by antibiotic resistant *Acinetobacter baumannii*, Carbapenemase-producing *Klebsiella pneumoniae* and multidrug resistant (MDR) *Pseudomonas aeruginosa,* among others. Challenges for implementation of phage therapy include the need for further development of therapeutic phage products in an appropriate manner for the current technology as well as clinical studies addressing gaps in the knowledge, especially in terms of dose, distribution, immune response, and legal framework, while drawing on the extensive history of phage used in other countries. Phage therapy has the potential to address a serious unmet need which has a significant economic impact, particularly the significant morbidity and mortality associated with antibiotic resistant infections.

## Introduction

The publication of this text comes at a time of exponential expansion in phage therapy. There is now a growing shift from phage therapy being seen as a quaint alternative medicine fairly widely practised in Eastern Europe and the former Soviet Bloc, to its consideration as a serious alternative for the treatment of antibiotic resistant infections at European, American, and Australian medical institutions. This shift has significantly progressed in the past five years. Phages, are viruses that only infect bacteria. Phages first appeared in descriptions by an English bacteriologist Ernest Hanbury Hankin in 1886. He noted that the Ganges and Jumna rivers in India had antibacterial properties against cholera and other gastrointestinal infections. Frederick Twort and Félix d'Hérelle independently discovered phage viruses in 1915 and 1917, respectively and characterised them into the early 1920s. Medical use of phages soon became quite popular in Europe and has remained a relevant part of modern medicine in post-Soviet Union countries such as Georgia, Poland, and Russia. However, the 1942 discovery of penicillin completely changed the pharmaceutical scene in most of the rest of the world. The period between the 1940s and 1970s can be viewed as the golden age of chemical antimicrobials; during this time, more than 40 antibiotics were discovered and introduced as *status quo* treatments. Today, with a drastic increase in multi-drug-resistant bacteria, the future of antibiotics seems less than promising. According to the Centers for Disease Control and Prevention (CDC), more than 2.8 million antibiotic resistant infections occur in the US annually. The CDC lists the following pathological agents as the most serious threats: *Carbapenem-resistant Acinetobacter*, *Candida auris*, *Clostridium difficile*, *Carbapenem-resistant Enterobacteriaceae* and *Drug-resistant Neisseria gonorrhoeae*. The UN Interagency Group on Antimicrobial Resistance warns us that by 2050 drug-resistant bacterial deaths could reach 10 million annually. The problem of "superbugs" is a continuing major health crisis which demands urgent attention. To find a solution, many scientists and physicians have begun to turn their attention to phages or phage therapy (PT) as a treatment for these infections. Phages are thus slowly shedding their reputation as an alternative medicine and establishing a seriously-considered position in the treatment of multi-drug-resistant bacteria. Gorski et al. (2018) explain that this transformation is in part due to increased communication among researchers, the exchange of articles, and phage and other microbiological conferences, as well as to the patients' needs and ability to search for alternative treatments.

The long-established phage therapy research centres in Georgia and Poland continue the application of phages for patients with antibiotic resistant infections and chronic infections (Kutter et al., 2016; Międzybrodzki, 2018). The Eliava Institute of Bacteriophages, Microbiology and Virology (IBMV) in Tbilisi isolated and characterised a wide range of individual phages and developed various phage cocktails that were used throughout the Soviet Union, adding major phage production facilities in Ufa, Alma Ata and Nizhniy Novgorod by 1980. It also added its own Eliava Phage Therapy Centre in 2010. This small outpatient clinic treats Georgian as well as foreign patients, carrying out research in the process, and complements the free-standing Phage Therapy Center established in 2005, which only caters to foreign patients. Most of the foreign patients who come to these centres have failed multiple rounds of antibiotic therapy. The most frequent conditions treated in the Eliava Institute's own clinic are urinary tract infections, infected surgical wounds, and both cystic fibrosis (CF) and non-CF lung infections. It treats far more Georgians than foreign patients, including many children. More information on the Eliava phage cocktails and phage therapy in Tbilisi is covered in other chapters in this book. In recent years, Russia has again greatly expanded its involvement in phage therapy, with most of the production made by their pharmacological giant Microgen, back in Ufa, Gorki and Nizhniy Novgorod and the development of several new very complex cocktails in addition to their own versions of Intestiphage and Pyophage, both of which are sold widely in pharmacies throughout the country. In 2016 and 2018, they held their own large phage therapy meetings, drawing a number of foreign scientists as well as researchers, physicians and phage producers from all over Russia.

The extensive use of phage therapy in these countries for 90 years attests to the perceived safety of phage (Chanishvili, 2012). However, despite the extensive historical documentation showing efficacy in both controlled and descriptive studies, there is still an urgent need for double-blind placebo-controlled trials for phages to become much more widely used.

Polish research on phage therapy dates to 1952, when Ludwik Hirszfeld established the Institute of Immunology and Experimental Therapy at the Polish Academy of Sciences in Wroclaw (Międzybrodzki et al., 2018; and see the relevant chapter in this book by Żaczek et al.,). The Hirszfeld Institute houses one of the largest phage libraries in the world. Between 1976 and 2018, the collection grew from 187 to 832 well-characterised phages. Polish researchers published about 36 articles between 2012 and 2017, exploring many aspects of phage therapy. Their teams also extensively explore bacteria-phage interactions, building on their large collection of plasmid and prophage-free hosts.

The Hirszfeld Institute research also concentrates strongly and uniquely on understanding the immunological responses to phage treatment along with many other aspects of immunology, measuring a wide variety of parameters. They have found that using highly purified phages corresponds with the highest antibody levels; however, 40% of patients who had high antibody levels still had an excellent clinical outcome (2018). They have expanded research to include areas such as phage in allergy, autoimmune liver disease, graft vs host disease and immune modulation (Gorski 2017, Gorski 2018a, Gorski 2018b, Gorski 2018c). The consideration of phages as a viable alternative for antibiotic resistant and refractory infections has only recently become more accepted by allopathic physicians, as can be seen by published descriptive studies several of which we discuss later in more detail.

## Early phage research in the United States

Before the advent of antibiotics, phage therapy was briefly used in the United States. Much of the disregard for phage in the US medical community came from this era where phage was not yet well understood in terms of mechanism of action and application.

In the US, an interest in phage therapy developed in the 1920s and the 1930s, and d'Hérelle actually taught at Yale for the 1928-1933 academic years before going back to focus on his phage therapy enterprises in Paris and Tbilisi. However, the US interest largely dwindled following the publication of a series of unfavourable reviews of most phage therapy by Eaton and Bayne-Jones, and by Krueger and Scribner in 1934 and 1941, respectively. One wonders what the results would have been having they included the French literature, where d'Hérelle maintained a strong influence in phage production and implementation and the reviews were generally positive, rather than limiting themselves to English-language papers. Quite extensive, very successful clinical work was continued only for *S. aureus* infections and for typhoid fever well into the 1940s (Kutter et al., 2010 and Abedon et al., 2011 and references therein).

Discovery of phages prompted vigorous research at the Pasteur Institute in the early 1920s. The Tbilisi IBMV, founded by George Eliava in 1923, also quickly became a leading organisation on phage research. Phage therapy became unpopular not because it was ineffective but because the discovery of penicillin quickly overshadowed phage therapy. It is very important to analyse the factors raised against phage therapy and address them, instead of erroneously considering phage therapy as inferior to other antibacterial treatments. Lack of molecular biology tools made it hard to understand the nature of phages as well as the characterisation of target bacteria.

Phage specificity demands *in vitro* analysis, before the therapeutic administration in patients. For therapeutic purposes, phages of high virulence should be used. During the initial stages of phage development, not much was known about the importance of the route of delivery, or additional factors such as gut pH. Advances in molecular biology, such as the development in the

1940s of ultra-high-speed centrifugation and the use of electron microscopes, helped with the purification and visualisation of these viruses. One-step growth curve experiments by Ellis and Delbrück and advancements made by Doermann, helped understand the interaction between lytic phages and bacteria, again showing that advancements in instrumental techniques are pivotal in fully comprehending the power of phage therapy. The creation of "Phage Group" in 1943, by Max Delbrück, Alfred Hershey, and Salvador Luria, greatly influenced the propagation of phage research in the United States.

Increasingly, we are becoming aware of very extensive research and clinical phage work that was carried out in France for much of the last century and in the US in the middle part of that century. However, since these topics have been quite extensively reviewed (cf. Kutter et al., 2010; Abedon et al., 2011), they will be mentioned here only briefly. In France, phage therapy was used on a larger scale until the early 1990s. Phages have been used in the treatment of a wide range of infections including orthopaedic infections, alimentary tract infections, skin infections, surgical infections, urinary tract infections, respiratory tract infections, otolaryngological infections, and septicaemia. The results of phage therapy conducted in France were published in a number of papers, documenting substantial success in most cases (Abedon et al., 2011, and references therein). Since the mid-1990s, when phage preparations stopped being produced in France, phage therapy has been practised by a few physicians who obtain phage preparations from Georgia or Russia.

## Current health crisis, the emergence of MDR

Antibiotic resistant infections are already a major crisis in medicine. The phenomenon of methicillin resistant *Staphylococcus aureus* (MRSA) gained such a degree of public awareness that its existence has become common knowledge. The emergence of multidrug resistant strains has continued and increasingly antibiotics are no longer effective treatments. Pan-resistant strains, ESKAPE pathogens (LIST), VRE, CRE among other resistant strains may not be as well known by the general public but are equally troublesome causing an estimated healthcare cost of 20 billion dollars per year (Dadgostar 2019).

Patients with such infections are forced to stay in hospitals, for a total of 8 million additional hospital days, with special isolation precautions. Even though many approaches are taken to limit the problem of bacterial resistance, such as the molecular study of resistant-bacterial strains, finding new agents mimicking properties of antibiotics, remodelling current antibiotics for efficacy, the problem is alarming and needs a viable fast solution (Taati, 2020). The time to take action in the fight against these pathogens has arrived. Serious measures should be taken to avoid a potential pandemic of antibiotic resistant infections. This is one reason why there is a renewed interest in phage therapy, reflected by the sharp increase in publications covering the topic coupled with the establishment of multiple commercially interested parties, many of whom are collaborating with academics within the field. Additionally, phage therapy has been the focus of numerous documentaries and television programs in spurts since the 1990s, including BBC, French TV5, German ZDF, ARTE TV, Dutch Zorg, VICE among others.

The first trial of phage therapy in humans was carried out in 1919 by Felix d'Hérelle, the co-discoverer of phages (Abedon et al., 2011*)*. However, d'Hérelle delayed the publication of his results pending further studies of the nature of phage, and the first report on the therapeutic use of phages in humans was published by Bruynoghe and Maisin (1921). Since then, a great number of studies have been conducted to evaluate the efficacy and safety of phage therapy and much clinical experience has been accumulated (for detailed reviews, see Alisky et al., 1998; Summers, 2001, Sulakvelidze et al., 2001; Sulakvelidze and Kutter, 2005; Gorski et al., 2009; Kutter et al., 2010, Abedon et al., 2011; Harper et al., 2011; Burrowes et al., 2011; Chanishvili, 2012). However, these earlier studies did not appear to meet the current rigorous standards for clinical trials, and many key aspects of data are unavailable. Only recently have the first double-blind, extensively controlled trials of phage therapy been instituted.

The first known phase I clinical trial meeting current rigorous standards was carried out in Britain by Exponential Biotherapies, Inc. under Richard Carlton, MD in 2000, and involved a single, sequenced phage targeting vancomycin-resistant *Enterococcus* (VRE) (R. Carlton, personal communication, 21 December 2012). The trial, approved and supervised by the institutional review board, is particularly noteworthy for having tested an intravenous application for safety and for pharmacokinetics in the bloodstream. On days 1, 3 and 9 of the trial, 12 healthy volunteers were each injected through i.v. with 5 x $10^6$ phage particles, an amount designed to give the same phage concentration as that of VRE found in VRE septicemia - about $10^3$ per mL. The only adverse event observed was a transient rash on one volunteer's arm, with no sequelae. After the first injection, the expected level of phage was observed. However, after the day-3 injection, the level was an order of magnitude lower, and only a few score phage per mL were detected after the day-9 injection. It thus appeared that the volunteers' immune systems were specifically activated by the initial doses to produce antibodies that inactivated later doses, even though the phage used had been selected by their proprietary method to be long-circulating. This made it unlikely that it could successfully treat VRE septicaemia, which often results from ongoing seeding into the bloodstream from VRE-colonized tissues, such as the skin or the gastrointestinal tract. Multiple doses of an anti infective agent would then be needed. Under the circumstances, they felt they could not raise sufficient funds for trials of other approaches of administration, and they did not proceed to the planned double-blind phase Ib/IIa trial, which would have involved larger numbers and actual patients, allowing them to also collect initial efficacy data. Instead, the company switched to other therapeutic targets, and never published this phase Ia trial.

By July 2013 the results of five randomized controlled trials (RCTs) of phage preparations were published in the peer reviewed literature (Bruttin and Briissow, 2005; Wright et al., 2009; Rhoads et al., 2009; Sarker et al., 2012; McCallin et *al.,* 20l3). A few other trials have since been conducted: the very extensive Bangladeshi phase II/III trial of the Nestle infant-diarrhoea product (www.clinicaltrials.gov/ct2/show/NCT00937274) and a series of compassionate-use studies under the Helsinki protocol at the Military Hospital in Brussels (Merabishvili et al., 2012). A pilot clinical study was also conducted to evaluate the safety of a phage preparation in burn wound patients, with the methodology of phage characterisation, preparation and administration discussed in great detail (Merabishvili et al., 2009). We here explore those trials in some depth, consider some of the challenges that have delayed the implementation of additional formal double-blind trials, and then explore some of the complementing broad data from the widespread clinical application of phage therapy over the last 90 years, largely in Poland and Georgia, particularly as related to issues of safety.

The first of these formal Western RCTs was conducted at the Nestle Research Centre in Lausanne, Switzerland (Bruttin and Brussow, 2005). It was a double-blinded randomised placebo controlled phase I trial whose main objective was to evaluate the safety of oral administration of phage. To that end, 15 healthy adult volunteers received 150 mL of mineral water containing a low dose of T4 phage ($10^3$ PFU/mL), a higher dose of the phage ($10^5$ PFU/mL), or placebo three times per day for two consecutive days. There was no formal separate control group in this trial; each subject served as an internal control and received all treatments in a random order (repeated measures design). This study showed no significant side-effects following the administration of the phage preparation. The levels of markers for liver toxicity were also measured; no increases in serum alanine aminotransferase or aspartate aminotransferase were observed. Although mild adverse events occurred for 5 volunteers over the course of the trial, these did not seem to be related to phage administration and no-one required any particular treatment. The authors also showed that oral administration of T4 did not disturb the balance of the *E. coli* gut population. Not even the higher dose reduced the total *E. coli* fecal counts. However, pre-trial 12 of 168 random colonies (7%) from the 15 volunteers were found to be susceptible to T4, while afterwards, only 1 of 116 random colonies was susceptible (0.9%). This apparent ability of phage to greatly reduce strains targeted by the specific phage used without shifting the overall species balance is encouraging, emphasising the potential usefulness of phage in maintaining non-pathogenic natural intestinal microflora while targeting pathogens.

Three main facts should be considered when interpreting and evaluating the results of the study. First, although the upper total phage dose of 1.5 x $10^7$ PFU employed by the investigators falls within the range of the standard doses of about $10^6$ to $10^8$ PFU reportedly administered to patients at both major centres of phage therapy, the HET and the Eliava Institute (Gorski et al., 2009; Chanishvili, 2012a), it was given for a shorter time. Second, there was no indication here that T4 phage particles were absorbed from the intestine to the blood, since neither T4 phage nor anti-phage antibodies were detected in the peripheral blood of the subjects. However, it should be noted that no anti-T4 antibodies could be expected within the time frame of this study. Third, T4, due to its narrow range on the common serious pathogenic strains of *E. coli,* is not likely to be useful in phage therapy. Nevertheless, it was used in this trial because it has been characterised at the genetic, molecular, and physiological levels in such great detail (see Miller et al., 2003) and because a large fraction of the therapeutic phages targeting serious *E. coli* pathogens are rather close relatives of T4 (in particular, all of the phages used in their later phase II trial in Bangladesh are T4-related, see below). In this context, it is also noteworthy that T4-like phages are known to have highly varied and overlapping host ranges, as well as to be able to recognise different receptors on different hosts. For example, on *E. coli* B, T4 itself uses B's unique lipopolysaccharide as its receptor, while on K12 strains it primarily targets outer-membrane protein OmpC. T4 itself hits 28/72 of the members of the *E. coli* collection of reference (ECOR), which is a higher fraction than that achieved by two-thirds of a tested set of 49 T4-like phages from around the world, with the highest hitting 47 strains, the lowest only four strains, and only seven targeting over 36 strains (Kutter, 2009). Every ECOR strain was hit by at least one of this set of phages.

The Nestle study compared the efficacy of T4-like coliphages, Russian Microgen phage cocktail, and placebo in Bangladeshi infants hospitalised with acute bacterial diarrhoea. The team had multiple goals, including understanding the safety and efficacy of phage treatment, as well as to determine total *E. coli* titres and their phage susceptibility in the stool. A prospective, single-centre, randomised, placebo-controlled, parallel-group clinical study recruited 400 infants, all between 6 to 24 months. Interim analysis was done only in the first 120 patients. Sarker et al. (2016) concluded that T4-like coliphages, as well as the Russian Microgen Phage cocktail, did not achieve treatment effects over placebo, they did not alter quantitative diarrhoea criteria. The team made an interesting point that when the *E .coli* pathogen is lysed by phage, the released endotoxin and enterotoxins can potentially cross into the circulation. This is true for infection where the barrier function of the gut mucosa in the small intestine is compromised. Since only 53% of planned patients were enrolled, it is essential to plan a study with higher enrollment to avoid-false negative outcomes. The study also underlines the importance of having sufficient titres of the targeted pathogen to study *in vivo* phage replication, since only a small percentage of children had a high female abundance *of E. coli* by molecular tests and stool titrations (Sarker, 2016). Sarker and Brussow explain that in comparison to low *E. coli* titres, *Streptococcal* bacterial titres were very high. Furthermore, symptoms of diarrhoea were correlated with increased titres of *Streptococcus* with low *Bifidobacterium*, while improvement was seen with an increase in *Bifidobacterium*.

**Compassionate use in the United States and beyond.**

Because of significant unmet therapeutic need, more physicians are willing to approach phage therapy for their patients. Compassionate treatment is defined as the use of unapproved medicines outside of clinical trials in circumstances where all other options have failed. This work is covered under the Helsinki declaration for human rights. We will outline a few of the cases where phage therapy was successfully used as a compassionate treatment. Most of these cases use phage as an augmentative therapy to antibiotics. Increasing numbers of patients travel to one of the longstanding phage therapy centres either in Georgia or Poland for treatment. The majority of these patients have antibiotic failure or other indications such as antibiotic allergies, and phage treatment is used as a monotherapy. Treating patients with phages beyond the current available approved framework is a reality some American physicians are accepting for patients without other viable alternatives.

**Current American case studies**

Many significant clinical studies have been documented by American researchers over the last few years. A recent very well-publicised case that has greatly stimulated awareness of and interest in phage therapy involves a professor from the University of California at San Diego, who became critically ill with an antibiotic resistant *Acinetobacter* infection during a trip to Egypt. The illness began abruptly with acute pancreatitis but quickly progressed to pseudocysts of the pancreas and disseminated infection. After developing severe nausea and vomiting as well as back pain, he was transferred to a local hospital. The patient's condition deteriorated quickly. Soon it became necessary to transfer him to Frankfurt, where he was diagnosed with a pancreatic pseudocyst. Bacterial culture revealed infection with *Acinetobacter baumannii*, which was sensitive to a limited number of antibiotics. By the time he was transferred to San Diego, this number had decreased even further. Running out of options, his wife Dr Steffanie Strathdee, an epidemiologist at the same university, advocated for her husband's health and began an intense search for any possible alternative antibacterial treatment. Through the PubMed search, she found an article about successful phage therapy of an *Acinetobacter* infection and began to search for a way to implement phage therapy on a compassionate use basis, sending out pleas on a worldwide basis for *Acinetobacter* phages that might work, none of which were among those being used in the Georgian or Polish phage centres. Phages targeting his strains were identified in collections from Australia, Texas, and the US navy - the latter reflecting the high levels of *Acinetobacter* acquired by servicemen fighting in Iraq. A very highly purified phage cocktail was put together by the UC San Diego investigational pharmacy unit with support from phage scientists from other American institutions and was delivered intravenously. Before phage treatment, her husband was in cardiopulmonary and renal failure. Three weeks post-treatment, he could already sit up in his bed. HA phage preparation was developed using a cocktail of well characterised phages. The treatment was successful and the patient has since fully recovered (Schooley 2018). This ordeal spurred the development of a centre based at the UC San Diego clinic which will focus on clinical trials for phage-based therapeutics. Dr Strathdee is co-director of the centre and also wrote a book about her husband's ordeal, The Perfect Predator:-A Scientist's Race to Save Her Husband from a Deadly Superbug (2019). This has become a best seller and has been a major force in creating further awareness about phage therapy.

### Aortic graft infection

Chan et al. (2018) describe a successful therapeutic use of phage OMKO1 in a 76-year old male who underwent aortic arch replacement surgery using a Dacron graft. Post-surgery the patient developed *P. aeruginosa* infection in the mediastinal and aortic grafts. Despite multiple debridement and washout procedures, blood and deep wound cultures did not become negative, and the infection persisted. During an extended period from July 2012 to late 2015, the patient entered the clinic multiple times for recurrent *P. aeruginosa* infections. Finally, the patient decided to explore an alternative antibacterial treatment. Phage OMKO1 was selected and granted approval as an emergency investigational new drug. 10 mL of phage OMKO1 and 0.2g/mL ceftazidime was introduced into the fistula under direct CT guidance in a sterile fashion. Four weeks post-procedure, the patient had to undergo emergency partial aorta and graft excision and replacement, and repair of a mediastinal fistula. The cultures taken during the surgery confirmed the growth of Candida from superficial chest wounds. After the surgery, the patient was taken off of antibiotics; he has not experienced any recurrence of *P. aeruginosa* infection.

Prosthetic vascular graft infections (PVGI) pose a significant threat to the health care system, putting a heavy economic strain on it; these infections are commonly caused by *P. aeruginosa*. Authors argue that with the increased number of MDR bacterial infections, it becomes vital to seek novel antimicrobial agents, while taking into consideration that further evolution and resistance development is unavoidable (Chan 2018).

### *Acientobacter baumannii* craniotomy infection

LaVergne et al. (2018) describe the use of phage therapy in a 77-year-old patient with multidrug-resistant *Acinetobacter baumannii* craniectomy site infection. The susceptibility testing was done using multiple antibiotics; only some isolates were sensitive to colistin. Fractional inhibitory concentration index (FICI) neared synergy for two combinations: colistin + rifampicin (FICI=0.53) and colistin + azithromycin (FICI=0,56). Colistin, rifampicin, and azithromycin were used together but without any clinical improvement. Due to failed antibiotic treatment, the team decided to seek alternative antibacterial therapy. The patient's isolate was matched against $10^4$ *A. baumannii* phage obtained from the Naval Medical Research Center-Frederick 's phage-Biolog system. The concentration of the phage was selected by depending on the endotoxin levels recommended by the FDA,endotoxin lipopolysaccharide limit of 5eu/kg body weight per hour. A total of 98 doses of phage were administered intravenously. The patient's family decided to withdraw care before the patient received the second phage cocktail, and he passed away on hospital day 20, he had received phage treatment for eight days prior. The researchers argue that if they had increased the concentration of phages from $2.14 \times 10^7$ PFU/mL, the treatment could have been more effective. It was also argued that the intrathecal route could guarantee better penetration in comparison to intravenous use. The authors also challenge the limitations of individualised therapy, since it is time-consuming, and suggest further exploration of the efficacy of phages with broader activity (LaVergne et al., 2018).

**Phage therapy in cystic fibrosis**

Cystic fibrosis is a genetic disease which results in dysfunction of chloride transport and results in thick mucus secretions and a predisposition to frequent bacterial infections. Cystic fibrosis associated lung infections are a logical target for phage therapy as these patients are heavily dependent on antibiotics and often have severe complications of infection. The Kutter lab at Evergreen State College started research on CF in the mid 1990s when a collection of 200 CF strains were donated by the Seattle Children's hospital. As part of undergraduate research, many Evergreen students have isolated phage from sewage against this CF collection, with more than 100 phages in the collection to date. Many *Pseudomonas* phages were isolated against this collection, some of which are being further studied in terms of pharmacokinetics and dynamics (Lin 2018). Phage treatment of patients with cystic fibrosis in Georgia started in 2007 with the treatment of eight Georgian patients in collaboration with the National CF Center. Infectious control against *Staphylococcus* and *Pseudomonas* was achieved with phage, and the patient's general condition improved. Further treatment of CF patients has continued at the Eliava Phage Therapy Center over the last decade with significant improvement in overall condition, reduction in antibiotic use and need for hospitalisation. Hoyle et al.

(2018) describe a case study of a CF patient with chronic *Achromobacter* infection which resulted in significant improvement of lung function over a three month period with periodic oral treatment and phage inhalation.

Researchers at Yale University began treatment of cystic fibrosis patients under emergent indication and have treated 13 patients. For some of these patient's infection is a factor which is preventing lung transplant. The work at Yale is an indication of an overall willingness of the medical community to try phage therapy for patients who have few, if any, alternatives. The results of these treatments have not yet been published.

A similar case was described by Law et al. (2019) who treated an MDR *Pseudomonas aeruginosa* infection in a 26-year-old Canadian girl. The patient was treated with an intravenous phage preparation, which was effective in preventing a recurrent infection for at least 100 days after the end of phage therapy. The patient successfully underwent double lung transplant nine months after treatment.

**Novel synthetic phage treatment against *Mycobacterium abscessus***

The treatment of intracellular bacteria with phage has not been possible until now. The first case was described by Dr Dedrick from the Great Ormond Hospital in London England where a 15-year-old cystic fibrosis patient was treated for a disseminated mycobacterial infection due to *Mycobacteria abscessus*, a non-tuberculosis mycobacteria (NTM) (Dedrick et al., 2019). The patient underwent a bilateral lung transplant after taking anti-NTM treatment for eight years prior. Unfortunately, the infection returned shortly after discontinuing anti-NTM treatment and the patient was sent home for palliative care with lung infection and skin lesions after seven months of treatment. However, the story does have a happy ending. A phage cocktail of genetically engineered phages was developed and applied both topically and by i.v. The patient responded well to the treatment, with significant improvement in her clinical condition. No Mycobacteria were detected in clinical samples after initiation of this second treatment (Dedrick et al., 2019).

The need for controlled clinical studies in CF is well recognised. The development of phage preparations which are active against wild type (mucoid) bacterial strains and strategies that take into consideration the presence of biofilm in the microbiota is essential for efficacy (Hraiech 2015, Rossitto 2018).

### DiGeorge *Pseudomonas* infection

Duplessis et al. (2018) describe a case report of a 2-year-old boy diagnosed with DiGeorge syndrome. Despite undergoing multiple antibiotic courses, the patient's *P. aeruginosa* infection persisted. Importantly, the patient developed Stevens-Johnson syndrome while being on cephalosporin antibiotics and fluoroquinolone. The patient's *P. aeruginosa* strain was matched against a proprietary lytic spectrum assay to identify a phage cocktail, which demonstrated efficacy towards the given bacterial strain. Phages were given for a limited 36 hours, a total of six doses of phage cocktails. Surveillance of blood cultures taken on days 4 and 5 post-therapy, were sterile. The therapy was stopped due to fear of the possibility of an anaphylactic reaction. Later in managing the patient, the team decided to put the patient back on phage therapy for eleven days. Unfortunately, the family had to withdraw care since the patient's health regressed significantly (Duplessis, 2018). The authors argue that for the future development of phage therapy, it is essential to create an extensive phage library, to work on phage cocktails to accommodate multivalence and reduce the chances of bacterial resistance. Concerns remain over the endotoxin content of phages prompting the development of an allergic reaction against proteinaceous organisms as a potential disadvantage of phage therapy (Duplessis, 2018). However, this phenomenon is also possible with the rapid killing of bacteria during antibiotic therapy. Widespread use of phage in countries with a long history of use have not reported this as a regular occurrence.

### Extensive applications since 2005 at a naturopathic and oriental medicine clinic in Portland, Oregon

In 1992, Dr. Joseph Pizzorno, president of Seattle's Bastyr University of Natural Medicine, asked Dr Kutter to write a chapter on phage therapy for the first edition of his Textbook of Natural Medicine; she had been a member of Bastyr's

Board of Directors since its founding in 1975. The 5th edition is now in press, complete with an updated chapter on phage therapy. Satya Ambrose, a student of Dr Kutter's in the early 1970's who had then spent seven years studying at an oriental medicine school, associated with Harvard, was brought to Portland by their longstanding National School of Natural Medicine to lead the development of a parallel School of Oriental Medicine there. She was so impressed with the focus and direction of the National Naturopathic students in classes she was teaching that she went about obtaining a Naturopathic degree herself. In 2014, she opened her own clinic and soon asked for help in adding the Intestiphage cocktails from Tbilisi to her arsenal, after getting approval from the Oriental Medicine licensing board to do so. In Oregon, such practitioners have the explicit right to use natural products that have been approved and extensively used in other countries. Here, she shares some of her nearly 1000 cases. Note that this was not implemented as a clinical trial of phage per se, but simply as an extra tool she could add and try when her other tools weren't sufficient - in the process collecting data on where the Intestiphage seemed to be most helpful. A wide variety of conditions were treated with phage, including chronic pulmonary infections, ENT infections such as otitis and sinusitis, skin infections including eczema, urinary tract infections like interstitial cystitis, and gastrointestinal problems ranging from functional disorders like IBS and SIBO to more severe inflammatory bowel disorders. The use of complex phage cocktails such as the Eliava Intestiphage in a manner similar to pre- or probiotics was carried out for many patients with functional gastrointestinal (g.i.) disorders. Retrospective analysis for these patients is planned and one of the Evergreen Phage Lab student researchers is currently collecting follow-up data for a substantial group of patients from recent years. The Naturopathic approach to disease is unique and looks holistically at causes of disease. One area of particular interest is the connection between pathogenesis, the microbiome, and the possibility of using phage to rebalance the microbiome, which has profound effects on the gut-associated lymphoid tissue (GALT), in turn strongly impacting immunity, neurotransmitters and neurofunction as well as nutrient absorption (Galland, 2014). This currently is a subject of study at many academic institutions.

### A phage therapy treatment that strongly made the news in Seattle

With the airing of television programs on phage therapy, there is frequently a large response from patients eager to see if their condition is one which can be treated with phage, as well as find ways to access the therapy. A very good example was a Seattle-area family whose 18 month old boy was very seriously ill with persistent pharyngeal dysphagia and aspiration, in addition to frequent aspiration pneumonia and *S. aureus* colonisation. The desperate family immediately sought the alternative therapy they had just heard about on a television program about phage therapy and undergraduate research at The Evergreen State College. They were particularly concerned that their Staph strain was methicillin resistant, and their child had already undergone multiple rounds of antibiotic treatment due to frequent lung infections. He had had twelve hospital visits in an eight month period, with three longer hospital stays. This had finally resulted in the doctors' decision to perform a bronchoscopy, through which a sizable laryngeal cleft was discovered and resolved with sclerosing therapy. There were clear signs of lung infection when the bronchoscopy was performed, with MRSA being isolated from samples taken via bronchoalveolar lavage.

The day after the phage therapy program, the whole family drove immediately down to Evergreen. We immediately supplied staph phage to the parents and two siblings, so they would not re-infect the child. Dr Kutter arranged for the patient to send samples to Tbilisi for phage sensitivity testing and to work with a local Naturopathic doctor to help facilitate the patient's treatment with phage. The patient's pediatric team was aware of the family's decision to try phage therapy and approved of it under the circumstances. However, they were not inclined to formally participate in the treatment without extensive involvement from their Institutional Review Board (IRB) and registration of an emergency IND with the FDA, and the situation seemed too dire for the delays that would involve. They did, however, stay involved in the other aspects of his care. Today, the FDA's involvement with phage has advanced far enough that we could and would have immediately filed an emergency IND with them, but we had not heard at the time that the option was required, or even possible.

The child was treated in collaboration with the Eliava Phage Therapy Center for seven days with 0.3 mL Eliava Staphylococcal phage intranasally and 1 mL diluted phage via nebuliser. The treatment was repeated after two weeks. The patient's family also performed MRSA decolonisation with phage intranasal application twice daily for two weeks. The patient's nasal cultures were negative after phage treatment. The ENT physician treating the boy could not exclude that his patient's clinical condition was due to the aspiration and dysphagia more so than the MRSA colonisation. His recommendation was for a diet of thickened liquids, which was also implemented until the age of five, when the final needed surgeries could be done. This was celebrated with a giant party, with nearly 100 people who had helped in various ways through the years to celebrate - and the KOMO TV crew again there to document and share this victory. Now, at nearly seven years of age, the boy is strong and healthy and has not had a recurrent infection since his early phage therapy (Kutter, personal communication 2020).

### Diabetic foot

The lifetime risk of people with diabetes to develop a foot ulcer is 34%. More than 50% of diabetic foot ulcers become infected, causing significant morbidity and cost (Armstrong, et al., 2017, Prompers et al., 2007). One in five foot ulcers ultimately leads to some level of amputation, and in the U.S., diabetes patients undergo about 80,000 of the 120,000 amputations carried out yearly (Walsh et al., 2016). The relative five year mortality rate after limb amputation is 68%. It is second only to lung cancer, at 86% (Armstrong, 2007). It behoves us, therefore, to treat diabetic foot ulcer infections with alacrity.

Podiatrist Randy Fish had been an Evergreen phage lab student in 1975 and then became a podiatrist. He was particularly frustrated by those of his elderly patients who developed toe ulcers that were totally recalcitrant to both local and intravenous antibiotics, leaving him no choice but to amputate. A few years ago, he serendipitously heard that Dr Kutter actually still had her phage lab, had established a Phagebiotics Research Foundation and was in fact working with the Eliava Institute people, was collaborating with a Naturopath to explore phage therapy, and had long been particularly interested in the idea of diabetic foot ulcers. Building on a longstanding collaboration with the Eliava Institute, supported by the Phagebiotics foundation, Dr Fish used the well characterised SB-1 phage (Kvachadze, 2011) to treat approximately ten patients with non-healing diabetic foot ulcers. Fish et al., (2018) reported a case series of infected diabetic toes where very small amounts of phages were introduced into the infection, with radiographic follow up demonstrating bone regeneration while the ulcers healed. All of the wounds closed, and remission was stable at after a one year follow up. Phage use on the diabetic foot model is one which is being further explored. Three representative cases are presented.

The foot ulcers treated were on patients having ulcers located at bony pressure points with culture proven *S aureus* infections. Circulation was evaluated and found to be poor. Several had bone exposed (Figure 1), others had infected bone within a toe or foot, with infection confirmed by MRI (Figure 2). While the ulcer healed quickly, bone re-ossification takes upwards of a year to complete.

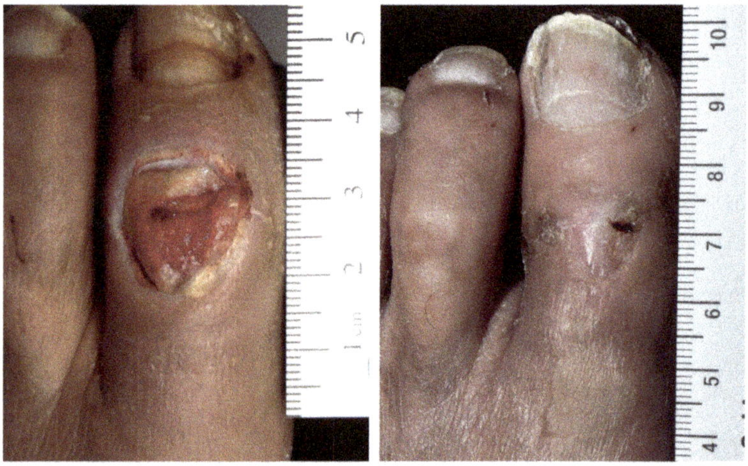

A (7th April 2014) B (18th August 2014)

**Figure 1. (A)** Diabetic foot ulcer in a 44-year-old male patient with a history of uncontrolled diabetes, morbid obesity, dyslipidemia, neuropathy, lymphedema, and sleep apnea. He was a former smoker. The Trans-Cutaneous Oxygen Pressure was 28 mm Hg. Note the open joint and bone in **(B)** after phage treatment (adapted from Fish et al., 2018).

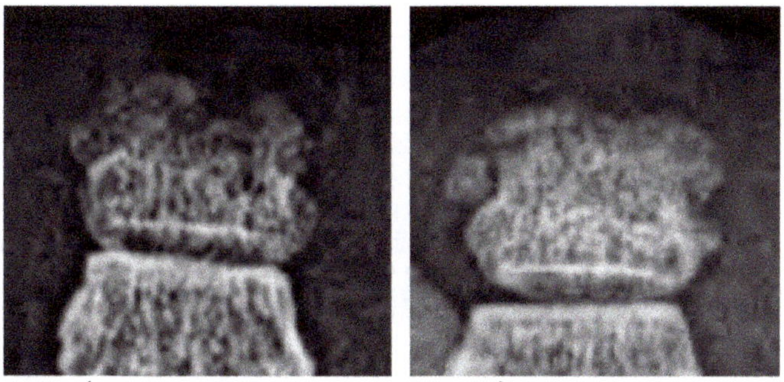

A (25th March 2015) B (22nd Sept 2015)

**Figure 2. (A)** MRI of the right second toe of a 63 year old diabetic female with osteomyelitis of the distal phalanx. **(B)** Note the re-ossification of the phalanx as the infection is controlled (adapted from Fish et al., 2018).

The result was a decrease in the localised inflammation and oedema, plus the initiation of the healing process. The idea of avoiding infection is another excellent use of phage. Antibiotics had been trialled, but unsuccessful in jumpstarting the healing process of all patients. The SB-1 phage solution was applied, once weekly, directly onto the open ulcer tissue or by injection into the tissue surrounding the infected bone or into the bone itself. One patient had prophylactic phage applied after extensive debridement, with the removal of gangrenous tissue and bone. The patient healed within seven weeks without infection (Figure 3).

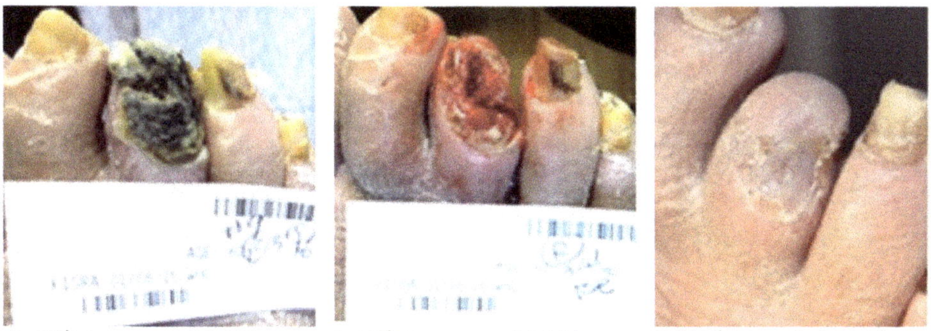

A (8th January 2015) B (8th January 2015) C (2nd March 2015)

**Figure 3.** The Patient is a 61 year old diabetic man with a history of hepatitis C, as well as heroin and alcohol abuse. He continued to smoke one pack per day. **(A)** His critical limb ischemia was reduced by stenting but still remained significant. **(B)** Gangrenous tissue and bone were removed from the middle and distal phalanges of the toe, and the remainder of the toe was treated with dehydrated human amniotic chorionic membrane that was rehydrated with the staphylococcal phage solution. **(C)** Although shortened, the toe completely healed (adapted from Fish et al., 2018).

In summation, the patients improved, saving limbs, and lives, as the healing potential following a minor amputation is poor and the morbidity is high. An unsuccessful minor amputation, by necessity, is followed by a major amputation, with a five-year mortality rate approaching 68%. Initial impressions of topical use of phage remain valid. It is noted that we see reduced erythema at the wound with reduced localised edema. Small amounts of phage seem to work well with localised cellulitis using the Infectious Disease Society of America's grading system of mild to moderate infections. Interesting work has been published by a Russian group in Novosibirsk, focusing on larger diabetic foot ulcers, utilising phage therapy with much success (Morozova 2018). While we are starting to see more descriptive cases in the literature, more work is necessary to better elucidate improved dosages and techniques.

## Phage treatment around the world

### *Phage treatment of musculoskeletal infections*

Onsea and colleagues (2019) report phage on therapy in four patients with complicated musculoskeletal (osteomyelitis) infections. A multidisciplinary team selected four patients with poor prognosis despite adequate medical and therapeutic intervention. The patient's wounds were surgically debrided and microbiological analysis and phage sensitivity testing was determined. Three of the patients were treated with BFC1 phage and one with Eliava Pyophage. The phage was applied during surgical intervention in addition to three times daily for ten days via a drainage system installed during surgery. This treatment was carried out as an adjunctive treatment; the patients were also given antibiotics based on sensitivity profiles. All of the patients had previously undergone extensive treatment for their condition before, but infection relapse had occurred on several occasions (Onsea 2019). No recurrence occurred and was seen in three out of four patients on an eight to sixteen month follow up. One of the patients had reinfection with a different strain. All of the patients' inflammatory markers decreased after phage therapy. A multidisciplinary team developed the protocol for phage therapy treatment design and application.

*Colistin-only-sensitive* Pseudomonas aeruginosa *septicemia*

A 61-year-old patient presented with a complicated medical history, including septicemia with colistin-only sensitive *P. aeruginosa*. The treatment was initially initiated with colistin; unfortunately, the patient developed nephrotoxicity, and the physicians had to find an alternative antimicrobial therapy. Purified phage cocktail BFC1 demonstrated *in vitro* activity against isolated *P. aeruginosa*. In addition to a six hour intravenous infusion, the wounds were irrigated with a 50 mL phage cocktail every eight hours for ten days. The team reported that following the treatment, blood cultures turned negative, CRP levels decreased, and the patient became afebrile. Unfortunately, the patient died four months after the therapy was started due to refractory cardiac arrest caused by *Klebsiella pneumoniae* sepsis (Jennes et al., 2017).

**Westmead clinical trial: the safety of phages in severe staphylococcal infections**

A recent single-arm non-comparative trial of thirteen patients with severe staphylococcal infections was conducted using a GMP phage cocktail. Selected patients had sepsis or prosthetic valve endocarditis and were critically ill but stable (death was not expected for at least 48 hours) The study was divided into two groups; phage sensitivity testing was conducted prior to administration in one group and in the other, phage was administered empirically while sensitivity testing was performed (Petrovic et al., 2020). Safety and tolerability were primary endpoints of the study and no symptoms such as fever, tachycardia, hypotension, diarrhoea or abdominal pain were observed. Efficacy was the secondary outcome: clinical improvement, inflammatory markers and mortality were assessed. Inflammatory markers decreased in eleven of thirteen patients soon after commencement of phage therapy. Pharmacokinetics was studied in detail. Distribution of phage twelve hours after dosing was detected in the blood, and phage DNA was detected in valve tissues after 14 days of treatment (Petrovic et al., 2020). This confirms the results of previous animal studies showing a wide distribution of phage after single doses (Dubos 1943, Abedon 2011). The authors found phage therapy to be safe and tolerable for patients with sepsis. Very high mortality due to severe staphylococcal infections such as the cases described here signals the importance of alternatives like phage as an adjunctive therapy to antibiotics. The use of intravenous phage against staphylococcal infections in sepsis and other severe infections is not a novelty. Some of the initial applications of phage involved using a phage lysate against *Staphylococcus* intravenously (D'Hérelle 1931).

## The Eliava experience with phage treatment for severe staphylococcal infections

Staphylococcal phage was used widely in Georgia from very early on. Phage preparations targeting *S. aureus* have always been of particular interest to Eliava scientists, reflecting its key roles in purulent wounds and hospital-acquired infections as well as the particularly high efficacy of such phages. In the 1970s, Eliava scientists developed a highly purified Intravenous Staph phage preparation, intended for the treatment of acute and chronic sepsis. After animal trials and tests on 20 volunteer patients with them showing no side-effects, permission was granted on 11 April 1979 to carry out clinical trials on a limited basis at three medical centres in Georgia and 3 in Moscow, involving 250 patients. The results were discussed in detail by Chanishvili (2012), drawing particularly on the very extensive compendium 'Results of Clinical Trials' (1983) from the Eliava Institute records. Further trials were conducted on 653 patients in two additional hospitals in Tbilisi and two in Moscow. A substantial fraction of the first batch of the preparation was commandeered by the Soviet Ministry of Health to treat a severe outbreak of sepsis involving hundreds of premature infants in a Moscow hospital. Altogether, 149 children (99 in Moscow and 50 in Tbilisi, 106 of them under one month of age, were treated with the intravenous phage, mostly in combination with antibiotics, while 98 children were in a control group and received only antibiotics. The infants receiving the phage recovered much more rapidly in terms of both physical symptoms and laboratory indices. Only one of the phage-treated infants died (0.67%), while eight infants from the control group died (8.2%).

This highly purified staphylococcal phage preparation was also used to treat peritonitis, osteomyelitis, mastitis, purulent arthritis, lung abscesses, chronic pneumonia and bronchitis, septic staphylodermia, purulent cysts and chronic pelvic inflammatory disease in adults. In the 1980s, clinical trials of intravenous and intra-arterial applications were performed in the Scientific-Research Institute of Clinical and Experimental Surgery of Georgia, Georgian Centre of Sepsis, Central Clinical Hospital, Scientific Centre of Traumatology and Orthopedics, and Scientific-Research Institute of Human Generative Functions. The route of administration of the phage depended on the nature and location of the infection. Intravenous and intra-arterial administration of the preparation was used in cases of traumatic osteomyelitis. For lung disease, the preparation was administered intravenously as well as by direct bronchoscopic introduction or intrathoracic injection into the site of the infection. These trials were performed on 653 patients (age 15–75; 355 men, 298 women) (Kutateladze and Adamia, 2008). The studied group consisted of 345 patients, 130 of whom were treated only with phage preparations, while 215 were administered phages and antibiotics. The control group included 308 patients who received no phages. The extent of total recoveries in the group treated only with phages, the group which received both phages and antibiotics, and the control group were 41.3%, 77.5%, and 11%, respectively.

Phage Sb-1, which is the main component of this anti-staph preparation, still reveals a broad spectrum of lytic activity *in vitro* against freshly isolated, genetically different clinical samples (including MRSA) from clinics in a wide range of different geographical areas. Analysis of the sequenced phage genome showed no integrase or potentially harmful genes such as those that might contribute to bacterial virulence. This phage is now again being prepared in a defined minimal medium and highly purified. It is legally approved for external applications, available in pharmacies and being successfully used for patient treatment (Kvachadze et al., 2011). A key aspect of the Georgian work to optimise phage therapy has always been the ongoing close collaboration of clinics and hospitals with phage producers. Eliava scientists worked closely with the Central Republican Hospital infectious surgery department (Dr G. Gvasalia and Academician N. Kipshidze), the Sepsis Center (Academician V. Bochorishvili), Gudushauri Hospital, Burn Center, Infectious Diseases Hospital and many others.

### Phage therapy in urology

Tsulukidze National Center of Urology, Tbilisi, Georgia carried out a randomised, double blind placebo-controlled study in collaboration between the Eliava Institute and several Swiss institutions. This trial used phage in 81 patients who were undergoing transurethral proctectomy. It was a three-arm study where one group was to be treated with commercially available pyophage adapted to the treated population, the second group was a placebo control and the last group antibiotic therapy. The study results are currently being analysed and publication is expected soon (Leitner 2017). There is a more detailed description of this study and of current phage therapy in Tbilisi in another chapter of this book by Chanishvili and co-authors.

### Phage against MDR carbapenemase-producing *Klebsiella pneumoniae*

Corbellino et al. present a case of personalised phage therapy against multidrug-resistant (MDR), carbapenemase-producing *K. pneumoniae* (*Kp*) (2020). A 57-year old patient diagnosed with Crohn's disease, at the age of 20, presented with multi-site colonisation of the gastrointestinal tract, urine, and ureteral stent with a MDR *Kp* strain. The patient had a long history of recurrent infections, which, despite antibiotic treatment, persisted. The isolated *Kp* strain belonged to the Sequence Type ST307 and was susceptible to only ceftazidime-avibactam (CZA), amikacin, fosfomycin, and tigecycline. Since the patient was at high risk for developing another infection, the physician deemed appropriate to find an alternative antimicrobial treatment. Five *Kp* isolates obtained from the urine, rectal swab, and ureteral stent were sent to the Eliava Institute of Bacteriophages in Tbilisi, Georgia, to check for sensitivity against phages. The institute developed a custom, lytic phage preparation, against the specific pathogen. Bacteriophage therapy (BT) was given on an outpatient basis; the patient did not report any adverse effects. The researchers report that 15 days after the beginning of phage therapy, the proximal end of the ureteral stent did not yield MDR *Kp*, which had been a continual problem after repeated antibiotic therapy. Molecular screening, using the Cepheid Xpert Carba-R Kit, also showed no presence of carbapenemase genes in the patient's rectal swabs. Importantly, since BT was initiated, swabs taken from the patient's stool samples, rectum, urine, and ureteral stents remained negative (Corbellino, 2020).

### Phage Therapy in a renal transplant patient

Kidney transplant cases are often complicated by chronic, relapsing UTIs caused by ESBL-positive *Klebsiella pneumoniae*. Dutch researchers present a case study of a 58 year old male patient, who tested positive for ESBL-producing *K. pneumoniae*, which was susceptible to meropenem and amikacin (Kupiers et al., 2019). Despite multiple attempts of antibiotic treatment, which included meropenem therapy seven times, the patient continued to develop episodes of urological infections. After sending culture to the Eliava Institute, a custom lytic phage suspension was made, which was used in combination with meropenem for eight weeks. Phage suspension was used per os and via intravesical administration. Kuipers et al. report that the symptoms of epididymitis resolved within a day and the urine cultures have remained negative 14 months post-treatment (2019).

### Phagoburn

Jault et al. (2018) aimed to compare the efficacy and tolerability of a cocktail of 12 natural lytic anti-*Pseudomonas aeruginosa* phages with 1% sulfadiazine silver emulsion cream. The randomised phase 1/2 trial was carried for seven days, with 14 days of follow-up, in patients with a confirmed wound infection. The team enrolled a total of 27 patients: 13 patients were designated in the phage treatment group, and 14 were given standard of care. One patient in the standard of care group was not exposed to treatment, while one patient in the PP1131 group did not have an infection at day 0, giving a final sample of PP1121 *n*=12 and standard of care n=13. Initially, phage cocktail PP1131 was selected with a concentration of 1 × $10^6$ PFU/mL; however, the titre decreased after manufacturing, and participants were given a lower concentration of 1×$10^2$ PFU/mL per daily dose. The team concluded that bacterial burden decreased at a slower pace with PP1131 treatment than standard of care, the research was stopped due to insufficient efficacy of phage cocktail against *P. aeruginosa* infection.

Despite the negative outcome of the study, it is vital to evaluate the role of cocktail concentration in wound infection treatment. Jault et al. (2018) said that further studies with increased phage concentration and a more significant sample number are necessary. This experience has been an important lesson in the significance of securing GMP manufactured phage products with high enough titres to adequately represent the "active ingredient" of the product.

**Promoting accessibility**

In order to improve the chances of phage becoming an accessible medicine in the United States and Europe some basic "rules" for standardising research should be agreed upon. Efforts toward consensus statements and general understanding have been attempted at several gatherings of experts (Rohde 2018, Expert panel 2016). The group identified three major areas where there is a gap in the knowledge; 1) bacterial phage resistance 2) phage training and 3) prophages in bacterial strains used for phage production. Co-evolution of bacteria and therapy. *In vitro* studies have shown that the development of phage resistance corresponds to less virulent bacteria (Friman, 2016, Oechslin, 2017). It is difficult to estimate the true rate of resistance to phage cocktails in a clinical setting. Some of the participants of the expert panel, including two biopharmaceutical researchers consider "it might be possible to develop resistance-proof therapeutic phage cocktails, using phages with a broad host range and targeting highly conserved structures that are essential for bacterial survival and/or infectivity. Two of them note that the phage resistance problem is not caused by the de novo emergence of phage resistant clones, but by the selection of naturally present phage resistant isolates harbouring antiviral mechanisms such as restriction modification systems and CRISPR/Cas systems" (Rodhe, 2018). Training phage is a longstanding method for therapeutic preparation. This method uses serial passaging to adapt the phage to its host *in vitro*, thus increasing its efficacy (Friman 2016, Rodhe 2018). d'Hérelle noted the importance of phage adaptation in his 1931 manuscript

> "Adaptability is an exclusive process of living beings and the bacteriophage possesses this character to a very high degree. There are, however, in this respect differences between different races for certain bacteriophages adapt themselves very readily,

> while others do so very slowly....When attacked by a powerful bacteriophage the bacterium succumbs, but if the potency of the bacteriophage is less the bacterium is capable of resistance, in which case it then contracts a true chronic disease accompanied by modifications in its characters." (d'Hérelle 1931).

Phage experts note the preference for phage production on bacterial strains that do not contain prophages. The presence of prophages in bacterial strains must be studied in more detail to understand its significance. There are many prophage containing bacteria in the human microbiome and their function is not completely understood (Rodhe, 2018).

Many consensus discussions have come out of phage therapy focused conferences such as the Phage Futures congress which have been held in both the USA and Europe. At the most recent meeting, much discussion on clinical applications and manufacturing issues took place. While a consensus statement has yet to be published from this gathering, some key points were made including exchange on lack of pre-clinical bacterial testing of the phage. Although testing occurs, the time involved in pre-use testing of the phage against the bacteria may be more than desired vs an active infection on an unstable patient. Against that thought was the idea to begin a phage cocktail which has very wide coverage, testing the bacteria against the phage and adjusting the phage as necessary. Some discussion was dedicated to the idea of inclusion of data from compassionate use applied to clinical trials which is currently not acceptable for the US FDA.

**Improving reporting**

Standardisation of reporting can be a significant support for the implementation of phage therapy. Various guidelines for this have been suggested (McCallin 2019, Abedon 2017). Abedon suggests including information about the sources of the phage, characterisation properties such as burst size, latent period, and adsorption rates, genomics: bioinformatic characterisation, host range, purification, dilution, stability, and toxicity among others (2017).

Such criteria will help the reliability of phage therapy, allowing researchers to build upon previous work, and for physicians to be able to apply reported knowledge into clinical practice (Abedon, 2017). Descriptive studies are excellent for spreading familiarity about phage therapy, they cannot however be used to draw conclusions on efficacy. For successful use of phage therapy as compassionate therapy, it is important for case studies to report about the pharmacodynamic and pharmacokinetic properties of phages in a detailed and authentic manner (McCallin, 2019).

**Developing therapeutic phage preparations**

Phages only target bacteria. They do not cause harm to the eukaryotic cell. Importantly, phages do not affect the normal microbiota of mammalian organisms. This is a significant advantage to antibiotics as it minimises the risk of developing antibiotic associated diarrhoea and infection with *C.difficile*. Phage can also be used as an adjunctive therapy to antibiotics. Numerous studies show significant synergistic action with some antibiotics (Abedon, 2019). It is time-consuming to find a phage for each isolated bacterial strain. It is important to select phages that are safe and effective. Authors suggest using phages that are lytic in nature and do not transduce bacterial virulence factor genes. Phage therapy is often scrutinised for the lack of double blinded Phase III trials. Importantly, authors suggest that this is oftentimes explained due to a lack of funding rather than underreporting failed attempts at demonstrating the efficacy of phage therapy. With advancements in biotechnics and characterisation strategies, the authors argue that genome annotation might be enough for regulatory approval, therefore dissipating the need for clinical trials (Chan 2013).

Phage cocktails are widely considered to be the best approach as a therapeutic as they have decreased risk of developing phage-resistance and are commercially more profitable. Instead of creating phage banks, Chan et al. (2013) suggest using phage cocktail banks. For the successful implementation of phage therapy, it is essential to create an appropriate phage bank and identification system. Others suggest phage banks of well characterised phage to use sequentially if necessary, but this method would need serious investigation (Rodhe, 2018). The composition of phage cocktails such as

Pyophage and Intestiphage, produced by the Eliava Institute, are updated bi-annually or annually based on the most prevalent pathogenic strains. This approach is similar to the development of seasonal vaccines. Any phage registration should have the ability to modify the preparation based on changes in clinical isolates. Developing phage cocktails which are effective and structured to allow for modification without new clinical trials for each iteration is a priority. There is debate on whether the model of "one size fits all"- or development of broad-spectrum phage cocktails is more efficacious than a tailor made- patient specific approach. Both approaches require well characterised phage, which are ultimately sensitive to the pathogenic bacterial strain causing the infection. Development of bacterial resistance towards phages is possible but could be avoided by the utilisation of "phage cocktails." It is also important to explore the impact of the synergic use of phages and antibiotics, to decrease the rates of bacterial resistance (Kutateladze 2010).

Metagenomic analysis of phage cocktails is a good way to evaluate them. Some of this work has been carried out on commercially produced Georgian phage cocktails such as Pyophage and Intestiphage, which were found to not contain any potentially harmful genes (Villarroel 2017, Zchach 2015). Similar metagenomic analysis and clinical safety testing were also carried out for a commercial therapeutic coliphage preparation called ColiProteus made by the Russian pharmaceutical company Microgen (McCallin et al., 2013) which was used as a third arm in the Bangladeshi clinical studies. When discussing phage as an alternative to antibiotics, one has to consider not only such advantages as a different mode of action, high specificity, and self-replication in the infection site and challenges such as the potential development of phage resistance, immunogenicity, varied bioavailability and complex pharmacokinetics. It is also important to take into consideration the ability of phage to replicate in the environment in which the trial is being held. For example, Bryan et al. (2016) showed that during infection in stationary phase T4 goes into a hibernation mode where the phage have replicated their DNA but do not make phage particles until fresh nutrients are added (casamino acids and glucose). This would help explain the low quantity of T4 phages isolated from patient samples in the Nestle trial. The Russian cocktail, by comparison, has a high amount of phages like T7, which does replicate well in stationary phase in the gut as long as there are sufficient bacteria (Weiss et al., 2009, Kutter et al., 2018). Clinical trials that have been conducted to date either have had an issue with dosing, quality of phage preparation (low titre by the time the study was run) or study design, which is not facilitative to phage therapy.

## Safety

Earlier therapeutic phage applications are important for further development of clinical phage therapy in multiple ways. Since phage therapy began in the 1900s, it is important to start the discussion of safety with the lack of reported cases of severe side effects in the scientific literature, despite it being used for more than 100 years in millions of patients worldwide. Importantly, like other antimicrobial reagents, phages are not recorded to cause anaphylaxis (Abedon et al., 2011). Phages are the most abundant biological entities on the planet, surpassing other viruses and bacteria by a factor of 10. It has been hypothesised that animals have been exposed to a large number of phage viruses throughout their evolution. This type of co-evolution could explain phage specificity, their target of bacterial pathogens, and avoidance of eukaryotic cells. Interestingly, in 1970 it was discovered that a large number of vaccines were contaminated by phages; however, an Executive Order was made to allow the use of vaccines, with no subsequent complaint of adverse reaction development (Henein, 2012). In addition to millions of doses of phage safely administered to patients throughout the past century, phage was also administered unintentionally to tens of thousands of patients via phage contaminated vaccines without adverse effect. Phages have been used for non-therapeutic reasons in developed countries. In 2006, the FDA approved a mixture of *Listeria monocytogenes* phages as having GRAS (generally recognised as safe) status for food production and processing (Lang 2006).

## Pharmacology

Malik et al. (2017) discusses the importance of determining phage pharmacokinetics and pharmacodynamics. The stability and formulation of phage therapeutics need the same level of attention as in any other pharmaceutical design. Efforts toward developing a stable and effective inhalation product have been undertaken by pharmacologists at the University of Sydney. Their group has studied *Pseudomonas* phage PEV2 and PEV40 in terms of stability using as a powder formulation as well as in various types of nebuliser (Lin 2018, Leung 2018, Leung 2019). Encapsulation of phage in lysosomes is a breakthrough technology which will significantly help with the distribution and bioavailability of phage (Chang, 2018). This technology can also be employed in the development of phage to treat intracellular bacteria. Studies are necessary to explore the parameters such as stability and formulation of all phages. These steps are vital in implementing phage therapy for therapeutic purposes (Malik, 2017). Further detailed pharmacokinetic and pharmacodynamic studies, dosing and toxicology, are essential for the implementation of phage products under Western regulatory body jurisdiction. For more comprehensive reading on encapsulation, see the chapter by Danish Malik in this book.

**Implementation challenges, goals**

We are at a point of critically evaluating roadblocks for the actualisation of phage therapy. One of which is clearly a flaw in the standard metrics for registration of new medicinal therapeutics. The interest of big "Pharma" is limited by the investability of its product particularly because of the high cost investment required to bring a new drug to market. Alternative pathways must be considered in the United States and the European Union to cater to the unique nature of phage in light of the serious unmet need of antibiotic resistant organisms.

Belgium has made significant gains in the legal framework for implementing phage therapy. Magistral products are the equivalent of an American compounding pharmacy which under the prescription of a medical doctor has the right to make a personalised product, as long as the formulation is part of the pharmacopoeia. Phage preparations have been included under this branch in Belgium with a few important stipulations including, "Phages should be delivered in the form of a magistral preparation to a specific nominal patient, magistral preparations should always be delivered under the direct responsibility of a medical doctor and a pharmacist, the relevant characteristics and qualities of the phage APIs should be defined in an internal monograph prepared by the supplier. Before the pharmacist can use the unlicensed material, he/she must ascertain—based on certificates of analysis issued by a Belgian Approved Laboratory—that the raw materials conform to the provisions of the internal monograph. Even if not legally required, it is recommended that the supplier submits the monograph for assessment by the FAMHP (Federal Agency for Medicines and Health Products)" (Pirnay 2018). This model is an example of working with the existing framework to make allowances which promote accessibility to phage therapy. Such type of progress is truly commendable and could be used as a model elsewhere. Limitations here are having an adequate phage library to address some of the issues delineated previously in this chapter.

**Phage Directory**

A new group was formed by young phage scientists motivated to help patients that required phage therapy. They started crowd-sourcing phages: for a productive exchange of information and, more importantly, for prompt isolation of active phage toward a specific bacterial strain. It is essential to have a shared space where patients/physicians/scientists will be able to exchange information. The recent creation of https://phage.directory has greatly increased the success rate of finding a phage with activity against the patient's isolate promptly.

The exchange of phages globally has limits because of the fact that the bacterial pathogen isolated from the patient should be shipped to a specific laboratory equipped with sensitivity testing as well as a phage library. Transfer of the pathogen across borders is not only time consuming, but also logistically challenging. This pushes the importance of creating a centralised stock of phages, with easy access and fast delivery (McCallin 2019).

**P.H.A.G.E**

P.H.A.G.E (Phages for Human Application Group Europe) is a nonprofit European advocacy group to promote phage therapy and support building a regulatory framework and the initiation of human trials in Europe (P-h-a-g-e.org). Their goals promote the involvement of phage experts in clinical study design, sharing of well characterised lytic phages and phage related data between partners, and advocacy for consensus and coordination within the EU to set a legal framework for phage therapy and phage prophylaxis. One of their current projects is establishing a therapeutic phage bank with the German DSMZ institute.

**Phagebiotics Research Foundation**

Is a nonprofit American foundation dedicated to research in phage therapy and conducting clinical trials in the United States, as well as assisting toward that end in Georgia and Europe and, more recently, helping support the biennial Evergreen International Phage Meetings. It was founded by Dr Elizabeth Kutter and Evergreen colleagues in 1996. The mission of the foundation is to further medical and veterinary applications of phages to treat bacterial disorders. Its goals include; 1) Supporting and providing phage therapy education for medical, scientific and general audiences 2) Supporting collaborative research that helps further clarify the safety and efficacy of phage therapy for key applications. 3) Working toward FDA approval of phage therapeutics. 4) Helping to develop the US infrastructure needed for the application of phage therapy. The Phagebiotics Research Foundation, working openly in collaboration with researchers here and abroad, has been trying to play a key role in developing models for proceeding with physician-initiated phage therapy trials in the US that can help move the field forward to FDA approval in a timely, cost-effective manner. This can best be accomplished by focusing particularly on the phages that have been most successful over the last 70 years: phages targeting *S. aureus.*, a major scourge in both hospital-acquired infections and the broader community. For example, one early stage study we have developed involves adding phage to the standard treatment of diabetic foot infections in cases resistant to all antibiotics. Preliminary small-scale case studies show rapid healing of previously intransigent wounds once phage are added, including an ability to resolve bone infections that all too commonly lead to amputation. Currently this work is led by Dr. Fish as reported earlier in this chapter. This is in agreement with the results long reported in the Republic of Georgia. The commercial Georgian phage preparation targeting *S. aureus* that is being used is fully characterised and sequenced but not amenable to patent protection due to restrictions on patent protection for prior art (in this case, long therapeutic use).

The equally well characterised public-domain staphylococcal phage used as one component in the P.H.A.G.E.-directed Belgian burn trial is also available to use in such diabetic ulcer studies. While corporations find it challenging to invest in the in-depth clinical trials for such non-patentable yet important phage products, the Veterans Administration and other governmental agencies that fund the care of many diabetic infections and amputees do have the incentive to invest in such a cost-saving addition to conventional care, as do wound-care centres. Further studies will need to be carried out to confirm the results and to optimise the protocol in the context of standard US wound-care clinical diabetic-foot protocols for use in the double-blind clinical trials being planned in several key venues. If this approach can indeed be validated, it would represent a major advance in the treatment of diabetic foot infections. The phage-aided cure of such bone infections could greatly reduce the rate of amputation, the associated disability, and the costs for surgery, rehabilitation, and prosthetics associated with limb loss. As these infections often take a very long time to heal, shortening the duration of infection and ulcer healing would also reduce disability and cut the cost of medical care for these patients.

The Phagebiotics Foundation is focused on supporting case studies of clinical applications of phage and building the human and other infrastructure necessary for carrying out physician-initiated controlled clinical trials both in the US and collaboratively with colleagues in foreign countries. Its purpose is to support high-quality research on the treatment of various key bacterial disorders with phage. As a nonprofit research foundation focused on science and community need, it provides the freedom and responsibility to support research that has great promise for meeting clinical needs and will make findings publicly available for further development of the field. Their goal is to help bridge the serious gap in funding through research that is focused on broad collaboration, and thus to help establish the desperately needed FDA recognized proof of principle through clinical trials of phage therapy. The trial on diabetic foot ulcers is focused on key clinical outcomes which reduce the cost of care and disability:

reducing treatment time, the need for amputation, the need for prosthetics and the need for rehabilitation. As a prelude to performing the studies needed to gain FDA approval for key applications of phage therapy, it is critical to develop the necessary infrastructure for performing such trials, including clinicians properly trained in phage application. The Phagebiotics Foundation also supported a cadre of young scientists in various labs at the Eliava Institute and support for training five sets of year-long training programs for surgical residents in phage therapy under Dr Guram Gvasalia, the head of the Surgical Infections division of the main medical school and a constant user in an effort to preserve the expertise of Georgian scientists and physicians.

Phages are slowly shedding their reputation as alternative medicine drugs and are establishing their position in the treatment of multidrug-resistant bacteria. Phage therapy is still largely unavailable in Western medicine. However, steps have been taken to implement PT as an experimental treatment. Gorski et al., (2018) suggest that active engagement of a dedicated industry that is willing to produce phages, with quality and safety controls, is key to propagating the future of phage therapy in pharmaceutical research and application. Phage therapy is not included in the five-year action plan of the European Commission against antibiotic resistance, which was updated in 2017. Authors argue that the distrust surrounding the PT stems from the principle of primum non nocere (first do no harm). Lack of Clinical Trials propagates concern around the widespread use of phages. Still, it should be underscored that many governments have become more accepting of using phages in emergency settings. The FDA gave phages the status of Emergency Investigational New Drug, in the case study at UC San Diego described above where phages were used against multidrug-resistant *Acinetobacter baumannii* (Schooley, 2018). Both the US Army and Navy have very substantial programs for phage targeting key pathogens of major interest to the military, including the second set of *Acinetobacter* phages used by Schooley in the case described above.

Brussow (2012) reviews current problems in phage therapy establishment based on the information presented in the following medical conferences: The First Phage meeting at Oxford University, the Banbury conference at Cold

Spring Harbor Laboratory, and the Brussels EMBO meeting on Viruses of Microbes. The pharmaceutical industry and the medical community were relatively absent from conferences covering PT. However, recent Evergreen International Phage meetings have hosted substantial numbers of presenters and participants from the pharmaceutical industry, multiple branches of the US military, NIH, and the FDA. At the Phage Futures meetings held in Washington D.C. in 2019 and 2020 there was a strong indication that the interest in phage therapy is increasing in the US, but most of the companies participating there were primarily interested in highly lucrative applications such as IV treatment rather than in trying to get more mundane applications like early-stage diabetic ulcers. There are only two topical phage clinical trials currently underway; one will use the diabetic foot model, and the other treatment of wounds Dr Fish is an experienced Podiatrist, and notes that many foot infections can be stopped early on a topical outpatient basis via the use of topical/local injections of phages as we have seen in the earlier described diabetic foot studies. More clinical trials are desired and needed-especially topical use trials. The idea of combined Phage/antibiotic trials was also discussed at Phage Futures in detail. This is a viable concept which meets the idea of reducing resistance by shortening the duration of antibiotics.

One clinical trial sponsored by Adaptive Phage Therapeutics INC. in collaboration with the US DoD is "A Phase I/II Study of Bacteriophage Therapy to Evaluate Safety, Tolerability, and Efficacy of Targeted "Personalised" Bacteriophage Treatments in Patients With Bacterial Infection of the Urinary Tract", which is due to start in 2020. The study will be a randomised placebo-controlled trial, which will use personalised therapy based on phage susceptibility testing via intravenous or intravesical administration. The primary outcome measures are to identify ideal phage treatment regimens based on improvements in disease control rates and microbiological eradication of target pathogen(s) identified at baseline (www.clinicaltrials.gov/NCT04287478)

One would hope this is an indication of movement in the right direction. A natural question is why pharmaceutical companies are reluctant about financing PT either due to possible competitiveness with antibiotics, lack of knowledge in the field, or possibly just the problem of familiarity. For now, phage therapy is used as a last resort for terminally ill patients. Since this group is relatively small, pharmaceutical companies do not invest in proper clinical studies: "The value of the individual human life is very high in civilised societies; it remains, however, "price-less" in economic terms" (Brussow 2012). "For the FDA, the phage product characterisation should comprise the DNA sequence of each phage, as well as sterility and stability of phages, should be analysed. For the acceptance of phage therapy, clinical trials are of paramount importance"(Brussow 2012).

We are restricted by the current regulations which get in the way of exploiting the full potential of phages or their registration for therapeutic use. Hopefully ongoing and future RCTs and deeper studies in the context of extensive current use will not only confirm the safety and efficacy of phage therapy, but also will outline a way to overcome regulatory difficulties and stimulate interest in phage therapy development, while our increasing knowledge of phage biology helps us choose the most efficacious and safe phages. Many experts in the field suggest phage therapy must become a target for public health agencies, where the proper study of phage for therapeutic use becomes a serious endeavour for the common good. It is already evident that there is an unmet need and the problem of antimicrobial resistance will continue to worsen if nothing is done. Serious work should be carried out to implement this therapeutic before it is too late.

## Acknowledgements

Thanks to the other authors on the first iteration of this work in 2014: Jan Borysowski, Ryszard Międzybrodzki, Andrzej Górski, Beata Weber-Dąbrowska, Mzia Kutateladze, Zemphira Alavidze and Marina Goderdzishvili. Thanks also to James Hoyle for support with editing this chapter.

## References

Abedon, S.T. (2017). Information phage therapy research should report. Pharmaceuticals (Basel, Switzerland). *10,* 43. https://doi.org/10.3390/ph10020043

Abedon, S. (2011). Phage therapy pharmacology: calculating phage dosing. Adv. Appl. Microbiol. 77, 1–40. doi: 10.1016/B978-0-12-387044-5.00001-7.

Abedon, S.T., Kuhl, S.J., Blasdel, B.G., and Kutter, E.M. (2011) Phage treatment of human infections. Bacteriophage. *1,* 66–85. doi:10.4161/bact.1.2.15845

Abedon, S.T. (2019). Phage-antibiotic combination treatments: antagonistic impacts of antibiotics on the pharmacodynamics of phage Therapy? Antibiotics (Basel). *8,* 182. doi:10.3390/antibiotics8040182

Alisky, J., Iczkowski, K., Rapoport, A., and Troitsky, N. (1998). Bacteriophages show promise as antimicrobial agents. J. Infect. *36,* 5–15.99 https://doi.org/10.1016/S0163-4453(98)92874-2

Armstrong, D., Boulton, A., and Bus, A. (2017). Diabetic foot ulcers and their recurrence. NEJM. *376,* 2367–2375. DOI: 10.1056/NEJMra1615439

Armstrong, D.G., Wrobel, J., and Robbins J.M. (2007). Guest Editorial: are diabetes-related wounds and amputations worse than cancer? Int Wound J. *4,* 286-287. DOI: 10.1111/j.1742-481X.2007.00392.x

Appelmans, R. (1921). Le bacteriophage dans l'organisme. C R Seances Soc Biol Fil *85,* 722-724

Brüssow, H. (2012). What is needed for phage therapy to become a reality in Western medicine?. Virology. *434,* 138–142. doi: 10.1016/j.virol.2012.09.015

Bruttin, A., and Brüssow, H. (2005). Human volunteers receiving *Escherichia coli* phage T4 orally: a safety test of phage therapy. Antimicrob. Agents. Chemother. *49,* 2874–2878. doi: 10.1128/AAC.49.7.2874-2878.2005

Bruynoghe, R., and Maisin, J. (1921). Essais de thérapeutique au moyen du bactériophage du Staphylocoque. J. Compt. Rend. Soc. Biol. *85*, 1120–1121.

Bryan, D., El-Shibiny, A., Hobbs, Z., Porter, J., & Kutter, E. M. (2016). Bacteriophage T4 infection of stationary phase *E. coli*: Life after log from a phage perspective. Frontiers in Microbiology. *7*, 1391. https://doi.org/10.3389/fmicb.2016.01391

Burrowes, B., Harper, D.R., Anderson, J., McConville, M., and Enright, M.C. (2011). Bacteriophage therapy: potential uses in the control of antibiotic-resistant pathogens. Expert Rev. Anti. Infect. Ther. *9,* 775–785. doi: 10.1586/eri.11.90.

Chan, B. K., Abedon, S. T., and Loc-Carrillo, C. (2013). Phage cocktails and the future of phage therapy. Future Microbiology. *8,* 769–783. doi: 10.2217/fmb.13.47.

Chan, B.K., Sistrom, M., Wertz, J.E., Kortright, K.E., Narayan, D., and Turner P.E. (2016). Phage selection restores antibiotic sensitivity in MDR *Pseudomonas aeruginosa*. Sci. Rep. *6,* 26717. doi: 10.1038/srep26717.

Chang, RYK., Wallin, M., Lin, Y., Leung, SSY., Wang, H., Morales, S., amd Chan, H.K. (2018). Phage therapy for respiratory infections. Adv Drug Deliv Rev. *133,* 76–86. doi:10.1016/j.addr.2018.08.001

Chanishvili, N. (2009). A literature review of the practical application of bacteriophage research. R. Sharp, ed. (Tbilisi, Georgia: Eliava Institute of Bacteriophage, Microbiology and Virology).

Chanishvili, N. (2012). Phage therapy - history from Twort and d'Hérelle through Soviet experience to current approaches. Adv. Virus Res. *83,* 3–40. doi: 10.1016/B978-0-12-394438-2.00001-3.Chaudhry, W.N., Concepción-Acevedo, J., Park, T., Andleeb, S., Bull, J.J., and Levin, B.R. (2017). Synergy and order effects of antibiotics and phages in killing *Pseudomonas aeruginosa* biofilms. PLoS One. *12,* e0168615. doi:10.1371/journal.pone.0168615

d'Hérelle F. Bacteriophage as a treatment in acute medical and surgical infections. (1931). Bull N Y Acad Med. *7,* 329–348.

d'Hérelle F. (1931). An address on bacteriophagy and recovery from infectious diseases. Can Med Assoc J. *24(5)*, 619–628.

Dadgostar, P. (2019). Antimicrobial Resistance: Implications and Costs. Infect Drug Resist. *12*, 3903–3910. doi:10.2147/IDR.S234610

Dedrick, R.M., Guerrero-Bustamante, C.A., Garlena, R.A., Russell, D.A., Ford, K., Harris, K., Gilmour, K.C., Soothill, J., Jacobs-Sera, D., Schooley, R.T., Hatfull, G.F., and Spencer, H. (2019). Engineered bacteriophages for treatment of a patient with a disseminated drug-resistant *Mycobacterium abscessus*. Nat Med. *25,* 730–733. doi:10.1038/s41591-019-0437-z

DeCorte, P., Rose, T., Jennes, S., Zizi, M., De Vos, D., and Vaneechoutte, M. (2009). Quality-controlled small-scale production of a well-defined bacteriophage cocktail for use in human clinical trials. PLoS One, *4*, e4944. doi: 10.1371/journal.pone.0004944

Duplessis, C., Biswas, B., Hanisch, B., Perkins, M., Henry, M., Quinones, J., and Hamilton, T. (2018). Refractory Pseudomonas Bacteremia in a 2-year-old Sterilized By Bacteriophage Therapy. Journal of the Pediatric Infectious Diseases Society. *7,* 253–256. doi: 10.1093/jpids/pix056.

Dubos, R., Straus, J.H., and Pierce, C. (1943) The multiplication of bacteriophage in vivo and its protective effects against an experimental infection with *Shigella dysenteriae*. J Exp Med. *78*, 161-168. doi: 10.1084/jem.78.3.161

Expert round table on acceptance and re-implementation of bacteriophage therapy. (2016). Silk route to the acceptance and re-implementation of bacteriophage therapy. Biotechnol J. *11*, 595–600. doi:10.1002/biot.201600023

Fish, R., Kutter, E., Wheat, G., Blasdel, B., Kutateladze, M., and Kuhl, S. (2018). Compassionate use of bacteriophage therapy for foot ulcer treatment as an effective step for moving toward clinical trials. Methods Mol Biol. *1693,* 159–170. doi:10.1007/978-1-4939-7395-8_14

Fish, R., Kutter, E., Bryan, D., Wheat, G., and Kuhl, S. (2018). Resolving digital staphylococcal osteomyelitis using bacteriophage - A case report. Antibiotics (Basel). *7,* 87. doi:10.3390/antibiotics7040087

Friman, V.P., Soanes-Brown, D., Sierocinski, P., Molin, S., Johansen, H.K., Merabishvili, M., Pirnay, J.P., De Vos, D., and Buckling, A. (2016). Pre-adapting parasitic phages to a pathogen leads to increased pathogen clearance and lowered resistance evolution with *Pseudomonas aeruginosa* cystic fibrosis bacterial isolates. J Evol Biol. *29,* 188–198. doi:10.1111/jeb.12774

Galland L. (2014). The gut microbiome and the brain. J Med Food. *17*(12), 1261–1272. https://doi.org/10.1089/jmf.2014.7000

Górski, A., Międzybrodzki, R., Łobocka, M., Głowacka-Rutkowska, A., Bednarek, A., Borysowski, J., Jończyk-Matysiak, E., Łusiak-Szelachowska, M., Weber-Dąbrowska, B., Bagińska, N., Letkiewicz, S., Dąbrowska, K., and Scheres, J. (2018). Phage therapy: What have we learned? Viruses. *10*, 288. https://doi.org/10.3390/v10060288

Górski, A., Dąbrowska, K., Międzybrodzki, R., Weber-Dabrowska, B., Lusiak-Szelachowska, M., Jonczyk-Matysiak, E., and Borysowski, J. (2017). Phages and immunomodulation. Future Microbiol. *12,* 905–914. doi:10.2217/fmb-2017-0049

Górski, A., Jończyk-Matysiak, E., Międzybrodzki, R., Weber-Dąbrowska, B., and Borysowski, J. (2018). Phage transplantation in allotransplantation": Possible

Treatment in Graft-Versus-Host Disease? Front Immunol. *9,* 941. doi:10.3389/ fimmu.2018.00941

Górski, A., Jończyk-Matysiak, E., Łusiak-Szelachowska, M., Weber-Dąbrowska, B., Międzybrodzki, R., and Borysowski, J. (2018). Therapeutic potential of phages in autoimmune liver diseases. Clin Exp Immunol. *192,*1–6. doi: 10.1111/cei.13092

Górski, A., Jończyk-Matysiak, E., Łusiak-Szelachowska, M., Międzybrodzki, R., Weber-Dąbrowska, B., and Borysowski, J. (2018). Phage therapy in allergic disorders?. Exp Biol Med (Maywood). *243*, 534–537. doi: 10.1177/1535370218755658

Górski, A., Międzybrodzki, R., Borysowski, J., Dąbrowska, K., Wierzbicki, P., Ohams, M., Korczak-Kowalska, G., Olszowska-Zaremba, N., Łusiak-Szelachowska, M., Kłak, M., Jończyk, E., Kaniuga, E., Gołaś, A., Purchla, S., Weber-Dąbrowska, B., Letkiewicz, S., Fortuna, W., Szufnarowski, K., Pawełczyk, Z., Rogóż, P. and Kłosowska, D. (2012). Phage as a modulator of immune responses: practical implications for phage therapy. Adv. Virus Res. *83,* 41–71. doi: 10.1016/B978-0-12-394438-2.00002-5.

Harper, D.R., Anderson, J., and Enright, M.C. (2011). Phage therapy: delivering on the promise. Ther. Deliv. *2,* 935–947.

Henein, A. (2013) What are the limitations on the wider therapeutic use of phage. Bacteriophage. 3. DOI: 10.4161/bact.24872

Hraiech, S., Brégeon, F., and Rolain, JM. (2015). Bacteriophage-based therapy in cystic fibrosis-associated *Pseudomonas aeruginosa* infections: rationale and current status. Drug Des Devel Ther. *9,* 3653–3663. doi:10.2147/ DDDT.S53123

Hoyle, N., Zhvaniya, P., Balarjishvili, N., Bolkvadze, N., Nadareishvili, L., Nizharadze, L., Wittmann, J., Rohde, C., and Kutateladze, M. (2018). Phage therapy against *Achromobacter xylosoxidans* lung infection in a patient with cystic fibrosis: a case report. Res Microbiol. *169,* 540–542. doi:10.1016/ j.resmic.2018.05.001

Jennes, S., Merabishvili, M., Soentjens, P., Pang, K. W., Rose, T., Keerseblick, E., Pirnay, J. P. (2017). Use of bacteriophages in the treatment of colistin-only-sensitive *Pseudomonas aeruginosa* septicemia in a patient with acute kidney injury-a case report. Critical Care. *21*, 1–3. doi: 10.1186/s13054-017-1709-y

Knouf, E.G., Ward, W.E., Reichle, P.A., Bower, A.G. and Hamilton, P.M. (1946) Treatment of typhoid fever With type specific bacteriophage. J Am Med Assoc. *132,* 134-138.

Kutateladze, M. and Adamia, R. (2008). Phage therapy experience at the Eliava Institute. Med. Mal. Infect. *38*, 426–430. doi: 10.1016/j.medmal.2008.06.023.

Kutateladze, M., and Adamia, R. (2010). Bacteriophages as potential new therapeutics to replace or supplement antibiotics. Trends Biotechnol. *28*, 591–595. doi: 10.1016/j.tibtech.2010.08.001.

Kutter, E., Bryan, D., Ray, G., Brewster, E., Blasdel, B., & Guttman, B. (2018). From host to phage metabolism: Hot tales of phage T4's takeover of *E. coli. Viruses*, *10*(7), 387. https://doi.org/10.3390/v10070387

Kvachadze, L., Balarjishvili, N., Meskhi, T., Tevdoradze, E., Skhirtladze, N., Pataridze, T., Adamia, R., Topuria, T., Kutter, E., Rohde, C., and Kutateladze, M. (2011). Evaluation of lytic activity of staphylococcal bacteriophage Sb-1 against freshly isolated clinical pathogens. Microb Biotechnol. *4,* 643-50. doi: 10.1111/j.1751-7915.2011.00259.x.

Lang, L.H. (2006). FDA approves use of bacteriophages to be added to meat and poultry products. Gastroenterology. *131,*1370. doi:10.1053/j.gastro. 2006.10.012

LaVergne, S., Hamilton, T., Biswas, B., Kumaraswamy, M., Schooley, R.T., and Wooten. D. (2018). Phage therapy for a multidrug-resistant *Acinetobacter baumannii* craniectomy site infection. Open Forum Infect Dis. *5,* ofy064. doi: 10.1093/ofid/ofy064

Law, N., Logan, C., Yung, G., Furr, C.L., Lehman, S.M., Morales, S., Rosas, F., Gaidamaka, A., Bilinsky, I., Grint, P., Schooley, R.T., and Aslam, S. (2019). Successful adjunctive use of bacteriophage therapy for treatment of multidrug-resistant *Pseudomonas aeruginosa* infection in a cystic fibrosis patient. Infection. *47,* 665–668. doi:10.1007/s15010-019-01319-0

Leitner, L., Sybesma, W., Chanishvili, N., Goderzishvili, M., Chkhotua, A., Ujmajuridze, A., Schneider, M.P., Sartori, A., Mehnert, U., Bachmann, L.M., and Kessler, T.M. (2017). Bacteriophages for treating urinary tract infections in patients undergoing transurethral resection of the prostate: a randomized, placebo-controlled, double-blind clinical trial. BMC Urol. *17,* 90. doi:10.1186/ s12894-017-0283-6

Leung, SSY., Carrigy, N.B., Vehring, R., Finlay, W.H., Morales, S., Carter, E.A., Britton, W.J., Kutter, E., and Chan H. (2019). Jet nebulization of bacteriophages with different tail morphologies - Structural effects. Int J Pharm. *554,* 322–326. doi:10.1016/j.ijpharm.2018.11.026

Leung, SSY., Morales, S., Britton, W., Kutter, E., and Chan, H.K. (2018). Microfluidic-assisted bacteriophage encapsulation into liposomes. Int J Pharm. *545*, 176–182. doi:10.1016/j.ijpharm.2018.04.063

Lin, Y., Chang, R.Y.K., Britton, W.J., Morales, S., Kutter, E., and Chan, H.K. (2018). Synergy of nebulized phage PEV20 and ciprofloxacin combination against *Pseudomonas aeruginosa.* Int J Pharm. *551*, 158–165. doi:10.1016/j.ijpharm.2018.09.024

Malik, D.J., Sokolov, I.J., Vinner, G.K., Mancuso, F., Cinquerrui, S., Vladisavljevic, G.T., Clokie, M.R.J., Garton, N.J., Stapley, A.G.F., and Kirpichnikova, A. (2017). Formulation, stabilisation and encapsulation of bacteriophage for phage therapy. Advances in Colloid and Interface Science. *249,* 100-133. https://doi.org/10.1016/j.cis.2017.05.014.

McCallin, S., Sacher, J.C., Zheng, J., and Chan, B. K. (2019). Current State of Compassionate Phage Therapy. Viruses. *11,* 343. https://doi.org/10.3390/v11040343

Merabishvili, M., Pirnay, J.P., Verbeken, G., Chanishvili, N., Tediashvili, M., Lashkhi, N., Glonti, T., Krylov, V., Mast, J., Van Parys, L., Lavigne, R., Volckaert, G., Mattheus, W., Verween, G.,

Merabishvili, M. (2012). The phage therapy experience of the Brussels burn wound centre [abstract]. In EuroPhages 2012: Bacteriophage in Medicine, Food and Biotechnology; Conference Handbook, 21.

Międzybrodzki, R., Hoyle, N., Zhvaniya, F., Lusiak-Szelachowska, M., Weber-Dabrowska, B., Labochka, M., Borysowski, J., Alavidze, Z., Kutter, E., Gorski, A. and Gogokhia, A. (2018). Current updates from the longstanding phage research centres in Georgia, Poland and Russia. In: Harper D, Abedon S, Burrowes B, et al., editors. Bacteriophages. Switzerland: Springer, Cham. 1–31. ISBN 978-3-319-40598-8.

Miller, E.S., Kutter, E., Mosig, G., Arisaka, F., Kunisawa, T., and Rüger, W. (2003). Bacteriophage T4 genome. Microbiol. Mol. Biol. Rev. *67,* 86–156.

Morozova, V. V., Kozlova, Y. N., Ganichev, D. A., & Tikunova, N. V. (2018). Bacteriophage treatment of infected diabetic foot ulcers. Methods Mol (Clifton, N.J.). 1693, 151–158. https://doi.org/10.1007/978-1-4939-7395-8_13

Oechslin, F., Piccardi, P., Mancini, S., Gabard, J., Moreillon, P., Entenza, J.M., Resch, G., and Yok-Ai, Q. (2017). Synergistic interaction between phage therapy and antibiotics clears *Pseudomonas aeruginosa* infection in endocarditis and reduces Virulence. J Infect Dis. *215,* 703–712. doi:10.1093/infdis/jiw632

Onsea, J., Soentjens, P., Djebara, S., Merabishvili, M., Depypere, M., Spriet, I., De Munter, P., Debaveye, Y., Nijs, S., Vanderschot, P., Wagemans, J., Pirnay, J.P., Lavigne, R., and Metsemakers, W.J. (2019). Bacteriophage application for difficult-to-treat musculoskeletal infections: Development of a Standardized Multidisciplinary Treatment Protocol. Viruses. *11,* 891. doi:10.3390/v11100891

Patey, O., McCallin, S., Mazure, H., Liddle, M., Smithyman, A., and Dublanchet, A. (2018). Clinical indications and compassionate use of phage therapy: personal experience and literature review with a focus on osteoarticular Infections. Viruses. *11,*18. doi:10.3390/v11010018

Petrovic, F. A., Lin, R.C.Y., Ho, J., Maddocks, S., Nouri, L., Zakour, B., Iredell, R.J., and Westmead Bacteriophage Therapy team (2020). Safety of bacteriophage therapy in severe *Staphylococcus aureus* infection [published correction appears in Nat Microbiol. *5,* 465–472. doi:10.1038/s41564-019-0634-z

Pirnay, J.P., Verbeken, G., Ceyssens, P.J., Huys, I., De Vos, D., Ameloot, C., and Fauconnier, A. (2018). The magistral phage. Viruses. *10,* 64. doi:10.3390/v10020064

Prompers, L., Huijberts, M., Apelqvist, J., Jude, E., Piaggesi, A., Bakker, K., Edmonds, M., Holstein, P., Jirkovska, A., Mauricio, D., Ragnarson, T.G., Reike, H., Spraul M., Uccioli, L., Urbanic, V., Van Acker, K., van Baal, J., van Merode, F., and Schaper, N. (2007). High prevalence of ischaemia, infection and serious comorbidity in patients with diabetic foot disease in Europe. Baseline results from the Eurodiale study. Diabetologia. *50,*18–25. doi:10.1007/s00125-006-0491-1

Rohde, C., Resch, G., Pirnay, J.P., Blasdel, B.G., Debarbieux, L., Gelman, D., Gorski, A., Hazan, R., Huys, I., Kakabadze, E., Lobocka, M., Maestri, A.,

Almedia, GMF., Makalatia, K., Malik, D.J., Maslanova, I., Merabishvili, M., Pantucek, R., Rose, T., Stverakova, D., Van Raemdonck, H., Verbeken, G., and Chanisvhili, N. (2018). Expert opinion on three phage therapy related topics: bacterial phage resistance, phage training and prophages in bacterial production strains. Viruses. *10,* 178. doi:10.3390/v10040178

Rhoads, D.D., Wolcott, R.D., Kuskowski, M.A., Wolcott, B.M., Ward, L.S., and Sulakvelidze, A. (2009). Bacteriophage therapy of venous leg ulcers in humans: results of a phase I safety trial. J. Wound Care. *18*, 237–238, 240–243.

Rossitto, M., Fiscarelli, E.V., and Rosati, P. (2018). Challenges and promises for planning future clinical research into bacteriophage therapy against *Pseudomonas aeruginosa* in cystic fibrosis. An argumentative review. Front Microbiol. *9,* 775. doi:10.3389/fmicb.2018.00775

Sarker, S.A., Sultana, S., Reuteler, G., Moine, D., Descombes, P., Charton, F., Bourdin, G., McCallin, S., Ngom-Bru, C., Neville, T., Akter, M., Huq, S., Qadir, F., Talukdar, K., Kassam, M., Delley, M., Loiseau, C., Deng, Y., El Aidy, S., Berger, B., and Brussow, H. (2016). Oral phage therapy of acute bacterial diarrhea with two coliphage preparations: A randomized trial in children from Bangladesh. EBioMedicine. *4,*124–137. doi:10.1016/j.ebiom.2015.12.023

Sarker, S.A., and Brüssow, H. (2016). From bench to bed and back again: phage therapy of childhood *Escherichia coli* diarrhea. Ann N Y Acad Sci. *1372,* 42–52. doi:10.1111/nyas.13087

Schooley, R.T., Biswas, B., Gill, J.J., Hernandez-Morales, A., Lancaster, J., Lessor, L., Barr, J.J., Reed, S.L., Rohwer, F., Benler, S., Segall, A.M., Taplitz, R., Smith, D.M., Kerr, K., Kumaraswamy, M., Nizet, V., Lin, L., McCauley, M.D., Strathdee, S.A., Benson, C.A., Pope, R.K., Leroux, B.M., Picel, A.C., Mateczun, A.J., Cliwa K.E., Regeimbal, J.M., Estrella, L.A., Wolfe, D.M., Henry, M.S., Quinones, J., Salka, S., Bishop-Lilly, K.A., Young, R., and Hamilton, T. (2017). Development and use of personalized bacteriophage-based therapeutic cocktails to treat a patient with a disseminated resistant *Acinetobacter baumannii* Infection [published correction appears in Antimicrob Agents Chemother. 2018 Nov 26;62(12):]. Antimicrob Agents Chemother. *61,* e00954-17. doi:10.1128/AAC.00954-17

Strathdee, S., and Patterson, T. ( 2019). The perfect predator - A scientist's race to save her husband from a deadly superbug. New York, NY: Hachette Books.

Summers, W.C. (2001). Bacteriophage therapy. Annu. Rev. Microbiol. *55*, 437–451.

Sulakvelidze, A., Alavidze, Z., and Morris, J.G. Jr. (2001). Bacteriophage therapy. Antimicrob. Agents Chemother. *45*, 649–659.

Sulakvelidze, A., and Kutter, E. (2005). Bacteriophage therapy in humans. In Bacteriophages: Biology and Application, E. Kutter and A. Sulakvelidze, eds. (Boca Raton, Florida: CRC Press), 381–436.

Taati Moghadam, M., Amirmozafari, N., Shariati, A., Hallajzadeh, M., Mirkalantari, S., Khoshbayan, A., and Masjedian Jazi, F. (2020). How phages overcome the challenges of drug resistant bacteria in clinical infections. Infection and drug resistance. *13*, 45–61. https://doi.org/10.2147/IDR.S234353

Villarroel, J., Larsen, M.V., Kilstrup, M., and Nielsen, M. (2017). Metagenomic analysis of therapeutic PYO phage cocktails from 1997 to 2014. Viruses. *9,* 328. doi:10.3390/v9110328

Walsh, J.W., Hoffstad, O.J., Sullivan, M.O., and Margolis, D.J. (2016). Association of diabetic foot ulcer and death in a population-based cohort from the United Kingdom. Diabet Med. *33,* 1493-1498.

Weiss, M., Denou, E., Bruttin, A., Serra-Moreno, R., Dillmann, M. L., & Brüssow, H. (2009). In vivo replication of T4 and T7 bacteriophages in germ-free mice colonized with *Escherichia coli. Virology*, *393*(1), 16–23. https://doi.org/10.1016/j.virol.2009.07.020

Wright, A., Hawkins, C.H., Anggård, E.E., and Harper, D.R. (2009). A controlled clinical trial of a therapeutic bacteriophage preparation in chronic otitis due to antibiotic-resistant *Pseudomonas aeruginosa*; a preliminary report of efficacy. Clin. Otolaryngol. *34*, 349–357.

Zschach, H., Joensen, K.G., Lindhard, B., Lund, O., Goderzishvili, M., Chkonia, I., Jgenti, G., Kvatadze, N., Alavidze, Z., Kutter, E.M., Hasman, H., and Larsen, M.V. (2015). What can we learn from a metagenomic analysis of a Georgian bacteriophage cocktail? Viruses. *7,* 6570–6589. doi:10.3390/v7122958

## Chapter 2

# Phage Therapy: The Pharmacology of Antibacterial Viruses

**Katarzyna Danis-Wlodarczyk[1], Krystyna Dąbrowska[2], Stephen T. Abedon[1*]**

[1] Department of Microbiology, The Ohio State University, Columbus, OH, USA
[2] Institute of Immunology and Experimental Therapy, Polish Academy of Sciences, Wrocław, Poland

*Corresponding author; email: abedon.1@osu.edu

DOI: https://doi.org/10.21775/9781913652517.02

**Abstract**

Pharmacology can be differentiated into two key aspects, pharmacodynamics and pharmacokinetics. Pharmacodynamics describes a drug's impact on the body while pharmacokinetics describes the body's impact on a drug. Another way of understanding these terms is that pharmacodynamics is a description of both the positive and negative consequences of drugs attaining certain concentrations in the body while pharmacokinetics is concerned with our ability to reach and then sustain those concentrations. Unlike the drugs for which these concepts were developed, including antibiotics, the bacteriophages (or 'phages') that we consider here are not chemotherapeutics but instead are the viruses of bacteria. Here we review the pharmacology of these viruses, particularly as they can be employed to combat bacterial infections (phage therapy). Overall, an improved pharmacological understanding of phage therapy should allow for more informed development of phages as antibacterial 'drugs', allow for more rational *post hoc* debugging of phage therapy experiments, and encourage improved design of phage therapy protocols. Contrasting with antibiotics, however, phages as viruses impact individual bacterial cells as single virions rather than as swarms of molecules, and while they are killing bacteria, bacteriophages also can amplify phage numbers, *in situ*. Explorations of phage therapy pharmacology consequently can often be informed as well by basic principles of the ecological interactions between phages and bacteria as by study of the pharmacology of drugs. Bacteriophages in phage therapy thus can display somewhat unique as well as more traditional pharmacological aspects.

## Introduction

The words 'drug' and 'chemotherapeutic' often can be used interchangeably. Traditionally this has been especially so when drugs were employed to combat roguely replicating entities such as cancer cells, pathogenic bacteria, and viruses. More generally, a drug can be viewed as a biologically active but non-food substance that provides some benefit to the body. What medical drugs and chemotherapeutics have in common is that both supply therapeutic value, while historically only chemotherapeutics strictly have been 'chemicals'. Drugs, especially defined more generally as medicaments, thus existed prior to the medical use of purified organic chemicals – such as Ehrlich's anti-syphilis chemotherapeutic, 'Salvarsan' (Lloyd et al., 2005; Zaffiri et al., 2012) – and the word 'drug' itself seems to have originated as a reference in French to medicaments sourced from nature, i.e., to dried herbs (Harper, 2020). Here we similarly define 'drug' broadly to include not only purified small molecules such as antibiotics but also less pure or more complex entities such as antibacterial viruses.

Bacteriophages, or phages, are the viruses of bacteria (Calendar and Abedon, 2006; Hyman and Abedon, 2012; Lehman, 2018). As phages can be biologically active within bodies, they also can be considered to be drug-like, i.e., as medicaments. This drug-like activity stems predominantly, though not exclusively, from phages infecting and subsequently killing and lysing body-associated bacteria, and the application of phages to bodies to combat bacterial infections is called phage therapy. With phages serving as drug-like entities, phage therapy therefore can be framed in terms of the science of pharmacology, which is the study of drug-body interactions. Phages as replicating entities, however, also are ecological in their interactions with bacteria (Abedon, 2014b; 2018a). Here therefore we review basic principles of pharmacology as applicable to phage antibacterial therapy but with emphasis as well on relevant aspects of phage-bacterial ecological interactions.

## Pharmacology

Pharmacology is the study of drug functionality. This includes issues of drug design or discovery as well as development of drug delivery strategies that are effective, convenient, minimally invasive, and relatively inexpensive (Goth, 1984; Mycek et al., 2000). As a basic premise, a drug typically will display greater functionality *in vivo* when its concentrations are higher, but at the same time also will display greater negative effects. The general goal then is to sustain drugs at concentrations *in vivo* that result in substantial positive effects while choosing drugs whose negative effects at those same concentrations are minor. In terms of drug choice, pharmacological considerations therefore can be greatly simplified if a drug displays both low toxicities and a relative lack of other side effects at otherwise efficacious *in situ* concentrations. A drug that has a toxic concentration that is substantially higher than its effective concentration in particular can be taken at higher doses and/or with longer intervals between doses than a drug that does not have this property. A drug for which this so-called therapeutic *window* or *index* is smaller, by contrast, may require constant patient monitoring, and particularly so if a drug's side effects are severe (Blix et al., 2010; Pereira and Kelley, 2011; Muller and Milton, 2012).

The science of pharmacology traditionally has been differentiated into two major components, pharmacodynamics and pharmacokinetics. Primary pharmacodynamic effects are the positive impacts of a drug, i.e., efficacy, such as the bactericidal activity of an antibacterial agent. Secondary pharmacodynamic effects instead include negative consequences, i.e., side effects or toxicities (Lees et al., 2004; Marino and Zito, 2020). These side effects often can be emergent properties, however (Curtright and Abedon, 2011). That is, because of the complex interactions between drugs and bodies, it can be difficult to predict the occurrence of drug side effects based solely on drug chemistry or even pre-clinical analysis. Drug development thus typically will necessitate extensive clinical safety determinations before such secondary pharmacodynamic effects can be either identified or ruled out. A great deal of pharmacological research consequently is devoted to a combination of identifying agents that can potentially deliver primary pharmacodynamic effects and then testing those agents for both efficacy and side effects.

Pharmacokinetics, contrasting pharmacodynamics, is the study of the reaching and sustaining of effective drug concentrations at sites of drug activity (Levison and Levison, 2009). Aspects of pharmacokinetics traditionally have been differentiated into four components, known as absorption, distribution, metabolism, and excretion. Particularly given systemic drug delivery, as opposed to topical application, the 'life cycle' of a drug is one of uptake into the systemic circulation of blood (*absorption*, meaning for the drug to become part of the body), movement out of the blood and into the tissues where it can then have an effect (*distribution*), reductions in blood densities as a consequence of *metabolism* (chemical breaking down of the drug, often by the liver), and *excretion* of otherwise intact drug, typically as effected by the kidneys. Absorption and distribution together describe especially the spatial delivery of a drug to sites in the body where it will display its primary pharmacodynamic effects. A summary of these concepts is provided in Figure 1.

Pharmacological development should be as applicable to phages, as phage-therapy effecting medicaments, as to drugs generally. At the same time, however, a number of characteristics of phages as antibacterial agents seem to have conspired to result in traditionally less formal pharmacological development for phages than has been the case for typical antibiotics. The reasons for this often-reduced phage therapy pharmacological emphasis include simply the long history of phage therapy, the typically low toxicity of phages as therapeutics, and likely as well the potential for phages to increase their numbers *in situ* in the course of effecting their antibacterial activities (Dąbrowska et al., 2018; Dąbrowska and Abedon, 2019). Notwithstanding these useful phage characteristics, approaching phage therapy development from a pharmacological perspective, as this chapter emphasizes, can be helpful toward improving upon phage therapy successes.

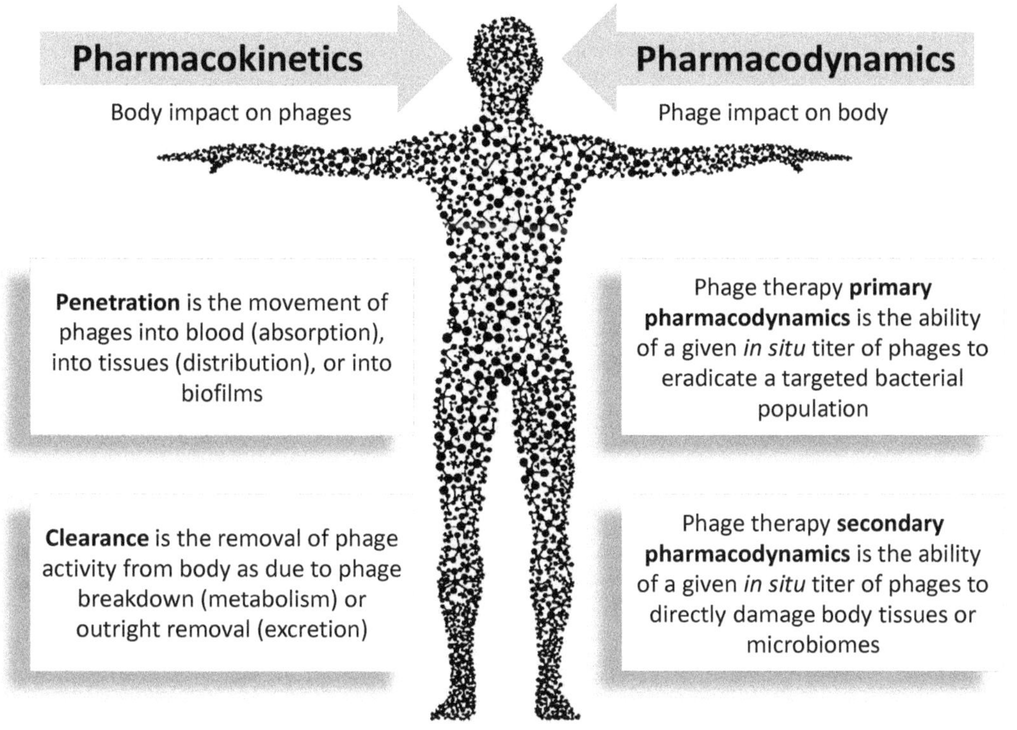


**Figure 1.** Phage therapy pharmacology. Pharmacokinetics traditionally consists of what are described as absorption, distribution, metabolism, and excretion. These are all things that contribute to drug *in situ* concentrations and are all consequences of a body's impact on a drug (with the concept of 'body' traditionally also including associated microorganisms). For phage therapy, we can also add the ability of phages to replicate, also *in situ*. Pharmacodynamics represents the impacts of a drug on the body, especially as a function of drug *in situ* concentration, and can be distinguished into primary and secondary effects. Primary effects are intended drug effects, such as phage bactericidal activity, while secondary effects are unintended and most notably include toxicities and side effects.

For discussion of advantages of employing phages as antibacterial agents, particularly in comparison to antibiotics, see (Sulakvelidze et al., 2001; Sulakvelidze and Kutter, 2005; Häusler, 2006; Kutter, 2008; Curtright and Abedon, 2011; Loc-Carrillo and Abedon, 2011; Nobrega et al., 2015; Wienhold et al., 2019). For review of the specifics of phage therapy pharmacokinetics, see (Dąbrowska, 2019; Dąbrowska and Abedon, 2019; Matsuzaki and Uchiyama, 2019). For an earlier version of this chapter, presenting additional topics as well as more mathematical detail, see Abedon (2014a) and for access especially to the older phage therapy literature, see Alves and Abedon (2017). Table 1 provides a glossary of terms.

**Table 1.** Concepts and terms relevant to phage therapy pharmacology.

| Term | Definition |
|---|---|
| Abortive infection | Phage infections that result in death of both the infected bacterium and the infecting phages, i.e., these are bactericidal infections that do not result in virion production; phage abortive infections are often also bacteriolytic but in principle need not be |
| Absorption | Movement of drug into the blood following dosing; not *ad*sorption |
| Active penetration | Layer-by-layer phage clearance of bacterial biofilms as mediated by active phage replication, or at least as associated with phage-mediated bacterial lysis |
| Active treatment | Clearance of bacterial populations *in situ* that is dependent upon auto dosing |
| Adsorption | Strictly the attachment of a phage to a bacterium but also typically describes as well the local movement of phages to the vicinity of targeted bacteria |
| Auto dosing | Drug dosing that is *not* a consequence of extrinsic application; phage *in situ* increases in titres that are a consequence of phage replication; a.k.a., self amplification |
| Bactericidal infection | Phage infection that results in bacterial death, and which also may result in bacteriolysis and virion production but doesn't necessarily |
| Bacteriolytic infection | Phage infection that results in bacterial death and phage-induced bacterial lysis though does not necessarily always also result in virion production, e.g., see also abortive infection |
| Bacteriophage translocation | Defined after the concept of bacterial translocation, describes the potential for phages to reach systemic circulation particularly starting from the gut |
| Body | Body tissues as well as associated microflora |
| Decimal reduction time (D-value; *D*) | The length of time required to reduce viability of target microorganisms by 90% as a result of antimicrobial application of a given type, density, and under a given set of conditions |
| Distribution | Drug movement out of systemic circulation and into other body tissues |
| Dosing | Mechanism initiating the elevation of drug concentrations *in situ*; here "application" and "delivery", both as resulting in dosing, are used synonymously |
| Efficacy (1) | Positive outcome of an antibacterial procedure, e.g., such as reductions in numbers of targeted bacteria; this is the definition of efficacy used here, and this is rather than definition (2)" |

| | |
|---|---|
| Efficacy (2) | Inherent ability of a drug to produce an effect following binding to its specific target, which for phages would be bacteria killing following adsorption; this definition of efficacy is *not* employed in this chapter other than in this table entry |
| Excretion | Drug removal from the body as typically mediated by kidneys |
| Free phage | Phage virions that are mature, functional, and found outside of bacteria, i.e., that state that phages are in when they are able to adsorb bacteria |
| Inundation threshold | Phage density required to exactly balance bacterial replication with bacterial deaths, the latter including as mediated by phages |
| Inundative densities | *In situ* phage titres sufficient to achieve adequate bacteria killing over reasonable time frames, though with 'adequate' and 'reasonable' not explicitly defined |
| Killing titre | Measure of phage density based on bacteria killing ability; assumes adsorption by approximately 100% of the adsorption-competent phage population that is present in the vicinity of a targeted bacterial population |
| Lytic phage | A phage that, at the end of its infection period, lyses its bacterial host, resulting in the death of its host bacterium and loss of continued infection productivity; as most temperate phages are also lytic phages, the concept of lytic phage contrasts instead with that of chronically (continuously) released phages, e.g., such as the filamentous phage M13 |
| Metabolism | Chemical modification of drugs typically resulting in drug inactivation though in the case of phages can instead result in both activation and auto dosing; inactivation for phages can include as due to immune system action |
| Minimum effective concentration | That drug density, as measured at the site of drug action, that achieves desired levels and rates of bacteria killing |
| Minimum toxic concentration | That drug density as measured *in situ* at which toxicity is apparent |
| Multiplicity of infection (MOI) | Strictly the ratio of *adsorbed* or infecting phages to target bacteria; too often used incorrectly as synonymous with the ratio of *added* phages to target bacteria; by either measure, MOI should never be employed as the sole dosing descriptor during phage therapy |
| $MOI_{actual}$ | Ratio of adsorbed phages to targeted bacteria; the historical meaning of MOI |

| | |
|---|---|
| $MOI_{input}$ | Ratio of added phages to target bacteria; under conditions of higher bacterial densities $MOI_{input}$ can approximate $MOI_{actual}$, but this is not the case at lower bacterial densities, or given slower rates of phage virion adsorption |
| Null | Phage-bacterial interactions that do not result in bactericidal activity |
| Passive treatment | Clearance of bacterial populations *in situ* that occurs without dependence on auto dosing |
| Penetration | Phage movement to the vicinity of bacteria including as mediated by bacterial lysis |
| Pharmacodynamics | Drug impact on bodies |
| Pharmacokinetics | Body impact on drugs; these are the factors impacting the attainment of drug densities necessary to effect primary pharmacodynamic effects |
| Pharmacologically emergent property | Especially toxicities and side effects associated with a drug that only become apparent upon clinical testing; pharmacologically emergent properties are a bane of drug development because by definition they only become obvious somewhat late in the drug development process; phages for phage therapy, unlike small-molecule drugs, tend to not display pharmacologically emergent properties |
| Primary pharmacodynamic effects | The positive impact of drugs on bodies, ideally resulting in efficacy (1) |
| Professionally lytic | Description of a phage that is both strictly lytic and which is not recently descended from a temperate phage; a virulent mutant of phage λ, for example, may be strictly lytic but would not be professionally lytic; to the extent that the encoding of bacterial virulence factor genes is associated only with temperate phages, then professionally lytic phages are not expected to encode bacterial virulence factor genes |
| Restrictive infection | Phage infections that result in death of the infecting phage but not in death of infected bacterium; these are infections are null rather than bactericidal, bacteriolytic, or virion productive |
| Secondary pharmacodynamic effects | Other than efficacy (2) effects, including the negative impact of drugs on bodies, i.e., toxicity, side effects, or destruction of normal flora bacteria |
| Strictly lytic | Description of a phage that both releases phage progeny via lysis and is unable to display lysogenic cycles; strictly lytic phages are not temperate, and can also be described, e.g., as obligately lytic. |

| | |
|---|---|
| Temperate phage | Phage that is able to initiate non-defective, ongoing, and metabolically active infections that produce no virion progeny but instead generate bacterial lysogens; temperate phages can carry bacterial virulence factor genes and bacterial lysogens typically display an acquired resistance to the associated temperate phage (or phages); note that the term "lysogenic phage" is a misnomer and should never be used |
| Titre | Phage density, usually as provided in per mL units; titres are the preferred phage dosing unit of measurement, with dosing ideally reported in terms of both volumes applied and titres of phages found within those volumes |
| Trough concentration | Minimum drug concentration experienced between dosing |
| Virion production | The maturation and release of progeny phage virions, which for lytic phages are products of bacteriolytic infections |

## Primary Pharmacodynamics

A drug's most salient feature is its primary pharmacodynamics, i.e., its positive effects. Positive effects will not come about, however, until sufficient drug concentrations are present in target tissues, which in the case of antibacterials are those tissues associated with targeted bacteria. With phages, these sufficient densities can be achieved by either of two approaches (Payne and Jansen, 2003; Cairns et al., 2009; Abedon and Thomas-Abedon, 2010; Letarov et al., 2010). These are (i) as following phage application to bodies (so-called passive treatment) or (ii) via a combination of phage application and *in situ* amplification of phage numbers (so-called active treatment). In this section, we discuss the basics of phage dosing in terms of impact on bacteria such as toward microbiologically positive outcomes. For extensive mathematical treatment of this issue, see Abedon (2011a). For less mathematically intensive explorations of these concepts, see Abedon (2012a; 2018b). Different routes of phage application are considered further below.

Another important consideration of phage therapy primary pharmacodynamics is that of phage host range and associated bacterial resistance (Dąbrowska and Abedon, 2019), though this issue is not addressed here. An additional consideration, also not further developed here, is the potential for phages to display synergistic, additive, or antagonistic interactions (Chaudhry et al., 2017) with other phages, or instead with other substances that phages may be delivered in association with, i.e., such as antibiotics (Chanishvili et al., 2001; Weber-Dąbrowska et al., 2003; Kutter, 2008; Chanishvili et al., 2009; Cooper et al., 2018; Torres-Barceló and Hochberg, 2016; Abedon, 2019a; Knezevic and Sabo, 2019; Segall et al., 2019; Tagliaferri et al., 2019). Commercially as well as practically, it is also highly relevant that bacterial resistance to antibiotics is not known to result generally in cross-resistance to phages, e.g., (Sulakvelidze, 2005; Międzybrodzki et al., 2012).

### *Phage therapy primary pharmacodynamic impacts*

The primary pharmacodynamic impact of phages, with only few exceptions, occurs during phage infection of targeted bacteria. Exceptions include where phages are employed more as drug delivery vehicles, e.g., (Yacoby and Benhar, 2008; Clark et al., 2012; Peng et al., 2020), in which case only phage adsorption but not infection is required. For the sake of avoiding ambiguity, we therefore note that the words 'adsorption' and 'infection' for phages should generally not be used interchangeably (Abedon, 2015a). Phage infection of bacteria thus begins with phage adsorption, but this is then followed by phage-genome translocation into the bacterial cytoplasm, and only then does phage infection proper start. Pharmacologically, four often not mutually exclusive outcomes therefore can result from phage application. These we described as null, bactericidal, bacteriolytic, and virion productive (Abedon, 2019a) (Figures 2 and 3), with the fourth category, especially, a pharmacokinetic rather than pharmacodynamic outcome.

#### Null phage impacts on targeted bacteria

Null, at its simplest, is a phage's inability to adsorb a targeted bacterium. Also in this category, however, are restrictive phage infections, defined here as infections that result in phage inactivation but not in bacterial death (Hyman and Abedon, 2010). Phage restriction occurs such as due to the action of bacterial restriction endonucleases (Korona et al., 1993) or instead CRISPR-Cas systems (Kumar et al., 2015; Strotskaya et al., 2017). Another sort of phage restriction is seen with secondary adsorption, which is the adsorption by phages to already phage-infected bacteria (Abedon, 2015a) and can result in superinfection immunity or superinfection exclusion (Abedon, 1994; Berngruber et al., 2010; Blasdel and Abedon, 2017; Domingo-Calap et al., 2020). Both are mechanisms that kill adsorbing phages and, if expressed by bacterial lysogens, are associated as well with survival of the phage-adsorbed bacterium. Alternatively is simply the initiation of lysogenic or pseudolysogenic infection cycles by infecting phages, i.e., reductive phage infections (Abedon, 2020b), as these at least in the near term also don't result in bacteria killing. Null phage infections generally are of little utility in terms of phage therapy primary pharmacodynamics.

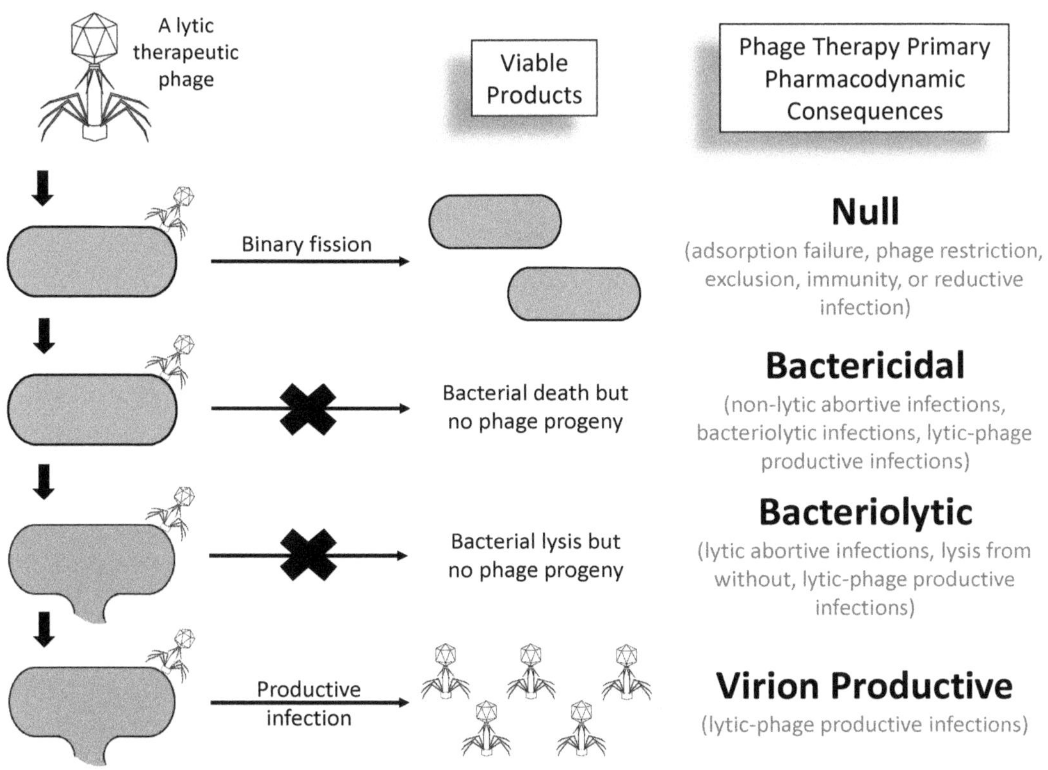


**Figure 2.** Categories of primary phage therapy pharmacodynamic effects. These are null, bactericidal, and bacteriolytic, which represent increasing phage impacts observed in the course of treatment of targeted bacterial populations. Virion production is shown as well, though this is a pharmacokinetic rather than pharmacodynamic effect.

Bactericidal phage impacts

Phage bactericidal activity results, of course, in bacterial death. This death may not always be accompanied by bacteriolytic activities, however, such as should abortive infection mechanisms interfere early with phage gene expression. More typically, though, phage bactericidal activity during phage therapy is associated with phage bacteriolytic activity, and also with the release of new phage virions. Notwithstanding these additional phage infection activities, phage bactericidal activities should be viewed as the minimum impact that phages must have on targeted bacteria to achieve primary phage therapy pharmacodynamic results, e.g., as associated with passive treatments.

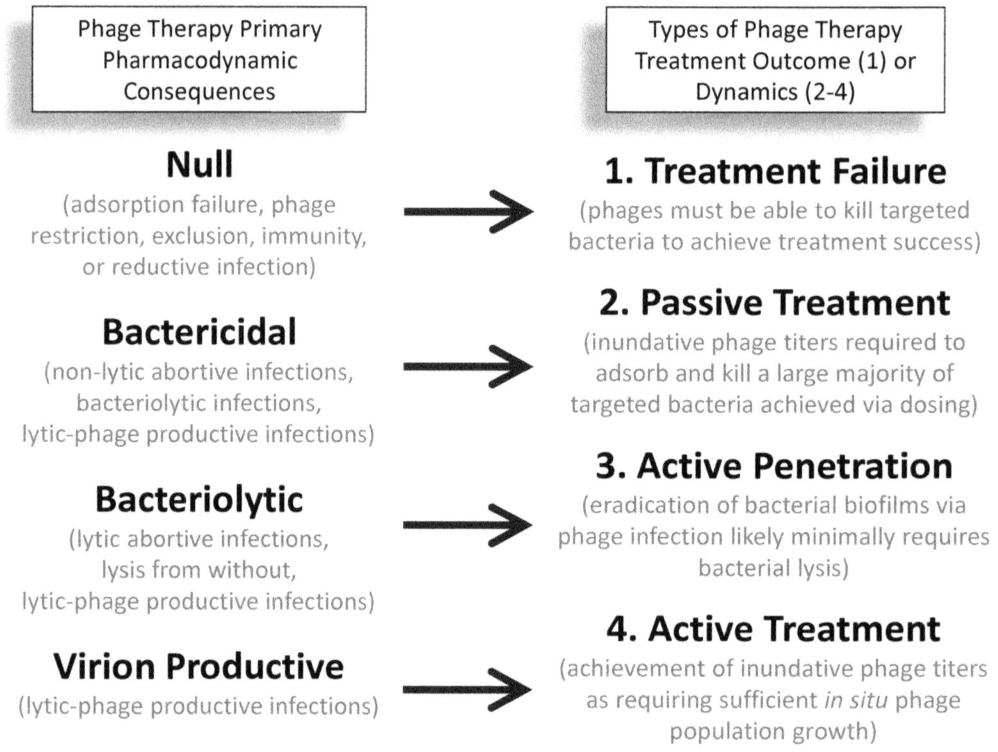


**Figure 3**. Various categories of phage treatment approaches or consequences.

### Bacteriolysis

Phage bacteriolytic activity under all circumstances will be associated as well with phage bactericidal activity against the phage-infected bacterium. The converse is not always true, however, as bacterial death can occur without accompanying bacterial lysis, e.g., such as when phages are engineered to kill but not lyse bacteria (Goodridge, 2010). Of course as highly familiar to those with interest in phage biology, bacteriolytic activity also often is associated as well with release of new phage virions from phage-infected bacteria. If lysis occurs but new virions are not released, then we again can use the term, abortive infection. Contrasting the more typical 'lysis from within' (Young, 1992), 'lysis from without' is also a phage bacteriolytic activity, in this case one which can be induced by high multiplicity phage adsorptions, but that also, by definition, is not associated with new phage virion production nor even necessarily with phage infection (Abedon, 2011d). In either case this lysis is a pharmacodynamic effect as it results in conversion of morphologically intact bacteria into no longer intact bacteria.

We can speculate that bacteriolysis may be especially important toward phage-mediated removal of bacterial biofilms (Abedon and Thomas-Abedon, 2010). This may be because lysed bacteria should be less susceptible to secondary phage adsorption than still-intact bacteria, thereby potentially allowing accelerated diffusion of phages to underlying layers of biofilm bacteria. Phage adsorption even to not-yet-phage-infected biofilm bacteria will likely also delay phage virion penetration into those same biofilms (Abedon, 2017d). We are not aware, however, of proof that bacteriolytic activity is essential for phage-mediated biofilm eradication, such as could possibly occur following high multiplicity application of engineered lysis-defective phages. Nevertheless, phages that retain bactericidal but not bacteriolytic activities possibly may be less suited to treatment of bacterial biofilms than phages retaining both. To the extent that the lysis of bacteria might aid in phage movement, such as penetration into biofilms, then this lysis can be viewed as contributing to phage therapy pharmacokinetics as well as pharmacodynamics.

Virion production

For lytic phages, virion release from phage-infected bacteria is always associated with bacteriolytic and thereby also bactericidal activities. Phage virion production thus can be used as a surrogate indicator of bactericidal and bacteriolytic phage activities, e.g., as may be assessed via one-step growth experiments (Hyman and Abedon, 2009; Kropinski, 2018). Notwithstanding these associations with bacteria killing and lysis, new virion production as occurring *in situ* should be viewed as a pharmacokinetic rather than a pharmacodynamic phage property, as it does not directly impact bacteria though does directly impact phage numbers found in the body. For further discussion of virion production as a pharmacokinetic concept, see “Metabolism and active treatment” as found under the heading of “Pharmacokinetics”, below.

*Using titres to describe phage potential to kill bacteria*

Most drugs are dosed in a manner that is intended to avoid exceeding toxic concentrations while at the same time, reaching or exceeding effective densities (Figure 4). For phages, the development of dosing strategies can be greatly simplified due to the low toxicity that is a typical phage characteristic (Abedon and Thomas-Abedon, 2010; Curtright and Abedon, 2011; Olszowska-Zaremba et al., 2012; Górski et al., 2018a; 2018b), meaning that efforts toward preventing phages from reaching relatively high concentrations *in situ* can be unnecessary. To a substantial extent, therefore, and other than in terms of purifying phages away from bacterial lysis products prior to use, emphasis during the design of many phage therapy protocols can be placed on enhancing antibacterial activities rather than on avoiding the induction of side effects. Specifically, to be an effective antibacterial agent, a phage must reach sufficient densities (titres) such that the rate of phage adsorption to bacteria is faster than the rate of bacterial replication and, ideally, quite a bit faster (see, that is, Inundation thresholds…, below). Bacteria, as a consequence of exposure to these sufficient phage densities, are killed, and this killing preferably occurs over relatively short intervals.

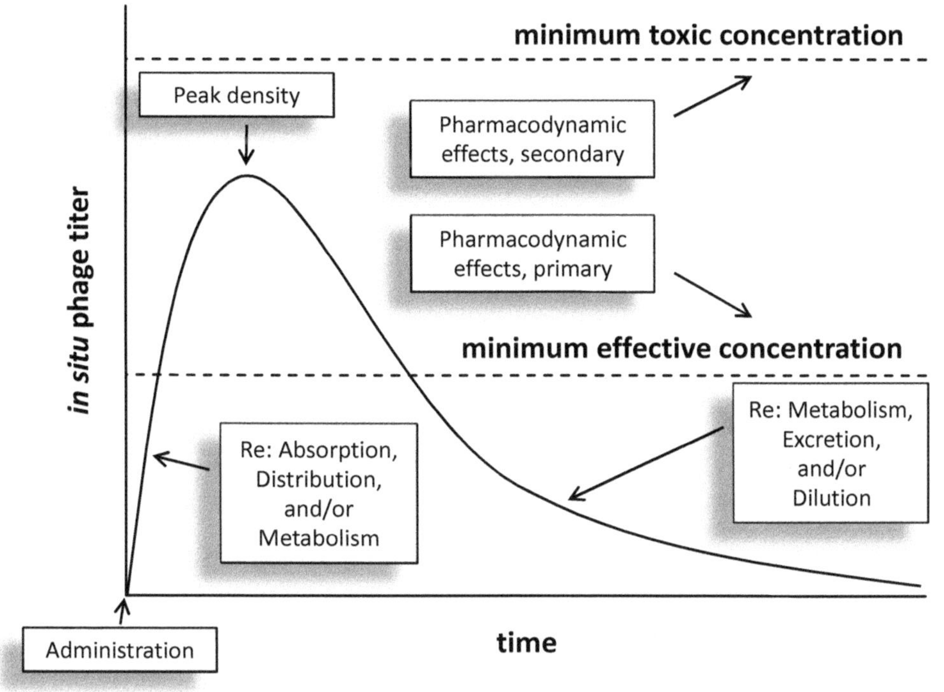


**Figure 4.** A representative drug time course highlighting pharmacodynamic issues (as explicitly indicated) and pharmacokinetic issues (absorption, distribution, metabolism, and excretion). Metabolism for phages, as indicated to the left (as resulting in 'activation'), is phage replication and is relevant particularly given active treatment. Metabolism, as indicated to the right (as resulting in 'inactivation'), is immune-system sequestration or inactivation of phage particles. Minimum effective concentrations and toxic concentrations refer to bacteria killing efficacy and side effects, respectively.

Most drugs also are dosed in a manner that is based on quantities of drug applied. This can be certain concentrations in combination with certain volumes (such as specific millilitres of milligrams per millilitre) or instead just absolute numbers, such as milligrams, though either often is dosed as a function of body mass as well. For phages we typically describe measures of phage concentration as titres, e.g., virions/mL, or operationally as plaque-forming units/mL. What is crucial, however, is the resulting phage titres as found *in situ*. For local phage delivery (Morozova et al., 2018b), such as the application of excess volumes of phage-containing buffer to the surface of wounds, just keeping track of what phage titres have been applied can be sufficient. Especially this can involve, for example, the soaking of gauze bandages or the application of phage-containing gels. For systemic delivery, however, *in situ* titres would be a function of numbers of phages applied but also the mass (weight) of the body that has been dosed. At a minimum and regardless, phage titres (and volumes) or instead absolute phage numbers as associated with dosing should always be reported as accurately and unambiguously as possible (Abedon, 2017c).

### $MOI_{actual}$

Assuming that there is a good bactericidal match between applied phages and target bacteria, then the extent of phage adsorption to individual bacteria should be the primary determinant of degrees of bacteria killing (Abedon, 2016). We can describe the extent of phage adsorption as an $MOI_{actual}$, which stands for *actual* multiplicity of phage infection ($MOI_{actual}$), or what more accurately can be described as a multiplicity of phage *adsorption*. Explicitly, this is the ratio of numbers of phages that have adsorbed to numbers of adsorbable bacteria. Thus, if on average 10 phages have *adsorbed* per available bacterium, then we would describe $MOI_{actual}$ as being equal to 10.

Note that $MOI_{actual}$ is not defined solely as a function of the titre of phages that have been *added* to targeted bacteria (i.e., so-called $MOI_{input}$), as time lags are expected to exist between the addition of phages to bacteria and the adsorption of those phages to bacteria. Nevertheless, what value of $MOI_{actual}$ may be achieved, as well as how quickly it may be achieved, should be an explicit function of phage titres. This dependence is due to the rate that individual bacteria become adsorbed by phages being a direct function of how many phages are present within a volume, i.e., the phage titre, and particularly as are found in a bacterium's immediate vicinity (Abedon, 2012b; 2012c; 2017a) (Table 2). The more phages which are present, that is, the higher the titre, then both the faster that a given $MOI_{actual}$ may be reached and the higher the $MOI_{actual}$ eventually achieved.

Titre-based descriptions of phage-mediated bacteria killing include what can be labeled as inundation thresholds and decimal reduction times (Payne and Jansen, 2001; Abedon and Thomas-Abedon, 2010; Abedon, 2011a; 2011b; 2013), i.e., as discussed below. These calculations – describing or based on what *in situ* phage titres have been achieved, whether those titres are theoretically estimated or actually measured – should be used as phage dosing guides. That is, it is useful to be aware of how many phages may be required, *in situ*, in the immediate vicinity of target bacteria, to achieve preferred bacteria-eradication goals.

**Table 2.** Some calculations regarding phage titres, multiplicities, and bacteria killing.

| *P*[1] | *D* (min)[2] | *D* (hours) | *M* (100 min)[3] | $N_t/N_0$[4] | *M* (100 min)[5] | $N_t/N_0$[6] | *M* (100 min)[7] | $N_t/N_0$ |
|---|---|---|---|---|---|---|---|---|
| $10^1$ | $10^8$ | ~$10^6$ | $2.5 \times 10^{-6}$ | 1 | $10^{-6}$ | 1 | $10^{-7}$ | 1 |
| $10^2$ | $10^7$ | ~$10^5$ | $2.5 \times 10^{-5}$ | 1 | $10^{-5}$ | 1 | $10^{-6}$ | 1 |
| $10^3$ | $10^6$ | 15000 | 0.00025 | 1 | 0.0001 | 1 | $10^{-5}$ | 1 |
| $10^4$ | $10^5$ | 1500 | 0.0025 | 1 | 0.001 | 1 | 0.0001 | 1 |
| $10^5$ | 10000 | 150 | 0.025 | 0.98 | 0.01 | 0.99 | 0.001 | 1 |
| $10^6$ | 1000 | 15 | 0.25 | 0.78 | 0.1 | 0.90 | 0.01 | 0.99 |
| $10^7$ | 100 | 1.5 | 2.5 | 0.082 | 1 | 0.37 | 0.1 | 0.90 |
| $10^8$ | 10 | 0.15 | 25 | $10^{-11}$ | 10 | 0.00005 | 1 | 0.37 |
| $10^9$ | 1 | 0.015 | 250 | 0 | 100 | 0 | 10 | 0.00005 |
| $10^{10}$ | 0.1 | 0.0015 | 2500 | 0 | 1000 | 0 | 100 | 0 |
| $10^{11}$ | 0.01 | 0.00015 | 25000 | 0 | 10000 | 0 | 1000 | 0 |

1. Phage titre, here per mL. A potentially inundative phage density is $10^8$ virions/mL.
2. Decimal reduction time, i.e., the time it takes to reduce the number of unadsorbed bacteria tenfold, here assuming a phage adsorption rate constant of $k = 2.5 \times 10^{-9}$ mL$^{-1}$ min$^{-1}$. *D* is calculated as $-\ln(0.1)/Pk$. Resulting numbers have been rounded for the sake of increasing clarity, where *P* is phage titre.
3. *M* meaning 'Multiplicity of adsorption' as equivalent to multiplicity of infection, actual, i.e., $MOI_{actual}$, which is the ratio of adsorbed phages to adsorbable bacteria. This is calculated in this column as seen after 100 min of phage adsorption, holding phage titres constant over time, assuming that bacteria do not lyse, and that adsorption is not saturable. Thus, *M* is calculated in this column as $Pkt = Pk100$ (Abedon, 2016), where *t* is time in minutes.

4. Fraction of unadsorbed bacteria for a given phage multiplicity of adsorption, $M$. Resulting reductions in bacterial numbers are as equivalent to the impact of a killing titre of value found one column to the left. This is equal to $e^{-M}$, as based on the $M$ value also as calculated to the left (thus, $e^{-Pk100}$). This then is the fraction of bacteria that are still alive after 100 min of phage adsorption given constant phage titres of $P$ (first column).
5. Calculations also can be made if not holding phage titres constant, instead allowing free phages to decline as a function of adsorption to bacteria present at a concentration of $N$. This is done using the formula, $P(1\text{-}e^{-Nkt})/N$, assuming that bacteria do not eventually lyse (Abedon, 2011a). Here we assume that $N = 10^7$ bacteria/mL. With lower bacterial concentrations, e.g., $10^6$/mL, then $M$ will be larger (for $N = 0$, see two columns to the left), whereas with larger bacterial concentrations then $M$ will be smaller (see two columns to the right).
6. This column also calculates the fraction of bacteria that remain unadsorbed after 100 min. These numbers are smaller than as previously indicated (two columns to the left) due to free phage losses stemming from adsorption to bacteria, resulting in reductions in $P$ over time. Bacterial lysis still is not assumed, however, as then calculations become somewhat more complicated. If lysis were allowed then phage adsorption could be assumed to occur to a greater extent as rates of free phage depletion over time would decline, assuming that phages cannot adsorb lysed bacteria. Thus, more bacteria would be adsorbed over time as more free phages at any given time would remain available to adsorb bacteria. Not considered as well is that phage-induced bacterial lysis usually would be associated with the release of additional, newly produced virions.
7. Same as two columns to the left except that $N$ is set as equal to $10^8$ bacteria/mL rather than $10^7$/mL.

### *Inundation thresholds and inundative densities*

For phage therapy, an 'inundation threshold' may be defined as the minimum phage density that can keep a bacterial infection from getting worse. More precisely, it is the phage density that can keep a bacterial culture from becoming more concentrated (Payne and Jansen, 2001; Abedon and Thomas-Abedon, 2010; Letarov et al., 2010). This, in turn, is equivalent to the phage density that achieves bacteria killing at a rate that equals and thereby offsets that of bacterial replication. In other words, and ecologically, when phage titres equal this inundation threshold, then co-located phage-susceptible bacteria neither increase in numbers, despite an intrinsic capacity to do so (i.e., bacteria are not in stationary phase), nor decrease in numbers despite the presences of phages.

Explicitly, when phage titres are equal to their inundation threshold, then for every new bacterium generated *in situ* by binary fission, one bacterium will be lost to phage adsorption. Though this on its face would appear to be a good thing, as bacteria therefore cannot increase in numbers, in practice this inundation threshold can be useful only toward understanding what phage titres would be highly *inadequate* to achieve effective phage therapy (Figure 5). That is, achievement of phage titres equal to the inundation threshold will be insufficient to result in effective phage therapy since not only do phages in phage therapy need to prevent bacteria from increasing in numbers, they also need to *reduce* numbers of already existing bacteria, and do so in a timely manner (Abedon, 2011a; 2017a). More generally, what this means is that the rate that phages will impact bacteria should be a function of phage titres, with relatively high phage titres resulting in net reductions in numbers of targeted bacteria, relatively low phage titres allowing net increases in numbers of bacteria, and the inundation threshold found exactly in between, allowing neither increases in numbers of bacteria nor causing decreases.

A phage therapy MIC?

The concept of an inundation threshold might be viewed as equivalent to minimum inhibitory concentrations (MICs) for antibiotics (Cairns and Payne, 2008). That is, an MIC is that concentration of an antibacterial agent that results in the maintenance of a constant bacterial concentration over time, neither reducing nor allowing an increase in bacterial concentration (Figure 5). This analogy between inundation thresholds and MICs is problematic, however, as stemming from a combination of irreversible single-hit killing by phages (Bull and Regoes, 2006; Abedon, 2011a; 2014a) and the potential for phages to replicate in the presence of targeted bacteria. In other words, while phages at inundation threshold titres should indeed represent the *minimum* concentration that will inhibit bacterial replication within a culture, in practice it is effectively impossible to hold phage titres at this density (Abedon, 2011a).

Inundation thresholds thus at best can only approximate MIC-type determinations (Wiegand et al., 2008). Such an approximation may be achieved if phage titres happen to greatly exceed bacterial densities (thereby with free phages not rapidly *irreversibly* lost to phage adsorption) and if the phages used are unable to replicate (and thereby are unable to increase in numbers over time). Furthermore, and especially if determined in liquid cultures, MIC-type determinations for phages will tend to give rise to false negatives if bacteria are able to readily mutate to phage resistance, as is often the case. Thus, inundation thresholds as an approximation of a phage MIC might have meaning in concept, but much less meaning in practice.

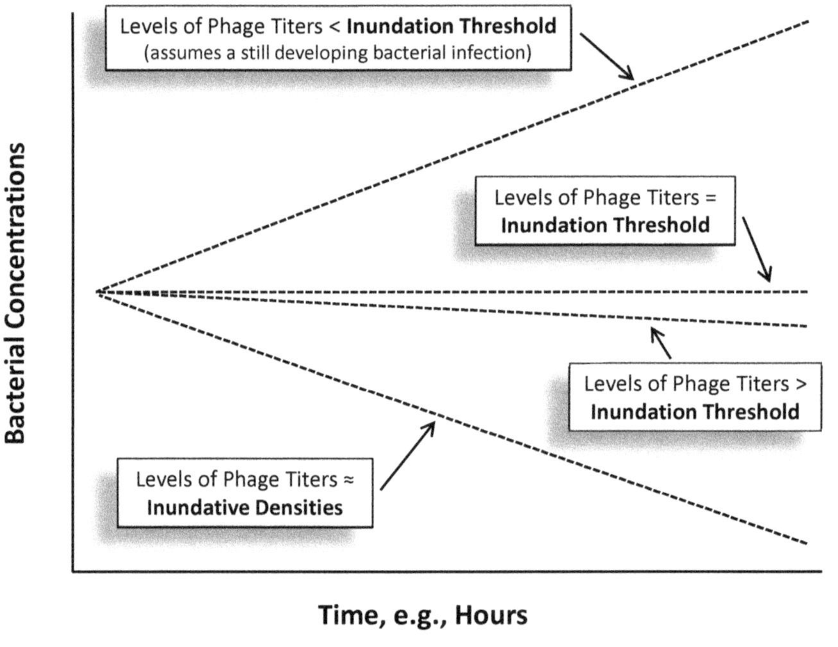


**Figure 5.** Inundation threshold and inundative density impacts on concentrations of target bacteria over time. Note that only inundative densities, here by definition, reduce bacterial densities substantial over time. In addition, the concept of inundation threshold is meaningless if bacterial concentrations within an infection have reached a steady state, i.e., in which bacterial numbers are no longer increasing independent of phage presence (contrasting the top curve where without treatment bacterial concentrations are assumed to instead be increasing as a function of time). Phage concentrations are described here in titres rather than multiplicities of infection, and the concept of multiplicity of infection is inappropriate to describe either phage inundation thresholds or phage inundative densities.

Not inundative densities

For a vigorously growing bacterial infection, $10^7$ phages/mL can be calculated as an inundation threshold, assuming a bacterial doubling time of 30 min and a phage adsorption rate constant of $2.5 \times 10^{-9}$ $mL^{-1}$ $min^{-1}$ (Abedon, 2014a). That is, given this bacterial doubling time, for a constant phage titre of about $10^7$ phages/mL [$\approx \ln(2)/(30 \times 2.5 \times 10^{-9})$], then every time a bacterial division occurs, so too will a phage adsorption occur, thereby keeping bacterial numbers constant over time.

As noted, a phage density that simply keeps bacterial cultures from increasing in size should be viewed as a product of *inadequate* phage dosing (see also Table 2), since by definition inundation thresholds are phage titres that can control but cannot reduce numbers of replicating bacteria. This furthermore would be the case even if the number of phages applied were to greatly exceed the number of bacteria present, i.e., as representing a high $MOI_{input}$ (Kasman et al., 2002; Abedon, 2016). For example, if $10^7$ phages/mL are applied to $10^5$ bacteria/mL, again assuming that bacteria are replicating with a doubling time of 30 min and phages are adsorbing at a per-virion rate of $2.5 \times 10^{-9}$ $mL^{-1}$ $min^{-1}$, then bacteria will not be expected to decline appreciably in number over time, and this is so even though the ratio of added phages to bacteria in this case would be equal to 100. Change the numbers instead to $10^7$ and $10^4$ and still the same conclusion would be reached, even though now the ratio would be 1,000-to-1 instead of 100-to-1.

To address this inadequacy of inundation thresholds as a guide to phage dosing, we introduce instead the concept of phage inundative densities (see Figure 5, and Table 2). Rather than simply keeping the bacterial culture constant in size, or indicating what phage titres must be exceeded to net reduce those numbers over some potentially very long span of time, an inundative density instead is a phage titre that not only substantially reduces bacterial numbers over time, but does so relatively rapidly. This concept we discuss in greater detail below, including positing that inundative phage densities generally would be equal to roughly $10^8$ phages/mL or more.

An additional consideration is that determination of inundation thresholds for phages during actual phage treatments is not straightforward. This is because this calculation explicitly is based on a combination of rates of bacterial replication and phage adsorption, neither of which is easily determined *in situ*. Rather than employing inundation thresholds or MICs toward better appreciating phage therapy pharmacodynamics, we suggest instead the use of killing titre and decimal reduction time determinations.

Killing titres are measures of the bactericidal densities of phages found in phage stocks or within formulated products (Stent, 1963; Carlson and Miller, 1994; Carlson, 2005; Abedon, 2011a; 2020a). Killing titres are a more practical as well as more direct measure of what phage densities can be required to adequately reduce bacterial numbers versus instead, e.g., inundation thresholds (Figure 5). This claim of greater utility comes with the caveat, however, and that is that killing titres don't take into account the issue of ongoing bacterial replication. Therefore, phage adsorption must be relatively fast for killing titre calculations to reasonably hold. That in turn means that densities of targeted bacteria must be fairly high, i.e., somewhat in excess of $10^7$ bacteria/mL, or instead that the numbers of phages that fail to adsorb following exposure to target bacteria must be explicitly taken into account.

Determining killing titres

To determine a killing titre, one mixes a phage population with a bacterial population in such a manner that, ideally, all or nearly all phages adsorb, and under conditions or over time spans in which bacterial replication is minimal. Measurement is made of bacterial viability both before and after phage adsorption. The ratio of still-viable bacteria present after phage adsorption ($N_t$) to that number present before phage exposure ($N_0$) is then assumed to be equal to $e^{-M}$, where $M$ is the ratio of adsorbed phages to bacteria present, i.e., $MOI_{actual}$. In other words, a Poisson distribution of phage adsorption to bacteria is assumed where the fraction of not adsorbed bacteria is equal to $e^{-M}$ (Abedon and Katsaounis, 2018).

Rearranging, $M$ = -ln($N_t$/$N_0$). From this calculation, it should be clear that the greater the fraction of bacteria that one wants to kill, i.e., the smaller that $N_t$ is relative to $N_0$, then the greater the number of phages that must adsorb, on average, to each bacterium that is present in a culture. Thus, in other words, the greater that $MOI_{actual}$ must be. For example, for $N_t$/$N_0$ = $10^{-1}$, $10^{-2}$, $10^{-3}$, $10^{-4}$, $10^{-5}$, or $10^{-6}$, the latter meaning, i.e., that only 1 in one-million bacteria survive, then $M$ must be equal to 2.3, 4.6, 6.9, 9.2, 11.5, or 13.8 adsorbed phages/bacterium, respectively. Be cautious, though, as at higher killing titres such as in excess of $M$ = 10, then phage-resistant bacterial mutants can be present in numbers that are equivalent to or higher than those of unadsorbed phage-sensitive bacteria, $N_t$. See Table 2 for additional killing titre calculations.

Insufficient phage penetration to bacteria?

Note that killing titre calculations assume equivalent rates of phage penetration to all targeted bacteria. That assumption is unlikely to hold, however, given environmental or bacterial infection heterogeneity. Making assumptions of equivalent rates of phage penetration to targeted bacteria – should equivalent penetration in fact not be the case – can be problematic as this can lead to underestimations of what ratios of adsorbed phages to bacteria, i.e., $MOI_{actual}$, may be necessary to achieve sufficient bacteria killing *in situ*. That is, longer phage adsorption times, as a proxy for slower phage penetration, will require higher phage titres to achieve similar ratios of adsorbed phages to adsorbable bacteria. To some degree this issue may be mitigated, though, if phage replication can generate these higher phage numbers, particularly in the immediate vicinity of otherwise less phage-available bacteria, e.g., as might occur during active phage penetration into bacterial biofilms (Abedon, 2017a).

Notwithstanding these various issues, killing titre calculations can be useful toward estimating what fraction of bacteria ought to be killed by the application of a given phage titre to a bacterial culture, including perhaps especially to bacterial biofilms (Abedon, 2011b; 2019a; 2020a). That is, you should be able to calculate how many bacteria should survive if all added phages adsorb (= $e^{-M}$). If the number of bacteria surviving is larger than as calculated, i.e., greater than $e^{-M}$, then not all added phages may be reaching targeted bacteria even if all phages are adsorbing. On the other hand, if the number of surviving bacteria is smaller than expected (< $e^{-M}$), then that may be a consequence of *in situ* phage replication generating more phages than were originally dosed with, i.e., as indicating active treatment.

A yet additional issue, related to that of phage penetration to bacteria and as already noted above, is that killing titre calculations do not explicitly take into account *rates* of phage impact on bacteria. This is because killing titres are determined by how many phages *adsorb* to bacteria, but phage adsorption itself generally takes time. To take into account this time issue, we suggest instead calculations of phage decimal reduction times.

### Decimal reduction time

The rate at which phage virions will acquire bacteria is a function of a combination of the intrinsic adsorption properties of individual phages (their adsorption rate constant), phage densities (titres), and the availability of bacteria to these phages, the latter as measurable as well in terms of phage adsorption rate constants (Abedon, 2020a). We thus focus in this section on phage adsorption properties and titres, assuming for convenience that all bacteria are equally available to all treatment phages. If that latter assumption is false, then that is a pharmacokinetic rather than pharmacodynamic concern, that is, it is an issue of the ability of phages to reach targeted bacteria rather than phage impact on those bacteria once they have been reached. Alternatively, phage adsorption rates can vary as a function of phage host range, a pharmacodynamic issue (Dąbrowska and Abedon, 2019).

Notwithstanding these latter considerations, one way of appreciating the impact of especially phage densities and intrinsic adsorption rates on phage therapy pharmacodynamics is in terms of what can be described as a decimal reduction time or D-value. D-value, or $D$, is the time that it takes to reduce the viability of a bacterial population by one log, here as a consequence of phage adsorption. For example, this could be from $10^6$ bacteria/mL down to $10^5$. Assuming a constancy in phage densities, and also that bacteria are not substantially replicating over the course of reduction times, then $D = -\ln(0.1)/Pk$, where $k$ is the phage adsorption rate constant, *P* is phage density (titre), and 0.1 represents the decimal reduction. *D*, again, is in units of time.

D-value example calculations

Employing the same phage adsorption rate constant used above to illustrate inundation thresholds, i.e., $2.5 \times 10^{-9}$ $mL^{-1}$ $min^{-1}$ (Stent, 1963), note that the titre of $10^7$ phages/mL would result in a D-value of approximately 92 min (as rounded to 100 in Table 2). Thus, it would take over 350 min, or about six hours under ideal conditions and without bacteria replicating, to reduce the viability of a bacterial culture approximately 10,000-fold. This reduction is found to be $e^{-9.2}$-fold when calculating killing titres, that is, where 9.2 = $M$ = $MOI_{actual}$. This in turn is approximately $M$ = 10, a multiplicity that we return to subsequently.

At $10^5$ phages/mL, by contrast, that same degree of reduction in bacterial viability would take instead approximately 25 days (not indicated in Table 2 but equal to about 37,000 min or ~600 hours), again assuming that phage titres remain constant and bacteria are not replicating. If faster reductions in bacterial densities are desired, if not all bacteria are equally available to phages (Barrow and Soothill, 1997), or if phage densities should decline over the course of treatment (Table 2), then supplying higher phage numbers would be necessary to achieve the same calculated rates of bacteria killing. See Abedon (2011a; 2011b) for additional discussion.

$10^8$ phages/mL as a potentially inundative density

From the above calculations, we can see that bacteria killing may not be sufficiently fast if phage densities are not sufficiently high. By 'sufficiently high' we mean phage densities that are near to or in excess of about $10^8$ phages/mL, contrasting the calculations of ~$10^7$ phages/mL as an inundation threshold under the same conditions. Derivation as well as experimental application of this idea – that it is phage density that determines rates of bacterial adsorption by phages along with a utility of phage application at titres in the vicinity of $10^8$/mL – can be found in a number of places, e.g., (Abedon, 1990; 1999; Payne and Jansen, 2001; Kasman et al., 2002; Goodridge, 2008; Bigwood et al., 2009; Hyman and Abedon, 2009; Abedon and Thomas-Abedon, 2010; Hagens and Loessner,

2010; Abedon, 2011a; 2011b; 2011c); see also (Spouge, 1994; Payne and Jansen, 2000; Abedon, 2018b). That is, for $k = 2.5 \times 10^{-9}$ $mL^{-1}$ $min^{-1}$, as above, then $10^8$ phages/mL corresponds to a D-value of approximately 10 min, while titres of $10^9$ phages/mL will instead have a D-value of 1 min, and as noted for $10^7$ phages/mL it instead is 100 min (Table 2).

In cases where excessively rapid bacterial lysis might be a concern, e.g., such as due to endotoxin release in the course of phage treatment of Gram-negative infections of the blood, then lower phage titres and therefore lower D values instead might be appropriate. That statement comes with the caveat, however, that it may not always be possible to control just what titres phages can attain *in situ* nor how rapidly cultures may be lysed by phages. Nevertheless, to achieve sufficient bacteria killing over reasonably short time frames, then sufficient phage titres must be present *in situ* to result in sufficiently short decimal reduction times (again, see Table 2). Such 'sufficient' phage titres are what we describe here as inundative, that is, inundative densities of phages, which we suggest can be in the range of at least $10^8$ phages/mL titres. What phage numbers need be supplied in the course of dosing, then, to result in inundative densities of phages *in situ*?

*Supplying sufficient phage titres*

As generally we are concerned not only with killing bacteria but also killing them over reasonable time frames, introduction of a time component to calculations of phage impact on bacteria is important. That is, bacteria typically will not be adsorbed instantaneously by phages. Nevertheless, for phage therapy sooner phage adsorption generally is preferable (minutes or at most hours) to greatly delayed phage adsorption (days or even weeks). It is this time component that is supplied by decimal reduction time calculations. Indeed, as illustrated by decimal reduction time calculations, those phage titres that can achieve adequate bacteria killing over reasonable time frames we have described as 'inundative'. Just what constitutes 'adequate' or 'reasonable' may be subject to debate, however, as well as vary with circumstances. With these considerations in mind, in this section, we discuss further the concept of inundative densities.

The problem of slow adsorption

Reduction of $10^6$ bacteria/mL to a density of $10^{-6}$ bacteria/mL – a drop in density to $10^{-12}$ of the original – will require a minimum of $3 \times 10^7$ phages/mL. Where does this number come from? To achieve that much bacteria killing then $MOI_{actual}$ must be equal to roughly 30, i.e., $M \approx 30$. That is, $N_t/N_0 = e^{-M}$, where $N_t$ is the ending bacterial concentration and $N_0$ is the bacterial concentration prior to phage addition. Thus, $N_t/N_0$ is the fraction of bacteria that survive following phage adsorption, which in the example is equal to $10^{-12}$, and $-\ln(10^{-12}) = 28 \approx 30$, i.e., where $30 \times 10^6 = 3 \times 10^7$. Therefore, $3 \times 10^7$ phages must fully adsorb $10^6$ bacteria to reduce that number of bacteria to $10^{-6}$.

Alternatively, such substantial bacteria overkill may not be necessary to the extent that phage-mediated disruption of bacterial infections is followed by immune system clearance (Levin and Bull, 2004; Leung and Weitz, 2017; Roach et al., 2017; Abedon, 2019b). This phage density of $3 \times 10^7$/mL, however, corresponds to an initial *D*-value, as seen prior to phage losses due to bacterial adsorption, of approximately 30 min, assuming again a phage adsorption rate constant of $2.5 \times 10^{-9}$ mL$^{-1}$ min$^{-1}$. This means that, if conditions for bacterial acquisition are poorer than anticipated, such that *D* instead, e.g., equals 1 hour, then this phage density, sufficient as it could be for bacteria overkill, may in fact *not* result in timely bacteria killing. That is, in this latter example, it would take 12 hours to reduce the bacterial density $10^{-12}$. That time frame may or may not be adequate for a given application.

Perhaps consistent with that concern of slowness of phage adsorption, note that some efforts to model phage therapy mathematically, as reviewed by Gill (2008), have employed adsorption rate constants of much lower magnitude than that used in the above calculations. For example, these appear to have been adsorption rate constants of ~$10^{-11}$ mL$^{-1}$ min$^{-1}$ or lower, which is ~40-fold smaller than $2.5 \times 10^{-9}$ mL$^{-1}$ min$^{-1}$. Confusingly, these values are often expressed in per hour rather than per min units, i.e., such that a constant of $10^{-9}$ mL$^{-1}$ h$^{-1}$ is equal to approximately $2 \times 10^{-11}$ mL$^{-1}$ min$^{-1}$. With a phage density of $3 \times 10^7$/mL and an adsorption rate constant of $10^{-11}$ mL$^{-1}$ min$^{-1}$, then the calculated D value is 5 days! Thus, in the above scenario the $10^{-12}$ reduction would require 60 days of constant phage adsorption as well as assumptions of a lack of compensating bacterial binary fission.

Consistently as well, phage densities of the reviewed simulations often reach approximately $10^{10}$/mL before substantial declines in bacterial densities are seen (Gill, 2008), which can serve as an example of how slower phage acquisition of, or penetration to individual bacterial cells can negatively impact phage therapy efficacy (Barrow and Soothill, 1997). Also consistently, Levin and Bull (1996) note in their model that with lower phage adsorption rates, bacteria are more likely to exceed their "lethal threshold", that is, to kill the patient prior to infections being brought under control following phage application. Thus, in short, a seemingly quite adequate $MOI_{input}$ such as $3 \times 10^7$/mL will not necessarily always translate into timely bacteria killing, and this explicitly is because it can take time, and sometimes substantial amounts of time, before $MOI_{input}$ comes to approximate the more pharmacodynamically useful $MOI_{actual}$ (Abedon, 2016) (Table 2).

Indeed, $MOI_{input}$, or even phage inundative densities, are more pharmacokinetic than pharmacodynamic descriptors, representing what numbers of phages have been made available toward bacteria killing rather than what number of phages are, at a given time, actually effecting that killing. It is the achievement instead of an $MOI_{actual}$ that is directly relevant to phage therapy primary pharmacodynamics. Nevertheless, we are defining here inundative densities as phage titres that over reasonable time frames are able to achieve adequate $MOI_{actual}$ values, such as 10 adsorbed phages per adsorbable bacterium. To achieve this, if working with phages displaying unusually slow intrinsic rates of phage adsorption (small adsorption rate constants), then phage titres representing inundative densities simply must be correspondingly higher, e.g., 40-fold higher given 40-fold slower adsorption.

The problem of free phage losses

A further concern is that D-value calculations are based on an assumption that phage titres will remain constant over time. This is rather than titres declining as a consequence of phage adsorption to bacteria or especially as due to bacteria-independent mechanisms of phage decay, the latter, e.g., such as due to the action of immune systems (Dąbrowska et al., 2018; Dąbrowska and Abedon, 2019). A utility of phage replication *in situ* thus is that it could contribute to a maintenance of phage titres in the face of these various mechanisms of titre decline. If bacterial densities are too low, however, then substantial increases in phage densities will not be able to proceed as a consequence of *in situ* phage replication. Thus, phage replication after dosing might help *in situ* phage titres to either reach or sustain inundative densities, though only if bacterial densities are sufficiently high to support that replication.

These statements are consistent with a scenario posited by Payne and Jansen (2000) in which too early phage application during the development of a bacterial infection, followed by substantial phage decay, could give rise to what they describe as a "failure threshold time". This is defined as a time before which numbers of targeted bacteria have grown to be sufficiently high to support adequate phage population growth. One can view this perspective, however, as also pointing to a problem of supplying too few phages rather than supplying those phages too early. Furthermore, it is an open question as to when, clinically, one in fact might catch a bacterial infection too early in its development to support active treatment, since bacterial infections often are fairly advanced before they are brought to a physician's attention. A possibly relevant scenario, though, could be during prophylactic phage treatment. Under such circumstances, an important consideration would be to supply as well as retain sufficient phage densities that newly forming bacterial infections are brought under control early in their development (Abedon, 2010; 2011b). Generally, such early control would require that phages have been supplied in sufficient numbers that inundative densities are retained despite free phage losses over time, as bacterial densities by definition in this case would be insufficient to support substantial phage replication especially early following the occurrence of bacterial contamination.

$10^8$ phages/mL again as a potentially inundative density

For these various reasons, striving for phage densities in the vicinity of $10^8$ phages/mL, or higher, and then sustaining those phage densities over the course of treatment – whether through passive or active means – should be the default dosing goal in phage therapy protocols. Lower densities should instead be employed or sought only to the extent that they are demonstrably and consistently effective in reducing bacterial densities to desirable extents. Substantially higher densities may also be employed should $10^8$ phages/mL be found to be inadequate in achieving desirable levels of bacteria killing, though with the caveat that an alternative explanation for poor efficacy, to that of too few phages, can be poor phage choice, e.g., such as stemming from insufficiently high adsorption rate constants.

Some of these issues are explored in Table 2. There we see that $10^8$ phages/mL may achieve a D value of 10 min, assuming no substantial reductions in phage numbers over time, whereas the D value given $10^7$ phages/mL of the same phages is 100 min. Perhaps more telling, after 100 min $MOI_{actual}$ is expected to reach 25 given a constant $10^8$ phages/mL, versus 2.5 for $10^7$ phages/mL. That means that at $10^8$ phages/mL an $MOI_{actual}$ of 10 would be reached in 40 min, versus 400 min given $10^7$ phages/mL. If phages decline in number over time, which as shown in Table 2 is due to phage adsorption to target bacteria, then so too does the $MOI_{actual}$ decline. Declines in phage numbers to adsorption, however, should only be a concern when employing phages that are unable to replicate *in situ*, as bacterial densities that are sufficient to substantially reduce phage numbers over time due to adsorption should also be sufficient to substantially support phage population growth as well. Nonetheless, for replication-competent phages, declines in phage numbers that are due to bacteria-independent mechanisms will still have an impact on what $MOI_{actual}$ can be achieved over a given time frame. So too, as noted, will smaller phage adsorption rate constants reduce the number of phages that will come to adsorb over a given time span. Thus, while achieving $10^7$ phages/mL *in situ* might appear under some circumstances to at least be adequate for a given phage treatment, that adequacy to a degree is dependent on phages finding and then adsorbing targeted bacteria quickly and virions declining in numbers only slowly. A more conservative dosing goal, and particularly so if phage penetration to some targeted bacteria is slower than it is for others, therefore could be one that is higher than $10^7$ phages/mL, e.g., such as $10^8$ phages/mL.

Keep in mind that it is also always important to monitor for toxicity during animal testing. This is particularly so given increased levels of dosing since the standard pharmacodynamic assumption is that greater drug densities will be associated with greater drug toxicity. This assumption, though, may be less relevant when using well-purified phage products given the generally low toxicity of phage therapy virions. Also, and as noted, what generally is key in terms of *in situ* phage titres are those titres found in the immediate vicinity of target bacteria rather than necessarily as found on average more systemically (Abedon, 2017a). Thus, phage titres of $10^8$/mL need not always be achieved more globally during phage treatments but instead only more locally, i.e., as may be readily achieved especially via more direct phage application to more localized bacterial infections (Table 3).

Striving for substantial killing of bacteria

A perhaps reasonable dividing line, following phage application, between substantial and insubstantial *reductions* in bacterial densities, whether *in vitro* or *in vivo*, is approximately 10,000-fold (Kasman et al., 2002). Experimentation that fails to achieve this level of bacteria killing might be deemed to be inadequate unless clinically satisfactory endpoints nonetheless are achieved, e.g., such as a consequence of concurrent immune-system action. Experimentation in which neither clinically positive outcomes nor substantial reductions in bacterial densities are observed, however, should not be used as evidence either for or against the potential for phage therapy to control a specific bacterial infection, unless higher phage density 'positive controls' are also attempted.

Examples of such bacteria-killing positive controls would be $10^9$, $10^{10}$, or even higher phage titres (phages/mL), as achieved within the vicinity of target bacteria *in situ*. In other words, one should not simply supply what may be phage doses that do not achieve inundative phage densities *in situ* and then from that declare that phage therapy generally has not been successful. Another way to state this is that given that phage treatments tend to display low toxicities, then especially during *in vitro* testing, and presumably during animal testing as well, it behooves researchers to establish dose-response relationships. Researchers particularly should not inadvertently or intentionally ignore higher phage dosing unless lower phage doses already provide desirable efficacies especially under more stringent or realistic conditions, e.g., such as toward the treatment of chronic bacterial infections (Abedon, 2019b).

### Pharmacokinetics

Pharmacokinetics, as noted, can be described as the body's impact on a drug. What that really means is that the concentration and location of a drug within a body will be impacted by the anatomy and physiology of that body. What that further means is that to properly function, a drug must become present in sufficient quantities within the immediate vicinity of target tissues. For antibacterial agents, this is in the immediate vicinity of target bacteria. The extent to which phages can attain as well as maintain those necessary quantities, or concentrations – what above we have described as inundative densities – is a function of their pharmacokinetics. This means that phages must become associated with the body (absorption), move within the body to the location of their target bacteria (distribution), and once reaching those bacteria change into a form that is explicitly antibacterial (metabolism). All the while they must avoiding becoming chemically inactivated (also metabolism) or otherwise being removed from the body (excretion) (for summaries, see the left side of Figure 1 and also Figure 4).

Given the diversity of phages, the diversity of dosing routes, and the diversity of infection properties, study of the pharmacokinetics of phages represents a vast undertaking. Specifically, just as the pharmacokinetics of every drug must be individually studied, including for every condition that the drug may be used to treat, so too the pharmacokinetics of every phage type that may be used for phage therapy might be individually studied for every dosing route and bacterial infection type. Ultimately, for treatment with phages as a whole, undertaking such an effort however becomes daunting. Therefore, general principles by necessity must be sought, some of which are our focus here.

In this section we start with metabolism especially in terms of phage 'activation', consider the ultimate product of pharmacokinetic processes, i.e., *in situ* phage titres, and then briefly address issues of phage absorption and distribution. For additional, recent reviews of phage therapy pharmacokinetics, see (Dąbrowska, 2019; Dąbrowska and Abedon, 2019; Matsuzaki and Uchiyama, 2019).

*Metabolism and active treatment*

Drug accumulation within body tissues involves diffusion and other movement processes. Pharmacokinetically, these are drug absorption and distribution (Figures 1 and 4). Alternatively, mechanisms exist by which certain drugs, such as the anti-tuberculosis drug, isoniazid (Then and Sahl, 2010), become *activated* in the course of pharmacokinetic metabolism. That is, isoniazid becomes chemically modified within the body – specifically by targeted bacteria – to its pharmacodynamically functional form. Thus, for some drugs it isn't good enough for them simply to have entered the body. Instead, to become active they must also be subject to pharmacokinetic metabolism.

Phage infection as a pharmacokinetic metabolism

Phages too can be chemically altered within the body. Most obviously, this can result in their inactivation, such as in the course of interacting with a body's immune system. Perhaps less obviously, the process of phage infection of bacteria itself represents a series of chemical alterations first of the phage particle and then as occurring over the course of phage gene expression as well.

As microbiomes pharmacologically are considered to be aspects of bodies, phage replication as a biochemical process – as occurs in association with microbiome bacteria – too should be considered to be a product of pharmacokinetic metabolism (Abedon, 2014b). In addition, however, so too are phage bactericidal activities products of pharmacokinetic metabolism as well, i.e., as also resulting from phage gene expression. While phage bactericidal activity nonetheless explicitly is pharmacodynamic, the point here is that its generation instead is pharmacokinetic, i.e., as a product of metabolism. For phage replication by contrast, both its generation (metabolism) and consequence (*in situ* changes in numbers of phages) are pharmacokinetic aspects (Figure 6). This *in situ* phage replication is explicitly required for so-called *active* phage treatments.

Bactericidal activity as a product of pharmacokinetic processes also is required for active treatment, but so too it is essential for passive treatment, the latter as considered above as well as discussed further in a subsequent section. See Figure 7 for a summary of the relationship between these various phage pharmacological activities and active vs. passive phage treatments.

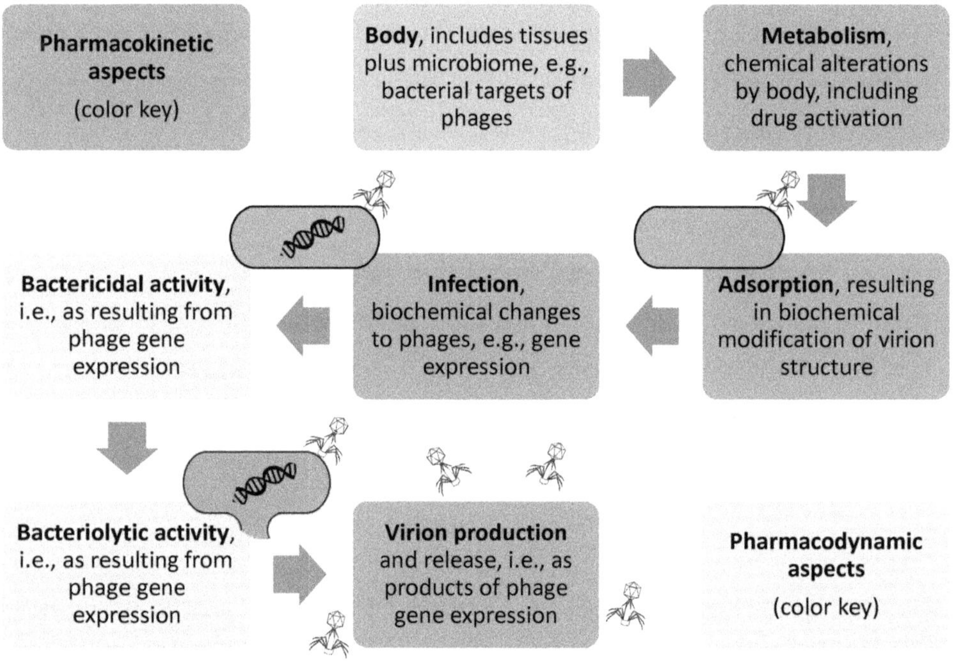


**Figure 6.** Phage infections and pharmacology. Target bacteria found within bodies pharmacologically are considered aspects of the body. Pharmacokinetic metabolism is the chemical alteration of pharmacological agents. Chemical and biochemical alterations of phages occur during their adsorption and infection in association with body bacteria. Bactericidal and bacteriolytic activities are primary pharmacodynamic consequences. Resulting virion production, as a contributor to changes in phage numbers instead is a pharmacokinetic product of this metabolism. If phages are inactivated upon infection (not shown), then that too would be a consequence of pharmacokinetic metabolism. Also, bacterial lysis, if it can contribute to phage penetration to bacteria, such as into bacterial biofilms, also would contribute to a phage therapy pharmacokinetic aspect, i.e., that of phage movement to target bacteria.

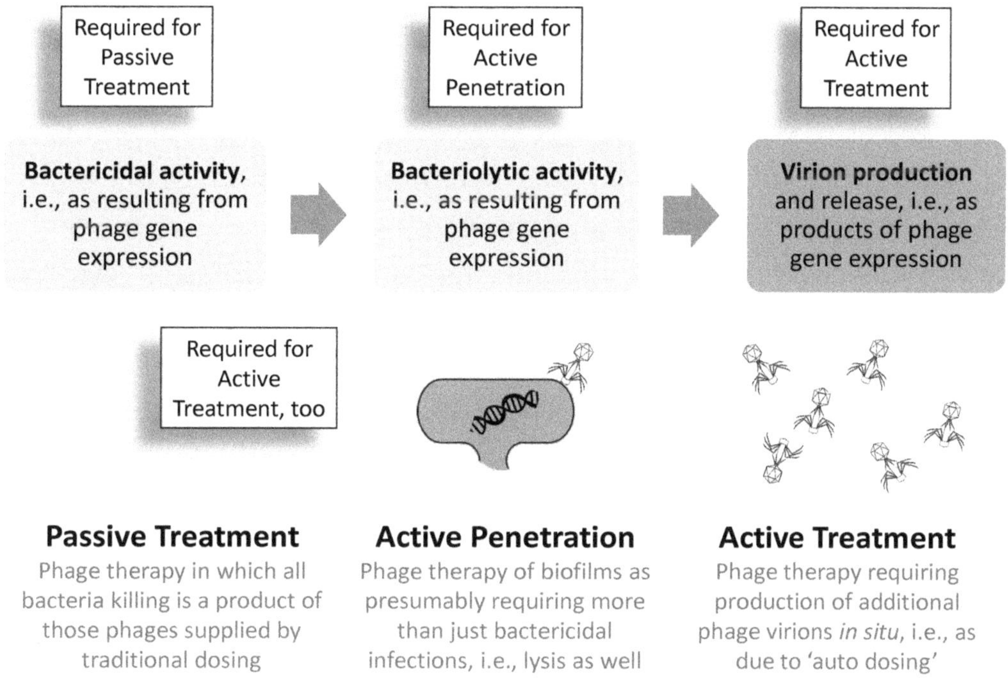


**Figure 7.** Restatement of relationship between different pharmacological phage activities and categories of phage treatment. While bactericidal and bacteriolytic activities are pharmacodynamic as phage impacts on bacteria (as an aspect of the body), virion production is pharmacokinetic, as bacteria impact on phage. As indicated, bactericidal activity is required for passive treatment, both bactericidal and virion production activities are required for active treatment, and bacteriolytic activity likely is either necessary or useful for phage penetration into bacterial biofilms, and also is required for the release aspect of virion production.

## A replicating drug but with resource requirements

Replication is decidedly unusual for a drug. Nonetheless, similar *in situ* drug amplification in densities can occur simply through dosing by conventional means. That is, by administering more drug, including via automatic dosing in response to *in situ* conditions, drugs too can increase in concentrations within the body. The increases in drug concentrations following conventional dosing, versus this phage "auto dosing" (Abedon and Thomas-Abedon, 2010; Loc-Carrillo and Abedon, 2011), can however result in changes in concentrations in distinctly different locations within the body. Phage self-amplification specifically will tend to occur precisely in the vicinity of phage antibacterial activity. That is, active treatments can allow the generation of inundative phage densities, especially where they may be most needed, rather than requiring that such densities be reached throughout the body (Abedon, 2017a). In addition, in those circumstances where phages are not delivered to a bacterial infection directly (e.g., locally) but rather indirectly (i.e., systemically), this ability of phages to increase their numbers though auto dosing can mean that phages need not pharmacokinetically penetrate to bacteria already possessing inundative densities for phage treatments to be successful.

Whether actively or passively supplied, ultimately phage therapy efficacy is dependent on the generation, in the vicinity of target bacteria, of peak phage concentrations that are sufficient, i.e., which are inundative, to result in both substantial and timely bacterial eradication (Abedon, 2011a; 2012a; 2017a). In terms of active treatments, strictly lytic phage populations explicitly will need to go through repeated rounds of bacterial adsorption, infection, and lysis to achieve necessary titres. Therefore, anything that speeds up adsorption, decreases the time spent infecting (shorter latent periods), or increases the yield from infections (larger burst sizes) should to a first approximation increase these rates of phage population growth (Levin and Bull, 2004; Abedon et al., 2009). Phage survival is also important (Abedon, 2009; Chan and Abedon, 2012), as discussed further below as well as touched upon above, and see also Abedon (2014a) for a listing of various complications on these processes. Rapid phage population growth, however, is only one component of active treatments.

Also as touched upon above, successful active treatment is also absolutely dependent on concentrations of target bacteria: If there are not enough bacteria present, less than the so-called proliferation threshold (Payne and Jansen, 2000; Payne and Jansen, 2001; 2003; Abedon and Thomas-Abedon, 2010; Abedon, 2011a) or what others refer to simply as a threshold (Brüssow, 2007; Gill, 2008; Letarov et al., 2010), then phage population growth and therefore active treatment simply will not work (Levin and Bull, 2004; Bigwood et al., 2009; Capparelli et al., 2010). Active treatments also require that phages replicate to sufficient titres. In particular, unless sufficient densities of target bacteria are present, then inundative phage densities will not be reached no matter how rapidly a phage population may replicate, and therefore active treatments again will not be successful. Nonetheless, what active treatment can supply to phage therapy protocols is the presentation of large numbers of phages to bacteria that otherwise are difficult to directly deliver phages to, though this is true only so long as there are sufficient numbers of bacteria present that physiologically are sufficiently able to support phage population growth.

*Phage* in situ *concentrations (titres)*

What phage titres, as found *in situ*, are sufficient for achieving phage therapy efficacy? Ultimately this is a somewhat unresolved question. Thus, as noted above, perhaps $10^8$ phages/mL could constitute a reasonable minimum target phage density under many circumstances (Abedon, 2018b), especially as achieved in the immediate vicinity of target bacteria (Abedon, 2017a). This titre might then be described as approximating a minimum effective phage concentration (Figure 4), i.e., as equivalent to an inundative phage density (above). What specific *in situ* phage titre constitutes a minimum effective concentration therefore should vary depending on the rate at which elimination of bacteria is desired, with killing over minutes requiring higher phage densities than killing that instead is envisaged to involve hours or days of treatment. Phage densities ideally would then remain at or higher than their minimum effective concentration throughout much of a course of treatment, again explicitly as found in the immediate vicinity of targeted bacteria (Abedon, 2017a). This section considers various aspects of the occurrence and maintenance of such phage *in situ* titres.

Passive treatment, with some consideration of active treatment

With passive treatment, phage adsorption need result only in the killing of target bacteria for therapeutic outcomes to be successful, that is, rather than also the production of additional phages (Figure 7). From Barrow and Soothill (1997), p. 271, "...sufficiently large dose of phage could be given that might be able to overwhelm the bacterial pathogens without any significant phage multiplication taking place." That is, with passive treatment, by definition (Payne and Jansen, 2000; 2001; 2003), doses supplied must be sufficiently large that a good approximation of all target bacteria become phage adsorbed without relying on *in situ* phage replication to bolster phage numbers. Phages must also adsorb bacteria reasonably quickly, i.e., see inundative density, above, with the more phages supplied then the faster bacteria will be adsorbed (Table 2). Passive treatment thus is dependent on four criteria: (i) phages reaching target bacteria in sufficient numbers, (ii) phages adsorbing the bacteria they reach, (iii) phages killing the bacteria they've adsorbed, and (iv) all of this occurring over reasonable time frames.

Though these criteria seemingly are at least modestly stringent, in fact they can be less strict in terms of phage properties than requirements for active treatment, since the latter by definition also requires adequate levels of phage population growth. Passive treatments thus can demand only relatively low levels of 'performance' on the part of the phages employed, i.e., just adsorption and killing. Furthermore, passive treatment can be effected by phages that have been engineered to kill bacteria without lysing them (Goodridge, 2010). On the other hand, the need for phages to reach targeted bacteria in adequate numbers can be *more* difficult with passive treatment than for active treatment (Abedon, 2017a). This is because with passive treatment every phage necessary to effect bacterial inundation must be supplied via external dosing rather than generated *in situ*.

Fortunately for the utility of phage therapy, naturally occurring phages often will retain a capacity to display auto dosing whether passive treatment is being attempted or not. Thus there usually will exist a possibility of near-immediate augmentation of passive dosing that is a consequence of active phage replication, presumably resulting in "a 'margin of safety' toward attaining phage therapy efficacy" (Abedon and Thomas-Abedon, 2010). These issues point to an importance of performing adequate *in vitro* growth as well as host range determinations with all potential therapeutic phages even if only passive treatment is envisaged, with determinations of phage growth characterization more relevant the more that active treatment is relied upon. For additional consideration of phage choice for phage therapy, see (Gill and Hyman, 2010; Łobocka et al., 2014).

Multiple or continuous dosing

Various factors can conspire to make it necessary to supply more than just that minimum number of phages which otherwise are thought to be adequate to achieve treatment success. One way of augmenting phage numbers, a means that should be assumed to be required unless it can be experimentally demonstrated to not be necessary (Abedon, 2012a), is to supply phages either in multiple doses (Summers, 2001; Capparelli et al., 2007) or continuously (Kutter et al., 2010). In Georgia and Poland for example, where phage therapy of humans has been routinely employed for decades (Kutter et al., 2010; Abedon et al., 2011), multiple or continuous dosing is typical. For example, treatment of intestinal ailments using the formulation Intestiphage involves ingestion of many tens of millilitres multiple times per day. Multiple dosing, similarly and elsewhere (Belgium) was found to be effective in a case of a human patient treated intravenously with phages every 6 hours due to a difficult sepsis (Jennes et al., 2017). See as well, e.g., (Fish et al., 2016; Schooley et al., 2017; Aslam et al., 2019b; Dedrick et al., 2019; Bao et al., 2020).

A further issue is the appropriate duration of multiple or continuous dosings, which in some instances, particularly against chronic infections, can take weeks (Ślopek et al., 1983; 1987; Duckworth and Gulig, 2002). Quoting d'Hérelle (d'Hérelle and Smith, 1930), p. 176:

> The number of administrations essential to induce a therapeutic effect varies with the type of condition under treatment; in acute infections or in infectious conditions that have not become fully chronic, a single administration or, perhaps, two administrations should overcome the organisms. In chronic conditions, on the other hand, it may be necessary to continue bacteriophage therapy over a relatively long period.

More generally, from Adembri and Novelli (2009), p. 523: "Microbiological and clinical success are less likely when suboptimal exposure and/or incorrect duration of antimicrobial therapy occur".

An exception to the suggestion that multiple or continuous dosing might always be assumed to be necessary can be found in the treatment of non-human subjects such as industrially raised chickens (Kosznik-Kwaśnicka et al., 2019). In this case limitations of convenience or of economic feasibility can be severe, resulting in greater impetus to design treatments in which a single dose is sufficient (Carvalho et al., 2010). Under those circumstances, dosing with higher phage titres also may be more feasible from a regulatory or safety perspective (Sulakvelidze and Pasternack, 2010; Kosznik-Kwaśnicka et al., 2019). Also, it is important to keep in mind that single doses *can* be effective even in the clinic (Wright et al., 2009; Chan et al., 2018), and potentially more effective than multiple dosing with too low phage titres (Jault et al., 2019). Nonetheless, single-dose phage therapy is not necessarily *always* more effective nor necessarily *always* more feasible than multiple or continuous dosing. As Lampert et al. put it (1935, p. 443), "We consider our successful results to be due in large measure to the technique of application which we employ routinely, and which is based upon direct contact of the bacteriophage with the infected tissues by means of generous daily applications." See Abedon (2011a; 2014a) for further consideration of phage frequency of dosing.

Peak and trough phage concentrations

For a standard drug, its peak concentration, or density, is a consequence of conventional dosing alone. Peak concentrations or titres for phages, however, can also be a consequence of *in situ* phage replication. While peak drug concentrations typically are referred to in terms of plasma concentrations, for phages peak concentrations often can be more relevantly described in terms of concentrations within other tissues. The reason for this distinction is that conventional drugs will typically achieve higher concentrations in blood than they will achieve following distribution outside of the blood, unless drugs are preferentially concentrated in specific tissues. Engineered phages that display tissue-targeting peptides or proteins too can potentially accumulate in tissues by affinity only (Górski et al., 2015), though not necessarily highly effectively (Hodyra-Stefaniak et al., 2019). Highest *in situ* concentrations for naturally occurring phages, by contrast, should be achieved especially at the site of infection due to the phage ability to replicate in association with target bacteria (Abedon, 2015b; 2017a).

Contrasting peak drug concentrations are instead trough concentrations (Czock et al., 2009). Specifically, the goal of multiple dosing should be to sustain drug concentrations at or above those required to achieve efficacy, and for at least some antibacterials, this should be as high as possible (Cerf et al., 2010). Thus, if there exists a minimum *effective* concentration of phages that is necessary to achieve a reasonable degree of bacteria killing at a reasonable rate (i.e., inundative densities), then the success or at least duration of phage therapy treatments likely will be a function of the length of time over which phage concentrations meet or exceed this density, however those densities are achieved. That is, trough phage densities, particularly as found in direct association with targeted bacteria, ideally will not fall substantially below minimum effective phage concentrations (Figure 4).

A related consideration is the question of what might contribute to an *upper* limit on peak phage densities. Should they, for example, be limited by safety concerns, regulatory considerations, economic issues, phage replication performance *in situ*, or even due to declines in antibacterial efficacy of ever greater levels of dosing? Generally for phages, however, the answer to this question is not known.

*Absorption and Distribution*

The ability of drugs to move between as well as within the different "compartments" making up bodies is crucial to their *in situ* activity. For many drugs this movement will consist of uptake into the blood stream (absorption) along with movement into other tissues (distribution). For phages, though, the terms translocation and penetration to a degree have been used synonymously with absorption and distribution, respectively. The phage ability to pass from compartment to compartment within bodies, such as from the gastrointestinal lumen into the blood, for example, has been described by Górski et al. (2006) in their review of the subject as "Bacteriophage translocation"; see also Olszowska-Zaremba et al. (2012). In a more narrow sense, phage translocation is a process of phage movement through an epithelial cell, as recently demonstrated *in vitro* by Nguyen et al. (2017) on cell layer cultures. Penetration can also be used to describe phage movement into localized infections, such as infiltration into biofilms, or into animal tissues in general (Dąbrowska et al., 2005). Distribution, or penetration, also can refer to movement from extracellular to intracellular compartments, though the phage potential to impact intracellular bacterial infections has not been substantially explored (Broxmeyer et al., 2002; Danelishvili et al., 2006; Capparelli et al., 2007). Issues of gastrointestinal (enteric) delivery of phages along with bacterial translocation will not be extensively addressed here, but see Dąbrowska and Abedon (2019) for review. We concentrate instead on a subset of topics pertaining to phage absorption and distribution, including examining the extent to which phage therapy researchers should even be concerned with these issues.

### Limitations to phage therapy pharmacokinetics

The pharmacokinetics of chemotherapeutics can differ among the species being treated, e.g., mouse vs. human, and also in terms of the health of the recipient (healthy volunteers vs. patients). Phage therapy so far has not been studied sufficiently robustly to greatly understand pharmacokinetic issues in even one system, however, much less comparatively. Indeed, such determinations are not fully worked out even for standard chemotherapeutic antibacterials (Adembri and Novelli, 2009; Czock et al., 2009). Furthermore, phage therapy may be unusual among antibacterial approaches in that there in fact appears to be a tradition of not always emphasizing especially pharmacokinetic analyses during the development of treatments (Dąbrowska et al., 2018; Dąbrowska and Abedon, 2019).

One example where even the most basic of phage therapy pharmacokinetic properties have not been well worked out is for *per os* delivery, especially that results in phage absorption into the blood. This appears to be possible (Górski et al., 2006; Olszowska-Zaremba et al., 2012; Nguyen et al., 2017), at least under some not yet well specified circumstances, but has not been found to be effective under other circumstances. For example, phage absorption following oral delivery is perhaps less effective for some phages such as certain T4-like and other large phages (Krylov, 2001; Bruttin and Brüssow, 2005; Denou et al., 2009). Some further generalizations also may yet exist, e.g., it appears that the higher the oral phage dose then the higher the probability of active phages coming to be found in blood. As with the pharmacokinetics of drugs more generally, the phage potential for translocation from the gut also might differ depending on the health state of the subject (Letarov et al., 2010). Nevertheless, our potential to predict the degree that a given oral dose of a given phage under a given set of circumstances will be absorbed by a patient is at best limited. A systematic review of such pharmacokinetic issues was recently presented by Dąbrowska (2019) and see also (Dąbrowska and Abedon, 2019; Matsuzaki and Uchiyama, 2019).

A related issue is that phages in phage therapy tend to be diverse, often differing greatly even among the phages sharing a particular target pathogen. Consequently, data collected for any one phage that impacts any one bacterial pathogen may not be greatly applicable even to other phages impacting that same pathogen. These same general issues also exist for antibiotics, however, i.e., an inability to exhaustively characterize their pharmacokinetics in all relevant systems. The number of antibiotics for which pharmacokinetic information is desired, though, is seemingly far smaller than is the case with phages (Harper and Kutter, 2009). The phage therapy literature, that is, is full of research groups isolating novel phage strains that are then relevantly characterized primarily in terms of their pharmacodynamics, i.e., efficacy but also toxicity. To fully appreciate phage therapy pharmacokinetics, however, all phage types used for treating a given pathogen and disease would need to be equivalently and comparatively studied across all relevant systems, e.g., mouse and human as well as healthy and diseased, though this is both unlikely to happen and, at least arguably, may not even be entirely necessary.

As a consequence of the diversity of phages that potentially could be used therapeutically, one could readily argue that performing detailed pharmacokinetic analyses of all phages of interest to phage therapy, under all relevant circumstances, would either represent a daunting task or instead be something that should represent a lower priority than pharmacodynamic determinations. The latter point may be particularly valid to the extent that concerns of phage absorption and distribution are less relevant given phage application directly to bacterial infections – such as topical application to wounds and lungs or application by injection directly to sites of infection – and so long as phages are capable of amplifying their numbers to effective densities once they have reached target bacteria. Thus, while at the same time pharmacokinetic issues may be more challenging to address for phage therapy as a whole, nonetheless under many treatment scenarios pharmacokinetic issues also may be less important to address than simply issues of efficacy and possible toxicity.

How much does phage inactivation really matter?

In this modern era of phage therapy research, roughly the mid-1990s and later (Abedon, 2017b), a fair amount of emphasis has been placed on the potential for bodies to reduce phage numbers *in situ*. Emphasis predominantly has been on what can be described pharmacokinetically as metabolism, and especially the impact of immune systems on continued phage presence. Though it has long been known that the adaptive immune system, especially antibodies, can inactivate phages (Bradley and Watson, 1963; Nelstrop et al., 1968; Ochs et al., 1971; Huff et al., 2010; Hodyra-Stefaniak et al., 2015), it was later found that the mononuclear-phagocyte system, a.k.a., the reticulo-endothelial system can be important as well (Inchley, 1969; Merril et al., 1996; Nilsson et al., 2004; Merril, 2008). Indeed, a standard concern outside of the phage therapy community seems to be that phages simply will be inactivated by the body's immune system prior to the achievement of efficacy. This potential, however, is inconsistent with numerous examples of seeming phage therapy efficacy found throughout the phage therapy literature (Chanishvili, 2012; Abedon, 2015c; 2018a; Morozova et al., 2018b; Abedon, 2019b; El Haddad et al., 2019; Melo et al., 2020), and indeed would constitute an argument against the use of non-human protein-based drugs generally. In addition, and importantly, there are conceptual as well as empirical arguments against the relevance of these concerns regarding phage therapy.

The conceptual arguments stem from issues of treatment duration and location. Adaptive immunity, at least for individuals who at first are naïve to the phages employed, should require up to weeks to fully develop, resulting in little expectation of substantial impact on efficacy if phage treatments are sufficiently short in duration. Alternatively, the mononuclear phagocyte system is an aspect of innate immunity so will not require substantial physiological development to have a negative impact on phage persistence. Nevertheless, this system is an aspect of microorganism removal from the blood, thus suggesting especially that topical or other more localized phage delivery (Table 3) should not be extensively impacted by this mechanism, nor have a substantial impact on those phages which have left the blood into other tissues, i.e., as following their pharmacokinetic distribution.

**Table 3.** Efficacy studies involving direct or nearly direct phage application to infections *in vivo*.[8]

| Treatment Target[9] | Model Animal-Treatment Studies | # (%) | Human-Treatment Studies | # (%) |
|---|---|---|---|---|
| Bacteremias | (Capparelli et al., 2007; Oduor et al., 2016a; Oduor et al., 2016b)[2] | 3 (4) | (Duplessis et al., 2018; Aslam et al., 2019b) | 2 (7) |
| Burn infections | (Kumari et al., 2011; Chadha et al., 2016) | 2 (3) | (Jikia et al., 2005; Marza et al., 2006; Rose et al., 2014; Jault et al., 2019; Patel et al., 2019) | 5 (17) |
| Ear infections | (Marza et al., 2006; Hawkins et al., 2010) | 2 (3) | (Weber-Dąbrowska et al., 2006; Wright et al., 2009) | 2 (7) |
| Eye infections (injection or corneal surface treatment) | (Fukuda et al., 2012; Furusawa et al., 2016; Kishimoto et al., 2019) | 3 (4) | (Fadlallah et al., 2015) | 1 (3) |
| Gastrointestinal delivery | (Tanji et al., 2005; Capparelli et al., 2006; Raya et al., 2006; Sheng et al., 2006; Niu et al., 2008; Rozema et al., 2009; Bach et al., 2009; Jamalludeen et al., 2009; Mai et al., 2010; Stanford et al., 2010; Maura et al., 2012; Jaiswal et al., 2013; Abdulamir et al., 2014; Jaiswal et al., 2014; Jun et al., 2014; Mai et al., 2015; Nale et al., 2015; Sarker et al., 2016; Nikkhahi et al., 2017; Yen et al., 2017; Zhao et al., 2017; Vahedi et al., 2018; Bhandare et al., 2019; Dallal et al., 2019; Dissanayake et al., 2019; Xue et al., 2020) | 26 (36) | (Sarker et al., 2017; Corbellino et al., 2019) | 2 (7) |
| Injection into or near local infections | (Wills et al., 2005; Gill et al., 2006; Capparelli et al., 2007; Bhowmick et al., 2009; Alam et al., 2011; Chhibber et al., 2013; Trigo et al., 2013; Basu et al., 2015; Kishor et al., 2016; Breyne et al., 2017; Chhibber et al., 2017; Yin et al., 2017; Albac et al., 2020; Geng et al., 2020)[3] | 14 (19) | (Chan et al., 2018; Ferry et al., 2018) | 2 (7) |

| | | | | |
|---|---|---|---|---|
| Instillation or irrigation in an infection's vicinity | | 0 | (Khawaldeh et al., 2011; Schooley et al., 2017; Bao et al., 2020) | 3 (10) |
| Lower respiratory tract delivery | (Debarbieux et al., 2010; Morello et al., 2011; Henry et al., 2013; Cao et al., 2015a; Cao et al., 2015b; Dufour et al., 2015; Yang et al., 2015; Gu et al., 2016; Pabary et al., 2016; Wang et al., 2016; Waters et al., 2017; Cha et al., 2018; Chang et al., 2018; Forti et al., 2018; Hua et al., 2018; Abd El-Aziz et al., 2019; Anand et al., 2019; Dufour et al., 2019; Jeon et al., 2019; Jeon and Yong, 2019; Ji et al., 2019) | 21 (29) | (Kutateladze and Adamia, 2008; Hoyle et al., 2018; Aslam et al., 2019a; Maddocks et al., 2019) | 4 (14) |
| Nasal or sinus treatment | (Drilling et al., 2014; Verstappen et al., 2016)[4] | 2 (3) | | 0 |
| Skin infection | | 1 (1) | (Zhvania et al., 2017) | 1 (3) |
| Wound infections (other than burn wounds) | (Mendes et al., 2013; Seth et al., 2013; Shivaswamy et al., 2015; Kusradze et al., 2016; Regeimbal et al., 2016; Sarhan and Azzazy, 2017; Yin et al., 2017; Chhibber et al., 2018) | 8 (11) | (Rhoads et al., 2009; Fish et al., 2016; Fish et al., 2018; Morozova et al., 2018a; Gupta et al., 2019; Patel et al., 2019; Tkhilaishvili et al., 2019) | 7 (24) |
| Totals | | 72 (100) | | 29 (100) |

8. Studies listed are limited to those published 2005 or later, in English, and in which mammals were treated.
9. Not explicitly shown is the study by Międzybrodzki *et al.*, 2009, in which phages were applied topically to humans, potentially in a variety though nonetheless not explicitly described contexts.
10. With Capparelli *et al.*, 2007, both phages and bacteria were supplied via i.v. Oduor *et al.*, 2016a and 2016b, were probably i.v. treated with phages but this is not stated explicitly
11. With Capparelli et al., 2007, it may be inferred that phages and bacteria were injected into similar locations toward abscess formation and treatment, respectively.
12. Verstappen et al., 2016, probably treated with phages intranasally but this is not stated explicitly.

Empirically, studies have looked for associations between the presence of neutralizing antibody in the blood and phage therapy outcomes. These studies, however, have failed to find much of a correlation to either presence or absence of phage therapy efficacy (Łusiak-Szelachowska et al., 2014; 2017). For recent reviews of phage-immune system interactions, see (Dąbrowska et al., 2018; Dąbrowska, 2019; Dąbrowska and Abedon, 2019; Van Belleghem et al., 2019; Żaczek et al., 2019).

## Phage movement into and out of the blood

Movement of phages into blood, i.e., absorption, is an important issue for a variety of reasons. First, it is a primary concern in the pharmacokinetic analysis of drugs in general. That is, essentially the default pathway that drugs take through the body in terms of systemic delivery can be broken down into application that is followed by absorption and only then distribution. Second, blood has the utility of being easily sampled, though plasma drug levels are not always the most meaningful measure of drug pharmacokinetics (Levison and Levison, 2009). Nonetheless, time courses of drug movement especially into and out of the blood can be easily obtained, particularly from larger animals and without animal sacrifice. The result not only is greatly reduced investigation expense but also ready determination of this aspect of drug pharmacokinetics in humans. Three, for targets that are limited to within systemic circulation, drug access to the blood is of course of primary concern. Lastly, to study drug distribution, that is, out of the blood and into other tissues, it is essential to first deliver a drug to the blood. For a general review of these various issues of phage movement within the body, see (Dąbrowska, 2019; Dąbrowska and Abedon, 2019). Nevertheless, absorption is relevant to phage therapy efficacy only to the extent that phages either cannot or are not delivered to sites of bacterial infection more directly.

There exist a variety of pathways by which absorption can be initiated, the most straightforward of which is intravenous (i.v.) delivery. Such parenteral delivery not only results in direct application to systemic circulation but also in principle can be applied at such a rate so as to achieve steady state plasma levels (Levison and Levison, 2009). I.v. application, especially, should be employed for the study of phage distribution – movement out of the blood into different locations – that is unless there is suspicion that alternative absorption protocols could provide qualitatively different phage distribution patterns. This point is also relevant for issues of decline in phage concentrations in the blood, where if phages are not first delivered directly to the blood then it may be difficult to distinguish slow phage movement into the blood from fast phage movement out of the blood (Uchiyama et al., 2009).

While i.v. delivery provides an obvious simplification of absorption, particularly in terms of the study of subsequent distribution, it is not necessarily always ideal for phage delivery, though it certainly can be useful when appropriate. Concerns include its overly invasive nature, the rapidity with which absorption is achieved (though nonetheless this can also be as slow as desired), concerns associated with carryover of contaminating bacterial toxins such as endotoxin directly into the blood (Carlton, 1999; Boratyński et al., 2004; Merril et al., 2006; Gill and Hyman, 2010), and a lack of convenience, particularly in limiting phage delivery to clinical settings. As a consequence, alternative approaches to phage absorption have been developed by various groups. See, though, Speck and Smithyman (2016) for arguments in favor of the use of i.v. delivery of phages for phage therapy and Table 4 for a listing of numerous phage therapy efficacy studies involving i.v., as well as other means of indirect phage dosing.

In terms of subsequent distribution, the more body compartments a drug must pass through to reach its site of activity, then the more steps during which drug concentration can be reduced (Levison and Levison, 2009). These reductions can be due simply to dilution or instead as a consequence of "translocation" inefficiencies. As a result, fewer phages may reach their target upon distribution than are required for treatment success, which in a worst-case scenario could mean no phages at all reaching target bacteria. In addition, effective concentrations such as for chemical antibiotics even following successful absorption can be difficult to achieve due poor blood circulation such as to burns (Ahmad, 2002), osteomyelitis, or diabetic ulcers (Stone, 2002). Therefore, the greater the number of phages that must be delivered to target bacteria to achieve desired levels of bacteria killing or control, then either the more directly phages must be applied to infections (Morozova et al., 2018b) (Table 3), the more phages that must be supplied to the patient overall, or both. Treatment failures, that is, can be due to insufficient numbers of phages encountering target bacteria rather than because of poor intrinsic phage potential to clear bacterial infections. It is crucial for researchers, therefore, to employ pharmacologically rigorous follow up should they observe disappointing phage therapy performance. In particular, though not necessarily easily determined, there can be a need to know whether adequate numbers of dosed phages have reached target bacteria or, instead, whether local phage replication might have been inadequate to compensate.

**Table 4.** Efficacy studies involving indirect phage application, i.e., as toward systemic delivery *in vivo*.[13]

| Application Approach[14] | Model Animal-Treatment Studies | # (%) | Human-Treatment Studies | # (%) |
|---|---|---|---|---|
| Intramuscular (i.m.) | (McVay et al., 2007; Bull et al., 2012; Ibrahim et al., 2016; Chhibber et al., 2017)[15] | 4 (7) | | 0 |
| Intraperitoneal (i.p.)[16] | (Matsuda et al., 2005; Vinodkumar et al., 2005; Wang et al., 2006a; Wang et al., 2006b; Capparelli et al., 2006; McVay et al., 2007; Chhibber et al., 2008; Nishikawa et al., 2008; Uchiyama et al., 2008; Vinodkumar et al., 2008; Kumari et al., 2009a; Kumari et al., 2009b; Zimecki et al., 2009; Carmody et al., 2010; Kumari et al., 2010; Sunagar et al., 2010; Zimecki et al., 2010; Hung et al., 2011; Tothova et al., 2011; Jun et al., 2014; Shivshetty et al., 2014; Takemura-Uchiyama et al., 2014; Pincus et al., 2015; Singla et al., 2015; Chai et al., 2016; Guang-Han et al., 2016; Regeimbal et al., 2016; Chadha et al., 2017; Cheng et al., 2017; Green et al., 2017; Yin et al., 2017; Can et al., 2018; Gelman et al., 2018; Jasim et al., 2018; Wang et al., 2018; Chen et al., 2019; Jeon and Yong, 2019; Kaabi and Musafer, 2019; Leshkasheli et al., 2019; Morris et al., 2019; Tang et al., 2019; Yuan et al., 2019; Alvi et al., 2020)[17] | 43 (80) | | 0 |
| Intrarectal | | 0 | (Letkiewicz et al., 2009; Corbellino et al., 2019) | 2 (12) |
| Intravenous (i.v.) | (Danelishvili et al., 2006; Oechslin et al., 2017; Prazak et al., 2019) | 3 (6) | (Fadlallah et al., 2015; Jennes et al., 2017; Schooley et al., 2017; Duplessis et al., 2018; Aslam et al., 2019b; Dedrick et al., 2019; Gilbey et al., 2019; Law et al., 2019; Maddocks et al., 2019; Nir-Paz et al., 2019) | 10 (59) |

| | | | | |
|---|---|---|---|---|
| Per os | (Hung et al., 2011) | 1 (2) | (Leszczyński et al., 2006; Weber-Dąbrowska et al., 2006; Międzybrodzki et al., 2009; Zhvania et al., 2017; Corbellino et al., 2019)[18] | 5 (29) |
| Subcutaneous (s.c.) | (McVay et al., 2007; Gupta and Prasad, 2010)[19] | 2 (4) | | 0 |
| Transdermal | (Rastogi et al., 2017) | 1 (2) | | 0 |
| Totals | | 54 (100) | | 17 (100) |

13. Studies listed are limited to those published 2005 or later, in English, and in which mammals were treated.
14. See Table 3 for a summary of the use of more direct phage application strategies.
15. Bull et al., 2012, specifically indicate that phages were inoculated into "opposite thighs" relative to application of the bacterial challenge
16. Studies are included in this list whether or not the bacterial challenge was also i.p. delivered, though that mostly was not the case. It also has not escaped our attention the extreme discrepancies seen between pre-clinical and clinical use of the i.p. route of phage application
17. Jeon and Yong, 2019, probably i.p. treated with phages but this is not stated explicitly.
18. While the Międzybrodzki et al., 2009, study is it presented as an efficacy study, it nevertheless piggybacked upon phage therapy treatments intended to be efficacious.
19. For McVay et al., 2007, it is not clear that phages were not applied in the immediate vicinity of the s.c. bacterial challenge, though the relatively poor efficacy especially in comparison with i.p. phage application is suggestive that such co-location of bacterial challenge and phage treatment was not the case.

## Conclusion

Pharmacology is the study of drug impact on bodies as well as body impact on drugs, where body is defined broadly to include normal microbiota and bacterial pathogens as well as normal body tissues. With phage therapy, however, pharmacology is complicated due to the potential for phages to replicate and also due to the large size and complexity of their virions. The latter, for example, makes the phage ability to translocate across certain tissues both somewhat surprising and less predictable. Though the phage potential to replicate *in situ* is a conceptual complication on their pharmacology, alternatively this ability can serve to reduce the importance of greatly understanding certain other aspects of their pharmacology, such as more quantitative measures of phage competence to move between body compartments. More important perhaps than even their ability to replicate, however, is that the use of phages appears to be fairly safe, usually generating at worst only minimal side effects. Notwithstanding this phage potential to amplify their numbers *in situ* and seeming ability to do so rather safely, it is desirable for the sake of improvement of efficacy, regulatory approval, and simply rigorous experimentation that pharmacological considerations not be ignored during phage therapy development. In short, it is important to know *in situ* and over time where in bodies phages are found, at what titres, and to what if any antibacterial or toxic effects. It is especially from a better understanding of these basic pharmacological measures that the effectiveness of phage therapy approaches may be improved.

## Acknowledgements

KD's work was supported by National Science Centre in Poland, grant UMO-2012/05/E/NZ6/03314. KDW's work was supported by The Ohio State University President's Postdoctoral Scholars Program.

## References

Abd El-Aziz, A.M., Elgaml, A., and Ali, Y.M. (2019). Bacteriophage therapy increases complement-mediated lysis of bacteria and enhances bacterial clearance after acute lung infection with multidrug-resistant *Pseudomonas aeruginosa*. J. Infect. Dis. *219*, 1439-1447.

Abdulamir, A.S., Jassim, S.A., and Abu, B.F. (2014). Novel approach of using a cocktail of designed bacteriophages against gut pathogenic *E. coli* for bacterial load biocontrol. Ann. Clin. Microbiol. Antimicrob. *13*, 39.

Abedon, S.T. (1990). Selection for lysis inhibition in bacteriophage. J. Theor. Biol. *146*, 501-511.

Abedon, S.T. (1994). Lysis and the interaction between free phages and infected cells. In The Molecular Biology of Bacteriophage T4, J.D. Karam, E. Kutter, K. Carlson, and B. Guttman, eds. (Washington, DC: ASM Press), pp. 397-405.

Abedon, S.T. (1999). Bacteriophage T4 resistance to lysis-inhibition collapse. Genet. Res. *74*, 1-11.

Abedon, S.T. (2009). Phage evolution and ecology. Adv. Appl. Microbiol. *67*, 1-45.

Abedon, S.T. (2010). Bacteriophages and biofilms. In Biofilms: Formation, Development and Properties, W.C.Bailey, ed. (Hauppauge, New York: Nova Science Publishers), pp. 1-58.

Abedon, S. (2011a). Phage therapy pharmacology: calculating phage dosing. Adv. Appl. Microbiol. *77*, 1-40.

Abedon, S.T. (2011b). Bacteriophages and Biofilms: Ecology, Phage Therapy, Plaques. (Hauppauge, New York: Nova Science Publishers).

Abedon, S.T. (2011c). Envisaging bacteria as phage targets. Bacteriophage *1*, 228-230.

Abedon, S.T. (2011d). Lysis from without. Bacteriophage *1*, 46-49.

Abedon, S.T. (2012a). Phage therapy best practices. In Bacteriophages in Health and Disease, P.Hyman and S.T.Abedon, eds. (Wallingford, UK: CABI Press), pp. 256-272.

Abedon, S.T. (2012b). Spatial vulnerability: bacterial arrangements, microcolonies, and biofilms as responses to low rather than high phage densities. Viruses *4*, 663-687.

Abedon, S.T. (2012c). Thinking about microcolonies as phage targets. Bacteriophage 2, 200-204.

Abedon, S.T. (2014a). Bacteriophages as drugs: the pharmacology of phage therapy. In Phage Therapy: Current Research and Applications, J. Borysowski, R. Międzybrodzki, and A. Górski, eds. (Norfolk, UK: Caister Academic Press), pp. 69-100.

Abedon, S.T. (2014b). Phage therapy: eco-physiological pharmacology. Scientifica *2014*, 581639.

Abedon, S.T. (2015a). Bacteriophage secondary infection. Virol. Sin. *30*, 3-10.

Abedon, S.T. (2015b). Ecology of anti-biofilm agents I. antibiotics versus bacteriophages. Pharmaceuticals. (Basel) *8*, 525-558.

Abedon, S.T. (2015c). Phage therapy of pulmonary infections. Bacteriophage *5*, e1020260.

Abedon, S.T. (2016). Phage therapy dosing: the problem(s) with multiplicity of infection (MOI). Bacteriophage *6*, e1220348.

Abedon, S.T. (2017a). Active bacteriophage biocontrol and therapy on sub-millimeter scales towards removal of unwanted bacteria from foods and microbiomes. AIMS Microbiol. *3*, 649-688.

Abedon, S.T. (2017b). Bacteriophage clinical use as antibactertial "drugs": utility, precedent. Microbiol. Spectr. *5*, BAD-0003-2016.

Abedon, S.T. (2017c). Information phage therapy research should report. Pharmaceuticals. (Basel) *10*, 43.

Abedon, S.T. (2017d). Phage "delay" towards enhancing bacterial escape from biofilms: a more comprehensive way of viewing resistance to bacteriophages. AIMS Microbiol. *3*, 186-226.

Abedon, S.T. (2018a). Bacteriophage-mediated biocontrol of wound infections, and ecological exploitation of biofilms by phages. In Recent Clinical Techniques, Results, and Research in Wounds, M.Shiffman, ed. Springer International Publishing AG).

Abedon, S.T. (2018b). Phage therapy: various perspectives on how to improve the art. Meth. Mol. Biol. *1734*, 113-127.

Abedon, S.T. (2019a). Phage-antibiotic combination treatments: antagonistic impacts of antibiotics on the pharmacodynamics of phage therapy? Antibiotics *8*, 182.

Abedon, S.T. (2019b). Use of phage therapy to treat long-standing, persistent, or chronic bacterial infections. Advanced Drug Delivery Reviews *145*, 18-39.

Abedon, S.T. (2020a). Phage therapy: killing titers, multiplicity of infection, adsorption theory, and passive versus active treatments. In Advances on the Applications of Bacteriophages, D.I.Kurtboke and R.Aminov, eds. (Hauppauge, New York: Nova Science Publishers).

Abedon, S.T. (2020b). Phage-phage, phage-bacteria, and phage-environment communication. In Biocommunication of Phages, G.Witzany, ed. Springer).

Abedon, S.T., Duffy, S., and Turner, P.E. (2009). Bacteriophage ecology. In Encyclopedia of Microbiology, M.Schaecter, ed. (Oxford: Elsevier), pp. 42-57.

Abedon, S.T. and Katsaounis, T.I. (2018). Basic phage mathematics. Meth. Mol. Biol. *1681*, 3-30.

Abedon, S.T., Kuhl, S.J., Blasdel, B.G., and Kutter, E.M. (2011). Phage treatment of human infections. Bacteriophage *1*, 66-85.

Abedon, S.T. and Thomas-Abedon, C. (2010). Phage therapy pharmacology. Curr. Pharm. Biotechnol. *11*, 28-47.

Abul-Hassan, H.S., El-Tahan k Massoud, B., and Gomaa, R. (1990). Bacteriophage therapy of *Pseudomonas* burn wound sepsis. Ann. Med. Burn Club *3*, 262-264.

Adembri, C. and Novelli, A. (2009). Pharmacokinetic and pharmacodynamic parameters of antimicrobials: potential for providing dosing regimens that are less vulnerable to resistance. Clin. Pharmacokinet. *48*, 517-528.

Ahmad, S.I. (2002). Treatment of post-burns bacterial infections by bacteriophages, specifically ubiquitous *Pseudomonas* spp. notoriously resistant to antibiotics. Med. Hypoth. *58*, 327-331.

Alam, M., Akhter, M.Z., Yasmin, M., Ahsan, C.R., and Nessa, J. (2011). Local bacteriophage isolates showed anti- *Escherichia coli* O157:H7 potency in an experimental ligated rabbit ileal loop model. Can. J. Microbiol. *57*, 408-415.

Albac, S., Medina, M., Labrousse, D., Hayez, D., Bonnot, D., Anzala, N., Laurent, F., Ferry, T., Dublanchet, A., Chavanet, P., Fevre, C., and Croisier, D. (2020). Efficacy of bacteriophages in a *Staphylococcus aureus* nondiabetic or diabetic foot infection murine model. Antimicrob. Agents Chemother. *64*, e01870-19.

Alves, D.R. and Abedon, S.T. (2017). An online phage therapy bibliography: separating under-indexed wheat from overly indexed chaff. AIMS Microbiol. *3*, 525-528.

Alvi, I.A., Asif, M., Tabassum, R., Aslam, R., Abbas, Z., and Rehman, S.U. (2020). RLP, a bacteriophage of the family *Podoviridae*, rescues mice from bacteremia caused by multi-drug-resistant *Pseudomonas aeruginosa*. Arch. Virol. *165*,1289-1297.

Anand, T., Virmani, N., Kumar, S., Kumar, M.A., Pavulraj, S., Ch, B.B., Vaid, R.K., Ahlawat, U., and Tripathi, B.N. (2019). Phage therapy for treatment of virulent *Klebsiella pneumoniae* infection in mouse model. J Glob. Antimicrob. Resist. *21*, 41.

Aslam, S., Courtwright, A.M., Koval, C., Lehman, S.M., Morales, S., Langlais Furr, C.L., Rosas, F., Brownstein, M.J., Fackler, J.R., Sisson, B.M., Biswas, B., Henry, M., Luu, T., Bivens, B.N., Hamilton, T., Duplessis, C., Logan, C., Law, N., Yung, G., Turowski, J., Anesi, J., Strathdee, S.A., and Schooley, R.T. (2019a). Early clinical experience of bacteriophage therapy in three lung transplant recipients. Am. J Transplant. *19*, 2631-2639.

Aslam, S., Pretorius, V., Lehman, S.M., Morales, S., and Schooley, R.T. (2019b). Novel bacteriophage therapy for treatment of left ventricular assist device infection. J Heart Lung Transplant. *38*, 475-476.

Bach, S.J., Johnson, R.P., Stanford, K., and McAllister, T.A. (2009). Bacteriophages reduce *Escherichia coli* O157:H7 levels in experimentally inoculated sheep. Can. J. Anim. Sci. *89*, 285-293.

Bao, J., Wu, N., Zeng, Y., Chen, L., Li, L., Yang, L., Zhang, Y., Guo, M., Li, L., Li, J., Tan, D., Cheng, M., Gu, J., Qin, J., Liu, J., Li, S., Pan, G., Jin, X., Yao, B., Guo, X., Zhu, T., and Le, S. (2020). Non-active antibiotic and bacteriophage synergism to successfully treat recurrent urinary tract infection caused by extensively drug-resistant *Klebsiella pneumoniae*. Emerg. Microbes. Infect. *9*, 771-774.

Barrow, P.A. and Soothill, J.S. (1997). Bacteriophage therapy and prophylaxis: rediscovery and renewed assessment of potential. Trends Microbiol. *5*, 268-271.

Basu, S., Agarwal, M., Kumar, B.S., Nath, G., and Kumar, S., V (2015). An *In vivo* wound model utilizing bacteriophage therapy of *Pseudomonas aeruginosa* biofilms. Ostomy. Wound. Manage. *61*, 16-23.

Berngruber, T.W., Weissing, F.J., and Gandon, S. (2010). Inhibition of superinfection and the evolution of viral latency. J. Virol. *84*, 10200-10208.

Bhandare, S., Colom, J., Baig, A., Ritchie, J.M., Bukhari, H., Shah, M.A., Sarkar, B.L., Su, J., Wren, B., Barrow, P., and Atterbury, R.J. (2019). Reviving phage therapy for the treatment of cholera. J. Infect. Dis. *219*, 786-794.

Bhatt, J.M. (2013). Treatment of pulmonary exacerbations in cystic fibrosis. European Respiratory Review *22*, 205-216.

Bhowmick, T.S., Koley, H., Das, M., Saha, D.R., and Sarkar, B.L. (2009). Pathogenic potential of vibriophages against an experimental infection with *Vibrio cholerae* O1 in the RITARD model. Int. J. Antimicrob. Agents *33*, 569-573.

Bigwood, T., Hudson, J.A., and Billington, C. (2009). Influence of host and bacteriophage concentrations on the inactivation of food-borne pathogenic bacteria by two phages. FEMS Microbiol. Lett. *291*, 59-64.

Blasdel, B.G. and Abedon, S.T. (2017). Superinfection immunity. In Reference Module in Life Sciences, (Elsevier). http://dx.doi.org/10.1016/B978-0-12-809633-8.07217-4

Blix, H.S., Viktil, K.K., Moger, T.A., and Reikvam, A. (2010). Drugs with narrow therapeutic index as indicators in the risk management of hospitalised patients. Pharm. Pract. (Granada. ) *8*, 50-55.

Boratyński, J., Syper, D., Weber-Dąbrowska, B., Łusiak-Szelachowska, M., Poźniak, G., and Górski, A. (2004). Preparation of endotoxin-free bacteriophages. Cell. Mol. Biol. Lett. *9*, 253-259.

Bradley, S.G. and Watson, D.W. (1963). Production of neutralizing antibody by mice injected with actinophage. J Immunol. *90*, 782-787.

Breyne, K., Honaker, R.W., Hobbs, Z., Richter, M., Żaczek, M., Spangler, T., Steenbrugge, J., Lu, R., Kinkhabwala, A., Marchon, B., Meyer, E., and Mokres, L. (2017). Efficacy and safety of a bovine-associated *Staphylococcus aureus* phage cocktail in a murine model of mastitis. Front. Microbiol. *8*, 2348.

Broxmeyer, L., Sosnowskai, D., Miltner, E., Chacón, O., Wagner, D., McGarvey, J., Barletta, R.G., and Bermudez, L.E. (2002). Killing of *Mycobacterium avium* and *Mycobacterium tuberculosis* by a mycobacteriophage delivered by a nonvirulent mycobacterium: A model for phage therapy of intracellular bacterial pathogens. J. Infect. Dis. *186*, 1155-1160.

Brüssow, H. (2007). Phage therapy: the Western perspective. In Bacteriophage: Genetics and Microbiology, S.Mc Grath and D.van Sinderen, eds. (Norfolk, UK: Caister Academic Press), pp. 159-192.

Bruttin, A. and Brüssow, H. (2005). Human volunteers receiving *Escherichia coli* phage T4 orally: a safety test of phage therapy. Antimicrob. Agents Chemother. *49*, 2874-2878.

Bull, J.J., Otto, G., and Molineux, I.J. (2012). *In vivo* growth rates are poorly correlated with phage therapy success in a mouse infection model. Antimicrob. Agents Chemother. *56*, 949-954.

Bull, J.J. and Regoes, R.R. (2006). Pharmacodynamics of non-replicating viruses, bacteriocins and lysins. Proc. R. Soc. Lond. B Biol. Sci. *273*, 2703-2712.

Cairns, B.J. and Payne, R.J.H. (2008). Bacteriophage therapy and the mutant selection window. Antimicrob. Agents Chemother. *52*, 4344-4350.

Cairns, B.J., Timms, A.R., Jansen, V.A., Connerton, I.F., and Payne, R.J. (2009). Quantitative models of in vitro bacteriophage-host dynamics and their application to phage therapy. PLoS Path. *5*, e1000253.

Calendar, R. and Abedon, S.T. (2006). The Bacteriophages. (Oxford: Oxford University Press).

Can, K., Aksu, U., and Yenen, O.S. (2018). Investigation of PhiKZ phage therapy against *Pseudomonas aeruginosa* in mouse pneumonia model. Turk. J. Med. Sci. *48*, 670-678.

Cao, F., Wang, X., Wang, L., Li, Z., Che, J., Wang, L., Li, X., Cao, Z., Zhang, J., Jin, L., and Xu, Y. (2015a). Evaluation of the efficacy of a bacteriophage in the treatment of pneumonia induced by multidrug resistance *Klebsiella pneumoniae* in mice. Biomed. Res. Int. *2015*, 752930.

Cao, Z., Zhang, J., Niu, Y.D., Cui, N., Ma, Y., Cao, F., Jin, L., Li, Z., and Xu, Y. (2015b). Isolation and characterization of a "phiKMV-like" bacteriophage and its therapeutic effect on mink hemorrhagic pneumonia. PLoS. One. *10*, e0116571.

Capparelli, R., Nocerino, N., Iannaccone, M., Ercolini, D., Parlato, M., Chiara, M., and Iannelli, D. (2010). Bacteriophage therapy of *Salmonella enterica*: a fresh appraisal of bacteriophage therapy. J. Infect. Dis. *201*, 52-61.

Capparelli, R., Ventimiglia, I., Roperto, S., Fenizia, D., and Iannelli, D. (2006). Selection of an *Escherichia coli* O157:H7 bacteriophage for persistence in the circulatory system of mice infected experimentally. Clin. Microbiol. Infect. *12*, 248-253.

Capparelli, R., Parlato, M., Borriello, G., Salvatore, P., and Iannelli, D. (2007). Experimental phage therapy against *Staphylococcus aureus* in mice. Antimicrob. Agents Chemother. *51*, 2765-2773.

Carlson, K. (2005). Working with bacteriophages: common techniques and methodological approaches. In Bacteriophages: Biology and Application, E.Kutter and A.Sulakvelidze, eds. (Boca Raton, Florida: CRC Press), pp. 437-494.

Carlson, K. and Miller, E.S. (1994). Enumerating phage: the plaque assay. In Molecular Biology of Bacteriophage T4, J.D.Karam, ed. (Washington, DC: ASM Press), pp. 427-429.

Carlton, R.M. (1999). Phage therapy: past history and future prospects. Arch. Immunol. Ther. Exp. *47*, 267-274.

Carmody, L.A., Gill, J.J., Summer, E.J., Sajjan, U.S., Gonzalez, C.F., Young, R.F., and LiPuma, J.J. (2010). Efficacy of bacteriophage therapy in a model of *Burkholderia cenocepacia* pulmonary infection. J. Infect. Dis. *201*, 264-271

Carvalho, C.M., Gannon, B.W., Halfhide, D.E., Santos, S.B., Hayes, C.M., Roe, J.M., and Azeredo, J. (2010). The *in vivo* efficacy of two administration routes of a phage cocktail to reduce numbers of *Campylobacter coli* and *Campylobacter jejuni* in chickens. BMC Microbiol. *10*, 232.

Cerf, O., Carpentier, B., and Sanders, P. (2010). Tests for determining in-use concentrations of antibiotics and disinfectants are based on entirely different concepts: "resistance" has different meanings. Int. J. Food Microbiol. *136*, 247-254.

Cerveny, K.E., Depaola, A., Duckworth, D.H., and Gulig, P.A. (2002). Phage therapy of local and systemic disease caused by *Vibrio vulnificus* in iron-dextran-treated mice. Infect. Immun. *70*, 6251-6262.

Cha, K., Oh, H.K., Jang, J.Y., Jo, Y., Kim, W.K., Ha, G.U., Ko, K.S., and Myung, H. (2018). Characterization of two novel bacteriophages infecting multidrug-resistant (MDR) *Acinetobacter baumannii* and evaluation of their therapeutic efficacy *in vivo*. Front. Microbiol. *9*, 696.

Chadha, P., Katare, O.P., and Chhibber, S. (2016). *In vivo* efficacy of single phage versus phage cocktail in resolving burn wound infection in BALB/c mice. Microb. Pathog. *99*, 68-77.

Chadha, P., Katare, O.P., and Chhibber, S. (2017). Liposome loaded phage cocktail: enhanced therapeutic potential in resolving *Klebsiella pneumoniae* mediated burn wound infections. Burns *43*, 1532-1543.

Chai, Q., Wu, D., and Liu, F. (2016). Therapy potential of tailless bacteriophage FHN161 and its ability in modulating inflammation caused by bacterial disease. Vet. Med. Open J. *1*, 36-42.

Chan, B.K. and Abedon, S.T. (2012). Bacteriophage adaptation, with particular attention to issues of phage host range. In Bacteriophages in Dairy Processing, A.Quiberoni and J.Reinheimer, eds. (Hauppauge, New York: Nova Science Publishers), pp. 25-52.

Chan, B.K., Turner, P.E., Kim, S., Mojibian, H.R., Elefteriades, J.A., and Narayan, D. (2018). Phage treatment of an aortic graft infected with *Pseudomonas aeruginosa*. Evol. Med. Pub. Health *1*, 60-66.

Chang, R.Y.K., Chen, K., Wang, J., Wallin, M., Britton, W., Morales, S., Kutter, E., Li, J., and Chan, H.K. (2018). Proof-of-principle study in a murine lung infection model of antipseudomonal activity of phage PEV20 in a dry-powder formulation. Antimicrob. Agents Chemother. *62*, e01714-e01717.

Chanishvili, N. (2012). A Literature Review of the Practical Application of Bacteriophage Research. (Hauppauge, New York: Nova Publishers).

Chanishvili, N., Chanishvili, T., Tediashvili, M., and Barrow, P.A. (2001). Phages and their application against drug-resistant bacteria. J. Chem. Technol. Biotechnol. *76*, 689-699.

Chanishvili, N., Khurtsia, N., and Malkhazova, I. (2009). Combined use of phages and antibiotics. In A Literature Review of the Practical Application of Bacteriophage Research, N.Chanishvili and R.Sharp, eds. (Tbilisi, Georgia: Eliava Institute), pp. 74-81.

Chaudhry, W.N., Concepcion-Acevedo, J., Park, T., Andleeb, S., Bull, J.J., and Levin, B.R. (2017). Synergy and order effects of antibiotics and phages in killing *Pseudomonas aeruginosa* biofilms. PLoS. One. *12*, e0168615.

Chen, Y., Guo, G., Sun, E., Song, J., Yang, L., Zhu, L., Liang, W., Hua, L., Peng, Z., Tang, X., Chen, H., and Wu, B. (2019). Isolation of a T7-like lytic *Pasteurella* bacteriophage vB_PmuP_PHB01 and its potential use in therapy against *Pasteurella multocida* infections. Viruses *11*, 86.

Cheng, M., Liang, J., Zhang, Y., Hu, L., Gong, P., Cai, R., Zhang, L., Zhang, H., Ge, J., Ji, Y., Guo, Z., Feng, X., Sun, C., Yang, Y., Lei, L., Han, W., and Gu, J. (2017). The bacteriophage EF-P29 efficiently protects against lethal vancomycin-resistant *Enterococcus faecalis* and alleviates gut microbiota imbalance in a murine bacteremia model. Front Microbiol *8*, 837.

Chhibber, S., Kaur, J., and Kaur, S. (2018). Liposome entrapment of bacteriophages improves wound healing in a diabetic mouse MRSA infection. Front. Microbiol. *9*, 561.

Chhibber, S., Kaur, T., and Sandeep, K. (2013). Co-therapy using lytic bacteriophage and linezolid: effective treatment in eliminating methicillin resistant *Staphylococcus aureus* (MRSA) from diabetic foot infections. PLoS. One. *8*, e56022.

Chhibber, S., Shukla, A., and Kaur, S. (2017). Transfersomal phage cocktail is an effective treatment against methicillin-resistant *Staphylococcus aureus*-mediated skin and soft tissue infections. Antimicrob. Agents Chemother. *61*, e02146-16.

Chhibber, S., Kaur, S., and Kumari, S. (2008). Therapeutic potential of bacteriophage in treating *Klebsiella pneumoniae* B5055-mediated lobar pneumonia in mice. J. Med. Microbiol. *57*, 1508-1513.

Cisło, M., Dąbrowski, M., Weber-Dąbrowska, B., and Woytoń, A. (1987). Bacteriophage treatment of suppurative skin infections. Arch. Immunol. Ther. Exp. *35*, 175-183.

Clark, J., Abedon, S.T., and Hyman, P. (2012). Phages as therapeutic delivery vehicles. In Bacteriophages in Health and Disease, P.Hyman and S.T.Abedon, eds. (Wallingford, UK: CABI Press), pp. 86-100.

Cooper, C.J., Koonjan, S., and Nilsson, A.S. (2018). Enhancing whole phage therapy and their derived antimicrobial enzymes through complex formulation. Pharmaceuticals. (Basel) *11*, 34.

Corbellino, M., Kieffer, N., Kutateladze, M., Balarjishvili, N., Leshkasheli, L., Askilashvili, L., Tsertsvadze, G., Rimoldi, S.G., Nizharadze, D., Hoyle, N., Nadareishvili, L., Antinori, S., Pagani, C., Scorza, D.G., Romano, A.L.L., Ardizzone, S., Danelli, P., Gismondo, M.R., Galli, M., Nordmann, P., and Poirel, L. (2019). Eradication of a multi-drug resistant, carbapenemase-producing *Klebsiella pneumoniae* isolate following oral and intra-rectal therapy with a custom-made, lytic bacteriophage preparation. Clin. Infect. Dis. *70*, 1998-2001.

Curtright, A.J. and Abedon, S.T. (2011). Phage therapy: emergent property pharmacology. J. Bioanalyt. Biomed. *S3*, 010.

Czock, D., Markert, C., Hartman, B., and Keller, F. (2009). Pharmacokinetics and pharmacodynamics of antimicrobial drugs. Expert Opin. Drug Metab. Toxicol. *5*, 475-487.

d'Hérelle, F. and Smith, G.H. (1930). The Bacteriophage and its Clinical Application. (Springfield, Illinois: Charles C. Thomas, Publisher).

Dąbrowska, K. (2019). Phage therapy: what factors shape phage pharmacokinetics and bioavailability? Systematic and critical review. Med. Res. Rev. *39*, 2000-2025.

Dąbrowska, K. and Abedon, S.T. (2019). Pharmacologically aware phage therapy: pharmacodynamic and pharmacokinetic obstacles to phage antibacterial action in animal and human bodies. Microbiol. Mol. Biol. Rev. *83, e00012-19.*

Dąbrowska, K., Górski, A., and Abedon, S.T. (2018). Bacteriophage pharmacology and immunology. In Bacteriophages: Biology, Technology, Therapy, D.Harper, S.T.Abedon, B.H.Burrowes, and M.McConville, eds. Springer International Publishing AG), p. https://link.springer.com/referenceworkentry/10.1007/978-3-319-40598-8_9-1.

Dąbrowska, K., Switała-Jeleń, K., Opolski, A., Weber-Dabrowska, B., and Gorski, A. (2005). Bacteriophage penetration in vertebrates. J. Appl. Microbiol. *98*, 7-13.

Dallal, M.M.S., Nikkhahi, F., Alimohammadi, M., Douraghi, M., Rajabi, Z., Foroushani, A.R., Azimi, A., and Fardsanei, F. (2019). Phage therapy as an approach to control *Salmonella enterica* serotype Enteritidis infection in mice. Rev. Soc. Bras. Med. Trop. *52*, e20190290.

Danelishvili, L., Young, L.S., and Bermudez, L.E. (2006). *In vivo* efficacy of phage therapy for *Mycobacterium avium* infection as delivered by a nonvirulent mycobacterium. Microb. Drug Resist. *12*, 1-6.

Debarbieux, L., Leduc, D., Maura, D., Morello, E., Criscuolo, A., Grossi, O., Balloy, V., and Touqui, L. (2010). Bacteriophages can treat and prevent *Pseudomonas aeruginosa* lung infections. J. Infect. Dis. *201*, 1096-1104.

Dedrick, R.M., Guerrero-Bustamante, C.A., Garlena, R.A., Russell, D.A., Ford, K., Harris, K., Gilmour, K.C., Soothill, J., Jacobs-Sera, D., Schooley, R.T., Hatfull, G.F., and Spencer, H. (2019). Engineered bacteriophages for treatment of a patient with a disseminated drug-resistant *Mycobacterium abscessus*. Nat. Med. *25*, 730-733.

Denou, E., Bruttin, A., Barretto, C., Ngom-Bru, C., Brüssow, H., and Zuber, S. (2009). T4 phages against *Escherichia coli* diarrhea: potential and problems. Virology *388*, 21-30.

Dissanayake, U., Ukhanova, M., Moye, Z.D., Sulakvelidze, A., and Mai, V. (2019). Bacteriophages reduce pathogenic *Escherichia coli* counts in mice without distorting gut microbiota. Front. Microbiol. *10*, 1984.

Domingo-Calap, P., Mora-Quilis, L., and Sanjuan, R. (2020). Social Bacteriophages. Microorganisms. *8, 533.*

Drilling, A., Morales, S., Boase, S., Jervis-Bardy, J., James, C., Jardeleza, C., Tan, N.C., Cleland, E., Speck, P., Vreugde, S., and Wormald, P.J. (2014). Safety and efficacy of topical bacteriophage and ethylenediaminetetraacetic acid treatment of *Staphylococcus aureus* infection in a sheep model of sinusitis. Int. Forum Allergy Rhinol. *4*, 176-186.

Duckworth, D.H. and Gulig, P.A. (2002). Bacteriophages: potential treatment for bacterial infections. BioDrugs *16*, 57-62.

Dufour, N., Debarbieux, L., Fromentin, M., and Ricard, J.D. (2015). Treatment of highly virulent extraintestinal pathogenic *Escherichia coli* pneumonia with bacteriophages. Crit. Care Medicine *43*, e190-e198.

Dufour, N., Delattre, R., Chevallereau, A., Ricard, J.D., and Debarbieux, L. (2019). Phage therapy of pneumonia is not associated with an overstimulation of the inflammatory response compared to antibiotic treatment in mice. Antimicrob. Agents Chemother. *63*, e00379-19.

Duplessis, C., Biswas, B., Hanisch, B., Perkins, M., Henry, M., Quinones, J., Wolfe, D., Estrella, L., and Hamilton, T. (2018). Refractory *Pseudomonas* bacteremia in a 2-year-old sterilized by bacteriophage therapy. J. Pediatric Infect. Dis. Soc. *7*, 253-256.

El Haddad, L., Harb, C.P., Gebara, M.A., Stibich, M.A., and Chemaly, R.F. (2019). A systematic and critical review of bacteriophage therapy against multi-drug resistant ESKAPE organisms in humans. Clin. Infect. Dis. *69*, 167-178.

Fadlallah, A., Chelala, E., and Legeais, J.M. (2015). Corneal infection therapy with topical bacteriophage administration. Open. Ophthalmol. J. *9*, 167-168.

Ferry, T., Leboucher, G., Fevre, C., Herry, Y., Conrad, A., Josse, J., Batailler, C., Chidiac, C., Medina, M., Lustig, S., and Laurent, F. (2018). Salvage debridement, antibiotics and implant retention ("DAIR") with local injection of a selected cocktail of bacteriophages: is it an option for an elderly patient with relapsing *Staphylococcus aureus* prosthetic-joint infection? Open Forum Infect. Dis. *5*, ofy269.

Fish, R., Kutter, E., Wheat, G., Blasdel, B., Kutateladze, M., and Kuhl, S. (2016). Bacteriophage treatment of intransigent diabetic toe ulcers: a case series. J Wound Care *25 Suppl 7*, S27-S33.

Fish, R., Kutter, E., Wheat, G., Blasdel, B., Kutateladze, M., and Kuhl, S. (2018). Bacteriophage therapy for foot ulcer treatment as an effective step for moving toward clinical trials. Meth. Mol. Biol. *1693*, 159-170.

Forti, F., Roach, D.R., Cafora, M., Pasini, M.E., Horner, D.S., Fiscarelli, E.V., Rossitto, M., Cariani, L., Briani, F., Debarbieux, L., and Ghisotti, D. (2018). Design of a broad-range bacteriophage cocktail that reduces *Pseudomonas aeruginosa* biofilms and treats acute infections in two animal models. Antimicrob. Agents Chemother. *62*, e02573-17.

Fukuda, K., Ishida, W., Uchiyama, J., Rashel, M., Kato, S., Morita, T., Muraoka, A., Sumi, T., Matsuzaki, S., Daibata, M., and Fukushima, A. (2012). *Pseudomonas aeruginosa* keratitis in mice: effects of topical bacteriophage KPP12 administration. PLoS. One. *7*, e47742.

Furusawa, T., Iwano, H., Hiyashimizu, Y., Matsubara, K., Higuchi, H., Nagahata, H., Niwa, H., Katayama, Y., Kinoshita, Y., Hagiwara, K., Iwasaki, T., Tanji, Y., Yokota, H., and Tamura, Y. (2016). Phage therapy is effective in a mouse model of bacterial equine keratitis. Appl. Environ. Microbiol. *82*, 5332-5339.

Gelman, D., Beyth, S., Lerer, V., Adler, K., Poradosu-Cohen, R., Coppenhagen-Glazer, S., and Hazan, R. (2018). Combined bacteriophages and antibiotics as an efficient therapy against VRE *Enterococcus faecalis* in a mouse model. Res. Microbiol. *169*, 531-539.

Geng, H., Zou, W., Zhang, M., Xu, L., Liu, F., Li, X., Wang, L., and Xu, Y. (2020). Evaluation of phage therapy in the treatment of *Staphylococcus aureus*-induced mastitis in mice. Folia Microbiol. *65*, 339-351.

Gilbey, T., Ho, J., Cooley, L.A., Petrovic, F.A., and Iredell, J.R. (2019). Adjunctive bacteriophage therapy for prosthetic valve endocarditis due to *Staphylococcus aureus*. Med. J Aust. *211*, 142-143.

Gill, J.J. (2008). Modeling of bacteriophage therapy. In Bacteriophage Ecology, S.T.Abedon, ed. (Cambridge, UK: Cambridge University Press), pp. 439-464.

Gill, J.J. and Hyman, P. (2010). Phage choice, isolation and preparation for phage therapy. Curr. Pharm. Biotechnol. *11*, 2-14.

Gill, J.J., Pacan, J.C., Carson, M.E., Leslie, K.E., Griffiths, M.W., and Sabour, P.M. (2006). Efficacy and pharmacokinetics of bacteriophage therapy in treatment of subclinical *Staphylococcus aureus* mastitis in lactating dairy cattle. Antimicrob. Agents Chemother. *50*, 2912-2918.

Goodridge, L.D. (2008). Phages, bacteria, and food. In Bacteriophage Ecology, S.T.Abedon, ed. (Cambridge, UK: Cambridge University Press), pp. 302-331.

Goodridge, L.D. (2010). Designing phage therapeutics. Current Pharmaceutical Biotechnology *11*, 15-27.

Górski, A., Dąbrowska, K., Hodyra-Stefaniak, K., Borysowski, J., Międzybrodzki, R., and Weber-Dąbrowska, B. (2015). Phages targeting infected tissues: novel approach to phage therapy. Future Microbiol. *10*, 199-204.

Górski, A., Wazna, E., Dąbrowska, B.-W., Switała-Jeleń, K., and Międzybrodzki, R. (2006). Bacteriophage translocation. FEMS Immunol. Med. Microbiol. *46*, 313-319.

Górski, A., Jończyk-Matysiak, E., Międzybrodzki, R., Weber-Dąbrowska, B., Łusiak-Szelachowska, M., Bagińska, N., Borysowski, J., Łobocka, M.B., Węgrzyn, A., and Węgrzyn, G. (2018a). Phage therapy: beyond antibacterial action. Frontiers in medicine *5*, 146.

Górski, A., Międzybrodzki, R., Łobocka, M., Głowacka-Rutkowska, A., Bednarek, A., Borysowski, J., Jończyk-Matysiak, E., Łusiak-Szelachowska, M., Weber-Dąbrowska, B., Bagińska, N., Letkiewicz, S., Dąbrowska, K., and Scheres, J. (2018b). Phage therapy: what have we learned? Viruses *10*, 288.

Goth, A. (1984). Medical Pharmacology. (Oxford: Blackwell Scientific).

Green, S.I., Kaelber, J.T., Ma, L., Trautner, B.W., Ramig, R.F., and Maresso, A.W. (2017). Bacteriophages from ExPEC reservoirs kill pandemic multidrug-resistant strains of clonal group ST131 in animal models of bacteremia. Sci. Rep. *7*, 46151.

Gu, J., Li, X., Yang, M., Du, C., Cui, Z., Gong, P., Xia, F., Song, J., Zhang, L., Li, J., Yu, C., Sun, C., Feng, X., Lei, L., and Han, W. (2016). Therapeutic effect of *Pseudomonas aeruginosa* phage YH30 on mink hemorrhagic pneumonia. Vet. Microbiol. *190*, 5-11.

Guang-Han, O., Leang-Chung, C., Vellasamy, K.M., Mariappan, V., Li-Yen, C., and Vadivelu, J. (2016). Experimental phage therapy for *Burkholderia pseudomallei* infection. PLoS. One. *11*, e0158213.

Gupta, P., Singh, H.S., Shukla, V.K., Nath, G., and Bhartiya, S.K. (2019). Bacteriophage therapy of chronic nonhealing wound: clinical study. Int. J Low Extrem. Wounds. *18*, 1534734619835115.

Gupta, R. and Prasad, Y. (2011). Efficacy of polyvalent bacteriophage P-27/HP to control multidrug resistant *Staphylococcus aureus* associated with human infections. Curr Microbiol. *62*, 255-260.

Hagens, S. and Loessner, M.J. (2010). Bacteriophage for biocontrol of foodborne pathogens: calculations and considerations. Curr. Pharm. Biotechnol. *11*, 58-68.

Harper, D.R. and Kutter, E. (2009). Bacteriophage: therapeutic use. In Encyclopedia of Life Sciences, John Wiley & Sons, Ltd.), pp. 1-7.

Hatfull, G.F. and Vehring, R. (2016). Respirable bacteriophage aerosols for the prevention and treatment of tuberculosis. In Delivery Systems for Tuberculosis Prevention and Treatment, A.J.Hickey, A.Misra, and P.B.Fourie, eds. (Chichester, UK: Wiley), pp. 277-292.

Häusler, T. (2006). Viruses vs. Superbugs: A Solution to the Antibiotics Crisis. (New York, NY: Macmillan).

Hawkins, C., Harper, D., Burch, D., Anggard, E., and Soothill, J. (2010). Topical treatment of *Pseudomonas aeruginosa* otitis of dogs with a bacteriophage mixture: A before/after clinical trial. Vet. Microbiol. *145*, 309-313.

Henry, M., Lavigne, R., and Debarbieux, L. (2013). Predicting *in vivo* efficacy of therapeutic bacteriophages used to treat pulmonary infections. Antimicrob. Agents Chemother. *57*, 5961-5968.

Hodyra-Stefaniak, K., Lahutta, K., Majewska, J., Kazmierczak, Z., Lecion, D., Harhala, M., Kęska, W., Owczarek, B., Jończyk-Matysiak, E., Kłopot, A., Miernikiewicz, P., Kula, D., Górski, A., and Dąbrowska, K. (2019). Bacteriophages engineered to display foreign peptides may become short-circulating phages. Microb. Biotechnol. *12*, 730-741.

Hodyra-Stefaniak, K., Miernikiewicz, P., Drapała, J., Drab, M., Jończyk-Matysiak, E., Lecion, D., Kazmierczak, Z., Beta, W., Majewska, J., Harhala, M., Bubak, B., Kłopot, A., Górski, A., and Dąbrowska, K. (2015). Mammalian host-versus-phage immune response determines phage fate *in vivo*. Sci. Rep. *5*, 14802.

Hoyle, N., Zhvaniya, P., Balarjishvili, N., Bolkvadze, D., Nadareishvili, L., Nizharadze, D., Wittmann, J., Rohde, C., and Kutateladze, M. (2018). Phage therapy against *Achromobacter xylosoxidans* lung infection in a patient with cystic fibrosis: a case report. Res. Microbiol. *169*, 540-542.

Hua, Y., Luo, T., Yang, Y., Dong, D., Wang, R., Wang, Y., Xu, M., Guo, X., Hu, F., and He, P. (2018). Phage therapy as a promising new treatment for lung infection caused by carbapenem-resistant *Acinetobacter baumannii* in mice. Front. Microbiol. *8*, 2659.

Huff, W.E., Huff, G.R., Rath, N.C., and Donoghue, A.M. (2010). Immune interference of bacteriophage efficacy when treating colibacillosis in poultry. Poult. Sci. *89*, 895-900.

Hung, C.H., Kuo, C.F., Wang, C.H., Wu, C.M., and Tsao, N. (2011). Experimental phage therapy in treating *Klebsiella pneumoniae*-mediated liver abscesses and bacteremia in mice. Antimicrob. Agents Chemother. *55*, 1358-1365.

Hyman, P. and Abedon, S.T. (2009). Practical methods for determining phage growth parameters. Meth. Mol. Biol. *501*, 175-202.

Hyman, P. and Abedon, S.T. (2010). Bacteriophage host range and bacterial resistance. Adv. Appl. Microbiol. *70*, 217-248.

Hyman, P. and Abedon, S.T. (2012). Smaller fleas: viruses of microorganisms. Scientifica *2012*, 734023.

Ibrahim, O.M.S., Sarhan, S.R., and Salih, S.I. (2016). Activity of isolated staphylococcal bacteriophage in treatment of experimentally induced chronic osteomyelitis in rabbits. Adv. Anim. Vet. Sci. *4*, 593-603.

Inchley, C.J. (1969). The activity of mouse Kuppfer cells following intravenous injection of T4 bacteriophage. Clin. Exp. Immunol. *5*, 173-187.

Jaiswal, A., Koley, H., Ghosh, A., Palit, A., and Sarkar, B. (2013). Efficacy of cocktail phage therapy in treating *Vibrio cholerae* infection in rabbit model. Microbes Infect. *15*, 152-156.

Jaiswal, A., Koley, H., Mitra, S., Saha, D.R., and Sarkar, B. (2014). Comparative analysis of different oral approaches to treat *Vibrio cholerae* infection in adult mice. Int. J. Med. Microbiol. *304*, 422-430.

Jamalludeen, N., Johnson, R.P., Shewen, P.E., and Gyles, C.L. (2009). Evaluation of bacteriophages for prevention and treatment of diarrhea due to experimental enterotoxigenic *Escherichia coli* O149 infection of pigs. Vet. Microbiol. *136*, 135-141.

Jasim, H.N., Hafidh, R.R., and Abdulamir, A.S. (2018). Formation of therapeutic phage cocktail and endolysin to highly multi-drug resistant *Acinetobacter baumannii*: *in vitro* and *in vivo* study. Iran J Basic Med. Sci *21*, 1100-1108.

Jault, P., Leclerc, T., Jennes, S., Pirnay, J.P., Que, Y.A., Resch, G., Rousseau, A.F., Ravat, F., Carsin, H., Le, F.R., Schaal, J.V., Soler, C., Fevre, C., Arnaud, I., Bretaudeau, L., and Gabard, J. (2019). Efficacy and tolerability of a cocktail of bacteriophages to treat burn wounds infected by *Pseudomonas aeruginosa* (PhagoBurn): a randomised, controlled, double-blind phase 1/2 trial. Lancet Infect. Dis. *19*, 35-45.

Jennes, S., Merabishvili, M., Soentjens, P., Pang, K.W., Rose, T., Keersebilck, E., Soete, O., Francois, P.M., Teodorescu, S., Verween, G., Verbeken, G., De, V.D., and Pirnay, J.P. (2017). Use of bacteriophages in the treatment of colistin-only-sensitive *Pseudomonas aeruginosa* septicaemia in a patient with acute kidney injury-a case report. Crit. Care *21*, 129.

Jeon, J., Park, J.H., and Yong, D. (2019). Efficacy of bacteriophage treatment against carbapenem-resistant *Acinetobacter baumannii* in *Galleria mellonella* larvae and a mouse model of acute pneumonia. BMC Microbiol. *19*, 70.

Jeon, J. and Yong, D. (2019). Two novel bacteriophages improve survival in *Galleria mellonella* infection and mouse acute pneumonia models infected with extensively-drug-resistant *Pseudomonas aeruginosa*. Appl. Environ. Microbiol. *85*, e02900-18.

Ji, Y., Cheng, M., Zhai, S., Xi, H., Cai, R., Wang, Z., Zhang, H., Wang, X., Xue, Y., Li, X., Sun, C., Feng, X., Lei, L., Ur, R.S., Han, W., and Gu, J. (2019). Preventive effect of the phage VB-SavM-JYL01 on rabbit necrotizing pneumonia caused by *Staphylococcus aureus*. Vet. Microbiol. *229*, 72-80.

Jikia, D., Chkhaidze, N., Imedashvili, E., Mgaloblishvili, I., Tsitlanadze, G., Katsarava, R., Glenn Morris, J.J., and Sulakvelidze, A. (2005). The use of a novel biodegradable preparation capable of the sustained release of bacteriophages and ciprofloxacin, in the complex treatment of multidrug-resistant *Staphylococcus aureus*-infected local radiation injuries caused by exposure to $Sr^{90}$. Clin. Exp. Dermatol. *30*, 23-26.

Jun, J.W., Shin, T.H., Kim, J.H., Shin, S.P., Han, J.E., Heo, G.J., De, Z.M., Shin, G.W., Chai, J.Y., and Park, S.C. (2014). Bacteriophage therapy of a *Vibrio parahaemolyticus* infection caused by a multiple antibiotic resistant O3:K6 pandemic clinical strain. J. Infect. Dis. *210*, 72-78.

Kaabi, S.A.G. and Musafer, H.K. (2019). An experimental mouse model for phage therapy of bacterial pathogens causing bacteremia. Microb. Pathog. *137*, 103770.

Kasman, L.M., Kasman, A., Westwater, C., Dolan, J., Schmidt, M.G., and Norris, J.S. (2002). Overcoming the phage replication threshold: a mathematical model with implications for phage therapy. J. Virol. *76*, 5557-5564.

Khawaldeh, A., Morales, S., Dillon, B., Alavidze, Z., Ginn, A.N., Thomas, L., Chapman, S.J., Dublanchet, A., Smithyman, A., and Iredell, J.R. (2011). Bacteriophage therapy for refractory *Pseudomonas aeruginosa* urinary tract infection. J Med. Microbiol. *60*, 1697-1700.

Kishimoto, T., Ishida, W., Fukuda, K., Nakajima, I., Suzuki, T., Uchiyama, J., Matsuzaki, S., Todokoro, D., Daibata, M., and Fukushima, A. (2019). Therapeutic effects of intravitreously administered bacteriophage in a mouse model of endophthalmitis caused by vancomycin-sensitive or -resistant *Enterococcus faecalis*. Antimicrob. Agents Chemother. *63*, e01088-19.

Kishor, C., Mishra, R.R., Saraf, S.K., Kumar, M., Srivastav, A.K., and Nath, G. (2016). Phage therapy of staphylococcal chronic osteomyelitis in experimental animal model. Indian J. Med. Res. *143*, 87-94.

Knezevic, P. and Sabo, V.A. (2019). Combining bacteriophages with other antibacterial agents to combat bacteria. In Phage Therapy: A Practical Approach, A.Górski, R.Międzybrodzki, and J.Borysowski, eds. Springer), pp. 257-293.

Korona, R., Korona, B., and Levin, B.R. (1993). Sensitivity of naturally occurring coliphages to type I and type II restriction and modification. J. Gen. Microbiol. *139*, 1283-1290.

Kosznik-Kwaśnicka, K., Topka, G., Dydecka, A., Necel, A., Nejman-Faleńczyk, B., Bloch, S., Węgrzyn, G., and Węgrzyn, A. (2019). The use of bacteriophages in animal health and food protection. In Phage Therapy: A

Practical Approach, A.Górski, R.Międzybrodzki, and J.Borysowski, eds. Springer), pp. 213-256.

Kropinski, A.M. (2018). Practical advice on the one-step growth curve. Meth. Mol. Biol. *1681*, 41-47.

Krylov, V.N. (2001). Phage therapy in terms of bacteriophage genetics: Hopes, prospects, safety, limitations. Rus. J. Genet. *37*, 715-730.

Kumar, M.S., Plotkin, J.B., and Hannenhalli, S. (2015). Regulated CRISPR Modules Exploit a Dual Defense Strategy of Restriction and Abortive Infection in a Model of Prokaryote-Phage Coevolution. PLoS Comput. Biol. *11*, e1004603.

Kumari, S., Harjai, K., and Chhibber, S. (2009a). Bacteriophage treatment of burn wound infection caused by *Pseudomonas aeruginosa* PAO in BALB/c mice. American Journal of Biomedical Sciences *1*, 385-394.

Kumari, S., Harjai, K., and Chhibber, S. (2010). Evidence to support the therapeutic potential of bacteriophage Kpn5 in burn wound infection caused by *Klebsiella pneumoniae* in BALB/c mice. J. Microbiol. Biotechnol. *20*, 935-941.

Kumari, S., Harjai, K., and Chhibber, S. (2011). Bacteriophage versus antimicrobial agents for the treatment of murine burn wound infection caused by *Klebsiella pneumoniae* B5055. J Med. Microbiol. *60*, 205-210.

Kumari, S., Harjai, K., and Chhibber, S. (2009b). Efficacy of bacteriophage treatment in murine burn wound infection induced by *Klebsiella pneumoniae*. J. Microbiol. Biotechnol. *19*, 622-628.

Kusradze, I., Karumidze, N., Rigvava, S., Dvalidze, T., Katsitadze, M., Amiranashvili, I., and Goderdzishvili, M. (2016). Characterization and testing the efficiency of *Acinetobacter baumannii* phage *vB-GEC_Ab-M-G7* as an antibacterial agent. Front. Microbiol. *7*, 1590.

Kutateladze, M. and Adamia, R. (2008). Phage therapy experience at the Eliava Institute. Med. Mal. Infect. *38*, 426-430.

Kutter, E. (2008). Phage therapy: bacteriophages as naturally occurring antimicrobials. In Practical Handbook of Microbiology, E.Goldman and L.H.Green, eds. (Boca Raton, FL: CRC Press), pp. 713-730.

Kutter, E., De Vos, D., Gvasalia, G., Alavidze, Z., Gogokhia, L., Kuhl, S., and Abedon, S.T. (2010). Phage therapy in clinical practice: treatment of human infections. Curr. Pharm. Biotechnol. *11*, 69-86.

Lampert, R., Boyce, F.F., and McFetridge, E.M. (1935). Bacteriophage therapy: A clinical study, with special reference to the technique of application. Am. J. Surgery *29*, 436-443.

Law, N., Logan, C., Yung, G., Furr, C.L.L., Lehman, S.M., Morales, S., Rosas, F., Galdamaka, A., Bilinsky, I., and Grint, P. (2019). Successful adjunctive use of bacteriophage therapy for treatment of multidrug-resistant *Pseudomonas aeruginosa* infection in a cystic fibrosis patient. Infection *47*, 665-668.

Lees, P., Cunningham, F.M., and Elliott, J. (2004). Principles of pharmacodynamics and their applications in veterinary pharmacology. J Vet. Pharmacol. Ther. *27*, 397-414.

Lehman, S.M. (2018). Bacteriophage diversity. In Viruses of Microorganisms, P.Hyman and S.T.Abedon, eds. (Caister Academic Press: Norwich, UK), pp. 145-165.

Leshkasheli, L., Kutateladze, M., Balarjishvili, N., Bolkvadze, D., Save, J., Oechslin, F., Que, Y.A., and Resch, G. (2019). Efficacy of newly isolated and highly potent bacteriophages in a mouse model of extensively drug-resistant *Acinetobacter baumannii* bacteraemia. J Glob. Antimicrob. Resist. *19*, 255-261.

Leszczyński, P., Weber-Dąbrowska, B., Kohutnicka, M., Łuczak, M., Górecki, A., and Górski, A. (2006). Successful eradication of methicillin-resistant *Staphylococcus aureus* (MRSA) intestinal carrier status in a healthcare worker - a case report. Folia Microbiol. *51*, 236-238.

Letarov, A.V., Golomidova, A.K., and Tarasyan, K.K. (2010). Ecological basis of rational phage therapy. Acta Naturae *2*, 60-71.

Letkiewicz, S., Międzybrodzki, R., Fortuna, W., Weber-Dąbrowska, B., and Górski, A. (2009). Eradication of *Enterococcus faecalis* by phage therapy in chronic bacterial prostatitis--case report. Folia Microbiol. *54*, 457-461.

Leung, C. Y. J. and Weitz, J. S. (2017). Modeling the synergistic elimination of bacteria by phage and the innate immune system. J. Theor. Biol. 429, 241-252.

Levin, B.R. and Bull, J.J. (1996). Phage therapy revisited: the population biology of a bacterial infection and its treatment with bacteriophage and antibiotics. Am. Nat. *147*, 881-898.

Levin, B.R. and Bull, J.J. (2004). Population and evolutionary dynamics of phage therapy. Nat. Rev. Microbiol. *2*, 166-173.

Levison, M.E. and Levison, J.H. (2009). Pharmacokinetics and pharmacodynamics of antibacterial agents. Infect. Dis. Clin. North Am. *23*, 791-815, vii.

Lloyd, N.C., Morgan, H.W., Nicholson, B.K., and Ronimus, R.S. (2005). The composition of Ehrlich's Salvarsan: resolution of a century-old debate. Angew. Chem. Int. Ed. Engl. *44*, 941-944.

Łobocka, M., Hejnowicz, M.S., Gągała, U., Weber-Dąbrowska, B., Węgrzyn, G., and Dadlez, M. (2014). The first step to bacteriophage therapy: how to choose the correct phage. In Phage Therapy: Current Research and Applications, J.Borysowski, R.Międzybrodzki, and A.Górski, eds. (Norfolk, UK: Caister Academic Press), pp. 23-67.

Loc-Carrillo, C. and Abedon, S.T. (2011). Pros and cons of phage therapy. Bacteriophage *1*, 111-114.

Łusiak-Szelachowska, M., Żaczek, M., Weber-Dąbrowska, B., Międzybrodzki, R., Kłak, M., Fortuna, W., Letkiewicz, S., Rogóż, P., Szufnarowski, K., Jończyk-Matysiak, E., Owczarek, B., and Górski, A. (2014). Phage neutralization by sera of patients receiving phage therapy. Viral Immunol. *27*, 295-304.

Łusiak-Szelachowska, M., Żaczek, M., Weber-Dąbrowska, B., Międzybrodzki, R., Letkiewicz, S., Fortuna, W., Rogóż, P., Szufnarowski, K., Jończyk-Matysiak, E., Olchawa, E., Walaszek, K.M., and Górski, A. (2017). Antiphage activity of sera during phage therapy in relation to its outcome. Future Microbiol. *12*, 109-117.

Maddocks, S., Fabijan, A.P., Ho, J., Lin, R.C.Y., Ben Zakour, N.L., Dugan, C., Kliman, I., Branston, S., Morales, S., and Iredell, J.R. (2019). Bacteriophage therapy of ventilator-associated pneumonia and empyema caused by *Pseudomonas aeruginosa*. Am. J Respir. Crit. Care Med. *200*, 1179-1181.

Mai, V., Ukhanova, M., Reinhard, M.K., Li, M., and Sulakvelidze, A. (2015). Bacteriophage administration significantly reduces *Shigella* colonization and shedding by *Shigella*-challenged mice without deleterious side effects and distortions in the gut microbiota. Bacteriophage *5*, e1088124.

Mai, V., Ukhanova, M., Visone, L., Abuladze, T., and Sulakvelidze, A. (2010). Bacteriophage administration reduces the concentration of *Listeria monocytogenes* in the gastrointestinal tract and its translocation to spleen and liver in experimentally infected mice. Int. J Microbiol. *2010*, 624234.

Marino, M. and Zito, P.M. (2020). Pharmacodynamics. (Treasure Island, FL: StatPearls Publishing).

Marza, J.A.S., Soothill, J.S., Boydell, P., and Collyns, T.A. (2006). Multiplication of therapeutically administered bacteriophages in *Pseudomonas aeruginosa* infected patients. Burns : journal of the International Society for Burn Injuries *32*, 644-646.

Matsuda, T., Freeman, T.A., Hilbert, D.W., Duff, M., Fuortes, M., Stapleton, P.P., and Daly, J.M. (2005). Lysis-deficient bacteriophage therapy decreases endotoxin and inflammatory mediator release and improves survival in a murine peritonitis model. Surgery *137*, 639-646.

Matsuzaki, S. and Uchiyama, J. (2019). Phage pharmacokinetics: relationship with administration route. In Phage Therapy: A Practical Approach, A.Górski, R.Międzybrodzki, and J.Borysowski, eds. Springer), pp. 43-57.

Maura, D., Galtier, M., Le, B.C., and Debarbieux, L. (2012). Virulent bacteriophages can target O104:H4 enteroaggregative *Escherichia coli* in the mouse intestine. Antimicrob. Agents Chemother. *56*, 6235-6242.

McVay, C.S., Velasquez, M., and Fralick, J.A. (2007). Phage therapy of *Pseudomonas aeruginosa* infection in a mouse burn wound model. Antimicrob. Agents Chemother. *51*, 1934-1938.

Melo, L.D.R., Oliveira, H., Pires, D.P., Dąbrowska, K., and Azeredo, J. (2020). Phage therapy efficacy: a review of the last 10 years of preclinical studies. Crit Rev. Microbiol. 1-22.

Mendes, J.J., Leandro, C., Corte-Real, S., Barbosa, R., Cavaco-Silva, P., Melo-Cristino, J., Górski, A., and Garcia, M. (2013). Wound healing potential of topical bacteriophage therapy on diabetic cutaneous wounds. Wound Repair Regen. *21*, 595-603.

Merril, C.R. (2008). Interaction of bacteriophages with animals. In Bacteriophage Ecology, S.T.Abedon, ed. (Cambridge, UK: Cambridge University Press), pp. 332-352.

Merril, C. R., Biswas, B., Carlton, R., Jensen, N. C., Creed, G. J., Zullo, S., and Adhya, S. (1996). Long-circulating bacteriophage as antibacterial agents. Proceedings of the National Academy of Sciences, USA *93*, 3188-3192

Merril, C.R., Scholl, D., and Adhya, S. (2006). Phage therapy. In The Bacteriophages, R.Calendar and S.T.Abedon, eds. (Oxford: Oxford University Press), pp. 725-741.

Międzybrodzki, R., Borysowski, J., Weber-Dąbrowska, B., Fortuna, W., Letkiewicz, S., Szufnarowski, K., Pawełczyk, Z., Rogóż, P., Kłak, M., Wojtasik, E., and Górski, A. (2012). Clinical aspects of phage therapy. Adv. Virus Res. *83*, 73-121.

Międzybrodzki, R., Fortuna, W., Weber-Dąbrowska, B., and Górski, A. (2009). A retrospective analysis of changes in inflammatory markers in patients treated with bacterial viruses. Clin. Exp. Med. *9*, 303-312.

Morello, E., Saussereau, E., Maura, D., Huerre, M., Touqui, L., and Debarbieux, L. (2011). Pulmonary bacteriophage therapy on *Pseudomonas aeruginosa* cystic fibrosis strains: first steps towards treatment and prevention. PLoS. One. *6*, e16963.

Morozova, V.V., Kozlova, Y.N., Ganichev, D.A., and Tikunova, N.V. (2018a). Bacteriophage treatment of infected diabetic foot ulcers. Meth. Mol. Biol. *1693*, 151-158.

Morozova, V.V., Vlassov, V.V., and Tikunova, N.V. (2018b). Applications of bacteriophages in the treatment of localized infections in humans. Front. Microbiol. *9*, 1696.

Morris, J.L., Letson, H.L., Elliott, L., Grant, A.L., Wilkinson, M., Hazratwala, K., and McEwen, P. (2019). Evaluation of bacteriophage as an adjunct therapy for treatment of peri-prosthetic joint infection caused by *Staphylococcus aureus*. PLoS. One. *14*, e0226574.

Muller, P.Y. and Milton, M.N. (2012). The determination and interpretation of the therapeutic index in drug development. Nat. Rev. Drug Discov. *11*, 751-761.

Mycek, M.J., Harvey, R.A., and Champe, P.C. (2000). Pharmacology. (Philadelphia, PA: Lippincott Williams & Wilkins).

Nale, J.Y., Spencer, J., Hargreaves, K.R., Buckley, A.M., Trzepinski, P., Douce, G.R., and Clokie, M.R. (2015). Bacteriophage combinations significantly reduce *Clostridium difficile* growth *in vitro* and proliferation *in vivo*. Antimicrob. Agents Chemother. *60*, 968-981.

Nelstrop, A.E., Taylor, G., and Collard, P. (1968). Studies on phagocytosis. II. In vitro phagocytosis by macrophages. Immunology *14*, 339-346.

Nguyen, S., Baker, K., Padman, B.S., Patwa, R., Dunstan, R.A., Weston, T.A., Schlosser, K., Bailey, B., Lithgow, T., Lazarou, M., Luque, A., Rohwer, F., Blumberg, R.S., and Barr, J.J. (2017). Bacteriophage transcytosis provides a mechanism to cross epithelial cell layers. MBio *8*, *e01874-17*.

Nikkhahi, F., Dallal, M.M.S., Alimohammadi, M., Rajabi, Z., Fardsanei, F., Imeni, S.M., and Bonab, P.T. (2017). Phage therapy: assessment of the efficacy of a bacteriophage isolated in the treatment of salmonellosis induced by *Salmonella enteritidis* in mice. Gastroenterology and Hepatology from bed to bench *10*, 131-136.

Nilsson, A. I., Kugelberg, E., Berg, O. G., Andersson, D. I. (2004) Experimental adaptation of *Salmonella typhimurium* to mice. Genetics *168*, 1119-1130.

Nir-Paz, R., Gelman, D., Khouri, A., Sisson, B.M., Fackler, J., Alkalay-Oren, S., Khalifa, L., Rimon, A., Yerushalmy, O., Bader, R., Amit, S., Coppenhagen-Glazer, S., Henry, M., Quinones, J., Malagon, F., Biswas, B., Moses, A.E., Merril, G., Schooley, R.T., Brownstein, M.J., Weil, Y.A., and Hazan, R. (2019). Successful treatment of antibiotic resistant poly-microbial bone infection with bacteriophages and antibiotics combination. Clin. Infect. Dis. *69*, 2015-2018.

Nishikawa, H., Yasuda, M., Uchiyama, J., Rashel, M., Maeda, Y., Takemura, I., Sugihara, S., Ujihara, T., Shimizu, Y., Shuin, T., and Matsuzaki, S. (2008). T-even-related bacteriophages as candidates for treatment of *Escherichia coli* urinary tract infections. Arch. Virol. *153*, 507-515.

Niu, Y.D., Xu, Y., McAllister, T.A., Rozema, E.A., Stephens, T.P., Bach, S.J., Johnson, R.P., and Stanford, K. (2008). Comparison of fecal versus rectoanal mucosal swab sampling for detecting *Escherichia coli* O157:H7 in experimentally inoculated cattle used in assessing bacteriophage as a mitigation strategy. Journal of Food Protection *71*, 691-698.

Nobrega, F.L., Costa, A.R., Kluskens, L.D., and Azeredo, J. (2015). Revisiting phage therapy: new applications for old resources. Trends Microbiol. *23*, 185-191.

Ochs, H.D., Davis, S.D., and Wedgwood, R.J. (1971). Immunologic response to bacteriophage ϕX 174 in immunodeficiency diseases. J. Clin. Invest. *50*, 2559-2568.

Oduor, J.M., Onkoba, N., Maloba, F., Arodi, W.O., and Nyachieo, A. (2016a). Efficacy of lytic *Staphylococcus aureus* bacteriophage against multidrug-resistant *Staphylococcus aureus* in mice. J. Infect. Dev. Ctries. *10*, 1208-1213.

Oduor, J.M.O., Onkoba, N., Maloba, F., and Nyachieo, A. (2016b). Experimental phage therapy against haematogenous multi-drug resistant *Staphylococcus aureus* pneumonia in mice. Afr. J. Lab Med. *5*, 435.

Oechslin, F., Piccardi, P., Mancini, S., Gabard, J., Moreillon, P., Entenza, J.M., Resch, G., and Que, Y.A. (2017). Synergistic interaction between phage therapy and antibiotics clears *Pseudomonas aeruginosa* infection in endocarditis and reduces virulence. J. Infect. Dis. *215*, 703-712.

Olszowska-Zaremba, N., Borysowski, J., Dąbrowska, K., and Górski, A. (2012). Phage translocation, safety, and immunomodulation. In Bacteriophages in Health and Disease, P.Hyman and S.T.Abedon, eds. (Wallingford, UK: CABI Press), pp. 168-184.

Pabary, R., Singh, C., Morales, S., Bush, A., Alshafi, K., Bilton, D., Alton, E.W., Smithyman, A., and Davies, J.C. (2016). Antipseudomonal bacteriophage reduces infective burden and inflammatory response in murine lung. Antimicrob. Agents Chemother. *60*, 744-751.

Patel, D.R., Bhartiya, S.K., Kumar, R., Shukla, V.K., and Nath, G. (2019). Use of customized bacteriophages in the treatment of chronic nonhealing wounds: a prospective study. Int. J Low Extrem. Wounds. 1534734619881076.

Payne, R.J.H. and Jansen, V.A.A. (2000). Phage therapy: The peculiar kinetics of self-replicating pharmaceuticals. Clin. Pharmacol. Ther. *68*, 225-230.

Payne, R.J.H. and Jansen, V.A.A. (2001). Understanding bacteriophage therapy as a density-dependent kinetic process. J. Theor. Biol. *208*, 37-48.

Payne, R.J.H. and Jansen, V.A.A. (2003). Pharmacokinetic principles of bacteriophage therapy. Clin. Pharmacokinet. *42*, 315-325.

Peng, H., Borg, R.E., Dow, L.P., Pruitt, B.L., and Chen, I.A. (2020). Controlled phage therapy by photothermal ablation of specific bacterial species using gold nanorods targeted by chimeric phages. Proc. Natl. Acad. Sci U. S. A *117*, 1951-1961.

Pereira, M.P. and Kelley, S.O. (2011). Maximizing the therapeutic window of an antimicrobial drug by imparting mitochondrial sequestration in human cells. J Am. Chem Soc. *133*, 3260-3263.

Pincus, N.B., Reckhow, J.D., Saleem, D., Jammeh, M.L., Datta, S.K., and Myles, I.A. (2015). Strain specific phage treatment for *Staphylococcus aureus* infection is influenced by host immunity and site of infection. PLoS. One. *10*, e0124280.

Prazak, J., Iten, M., Cameron, D.R., Save, J., Grandgirard, D., Resch, G., Goepfert, C., Leib, S.L., Takala, J., Jakob, S.M., Que, Y.A., and Haenggi, M. (2019). Bacteriophages improve outcomes in experimental *Staphylococcus aureus* ventilator-associated pneumonia. Am. J Respir. Crit Care Med. *200*, 1126-1133.

Rastogi, V., Yadav, P., Verma, A., and Pandit, J.K. (2017). *Ex vivo* and *in vivo* evaluation of microemulsion based transdermal delivery of *E. coli* specific T4 bacteriophage: A rationale approach to treat bacterial infection. Eur. J. Pharm. Sci. *107*, 168-182.

Raya, R.R., Varey, P., Oot, R.A., Dyen, M.R., Callaway, T.R., Edrington, T.S., Kutter, E.M., and Brabban, A.D. (2006). Isolation and characterization of a new T-even bacteriophage, CEV1, and determination of its potential to reduce *Escherichia coli* O157:H7 levels in sheep. Appl. Environ. Microbiol. *72*, 6405-6410.

Regeimbal, J.M., Jacobs, A.C., Corey, B.W., Henry, M.S., Thompson, M.G., Pavlicek, R.L., Quinones, J., Hannah, R.M., Ghebremedhin, M., Crane, N.J., Zurawski, D.V., Teneza-Mora, N.C., Biswas, B., and Hall, E.R. (2016). Personalized therapeutic cocktail of wild environmental phages rescues mice from *Acinetobacter baumannii* wound infections. Antimicrob. Agents Chemother. *60*, 5806-5816.

Rhoads, D.D., Wolcott, R.D., Kuskowski, M.A., Wolcott, B.M., Ward, L.S., and Sulakvelidze, A. (2009). Bacteriophage therapy of venous leg ulcers in humans: results of a phase I safety trial. J. Wound Care *18*, 237-244.

Roach, D.R., Leung, C.Y., Henry, M., Morello, E., Singh, D., Di Santo, J.P., Weitz, J.S., and Debarbieux, L. (2017). Synergy between the host immune system and bacteriophage is essential for successful phage therapy against an acute respiratory pathogen. Cell Host. Microbe *22*, 38-47.

Rose, T., Verbeken, G., Vos, D.D., Merabishvili, M., Vaneechoutte, M., Lavigne, R., Jennes, S., Zizi, M., and Pirnay, J.P. (2014). Experimental phage therapy of burn wound infection: difficult first steps. Int. J. Burns Trauma *4*, 66-73.

Rozema, E.A., Stephens, T.P., Bach, S.J., Okine, E.K., Johnson, R.P., Stanford, K., and McAllister, T.A. (2009). Oral and rectal administration of bacteriophages for control of *Escherichia coli* O157:H7 in feedlot cattle. J. Food Prot. *72*, 241-250.

Sarhan, W.A. and Azzazy, H.M. (2017). Apitherapeutics and phage-loaded nanofibers as wound dressings with enhanced wound healing and antibacterial activity. Nanomedicine (Lond) *12*, 2055-2067.

Sarker, S.A., Berger, B., Deng, Y., Kieser, S., Foata, F., Moine, D., Descombes, P., Sultana, S., Huq, S., Bardhan, P.K., Vuillet, V., Praplan, F., and Brüssow, H. (2017). Oral application of *Escherichia coli* bacteriophage: safety tests in healthy and diarrheal children from Bangladesh. Environ. Microbiol. *19*, 237-250.

Sarker, S.A., Sultana, S., Reuteler, G., Moine, D., Descombes, P., Charton, F., Bourdin, G., McCallin, S., Ngom-Bru, C., Neville, T., Akter, M., Huq, S., Qadri, F., Talukdar, K., Kassam, M., Delley, M., Loiseau, C., Deng, Y., El, A.S., Berger, B., and Brüssow, H. (2016). Oral phage therapy of acute bacterial diarrhea with two coliphage preparations: a randomized trial in children from Bangladesh. EBioMedicine *4*, 124-137.

Schooley, R.T., Biswas, B., Gill, J.J., Hernandez-Morales, A., Lancaster, J., Lessor, L., Barr, J.J., Reed, S.L., Rohwer, F., Benler, S., Segall, A.M., Taplitz, R., Smith, D.M., Kerr, K., Kumaraswamy, M., Nizet, V., Lin, L., McCauley, M.D., Strathdee, S.A., Benson, C.A., Pope, R.K., Leroux, B.M., Picel, A.C., Mateczun, A.J., Cilwa, K.E., Regeimbal, J.M., Estrella, L.A., Wolfe, D.M., Henry, M.S., Quinones, J., Salka, S., Bishop-Lilly, K.A., Young, R., and Hamilton, T. (2017). Development and use of personalized bacteriophage-based therapeutic cocktails to treat a patient with a disseminated resistant *Acinetobacter baumannii* infection. Antimicrob. Agents Chemother. *61*, e00954-17.

Segall, A.M., Roach, D.R., and Strathdee, S.A. (2019). Stronger together? Perspectives on phage-antibiotic synergy in clinical applications of phage therapy. Curr. Opin. Microbiol. *51*, 46-50.

Seth, A.K., Geringer, M.R., Nguyen, K.T., Agnew, S.P., Dumanian, Z., Galiano, R.D., Leung, K.P., Mustoe, T.A., and Hong, S.J. (2013). Bacteriophage therapy for *Staphylococcus aureus* biofilm-infected wounds: a new approach to chronic wound care. Plast. Reconstr. Surg. *131*, 225-234.

Sheng, H., Knecht, H.J., Kudva, I.T., and Hovde, C.J. (2006). Application of bacteriophages to control intestinal *Escherichia coli* O157:H7 levels in ruminants. Appl. Environ. Microbiol. *72*, 5359-5366.

Shivaswamy, V.C., Kalasuramath, S.B., Sadanand, C.K., Basavaraju, A.K., Ginnavaram, V., Bille, S., Ukken, S.S., and Pushparaj, U.N. (2015). Ability of bacteriophage in resolving wound infection caused by multidrug-resistant *Acinetobacter baumannii* in uncontrolled diabetic rats. Microb. Drug Resist. *21*, 171-177.

Shivshetty, N., Hosamani, R., Ahmed, L., Oli, A.K., Sannauallah, S., Sharanbassappa, S., Patil, S.A., and Kelmani, C.R. (2014). Experimental protection of diabetic mice against lethal *P. aeruginosa* infection by bacteriophage. Biomed. Res. Int. *2014*, 793242.

Singla, S., Harjai, K., Katare, O.P., and Chhibber, S. (2015). Bacteriophage-loaded nanostructured lipid carrier: Improved pharmacokinetics mediates effective resolution of *Klebsiella pneumoniae* induced lobar pneumonia. J. Infect. Dis. *212*, 325-334.

Ślopek, S., Durlakova, I., Weber-Dąbrowska, B., Kucharewicz-Krukowska, A., Dąbrowski, M., and Bisikiewicz, R. (1983). Results of bacteriophage treatment of suppurative bacterial infections. I. General evaluation of the results. Arch. Immunol. Ther. Exp. *31*, 267-291.

Ślopek, S., Weber-Dąbrowska, B., Dąbrowski, M., and Kucharewicz-Krukowska, A. (1987). Results of bacteriophage treatment of suppurative bacterial infections in the years 1981-1986. Arch. Immunol. Ther. Exp. *35*, 569-583.

Speck, P. and Smithyman, A. (2016). Safety and efficacy of phage therapy via the intravenous route. FEMS Microbiol. Lett. *363*, fnv242.

Spouge, J.L. (1994). Viral multiplicity of attachment and its implications for human immunodeficiency virus therapies. J. Virol. *68*, 1782-1789.

Stanford, K., McAllister, T.A., Niu, Y.D., Stephens, T.P., Mazzocco, A., Waddell, T.E., and Johnson, R.P. (2010). Oral delivery systems for encapsulated bacteriophages targeted at *Escherichia coli* O157:H7 in feedlot cattle. J. Food Prot. *73*, 1304-1312.

Stent, G.S. (1963). Molecular Biology of Bacterial Viruses. (San Francisco, CA: WH Freeman and Co.).

Stone, R. (2002). Bacteriophage therapy. Stalin's forgotten cure. Science (New York, N. Y. ) *298*, 728-731.

Strotskaya, A., Savitskaya, E., Metlitskaya, A., Morozova, N., Datsenko, K.A., Semenova, E., and Severinov, K. (2017). The action of *Escherichia coli* CRISPR-Cas system on lytic bacteriophages with different lifestyles and development strategies. Nucl. Acids Res. *45*, 1946-1957.

Sulakvelidze, A. (2005). Phage therapy: an attractive option for dealing with antibiotic-resistant bacterial infections. Drug Discov. Today *10*, 807-809.

Sulakvelidze, A., Alavidze, Z., and Morris, J.G., Jr. (2001). Bacteriophage therapy. Antimicrob. Agents Chemother. *45*, 649-659.

Sulakvelidze, A. and Kutter, E. (2005). Bacteriophage therapy in humans. In Bacteriophages: Biology and Application, E.Kutter and A.Sulakvelidze, eds. (Boca Raton, Florida: CRC Press), pp. 381-436.

Sulakvelidze, A. and Pasternack, G.R. (2010). Industrial and regulatory issues in bacteriophage applications in food production and processing. In Bacteriophages in the Control of Food- and Waterborne Pathogens, P.M.Sabour and M.W.Griffiths, eds. (Washington, DC: ASM Press), pp. 297-326.

Summers, W.C. (2001). Bacteriophage therapy. Ann. Rev. Microbiol. *55*, 437-451.

Sunagar, R., Patil, S.A., and Chandrakanth, R.K. (2010). Bacteriophage therapy for *Staphylococcus aureus* bacteremia in streptozotocin-induced diabetic mice. Res. Microbiol. *161*, 854-860.

Tagliaferri, T.L., Jansen, M., and Horz, H.P. (2019). Fighting pathogenic bacteria on two fronts: phages and antibiotics as combined strategy. Front Cell Infect. Microbiol. *9*, 22.

Takemura-Uchiyama, I., Uchiyama, J., Osanai, M., Morimoto, N., Asagiri, T., Ujihara, T., Daibata, M., Sugiura, T., and Matsuzaki, S. (2014). Experimental phage therapy against lethal lung-derived septicemia caused by *Staphylococcus aureus* in mice. Microbes Infect. *16*, 512-517.

Tang, F., Zhang, P., Zhang, Q., Xue, F., Ren, J., Sun, J., Qu, Z., Zhuge, X., Li, D., Wang, J., Jiang, M., and Dai, J. (2019). Isolation and characterization of a broad-spectrum phage of multiple drug resistant *Salmonella* and its therapeutic utility in mice. Microb. Pathog. *126*, 193-198.

Tanji, Y., Shimada, T., Fukudomi, H., Miyanaga, K., Nakai, Y., and Unno, H. (2005). Therapeutic use of phage cocktail for controlling *Escherichia coli* O157:H7 in gastrointestinal tract of mice. J. Biosci. Bioeng. *100*, 280-287.

Then, R.L. and Sahl, H.G. (2010). Anti-infective strategies of the future: is there room for species-specific antibacterial agents? Curr. Pharm. Des. *16*, 555-566.

Tkhilaishvili, T., Winkler, T., Muller, M., Perka, C., and Trampuz, A. (2019). Bacteriophages as adjuvant to antibiotics for the treatment of periprosthetic joint infection caused by multidrug-resistant *Pseudomonas aeruginosa*. Antimicrob. Agents Chemother. *64*, e00924-19.

Torres-Barceló, C. and Hochberg, M.E. (2016). Evolutionary rationale for phages as complements of antibiotics. Trends Microbiol. *24*, 249-256.

Tothova, L., Celec, P., Babickova, J., Gajdosova, J., Al-Alami, H., Kamodyova, N., Drahovska, H., Liptakova, A., Turna, J., and Hodosy, J. (2011). Phage therapy of *Cronobacter*-induced urinary tract infection in mice. Med. Sci. Mon. *17*, BR173-BR178.

Trigo, G., Martins, T.G., Fraga, A.G., Longatto-Filho, A., Castro, A.G., Azeredo, J., and Pedrosa, J. (2013). Phage therapy is effective against infection by *Mycobacterium ulcerans* in a murine footpad model. PLoS Negl. Trop. Dis. *7*, e2183.

Uchiyama, J., Maeda, Y., Takemura, I., Chess-Williams, R., Wakiguchi, H., and Matsuzaki, S. (2009). Blood kinetics of four intraperitoneally administered therapeutic candidate bacteriophages in healthy and neutropenic mice. Microbiol. Immunol. *53*, 301-304.

Uchiyama, J., Rashel, M., Takemura, I., Wakiguchi, H., and Matsuzaki, S. (2008). *In silico* and *in vivo* evaluation of bacteriophage fEF24C, a candidate for treatment of *Enterococcus faecalis* infections. Appl. Environ. Microbiol. *74*, 4149-4163.

Vahedi, A., Dallal, M.M.S., Douraghi, M., Nikkhahi, F., Rajabi, Z., Yousefi, M., and Mousavi, M. (2018). Isolation and identification of specific bacteriophage against enteropathogenic *Escherichia coli* (EPEC) and *in vitro* and *in vivo* characterization of bacteriophage. FEMS Microbiol. Lett. *365*, fny136.

Van Belleghem, J.D., Dąbrowska, K., Vaneechoutte, M., and Barr, J.J. (2019). Phage interaction with the mammalian immune system. In Phage Therapy: A Practical Approach, A.Górski, R.Międzybrodzki, and J.Borysowski, eds. Springer), pp. 91-122.

Verstappen, K.M., Tulinski, P., Duim, B., Fluit, A.C., Carney, J., van, N.A., and Wagenaar, J.A. (2016). The effectiveness of bacteriophages against methicillin-resistant *Staphylococcus aureus* ST398 nasal colonization in pigs. PLoS. One. *11*, e0160242.

Vinodkumar, C.S., Kalsurmath, S., and Neelagund, Y.F. (2008). Utility of lytic bacteriophage in the treatment of multidrug-resistant *Pseudomonas aeruginosa* septicemia in mice. Indian Journal Of Pathology & Microbiology *51*, 360-366.

Vinodkumar, C.S., Neelagund, Y.F., and Kalsurmath, S. (2005). Bacteriophage in the treatment of experimental septicemic mice from a clinical isolate of multidrug resistant *Klebsiella pneumoniae*. J. Commun. Dis. *37*, 18-29.

Wang, J., Hu, B., Xu, M., Yan, Q., Liu, S., Zhu, X., Sun, Z., Reed, E., Ding, L., Gong, J., Li, Q.Q., and Hu, J. (2006a). Use of bacteriophage in the treatment of experimental animal bacteremia from imipenem-resistant *Pseudomonas aeruginosa*. Int. J. Mol. Med. *17*, 309-317.

Wang, J.L., Kuo, C.F., Yeh, C.M., Chen, J.R., Cheng, M.F., and Hung, C.H. (2018). Efficacy of φkm18p phage therapy in a murine model of extensively drug-resistant *Acinetobacter baumannii* infection. Infect. Drug Resist. *11*, 2301-2310.

Wang, J., Hu, B., Xu, M., Yan, Q., Liu, S., Zhu, X., Sun, Z., Tao, D., Ding, L., Reed, E., Gong, J., Li, Q.Q., and Hu, J. (2006b). Therapeutic effectiveness of bacteriophages in the rescue of mice with extended spectrum beta-lactamase-producing *Escherichia coli* bacteremia. Int. J. Mol. Med. *17*, 347-355.

Wang, Y., Mi, Z., Niu, W., An, X., Yuan, X., Liu, H., Li, P., Liu, Y., Feng, Y., Huang, Y., Zhang, X., Zhang, Z., Fan, H., Peng, F., Tong, Y., and Bai, C. (2016). Intranasal treatment with bacteriophage rescues mice from *Acinetobacter baumannii*-mediated pneumonia. Future. Microbiol *11*, 631-641.

Waters, E.M., Neill, D.R., Kaman, B., Sahota, J.S., Clokie, M.R., Winstanley, C., and Kadioglu, A. (2017). Phage therapy is highly effective against chronic lung infections with *Pseudomonas aeruginosa*. Thorax *72*, 666-667.

Weber-Dąbrowska, B., Mulczyk, M., and Górski, A. (2003). Bacteriophages as an efficient therapy for antibiotic-resistant septicemia in man. Transplant. Proc. *35*, 1385-1386.

Weber-Dąbrowska, B., Zimecki, M., Kruzel, M., Kochanowska, I., and Łusiak-Szelachowska, M. (2006). Alternative therapies in antibiotic-resistant infection. Adv. Med. Sci. *51*, 242-244.

Wiegand, I., Hilpert, K., and Hancock, R.E. (2008). Agar and broth dilution methods to determine the minimal inhibitory concentration (MIC) of antimicrobial substances. Nat. Protoc. *3*, 163-175.

Wienhold, S.M., Brack, M.C., Nouailles, G., Suttorp, N.W., Seitz, C., Ross, A., Ziehr, H., Gurtner, C., Kershaw, O., and Dietert, K. (2018). Intratracheal phage therapy against *Acinetobacter baumannii* lung infection in mice. In Host Pathogen Interactions, American Thoracic Society), p. A7571.

Wienhold, S.M., Lienau, J., and Witzenrath, M. (2019). Towards inhaled phage therapy in Western Europe. Viruses. *11*.

Wills, Q.F., Kerrigan, C., and Soothill, J.S. (2005). Experimental bacteriophage protection against *Staphylococcus aureus* abscesses in a rabbit model. Antimicrob. Agents Chemother. *49*, 1220-1221.

Wright, A., Hawkins, C.H., Anggård, E.E., and Harper, D.R. (2009). A controlled clinical trial of a therapeutic bacteriophage preparation in chronic otitis due to antibiotic-resistant *Pseudomonas aeruginosa*; a preliminary report of efficacy. Clin. Otolaryng. *34*, 349-357.

Xue, Y., Zhai, S., Wang, Z., Ji, Y., Wang, G., Wang, T., Wang, X., Xi, H., Cai, R., Zhao, R., Zhang, H., Bi, L., Guan, Y., Guo, Z., Han, W., and Gu, J. (2020). The *Yersinia* phage X1 administered orally efficiently protects a murine chronic enteritis model against *Yersinia enterocolitica* Infection. Front. Microbiol. *11*, 351.

Yacoby, I. and Benhar, I. (2008). Targeted filamentous bacteriophages as therapeutic agents. Expert Opin. Drug Deliv. *5*, 321-329.

Yang, M., Du, C., Gong, P., Xia, F., Sun, C., Feng, X., Lei, L., Song, J., Zhang, L., Wang, B., Xiao, F., Yan, X., Cui, Z., Li, X., Gu, J., and Han, W. (2015).

Therapeutic effect of the YH6 phage in a murine hemorrhagic pneumonia model. Res. Microbiol. *166*, 633-643.

Yen, M., Cairns, L.S., and Camilli, A. (2017). A cocktail of three virulent bacteriophages prevents *Vibrio cholerae* infection in animal models. Nat. Commun. *8*, 14187.

Yin, S., Huang, G., Zhang, Y., Jiang, B., Yang, Z., Dong, Z., You, B., Yuan, Z., Hu, F., Zhao, Y., and Peng, Y. (2017). Phage Abp1 rescues human cells and mice from infection by pan-drug resistant *Acinetobacter baumannii*. Cell Physiol. Biochem. *44*, 2337-2345.

Young, R. (1992). Bacteriophage lysis: mechanisms and regulation. Microbiol. Rev. *56*, 430-481.

Yuan, Y., Wang, L., Li, X., Tan, D., Cong, C., and Xu, Y. (2019). Efficacy of a phage cocktail in controlling phage resistance development in multidrug resistant Acinetobacter baumannii. Virus Res. *272*, 197734.

Zaffiri, L., Gardner, J., and Toledo-Pereyra, L.H. (2012). History of antibiotics. From Salvarsan to cephalosporins. J. Invest. Surg. *25*, 67-77.

Zhao, J., Liu, Y., Xiao, C., He, S., Yao, H., and Bao, G. (2017). Efficacy of phage therapy in controlling rabbit colibacillosis and changes in cecal microbiota. Front Microbiol *8*, 957.

Zhvania, P., Hoyle, N.S., Nadareishvili, L., Nizharadze, D., and Kutateladze, M. (2017). Phage therapy in a 16-year-old boy with Netherton syndrome. Front Med. (Lausanne) *4*, 94.

Zimecki, M., Artym, J., Kocieba, M., Weber-Dąbrowska, B., Borysowski, J., and Górski, A. (2010). Prophylactic effect of bacteriophages on mice subjected to chemotherapy-induced immunosuppression and bone marrow transplant upon infection with *Staphylococcus aureus*. Med. Microbiol. Immunol. *199*, 71-79.

Zimecki, M., Artym, J., Kocieba, M., Weber-Dąbrowska, B., Borysowski, J., and Górski, A. (2009). Effects of prophylactic administration of bacteriophages to immunosuppressed mice infected with *Staphylococcus aureus*. BMC Microbiol. *9*, 169.

Żaczek, M., Łusiak-Szelachowska, M., Weber-Dąbrowska, B., Mi-Ödzybrodzki, R., Fortuna, W., Rogóż, P., Letkiewicz, S., and Górski, A. (2019). Humoral immune response to phage-based therapeutics. In Phage Therapy: A Practical Approach, A.Górski, R. Międzybrodzki, and J. Borysowski, eds. Springer), pp. 123-143.

# Chapter 3

# Human Gut Bacteriophages: Peacekeepers and Warriors at the Microbiota-Gut Interface

**Susan Mills[1], Colin Hill[1,2] and R. Paul Ross[1,2*]**

[1]APC Microbiome Ireland, University College Cork, Ireland
[2]School of Microbiology, University College Cork, Ireland

[*]Corresponding author : p.ross@ucc.ie

DOI: https://doi.org/10.21775/9781913652517.03

**Abstract**

The human gut phageome is not as well studied as its bacterial counterpart but studies to date strongly point to its role in generating microbiota diversity in early life and its contribution to microbiota fitness in adulthood. Kill-the-winner dynamics in the infant gut, governed by the bacteriophage (phage) lytic cycle, contribute to bacterial diversity and abundance levels, while piggy-back-the-winner dynamics in the adult gut is governed by the lysogenic cycle and contributes to bacterial fitness. However, increased phage abundance, presumably a result of prophage induction, has been associated with a number of diseases. The lysogenic-lytic switch appears to be a delicate balancing act and its full repertoire of triggers in this environment requires further investigation. Phages can also interact with host immunity and traverse the epithelial barrier. However, phage-bacteria-mammalian host interactions and phage-mammalian host interactions in the gut are only beginning to uncover how the phageome actually impacts health. In this review, we present the current state of the knowledge with regards to the cross-kingdom interactions that ensue. We also address the impact of the phageome on fecal microbiota transplantation and the potential of phage therapy for gut related diseases and its capacity to improve microbiota health. In evaluating the existing knowledge, we also shed light on gaps in the literature.

## Introduction

The number of virus-like particles (VLPs) estimated to exist globally is ~$10^{31}$, the majority of which are bacterial predators, known as bacteriophages (phages). In many microbial ecosystems, viral predation aids in controlling microbial numbers within the carrying capacity of a system and is said to contribute significantly to the ocean's carbon cycling (Rohwer et al., 2009). It was postulated that phages can outnumber their prokaryotic hosts by a factor of 10 (Wommack and Colwell, 2000), although this has been recently contested with the virus-to-microbial cell ratio proposed to vary pending on environmental location and to follow non-linear, power-law functions of microbial cell abundances (Wigington et al., 2016).

In terms of microbial load, the human gut is one of the most densely populated ecosystems on the planet with an estimated load of ~ $10^{13}$ microorganisms (Sender et al., 2016) belonging to Bacteria, Archaea and Eukarya in addition to prokaryotic and eukaryotic viruses. The bacterial component of the human gut microbiome is the most studied and it is now accepted that it plays an integral role in health by providing nutrients and bioactive metabolites many of which have far-reaching effects in the body, it protects the host against infection, sustains gut mucosal integrity and regulates gut immunity and energy homeostasis in the body (Mills et al., 2019). The viral component has received much less attention but counts to date suggest a viral load which is less than that of its microbial hosts, with ~ $10^9$-$10^{10}$ VLPs/g found in human feces, in the colonic mucosa and in cecal contents based on electron microscopy studies (LePage et al., 2008; Hoyles et al., 2014). These VLPs are composed mainly of tailed phages belonging to the order *Caudovirales* which have double-stranded (ds) DNA linear genomes and are of the families *Myoviridae* (short contractile tail), *Siphoviridae* (long non-contractile tail), and *Podoviridae* (short tail) (Breitbart et al., 2003). These families contain phages which infect the major bacterial phyla found in the human gut including *Firmicutes*, *Bacteroidetes*, *Proteobacteria* and *Actinobacteria* (Shkoporov and Hill, 2019). The largely lytic single-stranded (ss) DNA microphages (*Microviridae* family) have also been identified as major constituents of the human fecal virome (Kim et al., 2011) and generally infect enterobacteria. Collectively, these have been termed the 'phageome.' However, studying the phageome is more complicated than studying the bacterial component of the gut microbiome (Shkoporov and Hill, 2019). Many of the bacterial hosts are strict anaerobes which are difficult to cultivate in the laboratory and thus a limited number of gut phages have been isolated this way. Before the advent of high-throughput sequencing platforms, electron microscopy was utilised to visualise phages in gastrointestinal tract (GIT) samples and feces. Metagenomic sequencing studies have expanded our knowledge of the gut phageome to some degree but given that the vast majority of the sequencing data is novel, large gaps in our understanding of the phageome still loom. Indeed, the term 'viral dark matter' has been coined to represent the uncharacterised phages of the human gut (Aggarwala et al., 2017). Based on such studies the ratio of phages to bacteria in the gut has been estimated to approach 1:1 (Reyes et al., 2010), and a more recent estimate set the ratio as low as 0.1 (Shkoporov and Hill 2019).

The aim of this review is to examine the current state of the knowledge in relation to the gut phageome; its developmental path from infancy to adulthood and through to old age, its cross-kingdom influences, namely, dynamic interactions between phage, bacterium and mammalian host (Figure 1), and its structure and behaviour during disease. In light of what we now know about the phageome, we also review its impact on gut treatments, in particular, fecal microbiota transplantation (FMT), and the potential of phage therapy for gut related infections. In evaluating the existing knowledge, this review also sheds light on the gaps in the literature.

## Phage Lifecycles in the Gut

Different lifecycles or infection strategies are available to phages; in simple terms, phages and their lifecycles can be categorised as lytic or lysogenic/ temperate (Figure 2), the latter of which enables phages to 'piggyback' their hosts. Lytic phages infect the host cell and undergo phage replication and

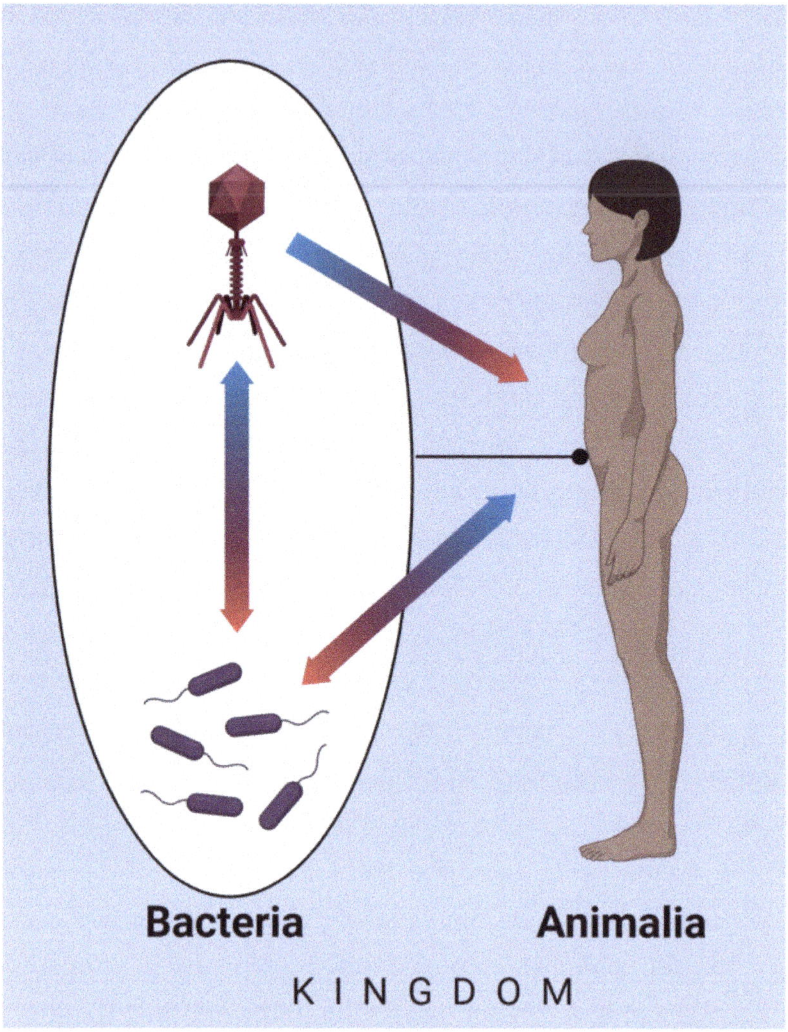


**Figure 1.** Cross-kingdom interactions in the gut. The mammalian host (Kingdom Animalia) influences the bacteriome (Kingdom Bacteria) which in turn influences and is influenced by the phageome, the latter of which also influences the mammalian host directly by interacting with host immunity. Created with BioRender

assembly after which the cell is lysed and free phage particles are released to continue the infection process. Lysogenic phages infect the host cell but their genome can become integrated into the bacterial chromosome where it exists as a prophage, and the host cell, the lysogen, remains intact until some event triggers prophage release from the chromosome and the host cell is killed as for the lytic cycle. Usual triggers include stress response pathways, DNA damage, reactive oxygen species and changes in host bacterial metabolism (Maurice,

2019), although a recent study reported that dietary fructose and microbiota-derived short chain fatty acids induced prophages from a gut symbiont (Oh et al., 2019), thus as research in this field continues we will undoubtedly uncover more lytic cycle triggers associated with the mammalian microbiome. The integrated prophage invariably protects the host cell from superinfection with other related phages (Bondy-Denomy and Davidson, 2014). Lysogeny is also important for horizontal gene transfer and driving genomic diversity. Furthermore, prophages may carry genes which are beneficial to the bacterial host providing it with distinct defence systems against phage attack, increased pathogenicity, improved fitness and metabolic capacities (Bondy-Denomy and Davidson, 2014). Pseudolysogeny is another phage-bacteria interaction which has been proposed to occur in the gut (Kai et al., 1985; Maura and Debarbieux, 2012). In this case, the phage exists as an episomal element in the cell but can enter either the lytic or lysogenic lifecycle when conditions permit (Manrique et al., 2017).

## Phagome Development in the Human Gut

### *Infants*

The human gut microbiota follows a particular trajectory over the course of a lifetime from birth through to old age with many factors known to influence its composition and functionality at various life stages (Mills et al., 2019). Much less is known about the phageome in terms of its development or the factors that may influence it although certain patterns are emerging as the phageome progresses from infancy to adulthood such as a contraction in phage abundance (Figure 3). Breitbart et al., (2008) presented the first description of the DNA viral community of an infant gut using metagenomic sequencing and epifluorescent microscopy. No viral particles were detected in the meconium based on direct epifluorescent microscope counts. By the first week of life, the viral community was dominated by phages which exhibited extremely low diversity and this community changed dramatically by the second week of age, although certain sequences were stable over the first three months of life. The most frequent hits were to phages that infect species such as *Bacillus*, *Lactobacillus*, *Lactococcus*, *Bacteroides*, *Listonella*, and *Staphylococcus*.

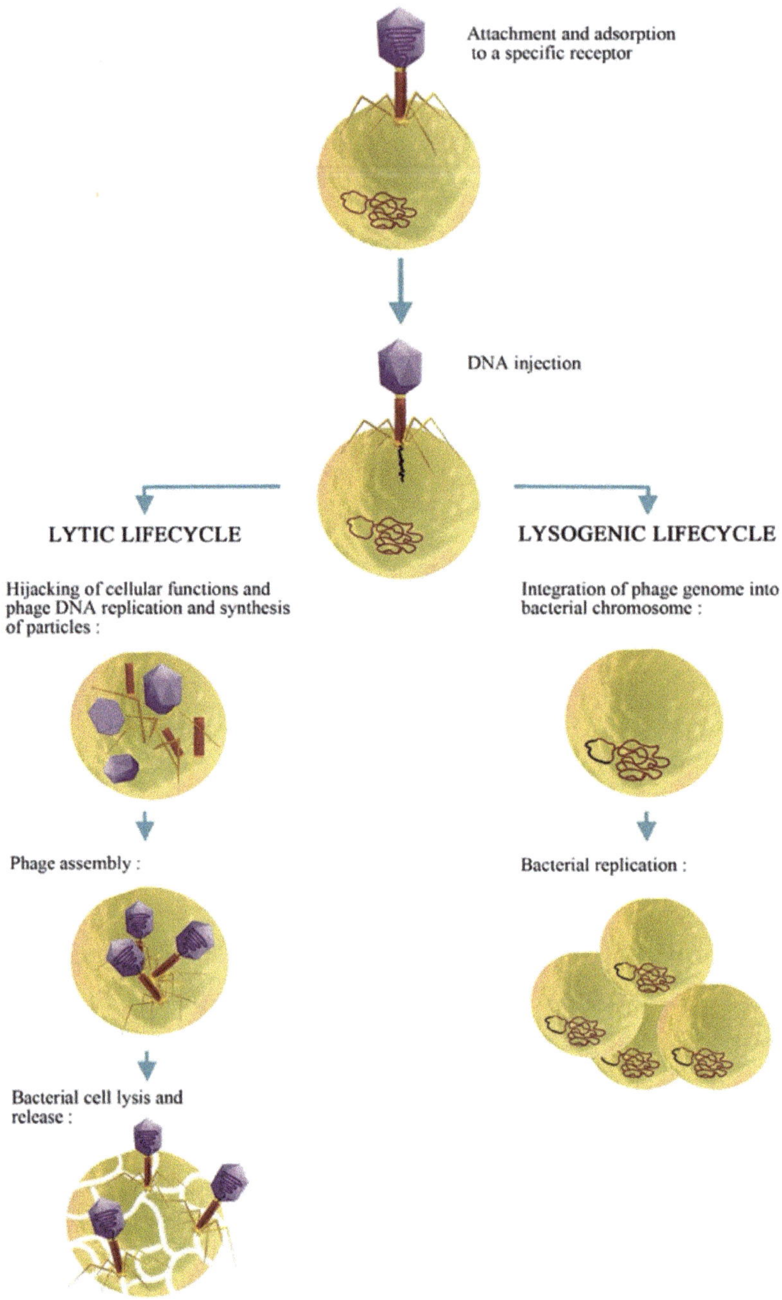


**Figure 2** Diagrammatic representation of the various stages of the lytic and lysogenic lifecycles of a phage.

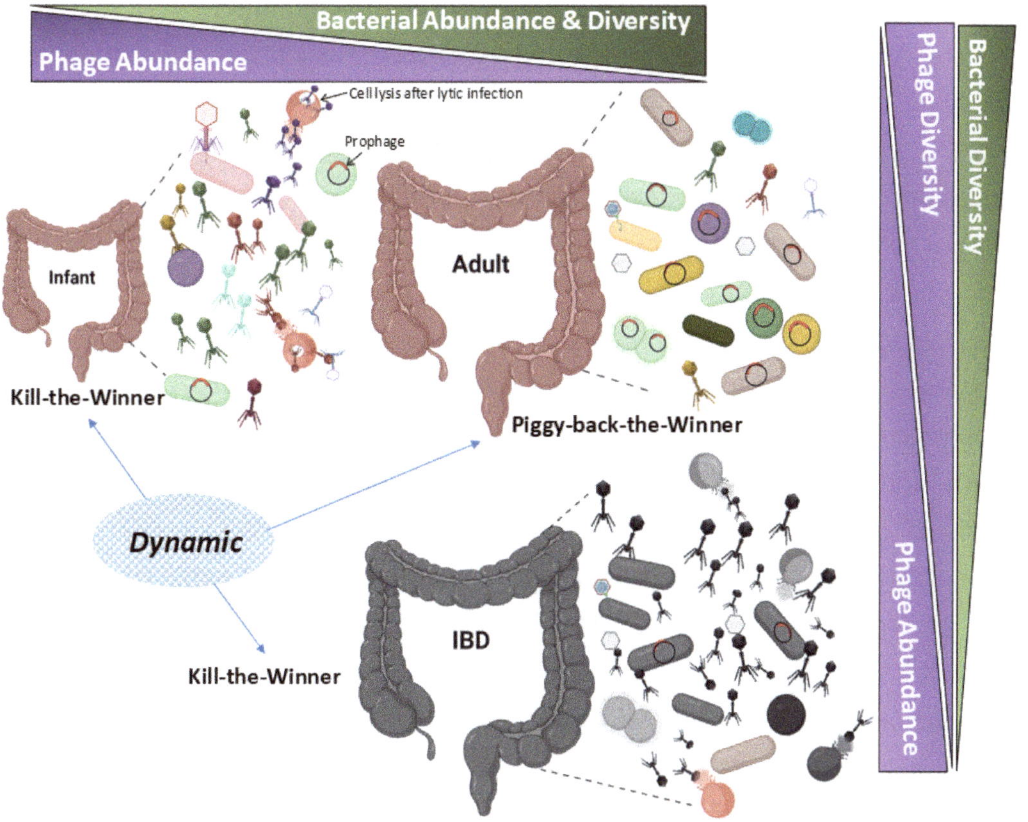


**Figure 3** Comparison of the infant gut phageome-bacteriome with that of the healthy adult, and comparison of the adult phageome-bacteriome with that which occurs during IBD (irritable bowel disease). The infant gut phageome-bacteriome is characterised by kill-the-winner dynamics. As the infant develops, the phageome shrinks and the bacteriome increases in diversity and abundance. At the adult stage, piggy-back-the-winner dynamics predominantly ensue where phages are integrated into the bacterial genome as prophages. During IBD, the phageome expands though phage diversity and richness can be reduced, at the same time bacterial diversity is also reduced. Created with BioRender.

By tracking species and strain level variations in the microbial communities in 11 fecal samples from a premature infant during the first month of life, Sharon et al., (2013) identified dramatic changes in the abundance of three particular strains of *Staphylococcus epidermidis* which correlated with changes in host-phage abundance. For example, the host:phage ratio for one of the strains increased from 1:1 to 10:1 possibly due to activation of phage resistance. Phage infection at this early stage may be vital to creating genetic diversity within a species. Indeed, replicating prophages of infant bifidobacterial species have been suggested to modulate infant gut bifidobacterial populations (Lugli et al., 2016). In this case, the microbiomes of 18 infant fecal samples collected from 5 days of age to 14 months revealed a correlation between the disappearance/ abundance reduction of a bifidobacterial species and the increase of specific phages which were prophages that had entered the lytic cycle. This dynamic is consistent with the 'kill-the-winner' hypothesis of the Lokta-Volterra "predator-prey" model which proposes that the dominant species will be targeted by phages (Rodriguez-Valera et al., 2009). A later study on *Bifidobacterium* prophages identified three distinct genetic groups which infect *Bifidobacterium breve* and *Bifidobacterium longum* hosts (Mavrich et al., 2018). Interestingly, many of these phages harbored a putative tyrosine DNA invertase-mediated phase variation shufflon that generates variation in the tail-associated receptor-binding protein thus potentially enabling the prophages to infect a broader range of host strains. Examination of the infant virome over 24 months noted a contraction in the phageome and an expansion in the bacterial microbiome (Lim et al., 2015). The authors, in this case, suggest that limited prey diversity in the infant gut in the early days of life results in a contraction of the phageome which is consistent with the reciprocal aspect of the Lokta-Volterra "predator-prey" model.

The source of phages in the new born infant gut has been proposed to come from non-dietary initial sources given the inability to identify viral sequences in breast milk or formula (Breitbart et al., 2008). However, later studies suggest vertical transmission of viruses from mother-to-infant since specific bifidophage sequences were common in fecal samples from mothers and corresponding infant pairs (Duranti et al., 2017; Milani et al., 2018). In addition, viromes are more similar between infant co-twins than between unrelated individuals (Lim et al., 2015; Reyes et al., 2015), although this similarity disappears for adult twins as discussed further on. Activated prophages from the developing bacterial community is also a source of phages (Lugli et al., 2016).

Interestingly, birth mode has been shown to influence fecal infant viral communities with those born by spontaneous vaginal delivery exhibiting greater viral and phage diversity than those born by caesarean section at one year of age, suggesting that the vertical transmission of viral communities from mother to child has an impact in shaping the early life microbiome (McCann et al., 2018). Furthermore, differences in phage populations along with eukaryotic viruses were capable of discriminating between twin pairs discordant and concordant for severe acute malnutrition over the first 30 months of life (Reyes et al., 2015).

These studies suggest that the infant phageome may play a fundamental role in the development of the gut microbiome thus understanding its composition and developmental path, and its significance to microbiome maturity is essential and may even aid the development of phage-based therapies to accompany bacterial therapies that endow all infants with the chance of an optimal microbiome from birth irrespective of birth mode, gestational age at birth, feeding regime etc.

*Adults*

While a limited number of microbiota studies have shown that the pre-adolescent and adolescent microbiome is still in a state of development (Mills et al 2019), phageome studies to date have focused on that of adults. The first metagenomic analysis of the adult viral community was performed on a fecal sample from a 33 year old male (Breitbart et al., 2003). It contained an estimated 1200 viral genotypes, and while recognisable viruses were predominantly siphophages most of the sequences were unrelated to anything in the databases. Sequencing the fecal viromes of four pairs of adult monozgotic twins and their mothers at three time points over a one year period revealed that viromes are unique to individuals regardless of their degree of genetic relatedness and the individual virome is highly stable with > 95% of virotypes retained over the sampling period (Reyes et al., 2010). Minot et al., (2011) also confirmed large interpersonal variation within the fecal viromes of healthy adults. The assembled contigs harbored functions required for lytic and lysogenic growth. A dietary intervention did cause a significant change in the virome community resulting in changes in the proportions of the pre-existing community and while virome compositions of individuals on the same intervention were shown to converge, they were not identical. Comparative analysis of the phageome on a more global-scale involving 64 healthy individuals from around the globe revealed a description of a healthy human gut phageome which is proposed to consist of a set of core phages shared among more than half of all individuals, a common set of phages found in 20-50% of individuals and a set of phages which are rarely shared or are unique to an individual (Manrique et al., 2016).

In order to understand the origin and evolution of the human gut virome, Minot et al., (2013) analysed the viral community of a single adult over 2.5 years. In total, 478 contigs were identified which consisted largely of previously unidentified phage genomes. Throughout the study period, > 80% of these contigs persisted with temperate phages showing lower mutation rates consistent with the accurate replication of bacterial DNA polymerases when integrated into the bacterial host chromosome. In contrast, the ssDNA lytic *Microviridae* showed high substitution rates of > $10^{-5}$ per nucleotide/day so that by the end of the 2.5 year period, sequence divergence was significant enough to distinguish new viral species. These longitudinal changes were also associated with diversity-generating retroelements (DGRs) and virus-encoded Clustered Regularly Interspaced Short Palindromic Repeats (CRISPRs). DGRs are reverse-transcriptase-based elements that introduce mutations at adenines in specific repeated sequences and can switch phage-host specificities (Medhekar and Miller, 2007). CRISPRs are phage resistance mechanisms described as innate immunity for prokaryotes (Mohanraju et al., 2016). In the arms-race between phages and prokaryotes, CRISPRs are hot-spots for genetic diversity. More recently, longitudinal metagenomic analysis of fecal viruses in ten healthy adults for at least a year revealed individual specificity, high temporal stability and correlation with the bacteriome (Shkoporov et al., 2019). Unlike other studies, the researchers used a database-independent approach that utilised most of the sequencing data as opposed to depending on incomplete viral sequence databases. They identified a highly prevalent, persistent, and individual-discriminatory consortia of viral genomes termed PPV that are mainly responsible for compositional virome stability and comprise the majority of VLP sequences. The PPV phages infect highly abundant and persistent inhabitants of the gut including *Prevotella*, *Faecalibacterium*, *Eubacterium*, *Parabacteroides* and *Bacteroides*. They also identified a less stable and non-individualised virome termed the transiently detected virome (TDV) which may be more shared between individuals although further studies with larger cohorts are required to confirm this. The TDV phages infect less abundant and more transient bacteria such as *Clostridium*, *Streptococcus*, *Akkermansia*, and *Listeria*. Importantly, temperate phages did not dominate the fecal virome thus the authors speculate that other mechanisms besides prophage integration enable phages to persist long-term in the human gut. Indeed, a dominant gut phage, namely crAss-phage (discussed further on), has been shown to proliferate with its host without significantly impacting the host (Shkoporov et al., 2018) but how this is achieved is currently unknown.

Microbiome diversity has been shown to correlate with virome diversity (Moreno-Gallego et al., 2019). An examination of the microbiomes and viromes of 21 adult monozygotic twins revealed that microbiome diversity and richness correlate to virome diversity and richness and vice versa. Earlier, De Sordi et al., (2017) reported that a phage was capable of adapting to infect an insensitive strain in the conventional mouse gut. However, this event did not occur in planktonic cultures or in dixenic mice and required the presence of an intermediate host. The event was associated with single amino acid substitutions and homologous intragenomic recombination of the phage genome but most importantly highlights how increased microbial diversity can promote phage diversification in the gut.

*Elderly*

The elderly (> 65 years) gut microbiome is characterised by a decline in microbiota diversity (Mills et al., 2019). In addition, residence location of the individual has been shown to influence gut microbiome diversity with healthy community dwellers having microbiotas that are significantly more diverse than those in long-stay care facilities (Claesson et al., 2012). The first and to date only description of the elderly gut virome did not report any differences in virome diversity between these two groups (Stockdale et al., 2018). The elderly virome was characterised by an abundance of *Microviridae* of the subfamily *Gokushovirinae*.

Further studies are clearly required that investigate the virome at different life stages. However, in order to gain an understanding of how the virome ultimately contributes to human health, studies should focus on the virome, the microbiota and the surrounding environment of the mammalian gut simultaneously. Such cross-kingdom studies should pave the way to a better understanding of how the virome and microbiome interact with each other and with the host, and how these interactions ultimately affect mammalian health.

## Phage Behaviour in Disease and During Antibiotic Treatment

### *Inflammatory Bowel Disease*

Inflammatory bowel disease (IBD) is a complex recurrent inflammatory disease of the intestine driven by genetic, immunological and environmental factors and collectively refers to ulcerative colitis (UC) and Crohn's disease (CD). Its primary symptoms include abdominal pain, rectal bleeding and weight loss. Alterations in the function and composition of the gut microbiota have been implicated in IBD and include reduced microbiota diversity, decreases in beneficial microbes (symbionts) and increases in pathogenic and non-beneficial microbes (pathobionts) (Nishida et al., 2018). The role of the gut virome in IBD is less studied but the analyses which have been conducted consistently report differences between healthy individuals and patients (Figure 3). The first such study used electron microscopy methods to compare the viral abundance of the gut mucosa of healthy individuals with CD patients and then compared ulcerated with non-ulcerated mucosa in the patient group (LePage et al., 2008). Significantly more VLPs were identified in CD patients ($2.9 \times 10^9$ VLPs/biopsy) compared with healthy individuals ($1.2 \times 10^8$ VLPs/biopsy) ($P = 0.024$). These VLPs were phages and were predominantly consistent with the morphotypes of the *Caudovirales* order. Ulcerated mucosa from CD patients harbored less VLPs ($2.1 \times 10^9$ VLPs/biopsy) compared with CD non-ulcerated mucosa ($4.1 \times 10^9$/ biopsy). The authors hypothesised that in non-ulcerated areas more phages are produced or can survive longer. Pérez-Brocal et al., (2013) performed the first metagenomic study of the viral and microbiome communities in fecal samples from CD patients. Compared with healthy individuals, the fecal virome and microbiome of CD patients showed lower diversity and greater levels of variation within the group compared with the healthy group. We have already mentioned the healthy gut phageome which consists of a core set of phages found in more than half of all people, notably 62% (Manrique et al., 2016). In the same study, the authors compared the prevalence of these core phages with those in CD and UC patients and noted significant reductions of 54% and 42% in CD and UC patients, respectively. While healthy subjects harbor 62% of the 23 core phages, CD and UC patients harbored significantly less at 30% and 37%, respectively, suggesting that the healthy gut phageome is markedly perturbed in IBD patients. An investigation of the viral load of ileal and colonic samples from paediatric CD patients and controls using metagenomic sequencing revealed that the largest number of viral hits was obtained from CD gut wash samples (n = 691), followed by CD ileal samples (n = 52) (Wagner et al., 2013) and the most abundant viral sequences belonged to *Caudovirales*. Norman et al., (2015) investigated the fecal viromes and microbiomes of CD and UC patients of three independent cohorts and used non-IBD household controls where possible thus controlling for confounders such as geographical location and diet and reported an expansion of *Caudovirales* phages in CD and UC samples. Furthermore, the viromes were disease- and cohort-specific. The expansion of *Caudovirales* in IBD patients was inversely related to the observed reduction in bacterial diversity and is suggestive of kill-the-winner dynamics. The authors suggest that the virome contributes to or is a biomarker of IBD. A recent study also reported an expansion of *Caudovirales* in rectal mucosal samples from patients with UC compared with healthy controls (Zuo et al., 2019). This expansion of phages was also accompanied by a decrease in *Caudovirales* diversity, richness and evenness compared with healthy controls. The altered mucosal virome in UC patients correlated with intestinal inflammation. Using whole virome analysis on the IBD cohort of Norman et al., (2015), as well as an in-house UC dataset, Clooney and colleagues interrogated both known and unknown viral sequences. Firstly, a core lytic virome lacking any identifiable temperate features was identified in healthy subjects. This was absent in IBD patients and replaced by an individual-specific shift toward induced temperate phages. However, no changes in viral richness were observed but the changes reflected alterations in bacterial composition.

Several hypotheses have been put forward as to the exact role of phages in IBD. Norman et al., (2015) suggest that the increased abundance of *Caudovirales* phages in IBD patients may have arisen from the induction of prophages from commensal bacteria. This seems to be supported by the observations of Clooney et al., 2019. We previously proposed a model whereby prophage induction from gut commensals may alter the ratio of symbionts to pathobionts thus providing an opportunity for pathobionts to expand (Mills et al., 2013). Gut inflammation in a murine *Salmonella* Typhimurium diarrhoea model was shown to trigger the SOS response in *Salmonella* lysogens which led to the expression of the phage anti-repressor protein and release of free phage particles that subsequently lysogenized a different *Salmonella* strain (Diard et al., 2017). Bacterial lysis results in the release of several different components including nucleic acids, lipids and proteins that can trigger inflammatory responses and ultimately cause tissue damage in the host (Norman et al., 2015; Zuo et al., 2019). Another important aspect is the interaction between the host immune system and phages themselves (Norman et al., 2015), a topic which is discussed in detail further on. However, administration of a phage cocktail to rats that targets gut inhabitants over a 10 day period was shown to induce a leaky gut which manifested as weight loss, messy hair and decreased activity – proposed symptoms of endotoxin translocation across the leaky mucosa (Tetz and Tetz, 2016). More recently, phages and phage DNA induced a gut immune response in germ-free mice by stimulating IFN-γ (Gogokhia et al., 2019). Increasing the phage levels exacerbated colitis via the same mechanism. Furthermore, phages from active UC patients induced more IFN-γ than phages from healthy individuals.

The studies to date do point to a role for the phageome in the onset of IBD. However, further studies are needed to thoroughly evaluate the extent of this role, and if indeed the phageome is an instigator of the microbial disruptions which are a recognised feature. The answers to these queries may aid in the development of preventative and therapeutic medicines but may also see the fecal phageome as a prognostic tool in the surveillance of IBD risk in humans.

*Colorectal Cancer*

Globally, colorectal cancer (CRC) is the third most diagnosed cancer in males and the second in females (Macrae, 2019). Specific microbial agents and decreased microbiota diversity have been identified as risk factors for CRC (Gao et al., 2017; Macrae, 2019).

Hannigan et al., (2018) analysed the dsDNA fecal virome in CRC patients and compared it to that of healthy individuals and reported that the virome of CRC patients was indeed altered; however, basic diversity metrics (alpha and beta) failed to identify community differences. Instead, the authors utilised an operational taxonomic unit (OTU) approach. The virome in CRC patients was composed mainly of phages. These phages did not necessarily infect the most influential bacteria as there was a lack of correlation between bacteria and phage abundances. Thus the authors postulated that the phages do not alter the influential bacteria but instead alter the greater bacterial community which indirectly influences the bacteria driving CRC progression. Essentially, through phage disruption of existing bacterial communities, niches become available for opportunistic bacteria such as *Fusobacterium nucleatum* which has been associated with CRC tumors (Castellarin et al., 2012) and leads to a cascade of events progressing towards the transformation of epithelial cells. However, further work is required to prove/disprove this hypothesis. Based on three different cohorts, Nakatsu et al., (2018) reported that the diversity of the fecal phage community was significantly increased in CRC patients compared with controls and 22 viral taxa were capable of discriminating cases from controls across the three cohorts. Furthermore, a subset of these viral taxa were enriched in CRC patients with poor survival outcomes. The study noted an enrichment of the *Inovirus* and *Tunalikevirus* phage species. *Inovirus* is a non-*Caudovirales* ssDNA phage that infects gram negative bacterial hosts including *bft*-positive enterotoxigenic *Bacteriodes fragilis*, *F. nucleatum* and *pks*-positive genotoxic *Escherichia coli* which have been associated with CRC development (Cuevas-Ramos et al., 2010; Castellarin et al., 2012; Boleij et al., 2015; Nakatsu et al., 2018). *Tunalikevirus* is a member of the *Caudovirales* order and *Siphoviridae* family. They are lysogenic and can infect commensal or lysing enteropathogenic *E. coli* (Niu et al., 2014; Nakatsu et al., 2018). It is unclear if the changes in the fecal virome were associated with changes in the mucosal microbiome. The authors suggest that phages may impact cancer cell behaviour by modulating the bacterial community or may influence the epithelial cells directly. Further work is required to provide mechanistic insights but the results to date clearly indicate that the virome in the context of CRC can no longer be ignored.

### *Type 1 Diabetes*

Type 1 diabetes is an autoimmune disease where the insulin-secreting beta cells of the pancreatic islets of Langerhans are attacked (Zhao et al., 2017). Studies have shown a link between the gut microbiota and the development of this disease (Needell and Zipris, 2016). For example, Kostic et al., (2015) noted a reduction in microbiome alpha diversity in seropositive children which occurred after the appearance of the autoantibodies that are predictors of the disease. In a follow-on study, the authors analysed the same fecal virome samples of these children which had been collected before and after the development of autoimmunity (Zhao et al., 2017). Of the 11 subjects (cases), six went on to develop the autoantibodies but did not progress to disease, while five developed the autoantibodies and progressed to type 1 diabetes. The viromes of these 11 cases were compared to healthy controls. The viromes of the case subjects were less diverse than those of controls and there was significant enrichment of particular eukaryotic viruses. In terms of phages, Shannon diversity and richness were lower in the cases compared to healthy controls and the changes in the virome over time differed between cases and controls. Furthermore, the authors identified disease-associated phage contigs that were linked to specific members of the bacteriome. Thus they concluded that in this cohort changes in the intestinal virome preceded the development of autoimmunity which could potentially modulate bacterial abundances resulting in dysfunctional alterations. It is proposed that the risk for autoimmunity in the case of type 1 diabetes is related to a combination of protective- versus disease risk-conferring effects of the virome. Further studies are required to determine if these effects are replicated in other cohorts and may open opportunities for virome targeting in at-risk groups to prevent progression to disease.

*Phage Behaviour during Antibiotic Treatment*

Antibiotics are known to impact the gut microbiota resulting in disturbances that have been implicated in disease development and aggravation and include loss of functional and taxonomic diversity along with diminished colonisation resistance (Lange et al., 2016). Nowadays, antibiotic resistance is a global crisis and novel alternatives are urgently required (Mills et al., 2017). However, given our knowledge of the microbiota damage caused by broad-spectrum antimicrobials, narrow-spectrum substitutes are the preferred option. For example, *Clostridium difficile*-associated diarrhoea is usually induced by antibiotic treatment (Theriot and Young, 2014). The impact of antibiotics on the phage community is worthy of consideration given that antibiotics have been shown to induce prophage induction which may be one of the mechanisms by which antibiotics induce the associated microbiome changes. For example, in-feed antibiotics were shown to induce prophage induction in swine fecal microbiomes (Allen et al., 2011). A subsequent decrease in the relative abundance of *Streptococcus* bacteria correlated with an increase in the relative abundance of *Streptococcus* phages indicating 'kill-the-winner' dynamics. A later study investigated the impact of the antibiotic carbadox on the swine fecal metatranscriptome and reported that the antibiotic induces transcription of phage genes involved in all stages of the lytic cycle just two days after antibiotic administration (Johnson et al., 2017). The viromes encoded antibiotic resistance genes but the frequency of these genes did not change over time or between antibiotic-treated and control groups. In a cohort of human subjects, long-term antibiotic administration resulted in characteristic changes in the oral and fecal bacterial communities but the changes in the fecal viral community were much less obvious and there was no significant impact on viral diversity (Abeles et al., 2015). However, the authors did note an expansion of antibiotic resistance genes in the fecal viromes following antibiotic therapy. While a similar phenomenon was reported in mice in that antibiotic treatment resulted in an enrichment of phage-encoded antibiotic resistance genes (Modi et al., 2013), a later independent inspection of the work concluded that the majority of such

genes were associated with bacterial DNA contamination and stated that phage genomes rarely encode antibiotic resistance genes (Enault et al., 2017). The technical difficulties which microbiologists face when extracting VLPs from biological samples and the short falls in subsequent data analyses owing largely to the fact that the majority of gut viral sequences are unknown, can lead to misinterpretations as discussed in some detail by Shkoporov and Hill (2019).
However, antibiotic treatment in severely malnourished diarrhea patients was associated with an expansion in fecal *E. coli* and coliphages (Kieser et al., 2015). The authors attributed the increase in coliphages to the increase in abundance of host cells. However, as the analysis was performed over a short time period (one day) the authors can't rule out if the expansion of coliphages could lead to a collapse of the *E. coli* hosts. However, the abundance of a particular *Escherichia* phage did negatively correlate with the abundance of its host.

Furthermore, recent studies of phage-host dynamics point to the potential of early-life antibiotic administration to increase opportunities for pathogenic lysogens to outcompete commensals at mucosal surfaces and become established infections. This has been proposed for lysogens of *Pseudomonas aeruginosa* in cystic fibrosis of the lungs (Silveira and Rohwer, 2016) and is discussed in further detail in the following section.

More studies are required to determine the full impact of antimicrobials on gut bacteria-phage interactions. While present-day antimicrobial screening studies aim to isolate narrow-spectrum antimicrobials, the impact of these on commensal lysogen growth is worthy of investigation to determine if exposure can ultimately alter whole bacterial community composition and functionality.

## Phage-Bacteria Dynamics in the Gut and Potential Consequences for the Mammalian Host

Until relatively recently, it was assumed that at high microbial cell densities when there is a plentiful supply of hosts, the phage lytic cycle dominates following the kill-the-winner model of predator-prey dynamics, while at low host densities, lysogeny is the favoured lifecycle. These assumptions were mainly based on investigations of microbial ecosystems in marine environments (Jiang and Paul, 1998; Payet and Suttle, 2013). But it soon became clear that in some of the most densely populated ecosystems, including the gut, scientists were unable to provide evidence of this dynamic. Knowles et al., (2016) noted this phenomenon in several densely populated microbial ecosystems including coral reefs and as a consequence proposed an extension of the kill-the-winner model called piggyback-the-winner where temperate dynamics dominate in ecosystems with high microbial cell density. The gut bacteriome and phageome generally follow this dynamic. Indeed, as viral and host densities increase, lysogeny is proposed to be the favoured scenario providing protection to the bacterial host against superinfection by closely related viruses (Bondy-Denomy et al., 2014; Thingstad et al., 2014), and the energetic costs of superinfection are deemed to be less than those incurred by mutations to protect the host against phage infection (Våge et al., 2013; Avrani and Lindell, 2015). Indeed, in the nutrient rich environment of the gut the energetic costs of carrying a prophage would potentially be minimal (Silveira and Rohwer, 2016). Lysogeny can also endow the host cell with other benefits (Howard-Varona et al., 2017). As an example, polymorphic MuF toxins were found to be abundant in temperate phages of the phylum Firmicutes, which are significant members of the human gut microbiota (Jamet et al., 2017). Such toxins can alter microbial community dynamics by inhibiting or killing competitors. Furthermore, the release of prophages as their lytic counterparts enables phage-killing of competitors. For example, prophage elements of the gut commensal *Enterococcus faecalis* V583 have been shown to confer a competitive advantage to the strain *in vitro* and *in vivo* as their release as lytic phages kills closely related strains enabling niche establishment and domination by *E. faecalis* V583 (Duerkop et al., 2012).

However, the anatomy of the gut also has an impact on phage-bacteria dynamics pending on location. The gut mucosal epithelium is a semi-permeable barrier that participates in immune-sensing and enables nutrient absorption while limiting/preventing the admission of harmful antigens and microorganisms (Vancamelbeke and Vermeire, 2017). Its first line of defence is the mucus layer which is synthesised and secreted by goblet cells and composed of highly glycosylated mucin proteins that prevent contact between bacteria and the epithelial cells (Cornick et al., 2015; Vancamelbeke and Vermeire, 2017). Following an analysis of a range of mucosal surfaces from cnidarians to humans, Barr et al., (2013) reported that at this location phage:bacteria ratios are increased relative to the adjacent environment. Based on cell culture studies the increase in phage numbers was proven to be mucin dependent and protected the eukaryotic cells from bacterial infection. Ig-like protein domains on phage capsids bind to glycan residues which enables phage enrichment in the mucus layer. The authors proposed the bacteriophage adherence to mucus (BAM) model as a non-host immunity mechanism against bacterial infection (Figure 4A). In a later study, subdiffusive motion of mucus-adherent phages in the mucus layer was shown to increase the frequency of bacterial encounters in the mucin mesh, while a non-mucus-adherent derivative phage, which accumulated and persisted at similar rates, was much less effective at reducing bacterial colonisation (< 4000 fold) in the mucus layer (Barr et al., 2015).

It is now postulated that BAM and the piggyback-the-winner dynamic are interrelated and play an important role in microbiome development (Silveira and Rohwer, 2016). Specifically, the lytic to lysogenic switch is spatially structured at mucosal surfaces owing to the mucin concentration gradient. In the top mucus layer where mucin concentration is low and phage abundance is thus low the abundance of commensal bacteria is higher. This favours lysogeny between bacteria and phages representing piggyback-the-winner. Moving into the mucin intermediate layer, phage abundance increases as mucin concentration rises which favour the lytic lifecycle. This system provides two steps of phage-mediated immunity at the mucosal layer: (1) competition with commensal lysogens which have increased fitness due to the integrated prophage; (2) lytic phage infection at the intermediate layer (Silveira and Rohwer, 2016).

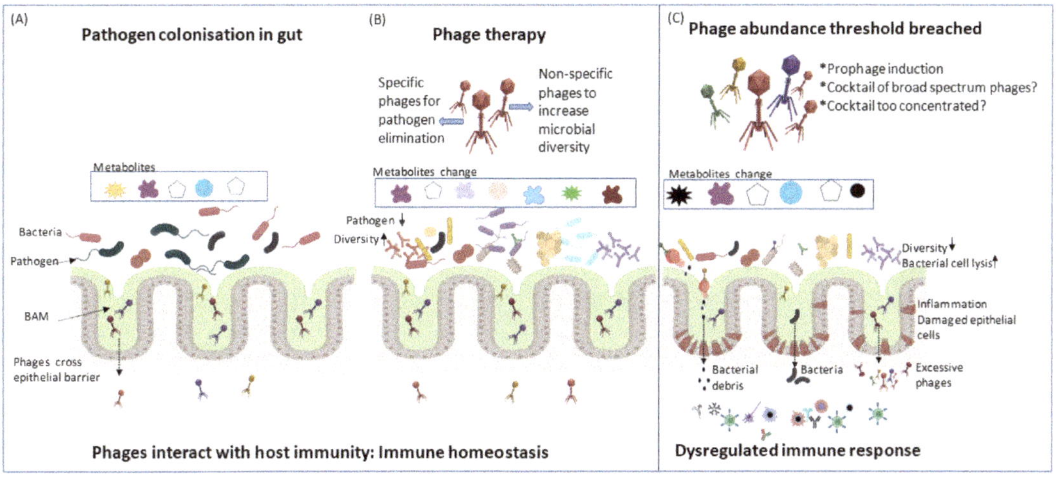


**Figure 4** (A) Pathogen colonisation of the gut. BAM (bacteriophage adherence to mucus) protects the epithelial cells from bacterial infection (Barr et al., 2015). (B) Phage therapy eradicates the pathogen (Hill et al., 2018) but can also increase bacterial diversity (Bao et al., 2018). The changes to the bacteriome subsequently change the metabolome (Hsu et al., 2019). (C) Hypothetical scenario representing what may happen when the phage abundance threshold is breached. The phage abundance threshold represents the maximum phage levels in the gut which do not detrimentally impact the bacteriome and subsequently the mammalian host. Above this level, phage lytic infection predominates, bacterial diversity decreases, bacterial cell lysis releases bacterial debris and excessive phages cross the epithelial barrier. The epithelial cells are damaged and inflammation ensues. Subsequently, bacterial debris and bacterial cells can cross the damaged gut barrier. Overall, this results in systemic immune dysregulation. The phage abundance threshold could be breached via prophage induction of resident gut bacteria, the use of broad-spectrum phage cocktails which infect a broad range of hosts or potentially by highly concentrated phage cocktails. Created with BioRend.

However, piggyback-the winner dynamics in opportunistic pathogens at mucosal surfaces represent a threat for the mammalian host (Silveira and Rohwer, 2016). For example, lysogenic conversion of the common cystic fibrosis pathogen, *Pseudomonas aeruginosa*, endowed it with a competitive advantage in cystic fibrosis rodent models (Winstanley et al., 2009; Davies et al., 2016). Recently, it was shown that immune cell uptake of *P. aeruginosa* containing an integrated prophage resulted in inappropriate antiviral immune responses following phage RNA production that impeded bacterial clearance, an event which did not occur with the prophage-free derivative (Sweere et al., 2019). Silveira and Rohwer (2016) propose that initial mucosal colonisation and external factors such as antibiotic administration will determine the dual role of piggyback-the winner dynamics in either conferring phage-mediated immunity or enhancing disease states. In terms of cystic fibrosis, opportunistic pathogen colonisation of the lungs occurs during the early years of life when antibiotic administration is a common treatment (Ratjen et al., 2001) which suppresses commensals thus providing opportunities for pathogenic lysogens to occupy niches and flourish (Han et al., 2012).

Piggyback-the winner dynamics in opportunistic pathogens in the gut could help explain the often observed inability of therapeutics such as probiotics and prebiotics to restore gut microbiota homeostasis in some individuals. Such lysogenic pathogens presumably have superior fitness and thus competitive advantage over commensals and introduced microorganisms (Silveira and Rohwer, 2016). Furthermore, given that phage host specificities can range from narrow to broad (de Jonge et al., 2019), the presence of broad-spectrum prophages in these lysogens can provide an additional competitive advantage by phage lytic killing of competitors (Silveira and Rohwer, 2016). Further insight into the occurrence of this dynamic in the gut may aid in the development of novel strategies such as probiotic lysogens with enhanced fitness capabilities for gut survival and pathogen takeover.

## The influence of Resident Phages on Host Immunity

The direct impact of resident gut phages on mammalian host immunity is largely unexplored. Given that the majority of gut phages reside in bacterial hosts as prophages during times of homeostasis, they undoubtedly exert indirect effects on mammalian immunity by shaping both the genetic and species diversity of the microbiota (Duerklop and Hooper, 2013). The human gut is exposed to free phage particles in the very early days of life (Breitbart et al., 2003; Lim et al., 2015) and these potentially play a role in the immune system education which takes place in the infant gut. Gorski et al., (2018) propose that resident phage particles have an immunomodulatory activity which targets the intestinal epithelial cells and contributes to immune homeostasis. Exposure of human peripheral blood monocytes to phages of *Staphylococcus aureus* and *Pseudomonas aeruginosa* induced the expression of immune-related genes and the responses were reproducible and predominantly anti-inflammatory (Van Belleghem et al., 2017). *Caudovirales* phages used to eliminate adherent invasive *Escherichia coli* in a mouse model of bacteria-aggravated colorectal cancer were shown to upregulate genes and pathways of both innate and adaptive immunity (Goghokia et al., 2019). Further experiments in the same study using purified phages and germ-free mice revealed that phages were capable of stimulating the expansion of IFN-γ-producing CD4+ T cells and CD8+ cells in the murine gut. The immune-stimulating ability of phages was further corroborated *in vitro* using CD4+ T cell-dendritic cell co-cultures and a range of *Caudovirales* phages purified from *Bacteriodes thetaiotaomicron*, *E. coli* and *Lactobacillus plantarum*, indicating that the response was not exclusive to the phages initially used to eliminate the pathogen. This phage activation of IFN-γ took place via toll-like receptor 9 and phages from active UC patients induced more IFN-γ than phages from healthy controls. Importantly, the phages were highly purified and free of endotoxins thus providing a strong indication that phages do directly interact with the mammalian immune system (Bollyky and Secor, 2019; Goghokia et al., 2019). In addition to this, *in vitro* studies have confirmed that phages can cross the epithelial cell layer with potential to spread throughout the body (Nguyen et al., 2017) and hence interact with immune cells at other sites and play a role in maintaining immunological homeostasis (Górski et al., 2017). Oral inoculation of mice with an eight-phage cocktail against *Salmonella* led to phage distribution in all organs and blood, as well as feces (Seo et al., 2018). In a recent review of the topic, Van Belleghum et al., (2019) state that phages can modulate innate immunity through phagocytosis and cytokine production and impact adaptive immunity by influencing antibody production and effector polarisation. While the overall responses are typically anti-inflammatory the authors state that this does not mean that phages are capable of suppressing innate immunity and the effects do not seem comparable to immunosuppressive drugs or agents.

When prophages enter the lytic cycle in the gut the abundance of free phage particles increases and intracellular bacterial components with immune-stimulating potential are released into the intestinal lumen. Increased phage abundance has been identified in certain diseases such as IBD as previously discussed. Gogokhia et al., (2019) set out to determine the consequence of this increased phage abundance for intestinal colitis using a mouse model. In this case, specific pathogen free mice received a phage cocktail (the target pathogen was absent from these mice) in the drinking water to artificially increase phage levels in the gut and were subsequently induced for dextran sodium sulfate colitis. The increased phage levels exacerbated colitis via TLR-9 and IFN-γ. These results suggest that increased phage abundance, as observed in certain diseases, may actually exacerbate intestinal inflammation and contribute to disease severity (Gogokhia et al., 2019). As stated by Bollyky and Secor (2019), this study raises several questions for future research such as, are the effects unique to *Caudovirales* phages? Would antibodies against these phages affect disease pathogenesis? And the ultimate question, do phages control intestinal homeostasis above the microbiota?

## Phages in Fecal Microbiota Transplantation

FMT describes the transfer of fecal microbiota from a healthy donor to an individual suffering from a gastrointestinal-related disorder (Manrique et al., 2017). Routes of administration include oral ingestion of capsules (Youngster et al., 2014; Khanna et al., 2016), endoscopic delivery (Mattila et al., 2012), retention enemas (Lee et al., 2014) or naso-intestinal tube delivery (Tian et al., 2017). It is a highly effective therapy for recurrent *C. difficile* infection (CDI) (Wilson et al., 2019). Indeed, a recent meta-analysis of randomized controlled trials revealed that FMT is significantly more effective than traditional methods, namely antibiotics, for treatment of recurrent CDI (Hui et al., 2019). FMT has been investigated for several other diseases associated with the intestinal microbiota including IBD but in such cases, its clinical efficacy varies as recently discussed by Wilson et al., (2019). Such limitations have been assigned to the complexity of these diseases, unlike CDI which results from the overgrowth of a specific pathogen. Donor microbial diversity is key and analyses of the gut microbiota of CDI patients before and after FMT have revealed that it is capable of increasing recipient diversity levels to donor levels (Song et al., 2013; Kelly et al., 2015; Khanna et al., 2017; Kellingray et al., 2018; Wilson et al., 2019). Only a limited number of studies have investigated the virome during FMT treatment but the results suggest that the virome may be an important component. Broecker and colleagues followed the progress of a CDI patient who successfully underwent FMT for up to 4.5 years post treatment, performing high-throughput sequencing on donor and recipient fecal samples (Broecker et al., 2013; 2016a, 2016b). Their results indicate that viruses (*Caudovirales* phages solely) were transferred from donor to recipient during FMT (Broecker et al., 2017) though the patient went on to establish a distinct virome from the donor (Broecker et al., 2016a). Interestingly, within the first seven months and beyond of FMT the recipient microbiota represented ongoing dysbiosis, varying from that of the donor before reaching donor similarity (Broecker et al., 2017). In contrast, during the same period, the virome (dsDNA free phages and integrated prophages were found to be indistinguishable using different methods) of donor and recipient exhibited comparable richness and diversity and proved to be stable. By purifying and sequencing viral particles from donor and recipient stools, Chehoud et al., (2016) demonstrated that viruses were transferred from donor to recipient during FMT to pediatric UC patients, with 32 different donor viral contigs appearing in a recipient sample. Interestingly, these contigs were identified as phages or suggestive of unstudied phages and the two most frequently transferred gene types were associated with temperate phage replication. Other studies have also provided evidence of phage transfer from donor to recipient during FMT (Kang et al., 2017; Draper et al., 2018) and long-term colonisation with donor phages (Draper et al., 2018).

Recently, Zuo et al., (2018) showed that phage transfer during FMT in CDI patients was associated with treatment outcome since cure after FMT was observed when donor-derived *Caudovirales* contigs occupied a greater fraction of the recipient enteric virome and especially when the richness of the donor virome was greater than that of the recipient. Indeed, prior to FMT, the recipient virome was associated with a significantly higher abundance of *Caudovirales* phages but with decreased diversity, richness and evenness compared with healthy household controls. The authors suggest that the donor virome should be an important aspect of donor selection. Similarly, in another study, successful FMT donors for recurrent CDI were shown to have higher phage alpha diversity and lower relative abundance compared to donors of failed FMT (Park et al., 2019). A sterile fecal filtrate containing a complex mixture of phages along with bacterial debris and metabolic products was sufficient to cure CDI in five patients (Ott et al., 2017). While the exact mechanism of action is unknown, the authors speculate that phages could be responsible. However, further studies are required to confirm this since cell wall components and DNA fragments may also have elicited the effect.

Finally, success of FMT in UC patients was associated with a lower relative abundance of *Caudovirales* phages in recipients at the time of transplant (Gogokhia et al., 2019). This suggests that phage levels of IBD patients could be used to predict responsiveness to FMT and patients with elevated phages compared with recipients may require further FMT treatment (Bollyky and Secor 2019; Gogokhia et al., 2019).

These studies strongly indicate that the phage component of both donor and recipient microbiota could be significant factors for the success of FMT. Indeed, the phageome may even provide the key to gastrointestinal diseases which have failed to consistently respond to FMT.

## Phage Therapy in the Gut

Phage therapy refers to the use of phages, preferably lytic, to kill disease-causing bacteria. It is practiced in certain regions of the globe including Russia, Poland and Georgia. Amid the ongoing antibiotic resistance crisis, phage therapy is gaining traction in research laboratories elsewhere with a view to identifying antibiotic alternatives and establishing phage therapy within evidence-based medicine through regulated clinical trials (for review of this topic see Hill et al., 2018). Phages offer several advantages over conventional antimicrobials; they multiply and increase in concentration at the site of infection, they tend to be highly strain specific and should have minimal impact on the surrounding microbiota, and they can be 'trained' to improve their infectivity and host range through basic laboratory procedures. However, current EU regulatory frameworks prevent phage therapy from being utilised to its maximum potential (Hill et al., 2018).

The complexity of the gut microbiota and the intricate relationships between gut bacteria, phages and the mammalian host might suggest that administering exogenous phages to target a specific pathogen in the gut would be akin to a *needle in a haystack* scenario. However, one of the key figures in phage discovery in the early twentieth century, namely Felix d'Herelle, initially used phages to treat dysentery (Abedon et al., 2011). More recently, phage therapy failed to treat diarrhoea in children in a phase II, randomised, placebo-controlled clinical trial (Sarker et al., 2016; Sarker and Brussow, 2016) and treatment efficacy in animal studies has varied from successful to complete inefficacy (Lourenço et al., 2018). The challenges of using phages to treat intestinal infections primarily relate to the different physiological states of bacteria in the gut owing to environmental stochasticity, a topic recently reviewed by Lourenço et al., (2018) who highlight the importance of bridging the gap between *in vitro* and *in vivo* studies to optimise phage applications. Furthermore, given the specificity of phages, identification of the aetiological agent prior to treatment is critical.

Phage therapy against *C. difficile* is of interest given the treatment failures and recurrent CDI associated with antibiotics (Surawicz et al., 2013). No strictly lytic phages against *C. difficile* have been isolated (Hargreaves and Clokie et al., 2014). Temperate phages have rendered promising results in *in vitro* (Meader et al., 2010; 2013; Nale et al., 2016b; 2018) and *in vivo* (Nale et al., 2016a; Nale et al., 2016b) studies. However, in an *in vitro* human colon model system, the specific phage ΦCD27 caused significant reductions in vegetative cells of *C. difficile* in 2 out of 3 replicates while *C. difficile* cells flourished in the remaining replicate when compared to control (Meader et al., 2013). This is presumably a consequence of lysogenisation of ΦCD27 thus imparting the pathogen with phage resistance. As well as superinfection immunity, prophages have been shown to encode distinct defence systems against closely related and unrelated lytic and temperate phages (Dedrick et al., 2017). However, Nale et al., (2016b) showed that specific phage combinations from a total of seven phages were capable of completely lysing *C. difficile in vitro* and preventing the emergence of lysogenic/resistant clones. In the same study, the optimised phage combinations reduced *C. difficile* colonisation in a hamster model 36 h post infection and delayed the onset of disease symptoms by 33 h in a challenge model of the disease. An optimised four phage cocktail against *C. difficile* caused a 6-log reduction in *C. difficile* numbers in a batch fecal fermentation model within 5 hours and complete eradication after 24 h but had no impact on commensal enterococci, lactobacilli, bifidobacteria, enterobacteria and total anaerobes (Nale et al., 2018). Interestingly, adding phages alone (without pathogen) caused an increase in specific commensals which could serve to promote colonisation resistance against the pathogen.

Likewise, a five-phage cocktail composed of lytic phages, termed ShigActive™, proved as effective as ampicillin for reducing *Shigella* colonization in challenged mice but its impact on the gut microbiota was significantly less than the antibiotic (Mai et al., 2015). Other studies have also demonstrated the ability of lytic phages to target specific gut pathogens but with minimal to no impact on the surrounding microbiota (Galtier et al., 2016; Cieplak et al., 2018; Hu et al., 2018; Seo et al., 2018). However, a recent study using gnotobiotic mice colonised with defined human gut commensal bacteria (ten culturable, characterised strains) and given the cognate lytic phages revealed that phage predation not only impacted the target bacteria but had cascading effects on other bacterial species via interbacterial interactions causing shifts across the microbiota in both low- and high-abundance species which had a direct impact on the gut metabolome with potential consequences for the mammalian host (Hsu et al., 2019). Thus the 'repressive influence' of phage(s) on the abundance of a target species could have a 'promoting influence' on another species (through niche and nutrient availability etc) which in turn could influence the abundance of another species and so on, enabling phages to modulate bacterial metabolite levels. As an example, the authors noted reductions in the concentration of the neurotransmitter tryptamine which coincided with reductions in *Clostridium sporogenes*. Even with next generation sequencing technologies, most other studies achieve a phylogenetic resolution to species level only but by creating a modest level of complexity this study was able to decipher the exact consequences of phage predation of specific targets on a bacterial community.

Phage dose is also an important factor, especially for broad-host range phages. As an example, a high dose ($10^9$ plaque forming units) of a broad spectrum phage against intraperitoneally administered *Enterococcus facecalis* not only cleared the vancomycin resistant *E. faecalis* from the blood of mice but also had an inhibitory effect on the *Enterococcus* genus in the gut and thus was capable of interfering with normal gut *E. faecalis* strains (Cheng et al., 2017). A single dose of a three-phage cocktail, capable of killing adherent invasive *E. coli*, reduced DSS-induced colitis symptoms in conventional mice colonised with the strain (Galtier et al., 2017).

Recently, a clinical trial was conducted to determine the safety and tolerability of consumption of a four-phage cocktail by healthy adults for 28 days (Gindin et al., 2019). The results revealed that therapeutic doses of the phage cocktail were safe and tolerable in the target human population. In another clinical trial, the researchers investigated the impact of a phage cocktail targeting *E. coli* on the gut microbiota and markers of intestinal and systemic inflammation in healthy human adults following phage cocktail consumption for 28 days (Febvre et al., 2019). The cocktail reduced fecal *E. coli* numbers. While there were no significant changes to alpha and beta diversity parameters, the phage cocktail caused specific population alterations such as increasing members of the butyrate-producing genera *Eubacterium* while decreasing taxa most closely related to *Clostridium perfringens* which is presumably a consequence of interbacterial interactions cascading from phage predation of susceptible bacteria as discussed above (Hsu et al., 2019). The phage cocktail had no impact on lipid metabolism, short chain fatty acid production or inflammatory markers, but a decrease in circulating interleukin-4 was detected (Febvre et al., 2019).

The ability of phages to alter gut microbiota composition has also been reported by Nale et al., (2018) when a phage cocktail against *C. difficile* caused an increase in specific commensals in a batch fecal fermentation model in the absence of the pathogen. Bao et al., (2018) investigated the impact of long-term (31 days) phage consumption (lytic or temperate) versus antibiotic administration on the gut microbiota in mice. The lytic phage, PA13076, had been originally isolated from chicken feces and was propagated on *Salmonella* Enteritidis and the temperate phage, BP96115, was induced from the gut pathogen *Salmonella* Pullorum. While the antibiotic, streptomycin, significantly reduced the diversity and richness of the gut microbiota, the temperate and lytic phages each increased species diversity and richness. All three treatments reduced the Bacteroidetes phylum and increased Firmicutes. The lytic and temperate phages increased bifidobacteria and lactobacilli species but the temperate phage also increased the abundance of *Escherichia*/*Shigella* compared to the control. Thus, while long-term lytic phage treatment promoted a beneficial ecosystem in the mouse gut, the temperate phage may actually promote disease-inducing conditions.

This study and others suggest that phages could be used to promote a healthy gut ecosystem (Figure 4B), strengthening colonisation resistance of the gut microbiota, and like probiotics could be taken as daily supplements, particularly after a course of antibiotics. As the phages are not targeting any particular pathogen, the limitations imposed by current regulations on regular phage therapy may not apply as the phage preparations may not have to be updated to kill phage resistant derivatives which can arise during the course of treatment. However, further research is warranted for this avenue of phage therapy. Firstly, lytic phages appear to be the most suitable candidates given the risk of lysogenisation of pathogens by temperate phages. A particular phage preparation should be tested on multiple mammalian hosts to determine if the responses are uniform or are dictated by the individual's microbiome and given the results of Hsu et al., (2019) the impact of all types of phage therapy on the gut metabolome is a significant consideration. Furthermore, the existence of a phage-abundance-threshold in the gut is a possibility and above this level phages may be detrimental to the mammalian host by expanding systemic phage levels, increasing bacterial debris from lysed cells, decreasing bacterial diversity, all of which negatively impacts host immunity and promotes inflammation (Figure 4C). In this regard, the impact of different concentrations of phage therapeutics on the gut microbiota, the gut metabolome and clinical markers of the mammalian host, including systemic immunity should be considered in tandem in order to define optimal phages and doses which can be used to promote a beneficial gut ecosystem, whether targeting specific pathogens or promoting microbiota diversity and richness.

## CrAssphage and Other Gut Phages that have Gained Attention

CrAssphage is an interesting gut inhabitant given its prevalence in the human population. It was first identified in 2014 when researchers noted its presence in the majority of published human fecal metagenomes (Dutilh et al., 2014). With a 97 kb genome, it comprised up to 90% of all reads in VLP-derived metagenomes and *Bacteroides* was predicted to be its host. Later, a diverse and expansive group of phages related to crAssphage was identified (Yutin et al., 2018). In the same study, protein functions were predicted for the majority of proteins and on account of their predicted short, stubby tails, this family of phages was suggested to belong to *Podoviridae*. This was subsequently confirmed in a study by Shkoporov and colleagues who successfully isolated a crAssphage (crAssphage001) from human fecal material and examined its morphology via electron microscopy (Shkoporov et al., 2018). CrAssphage001 has a genome size of 102 kb and lacks lysogeny genes yet it did not disrupt proliferation of its host *Bacteriodes intestinalis*. Indeed, infection of the exponentially growing cells did not cause complete lysis but delayed the onset of the stationary phase and the final density at stationary phase. The long-term interactions between host and phage were investigated over 23 days. While phage titres reached $10^{10}$ pfu/ml on the first two passages, they were reduced to between $10^6$ and $10^8$ pfu/ml thereafter and remained at that level for the duration of the experiment. At the final time point, 46% of tested colonies were phage resistant and the remainder were weakly or fully sensitive. Thus phage and host were capable of co-existing in a stable interaction over a prolonged period of time (Shkoporov et al., 2018). However, the exact nature of this interaction is yet to be deciphered in future studies which should go some way in aiding our understanding of how phages and their bacterial hosts coexist in the gut. Based on 702 human fecal virome/phageome samples, Guerin et al., (2018) presented 242 previously unrecognised crAss-like phages. Comparative genomics and taxonomic analysis resulted in the classification of human crAss-like phages into four candidate subfamilies and ten candidate genera. Using ex-vivo human fecal fermentations, the researchers were able to propagate crAss-like phages which were identified as *Podoviridae* by electron microscopy.

CrAssphage is acquired early in life and is not influenced by birth mode (McCann et al., 2018). A recent study investigating the origin, evolution and epidemiology of crAssphage revealed that crAssphages are temporarily and geographically clustered throughout the world indicating rapid evolution and local dispersion (Edwards et al., 2019). Fully collinear crAss-like phage genomes were found in both Old-World and New-World primates indicating that the phage has had a stable genome structure for millions of years and its association with primates is potentially millions of years old. In the same study, no significant associations were detected between crAssphages and health and disease. However, Draper et al., (2018) noted decreased levels of crAssphages in individuals with recurrent CDI and confirmed its transfer during FMT. Similarly, Cornault et al., (2018) noted increased levels of crAssphages in healthy controls compared with IBD patients.

Recently, another group of phages which are widespread in the gut microbiomes of humans and animals were identified and referred to as Lak phages (Devoto et al., 2019). With genomes > 540 kb in length, these are the largest phages reported in the human microbiome to date and are believed to infect *Prevotella* based on CRISPR targeting, a genus associated with humans and animals that consume high-fibre, low-fat diets (Gupta et al., 2017). Such megaphages can be overlooked due to genome fragmentation issues in metagenomics and as a result of their restricted mobility on plaque assay plates (Serwer et al., 2007; Devoto et al., 2019).

By examining complete prophage sequences in genomes of *Faecalibacterium prausnitzii*, Cornault et al., (2018) identified 18 different prophages which were grouped into eight new viral genera. These phages were highly diverse and a large proportion was found to contain DGRs. Furthermore, four of these phages were found to be significantly more abundant or more prevalent in stools of IBD patients compared with controls.

Given the difficulties associated with analyzing viral metagenomic data, further studies of this nature should shed some light on the viral dark matter within the gut microbiome and contribute to our understanding of gut phageome composition and function.

## Conclusion and Future Perspectives

Scientists have made admirable progress in establishing the significance of the gut microbiome to human health over the past 15 years. However, if these studies have taught us anything it's that the single, largest challenge facing gut microbiome scientists is grappling with the extensive level of complexity. This can be overcome to some degree by studying its components in isolation which is a common thread in microbiome studies and has exposed dynamic patterns of bacteriome evolution over the course of a lifetime. Using this approach, we now know that the gut phageome is an integral component of the microbiome with its own trajectory. However, knowledge of the interactions between phageome and bacteriome and the consequences of these for the mammalian host will provide the most insight into the microbiome and its significance to mammalian health. In this regard, we would like to draw attention to the study of Hsu et al., (2019) which investigated phage-bacteria interactions, along with the cascading interbacterial interactions and potential consequences for the mammalian host by creating a modest level of complexity in the murine gut using 10 bacterial strains and their cognate phages. Using such an approach along with strategic experimental designs should help scientists make reliable predictions of real-world scenarios and aid interpretation of real-world observations.

Indeed, it now appears that phage-bacteria interactions in the gut may serve as central regulators of microbiome development and stability by increasing and maintaining the general fitness of the microbiota from infancy to adulthood. Evidence of kill-the-winner dynamics in the infant gut suggests its importance in creating genetic diversity down to the strain level in early life, while piggy-back-the-winner dynamics in adulthood maintains genetic diversity and overall bacterial fitness. The phage: bacterial cell ratio in the infant gut may influence the switch from one dynamic to the other with low initial bacterial cell numbers (prey) causing a contraction in phages (predators) (Lim et al., 2015) and a pre-inclination for the lysogenic lifecycle. However, there is a dearth of studies investigating these interactions at different life stages and particularly in vulnerable populations such as infants and the elderly.

We now know that the healthy adult phageome is highly individualised and relatively stable but in certain diseased states, including IBD, the phage-bacteria dynamic changes with a reduction in phage and bacterial diversity but an increase in phage abundance. Given the ability of phages to interact with host immunity (Gogokhia et al., 2019; Van Belleghem et al., 2019) this may have far-reaching consequences for the mammalian host. It appears that the lysogenic-to-lytic switch in the gut is in a delicate balance. In this regard, more studies are required to determine what triggers the switch from one life cycle to the other in the gut environment and confirm if this alteration alone explains the increased abundance of phages associated with certain diseases and the reduction in bacterial diversity.

We have also seen how phage-bacteria interactions can impact the metabolome in the gut (Hsu et al., 2019). Thus, switching from lysogeny to lysis alters the gut environment in many ways; the metabolome changes as a result of bacterial alterations which could have dramatic consequences for the host, further, phage numbers expand, bacterial debris increases, all of which can impact host immunity and inflammation levels.

We hypothesise that homeostasis in the gut may be governed by a phage abundance threshold, above which is detrimental to both the bacteriome and the mammalian host (Figure 4C). Indeed, increased phage levels were found to exacerbate colitis in mice (Gogokhia et al., 2019) and a high titre of a broad spectrum phage interfered with normal gut inhabitants (Cheng et al., 2017). Furthermore, phage abundance and diversity in donors and recipients can impact the success of FMT (Goghokia et al., 2019; Park et al., 2019). While the existence of a phage abundance threshold is hypothetical, its establishment could have major consequences for screening tests for disease (increasing phage levels could predict disease susceptibility), for phage therapy (fine-tuning of treatments) and even disease prevention/treatment (using phage-specific antibodies to eliminate phages and prevent the cascading damage).

In this review, we also addressed the issue of pathogenic lysogens in the gut and their potential for superior fitness and competitive advantage over commensals. This highlights the importance of the correct development of the bacteriome at infancy with minimal opportunity for pathogenic lysogens to become established. Indeed, an understanding of prophage significance to pathogen survival in the gut may aid in the development of targeted therapies. For example, scientists may be able to identify species-specific switches of the lysogenic-to-lytic lifecycle, and accompanied with prophage-specific antibodies may be able to eradicate such lysogens and their released phages.

Finally, phage therapy may be on its way to becoming an evidence-based medical treatment with the advent of clinical trials investigating its efficacy. However, certain studies also show that administration of phages can actually increase bacteriome diversity amongst non- phage targets (Bao et al., 2018; Nale et al., 2018). This is an interesting angle of phage therapy that is worthy of further investigation that could see phages being used in a similar manner to probiotics and prebiotics and may even serve as a complementary therapy.

In conclusion, moving forward, microbiome scientists must now undertake more cross-kingdom analyses from phages right through to the mammalian host at the clinical level with the ultimate goal of understanding how these communities influence mammalian health and how they can be manipulated to avoid disease.

## References

Abedon, S.T., Kuhl, S.J., Blasdel, B.G., and Kutter, E.M. (2011). Phage treatment of human infections. Bacteriophage 1, 66–85. https://doi.org/ 10.4161/bact.1.2.15845 Abeles, S.R., Ly, M., Santiago-Rodriguez, T.M., and Pride, D.T. (2015). Effects of long-term antibiotic therapy on human oral and fecal viromes. PLoS ONE 10, e0134941. https://doi.org/10.1371/journal.pone. 0134941 Aggarwala, V., Liang, G., and Bushman, F.D. (2017). Viral communities of the human gut: metagenomic analysis of composition and dynamics. Mob. DNA 8, 12. https://doi.org/10.1186/s13100-017-0095-y

Allen, H.K., Looft, T., Bayles, D.O., Humphrey, S., Levine, U.Y., Alt, D., and Stanton, T.B. (2011). Antibiotics in feed induce prophage induction in swine fecal microbiomes. mBio 2, e00260-11. 10.1128/mBio.00260-11.

Avrani, S., and Lindell, D. (2015). Convergent evolution toward an improved growth rate and a reduced resistance range in *Prochlorococcus* strains resistant to phage. Proc. Natl. Acad. Sci. USA 112, E2191–E2200. 10.1073/pnas.1420347112.

Bao, H.D., Pang, M.D., Olaniran, A., Zhang, X.H., Zhang, H., Zhou, Y., Sun, L.C., Schmidt, S., and Wang, R. (2018). Alterations in the diversity and composition of mice gut microbiota by lytic or temperate gut phage treatment. Appl. Microbiol. Biotechnol. 102, 10219-10230. 10.1007/s00253-018-9378-6

Barr, J.J., Auro, R., Furlan, M., Whiteson, K.L., Erb, M.L., Pogliano, J., Stotland, A., Wolkowicz, R., Cutting, A.S., Doran, K.S., Salamon, P., Youle, M., and Rohwer, F. (2013). Bacteriophage adhering to mucus provide a non–host-derived immunity. Proc. Natl. Acad. Sci. USA 110, 10771-10776. 10.1073/pnas.1305923110

Barr, J.J., Auro, R., Sam-Soon N., Kassegne, S., Peters, G., Bonilla, N., Hatay, M., Mourtada, S., Bailey, B., Youle, M., Felts, B., Baljon, A., Nulton, J., Salamon, P., and Rohwer, F. (2015). Subdiffusive motion of bacteriophage in mucosal surfaces increases the frequency of bacterial encounters. Natl. Acad. Sci. USA 112, 13675-13680. https://doi.org/10.1073/pnas.1508355112

Boleij, A., Hechenbleikner, E.M., Goodwin, A.C., Badani, R., Stein, E.M., Lazarev, M.G., Ellis, B., Carroll, K.C., Albesiano, E., Wick, E.C., Platz, E.A., Pardoll, D.M., and Sears, C.L. (2015). The *Bacteroides fragilis* toxin gene is prevalent in the colon mucosa of colorectal cancer patients. Clin. Infect. Dis. 60, 208–215. 10.1093/cid/ciu787

Bollyky, P.L., and Secor, P.R. (2019). The innate sense of bacteriophages. Cell Host Microbe 25, 177-179. 10.1016/j.chom.2019.01.020.

Bondy-Denomy, J., and Davidson, A.R. (2014). When a virus is not a parasite: the beneficial effects of prophages on bacterial fitness. J. Microbiol. 52, 235-242. 10.1007/s12275-014-4083-3

Breitbart, M., Haynes, M., Kelley, S., Angly, F., Edwards, R.A., Felts, B., Mahaffy, J.M., Mueller, J., Nulton, J., Rayhawk, S., Rodriguez-Brito, B., Salamon, P., and Rohwer, F. (2008). Viral diversity and dynamics in an infant gut. Res. Microbiol. 159, 367–373. 10.1016/j.resmic.2008.04.006

Breitbart, M., Hewson, I., Felts, B., Mahaffy, J.M., Nulton, J., Salamon, P., and Rohwer, F. (2003). Metagenomic analysis of an uncultured viral community. J. Bacteriol. 185, 6220-6223. 10.1128/JB.185.20.6220-6223.2003

Broecker, F., Klumpp, J., and Moelling, K. (2016a). Long-term microbiota and virome in a Zürich patient after fecal transplantation against *Clostridium difficile* infection. Ann. N. Y. Acad. Sci. 1372, 29–41. doi: 10.1111/nyas.13100

Broecker, F., Klumpp, J., Schuppler, M., Russo, G., Biedermann, L., Hombach, M., Rogler, G., and Moelling, K. (2016b). Long-term changes of bacterial and viral compositions in the intestine of a recovered *Clostridium difficile* patient after fecal microbiota transplantation. Cold Spring Harb. Mol. Case Stud. 2, a000448. 10.1101/mcs.a000448

Broecker, F., Kube, M., Klumpp, J., Schuppler, M., Biedermann, L., Hecht, J., Hombach, M., Keller, P.M., Rogler, G., and Moelling, K. (2013). Analysis of the intestinal microbiome of a recovered *Clostridium difficile* patient after fecal transplantation. Digestion 88, 243–251. 10.1159/000355955

Broecker, F., Russo, G., Klumpp, J., and Moelling, K. (2017). Stable core virome despite variable microbiome after fecal transfer. Gut Microbes 8, 214-220. 10.1080/19490976.2016.1265196

Castellarin, M., Warren, R.L., Freeman, J.D., Dreolini, L., Krzywinski, M., Strauss, J., Barnes, R., Watson, P., Allen-Vercoe, E., Moore, R.A., and Holt, R.A. (2012). *Fusobacterium nucleatum* infection is prevalent in human colorectal carcinoma. Genome Res 22, 299–306. 10.1101/gr.126516.111

Chehoud, C., Dryga, A., Hwang, Y., Nagy-Szakal, D., Hollister, E.B., Luna, R.A., Versalovic, J., Kellermayer, R., and Bushman, F.D. (2016). Transfer of viral communities between human individuals during fecal microbiota transplantation. MBio 7, e00322. 10.1128/mBio.00322-16

Cheng, M., Liang, J., Zhang, Y., Hu, L., Gong, P., Cai, R., Zhang, L., Zhang, H., Ge, J., Ji, Y., Guo, Z., Feng, X., Sun, C., Yang, Y., Lei, L., Han, W., and Gu, J. (2017). The bacteriophage EF-P29 efficiently protects against lethal vancomycin-resistant *Enterococcus faecalis* and alleviates gut microbiota imbalance in a murine bacteremia model. Front. Microbiol. 8, 837. 10.3389/fmicb.2017.00837.

Cieplak, T., Soffer, N., Sulakvelidze, A., and Nielsen, S.D. (2018). A bacteriophage cocktail targeting *Escherichia coli* reduces *E. coli* in simulated gut conditions, while preserving a non-targeted commensal microbiota. Gut Microbes 9, 391-399. 10.1080/19490976.2018.1447291

Claesson, M.J., Jeffery, I.B., Conde, S., Power, S.E., O'Connor, E.M., Cusack, S., Harris, H.M, Coakley, M., Lakshminarayanan, B., O'Sullivan, O., Fitzgerald, G.F., Deane, J., O' Connor, M., Harnedy, N., O' Connor, K., O' Mahony, D., van Dinderen, D., Wallace, M., Brennan, L., Stanton, C., Marchesi, J.R., Fitzgerald, A.P., Shanahan, F., Hill, C., Ross, R.P., and O' Toole, P.W. (2012). Gut microbiota composition correlates with diet and health in the elderly. Nature 488, 178–184. 10.1038/nature11319

Clooney, A.G., Sutton, T.D.S., Shkoporov, A.N., Holohan, R.K., Daly, K.M., O' Regan, O., Ryan, F.J., Draper, L.A., Plevy, S.E., Ross, R.P., and Hill, C. (2019). Whole-virome analysis sheds light on viral dark matter in inflammatory bowel disease. Cell Host Microbe 26, 1-15. 10.1016/j.chom.2019.10.009

Cornault, J.K., Petit, M.-A., Mariadassou, M., Benevides, L., Moncaut, E., Langella, P., Sokol, H., and De Paepe, M. (2018). Phages infecting *Faecalibacterium prausnitzii* belong to novel viral genera that help to decipher intestinal viromes. Microbiome 6, 65. 10.1186/s40168-018-0452-1.

Cornick, S., Tawiah, A., and Chadee, K. (2015). Roles and regulation of the mucus barrier in the gut. Tissue Barriers 3, e982426. 10.4161/21688370.2014.982426

Cuevas-Ramos, G., Petit, C.R., Marcq, I., Boury, M., Oswald, E., and Nougayrède, J.P. (2010). *Escherichia coli* induces DNA damage *in vivo* and triggers genomic instability in mammalian cells. Proc. Natl. Acad. Sci. USA 107, 11537–11542. 10.1073/pnas.1001261107

Davies, E.V., James, C.E., Kukavica-Ibrulj, I., Levesque, R.C., Brockhurst, M.A., and Winstanley, C. (2016). Temperate phages enhance pathogen fitness in chronic lung infection. ISME J. 10, 2553-2555. 10.1038/ismej.2016.51

Dedrick R.M., Jacobs-Sera, D., Bustamante, C.A., Garlena, R.A., Mavrich, T.N., Pope, W.H., Reyes, J.C., Russell, D.A., Adair, T., Alvey, R., Bonilla, J.A., Bricker, J.S., Brown, B.R., Byrnes, D., Cresawn, S.G., Davis, W.B., Dickson,

L.A., Edgington, N.P., Findley, A.M., Golebiewska, U., Grose, J.H., Hayes, C.F., Hughes, L.E., Hutchison, K.W., Isern, S., Johnson, A.A., Kenna, M.A., Klyczek, K.K., Mageeney, C.M., Michael, S.F., Molloy, S.D., Montgomery, M.T., Neitzel, J., Page, S.T., Pizzorno, M.C., Poxleitner, M.K., Rinehart, C.A., Robinson, C.J., Rubin, M.R., Teyim, J.N., Vazquez, E., Ware, V.C., Washington, J., and Hatfull, G.E. (2017). Prophage-mediated defense against viral attack and viral counter-defense. Nat Microbiol. 2, 16251. 10.1038/nmicrobiol.2016.251

de Jonge, P.A., Nobrega, F.L., Brouns, S.J.J., and Dutilh, B.E. (2019). Molecular and evolutionary determinants of bacteriophage host range. Trends Microbiol. 27, 51-63. 10.1016/j.tim.2018.08.006

Devoto, A.E, Santinni, J.M., Olm, M.R., Anantharaman, K., Muck, P., Tung., J., Archie, E.A., Turnbaugh, P.J., Seed, K.D., Blekhman, R., Aarestrup, F.M., Thomas, B.C., and Banfield, J.F. (2019). Megaphages infect *Prevotella* and variants are widespread in gut microbiomes. Nat. Microbiol. 4, 693-700. 10.1038/s41564-018-0338-9oi

De Sordi, L.D., Khanna, V., and Debarbieux. L. (2017). The gut microbiota facilitates drifts in the genetic diversity and infectivity of bacterial viruses. Cell Host Microbe 22, 801-808. 10.1016/j.chom.2017.10.010

Diard, M., Bakkeren, E., Cornault, J.K., Moor, K., Hausmann, A., Sellin, M.E., Loverdo, C., Aertsen, A., Ackermann, M., De Paepe, M., Slack, E., and Hardt, W.D. (2017). Inflammation boosts bacteriophage transfer between *Salmonella* spp. Science 355, 1211-1215. 10.1126/science.aaf8451

Draper, L.A., Ryan, F.J., Smith, M.K., Jalanka, J., Mattila, E., Arkkila, P.A., Ross, R.P., Satokari, R., and Hill, C. (2018). Long-term colonisation with donor bacteriophages following successful faecal microbial transplantation. Microbiome 6, 220. 10.1186/s40168-018-0598-x

Duerkop, B.A., Clements, C.V., Rollins, D., Rodriguez, J.L.M., and Hooper, L.A. (2012). A composite bacteriophage alters colonisation by an intestinal commensal bacterium. Proc. Natl. Acad. Sci. USA 109, 17621-17626. 10.1073/pnas.1206136109

Duerkop, B.A., and Hooper, L.V. (2013). Resident viruses and their interactions with the immune system. Nat. Immunol. 14, 654-659. https://doi.org/10.1038/ni.2614

Duranti, S., Lugli, G.A., Mancabelli, L., Armanini, F., Turroni, F., James, K., Ferretti, P., Gorfer, V., Ferrario, C., Milani, C., Mangifesta, M., Anzalone, R., Zolfo, M., Viappini, A., Pasolli, E., Bariletti, I., Canto, R., Clementi, R., Cologna, M., Crifó, T., Cusumano, G., Fedi, S., Gottardi, S., Innamorati, C., Masè, C., Postai, D., Savoi, D., Soffiati, M., Tateo, S., Pedrotti, A., Segata, N., van Sinderen, D., and Ventura, M. (2017). Maternal inheritance of bifidobacterial communities and bifidophages in infants through vertical transmission. Microbiome 5, 66. https://doi.org/10.1186/s40168-017-0282-6

Dutilh, B.E., Cassman, N., McNair, K., Sanchez, S.E., Silva, G.G.Z., Boling, L., Barr, J.J., Speth, D.R., Seguritan, V., Aziz, R.K., Felts, B., Dinsdale, E.A., Mokili, J.L., and Edwards, R.A. (2014). A highly abundant bacteriophage discovered in the unknown sequences of human faecal metagenomes. Nat. Commun. 5, 4498. 10.1038/ncomms5498

Edwards, R.A., Vega, A.A., Norman, H.M., Ohaeri, M., Levi, K., Dinsadale, E.A., Cinek, O., Aziz, R.K., McNair, K., Barr, J.J., Bibby, K., Brounds, S.J.J., Cazares, A., de Jonge, P.A., Desnues, C., Díaz-Muñoz, S.L., Fineran, P.C., Kurilsjikov, A., Lavigne, R., Mazankova, K., McCarthy, D.T., Nobrega, F.L., Reyes Muñoz, A., Tapia, G., Trefault, N., Tyakht, A.V., Vinuesa, P., Wagemans, J., Zhernakova, A., Aarestrup, F.M., Ahmadov, G., Alassaf, A., Anton, J., Asangba, A., Billings, E.K., Cantu, V.A., Carlton, J.M., Cazares, D., Cho, G.S., Condeff, T., Cortés, P., Cranfield, M., Cuevas, D.A., De la Iglesia, R., Decewicz, P., Doane, M.P., Dominy, N.J., Dziewit, L., Elwasila, B.M., Eren, A.M., Franz, C., Fu, J., Garcia-Aljaro, C., Ghedin, E., Gulino, K.M., Haggerty, J.M., Head, S.R., Hendriksen, R.S., Hill, C., Hyöty, H., Ilina, E.N., Irwin, M.T., Jeffries, T.C., Jofre, J., Junge, R.E., Kelley, S.T., Khan Mirzaei, M., Kowalewski, M., Kumaresan, D., Leigh, S.R., Lipson, S.R., Lisitsyna, E.S., Llagosterra, M., Maritz, J.M., Marr, L.C., McCann, A., Molshanski-Mor, S., Monteiro, S., Moreira-Grez, B., Morris, M., Mugisha, L., Muniesa, M., Neve, H., Nguyen, N.P., Nigro, O.D., Nilsson, A.S., O' Connell, T., Odeh, R., Oliver, A., Piuri, M., Prussin li, A.J.,Qimron, U., Quan, Z.X., Rainetova, P., Ramírez-Rojas, A., Raya, R., Reasor, K., Rice, G.A.O., Rossi, A., Santos, R., Shimashita, J., Stachler, E.N., Stene, L.C., Strain, R., Stumpf, R., Torres, P.J., Twaddle, A., Ugochi Ibekwe, M., Villagra, N., Wandros, S., White, B., Whiteley, A., Whiteson, K.L., Wijmenga, C., Zambrano, M.M., Zschach, H., and Dutilh,

B.E. (2019). Global phylogeopraphy and ancient evolution of the widespread human gut virus crAssphage. Nat. Microbiol. 4, 1727-1736. 10.1038/s41564-019-0494-6

Enault, F., Briet, A., Bouteille, L., Roux, S., Sullivan, M.B., and Petit, M.-A. (2017). Phages rarely encode antibiotic resistance genes: a cautionary tale for virome analyses. ISME J. 11, 237–247. 10.1038/ismej.2016.90

Febvre, H.P., Rao, S., Gindin, M., Goodwin, N.D.M., Finer, E., Vivanco, J.S., Lu, S., Manter, D.K., Wallace, T.C., and Weir, T.L. (2019). PHAGE Study: Effects of supplemental bacteriophage intake on inflammation and gut microbiota in healthy adults. Nutrients 11, 666. 10.3390/nu11030666.

Gao, R., Gao, Z., Huang, L., and Qin, H. (2017). Gut microbiota and colorectal cancer. Eur. J. Clin. Microbiol. Infect. Dis. 36, 757-769. 10.1007/s10096-016-2881-8

Galtier, M., De Sordi, L., Maura, D., Arachchi, H., Volant, S., Dillies, M.A., and Debarbieux, L. (2016). Bacteriophages to reduce gut carriage of antibiotic resistant uropathogens with low impact on microbiota composition. Environ. Microbiol. 18, 2234-2245. 10.1111/1462-2920.13284

Galtier, M., De Sordi, L., Sivignon, A., de Vallée, A., Maura, D., Neut, C., Rahmouni, O., Wannerberger, K., Darfeuille-Michaud, A., Desreumaux, P., Barnich, N., and Debarbieux, L. (2017). Bacteriophages targeting adherent invasive *Escherichia coli* strains as a promising new treatment for Crohn's disease. J. Crohns Colitis 11, 840-847. 10.1093/ecco-jcc/jjw224

Gindin, M., Febvre, H.P., Rao, S., Wallace, T.C., and Weir, T.L. (2019). Bacteriophage for gastrointestinal health (PHAGE) study: Evaluating the safety and tolerability of supplemental bacteriophage consumption. J. Am. Coll. Nutr. 38, 68-75. 10.1080/07315724.2018.1483783

Gogokhia, L., Buhrke, K., Bell, R., Hoffman, B., Brown, D.G., Hanke-Gogokhia, C., Ajami, N.J., Wong, M.C., Ghazaryan, A., Valentine, J.F., Porter, N., Martens, E., O'Connell, R., Jacob, V., Scherl, E., Crawford, C., Stephens, W.Z., Casjens, S.R., Longmam, R.S., and Round, J.L. (2019). Expansion of bacteriophages is linked to aggravated intestinal inflammation and colitis. Cell Host Microbe 25, 285-299. 10.1016/j.chom.2019.01.008.

Górski, A., Dąbrowska, K., Międzybrodzki, R., Weber-Dąbrowska, B., Łusiak-Szelachowska, M., Jończyk-Matysiak, E., and Borysowski, J. (2017). Phages and immunomodulation. Future Microbiol. 12, 905-914. 10.2217/fmb-2017-0049

Górski, A., Jończyk-Matysiak, E., Łusiak-Szelachowska, M., Międzybrodzki, R., Weber-Dąbrowska, B., and Borysowski, J. (2018). Bacteriophages targeting intestinal epithelial cells: a potential novel form of immunotherapy. Cell. Mol. Life Sci. 75, 589-595. 10.1007/s00018-017-2715-6

Guerin, E., Shkoporov, A., Stockdale, S.R., Clooney, A.G., Ryan, F.J., Sutton, T.D.S., Draper, L.A., Gonzalez-Tortuero, E., Ross, R.P., and Hill, C. (2018). Biology and taxonomy of crAss-like bacteriophages, the most abundant virus in the human gut. Cell Host Microbe 24, 653-664. 10.1016/j.chom.2018.10.002

Gupta, V. K., Paul, S. and Dutta, C. (2017). Geography, ethnicity or subsistence-specific variations in human microbiome composition and diversity. Front. Microbiol. 8, 1162 10.3389/fmicb.2017.01162

Han, M.K., Huang, Y.J., Lipuma, J.J., Boushey, H.A.,Boucher, R.C., Cookson, W.O., Curtis, J.L., Erb-Downward, J., Lynch, S.V., Sethi, S., Toewa, G.B., Young, V.B., Wolfgang, M.C., Huffnagle, G.B., and Martinez, F.J. (2012). Significance of the microbiome in obstructive lung disease. Thorax 67, 456–464. 10.1136/thoraxjnl-2011-201183

Hannigan, G.D., Duhaime, M.B., Ruffin, IV M.T., Koumpouras, C.C., and Schloss, P.D. (2018). Diagnostic potential and interactive dynamics of the colorectal cancer virome. mBio 9, e02248-18. 10.1128/mBio.02248-18

Hargreaves, K,R., and Clokie, M.R.J. (2014). *Clostridium difficile* phages: still difficult? Front. Microbiol. 5, 184. 10.3389/fmicb.2014.00184

Hill, C., Mills, S., and Ross, R.P. (2018). Phages and antibiotic resistance: are the most abundant entities on earth ready for a comeback? Future Microbiol. 13, 711-726. 10.2217/fmb-2017-0261

Howard-Varona, C., Hargreaves, K.R., Solonenko, N.E., Markillie, L.M., White, R.A., Brewer, H.M., Ansong, C., Orr, G., Adkins, J.N., and Sullivan, M.B. (2018). Multiple mechanisms drive phage infection efficiency in nearly identical hosts. ISME J. 12, 1605-1618. 10.1038/s41396-018-0099-8

Hoyles, L., McCartney, A.L., Neve, H., Gibson, G.R., Sanderson, J.D., Heller, K.J., and van Sinderen, D. (2014). Characterization of virus-like particles associated with the human faecal and caecal microbiota. Res. Microbiol. 165, 803–812. 10.1016/j.resmic.2014.10.006

Hsu, B.B., Gibson, T.E., Yeliseyev, V., Liu, Q, Lyon, L., Bry, L., Silver, P.A., and Gerber GK. (2019). Dynamic modulation of the gut microbiota and metabolome by bacteriophages in a mouse model. Cell Host Microbe 25, 803-814. https://doi.org/10.1016/j.chom.2019.05.001

Hu, Y.O.O., Hugerth, L.W., Bengtsson, C., Alisjahbana, A., Seifert, M., Kamal, A., Sjöling, Å., Midtvedt, T., Norin, E., Du, J., and Engstrand L. (2018). Bacteriophages synergize with the gut microbial community to combat *Salmonella*. mSystems 3, e00119-18. 10.1128/mSystems.00119-18

Hui, W., Li, T., Liu, W., Zhou, C., and Gao, F. (2019). Fecal microbiota transplantation for treatment of recurrent *C. difficile* infection: An updated randomized controlled trial meta-analysis. PLoS ONE 14, e0210016. 10.1371/journal.pone.0210016

Jamet, A., Touchon, M., Ribeiro-Gonçalves, B., Carriço, J.A., Charbit, A., Nassif, X., Ramirez, M., and Rocha, E.P. (2017). A widespread family of polymorphic toxins encoded by temperate phages. BMC Biol. 15, 75. 10.1186/s12915-017-0415-1.

Jiang, S.C., and Paul, J. H. (1998). Significance of lysogeny in the marine environment: studies with isolates and a model of lysogenic phage production. Microb. Ecol. 35, 235–243

Johnson, T.A., Looft, T., Severin, A.J., Bayles, D.O., Nasko, D.J., Wommack, K.E., Howe, A., and Allen, H.K. (2017). The in-feed antibiotic carbadox induces phage gene transcription in the swine gut microbiome. mBio 8, e00709-17. 10.1128/mBio.00709-17

Kai, M., Watanabe, S., Furuse, K., and Ozawa, A. (1985). Bacteroides bacteriophages isolated from human feces. Microbiol. Immunol. 29, 895–899.

Kang, D.-W., Adams, J.B., Gregory, A.C., Borody, T., Chittick, L., Fasano, A., Khoruts, A., Geis, E., Maldonado, J., McDonough-Means, S., Pollard, E.L., Roux, S., Sadowsky, M.J., Schwarzberg Lipson, K., Sullivan, M.B., Caporaso, J.G., and Krajmalnik-Brown, R. (2017). Microbiota transfer therapy alters gut ecosystem and improves gastrointestinal and autism symptoms: an open label study. Microbiome 5, 10. https://doi.org/10.1186/s40168-016-0225-7

Kellingray, L., Gall, G. L., Defernez,M., Beales, I. L. P., Franslem-Elumogo, N., and Narbad, A. (2018). Microbial taxonomic and metabolic alterations during faecal microbiota transplantation to treat *Clostridium difficile* infection. J. Infect. 77, 107–118. 10.1016/j.jinf.2018.04.012

Kelly, C. R., Kahn, S., Kashyap, P., Laine, L., Rubin, D., Atreja, A., Moore, T., and Wu, G. (2015). Update on fecal microbiota transplantation : indications, methodologies, mechanisms, and outlook. Gastroenterology 149, 223–237. 10.1053/j.gastro.2015.05.008

Khanna, S., Pardi, D.S., Kelly, C.R., Kraft, C.S., Dhere, T., Henn, M.R., Lombardo, M.J., Vulic, M., Ohsumi, T., Winkler, J., Pindar, C., McGovern, B.H., Pomerantz, R.J., Aunins, J.G., Cook, D.N., and Hohmann, E.L. (2016). A novel microbiome therapeutic increases gut microbial diversity and prevents recurrent *Clostridium difficile* infection. J. Infect. Dis. 214, 173-181. 10.1093/infdis/jiv766.

Khanna, S., Vazquez-Baeza, Y., González, A., Weiss, S., Schmidt, B., Muñiz-Pedrogo, D. A., Rainey III, J.F., Kammer, P., Nelson, H., Sadowsky, M., Khoruts, A., Farrugia, S.L., Knight, R., Pardi, D., and Kashyap, P.C. (2017). Changes in microbial ecology after fecal microbiota transplantation for recurrent C. difficile infection affected by underlying inflammatory bowel disease. Microbiome 5, 55. https://doi.org/10.1186/s40168-017-0269-3

Kieser, S., Sarker, S.A., Berger, B., Sultana, S., Chisti, M.J., Islam, S.B., Foata, F., Porta, N., Betrisey, B., Fournier, C., Descombes, P., Mercenier, A., Sakwinska, O., and Brussöw H. (2015). Antibiotic treatment leads to fecal *Escherichia coli* and coliphage expansion in severely malnourished diarrhea patients. Cell. Mol. Gastroenterol. Hepatol. 5, 458-460. 10.1016/j.jcmgh.2017.11.014

Kim, M.-S., Park, E.-J., Roh, S.W., and Bae, J.W. (2011). Diversity and abundance of single-stranded DNA viruses in human feces. Appl. and Environ. Microbiol. 77, 8062-8070. 10.1128/AEM.06331-11

Knowles, B., Silveira, C.B., Bailey, B.A., Barott, K., Cantu, V.A., Cobián-Güemes, A.G., Coutinho, F.H., Dinsdale, E.A., Felts, B., Furby, K.A., George, E.E., Green, K.T., Gregoracci, G.B., Haas, A.F., Haggerty, J.M., Hester, E.R., Hisakawa, N., Kelly, L.W., Lim, Y.W., Little, M., Luque, A., McDole-Somera, T., McNair, K., de Oliveira, L.S., Quistad, S.D., Robinett, N.L., Sala, E., Salamon, P., Sanchez, S.E., Sandin, S., Silva, G.G.Z., Smith, J., Sullivan, C., Thompson, C., Vermeij, M.J.A., Youle, M., Young, C., Zgliczynski, B., Brainard, R.,

Edwards, A., Nulton, J., Thompson, F., and Rohwer, F. (2016). Lytic to temperate switching of viral communities. Nature 531, 466–470. https://doi.org/10.1038/nature17193

Kostic, A.D., Gevers, D., Siljander, H., Vatanen, T., Hyötyläinen, T., Hämäläinen, A.M., Peet, A., Tillmann, V., Pöhö, P., Mattila, I., Lähdesmäki, H., Franzosa, E.A., Vaarala, O., de Goffau, M., Harmsen, H., Ilonen, J., Virtanen, S.M., Clish, C.B., Orešič, M., Huttenhower, C., Knip, M., DIABIMMUNE Study Group, and Xavier, R.J. (2015). The dynamics of the human infant gut microbiome in development and in progression toward type 1 diabetes. Cell Host Microbe 17, 260–273. 10.1016/j.chom.2015.01.001

Lange, K., Buerger, M., Stallmach, A., and Bruns, T. (2016). Effects of antibiotics on gut microbiota. Dig. Dis. 34, 260-268. https://doi.org/10.1159/000443360

Lee, C. H., Belanger, J. E., Kassam, Z., Smieja, M., Higgins, D., Broukhanski, G., and Kim, T. (2014). The outcome and long-term follow-up of 94 patients with recurrent and refractory *Clostridium difficile* infection using single to multiple fecal microbiota transplantation via retention enema. Eur. J. Clin. Microbiol. Infect. Dis. 33, 1425–1428. 10.1007/s10096-014-2088-9.

LePage, P., Colombet, J., Marteau, P., Sime-Ngando, T., Doré, J., and Leclerc, M. (2008). Dysbiosis in inflammatory bowel disease: a role for bacteriophages? Gut 57, 424–425. 10.1136/gut.2007.134668

Lim, E.S., Zhou, Y., Zhao, G., Bauer, I.K., Droit, L., Ndao, I.M., Warner, B.B., Tarr, P.I., Wang, D., and Holtz, L.R. (2015). Early life dynamics of the human gut virome and bacterial microbiome in infants. Nat. Med. 21, 1228-1234. 10.1038/nm.3950

Lourenço, M., De Sordi, L., and Debarbieux, L. (2018). The diversity of bacterial lifestyles hampers bacteriophage tenacity. Viruses 10, 327. 10.3390/v10060327.

Lugli, G.A., Milani, C., Turroni, F., Tremblay, D., Ferrario, C., Mancabelli, L., Duranti, S., Ward, D.V., Ossiprandi, M.C., Moineau, S., van Sinderen, D., and Ventura, M. (2016). Prophages of the genus *Bifidobacterium* as modulating agents of the infant gut microbiota. Environ. Microbiol. 18, 2196-2213. 10.1111/1462-2920.13154.

Mai, V., Uhkanova, M., Reinhard, M.K., Li, M., and Sulakvelidze, A. (2015). Bacteriophage administration significantly reduces *Shigella* colonization and shedding by *Shigella*-challenged mice without deleterious side effects and distortions in the gut microbiota. Bacteriophage 5, e1088124. 10.1080/21597081.2015.1088124

Macrae, F.A. (2019). Colorectal cancer: Epidemiology, risk factors, and protective factors. Goldberg, R.M, and Seres, D., eds. UpToDate. (Waltham, MA: UpToDate Inc) https://www.uptodate.com (Accessed on September 10, 2019).

Manrique, P., Bolduc, B., Walk, S.T., van der Oost, J., de Vos, W.M., and Young, M.J. (2016). Human healthy gut phageome. Proc. Natl. Acad. Sci. USA 113, 10400-10405. https://doi.org/10.1073/pnas.1601060113

Manrique, P., Dills, M., and Young, M.J. (2017). The human gut phage community and its implications for health and disease. Viruses 9, 141. 10.3390/v9060141.

Mattila, E., Uusitalo-Seppälä, R.,Wuorela,M., Lehtola, L.,Nurmi, H., Ristikankare, M., Moilanen, V., Salminen, K., Seppälä, M., Mattila, P.S., Anttila, V.J., and Arkkila, P. (2012). Fecal transplantation, through colonoscopy, is effective therapy for recurrent *Clostridium difficile* infection. Gastroenterology 142, 490–496. 10.1053/j.gastro.2011.11.037

Maura, D., and Debarbieux, L. (2012). On the interactions between virulent bacteriophages and bacteria in the gut. Bacteriophage 2, 229–233. 10.4161/bact.23557

Maurice, C.F. (2019). Considering the other half of the gut microbiome: bacteriophages. mSystems 4, e00102-19. 10.1128/mSystems.00102-19

Mavrich, T.N., Casey, E., Oliveira, J., Bottacini, F., James, K., Franz, C.M.A.P., Lugli, G.A., Neve, H., Ventura, M., Hatfull, G.F., Mahony, J., and van Sinderen D. (2018). Characterization and induction of prophages in human gut-associated *Bifidobacterium* hosts. Sci. Rep. 8, 12772. 10.1038/s41598-018-31181-3.

McCann, A., Ryan, F.J., Stockdale, S.R., Dalmasso, M., Blake, T., Ryan, C.A., Stanton, C., Mills, S., Ross, P.R., and Hill, C. (2018). Viromes of one year old infants reveal the impact of birth mode on microbiome diversity. PeerJ 6, e4694. 10.7717/peerj.4694

Meader, E., Mayer, M.J., Gasson, M.J., Steverding, D., Carding, S.R., and Narbad, A. (2010). Bacteriophage treatment significantly reduces viable

*Clostridium difficile* and prevents toxin production in an *in vitro* model system. Anaerobe 16, 549-554. 10.1016/j.anaerobe.2010.08.006

Meader, E., Mayer, M.J., Steverding, D., Carding, S.R., and Narbad, A. (2013). Evaluation of bacteriophage therapy to control *Clostridium difficile* and toxin production in an *in vitro* human colon model system. Anaerobe 22, 25-30. 10.1016/j.anaerobe.2013.05.001.

Medhekar, B., and Miller, J.F. (2007). Diversity-generating retroelements. Curr. Opin. Microbiol. 10, 388-395. 10.1016/j.mib.2007.06.004

Milani, C., Casey, E., Lugli, G.A., Moore, R., Kaczorowska, J., Feehily, C., Magnifesta, M., Mancabelli, L., Duranti, S., Turroni, F., Bottacini, F., Mahony, J., Cotter, P.D., McAuliffe, F.M., van Sinderen, D., and Ventura, M. (2018). Tracing mother-infant transmission of bacteriophages by means of a novel analytical tool for shotgun metagenomic datasets: METAnnotatorX. Microbiome 6, 145. https://doi.org/10.1186/s40168-018-0527-z

Mills, S., Ross, R.P., and Hill, C. (2017). Bacteriocins and bacteriophage; a narrow-minded approach to food and gut microbiology. FEMS Microbiol. Rev. 41, S129-S153. 10.1093/femsre/fux022.

Mills, S., Stanton, C., Lane, J.A., Smith, G.J., and Ross, R.P. (2019). Precision nutrition and the microbiome, Part I: Current state of the science. Nutrients 11, 923. 10.3390/nu11040923

Mills, S., Shanahan, F., Stanton, C., Hill, C., Coffey, A., and Ross, R.P. (2013). Movers and Shakers: Influence of bacteriophages in shaping the mammalian gut microbiota. Gut Microbes 4, 1-16. 10.4161/gmic.22371

Minot, S., Bryson, A., Chehoud, C., Wu, G.D., Lewis, J.D., and Bushman, F.D. (2013). Rapid evolution of the human gut virome. Proc. Natl. Acad. Sci. USA 110, 12450-12455. 10.1073/pnas.1300833110

Minot, S., Sinha, R., Chen, J., Li, H., Keilbaugh, S.A., Wu, G.D., Lewis, J.D., and Bushman, F.D. (2011). The human gut virome: Inter-individual variation and dynamic response to diet. Genome Res. 21, 1616-1625. 10.1101/gr.122705.111

Modi, S.R., Lee, H.H., Spina, C.S., and Collins, J.J. (2013). Antibiotic treatment expands the resistance reservoir and ecological network of the phage metagenome. Nature 499, 219–222. 10.1038/nature12212

Mohanraju, P., Makarova, K.S., Zetsche, B., Zhang, F., Koonin, E.V., and van der Oost, J. (2016). Diverse evolutionary roots and mechanistic variations of the CRISPR-Cas systems. Science. 353, aad5147, 10.1126/science.aad5147.

Moreno-Gallego, J.L., Chou, S.-P., Di Rienzi, S.C., Goodrich, J.K., Spector, T.D., Bell, J.T., Youngblut, N.D., Hewson, I., Reyes, A., and Ley, R.E. (2019). Virome diversity correlates with intestinal microbiome diversity in adult monozygotic twins. Cell Host Microbe 25, 261-272. 10.1016/j.chom.2019.01.019

Nakatsu, G., Zhou, H., Wu, W.K.K., Wong, S.H., Coker, O.O., Dai, Z., Li, X., Szeto, C.-H., Sugimura, N., Lam, T.Y.-T., Yu, A.C.-S., Wang, X., Chen, Z., Wong, M.C.-S., Ng, S.C., Chan, M.T.V., Chan, P.K.S., Chan, F.K.L., Sung, J.J.-Y., and Yu, J. (2018). Alterations in enteric virome are associated with colorectal cancer outcome. Gastroenterology 155, 529-541. 10.1053/j.gastro.2018.04.018

Nale, J.Y., Chutia, M., Carr, P., Hickenbotham, P.T., and Clokie, M.R.J. (2016a). 'Get in Early'; Biofilm and wax moth (*Galleria mellonella*) models reveal new insights into the therapeutic potential of *Clostridium difficile* bacteriophages. Front. Microbiol. 7, 1383. 10.3389/fmicb.2016.01383

Nale, J.Y., Redgwell, T.A., Millard, A., and Clokie, M.R.J. (2018). Efficacy of an optimized bacteriophage cocktail to clear *Clostridium difficile* in a batch fermentation model. Antibiotics 7, 13. 10.3390/antibiotics7010013.

Nale, J.Y., Spender, J., Hargreaves, K.R., Buckley, A.M., Trzepiński, P., and Douce, G.R. (2016b). Bacteriophage combinations significantly reduce *Clostridium difficile* growth *in vitro* and proliferation *in vivo*. Antimicrob. Agents Chemother. 60, 968-981. 10.1128/AAC.01774-15

Needell, J.C., and Zipris, D. (2016). The role of the intestinal microbiome in type 1 diabetes pathogenesis. Curr. Diab. Rep. 16, 89. 10.1007/s11892-016-0781-z.

Nguyen, S., Baker, K., Padman, B.S., Patwa, R., Dunstan, R.A., Weston, T.A., Schlosser, K., Bailey, B., Lithgow, T., Lazarou, M., Luque, A., Rohwer, F., Blumberg, R.S., and Barr, J.J. (2017). Bacteriophage transcytosis provides a mechanism to cross epithelial cell layers. mBio 8, e01874-17. 10.1128/mBio.01874-17

Nishida, A., Inoue, R., Inatomi, O., Bamba, S., Naito, Y., and Andoh, A. (2018). Gut microbiota in the pathogenesis of inflammatory bowel disease. Clin. J. Gastroenterol. 11, 1-10. 10.1007/s12328-017-0813-5

Niu, Y.D., McAllister, T.A., Nash, J.H.E., Kropinski, A.M., and Stanford, K. (2014). Four *Escherichia coli* O157:H7 phages: a new bacteriophage genus and taxonomic classification of T1-like phages. PLoS ONE 9, e100426. 10.1371/journal.pone.0100426.

Norman, J.M., Handley, S.A., Baldridge, M.T., Droit, L., Liu, C.Y., Keller, B.C., Kambal, A., Monaco, C.L., Zhao, G., Fleshner, P., Stappenbeck, T.S., McGovern, D.P.B., Keshavarzian, A., Mutlu, E.A., Sauk, J., Gevers, D., Xavier, R.J., Wang, D., Parkes, M., and Virgin, H.W. (2015). Disease-specific alterations in enteric virome in inflammatory bowel disease. Cell 160, 447-460. 10.1016/j.cell.2015.01.002

Oh, J.-H., Alexander, L.M., Pan, M., Schueler, K.L., Keller, M.P., Attie, A.D., Walter, J., and van Pijkeren, J.-P. (2019). Dietary fructose and microbiota-derived short-chain fatty acids promote bacteriophage production in the gut symbiont *Lactobacillus reuteri.* Cell Host Microbe 25, 273-284. https://doi.org/10.1016/j.chom.2018.11.016

Ott, S.J., Waetzig, G.H., Rehman, A., Moltzau-Anderson, J., Bharti, R., Grasis, J.A., Cassidy, L., Tholey, A., Fickenscher, H., Seegert, D., Rosenstiel, P., and Scheriber, S. (2017). Efficacy of sterile fecal filtrate transfer for treating patients with *Clostridium difficile* infection. Gastroenterology 152, 799-811. 10.1053/j.gastro.2016.11.010

Park, H., Laffin, M.R., Jovel, J., Millan, B., Hyun, J.E., Hotte, N., Kao, D., and Madsen, K.L. (2019). The success of fecal microbial transplantation in *Clostridium difficile* infection correlates with bacteriophage relative abundance in the donor: a retrospective cohort study. Gut Microbes doi: 10.1080/19490976.2019.1586037

Payet, J., and Suttle, C.A. (2013).To kill or not to kill: the balance between lytic and lysogenic viral infection is driven by trophic status. Limnol. Oceanogr. 58, 465–474 https://doi.org/10.4319/lo.2013.58.2.0465

Pérez-Brocal, V., Garcia-Lopéz, R., Vázquez-Castellanos, J.F., Nos, P., Beltrán, B., Latorre, A., and Moya, A. (2013). Study of the viral and microbial communities associated with Crohn's disease: a metagenomic approach. Clin. Transl. Gastroenterol. 4, e36. 10.1038/ctg.2013.9

Ratjen, F., Döring, G., and Nikolaizik, W.H. (2001). Effect of inhaled tobramycin on early *Pseudomonas aeruginosa* colonisation in patients with cystic fibrosis. Lancet 358, 983–984 10.1016/S0140-6736(01)06124-4

Reyes, A., Blanton, L.V., Cao, S., Zhao, G., Manary, M., Trehan, I., Smith, M.I., Wang, D., Virgin, H.W., Rohwer, F., and Gordon, J.I. (2015). Gut DNA viromes of Malawian twins discordant for severe acute malnutrition. Proc. Natl. Acad. Sci. USA. 112, 11941-11946. https://doi.org/10.1073/pnas.1514285112

Reyes, A., Haynes, M., Hanson, N., Angly, F.E., Heath, A.C., Rohwer, F., and Gordon, J.I. (2010). Viruses in the faecal microbiota of monozygotic twins and their mothers. Nature 466, 334–338. https://doi.org/10.1038/nature09199

Rodriguez-Valera, F., Martin-Cuadrado, A.B., Rodriguez-Brito, B., Pasić, L., Thingstad, T.F., Rohwer, F., and Mira, A. (2009). Explaining microbial population genomics through phage predation. Nat. Rev. Microbiol. 7, 828–836. 10.1038/nrmicro2235.

Rohwer, F., Prangishvilli, D., and Lindell, D. (2009). Role of viruses in the environment. Environ. Microbiol. 11, 2771-2774. 10.1111/j. 1462-2920.2009.02101.x.

Sarker, S.A., and Brüssow, H. (2016). From bench to bed and back again: phage therapy of childhood *Escherichia coli* diarrhoea. Ann. N.Y. Acad. Sci. 1372, 42–52. 10.1111/nyas.13087

Sarker, S.A., Sultana, S., Reuteler, G., Moine, D., Descombes, P., Charton, F., Bourdin, G., McCallin, S., Ngom-Bru, C., Neville, T., Akter, M., Huq, S., Qadri, F., Talukdar, K., Kassam, M., Delley, M., Loiseau, C., Deng, Y., El Aidy, S., Berger, B., and Brüssow, H. (2016). Oral phage therapy of acute bacterial diarrhea with two coliphage preparations: a randomized trial in children from Bangladesh. *EBioMedicine* 4,124–37. 10.1016/j.ebiom.2015.12.023

Sender, R., Fuchs, S., and Milo, R. (2016). Are we really vastly outnumbered? Revisiting the ratio of bacterial to host cells in humans. Cell 164, 337–340. 10.1016/j.cell.2016.01.013.

Seo, B.-J., Song, E.-T., Lee, K., Kim, J.-W., Jeong, C.-G., Moon, S.-H., Son, J.S., Kang, S.H., Cho, H.-S., Jung, B.Y., and Kim, W.-I. (2018). Evaluation of the broad-spectrum lytic capability of bacteriophage cocktails against various

*Salmonella* serovars and their effects on weaned pigs infected with *Salmonella* Typhimurium. J.Vet. Med. Sci. 80, 851-860. 10.1292/jvms.17-0501

Serwer, P., Hayes, S.J., Thomas, J.A., and Hardies, S.C. (2007). Propagating the missing bacteriophages: a large bacteriophage in a new class. Virol. J. 4, 21. https://doi.org/10.1186/1743-422X-4-21

Sharon, I., Morowitz, M.J., Thomas, B.C., Costello, E.K., Relman, D.A., and Banfield, J.F. (2013). Time series community genomics analysis reveals rapid shifts in bacterial species, strains, and phage during infant gut colonization. Genome Res. 23, 111-120. 10.1101/gr.142315.112

Shkoporov, A.N., Clooney, A.G., Sutton, T.DS., Ryan, F.J., Daly, K.M., Nolan, J.A., McDonnell, S.A., Khokhlova, E.V., Draper, L.A., Forde, A., Guerin, E., Velayudhan, V., Ross, R.P, and Hill C. (2019). The human gut virome is highly diverse, stable, and individual specific. Cell Host Microbe 26, 527-541. https://doi.org/10.1016/j.chom.2019.09.009

Shkoporov, A.N., and Hill, C. (2019). Bacteriophages of the human gut: The "known unknown" of the microbiome. Cell Host Microbe 25, 195-209. 10.1016/j.chom.2019.01.017.

Shkoporov, A.N., Khokhlova, E.V., Fitzgerald, C.B., Stockdale, S.R., Draper, L.A., Ross, R.P., and Hill, C. (2018). ΦCrAss001 represents the most abundant bacteriophage family in the human gut and infects *Bacteroides intestinalis*. Nat. Commun. 9, 4781. https://doi.org/10.1038/s41467-018-07225-7

Silveira, C.B., and Rohwer, F.L. (2016). Piggyback-the-Winner in host-associated microbial communities. NPJ Biofilms Microbiomes 2, 16010. https://doi.org/10.1038/npjbiofilms.2016.10

Song, Y., Garg, S., Girotra, M., Maddox, C., von Rosenvinge, E. C., Dutta, A., Dutta, S., and Fricke, W.F. (2013). Microbiota dynamics in patients treated with fecal microbiota transplantation for recurrent *Clostridium difficile* infection. PLoS ONE. 8, e81330. https://doi.org/10.1371/journal.pone.0081330

Stockdale, S., Ryan, F., McCann, A., Dalmasso, M., Ross, P., and Hill, C. (2018). Viral dark matter in the gut virome of elderly humans. Preprints. https://doi.org/10.20944/preprints201807.0101.v1.

Surawicz, C.M., Brandt, L.J., Binion, D.G., Ananthakrishnan, A.N., Curry, S.R., Gilligan,P.H., McFarland, L.V., Mellow, M., Zuckerbraun, B.S. (2013).

Guidelines for diagnosis, treatment, and prevention of *Clostridium difficile* infections. Am. J. Gastroenterol. 108, 478–498. 10.1038/ajg.2013.4.

Sweere, J.M., Van Belleghem, J.D., Ishak, H., Bach, M.S., Popescu, M., Sunkari, V., Kaber, G., Manasherob, R., Suh, G.A., Cao, X., de Vries, C.R., Lam, D.N., Marshall, P.L., Birukova, M., Katznelson, E., Lazzareschi, D.V., Balaji, S., Keswani, S.G., Hawn, T.R., Secor, P.R., and Bollyky, P.L. (2019). Bacteriophage trigger antiviral immunity and prevent clearance of bacterial infection. Science 363, eaat9691. 10.1126/science.aat9691

Tetz, G., and Tetz, V. (2016). Bacteriophage infections of microbiota can lead to leaky gut in an experimental rodent model. Gut Pathogens 8, 33. https://doi.org/10.1186/s13099-016-0109-1

Theriot, C.M., and Young, V.B. (2014). Microbial and metabolic interactions between the gastrointestinal tract and *Clostridium difficile* infection. Gut Microbes 5, 86–95. 10.4161/gmic.27131

Thingstad, T. F., Våge, S., Storesund, J. E., Sandaa, R.-A., and Giske, J. (2014). A theoretical analysis of how strain-specific viruses can control microbial species diversity. Proc. Natl Acad. Sci. USA 111, 7813–7818. https://doi.org/10.1073/pnas.1400909111

Tian, H., Ge, X., Nie, Y., Yang, L., Ding, C., McFarland, L.V., Zhang, X., Chen, Q., Gong, J., and Li, N. (2017). Fecal microbiota transplantation in patients with slow-transit constipation: a randomized, clinical trial. PLoS ONE 12, e0171308. 10.1371/journal.pone.0171308

Våge, S., Storesund, J. E., and Thingstad, T.F. (2013). Adding a cost of resistance description extends the ability of virus-host model to explain observed patterns in structure and function of pelagic microbial communities. Environ. Microbiol. 15, 1842–1852. 10.1111/1462-2920.12077

Van Belleghem, J.D., Clement, F., Merabishvili, M., Lavigne, R., and Vaneechoutte, M. (2017). Pro- and anti-inflammatory responses of peripheral blood mononuclear cells induced by *Staphylococcus aureus* and *Pseudomonas aeruginosa* phages. Sci. Rep. 7, 8004. 10.1038/s41598-017-08336-9

Van Belleghem, J.D., Dąbrowska, K., Vaneechoutte. M., Barr, J.J., and Bollyky, P.L. (2019). Interactions between bacteriophage, bacteria, and the mammalian immune system. Viruses 11, E10. 10.3390/v11010010.

Vancamelbeke, M., and Vermeire, S. (2017). The intestinal barrier: a fundamental role in health and disease. Expert Rev. Gastroenterol. Hepatol. 11, 821-834. 10.1080/17474124.2017.1343143

Wagner, J., Maksimovic, J., Farries, G., Sim, W.H., Bishop, R.F., Cameron, D.J., Catto-Smith, A.G., and Kirkwood, C.D. (2013). Bacteriophages in gut samples from pediatric Crohn's disease patients: metagenomic analysis using 454 pyrosequencing. Inflamm. Bowel Dis. 19, 1598-1608. 10.1097/MIB. 0b013e318292477c.

Wigington, C.H., Sonderegger, D., Brussaard, C.P., Buchan, A., Finke, J.F., Fuhrman, J.A., Lennon, J.T., Middelboe, M., Suttle, C.A., Stock, C., Wilson, W.H., Wommack, K.E., Wilhelm, S.W., and Weitz, J.S. (2016). Re-examination of the relationship between marine virus and microbial cell abundances. Nat. Microbiol. 1, 15024. 10.1038/nmicrobiol.2015.24.

Wilson, B.C., Vatanen, T., Cutfield, W.S., and O' Sullivan, J.M. (2019). The super-donor phenomenon in fecal microbiota transplantation. Front. Cell. Infect. Microbiol. 9, 2. 10.3389/fcimb.2019.00002

Winstanley, C., Langille, M.G., Fothergill, J.L., Kukavica-Ibrulj, I., Paradis-Bleau, C., Sanschargin, F., Thomson, N.R., Winsor, G.L., Quail, M.A., Lennard, N., Bignell, A., Clarke, L., Seeger, K., Saunders, D., Harris, D., Parkhill, J., Hancock, R.E., Brinkman, F.S., and Levesque, R.C. (2009). Newly introduced genomic prophage islands are critical determinants of in vivo competitiveness in the Liverpool Epidemic Strain of *Pseudomonas aeruginosa*. Genome Res. 1, 12–23. 10.1101/gr.086082.108

Wommack, K.E., and Colwell, R.R. (2000). Virioplankton: viruses in aquatic ecosystems. Microbiol. Mol. Biol. Rev. 64, 69–114. 10.1128/mmbr. 64.1.69-114.2000

Youngster, I., Russell, G. H., Pindar, C., Ziv-Baran, T., Sauk, J., and Hohmann, E. L. (2014). Oral, capsulized, frozen fecal microbiota transplantation for relapsing *Clostridium difficile* infection. JAMA 312:1772. 10.1001/jama. 2014.13875.

Yutin, N., Makarova, K.S., Gussow, A.B., Krupovic, M., Segall, A., Edwards, R.A., and Koonin, E.G. (2018). Discovery of an expansive bacteriophage family that includes the most abundant viruses from the human gut. Nat. Microbiol. 3, 38-46. 10.1038/s41564-017-0053-y

Zhao, G., Vatanen, T., Droit, L., Park, A., Kostic, A.D., Poon, T.W., Vlamakis, H., Siljander, H., Härkönen, T., Hämäläinen, A.-M., Peet, A., Tillmann, V., Ilonen, J., Wang, D., Knip, M., Xavier, R.J., and Virgin, H.W. (2017). Intestinal virome changes precede autoimmunity in type 1 diabetes-susceptible children. Proc. Natl. Acad. Sci. USA. 114, E6166-E6175. 10.1073/pnas.1706359114

Zuo, T., Lu, X.-J., Zhang, Y., Cheung, C.P., Lam, S., Zhang, F., Tang, W., Ching, J.Y.L., Zhao, R., Chan, P.K.S., Sung, J.J.Y., Yu, J., Chan, F.K.L., Cao, Q., Sheng, J.-.Q, and Ng, S.C. (2019). Gut mucosal virome alterations in ulcerative colitis. Gut (ahead of print) 10.1136/gutjnl-2018-318131.

Zuo, T., Wong, S.H., Lam, K., Lui, R., Cheung, K., Tang, W., Ching, J.Y.L., Chan, P.K.S., Chan, M.C.W., Wu, J.C.Y., Chan, F.K.L., Yu, J., Sung, J.J.Y., and Ng, S.C. (2018). Bacteriophage transfer during fecal microbiota transplantation in *Clostridium difficile* infection is associated with treatment outcome. Gut 67, 634-643. 10.1136/gutjnl-2017-313952

## Chapter 4

# Polish Contribution to the Advancement of Phage Treatment in Humans

**Maciej Żaczek[1*], Beata Weber-Dąbrowska[1,2], Marzanna Łusiak-Szelachowska[1], Ryszard Międzybrodzki[1,2,3] and Andrzej Górski[1,2]**

[1]Bacteriophage Laboratory, Hirszfeld Institute of Immunology and Experimental Therapy, Polish Academy of Sciences (HIIET PAS), R. Weigla 12, 53-114 Wrocław, Poland
[2]Phage Therapy Unit, Hirszfeld Institute of Immunology and Experimental Therapy, Polish Academy of Sciences (HIIET PAS), R. Weigla 12, 53-114 Wrocław, Poland;
[3]Infant Jesus Hospital, The Medical University of Warsaw, ul. W. H. Lindleya 4, 02-005 Warsaw, Poland

*Corresponding author: maciej.zaczek@hirszfeld.pl

DOI: https://doi.org/10.21775/9781913652517.04

### Abstract

Polish research has helped to understand the significance of phage treatment in the new post-antibiotic reality. Its continuous growth over the decades distinguishes Poland from other countries, particularly from those located in the west of Europe and North America. Polish pioneering reports, although somewhat forgotten, remain one of the most compelling historical sources for researchers trying to piece together an incomplete picture of phage-related information. An ongoing phage renaissance constitutes an excellent opportunity to emphasize the significance of Polish efforts towards the development of phage therapy. Notably, the recent pandemic of COVID-19 demonstrates the importance of compassionate treatment in life-threatening circumstances. Such expanded access to phages has had a long tradition in Poland and shows robust insights into the clinical aspects of phage treatment in the last few decades.

## Introduction

Phage therapy is as old as the discovery of bacteriophages (phages). Thus, several nations can boast their history of treating people with bacterial viruses. Development of phage treatment has its roots dating back before the Second World War (WWII). In those early days, phages turned out to be a useful tool in fighting bacterial infections, particularly during early 20th-century warfare. However, the postwar period marked a rapid decline in the therapeutic application of phages. Difficulties in assessing the effectiveness of phage treatment *in vivo*, its narrow-spectrum of activity and the arising significance of newly discovered antibiotics eventually buried the chances for further progress on phage therapy. Indeed, there are not many countries where phage research survived these challenging times, which were full of controversy, scientific distrust and political disturbances which pushed phage therapy behind the Iron Curtain for over 50 years. In Poland, partially due to its above-mentioned geopolitical situation, the therapeutic use of phages has become an integral part of the Polish medical school which has lasted to the present day. The history of phage therapy, complicated and full of unexpected events, was even described as "strange" in scientific literature (Summers et al., 2012). But undoubtedly, the history would not be complete without the Polish contribution.

## Phage treatment before the Ludwik Hirszfeld era

One of the first Polish physicians interested in the therapeutic use of phages was Dr Jerzy Jasieński from the Jagiellonian University in Kraków. He was a pioneer of the topical application of phages and the use of phage preparations in patients with a single furuncle, which led him to the conclusion that phages are capable of destroying bacteria in purulent wounds. Jasieński focused on staphylococcal phage isolation from the wounds of patients. It is worth mentioning that he raised an important point about the possible induction of anti-phage antibodies, primarily with a prolonged treatment was conducted. He also noted the narrow spectrum of phage infection when he tried to find phages active against *Streptococcus* among his staphylococcal phage collection (Jasieński, 1927). He had moderate optimism with regards to the efficacy of phage treatment which led him to the conclusion that its topical application was a promising treatment, and should take a precedence over other therapeutic methods. He recommended the use of phage therapy for purulent inflammations, in particular staphylococcal inflammations of large joints.

Among other Polish scientists from the interwar period who were focused on phages, two female researchers, Bronisława Fejgin and Irena Lipska deserve special recognition. The scientific work of Bronisława Fejgin was tragically interrupted by her death in the Warsaw Ghetto in January of 1943. In 1927, in an article published in Polish she described "d'Hérelle's phenomenon" as the main driving force responsible for bacterial lysis (Fejgin, 1927). In her work, she describes phages as a "lytic factor" which clearly indicates how far back in phage history her experimental activity occurred. Nevertheless, her insight into the matter did not receive international recognition, although the majority of her work was published in French and is thus not widely accessible these days. Judging by the quality of these articles, we can assume she was a pioneer of her era (Fejgin, 1923; 1927; 1936). The second female researcher, Irena Lipska, survived WWII, becoming the only known Polish phage researcher active in both pre- and postwar periods. Her first work was published as early as the 1930s with more detailed articles published 20 years later, where Poland found itself in the Soviet sphere of domination when research on phages was almost completely abandoned elsewhere. The Polish journal "*Pediatria Polska* [Polish Journal of Paediatrics]" constitutes a valuable source of her work. In 1951, she published a summary of her previous phage experience, which also covered the war period (Lipska, 1951). This exhaustive narrative has given us an opportunity to get familiar with some aspect of phage treatment in postwar Poland. Of particular interest are her attempts to use phages prophylactically in newborns. According to her writings, about one hundred internal medicine specialists, urologists, paediatricians and surgeons used *Escherichia* and *Staphylococcus* phages during the German occupation. Unfortunately, due to difficulties in publishing their work, we know little about their methodology and their observed clinical outcomes of phage treatments from this period. Dr Lipska is an exception given that she provides a detailed description of her attempts in treating newborns with coliphages. However, in the case of her work, postwar disturbances made it impossible to track the health status of all her patients. We must remember that during the six-year course of WWII, Poland was in the narrow group of countries that suffered the most. The war's events completely rearranged political and economic systems in Europe. The post-war borders of Poland had dramatically changed, which was inevitably associated with human resettlement on a large scale. In 1942, Dr Lipska applied phage preparations to 20 children but was only able to follow-up on half of them. Significantly, all of those patients did not suffer from diarrhoea, even after eight years from the termination of phage treatment, despite the ubiquitous poverty and inferior sanitary conditions of those times. We learn from her articles of at least three more centres where coliphages were applied between 1947 and 1951. These are the Obstetric-Gynecological Clinic in Warsaw, where 129 newborns were treated by Dr Starkiewiczowa. The Warsaw Residential Care Center, where 93 children were treated under the supervision of Dr Morawska, as well as the 1st Health Care Center in Warsaw, where 126 children (up to two years old) received treatment. The majority of these patients finished treatment with a favourable clinical outcome. Obviously, such results should be considered more in a historical context rather than scientific data gathered in accordance with Evidence-Based Medicine, but they are certainly worthy of mention.

**Ludwik Hirszfeld Institute of Immunology and Experimental Therapy**

Following WWII, Poland retained its independence, at least at an official level, and never became a Soviet republic. Although Poland was dominated by the Soviet regime together with other so-called "satellite states", influences from the Western world continued to have a significant impact on Polish society. Of note, people's mentality has remained quite different from the official forced philosophy, which 40 years later led to the fall of communism in Poland. In this spirit, an internationally renowned immunologist Prof. Ludwik Hirszfeld came to Wrocław in 1945 to continue what was recognised to be a remarkable scientific journey. As a result of the postwar re-drawing of the Polish border, Wrocław along with other former German cities became incorporated into Poland. Notably, Wrocław was recognised as the most important German scientific, economic and cultural centre east of Berlin before WWII. Hirszfeld's appearance in Wrocław marks the re-initiation of phage research in Poland after WWII, and which continues to this day. Although Hirszfeld was famous for his achievements in the field of immunology and serology, he devoted the last years of his career to phages. He also became the founder, first director and later a patron of the Institute of Immunology and Experimental Therapy of the Polish Academy of Sciences which was founded in December 1952. His passion for phages was fully reflected in his work published after WWII (Hirszfeld, 1948a; Hirszfeld et al., 1948b; Hirszfeld, 1949).

In 1948, Prof. Ludwik Hirszfeld established the Bacteriophage Collection as a result of his work on *Salmonella* phages and cooperation with a Canadian phage researcher, Prof. James Craigie from Toronto. During the subsequent decades, the collection was expanded with new phages isolated by members of the Hirszfeld Institute's Bacteriophage Laboratory. These were isolated from different environmental samples including crude and purified city sewages, inland and sea waters as well as filtrates from biological samples all in accordance with the protocol developed at the Hirszfeld Institute. The summary of the institute's long-term experience regarding phage isolation, characterization and propagation for therapeutic purposes has been presented in our article (Weber-Dąbrowska et al., 2016). Back in 1976, 187 phages were deposited in the Hirszfeld Institute's collection (Międzybrodzki et al., 2018). Over 40 years later, the collection consists of nearly 900 phages against the most common bacterial pathogens, including over 300 *Escherichia coli* phages. Indeed, only during the last two years, 56 new *E. coli* phages have been isolated from human and environmental samples and incorporated into the collection. In the years 2000-2011 for the first time, we were able to isolate phages against *Enterococcus, Stenotrophomonas, Citrobacter, Morganella,* and *Streptococcus* strains. A detailed summary of the Hirszfeld Institute's collection is presented in Table 1. Importantly, the majority of antibiotic-resistant strains of *Staphylococcus aureus, Pseudomonas aeruginosa*, and *Enterococcus faecalis* isolated from patients diagnosed at the Hirszfeld Institute were sensitive to specific phages from the collection: 90% of the *S. aureus* strains, 72% of *P. aeruginosa* strains, and 77% of the *E. faecalis* strains. Crude city sewage from several municipal water stations proved to be the best reservoir of new therapeutic phages. It is worth mentioning that in recent years our team has collected nearly 300 environmental samples from 21 countries in order to isolate new potentially therapeutic phages. We expect that new phages active against *E. coli, Citrobacter freundii, Klebsiella pneumoniae* and *Morganella morganii* ESBL (+) strains isolated from internal hospital sewage will also be useful in the treatment of human infections.

**Table 1.** Current (March 2020) bacteriophage collection of the L. Hirszfeld Institute of Immunology and Experimental Therapy, Polish Academy of Sciences

| No. | Phages | Number of phages |
|---|---|---|
| 1 | *Staphylococcus aureus, S. epidermidis, S. haemolyticus* (including MSSA, MRSA, MRCNS) | 9 |
| 2 | *Enterococcus faecalis, E. faecium* (including HLAR, HLGR, VRE) | 105 |
| 3 | *Escherichia coli* (including ESBL+) | 303 |
| 4 | *Klebsiella pneumoniae, oxytoca* (including ESBL +) | 124 |
| 5 | *Enterobacter cloacae, E. aerogenes, E. sakazakii* (including ESBL+) | 51 |
| 6 | *Shigella flexneri, S. sonnei* | 39 |
| 7 | *Citrobacter freundii, C. koseri* | 32 |
| 8 | *Pseudomonas aeruginosa, P. fluorescens* | 55 |
| 9 | *Salmonella enteritidis, S. typhimurium* | 48 |
| 10 | *Stenotrophomonas maltophilia* (including ESBL+) | 21 |
| 11 | *Serratia marcescens, S. liquefaciens* (including ESBL+) | 16 |
| 12 | *Proteus mirabilis, P. vulgaris* (including ESBL+) | 26 |
| 13 | *Morganella morganii* (including ESBL+) | 16 |
| 14 | *Acinetobacter baumannii, A. lwoffii* | 18 |
| 15 | *Streptococcus sp.* | 10 |
| 16 | *Raoultella terrigena* | 7 |
| | **TOTAL** | 880 |

Furthermore, in recent years we established cooperation with the Institute of Biochemistry and Biophysics at the Polish Academy of Sciences in Warsaw to include prophage and plasmid-free (PPF) bacteria strains for the propagation of therapeutic phages and studies examining the molecular basis of bacteria-phage interactions. This step allowed us to obtain phage lysates deprived of bacterial plasmids and temperate phages which minimized the risk of uncontrolled spread of bacterial virulence or antibiotic-resistant genes (Międzybrodzki et al., 2018).

After Hirszfeld's death in September 1954, Prof. Stefan Ślopek was appointed the second director of the Institute. He served this role for 31 years and contributed significantly to the development of phage treatment. Prof. Ślopek initiated the fruitful cooperation with the Kraków-based serum and vaccine manufacturer BIOMED. In March 1972, BIOMED received an order from Prof. Ślopek and Prof. Mulczyk (Head of the Institute's Microbiology Department) for the production of a series of *Shigella flexneri* and *sonnei* phage cocktails for oral application a with a final titer no lower than $10^8$ pfu/mL (Figure. 1). This partnership has continued to the present day. Since 2006, some of the phage preparations used for experimental phage treatment have been produced by IBSS BIOMED SA, Kraków, Poland as sterile lysates under GLP conditions. Notably, phage cocktails against *S. flexneri* and *S. sonnei* manufactured in the 1970s were observed to have retained their lytic activity after 30 years of storage at 4°C (Międzybrodzki et al., 2018).

During Ślopek's era, one of the most notable activities was the development of phage preparations against *Shigella* species. Research on phage typing of over 5,000 *Shigella* strains performed by Ślopek`s team led to the creation of the international typing set for *Shigella*. Thanks to this set of phages a high percentage of *Shigella strains* (both *S. flexneri* and *S. sonnei*) isolated from various sources in different European countries was tested and shown to match an active phage (Ślopek et al., 1972; 1973). Those days were marked by

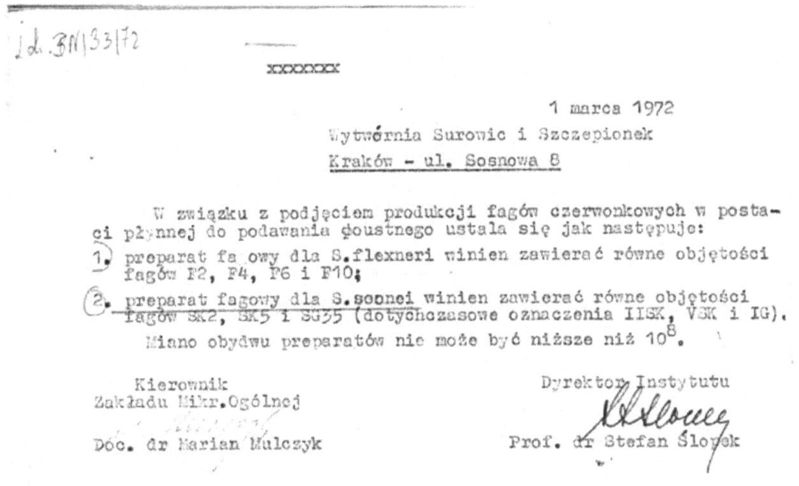

xxxxxxx

1 marca 1972

Wytwórnia Surowic i Szczepionek
Kraków - ul. Sosnowa 8

W związku z podjęciem produkcji fagów czerwonkowych w postaci płynnej do podawania doustnego ustala się jak następuje:

1. preparat fagowy dla S.flexneri winien zawierać równe objętości fagów F2, F4, F6 i F10;
2. preparat fagowy dla S.sonnei winien zawierać równe objętości fagów SK2, SK5 i SG35 (dotychczasowe oznaczenia IISK, VSK i IG).

Miano obydwu preparatów nie może być niższe niż $10^8$.

Kierownik
Zakładu Mikr.Ogólnej
Doc. dr Marian Mulczyk

Dyrektor Instytutu
Prof. dr Stefan Ślopek

**Figure 1.** A short note (in Polish) dated March 1, 1972, from Prof. Ślopek and Prof. Mulczyk to BIOMED with details regarding the production of *S. flexneri* and *S. sonnei* phage cocktails.

intensive international cooperation. It is known from various correspondences that in September 1967, Swedish scientists from the Statens Bakteriologiska Laboratorium in Stockholm visited Hirszfeld's Institute and signed an agreement with Prof. Ślopek regarding mutual work on *Shigella* phages. As a result of this partnership, Prof. Ślopek received a series of photos documenting phage plaque morphology and the lysis patterns of 28 tested *S. sonnei* strains from the Swedish collection along with a technical description of phage typing, recipes on phage agar and broth used in Stockholm as well as details on their titres and routine test dilutions (RTD). Prof. Lars Olof Kallings from Stockholm was responsible for distributing those phages among other research centres (including Hiszfeld's Institute). Prof. Ślopek's work on phage typing of *S. flexneri* was also recognized in such distant countries as Vietnam and Japan. In a letter dated May 1975, Dr Nguyen from the National Institute of Hygiene and Epidemiology in Hanoi expressed his interest in introducing phage typing of *S. flexneri* in Vietnam. He requested *S. flexneri* phages and propagating strains along with technical documentation. In the same year, Dr Torikai from the School of Medicine in Tokyo asked for 15 phages and reference *Klebsiella* strains used for propagation in order to classify more bacterial strains with the use of phages and elucidate any possible relations between phage groups, antibiotic resistance and pathogenesis. Of note, Dr Torikai referred to Prof. Ślopek's article published in 1967 by the Polish journal Archivum Immunologiae et Therapiae Experimentalis (Ślopek et al., 1967). In January 1982, exactly one month after the outbreak of martial law in Poland, Prof. Ślopek forwarded (free of charge) 13 *S. flexneri* phages and their propagating hosts to the Central Laboratory and Research Institute in Addis Abeba, Ethiopia at the request of local scientists (Figure. 2).

Clearly, the examples mentioned above show the significance and international recognition of Polish research despite Poland's isolation due to the geopolitical situation in those years. The archives of the Hirszfeld Institute possess many examples of such correspondence which paints a vivid picture of the unusual activity of Polish phage scientists at that time. Interestingly, such pioneering work was carried out in the "golden era" of antibiotics when research on phages was abandoned entirely in countries on the other side of the Iron Curtain.

**Phage therapy in the era of European Union regulations**

Poland's admission to the European Union (EU) in 2004 led to the creation of a new legislative system that was more advanced and thoroughly adapted to EU regulations. In this new reality, experimental treatment was expected to follow new guidelines that put human safety first and required a more thoughtful approach to experimental medicine. In this context, Hirszfeld's Institute has become the first ethically-approved phage treatment facility in the European Union. Since 2005 the Phage Therapy Unit (PTU), initiated by the efforts of Prof. Andrzej Górski, has admitted over 700 patients from around the globe. Phage treatment at the Hirszfeld Institute is strictly limited with clearly defined criteria of inclusion and exclusion. Patients are treated on an outpatient basis under the protocol approved by the Independent Bioethics Committee. Furthermore, circulation of phage therapeutics is under constant control with access only for

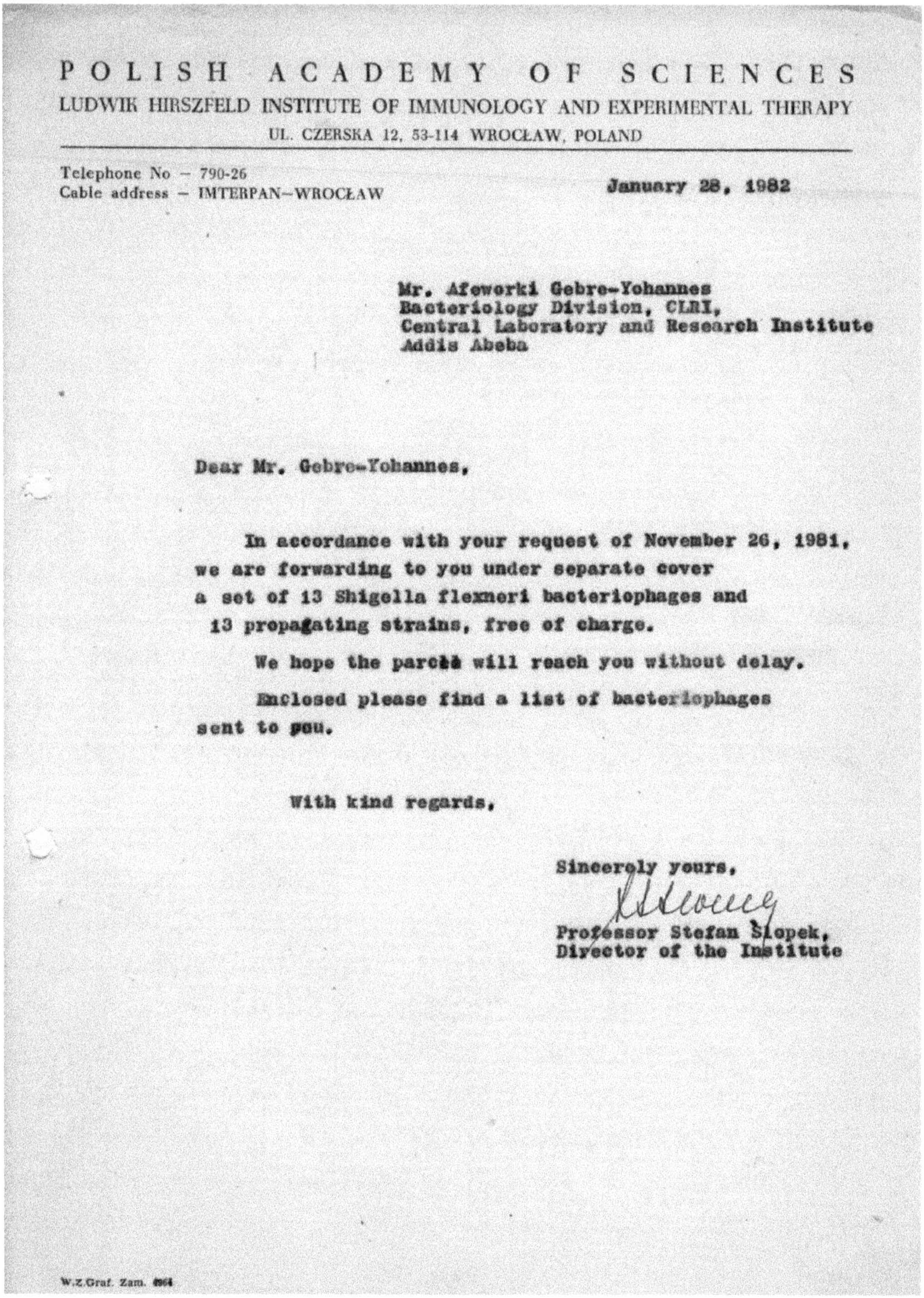

POLISH ACADEMY OF SCIENCES
LUDWIK HIRSZFELD INSTITUTE OF IMMUNOLOGY AND EXPERIMENTAL THERAPY
UL. CZERSKA 12, 53-114 WROCŁAW, POLAND

Telephone No – 790-26
Cable address – IMTERPAN–WROCŁAW

January 28, 1982

Mr. Afeworki Gebre-Yohannes
Bacteriology Division, CLRI,
Central Laboratory and Research Institute
Addis Abeba

Dear Mr. Gebre-Yohannes,

In accordance with your request of November 26, 1981, we are forwarding to you under separate cover a set of 13 Shigella flexneri bacteriophages and 13 propagating strains, free of charge.

We hope the parcel will reach you without delay.

Enclosed please find a list of bacteriophages sent to you.

With kind regards,

Sincerely yours,

Professor Stefan Ślopek,
Director of the Institute

W.Z.Graf. Zam. 8964

**Figure 2**. A letter dated January 28, 1982, from Prof. Ślopek to Mr Gebre-Yohannes from the Central Laboratory and Research Institute in Addis Abeba, Ethiopia informing about the shipment of requested 13 *S. flexneri* phages and their bacterial hosts.

patients enrolled in the experimental treatment. Such an approach does not seem to be the case in other Eastern European countries where phage therapy is currently conducted.

The results of phage therapy carried out at our Phage Therapy Unit have been presented in detail (Międzybrodzki et al., 2012; Górski et al., 2018; Międzybrodzki et al., 2018). Interestingly, the most favourable results have been achieved using purified staphylococcal phage preparations (good results in >50 % of cases). The therapy was well tolerated: mild side effects occurred in 12% of patients, and termination of treatment was necessary for 6% of patients. Our recent reports also highlight general and, most importantly, remote safety of phage therapy (Rogóż et al., 2019). Those data strongly suggest that the time has come to finally determine the efficacy of phage therapy in accord with Evidence-Based Medicine (controlled clinical trials).

In order to strengthen knowledge regarding the immune response to phages in patients undergoing experimental phage treatment, on the initiative of Prof. Górski, we started to investigate the anti-phage activity of patient's sera (AAS). In recent years, our team has published significant results in this research field (Górski et al., 2012; Łusiak-Szelachowska et al., 2014, 2016, 2017; Żaczek et al., 2016, 2019). Our preliminary results showed that the level of anti-phage activity of sera depends on the type of infection, the type of phage preparation (monovalent phage or phage cocktail), the dose of phage preparation, route of phage administration and duration of phage therapy. Our data suggests that even if high levels of anti-phage antibodies are induced throughout a phage therapy period, it does not necessarily exclude a positive outcome. Moreover, our analysis revealed that identical phages can induce different levels of anti-phage antibodies during phage therapy, which likely depends on a patient's immune status. The value of this research cannot be underestimated. Although studies performed on animals provide us with essential data about the anti-phage humoral immune response, those studies cannot be directly extrapolated to human models. Contrary to the tested animals, approximately 50% of patients undergoing phage treatment have different types of immunodeficiencies often caused by prolonged antibiotic therapy. (Łusiak-Szelachowska et al., 2017).

Interesting results for the future development of phage therapy was obtained in patients administered with purified phage preparations. Patients taking a purified staphylococcal phage cocktail showed the highest level of both AAS and antibody units (AU) measured in the ELISA test during treatment. These results indicate that titers of purified phage preparations (about four orders of magnitude higher compared to therapeutic phage lysates) could be responsible for the increased immune response. Considering the production costs of highly purified phage preparations, the ideal formulation for therapeutic applications unquestionably needs further attention (Żaczek et al., 2016).

In recent years our team published two books summarizing the current research on phages (Borysowski et al., 2014; Górski et al., 2019a). We had an excellent opportunity to work with numerous recognized phage scientists from different research centres who contributed to writing individual chapters. The latter position is probably the first guide in which experts from both the European Medicines Agency (EMA) and the U.S. Food and Drug Administration (FDA) shared their experience and opinions on experimental phage treatment and associated regulatory challenges. Evidently, both positions provide evidence of the unique position of the Hirszfeld Institute in the scientific world.

## Phage therapy in the era of drug repurposing

The current situation with the COVID-19 pandemic has been aggravated by the lack of drugs with clinically confirmed anti-coronavirus activity, which has resulted in the utilisation of the policy of "drug repurposing" (DR). This is due to the high costs associated with R&D and the slow pace of new drug development making DR an attractive alternative, by the utilisation of compounds with already known side effects, allowing shorter development timeslines (Pushpakom et al., 2019). As we have suggested in a series of recent publications, such options exist for DR for phage therapy of non-bacterial infections (Górski et al., 2019b) as well as autoimmune diseases, graft-versus-host disease, etc. Those effects are associated with diverse phage-specific influences on the immune system (Górski et al., 2019c). There is also data which suggests that phage therapy may be considered as an adjunct treatment for COVID-19 infections (Górski et al., 2020; submitted). Thus, we believe that the DR application of phages could be another significant Polish input into the history of phage therapy.

The COVID-19 pandemic has confirmed the compassionate use model as an accepted form of the development of new drugs by paving a way to clinical trials and, consequently, to determine the clinical value of the tested medicinal product (Grein et al., 2020). Evidently, Polish compassionate use of phage therapy has been following this principle for years.

## References

Borysowski, J., Międzybrodzki, R., and Górski, A., (2014) Phage Therapy: Current Research and Applications (Norfolk, UK: Caister Academic Press)

Jasieński, J. (1927) Próby zastosowania bakteriofagi w chirurgji. Polska Gazeta Lekarska. 4, 67-73. [article in Polish]

Fejgin, B. (1927) O zmienności bakteryj w związku ze zjawiskiem d'Hérellea. Warszawskie Czasopismo Lekarskie. 21, 694-697. [article in Polish]

Fejgin, B. (1923) Sur la nature du phenomene d'Hérelle. Compt. Rend. Soc. Biol. 89, 1383. [article in French]

Fejgin, B. (1936) La toxicite de bacilli de Shiga bacteriophage. Compt. Rend. Soc. Biol. 89, 1385. [article in French]

Górski. A., Bollyky, P.L., Przybylski, M., Borysowski, J., Międzybrodzki. R., Jończyk-Matysiak, E. and Weber-Dąbrowska, B. (2019b) Perspectives of Phage Therapy in Non-bacterial Infections. Front. Microbiol. 9, 3306. doi: 10.3389/fmicb.2018.03306

Górski, A., Międzybrodzki, R., and Borysowski, J. (2019a), Phage Therapy: A Practical Approach (Cham, Switzerland: Springer Nature Switzerland), doi: 10.1007/978-3-030-26736-0

Górski, A., Międzybrodzki, R., Borysowski, J., Dąbrowska, K., Wierzbicki, P., Ohams, M., Korczak-Kowalska, G., Olszowska-Zaremba, N., Łusiak-Szelachowska, M., Kłak, M., Jończyk, E., Kaniuga, E., Gołaś, A., Purchla, S., Weber-Dąbrowska, B., Letkiewicz, S., Fortuna, W., Szufnarowski, K., Pawełczyk, Z., Rogóż, P., and Kłosowska, D. (2012) Phage as a Modulator of Immune Responses: Practical Implications for Phage Therapy. Adv Virus Res 83:41–71. doi: 10.1016/B978-0-12-394438-2.00002-5

Górski, A., Międzybrodzki, R., Jończyk-Matysiak, E., Żaczek, M., and Borysowski, J. (2019) Phage-specific diverse effects of bacterial viruses on the immune system. Fut. Microbiol. 14, 1171-1174. doi: 10.2217/fmb-2019-0222

Górski, A., Międzybrodzki, R., Łobocka, M., Głowacka-Rutkowska, A., Bednarek, A., Borysowski, J., Jończyk-Matysiak, E., Łusiak-Szelachowska, M., Weber-Dąbrowska, B., Bagińska, N., Letkiewicz, S., Dąbrowska, K., and Scheres, J. (2018) Phage Therapy: What Have We Learned? Viruses. 10, pii: E288. doi: 10.3390/v10060288.

Górski, A., Międzybrodzki, R., Żaczek, M., and Borysowski, J. (2020) Phages in the fight against coronavirus (Cov)? Fut. Microbiol. (submitted).

Grein, J., Ohmagari, N., Shin, D., Diaz, G., Asperges, E., Castagna, A. et al. (2020) Compassionate Use of Remdesivir for Patients with Severe Covid-19. N Eng J Med. doi: 10.1056/NEJMoa2007016

Hirszfeld, L. (1948a) Proceedings of the Wrocław Scientific Society: The battle of the invisible with the imperceptible. State Publishing Institute. 3, 3-18.

Hirszfeld, L. (1949) Immunologia ogólna (Wrocław: Czytelnik). [book in Polish]

Hirszfeld, L., Galis-Malejczyk, A., Sembrat-Niewiadomska, Z., Zwierz C. (1948b) Bakteriofagi i ich rola w rozpoznawaniu duru brzusznego. Polski Tygodnik Lekarski. 14, 417-423. [article in Polish]

Lipska, I. (1951) Dotychczasowe spostrzeżenia nad leczniczym i zapobiegawczym stosowaniem bakteriofagów w Polsce. Pediatria Polska. 26, 449-453. [article in Polish]

Łusiak-Szelachowska, M., Żaczek, M., Weber-Dąbrowska, B., Kłak, M., Międzybrodzki, R., Fortuna, W., Rogóż, P., Szufnarowski, K., Jończyk-Matysiak, E., and Górski, A. (2016). Antiphage activity of sera from patients receiving staphylococcal phage preparations. In Microbes in the Spotlight: Recent Progress in the Understanding of Beneficial and Harmful Microorganisms, A. Méndez-Vilas ed. (Boca Raton, FL: BrownWalker Press), pp. 245–249.

Łusiak-Szelachowska, M., Żaczek, M., Weber-Dąbrowska, B., Międzybrodzki, R., Kłak, M., Fortuna, W., Letkiewicz, S., Rogóż, P., Szufnarowski, K., Jończyk-Matysiak, E., Owczarek, B., and Górski, A. (2014). Phage neutralization by sera of patients receiving phage therapy. Viral Immunol 27, 295–304. doi: 10.1089/vim.2013.0128

Łusiak-Szelachowska, M., Żaczek, M., Weber-Dąbrowska, B., Międzybrodzki, R., Letkiewicz, S., Fortuna, W., Rogóż, P., Szufnarowski, K., Jończyk-Matysiak, E., Olchawa, E., Walaszek, K. M., and Górski, A. (2017). Antiphage activity of sera during phage therapy in relations to its outcome. Future Microbiol. 12, 109–117. doi: 10.2217/fmb-2016-0156

Międzybrodzki, R., Borysowski, J., Weber-Dąbrowska, B., Fortuna, W., Letkiewicz, S., Szufnarowski, K., Pawełczyk, Z., Rogóż, P., Kłak, M., Wojtasik, E., and Górski, A. (2012) Clinical Aspects of Phage Therapy. Adv Virus Res 83, 73–121. doi: 10.1016/B978-0-12-394438-2.00003-7

Międzybrodzki, R., Hoyle, N., Zhvaniya, F., Łusiak-Szelachowska, M., Weber-Dąbrowska, B., Łobocka, M., Borysowski, J., Alavidze, Z., Kutter, E., Górski, A., and Gogokhia L. (2018) Current Updates from the Long-Standing Phage Research Centers in Georgia, Poland, and Russia. In Bacteriophages, D. Harper et al. eds. (Cham, Switzerland: Springer Nature Switzerland AG), pp. 1-31. https://doi.org/10.1007/978-3-319-40598-8_31-1

Pushpakom, S., Iorio, F., Eyers, P.A., Escott, K.J., Hopper, S., Wells, A., Doig, A., Guilliams, T., Latimer, J., McNamee, C., Norris, A., Sanseau, P., Cavalla, D., and Pirmohamed, M. (2019) Drug repurposing: progress, challenges and recommendations. Nat Rev Drug Discov. 18, 41-58. doi: 10.1038/nrd.2018.168

Rogóż, P., Amanatullah, D.F, Międzybrodzki, R., Manasherob, R., Tikunova, N.V., Weber-Dąbrowska, B., Fortuna, W., Letkiewicz, S., and Górski, A. (2019) Phage Therapy in Orthopaedic Implant-Associated Infections. In Phage Therapy: A Practical Approach, A. Górski et al. eds. (Cham, Switzerland: Springer Nature Switzerland AG), pp. 189-211. https://doi.org/10.1007/978-3-030-26736-0_8

Summers, W.C. (2012) The Strange History of Phage Therapy. Bacteriophage 2, 130-133. http://dx.doi.org/10.4161/bact.20757

Ślopek, S., Durlakowa, I., Kucharewicz-Krukowska, A., Krzywy, T., Ślopek, A., and Weber, B. (1972). Phage typing of *Shigella flexneri*. Arch. Immunol. et Ther. Exp. 20, 1-60.

Ślopek, S., Durlakowa, I., Kucharewicz-Krukowska, A., Krzywy, T., Ślopek, A., and Weber, B. (1973). Phage typing of *Shigella sonnei*. Arch. Immunol. et Ther. Exp. 21, 1-161.

Ślopek, S., Przondo-Hessek, A., Milch, H., Deak, S. (1967) A working scheme for bacteriophage typing of *Klebsiella bacilli*. Arch. Immunol. et Ther. Exp. 15, 589-599.

Weber-Dąbrowska, B., Jończyk-Matysiak, E., Żaczek, M., Łobocka, M., Łusiak-Szelachowska, M., and Górski, A. (2016) Bacteriophage Procurement for Therapeutic Purposes. Front. Microbiol. 7, 1177. doi: 10.3389/fmicb.2016.01177

Żaczek, M., Łusiak-Szelachowska, M., Jończyk-Matysiak, E., Weber-Dąbrowska, B., Międzybrodzki, R., Owczarek, B., Kopciuch, A., Fortuna, W., Rogóż, P., and Górski, A. (2016). Antibody production in response to staphylococcal MS-1 phage cocktail in patients undergoing phage therapy. Front Microbiol 7, 1681. doi: 10.3389/fmicb.2016.01681

Żaczek, M., Łusiak-Szelachowska, M., Weber-Dąbrowska, B., Międzybrodzki, R., Fortuna, W., Rogóż, P., Letkiewicz, S., and Górski, A. (2019) Humoral Immune Response to Phage-Based Therapeutics. In Phage Therapy: A Practical Approach, A. Górski et al. eds. (Cham, Switzerland: Springer Nature Switzerland AG), pp. 123-143. https://doi.org/10.1007/978-3-030-26736-0_5

# Chapter 5

# Commercial Products for Human Phage Therapy

**Nina Chanishvili* and Marina Goderdzishvili**

The Eliava Institute of Bacteriophage, Microbiology & Virology, 3 Gotua Street, Tbilisi 0160, Georgia.

*Corresponding author: nina.chanishvili@pha.ge

DOI: https://doi.org/10.21775/9781913652517.05

## Abstract

This article describes the development and distribution of early therapeutic bacteriophage (phage) products for the treatment of problematic bacterial infections before the advent of antibiotics. It details the beginnings of commercial phage production in Paris by Félix d'Hérelle in the 1920s leading subsequently to their production by L'Oréal, the manufacture of products in India to treat cholera, in Latin America to treat dysentery, and several in the USA by Ely Lilly to treat a range of infections up to the 1940s. It also discusses the various commercial products and the manufacturing approaches used to make them in the Republic of Georgia and throughout Russia from the Soviet period up to the present day, and details twenty-five commercial products for human phage therapy which are still officially registered there, most of which are widely purchased and used today. Genetic and metagenomic analysis of some of the latter products is discussed.

## Early phage products: France, India, Latin America and the USA

Commercially available therapeutic phage-based preparations associated with the Eliava Institute of Bacteriophage have a long history. Soon after the discovery of phages in the 1920s, Félix d'Hérelle founded a commercial laboratory in Paris which manufactured at least five phage preparations targeting a number of bacterial infections. These were called *Bacté-coli-phage, Bacté-rhino-phage, Bacté-intesti-phage, Bacté-pyo-phage,* and *Bacté-staphy-phage* (Figure 1). D'Hérelle's laboratory was later incorporated into the large French company L'Oréal (d'Hérelle, 1935; Summers, 1999; Sulakvelidze et al, 2001; Chanishvili, 2012). In his book published in Tbilisi in 1935, d'Hérelle mentioned that due to the results of his experiments in 1927 in Panjab in India (held in villages affected by a cholera epidemic), two centres for the industrial production of phages against cholera were established. One of these sites was directed by Prof. Igor Nikolaevich Asheshov, who was invited by the government of British India to take over d'Hérelle's research work on the cholera phages. This laboratory was providing phage preparations mainly to the Bihar province (located near Nepal and Bangladesh), which was the hotspot of the cholera epidemic. The other laboratory was established in Shilong at the Medical Research Institute of Assam (located between Bangladesh and Bhutan). Production at this site was led by Dr John Morrson who later became a director of the Pasteur Institute in Rangoon (Burma). D'Hérelle's writings refer to an interesting detail: according to the reports of the local healthcare agency, the cost of epidemic measures that were taken in 1931 in North Bihar using phages was only 25 rupees; while in 1930, when the same campaign had been carried out using vaccination and chemical disinfection of the water supply system the costs had come to 3118 rupees (d'Hérelle, 1935). During his stay in Soviet Georgia under Soviet rule in the early 1930s (described elsewhere in this book), d'Hérelle established further production of phages.

During the same general era in Latin America, the Oswaldo Cruz Institute which was founded in 1900 in Rio-de-Janeiro, had become important for sanitary campaigns, production of biological products, and medical research in Brazil. The institute started production of the anti-dysentery phages in 1924 to combat the disease in Latin American countries (Costa Cruz, 1924; d'Hérelle, 1935, Dublanchet and Bourne, 2007; Chanishvili, 2012; Almeida and Sundberg, 2020).

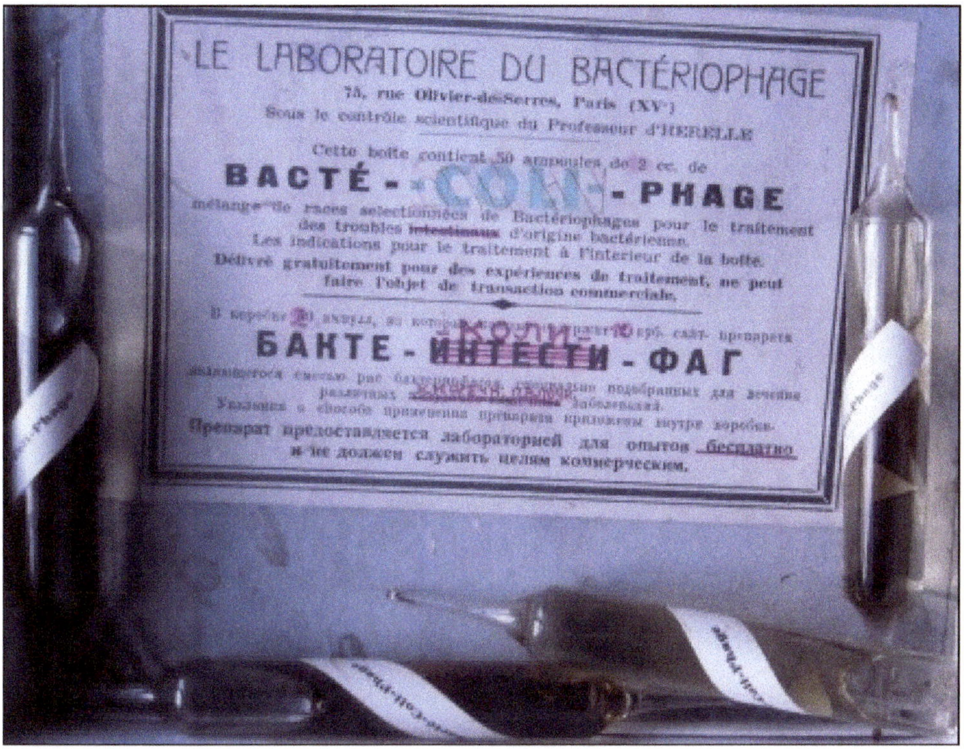


**Figure1.** A box of Intesti-phage produced by d'Hérelle's laboratory in Paris. The text is in French and Russian. Inscription on the box states that the box contains 20 ampoules of preparation composed of phages active against *E.coli.* At the bottom of each description on the label, there is a note stating: "This preparation is provided by the laboratory for experimental use for free (underlined) and should not serve commercial aims".

Within a year, the institute produced 10,000 vials of phages, which were sent to hospitals all over Brazil (d'Hérelle, 1935; Chanishvili, 2012; Almeida and Sundberg, 2020). The feedback led the Oswaldo Cruz Institute to recommend phage use, and it started to provide phages by request, under the name *Bacteriofagina disenterica*. This product was noted to be the first therapeutic phage preparation offered to the public on the basis of thorough observation (the first phage product that was tested in large human trials). *Bacteriofagina disenterica* was largely used during the 1924 revolution in São Paulo among government troops. Almeida and Sundberg assume that this was the first large scale human trial of phage therapy in history, preceding those done by d'Hérelle in India and those done in the Soviet Union (Almeida and Sundberg, 2020). In addition to the *Bacteriofagina disenterica* preparation, other phages have also been produced in Brazil, such as Estafilofagina (against *Staphylococcus*) and Colifagina. Such phage preparations were made by the Oswaldo Cruz Institute and distributed throughout the country, and commercial phage products were also made and sold by private companies, such as Raul Leite's laboratory at Rio de Janeiro, Brazil (Almeida and Sundberg, 2020).

Therapeutic phages were also produced in the United States. In the 1940s, the Eli Lilly Company (Indianapolis) produced seven phage products for human use, including preparations targeted against staphylococci, streptococci, *Escherichia coli*, and other bacterial pathogens (Summers, 1999; Sulakvelidze et al, 2001; Hausler, 2008; Chanishvili, 2012). These preparations consisted of phage-lysed, bacteriologically sterile broth cultures of the targeted bacteria and designated, for example, *Colo-lysate, Ento-lysate, Neiso-lysate*, and *Staphylo-lysate*) or the same preparations in a water-soluble jelly base, for example, *Colo-jel, Ento-jel*, and *Staphylo-jel*. They were used to treat various infections, including abscesses, suppurating wounds, vaginitis, acute and chronic infections of the upper respiratory tract, and mastoiditis infections (Summers, 1999; Chanishvili, 2012). However, the efficacy of such phage preparations was controversial (Chanishvili, 2012), and this could be explained by the fact that the nature and biological characteristics of phages were not well studied and understood. Also, the methods allowing the production of high titre and broad host range phage preparations had not yet been developed. As a result, with the advent of antibiotics, commercial production of therapeutic phages was brought to an end in most of the Western countries, while in the former Soviet Union the development of this field of science continued largely due to joint efforts of George Eliava and the above mentioned Felix d'Hérelle who together founded and developed the Bacteriophage Institute in Tbilisi, Georgia (Chanishvili, 2012; 2016). Figure 2 shows phage preparations produced in Tbilisi during the early antibiotic-era.

**Phage production in Georgia and Russia up to the present day**

From the very beginning of phage production almost up to the end of the 20th century, phage production was performed using natural associations of phages isolated mostly from water environments. Similarly, like mixed undefined

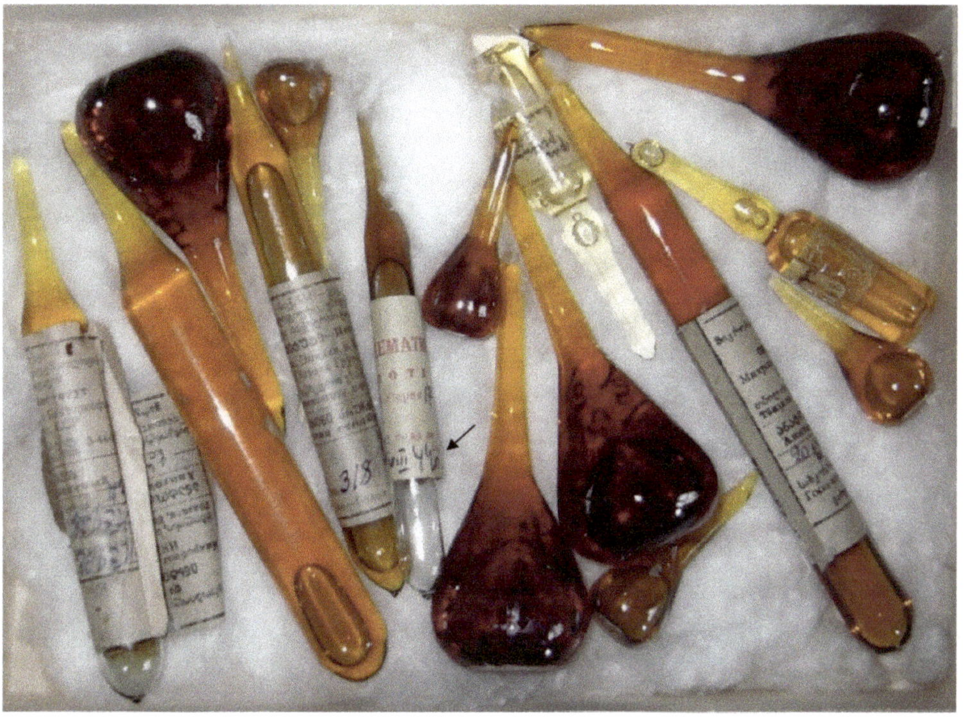

**Figure 2.** Early vials of phage preparations from Tbilisi, one of which (see arrow) was produced in August 1944.

fermented food starter cultures, the natural phage associations, which nowadays are wrongly termed as "cocktails", represent an ecosystem which is shaped by long term evolution under relatively defined environmental conditions and provides an interesting model to study co-evolution of phages and their hosts as well as the impact of diversity on microbial community stability. This tradition still dominates in many contemporary phage producing production plants. At present, commercial phage preparations are produced in Georgia and Russia, where they can be purchased at a regular pharmacy without a prescription. Unlike other drugs with the exact chemical formula and concentration of active ingredients, the phage preparations are non-standard solutions containing biologically active viral particles. It is noteworthy that a phage product with the same name, but produced by different enterprises or at different times, may contain different combinations and/or proportions of constituent phages.

All these differences in content are due to the remarkable specificities of the phage selection procedure and their production. Phages are selected according to their ability to lyse specific bacterial isolates, then a mixture of phages is grown on a given bacterial cultures, and then they are "launched" into production, where phages are propagated in large containers (or fermenters) using plasmid-free so-called "production" bacterial strains.

The therapeutic and prophylactic phage preparations present on the Georgian and Russian markets are composed of the natural associations of virulent phages targeting a broad range of species and having overlapping host ranges, which are equally destructive to antibiotic-susceptible and antibiotic-resistant bacteria. The phage preparations are classified as either mono- and polyvalent (Figure 3). The monovalent phage preparations are active against one particular species of bacterial pathogen, e.g. *E. coli*, while polyvalent preparations such as Intesti-phage are characterized with a wide and simultaneous coverage of several bacterial genera and species, e.g. *E. coli*, *Salmonella*, *Shigella* spp., *Enterococcus* spp. and other enteric bacteria. In most cases, it is recommended to use polyclonal, polyvalent phage preparations (Krasilnikov et al., 2011).

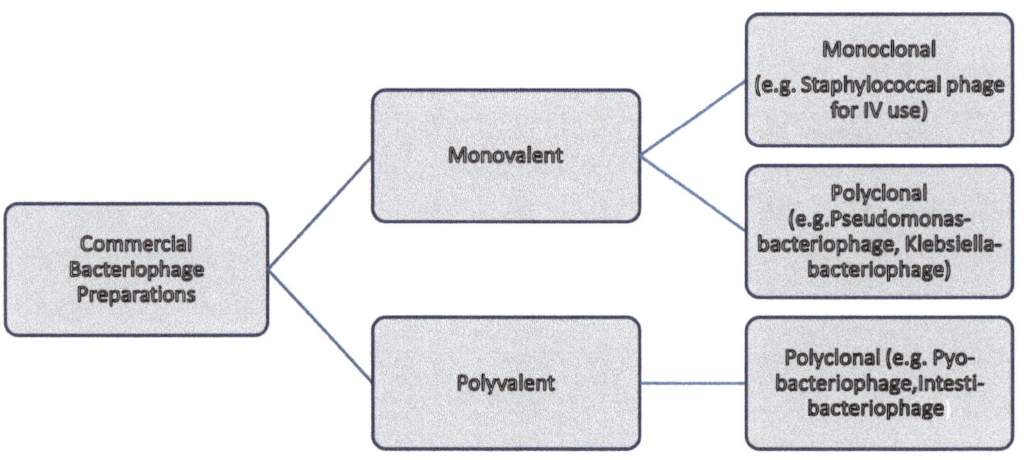


**Figure 3.** Classification of the phage preparations according to their clonal composition.

Phage preparations are nowadays used for treatment and prophylaxis of the following:

(a) gastrointestinal infections (dysentery, typhoid fever, salmonellosis, dysbiosis, gastro-enterocolitis, cholecystitis colitis, dyspepsia, etc.);
(b) purulent-inflammatory diseases of eyes, nasal and oral cavities, throat, lungs (conjunctivitis, otitis, tonsillitis, angina, pharyngitis, stomatitis, periodontitis, highmoritis, sinusitis, frontitis, pneumonia, etc.)
(c) surgical infections (abscesses, phlegmon, osteomyelitis, peritonitis); post-operational complications;
(d) dermatology problems (furunculosis, carbunculosis, hidradenitis, acne, etc.);
(e) burn wound infections;
(f) urogenital infections (urethritis, cystitis, pyelonephritis, vulvitis, colpitis, endometritis, female infertility, etc.) (Leitner et al. L., 2017; 2020)

The main requirements for phage preparations are:
(a) The titre of the commercial phage preparations may not be lower than $10^{-4}$ – $10^{-5}$ according to the system of Appelmans (1921), which is corresponding to $10^7$ to $10^9$ pfu/mL by Gratia (1939)
(b) They should include only virulent phages with a broad lytic spectrum against the bacterial pathogen. The phages composing the preparation should have an overlapping host range and differ according to the mechanisms of their interaction with the host cells. At the same time, these phages should not be active to the inhabitants of normal micro-flora.
(c) The phages composing the phage preparation/mixture should be characterized with a high yield of their progeny per host cell.
(d) Lytic activity of the preparation should remain stable during its prolonged storage.

The use of such phages for treatment or prophylaxis of bacterial infections limits the emergence of phage-resistant among the pathogen population (Krasilnikov et al., 2011).
The first commercial phage preparations designed by Felix d'Hérelle, were named Pyo- and Intesti-bacteriophage (Figure 3 and 4). They are still produced today under the same names in Russia and Georgia by the Eliava

BioPreparations Ltd (EBP), which was established by the Eliava Foundation for Bacteriophage Research in 2008. At the moment EPB produces six phage-based preparations: Pyo-bacteriophage, Intesti-bacteriophage, Staphylococcus-bacteriophage, Ses-bacteriophage, ENCO-bacteriophage and Fersis-bacteriophage. These products were developed at the Eliava Institute of Bacteriophage, Microbiology & Virology at Tbilsii, Georgia, which is a world known centre for excellence in phage therapy research. Together with the manufacturing process, the EBP performs significant R&D activities to maintain and improve quality of its products continuously. EBP is collaborating with several diagnostic centres to collect samples from the field to ensure high activity and a wide spectrum of its phages. The company routinely maintains and grows its collection of bacterial strains and phages in collaboration with the Eliava Institute. With the support of the Eliava Institute and EU funding, the EBP company is renovating and updating its production spaces to comply with GMP standards.

Another private Georgian company, BioChimPharm (formed in the 1990s on the basis of one of the industrial plants of the Eliava Institute of Bacteriophage, Microbiology & Virology) is manufacturing the following phage preparations: Travel-phage, (produced in liquid and tablet forms also under the trade name Septaphage), Phagio (an analogue of Pyo-bacteriophage) and Phagosept (liquid) containing a combination of phages for treatment of *Staphylococcus* infections.

Microgen Bioproducts is a leading Russian state company producing a variety of biological preparations. The enterprise known as “The Scientific and Production Association Microgen” was established in May 2003 as a result of the merger of state-owned enterprises producing medical immunobiological products and other medical drugs. Because of the special status of the company, its work is controlled personally by the Prime Minister of Russia. Microgen possesses nine production platforms located in nine cities across Russia, with its major production sites located in Stavropol, Irkutsk, Nizhny Novgorod, Moscow, Tomsk, Perm and Ufa. Microgen produces a range of over 120

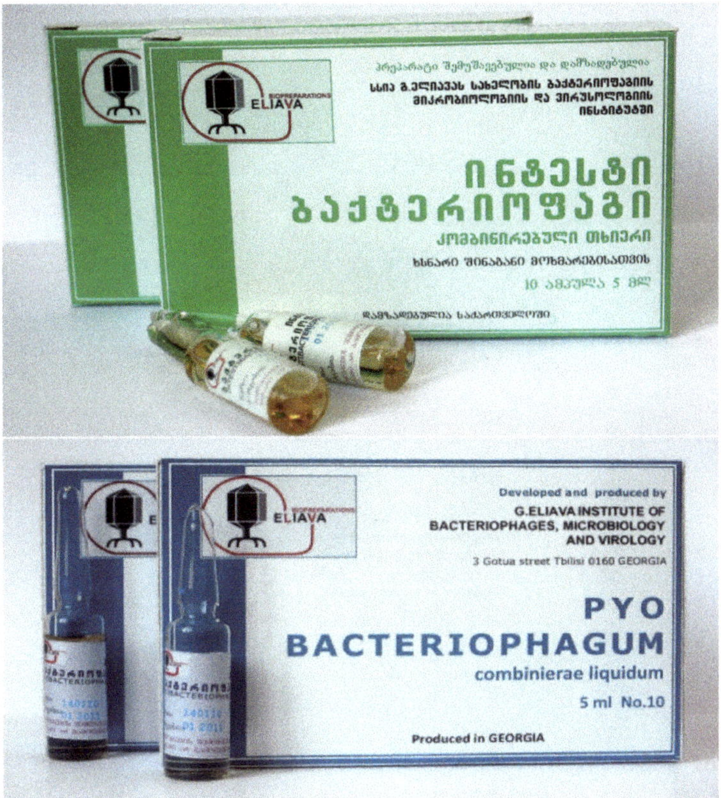


**Figure 4.** Products of the Eliava BioPreparations Ltd., Tbilisi, Georgia: Intesti-bacteriophage - labelled in Georgian script (upper image), Pyo-bacteriophage (lower image).

immunobiological products including vaccines, sera, specific and polyvalent immunoglobulins, nutrient media, allergens, probiotics etc. It is the only enterprise in Russia producing phages on an industrial scale. Phages are produced at the facilities located in Nizhny Novgorod ("ImBio"), Perm ("BioMed") and Ufa ("Immugen") and the website of Microgen is also presented in English (www.microgen.ru). The company is planning to develop commercially available phages targeting *Enterobacter* spp. and *Acinetobacter baumanii*. In 2020, phage preparations were included in the national clinical guidelines for antimicrobial therapy in urology, and the information is available on their website (https://www.bacteriofag.ru/press/) presented in English as well as Russian. The preparations registered in Georgia and Russia are listed in Table 1. The total number of phage products is 25 although some, in particular those that are targeting single pathogens, are not currently in production (e.g.

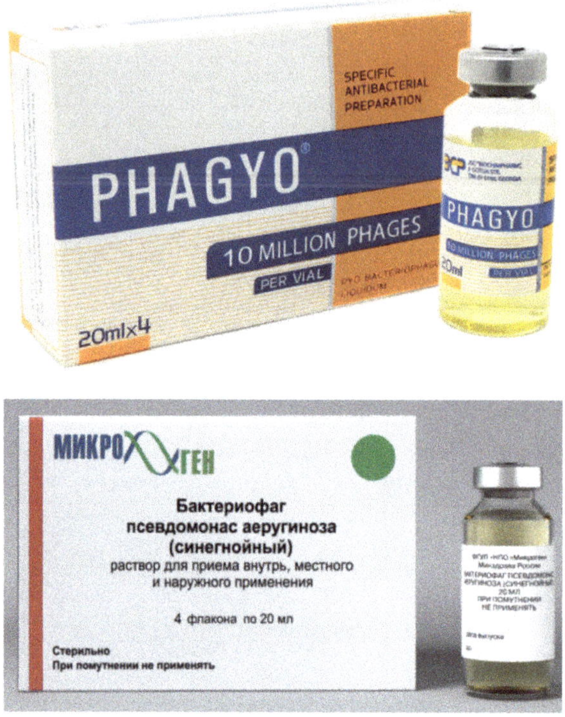


**Figure 5.** Phagio (analogue of Pyo-bacteriophage), one of the products of BioChimPharam, Tbilisi, Georgia (upper image); Pyo-bacteriophage, one of several products of Microgen, Russia (lower image).

Coli-bacteriophage, Proteus-bacteriophage). The complex preparations such as Pyo- and Intesti- bacteriophages and their modifications with broad coverage are manufactured. The largest volume phage product mass-produced in Russia is the Pyo-bacteriophage, which is manufactured by three branches of Microgen (in Nizhniy Novgorod, Perm and Ufa).

It is necessary to underline that development of phage preparations for medicinal use across the Soviet Union was first coordinated by the All-Union Ministry of Health, which was formed in the 1940s and later in 1980s by the All-Union Ministry for the production of bacterial and biological preparations. The information, documentation (regulations, etc.) and so called “mother strains” for production of various versions of microbiological (including phages) and immunological products were controlled, stored and, if necessary, distributed within the network of the scientific–industrial units by the Tarasevich State

Control Institute, which was a core organization of this association. Prior to the collapse of the Soviet Union, the phage production was held in Tbilisi (Georgia), considered a lead organization for phage research, and in the Russian cities: Gorkyi (nowadays Nizhnii Novgorod), Ufa and Perm, as they were united under the umbrella of the same ministry. Since the collapse of the Soviet Union, the contemporary production plants have not shared trade names, manufacturing traditions, and often the phage composition of their products. Differences in phage preparations among Georgian and Russian facilities have been highlighted by metagenomic analysis, which showed variation in the composition of their phage products (McCallin et al., 2013. 2018).

**Table 1.** Phage-based preparations registered in Georgia and Russia

| No. | Trade name of the phage preparation | Type of preparation | Composition/phages active against: | Producer company | Country |
|---|---|---|---|---|---|
| 1 | Pyo-bacteriophage | Polyvalent, polyclonal, liquid | *Staphylococcus* sp., *Streptococcus* sp, *P. aeruginosa, Proteus mirabilis, Proteus vulgaris, E.coli* (different serovars) | Eliava BioPreparations Ltd | Tbilisi, Georgia |
| 2 | Intesti-bacteriophage | Polyvalent, polyclonal, liquid | *Shigella* sp, *Salmonella* sp., *E.coli, Staphylococcus* sp., *Enterococcus* sp., *Proteus* sp., *Pseudomonas aeruginosa* | Eliava BioPreparations Ltd | Tbilisi, Georgia |
| 3 | Staphylococcal bacteriophage | Monovalent, monoclonal, liquid | *Staphylococcus aureus* | Eliava BioPreparations Ltd | Tbilisi, Georgia |
| 4 | Enco-bacteriophage | Polyvalent, polyclonal, liquid | *Salmonella* sp, *Shigella* spp., different serovars of *E.coli* and *Staphylococcus* spp | Eliava BioPreparations Ltd | Tbilisi, Georgia |

| | | | | | |
|---|---|---|---|---|---|
| 5 | SES-bacteriophage | Polyvalent, polyclonal, liquid | *Staphylococcus* spp, *Streptococcus* spp. and different servars of *E.coli* | Eliava BioPreparations Ltd | Tbilisi, Georgia |
| 6 | Fersis - bacteriophage | Polyvalent, polyclonal, liquid | *Staphylococcus* spp (*S. aureus, S. epidermidis,* and *S. saprophyticus*) and *Streptococcus* spp. (*S. pyogenes, S. sanguis, S. salivarius* and *S. agalactiae).* | Eliava BioPreparations Ltd | Tbilisi, Georgia |
| 7 | Travel-phage | Polyvalent, polyclonal, ltablets | *Shigella* spp., *Salmonella* spp., *E.coli, Proteus* spp., *Staphylococcus* spp., *Pseudomonas* spp., *Enterococus* spp | BioChimPharm Ltd | Tbilisi, Georgia |
| 8 | Septaphage | Polyvalent, polyclonal, liquid, tablets | *Shigella* spp., *Salmonella* spp., *E.coli, Proteus* spp., *Staphylococcus* spp., *Pseudomonas* spp., *Enterococus* spp | BioChimPharm Ltd | Tbilisi, Georgia |
| 9 | Phagio (analogue of Pyo-bacteriophage) | Polyvalent, polyclonal, liquid | *Staphylococcus* spp., *Streptococcus* spp., *Pseudomonas* spp., *Proteus* spp, *E.coli* | BioChimPharm Ltd | Tbilisi, Georgia |
| 10 | Phagosept | Polyvalent, polyclonal, liquid | *Staphylococcus* spp. | BioChimPharm Ltd | Tbilisi, Georgia |
| 11 | Staphylococcus-bacteriopahge | Monovalent, polyclonal | *Staphylococcus* spp. | Microgen-BioMed, | Perm, Russia |
| 12 | Streptococcus-bacteriophage | Monovalent, polyclonal | *Staphylococcus* spp. (including *Enterococcus* spp.) | Microgen-BioMed, | Perm, Russia |
| 13 | Pseudomonas aeruginosa-phage | Monovalent, polyclonal | *Pseudomonas aeruginosa* | Microgen-ImBio | Nizhnii Novgorod, Russia |

| | | | | | |
|---|---|---|---|---|---|
| 14 | Coli-bacteriophage | Monovalent, polyclonal | *E.coli* | - | - |
| 15 | Proteus-bacteriophage | Polyvalent, polyclonal | *Proteus mirabilis, Proteus vulgaris* | - | - |
| 16 | Dysenterial-bacteriophage | Polyvalent, polyclonal, liquid | *Shigella flexneri* serovars 1,2,3,4,6 & *Shigella sonnei* | Microgen-BioMed | Perm, Russia |
| 17 | Typhoid-bacteriophage | Polyvalent, polyclonal, liquid | *Salmonella* Typhi | - | - |
| 18 | Salmonella – bacteriophage | Polyvalent, polyclonal, liquid | *S.* Typhimurium, *S.*paratyphi A & B, S. Heidelberg, *S.* Newport, *S.* Cholerasuis, *S.*Oranienburg, *S.* Infantis, *S.* Dublin, *S.* Enterritidis, *S.*Newlands | Microgen-BioMed | Perm, Russia |
| 19 | Klebsiella – bacteriophage | Monovalent | Klebsiella pneumoniae | Microgen-BioMed | Ufa, Russia |
| 20 | Coli-Proteus bacteriophage | Polyclonal, polyclonal, liquid | *E.coli, P. mirabilis, P.vulgarris* | Microgen-ImBio | Nizhnii Novgorod, Russia |
| 21 | Klebsilella-bacteriophage | Polyvalent, polyclonal, liquid | *K. pneumonae, K.ozaenae, K.rhinoscleromatis* | - | - |
| 22 | Pyo-bacteriophage, purified | Polyvalent, polyclonal, liquid | *Staphylococcus* sp., *Streptococcus* sp, *P. aeruginosa, Proteus mirabilis, Proteus vulgaris, P. mirabilis, K.pneumoniae* | Microgen-ImBio & Imunogen – BioMed. | Nizhnii Novgorod, & Ufa, Russia |
| 23 | Sexsta-phage | Polyvalent, polyclonal, liquid | *Staphylococcus* sp., *Streptococcus* sp, *P. aeruginosa, Proteus mirabilis, Proteus vulgaris, P. mirabilis, K.pneumoniae, E.coli* | Microgen-Immunogen | Perm, Russia |

| | | | | | |
|---|---|---|---|---|---|
| 24 | Pyo-bacteriophage, complex | Polyvalent, polyclonal, liquid | *Staphylococcus* sp., *Streptococcus* sp, *Enterococcus* spp., *P. aeruginosa, Proteus mirabilis, Proteus vulgaris, P. mirabilis, E.coli, K.pneumoniae, K.oxitoca* | Microgen-ImBlo, Microgen-Immunogen, Microgen-BioMed | Nizhnii Novgorod, Ufa & Perm, Russia |
| 25 | Intesti-bacteriophage | Polyvalent, polyclonal, liquid | *Shigella flexneri* serovars 1,2,3,4,6 & *Shigella sonnei, S.* Typhimurium, *S.*paratyphi A & B, S. Heidelberg, *S.* Newport, *S.* Cholerasuis, *S.*Oranienburg, *S.* Infantis, *S.* Dublin*, S.* Enterritidis, *S.*Newlands, *E.coli, Enetrococcus* spp., *P. aeruginosa, P.mirabilisi, P. vulgaris, Staphylococcus* spp. | Microgen - ImBio & Microgen-Imunogen | Nizhnii Novgorod & Perm, Russia |

## Genetic and metagenomic analysis of commercial phage products

Studies on the genetic composition of the commercial phages started several decades ago, but even today, we are far from the complete decoding of these natural combinations. For example, comparison of the compositions of three PYO cocktails produced in 1997 (PYO97), 2000 (PYO2000) and 2014 (PYO2014) performed by Villarroel et al. (2017) (using next-generation sequencing, de-novo assembly of contigs into draft genomes based on tetranucleotide distance) allowed the authors to predict thirty and twenty-nine phage draft genomes in PYO97 and PYO2014, with thirteen and fifteen sharing high similarity to previously known phages, respectively. Eleven draft genomes were found to be common to both cocktails. One of these showed no similarity to publicly available phage genomes. Representatives of phages targeting *E. faecalis, E. faecium, E. coli, Proteus, P. aeruginosa* and *S. aureus* were found in both cocktails. Finally, the authors estimated larger overlap of the PYO2000 cocktail to PYO97 compared to PYO2014. Thus, the authors concluded that even though the cocktail composition is upgraded every six months, it had remained relatively stable over 17 years (Villarroel, et al., 2017).

Another article investigated the composition of the Coli-Proteus bacteriophages produced by Microgen, Russia (McCallin et al., 2013). Electron microscopy (EM) identified six phage types, with T7-like phages dominating numerically over T4-like phages. A metagenomic approach using taxonomic classification, reference mapping and de-novo assembly identified 18 distinct phage types, including seven genera of *Podoviridae*, two proposed genera of *Myoviridae*, and two genera of *Siphoviridae*. De-novo assembly yielded seven contigs greater than 30kb, including a 147 kb Myovirus genome and a 42 kb genome of a potentially new phage. Due to thorough reference mapping and de-novo assembly with the nucleotide reads from the Microgen cocktail, the authors were able to identify the presence of 17 different groups of phages. They analyze the full set of reads in order to gain insight into their relative proportions within the Microgen cocktail (anticipating no PCR amplification bias). This analysis identified T7-like phages as the most abundant ingredient of the Microgen phage cocktail, followed by T4-like phages, which was consistent with the EM observations. Nine further phage groups made important contributions to the cocktail that clustered, with the exception of two 140-kb and one 70-kb phage genomes and one with a genome size of 40kb. The observed contribution of phage ColProt2 to the mix was surprising. Despite the fact that it showed no homology over 75% of its length it appeared to represent (second to the T7-like phages), one of the most important contributors to the Coli-Proteus cocktail. It is important to state that bioinformatic analysis did not reveal undesired genes and a small human volunteer trial did not associate adverse effects with oral phage exposure (McCallin et al., 2013).

McCallin et al. (2018) also performed a comparison of the metagenomic composition of two Pyophage products from Georgia and Russia. The results revealed substantial differences in phage-types targeting *Escherichia, Enterococcus, Salmonella, Pseudomonas aeruginosa* and *Proteus*. The authors concluded that this suggested diverse technical strategies for composing phage cocktails against these bacteria.

Nevertheless, across all products, the preparations shared phage components targetting *S. aureus*, in that they contained closely-related phages of the genus *Kayvirus*, apart from the inclusion of a secondary *S. aureus Podovirus* in one Microgen cocktail. As well as reflecting differences in preparation protocols, the differences in phage composition in Pyo-bacteriophage batches from Georgia and Russia also reflects distinct pathogen epidemiology in these regions, as phage products are regularly updated for efficiency against newly emerging pathogen variants within a region. Comparison of Pyophage products from before USSR dissolution in 1991, with the Pyophage products from the two manufacturers operating independently since then, revealed that few phage-types remained as shared components. Nevertheless, one widely shared phage component between many commercial Eastern phage cocktails are phages belonging to the genus *Kayvirus* infecting *S. aureus*. Phages with near identical genome sequences were found in eight commercial phage products, suggesting a high level of genetic stability of members of this phage genus over decades of propagation. The maintenance of *Kayvirus* phages in all commercial products from commercial Eastern phage cocktails is also explained by the exceptionally broad host range of such phages, which achieves 60% to >95% coverage when tested against different collections of *S. aureus* clinical isolates. The Microgen Pyophage cocktail contains a second *S. aureus* phage type, belonging to the group of P68-like phages of the subfamily *Picovirinae*. The metagenomic analysis also revealed the presence of several probable prophage sequences but detected no genetic safety risks in terms of virulence factors or antibiotic resistance genes (McCallin et al., 2018).

Another research group (Zschach et al., 2018) performed the sequencing of the full Intesti cocktail (produced by the Eliava BioPreparations Ltd, Tbilisi, Georgia) which resulted in 440,392 reads with an average read length of 174.9 bp. De-novo assembly yielded 420 contigs ranging in size from 500 - 134,226 bp.and a total assembly size of 2,041 kb. Within the cocktail, 22 phage clusters were recovered by grouping using BLAST hit profiles; plus one novel *Proteus* phage was cluster identified. Overall, clusters ranged in size from 13.4 to 212 kb and were composed of between one and 56 contigs. 70 contigs, which together make up 217 kb of sequence or 10.6 % of the total assembly size, had no significant hits to NCBI's nr nucleotide database. They could therefore not be assigned to a phage cluster. Eighteen clusters were formed and compared to the known database. For the majority of the phage clusters there were several references with a comparable query coverage and percent identity. They were divided into two groups: Clusters with several highly similar references (query coverage and percentage identity >90%), and clusters with similar medium references (query coverage and percentage identity between 90% and 70%). The authors concluded that within 11 clusters with several highly similar references they identified close relatives of *Staphylococcus* phage Team1 (KC012913.1) and *Pseudomonas* phage PAK_P5 (KC862301.1) in the Intesti phage cocktail. The second group of clusters with a slightly lower but still apparent similarity to their references contained the parts that differed from their references, either because they were acquired from other phage types or because they are novel. The authors looked also at the abundances of the phage families by summing the reads. They found 41% *Myoviridae*, 16% *Podoviridae*, 29% *Siphoviridae* and 14% belonging to an unknown family. The authors suggested that their study in comparison with previous research done by McCallin et al. (2013; 2015) had its focus on bioinformatics and specifically sequence analysis tools. This approach, based on sequencing first and employing bioinformatics prior to further lab work, allows increased insight and leads to designing experiments with more efficiency. In general, they concluded that with careful interpretation, a great amount of information could be gained from using bioinformatic tools to investigate a phage cocktail (Zschach et al., 2018).

According to the acting Russian Pharmacopea (http://pharmacopoeia.ru/en/) for construction of the commercial preparations, the phages are isolated from natural sources - clinical material, wastewater, soil, etc. Usually, these natural associations/combinations of phages undergo passaging (sequential propagation) on the strains of homologous bacteria that are freshly isolated as well as industrial (domesticated). Highly active strains/clones of phages are selected that are lysing bacterial strains previously identified as phage-resistant. The replenishment of production phage strains/clones with various isolates from natural sources allows the initial phage resistance of pathogens to be overcome. The requirements with "mother strains" includes to their check for lytic activity: the phages should be strictly lytic, they should have a broad host range, they should be characterised to possess long stable lysis over 24-48 h duration (checked using Applemans method (1921) of titration in liquid), and they should have a high yield of progeny phage from a single cell. EM imaging is also desirable. However, phages that are produced from "mother strains" in a low titer may not be sufficiently concentrated be visible in EM analysis. During "passaging" of the mother strains of phages through a number of clinical isolates, it plausible that the addition of new phages will result in those already established to be overtaken, causing such phages to represent a lower proportion or vanish over time in subsequent preparations.

This fact was confirmed by McCallin et al. (2018) who made a puzzling observation during metagenomic analysis that only half of the bins detected in Pyo phage preparations were the major contributors to the overall metagenomic composition. Many phage genomes were detected with low coverage raising the question as to whether they are intended phage ingredients of the commercial phage preparation. If multiple phages are amplified in a batch culture, McCallin et al. (2018) drew an assumption that there is a possibility that a 'winner' phage is amplified more efficiently and predominates the lysate, with less-efficiently-replicating phages representing low the coverage phages. Regular enrichment of the bacterial collection with new clinical isolates allows adaptation of phages to pathogens circulating at certain times in particular geographical areas and thus extend their lytic activity. The adaptation cycles, which can be compared with the training of phages to new environments/hosts, are performed at least once per 6 months. These cycles are the driving forces for human-made phage co-evolution, in which some phages win, and the others lose.

## Conclusion

The major issue for acceptance of phage cocktail products in western countries is that their composition is not standard and stable. If we assume that in future we will build a phage preparation using the "cocktail" principle with only well-characterised sequenced phages, are we sure that their combinations will remain stable? Are we sure that these phages in this mixture will not compete with each other for domination? Actually how much do we know about phage-phage interaction in such phage mixtures? Of course, contemporary research is developing, and new technologies are emerging that allow us to synthesize phages, but once synthesized would these phages remain effective for a long period of time? Would it be necessary to adapt it using gene engineering approaches? How often would it be necessary to use "gene fixing," i.e. adaptation to improve the quality of the phage product? These are the questions that require proper answers. For future solutions, sometimes it is useful to look back, analyze the previous old fashioned experiences and find a rational answer there.

## References


Almeida, G.M.F. and Sundberg, L.R. (2020). The forgotten tale of Brazilian phage therapy. Lancet Infect Dis. pii: S1473-3099(20)30060-8. doi: 10.1016/S1473-3099(20)30060-8. [Epub ahead of print]

Appelmans, R. (1921). Le dosage du bacteriophage. C.R. Soc. Biol. 85, 1098–1099.

Gratia, A. (1936). Des relations numeriques entre bactéries lysogenes et particules de bactériophage. Annales de l'Institut Pasteur *57*:652–676.

Chanishvili, N. (2012). Phage therapy - History from Twort and d'Hérelle through Soviet experience to current approaches. Adv Virus Res 83, 3–40.

Chanishvili, N. (2016). Bacteriophages as therapeutic and prophylactic means: summary of the Soviet and post Soviet experiences. Curr Drug Deliv. 13(3), 309-323. DOI:10.2174/156720181303160520193946

Costa Cruz, J. (1924). Le traitement des dysenteries bacillaires par le bacteriophages. C.R. Soc. Biol. 91, 845.

Dublanchet, A. and Bourne, S. (2007). The epic of phage therapy. Can. J. Infect. Dis. Med. Microbiol, 18, 15–18.

D'Hérelle, F. (1935). Bacteriophage and phenomenon of recovery TSU Press, Tbilisi State University, Tbilisi, Georgia.

Hausler, T. (2008). Viruses vs. Superbugs; (MacMillan, New York).

Krasilnikov, I.V., Lysko, K.A., Otrashevskaya, E.V. and Lobastova, A.K. Bacteriophage based preparations: a brief survey of current state and future development.( 2011) *Siberian Medical Journal*, *26*(2/2), 33–36.

Leitner, L., Sybesma, W., Chanishvili, N., Goderdzishvili, M., Chkhotua, A., Ujmajuridze, A., Schneider, M.P., Sartori, A., Mehnert, U., Bachmann, L.M. and Kessler, T.M. (2017). Bacteriophages for treating urinary tract infections in patients undergoing transurethral resection of the prostate: a randomized, placebo-controlled, double-blind clinical trial. BCM Urology, 17(1), 90. doi: 10.1186/s12894-017-0283-6.

Leitner, L., Ujmajuridze, A., Chanishvili, N., Goderdzishvili, M., Chkonia, I., Rigvava, S., Chkhotua, A., Changashvili, G., Schneider, M.P., Liechti M.D., Mehnert, U., Bachmann, L.M., Sybesma, W., Kessler, T.M. (2020). Intravesical bacteriophages for treating urinary tract infections in patients undergoing transurethral resection of the prostate: A randomised, placebo-controlled, double-blind clinical trial. Lancet Infect Dis.

McCallin, S., Sarker, S.A., Barretto, C., Sultana, S., Berger, B., Huq, S., Krause, L., Bibiloni, R., Schmitt, B., Reuteler, G. and Brüssow, H. (2013). Safety analysis of a Russian phage cocktail: From metagenomic analysis to oral application in healthy human subjects. Virology 443,187–196.

McCallin, S., Sarker, S.A., Sultana, S., Oechslin, F. and Brüssow, H. (2018). Metagenome analysis of Russian and Georgian Pyophage cocktails and a placebo-controlled safety trial of single phage versus phage cocktail in healthy *Staphylococcus aureus* carriers. Environ Microbiol. 20(9), 3278-3293. doi: 10.1111/1462-2920.14310.

Summers, W.C. (1999). Felix d'Hérelle and the Origins of Molecular Biology Yale University Press: New Haven, CT.

Sulakvelidze, A., Alavidze, Z. and Morris, J.G. (2001). Bacteriophage therapy. Antimicrob. Agents Chemother. 45, 649–659.

Villarroel, J., Larsen, M.L., Kilstrup, M., and Nielsen, M., (2017) Metagenomic analysis of therapeutic PYO phage cocktails from 1997 to 2014, Viruses 9(11), pii: E328. doi: 10.3390/v9110328.

Zschach, H., Joensen, K.G., Lindhard, B., Lund, O., Goderdzishvili, M., Chkonia, I., Jgenti, G., Kvatadze, N., Alavidze, Z., Kutter, E.M., Hasman, H. and Larsen M.V. (2015). What can we learn from a metagenomic analysis of a Georgian bacteriophage cocktail? Viruses 7(12), 6570–6589. https://doi.org/10.3390/v7122958

Chapter 6

# Application of Bacteriophages in Human Therapy: Recent Advances at the George Eliava Institute

**Nina Chanishvili[1]*, Lia Nadareishvili[2], Elisabed Zaldastanishvili[1], Nana Balarjishvili[1] and Mzia Kutateladze[1]**

[1]George Eliava Institute of Bacteriophage, Microbiology & Virology, 3 Gotua Street, Tbilisi 0160, Georgia
[2]Eliava Phage Therapy Center, 3 Gotua Street, Tbilisi 0160, Georgia

*Corresponding author: nina.chanishvili@pha.ge

DOI: https://doi.org/10.21775/9781913652517.06

**Abstract**
This chapter describes a variety of recent compassionate-case phage therapy interventions performed using the expertise of the staff at the George Eliava Institute. The cases described include a variety of complex problematic infections involving various skin locations, those with cystic fibrosis, physiologically damaged diabetic feet, post-transplant conditions and urinary tract complications. The most remarkable developments are the positive findings following the execution of a small-scale human clinical trial, one which leads to the successful completion of a large-scale double-blind, randomised controlled human clinical trial aimed at eliminating problematic urinary tract infection in patients intended for transurethral resection of the prostate. The necessity for pre-adaption of existing phage cocktails to target a particular spectrum of infectious pathogens is also discussed.

## Introduction

The present George Eliava Institute of Bacteriophage, Microbiology and Virology is one of the oldest research organisations in Georgia. It was founded in 1923 due to an initiative of professor George Eliava, who by coincidence met Félix d'Hérelle in Paris during his stay at the Pasteur Institute in the early 1920s. By this time, Eliava already had some experience observing spontaneous lysis of bacteria, which he could not explain. Hearing within the walls of the Pasteur Institute about a strange person claiming the existence of living creatures eating bacteria, which could be used for the treatment of bacterial diseases, he immediately recalled his observations and became interested in d'Hérelle's experiments. Professor Emile Ru, who was a director of the Pasteur Institute at that time, allowed Eliava to reproduce d'Hérelle's findings, which were confirmed by the young Georgian scientist. This fact laid the foundation for their friendship. In the 1930s, George Eliava invited d'Hérelle to visit Georgia. This invitation was readily accepted, because at that time, d'Hérelle experienced an often critical attitude to his ideas while living in Paris. At that time an intense scientific debate about the nature of bacteriophages (phages) was ongoing, and many researchers were inclined to believe that these were not living things as d'Hérelle believed, but only enzymes secreted by bacteria themselves (Eaton and Bayne-Jones, 1934; Northrop, 1937; Northrop, 1938).

D'Hérelle stayed in Georgia for about two years, during which the two of them worked out a joint plan to create a world centre for bacteriophage research, which should have a scientific department, a production centre and a diagnostic centre, which would make it the most modern clinic at that time where patients could be treated with phages (Figure 1). Eliava enthusiastically began to implement this plan and secured from the government the allocation of 17 hectares of land for the construction of the centre. Unfortunately, by the time the first buildings were built, Eliava was already arrested and executed in 1937 on false charges as an "Enemy of the People". When d'Hérelle left Georgia in 1934, he had been intending to return to settle permanently, but ultimately, he never did return. The grand plan developed by Eliava and d'Hérelle together was only partially implemented.

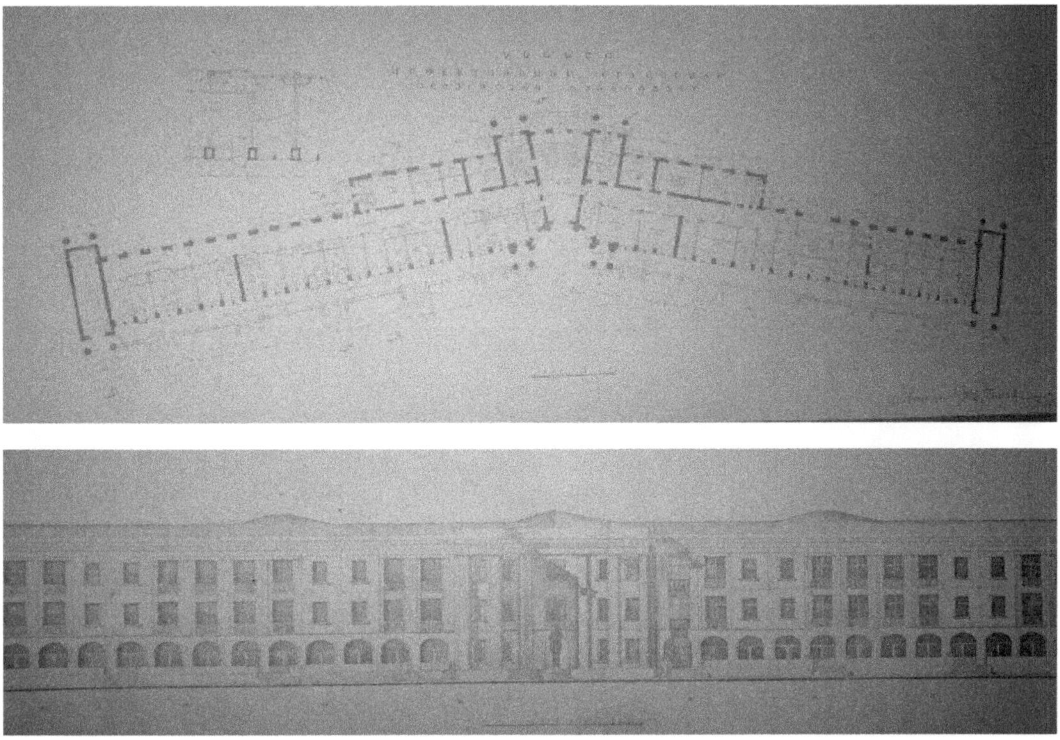

**Figure 1.** Plan and front elevation of the Infectious Disease Hospital of the All Union Scientific-Research Institute of Bacteriophage (from the blueprints of the institute, dated 1934).

Several buildings from the grand plan have been built including research and diagnostic laboratories, manufacturing premises, a hostel for the personnel, utility rooms, storage houses, garage, workshops, vivarium and administrative buildings, and even a residence house for d"Herelle and Eliava. However, the Infectious Disease Hospital element, which was envisaged to provide 500-600 beds and to be equipped according to the latest medical technology, was not constructed (Figure 1).

Only after many years in 2012 did the Eliava Institute manage to establish a small outpatient clinic (Eliava Phage Therapy Center), a diagnostic (Eliava Diagnostic Center), a phage production centre (Eliava BioPreparations Ltd) and an authorised pharmacy (Eliava Pharmacy) (Figure 2). Thus, the dream of d'Hérelle and Eliava was indeed realised - although on a much smaller scale. With the above entities, the Eliava Institute works to achieve the main goal - the treatment of patients with chronic bacterial infections, who in many cases arrive in Georgia in a desperate condition after long term unsuccessful antibiotic therapies. The Eliava Phage Therapy Center (EPTC) provides services in surgery, internal medicine, urology, gynaecology, paediatrics, dermatology and for ear, nose and throat (ENT) infections.

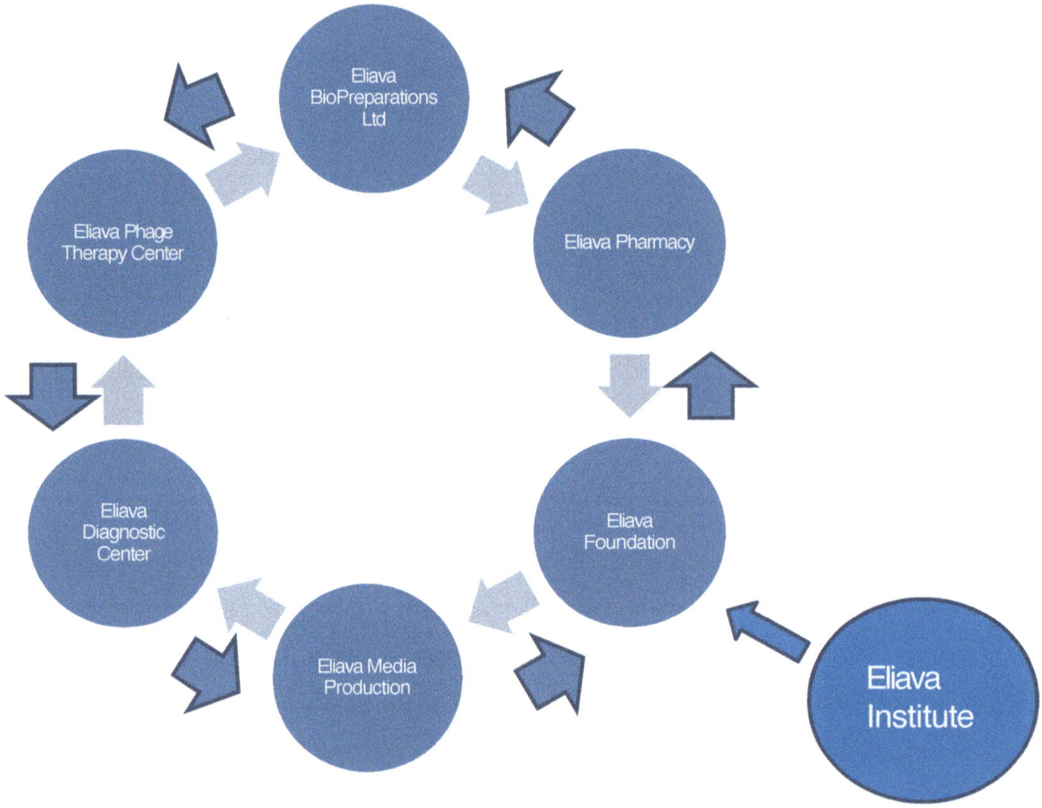


**Figure 2.** Scheme of the Eliava Hub including the Eliava Bacteriophage Institute, Eliava Foundation and the spin-off companies.

## Statistics

From 2012 to 2019, the Eliava Phage Therapy International Center has received 10,160 patients. Over 1500 patients were from foreign countries in Europe, Asia,

Australia, North and South America and Africa (altogether 71 countries). Sometimes, typically upon receiving a request from patients who cannot travel to Georgia, the personalised and commercial phage preparations are sent to them. Frequently occurring diagnoses treated with phages are summarised in Table 1. Recently, an analysis was performed on infections treated at the EPTC during the years 2018 and 2019. During that period, 1132 patients presented, 37% of whom were recurrent patients, who had previously applied more than once. Altogether, 3605 bacterial strains representing 88 species were identified - and among them, Gram negative species predominated (Figure 3).

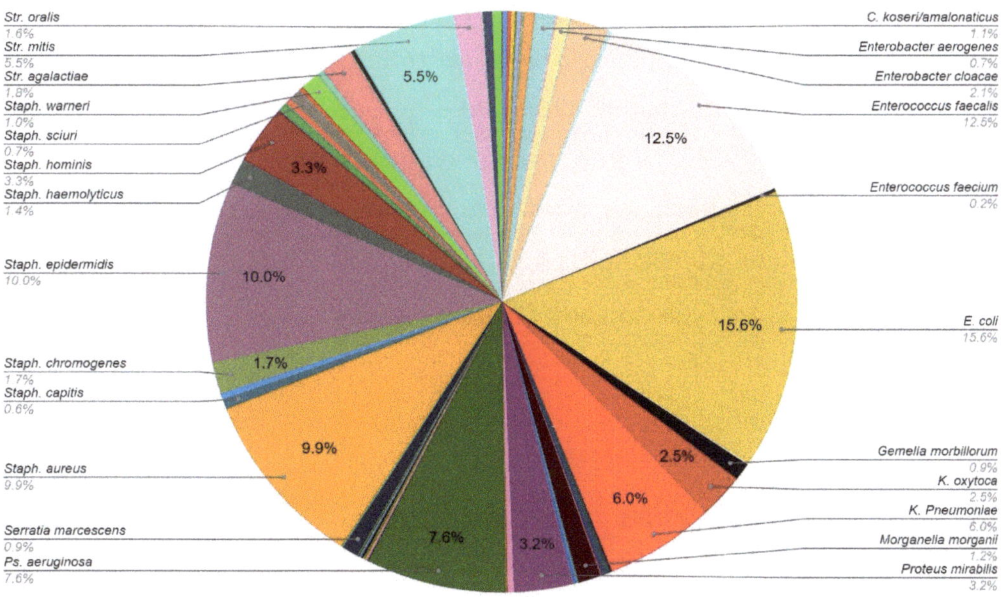


**Figure 3**. Diversity of clinical strains isolated from bacterial infections presenting for phage therapy in 2018 and 2019. A total of 3605 strains representing 88 different species are grouped in colour according to prevalence.

Table 1. Frequently occurring conditions treated with phage therapy

| Type of diseases | Diagnosis |
|---|---|
| Urological and gynecological diseases | Prostatitis |
| | Uretritis |
| | Vaginistis |
| | Cystitis, other inflammatory diseases of urinary tract |
| Surgical infections | Chronic wounds |
| | Diabetic foot |
| | Prosthetic associated infections |
| Internal medicine, ENT (ear, nose, throat), pediatrics | Gastrointestinal tract diseases: antibiotic associated diarrhea, irritable bowel syndrome, small intestine overgrowth syndrome and infectious diarrhoea |
| | Respiratory system diseases: sinusitis, otitis, tonsillitis, bronchitis, bronchiectasis, pneumonia |
| | Cystic fibrosis |
| | Skin and soft tissue diseases |

Bacterial pathogens frequently treated by the EPTC include: *Staphylococcus* spp., *Escherichia coli, Pseudomonas aeruginous, Enterobacter* spp., *Klebsiella* spp., *Proteus* spp., *Klebsiella* spp. etc. Many of these infectious pathogens are multidrug resistant and are of nosocomial origin.

Treatment of a typical patient begins with a study of the medical history based on a detailed description of current status and previous treatments. Diagnosis is based on blood, biochemical, bacteriological and other with instrumental based investigations. After identification of the bacterial causative agent of the infection, the organism is tested for susceptibility to phages. Figure 4 shows the distribution of the phage-resistant strains. Those that were not lysed by any of six therapeutic phage preparations were designated as "resistant", while the strains that responded to the above phages with intermediate or complete lysis were designated as sensitive. The lowest percent of resistant strains was detected for *Staphylococcus aureus* strains (9.8%), while *Streptococcus agalactae* strains showed the highest resistance rate (89.06%) to the commercial phage preparations.

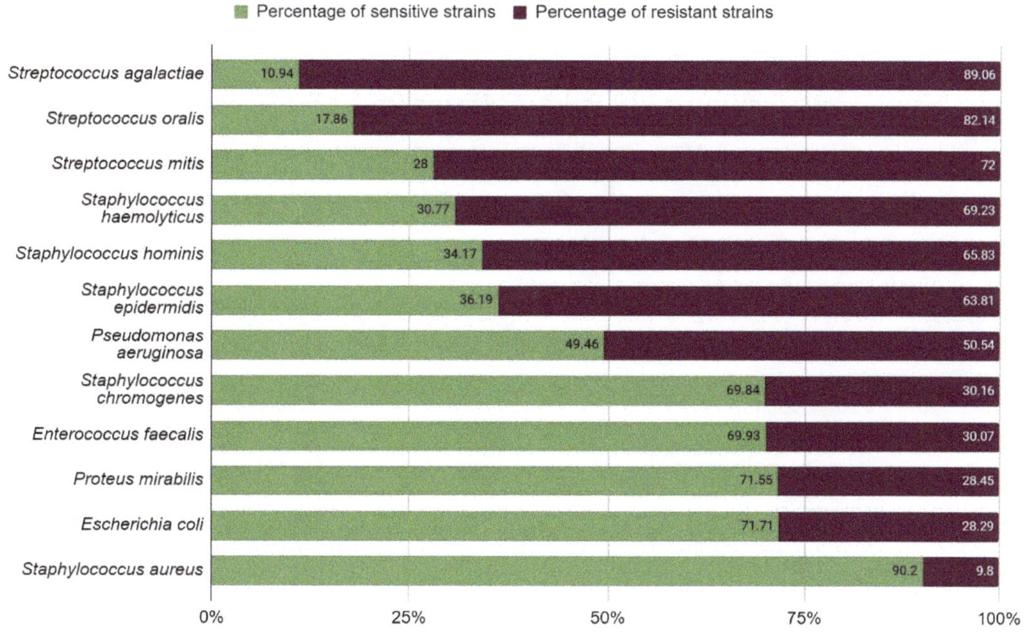


**Figure 4.** Distribution of phage-resistant strains among the bacterial species represented by at least 50 strains over the two-year period (2018 and 2019). The fraction of phage-sensitive strains is shown in green with phage-resistant strains in purple. The numbers of the strains used to calculate the percentages were: *Strep. agalactiae* - 64, *Strep. oralis* - 56, *Strep. mitis* - 200, *Staph. haemolyticus* - 52, *Staph. hominis* - 120, *Staph. epidermidis* - 362, *Pseud. aeruginosa* - 275, *Staph. chromogenes* - 63, *Ent. faecalis* - 449, *Prot. mirabilis* - 116, *E. coli* - 562, *Staph. aureus* - 357

If the commercial preparations are weakly efficient or not efficient at all, a personalised treatment is designed. Often treatment strategy is complex and may include spasmolytics (muscle relaxants), probiotics, antibiotics and other agents. The standard duration of phage treatment is twenty days, which may be repeated after a fourteen-day break. For example, in cases of irritable bowl syndrome (IBS) patients with bowel infections, complete eradication of the pathogenic bacterium was observed in 85% of cases (with rapid improvement in 75%) and no effect observed only in 15%. Among cystic fibrosis patients, outcomes of phage therapy objectively were expressed in terms of improvement of patient's appetite, weight gain, enhancement of patient's general and psycho-emotional condition, with the frequency of exacerbation of the disease decreased. The oxygen saturation of CF patients increased up to 94% - 95%. The main advantage of phage treatment was expressed in minimising the use of antibiotics in the complex therapy and prolongation of remission periods (Kutateladze, 2015). Among the patients receiving personalised/custom phages, the infection recurrence rate is rather low, typically between 10% and 25% as shown in Figure 5. This data demonstrates the precise efficacy of the customised phage therapeutic approach.

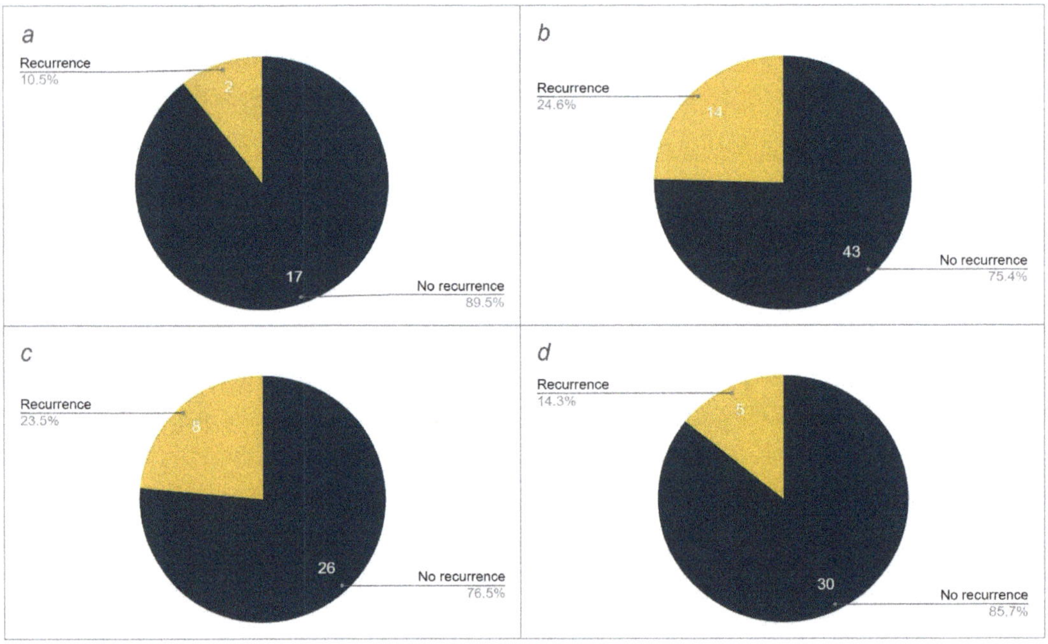


**Figure 5.** Infection recurrence rates in patients who received customised phage preparations for infections caused by (a) *Klebsiella oxytoca,* (b) *Klebsiella pneumonia* ssp. *pneumoniae* (c), *Escherichia coli* and (d) *Pseudomonas aeruginosa*. The numbers (and percentages) of patients with a recurring infection after therapy are shown in yellow. Those with no report of recurring infection after therapy are shown in blue.

In the remainder of this article, we will describe some recent clinical cases treated with phages.

**Skin diseases**

Among many successful cases, it is interesting to draw attention to a particular case with Netherton Syndrome (NS), a rare genetic disease known to be an autosomal recessive disorder. Manifestations of the disease are congenital ichthyosiform erythroderma, trichorrhexis invaginata, and atopic diathesis, which is often associated with bacterial infections. The publication by Zhvania et al., (2017) describes a case of a 16-year-old male with all of the typical manifestations of NS, including atopic diathesis and ongoing serious long-lasting staphylococcal infections for the previous four years at least, and allergy to multiple antibiotic groups including beta-lactams, aminoglycosides, macrolides, sulfonamides, cephalosporins and also allergy to various kinds of food. Treatment of the patient was complicated by the existence of antibiotic resistance of the infectious bacteria. This patient had been hospitalised every two months due to infectious complications. After testing susceptibility of *S. aureus* to a number of commercial phage preparations, the two mentioned above were selected for treatment, namely *Staphylococcus* and *Pyo-phage*. Both preparations were used topically as well as orally. Because of the patient's allergic predisposition, an allergy test with the phage preparations was first performed on a 1 $cm^2$ area and observed for 2 hours. After confirmation that no adverse reactions had developed, the following daily procedures were carried out. First, the patient's limbs were wrapped with sterile gauze (Figure 6) and liquid *Pyo-phage* was sprayed on with a sterile syringe to soak the gauze. After 20 min, the gauze was removed and a thin layer of *Staphylococcus* phage cream was applied, which was specifically prepared for this purpose by the Eliava Authorized Compounding Pharmacy using a hypoallergenic cream base to which the patient did not have an allergic reaction. The regimen for oral treatment included 10 mL of *Staphylococcus* phage preparation and 10 mL of *Pyo-phage* administered once a day. While the complete duration of both topical and oral treatments was 20 days, an obvious improvement was observed by the $7^{th}$ day. The infiltrated, hyperemic areas were reduced, the thickness of the yellowish film layer decreased, and mobility improved in the joints with areas of normal skin reappearing. No allergic reaction to the phage preparations were recorded. After the first course, the patient took a two weeks pause and then continued treatment for another 20 days with the same phage preparations. After three months, the repeated bacteriological testing showed the presence of *Staphylococcus* and *Streptococcus* species, which appeared to be resistant to both phage preparations initially used. Therefore a third commercial preparation, namely *FERSISI* (composed of phages active against *Staphylococcus aureus* and *epidermidis,* and *Streptococcus pyogenes, sanguis, salivarius* and *agalactiae),* which on testing had been observed to efficiently lyse the persisting bacteria, was selected for further treatment. During the next three month period, the patient took this preparation for two-week periods, with a two week breaks inbetween treatment courses. At a six month follow-up, the patient's condition remained stable with a visually apparent improvement of his skin condition. Since therapy, he had not required hospitalisation for generalised infection. This was the first documented case of phage therapy with Netherton Syndrome.

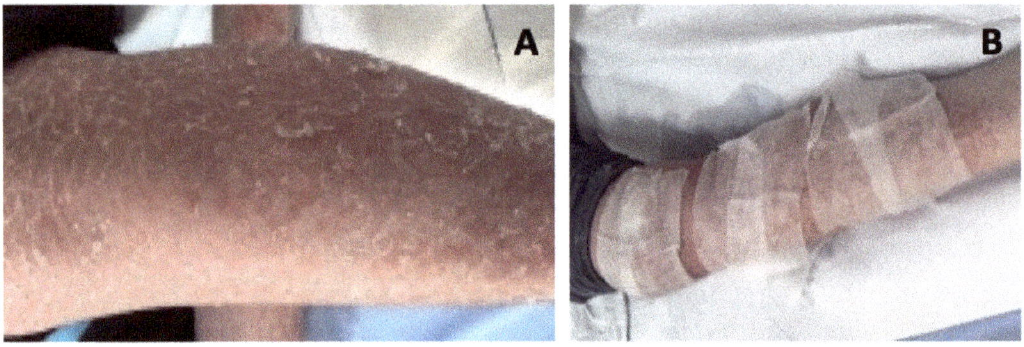


**Figure 6.** Chronic staphylococcal skin infection in a case of Netherton Syndrome (A) treated by local application of liquid phage preparations on gauze (B).

## Pulmonary diseases

Among pulmonary cases, one example of the treatment of bronchiectasis is notable. In September 2017, a 43-year-old male patient (designated V.S.) who was diagnosed with bronchiectasis in early childhood, applied to the EPTC. He had frequent respiratory tract infections caused by *Pseudomonas aeruginosa* and had undergone long-term ineffective antibiotic treatments. At the time of application to the EPTC, he complained of shortness of breath, a productive cough with yellow/greenish sputum, fatigue on exertion, as well as flatulence and frequent diarrhoea. Objective examination showed no pathological changes in skin and mucosa, with other health parameters being normal. However, pulmonary auscultation revealed multiple dry and wet crackles and wheezing on both sides, with a respiratory rate (RR) of 22 to 24 breaths per minute; His peripheral capillary oxygen saturation level ($SpO_2$) was 93%. The abdomen was soft, liver and spleen were not palpable. Serological analysis of blood samples showed increased levels of anti-staphylococcal and anti-streptococcal antibodies and antibodies to staphylococcal α toxin. Bacteriological analysis of stools demonstrated Grade IV bacterial imbalance, which was marked by decreased numbers of *Escherichia coli, Lactobacillus* spp., *Bifidobacterium* spp. and *Enterococcus* spp with a domination of haemolytic *Escherichia coli*. The latter organism appeared to be sensitive to all standard commercial phages. Bacteriological analysis of sputum revealed *Pseudomonas aeruginosa* (at $5x10^5$ CFU/mL), which demonstrated low sensitivity to standard phages. A recommendation to perform personalised treatment with the phages against *P. aeruginosa* was given.

The first phase of treatment started in September 2017 and was performed by using oral doses of *Intesti-* and *Pyo-phage* at 10 mL, once per day, and using 2 mL of the same preparation for inhalation through a nebuliser once per day. The treatment continued for 20 days, followed by a 14-day pause. This treatment regimen was repeated three times. After the first course, the patient applied to his pulmonologist in his home country, who performed computer tomography and reported improvements in bronchiectasis status. The second time the patient applied to the EPTC was in February 2018. His sputum sample showed the presence of *P. aeruginosa* at $10^5$ CFU/mL. A personalised phage preparation against *P. aeruginosa* was developed and sent out to the patient in his home country, where he continued applying phages according to the following regimen: 10 mL phage orally twice per day and 2 mL of phage for inhalation for 20 days with 14 days of pause. The patient again applied to the EPTC again in January 2019. His sputum sample showed the presence of a new strain of *P.*

*aeruginosa*, which was not sensitive to the previously personalised phage batch. Because of this, a new personalised phage active against the new strain of *P. aeruginisa* was prepared and sent to the patient. The third phase of treatment was performed according to the same regimen as described above.

It is noteworthy that during the phage therapy courses, and afterwards, the patient did not take antibiotics. During the periods of 5-11 months between the phage therapy courses, the patient did not experience any exacerbations and lived a normal life. A total cure of chronic diseases such as bronchiectasis is not expected. It is well known that mucosal surfaces are the main entry point for pathogens and the principal sites of defence against infection. Both bacteria and phage are associated with this mucus, which causes difficulties in treatment. Recent research has highlighted the diversity of phage populations associated with respiratory bacterial communities, a potentially important component of the airway microbiome due to their ability to profoundly influence bacterial behaviour (Hargreaves et al., 2014; Ganeshan and Hosseinidoust, 2019). In this example, we see a positive role of phage therapy in the improvement of a bronchiectasis status.

Another case of a 17-year-old female patient (designated S.J.) involved phage treatment of cystic fibrosis. She was diagnosed with the homozygous ΔF508 mutation in the cystic fibrosis transmembrane conductance regulator gene (CFTR). The patient applied to the EPTC in March 2016. Her history of respiratory infections began at the age of 11 with an initial infection caused by *Burkholderia cepacia*, and then since the age of 12 an infection caused by *Achromobacter xylosoxidans*, which was unsuccessfully treated with many courses of antibiotics. She reported a persistent cough, increased dyspnea, and moderate sputum production. On examination, she appeared dyspneic, and inspiratory crackles were heard in both upper lobes. Oxygen saturation level was $SpO_2$ 95% while breathing ambient air. A specimen of sputum yielded a multi-drug-resistant *Achromobacter,* with intermediate sensitivity only to piperacillin. A personalised phage reparation active against *Achromobacter* was developed and prescribed to the patient 10 mL as a daily oral dose and 2 mL ($3x10^8$ PFU/mL) diluted using 2-3 mL of sterile 0.9% NaCl solution with half of this (slightly over 2 mL) inhaled twice per day. The treatment was performed for 20 days. This course was repeated a total of four times; at the first, third, sixth, and twelfth months after the start of initial treatment.

Following the above phage therapy, the patient's subjective conditions improved significantly with dyspnea resolved, and cough reduced. Her lung function, FEV1 (Forced Expiratory Volume, a measurement taken from a pulmonary function test) increased from an initial 1.83 L (54%) to 1.88 L (62%) after three months of treatment. The patient underwent her normally scheduled intravenous antibiotic therapy for six months after phage therapy with piperacillin/ tazobactam, followed by a final course of *Achromobacte*r phages after which the lung function further improved to (3.33 L) 84%.

Another case relating to phage therapy to a Cystic Fibrosis (CF) infection, was a 44 year old male (designated L.M.) who applied to the EPTC in January 2017. He had been diagnosed with CF, gene-type Δ F508, and was suffering from persistent *P. aeruginisa* infections from his early teenage years. He had undergone several courses of antibiotic treatment without any effect. His main complaints were a productive cough and fatigue on exertion. In addition to CF, he had also been diagnosed with insulin-dependent diabetes (at the age of 2) and exocrine pancreatic insufficiency. His examination showed a respiratory rate (RR) of 26 to 28, a peripheral capillary oxygen saturation level ($SpO_2$) of 95-97%, and a pulmo-auscultation demonstrated multiple dry and wet crackles and wheezing on both sides. His pharyngeal mucosa was hyperaemic and slightly swollen. Blood diagnostics showed increased levels of antibodies against *Staphylococcus* and staphylococcal α toxins. Bacteriological analysis of a sputum sample revealed infection by *Pseudomonas aeruginosa* ($5x10^5$ CFU/ mL), which was sensitive to the standard commercial preparations *Intesti-phage* and *Pyo-phage*. The patient was recommended to use these preparations according to the following regimen: a mixture of *Intesti-* and *Pyo-phage* 10 mL each as an oral dose twice per day and 2 mL of *Pyo-phage* for inhalations. The recommended duration of phage therapy was 20 days, with 14 days of pause. In addition to phage therapy, the patient was prescribed oral acetylcysteine (600 mg) once per day and Medrol (4 mg) once per day for ten days. Subjective and objective improvements were obvious after the first course of treatment.

Typically, patients spend 2-3 weeks in Georgia after which monitoring is continued at a distance after they return home. Continued analyses of chronic patients with lung diseases are performed regularly every 2-3 months. If there are problems, the patients send their sputum samples for bacteriological analysis and testing for phage susceptibility, after which new recommendations are issued, and if necessary, custom phages are produced and sent out with corresponding instructions. The main achievement of phage therapy in such chronic cases is that the patients have long remission periods and do not require antibiotic treatment.

**Diabetic foot**

Among the complications of diabetes mellitus such as cardiovascular disease, neuropathy, nephropathy, retinopathy, foot damage is rather common. Foot problems most often are associated with nerve damage resulting in skin changes. Typically, the skin becomes very dry, easily peels and cracks. Calluses occur more frequently and build up faster, get very thick, break down, and turn into ulcers. Poor circulation makes feet susceptible to infections that are slow to heal. Bacterial infections of diabetic foot ulcers (DFU) are a major well-being issue for these patients, which complicates ulcer treatment. Additionally, osteomyelitis in an ulcerated foot substantially challenges effective treatment and antibiotic resistance of the infecting bacterium complicates the treatment. Failure of standard treatments with these conditions generally leads to amputation of the foot.

A 69-year-old male patient (designated T.H.J.) with Type 2 diabetes mellitus applied to the EPTC in July 2017, with a diabetic foot ulcer that had developed several months prior on the lower side of the left foot near the first toe (Figure 7). The wound was 5 cm in diameter, which was characterised by purulent discharge. By the time of his application to the EPTC, the patient had previously undergone a long-term unsuccessful treatment with antibiotics. The X-ray examination of the foot revealed degenerative changes in the bones.

Bacteriological analysis of the discharge showed a mixed infection caused by *Burkholderia cepacia* (abundant growth), *Staphylococcus aureus* and *Enterococcus faecalis.* Isolates of the latter two pathogens were shown to be sensitive to all standard phage preparations.

The patient was prescribed with oral *Staphylo-phage* (20 mL) and oral *Intesti-phage* (20 mL), each to be administered every alternate day, while 10 mL of *Intesti-phage* was applied locally every day on sterile gauze followed by bandaging. The treatment was performed for 20 days, with a two week break. After one month, a granulation process (formation of new connective tissue and microscopic blood vessels) became obvious. The therapy was repeated three times after which the wound was completely healed (Figure 7). Interestingly, this result was achieved without the use of *Burkholderia cepacia* phages.

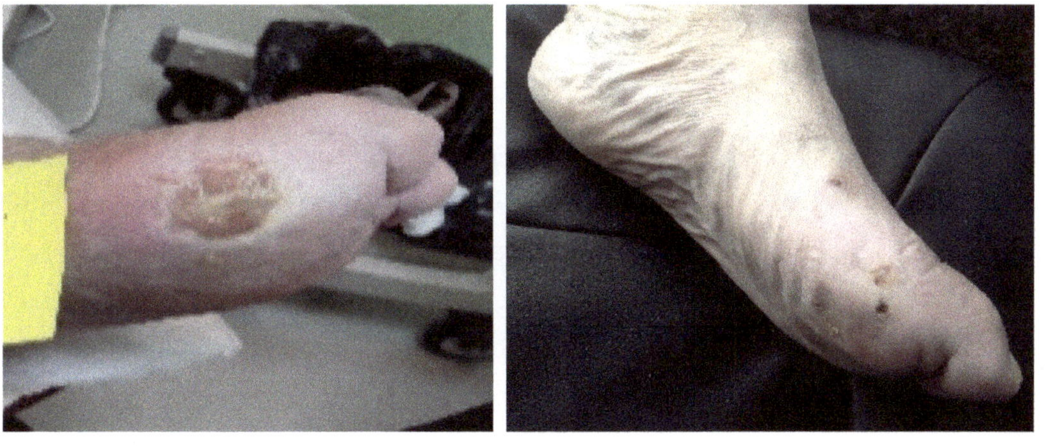

**Figure 7.** Left: Infected diabetic foot ulcer before treatment with *Staphylo-phage* and *Intesti-phage*; Right: The wound after three months of phage therapy.

The experience of the EPTC was also shared by foreign doctors who used the phage preparations produced by the Eliava BioPreparations Ltd. (a spin-off organisation of the Eliava Institute). For example, Fish et al., (2016) reported on a continuous case series of patients in assessing the efficacy of phage therapy for treating infected, poorly-vascularised toe ulcers with exposed bone. These were compassionate-use diabetic cases where the toe ulcers, bone and soft tissue were culture-proven to be infected with *Staphylococcus aureus*. In all cases, there was a poor response to conventional antibiotic therapy and the only remaining option in each case was toe amputation. Exposed portions of the infected phalanges were removed in three cases and left in place in two cases. The phage used was the above mentioned staphylococcal monoclonal phage Sb-1, which lyses almost all *Staphylococcus aureus*, including most methicillin-resistant (MRSA) strains (Kvachadze et al., 2011; Sergueev et al., 2019). The phage solution was applied topically to the ulcerations once weekly, following standard good wound care. The amount of phage solution applied varied from 0.1 to 0.5 mL depending on the volume and area of the ulceration. All infections responded to the phage applications and with debridement (removal) of infected bone material, the ulcers healed in an average of seven weeks. One ulcer, where vascularity was extremely poor, and bone was not removed to preserve hallux function, required 18 weeks of treatment. One additional case had presented as a micro-clot induced gangrene following vascular stenting. Here, following the phage therapy, the toe was salvaged and healed in seven weeks without complications (Fish et al., 2016).

In another study Fish et al., (2018a) reported on a 63 year-old type-2 diabetic female with distal phalangeal osteomyelitis with underlying diabetic neuropathy, COPD, hypertension, hyperlipidemia, anxiety disorder, and depression, who had developed an infected ulcer that over a period of three months did not respond to topically-applied over-the-counter antibiotic ointment with normal hygiene. She was morbidly obese and her legs demonstrated 2 mm pitting oedema. Orthopedically, she had a contraction deformity of all lesser toes, without bunion deformity. X-rays of the foot revealed osteolysis of the distal phalanx of the right second toe, reading as positive for osteomyelitis. On presentation, she demonstrated an edematous, erythematous, contracted right second mallet toe with ulceration at the distal tip measuring 0.6 x 0.9 x 0.3 cm, probing to the bone. Microbiological culturing from the infected tissue revealed a *Staphylococcus aureus*, sensitive to all antibiotics except for penicillin. As the patient refused to undergo the conventional treatment options, which would include excision of the distal phalanx along with a standard follow-up six-week intravenous antibiotic course, phage therapy was explored. In contrast with the previous cases, the phage was administered into the soft tissue surrounding the distal phalanx, inserted directly into the bone as much as possible. The treatment was supplemented with a toe crest and standard good wound care. The phage treatment consisted of 0.7 mL of the staphylococcal phage once weekly for seven weeks, to a total of 4.9 mL. Serial X-rays demonstrated the re-ossification of the distal phalanx over time. The erythema and oedema slowly decreased over the period of treatment and continued to decrease after the phage injection treatment was discontinued. Following completion of the phage therapy, the patient returned for observations over the following two years. While new toe ulcers had developed, these were resolved by using a simple toe crest to stop the pressure. The medical doctor authors suggested that adding phage to the standard care of diabetic foot ulcers with osteomyelitis can help to heal diabetic wounds more quickly and effectively while reducing the harm associated with antibiotic therapy (Fish et al., 2018a), recommending further clinical trials to prove the benefits of phage therapy for the treatment of diabetic foot ulcers (Fish et al., 2018b).

## Surgery

A 44-year-old Georgian male (designated V.J.) presented to the EPTC with a hospital-acquired multi-drug resistant *Pseudomonas aeruginosa* infection following liver transplant surgery, which had been performed outside the country. The patient had also had an ileostomy with drainages at the pleural cavity (Figure 8) and biliary duct (Figure 9) and also signs of kidney failure were emerging. The infection had spread into the biliary duct, pleural cavity and was present as a surgical non-healing wound running along the *linea alba* (Figure 10). Prior to application to the EPTC, the patient had been treated with a number of intravenous (IV) antibiotics (broad-spectrum carbapenems: meropenem, imipenem, etc.) without improvement. The patient was in a critical condition on the presentation for phage therapy, where the aim was to eradicate the infection and remove inserted drainages, despite the threat of systemic disease such as sepsis. The *P. aeruginosa* strain isolated from the wound showed sensitivity to both *Pyo-* and *Intesti-phage*, which were chosen for treatment. They were used alternately, with *Pyo-phage* on one day and *Intesti-phage* on the next with both introduced via the existing drainage tubes (Figures

5 and 6) directly into the common biliary duct where 5 mL was administered and the pleura where 2 mL was administered. The surface wound sites were also sprayed with 5 mL of each of these phage preparations at each site and after spraying, the wound was covered with a bandage soaked with the phages and left for 18-24 hours. The patient was also given 10 mL oral doses of *Pyo-phage* each morning and *Intesti-phage* each evening and a rectal dose of *Intesti-phage* 20 mL diluted with 20 mL of saline solution. The treatment continued for 20 days. After this, the pleural cavity was cleared from infection and the drainage duct was removed from that location. However, the other wound site remained infected with *P. aeruginosa*, which showed resistance to the commercial phage preparations. This problem required the production of a custom phage preparation, a process typically requiring 8-12 weeks. During this time, the 2nd phase of phage therapy treatment, which included the same regimen as described above was performed. During this period also, imipenem was administered intravenously to help combat the phage resistant strain. A 3rd phase of phage therapy involved the custom phage being administered into the biliary drainage duct (2 mL) and also being sprayed onto the wound surface (5 mL). Simultaneously the patient was getting rectal doses of *Intesti-phage* as described above. The entire *Pseudmonas* infection was successfully eradicated after six months of treatment.

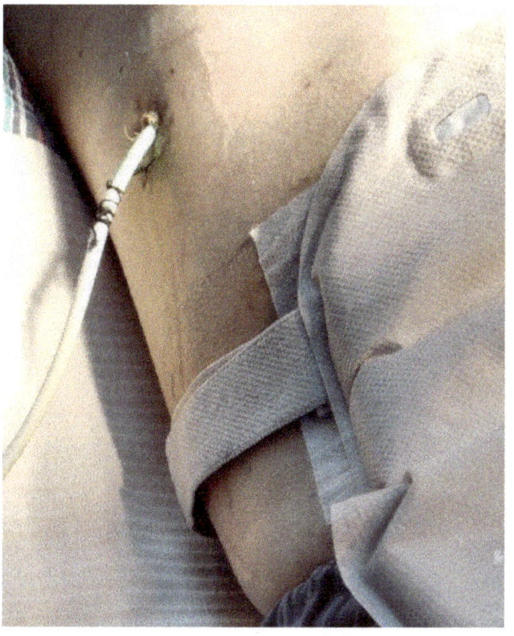

**Figure 8**. Pleural cavity drainage duct, which was also used for administration of phages to eliminate the *Pseudomonas* infection in a patient following liver transplant and ileostomy surgery.

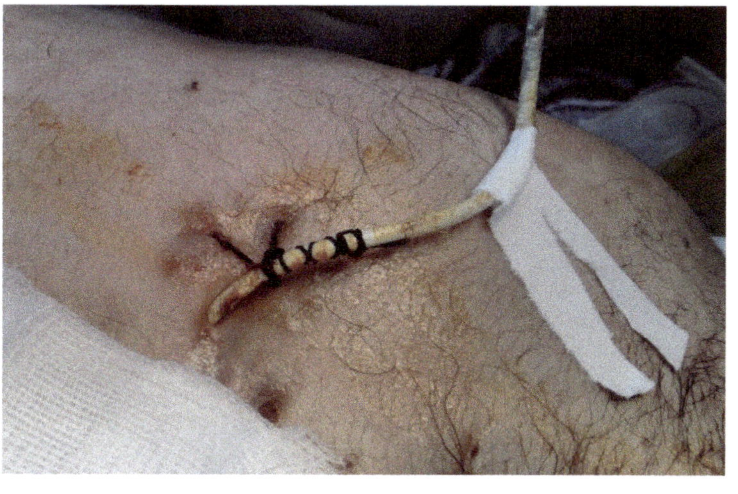

**Figure 9**. Biliary drainage duct, which was also used for administration of phages to eliminate the *Pseudomonas* infection in a patient following liver transplant and ileostomy surgery.

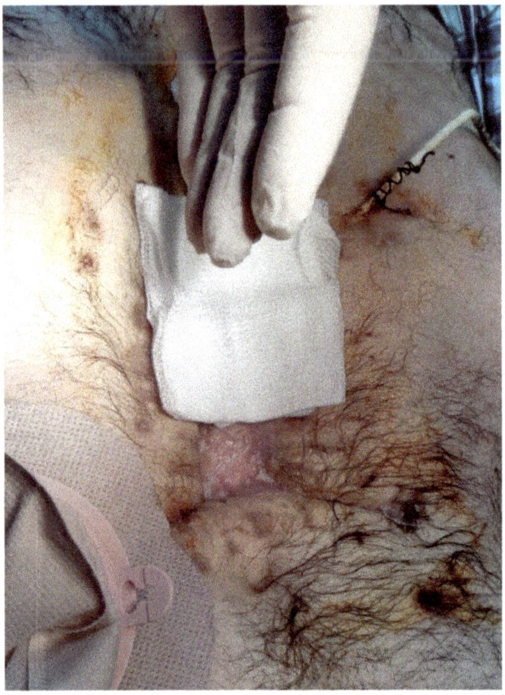

**Figure 10**. Phage therapy of wound running along the *linea alba*.

A 67-year-old male patient (designated A.G.) with laryngeal carcinoma required a skin graft for reconstructive surgery following a tracheostomy. The source of the graft had become infected with the nosocomial *P. aeruginosa*. Multiple antibiotic therapies had been conducted, including intravenous and oral antibiotics. The strain of *P. aeruginosa* showed sensitivity to *Pyo*- and *Intesti-phage*, and accordingly, each was orally administered (10 mL) alternately every second day. A topical application was also conducted with phage soaked gauze. The results of treatment after the first transplantation are illustrated in Figure 11.

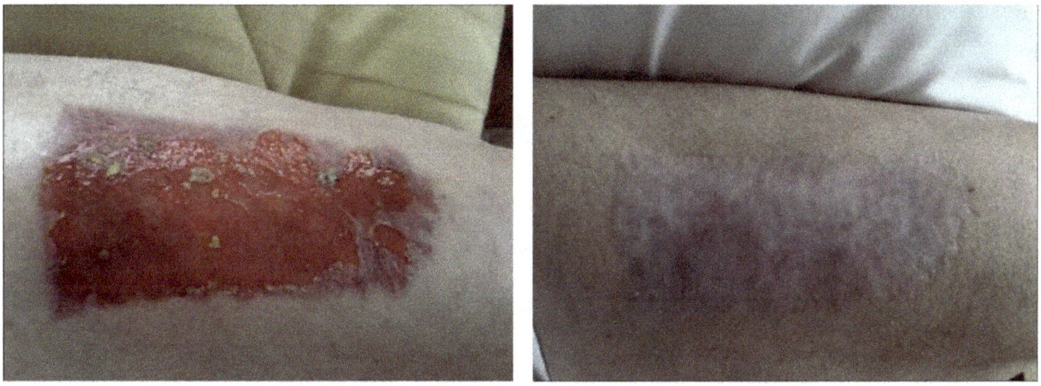

**Figure 11.** Left: Initial status of the *Pseudomonas aeruginosa*-infected skin graft wound; Right: Status of the wound after 12 weeks of phage therapy.

The same patient underwent a second skin graft, which also resulted in an infection. In this case, the bacteriological analysis revealed the presence of *S. aureus* and also *Serratia marcescens.* Initial treatment was performed using *Pyo-phage* active against *Staphylococcus* spp. While no phages against *Serratia* were present nor available, a positive outcome occurred by the eighth week of therapy as shown in Figure 12.

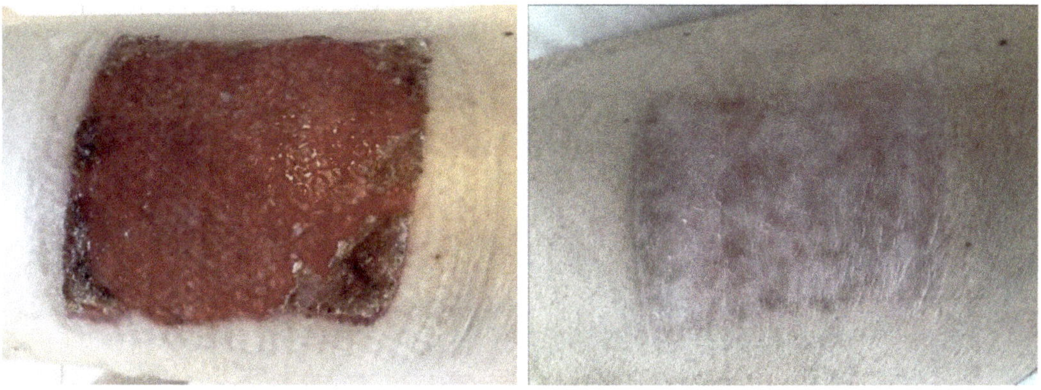

**Figure 12.** Left: Initial status of the *S. aureus* and *Serratia marcescens*-infected skin graft wound; B. Status of the wound after eight weeks of phage therapy.

A 42 year old male (designated P.H.) presented with an infected fistula that had formed after hip replacement surgery some years prior. The fistula had developed on the scar area and the wound was characterised by purulent discharge. The patient had previously been treated with antibiotics to eliminate the infection, but without any success. By the time of arrival to the EPTC, the fistula was noted on the lateral surface of the upper third of the right thigh, with measurements 2.5 x 1.0 cm (Figure 13A). Bacteriological analysis confirmed infection by *S. aureus*, which was susceptible to standard commercial phages. The *Staphylococcus* phages were applied via instillations directly into the fistula (Figure 13B). This application was performed once per day for 20 days. Simultaneously, the patient was given 10 mL oral doses of each of *Pyo-, Intesti-* and *Staphylo- phage* each day. Duration of treatment was 20 days after which the fistula had healed with the infection eliminated (Figure 13C).

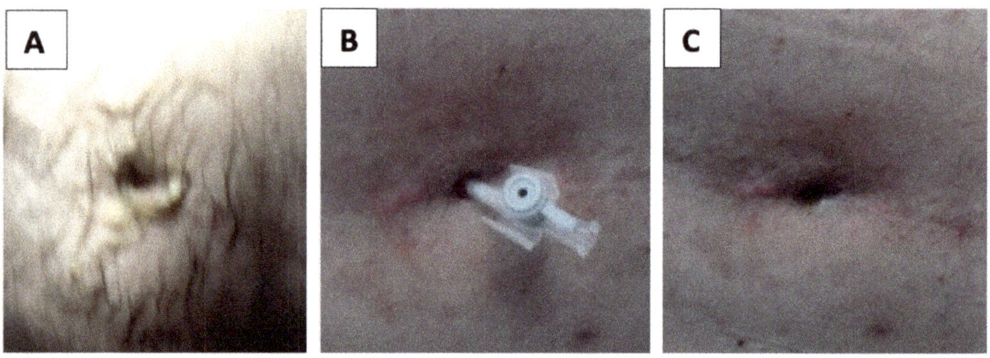


**Figure 13. A.** The initial status of the *S. aureus* infected fistula; **B.** Administration of staphylococcal phages directly into the fistula; **C.** Status of the fistula following 20 days of local and oral phage treatment.

## Phage therapy in implant-related infection

Biofilm formation on orthopaedic implants is a common problem. Accordingly, pre-treatment of orthopaedic implants with phages before surgery is considered worthy of investigation. An experimental study performed by the Georgian scientists in collaboration with colleagues in Turkey (Yilmaz et al., 2013) demonstrated a preventive effect of phage treatment on biofilm development on the surface of orthopaedic implants. The demonstration was performed in animal models that involved 96 rats, half of which were infected with MRSA, while another half were infected with *P. aeruginosa*. These two major groups were divided into four subgroups, with 12 individuals in each group. The first set of subgroups from each major group received 0.1 mL of phages with a titre of $10^7$ PFU/mL (*Staphlococcus* phage Sb-1 in case of MRSA infection group and *Pseudomonas* phage PAT14 in case of *Pseudomonas* infection group), the second set of subgroups was treated with antibiotics (20 mg/kg/day teicoplanin for the MRSA group; 20 mg/kg/day imipenem & cilastatin and 25 mg/kg/day amikacin for the *P. aeruginosa* group), the third set of subgroups were treated with the combination of corresponding phages and antibiotics, while the fourth set of subgroups were a control. Phages were administered as local injections during three consecutive days. On the fifteenth day after initiation of the experiment, the animals were culled and their tibias were excised and examined. One third of the implant with the surrounding bone was reserved for examination by electron microscopy for assessment of biofilm formation, another one third for examination by light microscopy for detection of cytological changes, and the remaining one third for microbiological assessment. The authors concluded that the combined treatment with phages along with antibiotics was the most efficient for the disruption of biofilms (the bacterial density score was measured with the use of electron microscopy). However, this effect was more significant in the MRSA infected animals than those infected with *P. aeruginosa* (Yilmaz et al., 2013).

Other collaborative research implemented jointly by the UK and Georgian scientists (Issa et al., 2019) is in tune with the previous article. Environmentally-sourced phages were studied for their effect on *P. aeruginosa* biofilms. The bacterial target strains were typed for their ability to produce the quorum sensing-controlled virulence factor pyocyanin, and then tested for phage susceptibility using the spot test method. From a total of forty phage isolates derived from environmental surface waters, the majority were only effective against pyocyanin-deficient strains of *P. aeruginosa*. These phages were then co-administered with ciprofloxacin in order to determine whether the phages would demonstrate synergistic or antagonistic behaviour to the antibiotic *in vitro*.

The results observed suggested a potential relationship between phage plaque size and biofilm inhibition, in that those phages producing smaller plaques appeared to be more effective at reducing bacterial biofilm formation. The authors suggested that this phenomenon may be explained by a high adsorption rate leading to the rapid formation of smaller plaques, and greater biofilm reduction associated with the loss of viable bacterial cells before the cells can adhere to the surface and develop the biofilm. Results from the co-administration of phage with ciprofloxacin suggested that the two agents were working synergistically in countering the *P. aeruginosa* biofilms. The data indicated that the efficacy of ciprofloxacin was enhanced by ≥50% when phages were present, with the biofilm analysis suggesting a reduction in cellular adherence, and in turn, biofilm formation. The authors concluded that combining phage therapy with antibiotic could offer a worth-while strategy for targeting such antibiotic-resistant infections (Issa et al., 2019).

**Urology**

A 65 year old female (designated B.R.) applied to the EPTC with chronic bacterial cystitis combined with chronic bacterial vaginitis caused by an antibiotic-resistant strain of *K. pneumonia*. The patient was suffering from frequent exacerbations during which she was treated with multiple courses of antibiotics, however, without any significant results. Several years prior (in 1992), the patient had undergone a hysterectomy. On presentation to the EPTC, the patient complained of severe periodic pains in the suprapubic area, frequent urination, with burning and pain on urination. Vaginal bacteriological analysis showed the presence of *Staphylococcus* spp., while the *K. pneumonia* was isolated from a urine sample. The *Staphylococcus* strain appeared to be susceptible to commercial *Pyo-phage*, while a custom phage preparation was manufactured to treat the *Klebsiella* infection. The *Pyo-phage* was applied as instillations into the vagina 10 mL per day and as vaginal suppositories once per day. The custom-generated *Klebsiella* phage was applied orally 10 mL twice per day and as instillations into the bladder 10 mL once per day. The patient's condition improved within 20 days of treatment in that the pain was decreased, and the urge of urination was less frequent. Also, after the 20 days of treatment *K. pneumonia* was no longer present in urine samples. As a precaution, after 14 days of pause, the phage therapy course was repeated. Follow-on bacteriological analysis showed that *K. pneumonia* had been completely eradicated.

Another 65-year-old female patient (designated F.G.) presented to the EPTC having been diagnosed with kidney failure. One of her kidneys had been removed, and a pathological procedure was ongoing in the remaining kidney. There was also a stent in the ureter. The patient was preparing for kidney transplantation, which was being prevented by a *K. pneumonia* infection. A custom *Klebsiella* phage was manufactured, which was used according to the following regimen: 10 mL two times per day administered orally and locally via instillations directly into the stent performed daily. It is notable that during the first days of phage treatment, the patient showed a temperature reaction ($38^{O}C$), although the treatment continued and her condition improved dramatically in a short period of time. The last bacteriological analysis was performed 45 days after the start of treatment, and it showed that the *Klebsilella* infection was completely eradicated. The short episode with a rise of temperature could be explained by an intensive release of *Klebsilella* endotoxins from the biofilm present in the stent. Usually, oral administration of phage preparations is absolutely safe. It should also be noted that all custom preparations are super-pure and do not contain LPS fractions.

**Clinical trial – Urology**

Recently, the Eliava Institute together with the doctors from the Tsulukidze National Urology Center (Tbilisi) and collaborators at the University of Zurich, Switzerland (the Department of Neuro-Urology, Balgrist University Hospital) finalised the implementation of a project titled: "Bacteriophages for treating urinary tract infections in patients undergoing transurethral resection of the prostate: a randomised, placebo-controlled, double-blind clinical trial" (funded by the Swiss National Science Foundation). This was one of 14 officially registered (registration number NCT03140085) clinical trials utilising phages for treatment of various bacterial infections. The commercial phage preparation *Pyo-phage* was selected for the clinical trial. The choice was based on the vast clinical and manufacturing experience related to this preparation and the composition of phages which target the most frequent urinary tract infection (UTI) causative agents, such as *E. coli, Staphylococcus* spp, *Streptococcus* spp (including group D streptococci, nowadays known as *Enterococcus*), *Pseudomonas* spp and *Proteus* spp. In cases involving personalised treatment, the therapeutic phages usually are pre-adapted to the patient's particular strain causing the infection, with the efficacy of such an adaptation approach being previously demonstrated by Sybesma et al., (2016). In that study, an adaptation of *Pyo-phage* against 41 *E. coli* strains isolated from UTI patients showed that with adaptation, coverage of the strains was expanded from 66% to 93%. Therefore, the phages of *Pyo-phage* selected for the clinical trial were adapted for clinical strains collected during the first year of the project. In all, 130 strains were collected from urine samples of patients planned for transurethral resection of the prostate, although only 118 strains were ultimately used for the pre-adaptation of *Pyo-phage*. These 118 isolates were those that could be defined as uropathogens (*S. aureus, E. coli, Streptococcus spp., P. aeruginosa* and *Proteus mirabilis*) present in the urine cultures at ≥$10^4$ CFU/mL. The initial screening with the *Pyo-phage* gave 41% (48/118) efficiency towards the selected uropathogens, and after six adaptation cycles, this improved to 75% (88/118). Figure 15 shows that prior to adaptation, the percentage of sensitive (S) and intermediary (I) strains was 24% and 16%, respectively (i.e. a total 40% of the test strains were lysed). After six adaptation cycles, these fractions changed to 41% and 34%, respectively (i.e. a total 75% of the test strains were lysed). Thus, at the beginning of the study, the fraction of resistant strains (R) strains was 60%, after the adaptation, the resistant fraction was reduced to 25% (i.e. 2.4 fold).

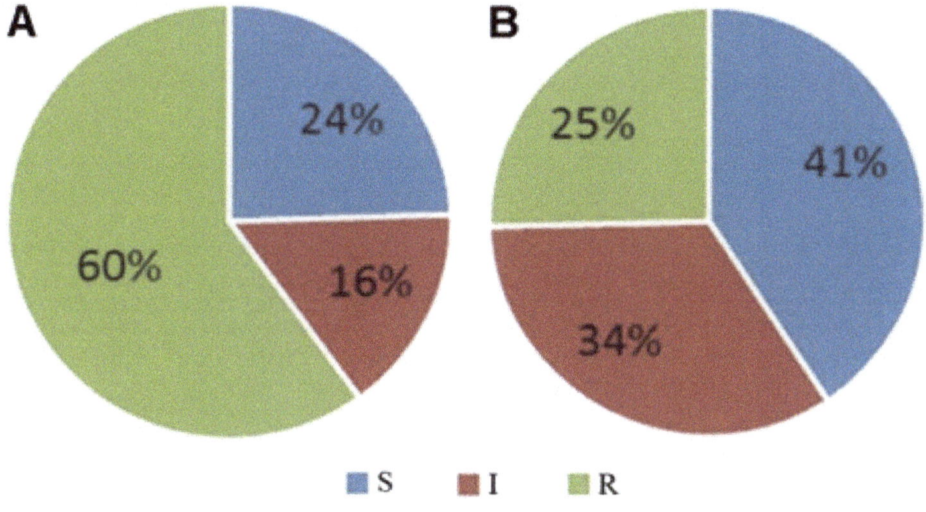


**Figure 15.** The proportional distribution of the sensitive (S), intermediate (I), and resistant (R) strains among 118 clinically isolated uropathogens samples before **(A)** and after **(B)** the adaptation cycles of the *Pyo-phage* cocktail (Ujmajuridze et al., 2018). Initially, 60% of strains were categorised as resistant; and the phage adaptation cycles reduced this fraction to 25%.

This achievement allowed the project team to perform a pilot clinical study involving nine patients. Six out of the nine patients showed between a 1-log and 5-log decrease of bacterial counts in urine samples. No adverse events were registered, although one patient developed fever due to *P. aeruginosa* infection, but with the relief of symptoms following antibiotic treatment. (Table 2) (Ujmajuridze et al., 2018).

**Table 2**. Summary of results of intravesical *Pyo-phage* treatment of urinary tract infections in nine patients undergoing transurethral resection of the prostate (Ujmajuridze et al., 2018).

| Case | Age (yrs) | Pre-treatment | | Post-treatment | | Prostate size (g) | PSA value (ng/mL) | Adverse effects / Remarks |
|---|---|---|---|---|---|---|---|---|
| | | Culture(s) | Ttitre (CFU/ml) | Cultures | Titre (CFU/ml) | | | |
| 1 | 62 | *E. coli* | 10^7 | *E. coli* | 10^5 | 70 | 1 | No side effects |
| 2 | 82 | *E. coli* | 10^7 | *Non pathogenic microflora* | N/a | 140 | 1 | No side effects |

| 3 | 58 | *E. coli* | 10^4 | *E. coli* | 10^4 | 65 | 0.3 | No side effects |
|---|---|---|---|---|---|---|---|---|
| 4 | 85 | *E. coli* | 10^7 | *Non pathogenic microflora* | N/a | 90 | 3.2 | No side effects |
| 5 | 56 | *Streptococcus* | 10^7 | *Streptococcus* | 10^6 | 130 | 0.8 | No side effects |
| 6 | 85 | *Streptococcus* | 10^5 | sterile | 0 | 29 | 2.3 | No side effects |
| 7 | 69 | *Enterococcus* | 10^6 | *E. coli* | 10^7 | 45 | 1 | Secondary infection (?) |
| 8 | 69 | *Enterococcus* | 10^6 | *Non pathogenic microflora* | N/a | 60 | 1.3 | No side effects |
| 9 | 58 | *Pseudomonas* | 10^6 | No data | No data | 60 | 1.3 | Catheter removed on the 2nd day, fever and chills started on the 3rd day |

The above study was designed to prove feasibility, tolerability, and safety of phage therapy and to evaluate clinical/microbiological outcomes of commercially available adapted *Pyo-phage* preceding a larger clinical placebo-controlled, double-blind randomised controlled trial (Ujmajuridze et al., 2018). These findings suggest that phage therapy might be effective and safe for treating UTIs.

Therefore, a clinical trial design was developed, discussed in detail in Leitner et al., (2017) (Figure 16). According to this design, the study was conducted at the Tzulukidze National Center of Urology, Tbilisi, Georgia. The phage preparations were the *Pyo-phage* solution: (commercially available and registered in Georgia) with continuous adaptation cycles to enhance their efficacy performed at the Eliava Institute of Bacteriophages, Microbiology, and Virology in Tbilisi during the course of the study. Patients planned for transurethral resection of the prostate were screened for UTIs and enrolled if urine culturing of eligible microorganisms demonstrated $\geq 10^4$ CFU/ml. Patients were randomised in a double-blind fashion to three treatment groups in a 1:1:1 ratio to receive either: (a) the *Pyo-phage* preparation, (b) a placebo solution with both (a) and (b) administered twice per day directly into the bladder through the catheter, or (c) antibiotic treatment administered orally in accordance with the antibiotic sensitivity pattern. All treatments continued for seven days with no antibiotic prophylaxes permitted for groups (a) and (b) during that time period. The *Pyo-phage* cocktail was subjected to periodic adaptation cycles during the study, which was aimed at having regular enrichment with new phages targeting those strains that had emerged over time and which scored resistant in the *in vitro* analyses. Urinalysis, urine culture, bladder and pain diary, and an IPSS questionnaire (International Prostate Symptom Score) were completed prior to and at the end of the 7-day treatment (or withdrawal from the study). Patients with persistent urinary tract infections following the trial underwent follow-up antibiotic treatment according to antibiotic sensitivity pattern. The study was registered at clinicaltrials.gov (Leitner et al., 2017).

Briefly, in this randomised placebo-controlled double-blind trial, following the inclusion criteria only 97 patients undergoing transurethral resection of the prostate presenting with UTI were assigned to receive intravesical *Pyo-phage* (*n*=28), intravesical placebo solution (*n*=32) or oral antibiotic treatment (*n*=37). The primary outcome of the trial was clinical and microbiological treatment response. Seven days after surgery, the urine samples were collected and bacteriologically studied. The microbiological outcome was based on quantitative microbiological urine assessment, i.e. a reduction of CFU ≥10/mL. Along with this data, successful treatment was defined as no signs of local UTI symptoms (urgency, frequency, and/or dysuria) that the patient had presented preoperatively and no signs for a systemic (fever >38°C, CRP >100mg/L) infection within the treatment period, which were based on the investigator's evaluation of patient's clinical signs and symptoms upon response to the treatment i.e. improvement or deterioration of UTI symptoms compared to the status presented at study entry and absence of new symptoms. Secondary outcomes included safety parameters, quantitative microbiological urine assessments and improvement or deterioration of UTI symptoms. Efficacy and safety analysis was done using statistical assessment with $p<0.05$ considered statistically significant (StataCorp. 2015. Stata Statistical Software: Release 14. College Station, TX: StataCorp LP). At treatment day seven, successful therapy was achieved in 27/97 (28%) of the intention-to-treat population. Compared to the *Pyo-phage* (5/28 (18%)), the placebo group had 9/32 (28%) (Odds Ratio (OR) 1.8, 95% confidence interval (CI) 0.52-6.2; p= 0.352) and the antibiotic group had 13/37 (35%) (OR 2.49, 95% CI 0.77-8.1; p=0.129) successes. Regarding the microbiological outcomes, a reduction of CFU could be found in 7/28 (25%), 12/32 (38%), and 18/37 (49%) of patients treated with *Pyo-phage*, placebo, or antibiotics, respectively. Overall, adverse events were reported in 22/97 (23%) of patients. Sudden onset of temperature >38°C and concomitant use of antibiotics accounted for most adverse events. Adverse events were similar between the three groups. They were seen in 6/28 (21%) patients in the *Pyo-phage* but in 13/32 (41%) in the placebo group (OR 0.4, 95% CI 0.13-1.25; p= 0.116) and in 11/37 (30%) in the antibiotic group (OR 0.65, 95% CI 0.21-2.03; p=0.452).

A local application (i.e. intravesical instillation) of *Pyo-phage* was decided with the purpose to refrain from systemic application to avoid potentially more frequent and relevant side effects. The treatment success rate was unexpectedly low for both active treatment groups, i.e. antibiotics and phages. Unexpectedly, twice-a-day intravesical instillation of *Pyo-phage* and placebo solution showed similar efficacy. The authors hypothesise that the considered placebo treatment had a therapeutic effect due to a mechanical reduction of the bacterial load following bladder irrigation. The effect of mechanical reduction would also be expected for the *Pyo-phage* solution. In addition, the incidence of fever was similar within all groups. At the same time, *Pyo-phage* therapy showed a favourable safety profile. No *Pyo-phage* related diarrhoea, nausea, or allergic reaction, i.e. adverse events typically seen in systemic antibiotic therapies were observed (Leiter et al., 2020).

The authors believe that this trial will significantly influence the future management of UTIs and form a basis for further studies involving phages for treating different bacterial infections. Moreover, the findings of this research project will provide further stimuli for competent authorities and physicians to use phages, as complimentary tools, in the prevention and treatment of otherwise virtually untreatable infections. The authors hope that this study will raise the acceptance to use phages in western civilisations and accelerate the necessary adaptions to regulatory processes.

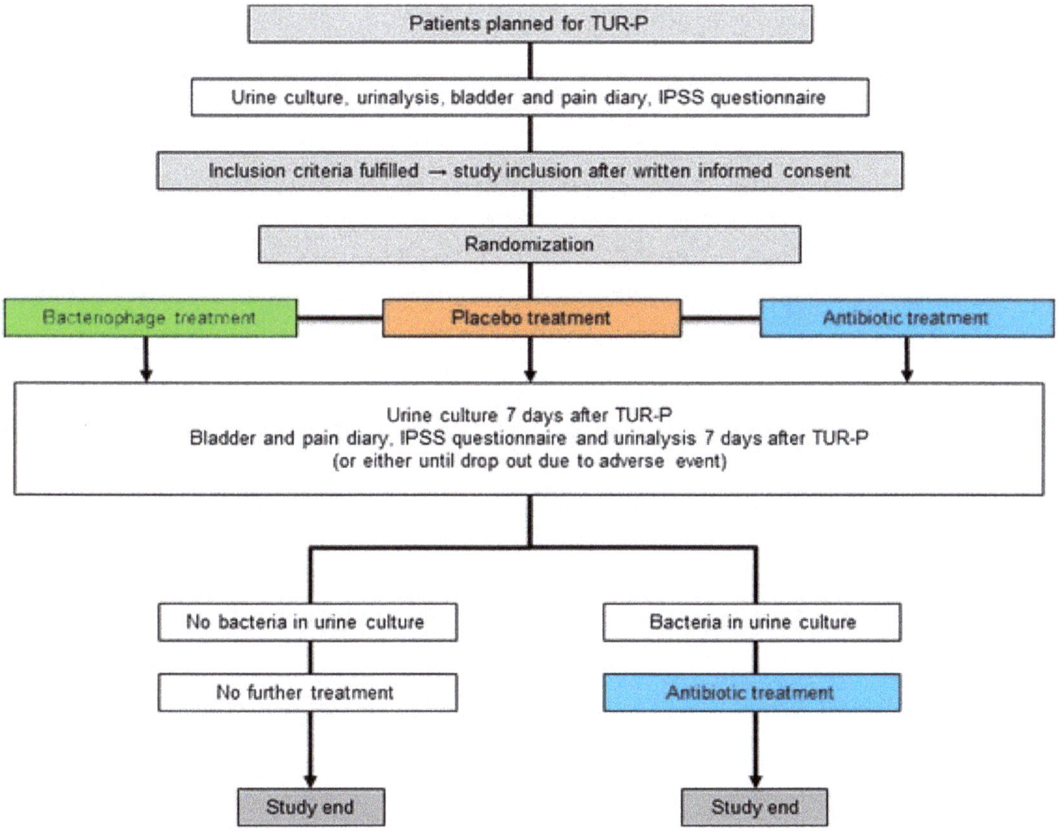


**Figure 16.** Flow chart of the procedures implemented for the clinical randomised controlled trial (RCT) involving patients intended for TUR-P (transurethral resection of the prostate) (Leitner et al., 2017).

## Conclusion

In summary, it is possible to conclude that phage therapy is effective for the treatment of various chronic infection conditions, especially those that are caused by multi-drug-resistant bacteria that do not respond to conventional antibiotic therapy. In cases of chronic pulmonary diseases, phage therapy prolongs the period between exacerbations, allowing patients to live their normal lives without permanent use of antibiotics. Phages show significant potential as prophylactic and therapeutic means in the management of diabetic foot infections, untreatable wounds, skin problems, post-operational bacterial complications, etc. Interestingly, in some cases phage therapy demonstrates efficacy even in those cases when non-specific phages are used for the treatment of mixed infections (e.g. a case of mixed infection with *Bukhalderial cepacia* treated with *Staphylococcus* and *Enterococcus* phages). It is of particular interest that after the use of phage therapy for an antibiotic-resistant bacterial infection, the pathogen population frequently has its antibiotic sensitivity restored to a significant degree. More insight into the mechanisms of such observations should be uncovered by collaborative efforts of bacteriologists, molecular biologists, geneticists, immunologists and medical practitioners. We still think that there is scope for optimisation of the doses, regimens and durations of phage therapy for each specific diagnosis in a particular field of medicine. For example, one may ask how efficient is an oral administration of phage-based medications for the treatment of a lung or a wound infection; or is there a strong accepted argument for a particular route of administration? Elucidation of these gaps in our knowledge, defining phage mechanisms of action should become future research topics.

One of the most important recent achievements of the Eliava Institute is the completion of the double-blind, randomised controlled trial, which will not only disseminate our experience of phage therapy but will also enrich our knowledge and stimulate further clinical trials. These will build trust in the medical profession and facilitate the establishment of phage therapy in clinical practice, either alone or in combination with antibiotics.

## References

Eaton, M.D. and Bayne-Jones, S. (1934). Bacteriophage therapy: review of the principles and results of the use of bacteriophage in the treatment of infections. JAMA 103, 1769-1776.

Fish, R., Kutter, E., Bryan, D., Wheat, G. and Kuhl, S. (2018a). Resolving digital staphylococcal osteomyelitis using bacteriophage - a case report. Antibiotics, 7: 87; doi:10.3390/antibiotics7040087

Fish, R., Kutter, E., Wheat, G., Blasdel, B., Kutateladze, M. and Kuhl, S. (2016) Bacteriophage treatment of intransigent diabetic toe ulcers: a case series. Journal of Wound Care. 25(Sup7):S27-S33

Fish, R., Kutter, E., Wheat, G., Blasdel, B., Kutateladze, M. and Kuhl, S. (2018b). Compassionate use of bacteriophage therapy for foot ulcer treatment as an effective step for moving toward clinical trials. In: Bacteriophage Therapy: From Lab to Clinical Practice. J.Azeredo & S. Sillankorva (Eds). Springer Protocols: 159-170.

Ganeshan, D, S., and Hosseinidoust, Z. (2019). Phage Therapy with a Focus on the Human Microbiota. Antibiotics (Basel, Switzerland). 8(3), 131. https://doi.org/10.3390/antibiotics8030131

Hargreaves, K. R., Kropinski, A. M., and Clokie, M. R. J. (2014). Bacteriophage behavioural ecology. Bacteriophage, 4(3) doi 10.4161/bact.29866.

Kutateladze, M., (2015). Experience of the Eliava Institute. Virol Sin. 30(1), 80-81. doi: 10.1007/s12250-014-3557-0

Kvachadze, L., Balarjishvili, N., Meskhi, T., Tevdoradze, E., Skhirtladze, N., Pataridze, T., Adamia, R., Topuria, T., Kutter, E., Rohde, C., Kutateladze, M.E. (2011). valuation of lytic activity of staphylococcal bacteriophage Sb-1 against freshly isolated clinical pathogens. Microb. Biotechnol. 4(5), 643-650. DOI: 10.1111/j.1751-7915.2011.00259.x

Leitner, L., Sybesma, W., Chanishvili, N., Goderdzishvili, M., Chkhotua, A., Ujmajuridze, A., Schneider, M.P., Sartori, A., Mehnert, U., Bachmann, L.M. and Kessler, T.M. (2017). Bacteriophages for treating urinary tract infections in patients undergoing transurethral resection of the prostate: a randomized, placebo-controlled, double-blind clinical trial. BCM Urology, 17: 90. doi: 10.1186/s12894-017-0283-6

Leitner, L., Ujmajuridze, A., Chanishvili, N., Goderdzishvili, M., Chkonia, I., Rigvava, S., Chkhotua, A., Changashvili, G., Schneider, M.P., Liechti M.D., Mehnert, U., Bachmann, L.M., Sybesma, W., Kessler, T.M. (2020). Intravesical bacteriophages for treating urinary tract infections in patients undergoing transurethral resection of the prostate: A randomised, placebo-controlled, double-blind clinical trial. Lancet Infect Dis (In Press).

Northrop, J. (1937). Chemical nature and mode of formation of pepsin, trypsin and bacteriophage. Science 86, 479-483.

Northrop, J. (1938). Concentration and purification of bacteriophage. J. General Philosophy 21(3), 335-366.

Sergueev, K. V., Filippov, A. A., Farlow, J., Su, W., Kvachadze L., Balarjishvili, N., Kutateladze, M. and Nikolich, M. P. (2019). Correlation of host range expansion of therapeutic bacteriophage Sb-1 with allele state at a hypervariable repeat locus. Appl. Environ. Microbiol. Appl. Environ. Microbiol. 85(22). Pii: e01209-19 DOI: 10.1128/AEM.01209-19

Sybesma W., Zbinden R., Chanishvili N., Kutateladze M., Chkhotua A., Ujmajuridze A., Mehnert U. and Kessler T. (2016). Bacteriophages as potential treatment for urinary tract infections. Front. Microbiol. 7, 465 doi: 10.3389/fmicb.2016.00465

Ujmajuridze A., Chanishvili, N., Goderdzishvili, M., Leitner, L., Mehnert, U., Chkhotua, A., Kessler, T.M. and Sybesma, W. (2018). Adapted bacteriophages for treating urinary tract infections. Front. Microbiol. 9. 1832. doi: 10.3389/ fmicb.2018.01832

Yilmaz, C., Colak, M., Yilmaz, B.C., Ersoz, G., Kutateladze, M. and Gozlugol, M. (2013). Bacteriophage therapy in implant-related infections: an experimental study. J. Bone Jt. Surg. 95(2), 117-125. doi: 10.2106/JBJS.K.01135

Zhvania, P., Sulinger Hoyle, N., Nadareishvili, L., Nizharadze, D. and Kutateladze, M. (2017). Phage therapy in a 16-year-old boy with Netherton syndrome. Front. Med. 4, 94. doi: 10.3389/fmed.2017.00094

# Chapter 7

# Considerations for Using Bacteriophages in Plant Pathosystems

**Aleksa Obradović[1*], Jeffrey B. Jones[2], Botond Balogh[3] and Katarina Gašić[4]**

[1]University of Belgrade, Faculty of Agriculture, Belgrade, Serbia
[2]University of Florida, Plant Pathology Department, Gainesville, FL, USA
[3]Nichino America Inc., Wilmington, DE, USA
[4]Institute for Plant Protection and Environment, Belgrade, Serbia

*Corresponding author: aleksao@agrif.bg.ac.rs

DOI: https://doi.org/10.21775/9781913652517.07

**Abstract**

Lytic bacteriophages (phages) have the potential for controlling susceptible bacteria in the rhizosphere or phyllosphere. The success of phage application in plant disease control requires that high populations of both phage and bacterium exist in order to initiate a chain reaction of bacterial lysis. Physical factors in natural environments such as the presence of biofilms that trap phages, low soil pH which inactivates phages, low rates of diffusion of phages in soil that prevent contact with target bacteria, and inactivation of phages upon exposure to ultraviolet (UV) light and desiccation, all impact successful use of phages in control of plant pathogenic bacteria. Other considerations relate to the bacterial strains which exist in nature. The bacterial species may have a low or high degree of variation in sensitivity to phages. Therefore, phage selection for field use requires careful monitoring of the targeted bacterial strains in the field due to the potential for strain variation and the likelihood for development of resistance to the deployed phages. Application timing has also been shown to be an important factor in improving the efficacy of phages. For instance, UV light is deleterious to phages and upon exposure phage populations plummet; therefore, evening applications of phages result in persistence on leaf surfaces for longer periods of time and may result in improved disease control. Extending the period of time that phages persist in the phyllosphere has been a major hurdle. Formulations have been identified which improve the persistence of phages on leaf surfaces; however, there is a need to identify superior formulations that extend the life on leaf surfaces from hours to days. Another strategy for maintaining high populations of phages has been to use non-pathogenic bacterial strains that are sensitive to the phage(s) or a closely related organism that does not cause disease on the plant host. Finally, phages may have value as part of an integrated management strategy.

## Introduction

Bacterial plant pathogens account for significant losses to agricultural production systems. Disease control for many bacterial incited ailments is a major challenge affected by various factors including: pathogen variation; ability to overcome plant genetic resistance as a result of mutation or loss of plasmid; the lack of effective bactericides as a result of strains developing tolerance; and the pathogen's ability to reach high populations in a relatively short period of time when conditions are favourable for disease development. Since control of bacterial incited diseases is such a major challenge, integrated pest management strategies have been critical to reducing the impact from bacterial diseases (Obradovic et al., 2004, 2005, Šević et al., 2019). A major component of integrated pest management strategies has been the use of bactericides; however, control with bactericides has been extremely difficult because few effective bactericides are available. Antibiotics and copper-based bactericides have been the principal chemicals used for disease control. Copper has been the most extensively used bactericide; however, copper resistance is present in many plant pathogenic bacteria and is associated with plasmids and in rare situations the chromosome (Bender and Cooksey, 1986, 1987, Stall et al., 1986, Bender et al., 1990, Lee et al., 1994, Basim et al., 1999, Canteros, 1999, Gutierrez-Barranquero et al., 2013). Transfer of this resistance is readily achieved (Stall et al., 1986). Antibiotics have also been used as part of a management strategy for various bacterial diseases, since the 1950s. The aminoglycoside antibiotic streptomycin was used extensively for control of bacterial diseases. Due to the extensive use, streptomycin-resistant strains became prevalent, resulting in reduced disease control efficacy of bacterial spot of tomato and pepper (Thayer and Stall, 1961), fire blight of apple and pear (Manulis et al., 1998), and many other bacterial plant pathogens (Cooksey, 1990). Systemic acquired resistance (SAR) compounds have provided a level of control against tomato and pepper bacterial diseases (Louws et al., 2001; Romero et al., 2001; Obradovic et al., 2004; Huang et al., 2012; Šević et al., 2019), *Xanthomonas* leaf blight on onion (Gent and Schwartz, 2005) and fire blight on apple (Maxson-Stein et al., 2002); however these inducers may have negative physiological effects on plant growth and yield (Romero et al., 2001; Gent and Schwartz, 2005).

Phage therapy has gained significant interest recently (Balogh et al., 2003, 2008, 2018; Obradovic et al., 2004, 2005; Gent and Schwartz, 2005; Jones et al., 2007, 2012, 2018; Lang et al., 2007; Fujiwara et al., 2011; Bae et al., 2012; Frampton et al., 2012; Iriarte et al., 2012; Lim et al., 2013) as an alternative to the use of conventional bactericides, although its use goes back to the early days following the discovery of phages in the early 1900s (Twort, 1915; d'Herelle, 1917). In 1924, a phage was isolated from black rot infected cabbage and was shown to inhibit *in vitro* growth of the pathogen, *Xanthomonas campestris* pv. *campestris* (Mallmann and Hemstreet, 1924). Shortly thereafter phages were isolated from soil with activity against *Pectobacterium atrosepticum*, the causal agent of blackleg disease of potato (Kotila and Coons, 1925); in growth chamber experiments co-inoculation of *P. atrosepticum* with phage prevented rotting of potato tubers (Kotila and Coons, 1925). Thomas (1935) treated corn seeds with a crude phage suspension isolated from Stewart's Wilt infected plant material and upon planting of the treated seed disease incidence was reduced from 18% to 1.4%.

More recently, Civerolo and Keil (1969) applied phage to peach foliage prior to *Xanthomonas campestris* pv. *pruni* (Xcp) and reduced disease incidence from 58% on control plants to 22% on phage treated plants. In another study, Civerolo (1973) reported that treatment of peach seedlings with crude lysates of a phage mixture resulted in fewer infected leaves and a significant reduction of disease compared to controls. Civerolo speculated that phages may regulate populations of Xcp and in so doing reduce disease. In other studies with this bacterial pathogen, Saccardi et al. (1993) and Zaccardelli et al. (1992) reduced fruit spot incidence on peaches with biweekly spray applications of phage suspensions active against the target bacterium, Xcp. In their work, they isolated eight phages active against the pathogen, screened them for host range and lytic ability, and selected a lytic phage strain with the broadest host range for disease control.

Although the above studies and others provided positive results using phage, phage therapy was not considered a good strategy for controlling plant pathogenic bacteria because of its unreliability. This is clearly the opinion in a review article in 1963 (Okabe and Goto, 1963). In a more recent review article, it was stated that the practical use of phages for control of bacterial plant disease in the field was unsuccessful (Goto, 1992). Furthermore, because of phages' narrow spectrum of activity, they were considered less likely to succeed compared to antibiotics (Summers, 2005). Therefore, phages were not considered a practical control strategy for bacterial plant diseases, with antibiotics and copper compounds becoming the principal control agents.

Nonetheless, in recent years several papers have been published on the use of phages against different plant pathogens. These include using phages targeted for killing *Pectobacterium carotovorum* subsp. *carotovorum* (formerly designated *E. carotovora* subsp. *carotovora)* (Lim et al., 2013), *Ralstonia solanacearum* (Fujiwara et al., 2011; Iriarte et al., 2012; Bae et al., 2012), *Xanthomonas arboricola* pv. *juglandis* (Romero-Suarez et al., 2012) and *Erwinia amylovora* (Born et al., 2011; Svircev et al., 2006); these also hold promise for using as part of plant disease management strategies.

## Impediments in using phage therapy in plant pathology

### *Development of resistance in bacterial strains*

The high probability of developing bacterial strains resistant to a phage is a real concern (Fujiwara et al., 2011). It was calculated based on the classic experiments of Luria and Delbrück (1943) that the mutation rate of *E. coli* to resistance to phage T1 is $3.4 \times 10^{-9}$ (Birge, 2000). Early on in studies utilizing phage, the concern of host resistance was addressed by Katznelson (1937) and later as a concern in a review article (Okabe and Goto, 1963) and then more demonstratively as a significant impediment in the implementation of this strategy for control (Vidaver, 1976).

*Phage selection*

An important and oftentimes neglected aspect of phage-based biological control is pre-screening phages for their biocontrol value before application; that is, identifying specific phages with particular characteristics that may prove effective in control rather than arbitrarily selecting them based strictly on lytic activity for disease control studies. In two studies (Zaccardelli et al., 1992, Saccardi et al., 1993) the phage selected from a collection of eight phages for biological control experiments was a lytic phage with the broadest host range. Balogh (2006) observed that phages vary in their ability to interact with and multiply on their host bacterium on the grapefruit leaves, an ability that correlated with their disease control efficacy against citrus canker. However, he found no *in vitro* characteristics that predicted disease control ability against bacterial spot on tomato (Balogh et al., 2018).

*Application timing*

Timing of phage applications relative to the arrival of the pathogen is a critical concern. Given the leaf surface is an extremely harsh environment for survival of phages (Gill and Abedon, 2003; Iriarte et al., 2007) and the drastic reduction in phage populations on leaf surfaces as a result of ultraviolet light (Iriarte et al., 2007), the timing of phage applications is essential to optimize that high phage populations be present in close proximity to the target bacterium in order to maximize phage therapy efficacy. This has been shown to influence the efficacy of disease control in several instances. Civerolo and Keil (1969) achieved a marked reduction of peach bacterial spot only if phage treatment was applied one hour or one day before inoculation with the pathogen. There was a slight disease reduction when phage was applied one hour after inoculation and no effect if applied one day later. Civerolo (1972) suggested that bacteria were inaccessible to phage in the intercellular spaces, or there were not enough phages reaching the pathogen. Schnabel et al. (1999) achieved a significant reduction of fire blight on apple blossoms when the phage mixture was applied at the same time as the pathogen, *Erwinia amylovora*. In contrast, disease reduction was not significant when phages were applied a day before inoculation. Likewise, Bae et al. (2012) and Iriarte et al. (2012) showed that phages were most effective when applied at the time of plant inoculation; use of phages up to a few days before or after pathogen inoculation resulted in a lower, or even no biocontrol efficacy. Bergamin Filho and Kimati (1981) investigated the effect of timing on the efficacy of phage treatment in greenhouse trials with two pathosystems: black rot of cabbage, caused by *Xanthomonas campestris* pv. *campestris* and bacterial spot of pepper, caused by *Xanthomonas campestris* pv. *vesicatoria*. Phage treatment was applied once varying from 7 days before to 4 days after pathogen inoculation. On cabbage significant disease reduction was achieved if the phage treatment was applied 3 days before to 1 day after inoculation, whereas on pepper from 3 days before to the day of inoculation. The greatest disease reduction occurred with the application of phages the same day as inoculation in both pathosystems. Recently, Das et al. (2015) have used phage cocktail to control Pierce's Disease of grapevine, caused by Xylella *fastidiosa* subsp. *fastidiosa*. They demonstrated that both prophylactic (3 weeks before the pathogen inoculation) and therapeutic (3 weeks after the pathogen inoculation) treatments significantly reduced the pathogen level in grapevines and prevented symptom development.

In terms of disease control in the field, the time of day when phages are applied also affects efficacy. Sunlight irradiation was identified as the single most detrimental factor in reducing phage persistence in the tomato canopy (Iriarte et al., 2007). Consequently, applying phages when they are not exposed immediately to direct sunlight (before dawn or after sunset) prolonged their residual activity (Iriarte et al., 2007) and resulted in more effective control of tomato bacterial spot, in comparison with daytime treatment applications (Balogh et al., 2003).

*Strain variation*

Better understanding of bacterial strains associated with particular plant species has shown that significantly greater genetic diversity may exist than considered in earlier studies (Guy et al., 2013). Much of our knowledge is based on experience with *Xanthomonas* spp. associated with tomato and pepper. Bouzar et al. (1999) used 26 phages to type approximately 100 *X. euvesicatoria* strains isolated from various countries in the Caribbean, including Central America. There was a very high level of diversity in which at least 26 different reaction patterns were observed. Beyond the variation within a particular bacterial species, plant species may be infected by more than one bacterial species within the same genus and as a result in all likelihood will require different phages. For instance, four different *Xanthomonas* species infect tomato (Jones et al., 2004), and as a result, require the use of different phages for disease control. In general, the phages that infect bacterial strains within one bacterial species do not infect strains within the other bacterial species.

*Phage utilization in the rhizosphere*

Phages have the potential for controlling plant pathogens in the rhizosphere or phyllosphere. However, Gill and Abedon (2003) identified factors that can hinder the success of disease control in the rhizosphere. The relatively low diffusion rate of phages through heterogeneous soil matrices changes as a function of available free water. Biofilms can trap phages (Storey and Ashbolt, 2001), soil clay particles can reversibly adsorb phages (Williams et al., 1987), and low soil pH can inactivate phages (Sykes et al., 1981; Bae et al., 2012). In natural environments, as a result of low rates of phage diffusion and high rates of phage inactivation, low numbers of viable phages are available to lyse target bacteria (Gill and Abedon, 2003). One additional factor needed for a high degree of success is that high populations of both phage and bacterium exist in order to initiate a chain reaction of bacterial lysis (Gill and Abedon, 2003). However, in spite of the above environmental conditions that may reduce the number of effective phage particles and limit their stability and efficacy in the rhizosphere, some phages were found to be fairly stable in soil (Fujiwara et al., 2011). Iriarte et al. (2012) proved that phages, applied as a rhizosphere drench treatment, could be translocated through the roots into above ground parts of the treated tomato plants.

*Persistence of phages on leaf surfaces*

Phages encounter deleterious factors in the phyllosphere, with their populations plummeting below the threshold level for activity as a biocontrol agent. Biocontrol efficacy is influenced by both the biocontrol agent and target pathogen densities (Johnson, 1994). Maintaining high populations of biocontrol agents in the target environment in close proximity to the target bacterium is of critical importance. Phage therapy requires high populations of phage and target bacterium (Gill and Abedon, 2003). Clearly, for good control, phages must exist above a threshold concentration; below that threshold, phages will have a minor effect on bacterial populations. In support of this threshold effect hypothesis, Balogh (2002) observed that phage mixtures applied at $10^6$ or $10^8$ PFU/mL concentration provided similar levels of control of tomato bacterial spot, but at $10^4$ PFU/mL was ineffective.

In the phyllosphere, a harsh environment exists and phages in this location degrade extremely rapidly (Civerolo and Keil, 1969; McNeil et al., 2001; Balogh, 2002; Balogh et al., 2003; Iriarte et al., 2007; Jones et al., 2012). This short-lived persistence on plant leaf surfaces is a limiting factor of phage treatment. Infield and laboratory studies, insect viruses were shown to be inactivated by high temperatures, high and low pH and sunlight, and were readily dislodged by rain (Ignoffo et al., 1989, Ignoffo and Garcia, 1992). Other factors that may contribute to phage inactivation include desiccation and exposure to certain chemical pesticides, such as copper bactericides commonly used in bacterial plant disease management strategies (Iriarte et al., 2007). Under field conditions, sunlight irradiation is the single major deleterious component affecting phage survival (Iriarte et al., 2007). In particular, UV-A and UV-B spectra of sunlight were determined to be the environmental factor most destructive to viruses (Ignoffo and Garcia, 1994). Initial phage studies on tomato showed that phage applied in the mid-morning was not effective in controlling bacterial spot of tomato (Jones, unpublished results).

Iriarte et al. (2007) demonstrated that phage populations plummeted precipitously during the early afternoon hours when UV radiation was high. However, if the phages were applied in the late afternoon or early evening close to sunset, phage populations did not plummet overnight, and thus high phage concentrations persisted overnight and potentially could interact with bacterial strains on the leaf surface. Phage survival was highly correlated with the intensity of sunlight UV irradiation. Furthermore, phage persistence varied depending on the ambient temperature (Iriarte et al., 2007).

**Future considerations for improving phage efficacy**

As stated earlier biological control using phages is dependent on the ability of the biological control agent to persist at high levels in close proximity to the target bacterium (Johnson, 1994). Furthermore, the phages must reach and attach to their hosts before themselves being destroyed by adverse factors (Goodridge, 2004). Several considerations exist with regard to improving phage-bacterium interactions: concentration and accessibility of target bacteria; timing of phage application to optimize efficacy; phage infection and replication ability in the target environment; phage concentration at the site of interaction (i.e., phyllosphere or rhizosphere); rates of virion degradation (phage vary in degradative properties); and the presence of adequate moisture as a medium for phage diffusion (Gill and Abedon, 2003). Below, we will address several of the key issues which must be considered for improving phage efficacy.

*Development of resistance in bacterial strains*

As a result of the high probability of bacterial strains developing resistance to the phage, Jackson (1989) developed a strategy to minimize the occurrence of phage-resistant mutants. This involved preparing mixtures of phages that lyse bacterial strains that are resistant to the parent phage (Adams, 1959) while maintaining the ability to lyse the wild-type bacterium. Using this strategy, a mixture of four phages including wild-type and host-range mutant (h-mutant) phages were applied twice weekly and provided significantly better disease control and produced greater yield of extra-large fruits than the standard copper-mancozeb (Flaherty et al., 2000).

A second strategy relies on constant monitoring of phage efficacy. Phage mixtures are applied to plants in the target field, and bacteria are regularly isolated from phage-treated plants during the treatment period. The isolates are evaluated for their sensitivity to the utilised phages. If resistance is observed, the phage in question is removed from the mix and replaced with a new one. This strategy relies on the existence of phage-banks (i.e., large collection of phages specific to the target bacterium).

*Phage selection*

The proper assay for phage selection is a critical and often overlooked factor in ensuring the success of phage therapy in agriculture. Although *in vitro* assays are frequently used in the selection process for phages, these may not be good predictors of biological control ability. These assays provide optimal conditions for phage infections - such as exponentially growing susceptible bacterial culture, controlled constant temperature, constant level of free moisture, more or less constant pH, the availability of a wide range of nutrients, and protection from sunlight exposure - and as such, does not represent the “reality of life” in the plant environment, where nutrients and water are scarce, and the environmental conditions are in constant limbo. Balogh (2006) found that two of three phages, which were active on *Xanthomonas citri* pv. *citri* in plate assays, were unable to lyse the bacterium on the grapefruit foliage. These two phages, not surprisingly, were not able to suppress citrus canker in greenhouse trials. Balogh et al. (2018) evaluated eight phages that were active against *Xanthomonas perforans* for a number of *in vitro* characteristics, such as the efficiency of plating (EOP), plaque size, antibacterial activity or phage multiplication rate, and found no correlations between these attributes and actual disease control efficacy. Based on these results, actual bioassays are unavoidable in order to gauge biocontrol activity. Such bioassays should evaluate the phages capacity to persist and replicate on the plant surface, and their effect on disease development. However, they demonstrated that EOP is not the only factor that is required for determining the ability of phage to persist on leaf surfaces when the bacterial host is present. Predicting the ability of phage to multiply on leaf surfaces cannot be determined strictly based on EOP as it is only a good predictor for those phages-bacterial host combinations where the EOP is low (Balogh et al., 2018).

*Application timing*

Given the difficulty in maintaining high phage populations in the phyllosphere or other aboveground plant parts, optimizing an application strategy that results in improved disease control from phage applications is necessary to essential. Using the bacterial spot of tomato system as a model, Balogh et al. (2003) and Iriarte et al. (2007) noted that in the phyllosphere, phage populations are drastically reduced when phages are applied in the morning or early afternoon. As a result, very low phage populations are present on the leaf surface to interact with the target bacterium. Given that phage persistence on leaf surfaces is very poor over a 24 hour period, frequent applications in the evening hours to maximize exposure of the target bacterium to high concentrations of phages are necessary to maximize disease control. Another factor one might consider when choosing when to apply (and what application volume to use), is the availability of free moisture on the leaves. Phages will only interact with their target bacterium if the leaf surface is wet, and consequently it seems advisable to apply them when free moisture is expected to stay on the leaves for an extended time period in order to ensure longer exposure time of the target bacterium.

*Persistence of phages on leaf surfaces*

Persistence on leaf surfaces is a major limiting factor in using phage therapy for disease control in the phyllosphere. Several strategies have been evaluated for increasing phage persistence, including the use of protective formulations, application scheduling for sunlight avoidance and co-application of bacterial hosts for *in vivo* phage propagation. In several studies, solar protectants were identified that increased biocontrol efficacy not only for phages but also for entomopathogenic viruses and proteinaceous biopesticides (Behle et al., 1996; Ignoffo et al., 1997; Balogh et al., 2003; Arthurs et al., 2006; Reddy et al., 2008; Behle and Popham, 2012). Balogh (2002) identified compounds that, when mixed with phage, extended the persistence of phage on the phyllosphere. Balogh et al. (2003) enhanced the efficacy of phage treatment with protective formulations that increased phage persistence on tomato foliage. These authors determined that phage populations persisted at significantly higher concentrations on leaf surfaces when the phage suspensions were applied in combination with skim milk alone or in combination with sucrose. Iriarte et al. (2007) corroborated that the combination of phages with skim milk was important for improved persistence on leaf surfaces, even under intense UV irradiation, although phage populations did drop to very low levels even with the formulated phage. Balogh et al. (2003) determined that the addition of several formulations to the phage suspension led to enhanced disease control efficacy (Balogh et al., 2003). In other field studies treatment of plants with formulated phages resulted in reduced disease and increased yield (Obradovic et al., 2004, 2005). Although Balogh et al. (2003) and Iriarte et al. (2007) demonstrated that the addition of skim milk to phage suspensions improved the ability of phages to persist on leaf surfaces, there is considerable need for identifying formulations that are superior to skim milk. In terms of other approaches, Born et al. (2015) reported on some natural extracts from carrot, red pepper and beetroot, as well as casein, soy peptone and purified aromatic amino acids, astaxanthin and Tween 80, protected phages from detrimental UV light *in vitro*.

A second approach for improving persistence is based on the unique advantage of phages over chemical pesticides in the ability to increase their populations by multiplying on the target bacterial host. This ability could potentially be used if phages are applied into an environment where a phage-sensitive bacterium is present, or where the phages and bacterial host are delivered together (Bae et al., 2012; Iriarte et al., 2012). On leaf surfaces, where high host populations persist, phages persist at significantly higher levels than on surfaces without the host (Balogh, 2006). This approach was investigated for the soilborne bacterium, *Ralstonia solanacearum* in which Tanaka et al. (1990) used an avirulent strain of *Ralstonia solanacearum* and its phage that was active against both the virulent and avirulent strains to reduce tobacco bacterial wilt incidence. Although application of the avirulent strain alone caused significant disease control, co-application of phage with the avirulent strain significantly improved disease control beyond the avirulent strain alone. Svircev et al. (2006) employed a similar strategy for controlling fire blight of pear, by selecting phages based on the ability to lyse both the target organism, the pathogen *Erwinia amylovora* and also an antagonistic phyllosphere bacterium, *Pantoea agglomerans*. *P. agglomerans*, a biological control agent for *E. amylovora*, served as well as a phage carrier, and as a vehicle propagator of phage on the inoculation sites. While *P. agglomerans* significantly reduced disease, combining it with phage resulted in significantly better disease control; the combination resulted in disease control comparable to streptomycin treatment.

Various approaches using non-pathogenic or attenuated bacterial strains as hosts could be used for propagating phage on plants. The main requirements are that they persist in the plant environment relatively well and are hosts to the phages used. This would mean in practice that they have to be closely related to the target pathogen. They can be epiphytes, such as the *Pantoea agglomerans* used by Svircev et al. (2006) for control of fire blight, avirulent strains of the pathogen present in rhyzosphere, such as the *Ralstonia solanacearum* strain utilized by Tanaka et al. (1990) for control of bacterial wilt, or even attenuated strains of the pathogen for control of bacterial spot of tomato (Iriarte et al., 2012). Such pathogenically attenuated strains are under consideration for *Xanthomonas* phages (Iriarte et al., 2012), as avirulent *Xanthomonas* strains are very weak epiphytes (Jones, unpublished results). The use of attenuated strains carries an inherent risk, since under favourable conditions, these strains will potentially cause a considerable level of disease.

**Phages as part of an Integrated management strategy**

In contrast to the long list of antifungal products, the search for bactericides suitable for crop protection has turned out to be exceedingly difficult. In order to reduce losses in yield and quality of agricultural crops, continuous efforts have been made to identify chemicals and other strategies that could be effectively used for controlling plant pathogenic bacteria. Chemical control of plant bacterial diseases mostly relied on the application of the antibiotic streptomycin and copper-based compounds. However, efficacy of these compounds in plant protection was compromised by occurrence of resistant strains of pathogenic bacteria. Recently, plant activators, triggering natural plant defence mechanisms, have been introduced in disease management programs (Qui et al., 1997; Louws et al., 2001; Du et al., 2012). One of them, acibenzolar-S-methyl (ASM), showed excellent potential as a part of an integrated strategy for control of tomato and pepper bacterial spot (Louws et al., 2001, Obradovic et al., 2004, 2005, 2008, Šević et al., 2019). However, some results showed adverse effects of this compound on tomato and pepper growth and yield (Louws et al., 2001). Due to lack of an efficient synthetic bactericide the most frequently implemented approaches in control of plant pathogenic bacteria were use of pathogen-free seed or planting material, genetic host resistance, cultural and sanitation practices, and biological control agents. In practice, none of the above non-chemical treatments have been able to stand alone and provide efficient control.

Multifaceted, integrated strategies carry the promise for effective, reliable and sustainable management of bacterial plant diseases. Phages offer an alternative to conventional management strategies for controlling plant bacterial diseases (Jones et al., 2007, 2018). Several approaches have been explored for using phage-treatment within an integrated management strategy. Tanaka et al., (1990) reduced tobacco bacterial wilt by co-application of an avirulent strain of the pathogen, *R. solanacearum*, with a phage that was active against both virulent and avirulent strains. Using a similar approach, Svircev et al. (2006) reduced fire blight of pear with co-application of an antagonistic epiphyte, *Pantoea agglomerans* and a phage that lysed both the antagonist and the pathogen, *Erwinia amylovora*.

Lang et al. (2007) evaluated phage treatment in combination with ASM or with copper-mancozeb for the control of *Xanthomonas* leaf blight of onion and found that both combinations resulted in enhanced disease control. Gašić et al. (2018) found that integrated application of the *Xanthomonas euvesicatoria* specific phage KΦ1 and copper-hydroxide resulted in increased efficacy of the treatments in control of pepper bacterial spot compared to the copper-hydroxide alone. To the contrary, Balogh et al. (2008) observed no improvement in control of citrus canker or citrus bacterial spot with the combination of phages with copper-mancozeb.

However, most of our knowledge is based on experience from several experiments of tomato bacterial spot control. In an effort to develop more sustainable strategies for reducing bacterial spot severity on tomato, Obradovic et al. (2004, 2005, 2008) compared efficacy of various combinations of unformulated phages, biocontrol agents, including strains of PGPR, bacterial antagonists, SAR inducers (harpin, ASM) and copper hydroxide in greenhouse experiments. The intention was to integrate some of these practices, optimizing their benefits in control of tomato bacterial spot in the greenhouse, aiming at developing a comprehensive phage-based integrated management strategy for disease control in commercial tomato fields. Several combinations of treatments that effectively controlled tomato bacterial spot in the greenhouse were tested in field trials carried out in north and central Florida during three consecutive seasons. Although copper sensitive strains were used, providing more effective control of bacterial spot with copper bactericides, application of formulated phages twice a week was either more effective or equally effective compared with the standard copper-mancozeb treatment. However, integration of phage treatments and ASM provided an additional reduction in disease pressure and resulted in more efficient foliar disease control than ASM, phage, or copper-mancozeb alone (Obradovic et al., 2004, 2008).

A similar strategy has been recently applied for control of pepper bacterial spot. In order to develop an efficient integrated disease management program, Šević et al. (2019) studied the efficacy of *X. euvesicatoria* specific phage KΦ1 treatment, applied alone or in combination with different biocontrol agents and chemicals. Although the phage KΦ1 treatment significantly reduced symptom development (62-77%) compared to the untreated control, its integration with ASM and copper-hydroxide provided more efficient disease control (96-98%). Different mode of action of phages, ASM and copper compound confronted the pathogen more efficiently and contributed to the sustainability of the disease management.

## Commercialization of phages in agriculture

Phages have been available commercially for a relatively short time. OmniLytics, Inc. (Salt Lake City, UT), the U.S. company, has received the first

EPA permit (EPA Reg. No. 67986-1) for the application of phages for control of bacterial spot on pepper and tomato and bacterial speck on pepper. Currently, this company extended its product line by biopesticides containing phages against tomato bacterial canker, apple and pear fire blight, and citrus canker. In Europe, the first biopesticide named Erwiphage, containing phages for control of fire blight of apple, was registered by Hungarian company Enviroinvest (Enviroinvest, n.d.). Moreover, the Scottish company, APS Biocontrol developed a product named Biolyse, containing phages for control of potato soft rot disease during storage (APS Biocontrol Ltd, n.d.).

## Conclusions

The effectiveness of phage therapy in the control of a number of bacterial plant diseases is under intense evaluation lately. The potential benefits are substantial: phage therapy would provide an environmentally friendly disease control strategy against a group of plant pathogens, bacteria, which are chronically hard to control. However, a number of hurdles exist such as the short residual activity and narrow host range of these agents, the considerable effect of the environment and development of resistance to the agent in the target bacteria. By now, most of these issues have been addressed, and it appears that phages do have a place in disease management as part of an integrated approach.

Phages have significant potential for controlling plant pathogens in the phyllosphere. Success requires that high populations of phage exist in order to guarantee that bacterial lysis occurs. The single most important physical factor that affects the efficacy of phages in natural environments is the UV spectrum of sunlight. Therefore, designing strategies for reducing exposure of phages to UV is critical to success. The timing of application can be altered to maintain high concentrations of the phage on the leaf surface. Protective formulations have been used to prolong phage activity in the field; however, considerably more work is needed in this area to identify compounds that can extend persistence on leaf surfaces. Another strategy for maintaining high populations of phages has been to use non-pathogenic bacterial strains that are sensitive to the phage(s) or a closely related organism that do not cause disease on the plant host. Selecting the correct phages for disease control is critical. This requires careful monitoring of bacterial strains in the field to minimize the development of bacterial strains with resistance to the deployed phages or the proliferation of wild-type strains that are not sensitive to the phages that are used for field application. Considering the present knowledge, phages can be effective in bacterial disease control but the best effectiveness is achieved if they are used as part of an integrated management strategy.

## Acknowledgements

The first and fourth author were supported by the Government of Serbia, Ministry of science, education and technological development, project III46008 and the contract No. 451-03-68/2020-14/200116.

## References

Adams, M.H. (1959). Bacteriophages. New York: Interscience Publishers.

Arthurs, S.P., Lacey, L.A., and Behle, R.W. (2006). Evaluation of spray-dried lignin-based formulations and adjuvants as solar protectants for the granulovirus of the codling moth, *Cydia pomonella* (L). Journal of Invertebrate Pathology 93, 88–95.

Bae, J.Y., Wu, J., Lee, H.J., Jo, E.J., Murugaiyan, S., Chung, E., and Lee, S.W. (2012). Biocontrol potential of a lytic bacteriophage PE204 against bacterial wilt of tomato. J. Microbiol. Biotechnol. 22, 1613-1620.

Balogh, B. (2002). Strategies of improving the efficacy of bacteriophages for controlling bacterial spot of tomato. MS Thesis, Gainesville, FL: University of Florida.

Balogh, B. (2006). Characterization and use of bacteriophages associated with citrus bacterial pathogens for disease control. PhD Dissertation, Gainesville, FL: University of Florida.

Balogh, B., Jones, J.B., Momol, M.T., Olson, S.M., Obradovic, A., King, P., and Jackson, L.E. (2003). Improved effi cacy of newly formulated bacteriophages for management of bacterial spot on tomato. Plant Disease 87, 949–954.

Balogh, B., Canteros, B.I., Stall, R.E., and Jones, J.B. (2008). Control of Citrus Canker and Citrus Bacterial Spot with Bacteriophages. *Plant Disease 92,* 1048-1052.

Balogh, B., Nga, N.T.T. and Jones, J.B. (2018). Relative Level of Bacteriophage Multiplication *in vitro* or in Phyllosphere May Not Predict *in planta* Efficacy for Controlling Bacterial Leaf Spot on Tomato Caused by *Xanthomonas perforans*. *Front. Microbiol.* 9:2176. doi: 10.3389/fmicb.2018.02176

Basim, H., Stall, R. E., Minsavage, G. V., and Jones, J. B. (1999). Chromosomal gene transfer by conjugation in the plant pathogen *Xanthomonas axonopodis* pv. *vesicatoria*. Phytopathology 1999, 1044–1049. doi: 10.1094/PHYTO. 1999.89.11.1044.

Behle, R.W, and Popham, H.J. (2012). Laboratory and field evaluations of the efficacy of a fast-killing baculovirus isolate from *Spodoptera frugiperda*. J. Inverteb. Pathol. *109,* 194-200.

Behle, R.W., McGuire, M.R., and Shasha, B.S. (1996). Extending the residual toxicity of *Bacillus thuringiensis* with casein-based formulations. Journal of Economic Entomology 89, 1399–1405.

Bender, C.L., and Cooksey, D.A. (1986). Indigenous plasmids in *Pseudomonas syringae* pv. *tomato*: Conjugative transfer and role in copper resistance. Journal of Bacteriology 165, 534–541.

Bender, C.L., and Cooksey D.A. (1987). Molecular cloning of copper resistance genes from *Pseudomonas syringae* pv. *tomato*. Journal of Bacteriology 169, 470–474.

Bender, C.L., Malvick, D.K., Conway, K.E., George, S., and Cooksey D.A. (1990). Characterization of pXv10A, a copper resistance plasmid in *Xanthomonas campestris* pv. *vesicatoria*. Applied and Environmental Microbiology 56, 170–175.

Bergamin Filho, A., and Kimati, H. (1981). Estudos sobre um bacteriofago isolado de *Xanthomonas campestris*. II. Seu emprego no controle de *X. campestris* e *X. vesicatoria*. Summa Phytopathologica 7, 35–43.

Birge, E.A. (2000). Bacterial and bacteriophage genetics, 4th ed. Springler-Verlag New York Inc., New York, NY. Page 74.

Born, Y., Fieseler, L., Marazzi, J., Lurz, R., Duffy, B., and Loessner, M.J. (2011). Novel virulent and broad-hostrange *Erwinia amylovora* bacteriophages reveal a high degree of mosaicism and a relationship to Enterobacteriaceae phages. Appl. Environ. Microbiol. 77, 5945-5954.

Born, Y., Bosshard, L., Duffy, B., Loessner, M.J., and Fieseler, L. (2015). Protection of *Erwinia amylovora* bacteriophage Y2 from UV-induced damage by natural compounds. *Bacteriophage* 5:e1074330. doi: 10.1080/21597081.2015.1074330

Bouzar, H., Jones, J.B., Stall, R.E., Louws, F.J., Schneider, M., Rademaker, J.L.W., de Bruijn, F. J., and Jackson, L.E. (1999). Multiphasic analysis of xanthomonads causing bacterial spot disease on tomato and pepper in the Caribbean and Central America: Evidence for common lineages within and between countries. Phytopathology 89, 328-335.

Canteros, B.I. (1999). Copper resistance in *Xanthomonas campestris* pv. *citri*. In: Mahadevan A, ed. Plant Pathogenic Bacteria – Proceedings of the 9th International Conference, 26–29 August, 1996. Chennai, India: Centre for Advanced Study in Botany, University of Madras, 455–459.

Civerolo, E.L. (1972). Interaction between bacteria and bacteriophages on plant surfaces and in plant tissues. In: Maas Geesteranus HP, ed. Proceedings of the Third International Conference of Plant Pathogenic Bacteria, April 14–21, 1971. Wageningen: Centre for Agricultural Publishing and Documentation, 25–37.

Civerolo, E.L. (1973). Relationship of *Xanthomonas pruni* bacteriophages to bacterial spot disease in prunus. Phytopathology 63, 1279–1284.

Civerolo, E.L., and Keil, H.L. (1969). Inhibition of bacterial spot of peach foliage by *Xanthomonas pruni* bacteriophage. Phytopathology 59, 1966–1967.

Cooksey, D.A. (1990). Genetics of bactericide resistance in plant pathogenic bacteria. Annual Review of Phytopathology 28, 201–219.

Das, M., Bhowmick, T.S., Ahern, S.J., Young, R., and Gonzalez, C.F. (2015). Control of pierce's disease by phage. *PLoS ONE* 10:e0128902. doi: 10.1371/journal.pone.0128902

D'Herelle, F. (1917). An invisible microbe that is antagonistic to the dysentery bacillus. *Comptes rendus Acad. Sciences* **165:** 373-375.

Du, Q., Zhu, W., Zhao, Z., Q!an, X., and Xu, Y. (2012). Novel benzo-1,2,3-thiadiazole-7 -carboxylate derivatives as plant activators and the development of their agricultural applications. J. Agric. Food Chern. 60, 346-353.

Frampton, R.A., Pitman, A.R., and Fineran, P.C. (2012). Advances in bacteriophage-mediated control of plant pathogens. Int. J. Microbiol. 2012,326452.

Flaherty, J.E., Jones, J.B., Harbaugh, B.K., Somodi, G.C., and Jackson, L.E. (2000). Control of bacterial spot on tomato in the greenhouse and fi eld with H-mutant bacteriophages. HortScience 35, 882–884.

Fujiwara, A., Fujisawa, M., Hamasaki, R., Kawasaki, T., Fujie, M., and Yamada, T. (2011). Biocontrol of Ralstonia solanacearum by treatment with lytic bacteriophages. Appl. Environ. Micro bioi. 77,4155-4162.

Gašić, K., Kuzmanović, N., Ivanović, M., Prokić, A., Sević, M., and Obradović, A. (2018). Complete genome of the *Xanthomonas euvesicatoria* specific bacteriophage KΦ1, its survival and potential in control of pepper bacterial spot. Front. Microbiol. 9, 2021.

Gent, D.H., and Schwartz, H.F. (2005). Management of xanthomonas leaf blight of onion with a plant activator, biological control agents, and copper bactericides. Plant Disease 89, 631–639.

Gill, J. J., and Abedon, S. T. (2003). Bacteriophage Ecology and Plants. doi: 10.1094/APSnetFeature-2003-1103.

Goodridge, L.D. (2004). Bacteriophage biocontrol of plant pathogens: Fact or fiction? Trends in Biotechnology 22, 384–385.

Goto, M. (1992). Fundamentals of Bacterial Plant Pathology. San Diego, CA: Academic Press.

Gutierrez-Barranquero, J.A., de Vicente, A., Carrion, V.J., Sundin, G.W., and Cazorla, F.M. (2013). Recruitment and rearrangement of three different genetic determinants into a conjugative plasmid increase copper resistance in Pseudomonas syringae. Appl. Environ. Microbiol. *79,* 1028-1033.

Guy, E., Genissel, A., Hajri, A., Chabannes, M., David, P., Carrere, S., Lautier, M., Roux, B., Boureau, T., Arlat, M., et al. (2013). Natural genetic variation of Xanthomonas campestris pv. campestris pathogenicity on arabidopsis revealed by association and reverse genetics. MBio. *4,* e00538-12.

Huang, Ch.-H., Vallad, G.E., Zhang, S., Wen, A., Balogh, B., Figueiredo, J.F.L., Behlau, F., Jones, J.B., Momol, M.T., and Olson, S.M. (2012). Effect of application frequency and reduced rates of acibenzolar-s-methyl on the field efficacy of induced resistance against bacterial spot on tomato. Plant Dis. 96,221-227.

Ignoffo, C.M., and Garcia. (1992). Combinations of environmental factors and simulated sunlight affecting activity of inclusion bodies of the *Heliothis* (*Lepidoptera: Noctuidae*) nucleopolyhedrosis virus. Environmental Entomology 21, 210–213.

Ignoffo, C.M., and Garcia, C. (1994). Antioxidant and oxidative enzyme effects on the inactivation of inclusion bodies of the *Heliothis* baculovirus by simulated sunlight- UV. Environmental Entomology 23, 1025–1029.

Ignoffo, C.M., Garcia, C., and Saathoff, S.G. (1997). Sunlight stability and rain-fastness of formulations of *Baculovirus heliothis*. Environmental Entomology 26, 1470–1474.

Ignoffo, C.M., Rice, W.C., and McIntosh, A.H. (1989). Inactivation of nonoccluded and occluded baculoviruses and baculovirus-DNA exposed to simulated sunlight. Environmental Entomology 18, 177–183.

Iriarte, F.B., Balogh, B., Momol, M.T., Smith, L.M., Wilson, M., and Jones, J.B. (2007). Factors affecting survival of bacteriophage on tomato leaf surfaces. Applied and Environmental Microbiology 73, 1704–1711.

Iriarte, F. B., Obradović, A., Wernsing, M.H., Jackson, L.E., Balogh, B., Hong, J.A., Momol, M.T., Jones, J.B., and Vallad, G.E. (2012). Soil-based systemic delivery and phyllosphere *in vivo* propagation of bacteriophages: Two possible strategies for improving bacteriophage persistence for plant disease control. Bacteriophage. 2, 215-224.

Jackson, L.E. (1989). Bacteriophage prevention and control of harmful plant bacteria. US Patent No.4828999.

Johnson, K.B. (1994). Dose–response relationships and inundative biological control. Phytopathology 84, 780–784.

Jones, J.B., Lacy, G.H., Bouzar, H., Stall, R.E., and Schaad, N.W. (2004). Reclassification of the xanthomonads associated with bacterial spot disease of tomato and pepper. Syst. Appl. Micro bioi. 27,755-762.

Jones, J.B., Jackson, L.E.,Balogh, B., Obradovic, A., Iriarte, F.B., and Momol, M.T. (2007). Bacteriophages or plant disease control. Annu. Rev. of Phytopathol. 45:7.1–7.18. Annual Reviews Inc. Palo Alto.

Jones, J.B., Vallad, G.E., Iriarte, F.B., Obradovic, A., Wernsing, M.H., Jackson, L.E., Balogh, B., Hong, J.C., and Momol, M.T. (2012). Considerations for using bacteriophages for plant disease control. Bacteriophage. 2, 208-214.

Jones J.B., Svircev A.M., Obradović A.Ž. (2018). Crop Use of Bacteriophages. In: Harper D., Abedon S., Burrowes B., McConville M. (eds) Bacteriophages. Springer, Cham. Pp 1-18. https://doi.org/10.1007/978-3-319-40598-8_28-1. Online ISBN 978-3-319-40598-8

Katznelson, H. (1937). Bacteriophage in relation to plant diseases. The Botanical Review 3, 499–521.

Kotila, J.E., and Coons, G.H. (1925). Investigations on the blackleg disease of potato. Michigan Agricultural Experimental Station Technical Bulletin 67, 3–29.

Lang, J.M., Gent, D.H., and Schwartz, H.F. (2007). Management of *Xanthomonas* leaf blight of onion with bacteriophages and a plant activator. Plant Disease 91, 871–878.

Lee, Y., Hendson, M., Ponopoulos, N.J., and Schroth, M.N. (1994). Molecular cloning, chromosomal mapping, and sequence analysis of copper resistance genes from *Xanthomonas campestris* pv. *juglandis*: Homology with small blue copper proteins and multicopper oxidase. Journal of Bacteriology 176, 173–188.

Lim, J.A., Jee, S., Lee, D.H., Roh, E., Jung, K., Oh, C., and Heu, S. ( 2013). Biocontrol of *Pectobacterium carotovorum* subsp. *carotovorum* using bacteriophage PP I. J. Microbiol. Biotechnol. 23, 1147-1153.

Louws, FJ., Wilson, M., Campbell, H.L., Cuppels, D.A., Jones, J.B., Shoemaker, P.B., Sahin, F., and Miller, S.A. (2001). Field control of bacterial spot and bacterial speck of tomato using a plant activator. Plant Disease 85, 481–488.

Luria, S.E., and Delbrück, M. (1943). Mutations of bacteria from virus sensitivity to virus resistance. Genetics 28, 491-511.

Mallmann, W.L., and Hemstreet, C.J. (1924). Isolation of an inhibitory substance from plants. Agricultural Research, 28, 599–602.

Manulis, S., Zutra, D., Kleitman, F., Dror, O., David, I., Zilberstaine, M., and Shabi, E. (1998). Distribution of streptomycin-resistant strains of *Erwinia amylovora* in Israel and occurrence of blossom blight in the autumn. Phytoparasitica 26, 223–230.

Maxson-Stein, K., He, S.Y., Hammerschmidt, R., and Jones, AL. (2002). Effect of treating apple trees with acibenzolar-*S*-methyl on fi re blight and expression of pathogenesis-related protein genes. Plant Disease 86, 785–790.

McNeil, D.L., Romero, S., Kandula, J., Stark, C., Stewart, A., and Larsen, S. (2001). Bacteriophages: A potential biocontrol agent against walnut blight (*Xanthomonas campestris* pv. *juglandis*). New Zealand Plant Protection 54, 220–224.

Obradovic, A., Jones, J.B., Momol, M.T., Balogh, B., and Olson, S.M. (2004). Management of tomato bacterial spot in the fi eld by foliar applications of bacteriophages and SAR inducers. Plant Disease 88, 736–740.

Obradovic, A., Jones, J.B., Momol, M.T., Olson, S.M., Jackson, L.E., Balogh, B., Guven, K., and Iriarte, F.B. (2005). Integration of biological control agents and systemic acquired resistance inducers against bacterial spot on tomato. Plant Disease 89, 712–716.

Obradović A., Jones J.B., Balogh B., Momol M.T. (2008). Integrated management of tomato bacterial spot. In: Integrated Management of Plant Diseases Caused by Fungi, Phytoplasma and Bacteria. Edited by A. Ciancio and K. G. Mukerji. Springer Science + Business Media B. V., 211-223.

Okabe, N., and Goto, M. (1963). Bacteriophages of plant pathogens. *Annual Review of* Phytopathology 1, 397–418.

Qui, D., Wei, Z. M., Bauer, D. W., and Beer, S. V. (1997). Treatment of tomato seed with harpin enhances germination and growth and induces resistance to *Ralstonia solanacearum*. Phytopathology, 87, S80.

Reddy, N.P., A Khan, P.A., Devi, K.U., Victor, J.S., and Sharma, H.C. (2008). Assessment of the suitability of Tinopal as an enhancing adjuvant in formulations of the insect pathogenic fungus Beauveria bassiana (Bals.) Vuillemin. Pest Manag. Sci. *64,* 909-915.

Romero, A.M., Kousik, C.S., and Ritchie, D.F. (2001). Resistance to bacterial spot in bell pepper induced by acibenzolar-*S*-methyl. Plant Disease 85, 189–194.

Romero-Suarez, S., Jordan, B., and Heinemann, J.A. (2012). Isolation and characterization of bacteriophages infecting *Xanthomonas arboricola* pv. *juglandis*, the causal agent of walnut blight disease. World J. Microbiol. Biotechnol. *28,* 1917-1927.

Saccardi, A., Gambin, E., Zaccardelli, M., Barone, G., Mazzucchi, U. (1993). *Xanthomonas campestris* pv. *pruni* control trials with phage treatments on peaches in the orchard. Phytopathologia Mediterranea 32, 206–210.

Schnabel, E.L., Fernando, W.G.D., Meyer, M.P., Jones, A.L., and Jackson, L.E. (1999). Bacteriophage of *Erwinia amylovora* and their potential for biocontrol. Acta Horticulturae 489, 649–654.

Stall, R.E., Loschke, D.C., and Jones, J.B. (1986). Linkage of copper resistance and avirulence loci on a self-transmissible plasmid in *Xanthomonas campestris* pv. *vesicatoria*. Phytopathology 76, 240–243.

Storey, M.V., and Ashbolt, N.J. (2001). Persistence of two model enteric viruses (B40–8 and MS- 2 bacteriophages) in water distribution pipe biofi lms. Water Science and Technology 43, 133–138.

Summers, W.C. (2005). Bacteriophage research: Early history. In: Kutter E, Sulakvelidze A, eds. Bacteriophages: Biology and Applications. Boca Raton, FL: CRC Press, 5–27.

Svircev, A.M., Lehman, S.M., Kim,W.S., Barszcz, E., Schneider, K.E., and Castle, A.J. (2006). Control of the fire blight pathogen with bacteriophages. In: Zeller W, Ullrich C, eds. Proceedings of the 1st International Symposium on Biological Control of Bacterial Plant Diseases, October 23–26, 2005. Berlin, Germany: Die Deutsche Bibliothek – CIP-Einheitsaufnahme, 259–261.

Sykes, I.K., Lanning, S., and Williams, S.T. (1981). The effect of pH on soil actinophage. Journal of General Microbiology 122, 271–280.

Šević M., Gašić, K., Ignjatov M., Mijatović M., Prokić, A., Obradović, A. (2019). Integration of biological and conventional treatments in control of pepper bacterial spot. Crop Protection, 119: 46-51.

Tanaka, H., Negishi, H., .and Maeda, H. (1990). Control of tobacco bacterial wilt by an avirulent strain of *Pseudomonas solanacearum* M4S and its bacteriophage. Annals of the Phytopathological Society of Japan 56, 243–246.

Thayer, P.L., and Stall, R.E. (1961). A survey of *Xanthomonas vesicatoria* resistance to streptomycin. Proceedings of the Florida State Horticultural Society 75, 163–165.

Thomas, R.C. (1935). A bacteriophage in relation to Stewart's disease of corn. Phytopathology, 25, 371–372.

Twort, F.W. (1915). An investigation on the nature of the ultra-microscopic viruses. Lancet, 189, 241–243.

Vidaver, A.K. (1976). Prospects for control of phytopathogenic bacteria by bacteriophages and bacteriocins. Annual Review of Phytopathology 14, 451–465.

Williams, S.T., Mortimer, A.M., and Manchester, L. (1987). Ecology of soil bacteriophages. In: Goyal SM, Gerba CP, Bitton G, eds. Phage Ecology. New York: John Wiley and Sons, 157–179.

Zaccardelli, M., Saccardi, A., Gambin, E., and Mazzucchi, U. (1992). *Xanthomonas campestris* pv. *pruni* bacteriophages on peach trees and their potential use for biological control. Phytopathologia Mediterranea 31, 133–140.

# Chapter 8

# Phage Biocontrol Applications in Food Production and Processing

**Amit Vikram*, Joelle Woolston and Alexander Sulakvelidze**

Intralytix, Inc., Columbia, Maryland 21046, U.S.A.

*Corresponding author: avikram@intralytix.com

DOI: https://doi.org/10.21775/9781913652517.08

**Abstract**

Bacteriophages, or phages, are one of the most – if not the most – ubiquitous organisms on Earth. Interest in various practical applications of bacteriophages has been gaining momentum recently, with perhaps the most attention (and most regulatory approvals) focused on their use to improve food safety. This approach, termed "phage biocontrol" or "bacteriophage biocontrol," includes both pre- and post-harvest application of phages as well as decontamination of the food contact surfaces in food processing facilities. This chapter focuses on post-harvest applications of phage biocontrol, currently the most commonly used type of phage mediation. We also briefly describe various commercially available phage preparations and discuss the challenges still facing this novel yet promising approach.

## Phages are Ancient and Abundant in Nature

Bacteriophages are the viruses that infect bacteria. They were discovered in 1917 by Félix d'Hérelle in 1917 (Salmond and Fineran, 2015; Sulakvelidze et al., 2001), who also coined the term "bacteriophage," derived from "bacteria" and the Greek φαγεῖν (*phagein*) meaning "to eat" or "to devour" bacteria. Numerous studies, including recent metagenomic surveys, suggest that phages are (i) arguably the oldest microorganisms on this planet that likely originated approximately 3 billion years ago (Brüssow, 2007), and (ii) likely the most ubiquitous organisms on Earth, abundant in all life-supporting environments including all natural untreated foods. To give a few examples: (1) There are an estimated $1.5 \times 10^8$ phage particles per gram of agricultural soil (Ashelford et al., 2003; Williamson et al., 2003); (2) There are an estimated 7 to $15 \times 10^6$ phages per mL in fresh water lakes (Mohiuddin and Schellhorn, 2015) and $10^6$ to $10^9$ particles per mL in sea water (Bergh et al., 1989); (3) Bacteriophages are likely present in 100% of fresh unprocessed foods and have been isolated from various food products such as beef, pork, chicken, fresh produce, dairy, and fermented foods (Aw et al., 2016; Park et al., 2011; Zhang et al., 2014); (4) Humans are constantly exposed to phages in their daily lives as highly diverse and abundant phage populations are present on various human organs such as the skin, vagina, etc. – with an estimated $10^{15}$ phages present in our gastrointestinal tract (Dalmasso et al., 2014; Hannigan et al., 2015; Shkoporov et al., 2019); (5) Similarly, animals carry a diverse and prolific phage population (Dalmasso et al., 2014; Delwart, 2012; Hannigan et al., 2015; Shkoporov et al., 2019). Altogether, phages are a significant part of the living biosphere and humans continuously interact with a multitude of phages, through food, water, and the general environment. The phage population is increasingly recognized as an active part of the mammalian microbiome and a contributor towards the health of their host (Cadwell, 2015).

Almost immediately after their discovery, the ability of phages to infect and kill bacteria led to the exploration of their therapeutic potential against bacterial pathogens, in a clinical approach known as "bacteriophage therapy" or "phage therapy" – with the first therapeutic use in humans described in 1919, just two years after their discovery by d'Herelle (Sulakvelidze and Kutter, 2005; Summers, 2001). Although the exact number of people treated therapeutically with phages is difficult to estimate, in all likelihood tens of thousands of people have been administered phages therapeutically since that initial clinical use, with no serious phage-related adverse effects ever reported (Kutter and Sulakvelidze, 2004). However, the discovery and increased use of antibiotics during the 1940s and 1950s, coupled with an incomplete understanding of phage biology, as well as several other factors, led to a decline in the clinical use of phages in Western Europe and North America. In contrast, phage therapy continued to be utilized in the former Soviet Union and some Eastern European countries (e.g., Poland) where therapeutic phage preparations are still readily available in pharmacies today (Sulakvelidze et al., 2001; Sulakvelidze and Kutter, 2005; Summers, 2012). The situation has changed recently as phage therapy and other phage-related technologies – including phage biocontrol – have been gaining a renewed interest in the West.

Two major developments played a significant role in rekindling the interest in phages as antimicrobial agents. First, the emergence and widespread distribution of antimicrobial resistant (AMR) bacterial pathogens have limited our therapeutic options. The situation is further exacerbated by the limited number of new antimicrobial drugs entering the market; also, most new antibiotics approved in recent years are modifications of existing drugs and, therefore, are at higher risk of being rendered ineffective in a short period (Talbot et al., 2019). Secondly, there has been an increased appreciation of the damage broad-spectrum antibiotics can inflict upon the microbiome. In fact, antibiotic mediated microbiome perturbation is now believed to contribute to several non-communicable and chronic diseases (Dietert and Dietert, 2015). Also, due to the rapidly growing field of microbiome research, connections are increasingly being made and/or hypothesized between microbiome composition and various health-associated conditions, ranging from obesity to Alzheimer to certain forms of cancer. If these links are substantiated through further research, there would be a need for a modality(ies) capable of the targeted elimination of certain bacteria in our microbiome which would not deleteriously impact other, beneficial bacterial species (i.e., a modality to fine-tuning the microbiome). As a result, there has been an increased interest in targeted antimicrobials, such as phages, which are capable of killing disease-causing bacteria in the gastrointestinal tract without altering the normal microflora. Lytic phages demonstrate remarkable bactericidal activity against specific bacterial strains and can be employed to treat and manage AMR as well as non-AMR bacterial infections either alone or in conjunction with existing antimicrobials (Hesse and Adhya, 2019). Since lytic phages exert their bactericidal activity through very different mechanisms than antibiotics, they demonstrate similar efficacy for both AMR and non-AMR bacterial pathogens (Hesse and Adhya, 2019; Kortright et al., 2019; Sulakvelidze and Morris, 2015). In addition, the mechanisms of bacterial resistance to phages differ from those for antibiotics and the strains that develop resistance to antibiotics typically remain susceptible and can be killed by bacteriophages (Chen et al., 2017). Therefore, phages may help to manage various bacterial infections, including those caused by pandrug-resistant bacterial strains (Sulakvelidze and Morris, 2015). The high specificity towards the target bacterial species provides an additional benefit, the preservation of microbial diversity and function (Cieplak et al., 2018). Thus, developing phage preparations for human clinical applications appears to be a logical fit for lytic bacteriophages. Yet, most commercial phage developments in the United States thus far have focused not on human therapeutic applications or microbiome modulations (although efforts in those areas are underway, including several clinical trials), but on using bacteriophages for improving the safety of our food products (Hesse and Adhya, 2019).

## Foodborne Pathogens – A Growing Challenge

Foodborne illnesses of microbial origin continue to be a serious food safety problem worldwide. In addition to being of significant public health importance, the economic impact of foodborne bacterial infections is considerable. For example, in the US alone, the average incidence of foodborne illness is estimated to cost ~$1500/person, with the total annual estimated cost of these foodborne diseases reaching over $75 billion (Scharff, 2012). In addition, substantive costs to the food industry are incurred due to product loss and brand-damaging negative publicity that is associated with the recall of products contaminated with pathogenic bacteria. Thus, there are both strong public health and economic incentives to develop novel approaches for managing contamination of a broad range of foods by specific foodborne bacterial pathogens.

Contamination prevention and the provision of safe food supply is one of the high priority areas to control and limit the foodborne pathogen outbreaks under the “One Health” approach (USDA, 2016). However, the challenge is not a straightforward one to solve rapidly or easily. The epidemiology of foodborne pathogens is complex and involves multiple routes of transmission from food animals and agricultural produce to consumers. For example, animals are often asymptomatic carriers of Shiga-toxin producing *Escherichia coli* (STEC), *Salmonella*, *Campylobacter*, and *Listeria* and they can spread the pathogens to other food animals, crops, slaughter facilities, and, in some instances, directly to humans. Food processing facilities can also harbour foodborne pathogens as biofilms, which can potentially transfer to food products and reach consumers. According to estimates published by the Foodborne Disease Burden Epidemiology Reference Group (FERG) of the World Health Organization (WHO), pathogenic bacteria were responsible for almost 350 million illnesses and 187,000 deaths globally in 2010 (Havelaar et al., 2015). Four bacterial pathogens, *Escherichia coli* (including Enterotoxigenic, Enteropathogenic, and Shiga-toxin producing *E. coli*), *Campylobacter* spp., non-typhoidal *Salmonella enterica*, and *Shigella* spp. were responsible for 336 million (96%) of the illnesses. While these four pathogens cause diarrheal diseases that are generally self-limiting, there are still over 159,000 deaths, attributed to them yearly. Children under five years old are disproportionally impacted, due to under-developed immune systems, and a staggering 45% of those deaths were in children under 5 years of age (Havelaar et al., 2015).

Food processors routinely employ antimicrobial interventions to reduce contamination of the foods with foodborne pathogens. Also, in an effort to improve the safety of foods, FSIS regulations and the Food Safety and Modernization Act have mandated implementation of hazard analysis and critical control point systems (HACCP) to manage pathogens by all US establishments processing red meats, fresh fruit and vegetables, dairy, and other food products (Food and Drug Administration, 2015; Food Safety and Inspection Service, 2012; U. S. Department of Agriculture, 2012). Currently, the conventional pathogen decontamination protocols in food processing facilities focus primarily on using chemicals, physical disruption techniques, and irradiation to reduce the microbial burden from those facilities and from the foods produced in them (Gomez-Lopez, 2012; Hui, 2003; Maukonen et al., 2003). For example, various harsh chemical sanitizers, such as chlorine and peracetic acid (PAA), are commonly utilized to reduce microbial contaminants of many fresh fruits and vegetables as well as Ready-To-Eat (RTE) food products (Beuchat and Ryu, 1997; Sohaib et al., 2016). Heat pasteurization is often used to reduce bacterial numbers, generally in liquids and dairy items, such as milk. High Pressure Processing (HPP) is also used to successfully reduce pathogens in liquid products, as well as pre-cooked, meant to be frozen meals (Bajovic et al., 2012; Wolbang et al., 2008). This technique exposes foods to high pressure to inactivate microbes. Ionizing radiation (i.e., irradiation) has been approved as a means for reducing the burden of pathogenic organisms in foods since 1997 (Food and Drug Administration, 1997). However, no single approach is 100% effective, and the above-mentioned approaches also have some significant drawbacks. For instance, many chemical sanitizers corrode and damage food processing equipment (Fatica and Schneider, 2009; Moye et al., 2018) and could have toxic chemical residues that may harm the environment (i.e., they are not environmentally friendly). Pasteurization and HPP are not suitable for fresh produce and meat products as they can adversely affect the organoleptic properties and/or the nutritional content of some foods (Bajovic et al., 2012; Wolbang et al., 2008). Irradiation, which can deleteriously affect the appearance of some foods, also has low customer acceptance, which is compounded by a labelling requirement for many food items treated with radiation (Suklim et al., 2014; Wheeler et al., 1999). As a result, the recent trend has been towards identifying alternative non-chemical, environmentally friendly (aka green) antimicrobial approaches. One such approach is phage biocontrol.

The concept of phage biocontrol is to apply lytic bacteriophages, with strong lytic potency against one or more foodborne bacterial pathogens, onto the foods at high risk of contamination. The phages can lyse the targeted contaminating bacteria and significantly reduce (or eradicate) the foodborne pathogen(s), thus making the foods safe for consumption. Phage biocontrol is increasingly being accepted as a natural and green technology, effective at specifically targeting bacterial pathogens in various foods (Table 1) and the development and commercialization of bacteriophage products is now an emerging worldwide industry (Sulakvelidze, 2013) (Table 2). Phage biocontrol addresses many of the challenges facing the traditional chemical- or irradiation-based approaches

| **Table 1.** A summary of studies of direct phage application onto a variety of foods† | | | |
|---|---|---|---|
| **Bacterium** | **Phages** | **Notes** | **Reference** |
| *Bacillus cereus* | BCP1-1 | *Bacillus cereus* counts were decreased after treatment with a single phage in fermented soya bean paste without affecting *Bacillus subtilis*, a critical component of the fermentation process. | (Bandara et al., 2012) |
| *Campylobacter jejuni* | Φ2 | Counts of *Campylobacter* were reduced by ~1 log on the surface of chicken skin stored at 4°C after the application of a single phage. | (Atterbury et al., 2003) |
| *Campylobacter jejuni*; *Salmonella* spp. | *C. jejuni* typing phage 12673, P22, 29C; *Salmonella* typing phage 12 | *C. jejuni* levels were decreased ~2 logs on experimentally-contaminated chicken skin after application of phage at an MOI of 100:1 or 1,000:1. *Salmonella* levels were reduced by ~2 logs on chicken skin treated with phage at an MOI of 100:1 or 1,000:1 and stored for 48 h, and bacterial counts were reduced below the limit of detection when lower levels of bacteria were used to contaminate the chicken. | (Goode et al., 2003) |
| *Campylobacter jejuni*; *Salmonella* spp. | Cj6; P7 | *Campylobacter* levels were significantly decreased in beef after application of the phage Cj6 and decrease in bacterial levels were not significant at low levels of bacterial contamination (~100 CFU/cm$^2$). *Salmonella* counts were decreased ~2-3 logs at 5°C and >5.9 logs at 24°C in raw and cooked beef after P7 phage application. Surviving *Salmonella* colonies were still sensitive to P7. For both phages, the killing of bacteria was higher at an MOI of 10,000:1 and ~10,000 CFU/cm$^2$ of bacteria. | (Bigwood et al., 2008) |
| *Cronobacter sakazakii* | ESP 1-3, ESP 7321 | In infant formula, *Cronobacter sakazakii* (formerly *Enterobacter sakazakii*) levels were decreased after phage addition. The reduction was dependent on the phage concentration, and the phages were more effective at 24°C than 37°C or 12°C. | (Kim et al., 2007) |
| *Cronobacter sakazakii* | Five phages | Growth of 36 of 40 test strains was inhibited by a phage cocktail tested in infant formula experimentally contaminated with *C. sakazakii*. Further, both high and low concentrations ($10^6$ and $10^2$ CFU/mL) of bacteria were eliminated from the liquid culture medium treated with the individual phage ($10^8$ PFU/mL). | (Zuber et al., 2008) |
| *Escherichia coli* O157:H7 | e11/2, pp01, e4/1c | After incubation at 37°C, a three-phage cocktail used to treat the surface of beef that was contaminated ($10^3$ CFU/g) with *E. coli* O157:H7 eliminated the bacterium from ~78% of the treated specimens. | (O'Flynn et al., 2004) |
| *Escherichia coli* O157:H7 | EcoShield™ (formerly ECP-100) | *E. coli* 0157:H7 levels were decreased by ~1-3 logs or reduced below the limits of detection, on tomatoes, broccoli or spinach after treatment with a phage cocktail while *E. coli* O157:H7 levels were decreased by ~1 log when the phages were applied to ground beef. | (Abuladze et al., 2008) |
| *Escherichia coli* O157:H7 | EcoShield™ | A phage cocktail applied to experimentally contaminated lettuce and cut cantaloupe significantly reduced *E. coli* O157:H7 levels by up to 1.9 and 2.5 logs, respectively. | (Sharma et al., 2009) |
| *Escherichia coli* O157:H7 | Cocktail BEC8 | At various temperatures (4, 8, 23 and 37°C), the phage cocktail significantly reduced the level of *E. coli* O157:H7 on leafy green vegetables by ~2-4 logs. The inclusion of essential oil increased this effect. | (Viazis et al., 2011) |

| | | | |
|---|---|---|---|
| *Escherichia coli* O157:H7 | EcoShield™ | The levels of *E. coli* O157:H7 were reduced by ≥ 94% and ~87% on the surface of experimentally contaminated beef and lettuce respectively after addition of the phage cocktail; however, the single treatment did not protect foods after recontamination with the same bacteria (i.e., phage biocontrol had no continued technical effect on the foods). | (Carter et al., 2012) |
| *Escherichia coli* O157:H7 | EcoShield™ | After a 30 mins phage treatment at both 4 and 10°C, levels of *E. coli* O157:H7 were decreased by >2 logs on leafy greens under both ambient and modified atmosphere packaging storage. | (Boyacioglu et al., 2013) |
| *Escherichia coli* | FAHEc1 | Contamination of raw and cooked beef was decreased by 2-4 logs at 5, 24 and 37°C in a concentration dependent manner after phage application. The *E. coli* displayed regrowth at higher temperatures. | (Hudson et al., 2013b) |
| *Escherichia coli* O157:H7 | EcoShield™ | A phage cocktail was applied to lettuce by spraying and dipping. A larger initial reduction (~0.8-1.3 logs) in *E. coli* O157:H7 counts was observed after spraying. Dipping required submerging the lettuce for as long as 2 mins, and the initial reductions were not significant. After 1 day of storage at 4°C, dipping in the highest concentration of the phage cocktail reduced *E. coli* by ~0.7 log. | (Ferguson et al., 2013) |
| *Escherichia coli* | EC6, EC9, EC11 | Two *E. coli* strains were eradicated from raw and ultra-high temperature processed (UHT) milk after treatment with a three-phage cocktail at 5-9°C and 25°C. For a third *E. coli* strain, phage treatment eliminated the bacteria from UHT milk; however, after an initial reduction, regrowth occurred in the raw milk after 144 or 9 h at 5-9°C and 25°C storage, respectively. | (McLean et al., 2013) |
| *Escherichia coli* O157:H7 | EcoShield PX™ | Application, via spraying, of the phage cocktail reduced the *E. coli* O157:H7 levels by as much as 97% on various food products. In addition, the phage cocktail reduced the occurrence of *E. coli* O157:H7 on chuck roast beef by ≥80 %. | (Vikram et al., 2020) |
| *Escherichia coli*, *Salmonella*, *Shigella* | EcoShield™, SalmoFresh™, ShigActive™ | Phage cocktails were as effective or more effective than chlorine wash at reducing targeted pathogenic bacteria from broccoli, cantaloupe and strawberries in samples containing a large amount of organic content. Combination of the phage cocktail and a produce wash produced a synergistic effect, i.e., higher reductions of bacteria. | (Magnone et al., 2013) |
| *Listeria monocytogenes* | ListShield™ (formerly LMP-102) | *Listeria* counts were decreased ~2 logs and ~0.4 logs after application of a phage cocktail on melon and apple slices respectively; a synergistic effect was observed when phage and nisin were used, decreasing levels of *Listeria* on the fruit ~5.7 logs and ~2.3 logs, respectively. | (Leverentz et al., 2003) |
| *Listeria monocytogenes* | ListShield™ | Application of a phage cocktail 1, 0.5, or 0 h before honeydew melon tissue were contaminated with the bacterium was most effective at reducing *Listeria* counts. This effect was depended on the concentration of phage applied. *Listeria* counts decreased ~5-7 logs after 7 days when the phages were applied at the times described above. | (Leverentz et al., 2004) |

| | | | |
|---|---|---|---|
| *Listeria monocytogenes* | PhageGuard Listex™ (formerly Listex™; P100) | Levels of *L. monocytogenes* were reduced at least 3.5 logs after a single phage was administered to the surface of ripened red-smear soft cheese. The surviving *L. monocytogenes* colonies isolated from the cheese after phage treatment were not resistant to the phage. | (Carlton et al., 2005) |
| *Listeria monocytogenes* | A511, PhageGuard Listex™ | Levels of *L. monocytogenes* in experimentally contaminated chocolate milk and mozzarella cheese brine were eradicated after phage treatment at 6°C. When the phage cocktail was applied to various solid foods, including sliced cabbage, iceberg lettuce leaves, smoked salmon, mixed seafood, hot dogs, and sliced turkey meat, a reduction of *Listeria* of up to 5 logs was observed. | (Guenther et al., 2009) |
| *Listeria monocytogenes* | PhageGuard Listex™ | *L. monocytogenes* counts were decreased between 1.8-3.5 logs after application of a single phage at ~$10^8$ PFU/g to the surface of raw salmon fillets that were stored at 4°C or 22°C. | (Soni and Nannapaneni, 2010) |
| *Listeria monocytogenes* | PhageGuard Listex™ | Levels of *L. monocytogenes* were decreased 1.4-2.0 logs CFU/g at 4°C, 1.7-2.1 logs CFU/g at 10°C, and 1.6-2.3 logs CFU/g at room temperature (22°C) after application a single phage to the surface of raw catfish fillets. Regrowth was not observed after ten days of storage at either 4°C or 10°C. | (Soni et al., 2010) |
| *Listeria monocytogenes* | A511 | The natural microbial community on soft cheese was maintained after the addition of the phage. Levels of *Listeria* on experimentally contaminated cheese were decreased 2 logs and additional phage administrations did not improve the reduction of *Listeria*. | (Guenther and Loessner, 2011) |
| *Listeria monocytogenes* | FWLLm1 | *Listeria* levels were decreased 1-2 logs on the surface of experimentally contaminated chicken stored in vacuum packages at 4°C or 30°C. Subsequent regrowth of *Listeria* was observed at 30°C but not at 4°C. | (Bigot et al., 2011) |
| *Listeria monocytogenes* | PhageGuard Listex™ | Counts of *Listeria* decreased ~3 logs in experimentally contaminated queso fresco cheese after the addition of a single phage; however, subsequent growth was observed. Regrowth was prevented, and a similar log reduction was observed when PL + SD were included with the phage. Reduction of *Listeria* was lower, and regrowth occurred when LAE was included with phage. | (Soni et al., 2012) |
| *Listeria monocytogenes* | PhageGuard Listex™ | Compared to PL or PL + SD, a single phage was most effective at decreasing *Listeria* levels on RTE roast beef and turkey after storage at 4°C or 10°C, and subsequent bacterial growth was observed at both temperatures. Similar log reductions occurred when PL or PL + SD were used in conjunction with the phage, and regrowth was prevented or diminished at both 4°C and 10°C. | (Chibeu et al., 2013) |
| *Listeria monocytogenes* | PhageGuard Listex™ | Counts of *Listeria* were decreased by ~1.5 logs on experimentally contaminated melon and pear slices but not apple slices after two days at 10°C. Additionally, treatment with phage did not impact *Listeria* levels in apple juice but decreased bacterial contamination by ~4 and ~2.5 logs in melon and pear juice respectively. | (Oliveira et al., 2014) |

| | | | |
|---|---|---|---|
| *Listeria monocytogenes* | PhageGuard Listex™ | *Listeria* levels on soft cheese were decreased ~2-3 logs after 30 mins and ~0.8-1 log after storage for 7 days at 10°C. | (Silva et al., 2014) |
| *Listeria monocytogenes* | ListShield™ | Counts of *L. monocytogenes* were decreased 0.7 and 1.1 logs on experimentally contaminated cheese and lettuce respectively after a 5 min treatment with phage and decreased the bacteria 1.1 logs on the surface of apple slices after 24 h when combined with an anti-browning solution. The phage cocktail also eliminated *L. monocytogenes* from experimentally contaminated frozen entrees that were frozen and thawed after treatment. It was also effective in eliminating environmental contamination by *L. monocytogenes* in a smoked salmon processing plant. | (Perera et al., 2015) |
| *Listeria monocytogenes* | PhageGuard Listex™ | When applied to the surface of experimentally contaminated sliced pork ham, the phage reduced *Listeria* counts below the limit of detection after 72 h and performed better than nisin, sodium lactate or combinations of these antibacterial measures. | (Figueiredo and Almeida, 2017) |
| *Mycobacterium smegmatis* | Six phages | *M. smegmatis* counts were reduced below the limit of detection in milk treated with a six-phage cocktail or each component phage. Subsequent bacterial growth occurred when the component phages were used, but no growth was observed after 96 h at 37°C when the cocktail was applied. | (Endersen et al., 2013) |
| *Salmonella* spp. | SJ2 | *Salmonella* levels were reduced by 1-2 logs in raw and pasteurized cheeses created using milk that was treated with phage, while cheese made from milk without phage saw *Salmonella* counts rise ~ 1 log. | (Modi et al., 2001) |
| *Salmonella* spp. | SCPLX-1 (an early version of SalmoFresh™) | Counts of *Salmonella* were decreased by ~3.5 logs at 5 and 10°C and ~2.5 logs at 20°C on melon slices after application of a four-phage cocktail; treatment of apple slices with phage showed no reduction of bacteria. | (Leverentz et al., 2001) |
| *Salmonella* spp. | Felix-O1 | *Salmonella* counts were decreased by 1.8-2.1 logs after phage application to chicken frankfurters. | (Whichard et al., 2003) |
| *Salmonella* spp. | PHL4 | The levels of *Salmonella* recovered from experimentally contaminated broiler and naturally contaminated turkey carcasses were reduced by as high as 100% or 60% respectively after phage administration. | (Higgins et al., 2005) |
| *Salmonella* spp. | | Levels of *Salmonella* were decreased ~3 logs after application of a phage cocktail to sprouts; addition of an antagonistic bacteria to the phage cocktail increased this reduction to ~6 logs. | (Ye et al., 2010) |
| *Salmonella* spp. | FO1-E2 | In chocolate milk and mixed seafood, *Salmonella* levels were reduced to undetectable after phage treatment and storage for 24 h at 8°C and remained below the limit of detection. When foods were treated with phage and stored at 15°C, *Salmonella* counts were reduced to undetectable levels within 24-48 h for hot dogs, sliced turkey breast, and chocolate milk, but regrowth occurred after 5 days. *Salmonella* levels were initially inhibited ~0.5-2 logs and ~1-3 logs in egg yolk and mixed seafood respectively after phage addition; but bacterial recovery matched controls in egg yolks after two days, while the log reduction was maintained in seafood. | (Guenther et al., 2012) |

| | | | |
|---|---|---|---|
| *Salmonella* spp. | UAB_Phi 20, UAB_Phi78, UAB_Phi87 | *Salmonella* counts were decreased by ~1 log on the shells of fresh eggs and 2-4 logs on lettuce 60 mins after application of the phage. After an initial reduction of 1-2 logs when chicken breasts were dipped in a phage cocktail, no further decrease in the bacterial counts was observed over the next seven days at 4°C. The levels of *Salmonella* were reduced 2-4 logs on pig skin after phage application and storage for 6 h at 33°C. | (Spricigo et al., 2013) |
| *Salmonella* spp. | wksl3 | *Salmonella* counts were decreased by ~3 logs on chicken skin after application of a single phage, and no further decrease in bacterial levels was observed over the next seven days at 8°C. Phage cocktail was also administered to mice to test for safety. Mice received a single dose of phage orally and displayed no adverse effects. | (Kang et al., 2013) |
| *Salmonella* spp. | Five phages | The levels of *Salmonella* were decreased by ~1 log on chicken skin after application of a five-phage cocktail comprised of closely related phages. The reduction was comparable to treatment with 200 ppm dichloroisocyanurate, 10 ppm peroxyacetic acid, and 2 % lactic acid. | (Hungaro et al., 2013) |
| *Salmonella* spp. | P22 | After the administration of a single temperate phage and storage at 4°C, levels of *Salmonella* decreased by 0.5-2 logs on chicken samples; below the limits of detection in whole and skimmed milk; by ~3 logs in apple juice; by ~2 logs in liquid egg; and by ~2 logs in an energy drink. | (Zinno et al., 2014) |
| *Salmonella* Enteritidis | Five phages | The levels of *S.* Enteritidis were reduced by as much as 3.2 logs and 2.8 logs after 10 days storage at 18°C and 4°C, respectively in raw salmon fillets. The phage treatment of smoked salmon fillets resulted in 1.9 logs and 1.2 logs reduction in Salmonella after 10 days of storage at 18°C and 4°C, respectively. | (Galarce et al., 2014) |
| *Salmonella* spp. | SalmoFresh™ | The stability of a *Salmonella*-specific phage preparation was determined in various chemical antimicrobials. Treatment of chicken breast fillets with a combination of phage and chemical antimicrobials did not produce a synergistic effect on the reduction of *Salmonella*; however, application of chlorine or PAA followed by spraying with phage significantly reduced *Salmonella* from chicken skin by up to 2.5 logs, compared to use of chlorine, low levels of PAA, or phage alone (0.5-1.5 logs). | (Sukumaran et al., 2015) |
| *Salmonella* spp. | SalmoFresh™ | Treatment of chicken breast fillets by dipping or surface application of a *Salmonella*-specific bacteriophage preparation and storage at 4°C significantly reduced *Salmonella* contamination by up to 0.9 logs; further, storing the meat in modified atmospheric packaging after surface application produced a higher reduction in bacterial counts (up to 1.2 logs). | (Sukumaran et al., 2016) |
| *Salmonella* spp. | SalmoLyse® | A phage cocktail was sprayed onto experimentally contaminated raw pet food ingredients, including chicken, tuna, turkey, cantaloupe, and lettuce, and reduced the levels of the targeted bacteria by 0.4-1.1 logs. | (Soffer et al., 2016) |

| | | | |
|---|---|---|---|
| *Salmonella* spp. | SJ2 | Application of the phage SJ2 significantly reduced *Salmonella* in experimentally contaminated ground pork and eggs with a larger reduction observed at room temperature compared to 4°C. After treatment, *Salmonella* colonies were screened for phage resistance, and more phage-resistant *Salmonella* isolates were recovered from eggs compared with ground pork. | (Hong et al., 2016) |
| *Salmonella* spp. | PhageGuard S™ (formerly Salmonelex™) | Boneless chicken thighs and legs were experimentally contaminated with *Salmonella* and treated with phage solution. A larger reduction of *Salmonella* was achieved when the bacteriophage preparation was diluted in tap water compared to filtered water prior to application, and the phage cocktail was more effective against *Salmonella* isolated from other sources compared to those from ground chicken. | (Grant et al., 2017) |
| *Salmonella* spp. | PhageGuard S™ | Treatment with a bacteriophage cocktail or irradiation significantly reduced (~1 log) the level of *Salmonella* on experimentally contaminated ground beef trim, and a combination of these methods was even more effective and decreased bacterial contamination by ~2 logs. | (Yeh et al., 2018) |
| *Salmonella* spp. | PhageGuard S™ | Bacteriophage application reduced the *Salmonella* levels >1 log on skinless and skin-on poultry products. | (Hagens et al., 2018) |
| *Salmonella* spp. | BSPM4, BSP101, BSP22A | The phage cocktail treatment achieved a reduction of 4.7-5.8 logs of *Salmonella* on lettuce and cucumber. | (Bai et al., 2019) |
| *Salmonella* spp. | LPSTLL, LPST94, LPST153 | Application of phage cocktail reduced 3.0 log *Salmonella* inoculum to below detectable limits on chicken breast and in milk. Phage cocktail was effective against *Salmonella* biofilm grown for 72 h on microtiter plates and steel chips, resulting in >5.23 log reduction in *Salmonella* viable cells. | (Islam et al., 2019) |
| *Salmonella* spp. | SalmoFresh™ | Phage biocontrol reduced *Salmonella* by 2-3 log on lettuce and sprouts. The pairing of the phage cocktail with chlorinated water resulted in 2.7-3.8 log reduction in viable *Salmonella* counts. | (Zhang et al., 2019) |
| *Shigella* spp. | SD-11, SF-A2, SS-92 | *Shigella* counts were reduced by ~1-4 logs in spiced chicken after application of a phage cocktail or each of the component phages and storage at 4°C. | (Zhang et al., 2013) |
| *Shigella* spp. | ShigaShield™ | Application of a five phage *Shigella*-specific cocktail to various RTE foods (including lettuce, melon, smoked salmon, corned beef, and pre-cooked chicken) reduced the levels of *Shigella* ~1.0-1.4 logs compared to control. | (Soffer et al., 2017) |
| *Staphylococcus aureus* | Φ88, Φ35 | *S. aureus* levels were decreased below the limit of detection in experimentally contaminated whole milk after treatment with a two-phage cocktail and storage at 37°C. After phage treatment, *S. aureus* was not recovered from the acid curd after storage for 4 h at 25°C and was eliminated from the renneted curd after 1 h at 30°C. | (Garcia et al., 2007) |
| *Staphylococcus aureus* | vB_SauS-phi-IPLA35, vB_SauS-phi-SauS-IPLA88 | Counts of *S. aureus* were significantly decreased in cheese made using milk treated with phage compared to milk made without the addition of phage. The microbiota of the cheese was not impacted by the addition of the phage. | (Bueno et al., 2012) |

† Modified from (Moye et al., 2018)

**Table 2.** Phage products approved for food safety applications†

| Company | Phage Product | Target Organism(s) | Regulatory | Certifications | References |
|---|---|---|---|---|---|
| FINK TEC GmbH | Secure Shield E1 | *E. coli* | FDA, GRN 724; USDA, FSIS Directive 7120.1 | | |
| Intralytix, Inc. | Ecolicide® (Ecolicide PX™) | *E. coli* O157:H7 | USDA, FSIS Directive 7120.1 | | |
| | EcoShield PX™ | *E. coli* O157:H7 and other STECs | FDA, GRN 834; USDA FSIS Directive 7120.1; FCN No. 1018 | | (Vikram et al., 2020) |
| | EcoShield™ | *E. coli* O157:H7 | FDA, FCN 1018; Israel Ministry of Health; Health Canada | Kosher; Halal | (Abuladze et al., 2008; Boyacioglu et al., 2013; Carter et al., 2012; Ferguson et al., 2013; Magnone et al., 2013; Sharma et al., 2009) |
| | ListShield™ | *L. monocytogenes* | FDA, 21 CFR 172.785; FDA, GRN 528; EPA Reg. No. 74234-1; National Food Service of Israel approved as a food processing aid for the treatment of ready-to-eat meat and poultry products (Ref: 70275202); Health Canada (iLONO) | Kosher; Halal; OMRI | (Leverentz et al., 2003; Leverentz et al., 2004; Perera et al., 2015) |
| | SalmoFresh™ | *Salmonella* spp. | FDA, GRN 435; USDA, FSIS Directive 7120.1; Israel Ministry of Health; Health Canada | Kosher; Halal; OMRI | (Sukumaran et al., 2015, 2016) |
| | ShigaShield™ (ShigActive™) | *Shigella* spp. | FDA, GRN 672 | | (Mai et al., 2015; Soffer et al., 2017) |
| Micreos Food Safety | PhageGuard Listex™ | *L. monocytogenes* | FDA, GRN 198/218; FSANZ; EFSA; Swiss BAG; Israel Ministry of Health; Health Canada | Kosher; Halal; OMRI; SKAL | (Carlton et al., 2005; Chibeu et al., 2013; Figueiredo and Almeida, 2017; Guenther et al., 2009; Oliveira et al., 2014; Silva et al., 2014; Soni et al., 2012; Soni and Nannapaneni, 2010; Soni et al., 2010) |
| | PhageGuard S™ | *Salmonella* spp. | FDA, GRN 468; USDA, FSIS Directive 7120.1; FSANZ; Swiss BAG; Israel Ministry of Health; Health Canada | Kosher; Halal; OMRI; SKAL | (Grant et al., 2017; Yeh et al., 2018) |
| | | *E. coli* O157:H7 | FDA, GRN 757 | | |
| Passport Food Safety Solutions | Finalyse® | *E. coli* O157:H7 | USDA, FSIS Directive 7120.1 | | |

| Phagelux | SalmoPro® | *Salmonella* spp. | FDA, GRN 603; USDA, FSIS Directive 7120.1 | | |
|---|---|---|---|---|---|
| | | *Salmonella* spp. | FDA, GRN 752; USDA, FSIS Directive 7120.1 | | |

† Modified from (Moye et al., 2018)

briefly described above, as well as many of the concerns voiced by consumers. For example, the traditional decontamination methods are broad-spectrum, killing not only the pathogen of concern but also the natural microflora of the foods, which are often beneficial. In contrast, phage biocontrol, due to the specificity of bacteriophages, enables targeted elimination of the foodborne bacteria in the foods, while maintaining the natural microbial population and preserving the nutritional composition/value of those foods (Moye et al., 2020).

As bacteriophages are the natural enemy of the pathogenic bacteria, phage biocontrol could be considered the most natural, environment-friendly antimicrobial intervention available today. The application rates for typical phage biocontrol interventions are low and are expected to have little if any, environmental impact. For instance, the US meat and poultry industry, the largest US agricultural sector, produced 52 billion pounds of meat and 48 billion pounds of poultry in 2017 (NAMI, 2020). If phages were applied at the maximum approved amount ($10^9$ PFU/g for one phage product, all other current approvals are for up to $10^7$-$10^8$ PFU/g) to all 100 billion pounds of meat and poultry produced in one year, then the total phages applied would be ~4.5 x $10^{22}$ PFU, which is just 0.000005% of the minimum estimate of 1 x $10^{30}$ PFU present in the world. This calculation is a gross overestimate, as it assumes the maximum approved amount of phage is applied, especially considering that most GRAS approvals only permit an application of up to $10^8$ PFU/g food (reducing the phage estimate by a factor of 10). Moreover, this estimate assumes bacteriophage biocontrol is universally used by all relevant food industries in the United States. If both assumptions were true, theoretical sales for the phage industry would exceed $650 billion per year. Additionally, our estimate envisions a worst-case scenario, wherein all the phages applied to the foods eventually end up in the environment. Again improbable, as they will likely be inactivated prior to reaching the environment, either through cooking, consumption (and subsequent inactivation in the GI tract) or due to other environmental factors. In short, the number of phages added to the environment as a result of phage biocontrol is less than negligible, especially when compared to naturally present phage populations.

There are several other characteristics of commercial phage biocontrol preparations that make them consumer friendly. For example, many of them do not contain any additives or preservatives, and several are certified Kosher, Halal, and "Organic" – that is to say, suitable for use in organic foods (e.g., OMRI-listed in the USA; SKAL in EU) (Table 2). Phage biocontrol may also provide some cost benefits. For example, the costs for some of the currently implemented non-phage interventions (e.g., irradiation and HPP) range from 10-30 cents per pound of treated food (Viator et al., 2017). In contrast, phage biocontrol costs range between 1-4 cents per pound, similar to the cost range of harsher chemical sanitizers. Therefore, phage biocontrol provides a 'green' and safe alternate method to control foodborne pathogens, which may demonstrate superior biocontrol compared to other interventions while preserving the environment and encouraging sustainability.

While the biological properties of lytic bacteriophages provide advantages for improving food safety, these properties also lead to some of the limitations and drawbacks to phage biocontrol. As mentioned previously, bacteriophages are highly specific and as such, they are only effective against the pathogen of interest. Still, if foods are contaminated with multiple pathogens, a combination of phage biocontrol products could be used to target more than one pathogen. Since bacteriophages themselves are also microorganisms, commonly used disinfectants or chemicals could inactivate them, so their use needs to be carefully coordinated within the processing line. In addition, the currently marketed phage preparations require refrigerated storage (typically 28°C). Thus, a good understanding of the biological properties of bacteriophages and designing optimal application regimens that consider those properties is essential for the best possible efficacy of phage biocontrol intervention. The main *Pros* and *Cons* of phage biocontrol are summarized in Table 3 and are also discussed in more detail later in this chapter.

**Table 3.** Pros and Cons of phage biocontrol

| ***Pros*** | ***Cons*** |
|---|---|
| Phages are a natural product | Not all phages make a good biocontrol agent, e.g. temperate phages. These *pros* refer to wild-type lytic phages. |
| Specific, only targets problem foodborne bacterial pathogens | May not ensure full safety of foods if the foods are contaminated by a different foodborne pathogen (e.g. one not targeted by the phage biocontrol) |
| Effective in killing targeted bacteria | Narrow host range limits their usage in theory; this shortcoming can be overcome by using a "cocktail" or a combination of phages |
| Single dose application | Residual activity has not been observed on foods despite the ability of lytic phages to infect new hosts; possibly due to physical inaccessibility on food surfaces |
| Narrow potential for resistance and lack of cross resistance with antibiotics | Phage resistant bacterial strains can emerge, but using a cocktail is shown to reduce the phage resistance |
| Rapid discovery and relative ease for formulation and application | |
| Low inherent toxicity and no adverse environmental impact | |
| Does not impact organoleptic, nutritional, and rheological properties of the food | |

For food safety applications, lytic bacteriophages can be used for both pre- and post-harvest interventions. When applied to live animals, phages can be administered via animal feed or spray-applied to hides or feathers prior to slaughter. For post-harvest applications, the phage preparations are generally applied directly to food surfaces, either via direct spraying, through the packaging materials, or by some other means (Lone et al., 2016; Sulakvelidze,

2013). Phage biocontrol could also be utilized as a disinfectant for surfaces within the food processing facility (Abuladze et al., 2008; Woolston et al., 2013). Several earlier reviews of those approaches are available (Endersen et al., 2014; Greer, 2005; Sulakvelidze, 2013; Woolston and Sulakvelidze, 2015). Other phage-related methods such as the use of phage endolysins or using bacteriophages to manage food spoilage have also been discussed in various earlier reviews (Greer, 2005; Schmelcher and Loessner, 2016). This chapter provides an updated overview of studies describing phage biocontrol predominantly in post-harvest applications, the segment that currently appears to be gaining the most momentum (Table 1).

## Phage Biocontrol of Foodborne Pathogenic Bacteria

### *Listeria monocytogenes*

Listeriosis, an infection caused by *Listeria monocytogenes*, results in a range of clinical symptoms including mild febrile gastroenteritis to more severe sepsis, meningitis, rhombencephalitis, perinatal infections, and abortions (Allerberger and Wagner, 2010). *L. monocytogenes* is generally transmitted when food is handled in environments contaminated with the pathogen and is usually contracted by consuming improperly processed or cooked meat and milk. *L. monocytogenes* can proliferate at the lower temperatures generally used to refrigerate (2-8°C) foods in households and transportation, making it a serious food safety threat. Therefore, the detection and control of *L. monocytogenes* is critically important for ensuring the safety of the food chain, especially in RTE foods. *L. monocytogenes* infections appear to have a very high mortality rate. For instance, Havelaar et. al. estimated that of the 14,000 *L. monocytogenes* global infections, recorded in 2010, approximately 22% of infections resulted in death (Havelaar et al., 2015). According to the CDC, *L. monocytogenes* infections result in approximately 1,600 illnesses and 320 mortalities each year in the US alone.

*L. monocytogenes* was the first foodborne pathogen for which a commercial phage biocontrol preparation was developed. That preparation, called ListShield™ (originally LMP-102) (Figure 1), was developed and is currently marketed by Intralytix, Inc. (Columbia, MD, USA) as either a Food Additive or

GRAS (Generally Recognized as Safe) food processing aid (Moye et al., 2018). Several studies have reported that the ListShield™ is highly efficacious in significantly reducing or eliminating *L. monocytogenes* contamination in a variety of foods, including RTE foods (Table 1) such as fruits and vegetables, cheese, and smoked fish (Perera et al., 2015). In addition, ListShield™ has been shown to reduce *L. monocytogenes* by approximately 2.2 logs in prepackaged frozen foods, such as those usually served in-flight (Perera et al., 2015). Similar efficacy has been reported for Listex™, another commercial monophage preparation. Listex™ reduced *L. monocytogenes* on pork ham by ca. 2.8 logs to undetectable levels; in comparison, nisin (a polycyclic antibacterial peptide) and sodium lactate resulted in 2 logs and <0.5 log reduction in *L. monocytogenes* levels, respectively (Figueiredo and Almeida, 2017). Only the phage mediated reductions were sustained during storage of the ham for 72 h at 6-8°C, suggesting phage biocontrol was more effective than antimicrobial treatments (Figueiredo and Almeida, 2017). The same monophage preparation was also shown to reduce *L. monocytogenes* on the surface of other deli meats (cooked sliced turkey and roast beef) stored at 4°C and 10°C (Chibeu et al., 2013). Additionally, Listex™ demonstrated synergistic activity against *L. monocytogenes*, with other antimicrobials such as sodium diacetate or potassium lactate (Chibeu et al., 2013).

**Figure 1.** *ListShield™*, the first phage-based product ever to be cleared by the FDA for food safety applications

*Salmonella* spp.

Non-typhoidal *Salmonella enterica* was estimated to account for 95.1 million cases of enterocolitis and 50,771 fatalities in 2017, worldwide (Stanaway et al., 2019). *Salmonella* infections are often self-limiting, with symptoms typically including stomach cramps, fever, nausea, and diarrhoea, but life-threatening instances can occur in cases of dehydration and when the bacteria invade the internal organs. According to the CDC estimates, non-typhoidal *Salmonella* causes about 1.35 million illnesses, 26,500 hospitalizations, and 420 deaths yearly in the US (Centers for Disease Control and Prevention, 2019c). While *Salmonellae* are typically associated with poultry products, the past several years have seen outbreaks caused by a variety of foods, including fruits and vegetables, nuts, and fish. In 2019, the CDC reported three fruit-related outbreaks, which sickened over 300 people in at least 11 states (Centers for Disease Control and Prevention, 2019b, 2020).

Phage biocontrol of *Salmonella* has been evaluated by several investigators (Table 1). In the US, two phage preparations are currently available commercially (Table 2). In general, both commercial and non-commercial phage-preparations have been shown to be an efficacious approach to control *Salmonella* contamination of various foods. For example, a recent study demonstrated an average of ca. 5 log reduction on lettuce and 0.8 log reduction on sprouts in *Salmonella* levels following application of the commercial six-phage cocktail SalmoFresh™ (Zhang et al., 2019). Another study showed ca. 2 log reduction in *Salmonella* population by the same phage cocktail on whole cucumbers (Sharma et al., 2017). The same phage cocktail was also reported to be effective in reducing *Salmonella* concentration on poultry products, with reductions up to 1.2 logs and 1.3 logs recorded in chicken breast and turkey breast, respectively (Sharma et al., 2015; Sukumaran et al., 2016). Finally, reports from our laboratory have shown that SalmoFresh™ can also be effective in reducing *Salmonella* contamination of dry pet food as well as raw pet food-ingredients (Soffer et al., 2016) and glass and stainless steel surfaces (Woolston et al., 2013). Raw pet foods present a high risk to not only the pets but also to the unsuspecting owners (Soffer et al., 2016).

Other *Salmonella*-specific phages with good efficacy were also reported for controlling *Salmonella* in various food applications. For example, Salmonelex™, a cocktail of the two phages Fo1a and S16, reduced a mixture of three *Salmonella* serovars by 0.4 logs and 0.7 logs after 30 min and 8 h treatment in ground chicken (Grant et al., 2017). The three serovars, *S.* Typhimurium, *S.* Newport, and *S.* Thompson, used in this study were isolated from retail ground chicken products and represented the real-life contamination of chicken products. Additionally, the same study also demonstrated that Salmonelex™ was effective against a combined challenge of *Salmonella* serovars Heidelberg, Enteritidis, and Typhimurium obtained from ATCC (Grant et al., 2017). A monophage preparation, consisting of SJ2, was able to significantly reduce *Salmonella* in both liquid egg and ground pork during storage at 4°C and 21°C (Hong et al., 2016).

*Shiga toxin-producing Escherichia coli*

Shiga toxin-producing *Escherichia coli* (STEC, also known as enterohemorrhagic *E. coli* or EHEC) are important foodborne pathogens that cause more than 2.5 million illnesses globally each year, resulting in 3,330 haemolytic uremic syndrome (HUS) cases and 269 deaths (Kirk et al., 2015). According to the CDC's National STEC Surveillance Program, an estimated 5,441 culture confirmed STEC infections were recorded in the US in 2016, which was a 26% increase over the previous year (Centers for Disease Control and Preventions, 2018). The clinical manifestations of STEC infections range from mild diarrhoea to hemorrhagic colitis and potentially fatal HUS (Besser et al., 1999). Beef and fresh produce are the two most common sources of STEC infections, implicated in about 75% of all STEC-related outbreaks in the US (Centers for Disease Control and Preventions, 2019; Interagency Food Safety Analytics Collaboration, 2018). Lately, poultry products have also been increasingly linked to STEC outbreaks (Chen et al., 2018; Mathusa et al., 2010). Additionally, STEC, in particular, *E. coli* O157:H7 has been identified in amphibian, fish, and invertebrate carriers (Ferens and Hovde, 2011; Sanath Kumar et al., 2001).

Phage biocontrol of STEC, particularly *E. coli* O157:H7, using commercial and noncommercial phage preparations has been reported by several investigators. As beef products are inherently at high risk for EHEC contamination, initial studies primarily focused on controlling EHEC contamination in beef products (Abuladze et al., 2008; Hudson et al., 2013a). For example, a single phage FAHEc1, isolated from raw sewage, was shown to reduce the STEC counts on beef slices by approximately 2 logs under conditions simulating hot boning and conventional carcass cooling (Hudson et al., 2013a). In another study, a commercial three phage preparation was shown to reduce *E. coli* O157:H7 *levels* by 1.2 logs in ground beef (Abuladze et al., 2008), and a different three phage preparation was reported to reduce the *occurrence* of *E. coli* O157:H7 on beef by approximately 77% (O'Flynn et al., 2004).

In addition to phage biocontrol of STEC on beef, several studies have also demonstrated the effectiveness of phage biocontrol in reducing *E. coli* O157:H7 contamination of other foods, such as fresh produce. A single bacteriophage, OSY-SP, that was isolated from sewage and livestock manure, was shown to reduce *E. coli* O157:H7 by 1-4 logs on cut green peppers and baby spinach (Snyder et al., 2016). Storage at 4°C for 72 h showed a sustained activity while at 25°C some bacterial regrowth was observed after 72 h (Snyder et al., 2016). Another study demonstrated the effectiveness of a phage cocktail on reducing *E. coli* O157:H7 on fresh-cut cantaloupes and lettuce (Snyder et al., 2016). The phage treatment resulted in ca. 2 logs reduction on lettuce and 2-3 log reduction in *E. coli* O157:H7 levels on cantaloupe, clearly demonstrating the effectiveness of phage biocontrol on fresh produce (Snyder et al., 2016).

Initially, most phage preparations, and therefore studies, were primarily focused on targeting *E. coli* O157:H7. However, with the increased frequency of non-O157 STEC associated disease, more recent research has focused on phage biocontrol that targets STEC in general. For example, the phage cocktail EcoShield PX™ was granted GRAS affirmation from the FDA (GRN 834) in the winter of 2019 and is effective against a broader range of STEC pathogens (Vikram et al., 2020). In a study using non-O157 strains, application of a three phage cocktail completely inhibited the *E. coli* strains ATCC 25922 and O127:H6 in ultra-high temperature treated (UHT) milk and raw milk at 4°C and 25°C (McLean et al., 2013). Furthermore, no regrowth of the two strains was observed following the phage treatment and storage at either 4°C or 25°C. In separate experiments, the authors demonstrated that, initially, a two-phage cocktail completely inhibited *E. coli* O5:H, an enterohemorrhagic strain, in UHT milk at 4°C and 25°C, but, in this case, regrowth of *E. coli* O5:H- was observed during storage at both temperatures. While multiple factors may contribute to the observed regrowth, two plausible causes could be strain-specific differences and use of three vs two phage cocktail. It is likely that a three phage cocktail may provide better activity and suppress the emergence of resistant mutants, as discussed later (Örmälä and Jalasvuori, 2013; Woolston et al., 2013; Yuan et al., 2019).

*Shigella* spp.

*Shigella* species are major foodborne and waterborne pathogens (Tack et al., 2019). Shigellosis, the infection caused by the bacteria, usually results in diarrheal disease, with symptoms ranging from mild stomach cramps to vomiting and bloody diarrhoea. While the infection is self-limiting and generally clears within 5-7 days, people with compromised immune systems may suffer from more severe and debilitating illnesses. In addition, developing countries have a very high incidence of shigellosis and children under the age of 5 are disproportionately affected (Havelaar et al., 2015; Kotloff et al., 1999). In the US, approximately 13,000 culture confirmed cases of shigellosis were reported in 2016 (Enteric Disease Laboratory Branch, 2016), with one species *S. sonnei,* accounting for the largest (80.5%) percentage of infections (Enteric Disease Laboratory Branch, 2016). *Shigella* is not a frequent cause of foodborne outbreaks in the US, but two recent outbreaks have been linked to contaminated asparagus (2018) and raw oysters (2019) (Flynn, 2019; Food and Drug Administration, 2019). *Shigella* is also an issue for the US military, as travel to developing countries puts military members at an elevated risk for shigellosis (Magnone et al., 2013).

While several groups have evaluated phage biocontrol of *Shigella*, currently there is only one commercially available phage preparation, ShigaShield™. This five-phage cocktail is GRAS-listed (GRN 672; Table 2) and has been shown to reduce the levels of *Shigella* by approximately 1 log in a variety of foods, including melons, lettuce, yoghurt, deli corned beef, smoked salmon, and chicken breast meat (Soffer et al., 2017). Notably, ShigaShield™ is also one of the few commercial phage biocontrol preparations that have been examined for its impact on the normal gut microflora in mice. In an article from the University of Florida, ShigaShield was reported to be well tolerated when administered to mice and, in contrast to an antibiotic (ampicillin), to have significantly less impact on the normal gut microflora (Mai et al., 2015). Other studies have also demonstrated the effectiveness of phage biocontrol in managing *Shigella* contamination of foods. For example, treatment with a phage, designated vB_SflS-ISF001, resulted in nearly a 2 log reduction of *S. flexneri* counts on raw and cooked chicken breast samples (Shahin and Bouzari, 2018). The same group examined vB_SflS-ISF001 and a second phage, vB_SsoSISF002, against a panel of *Shigella* that had been isolated directly from foods (Shahin et al., 2019). Combining the two phages showed increased efficacy in reducing the *Shigella* than when either phage was used alone. Phage biocontrol has also been shown to effectively control *Shigella* in contaminated water (Jun et al., 2016).

*Campylobacter jejuni*

Infections due to *Campylobacter* spp. are one of the most common causes of gastroenteritis worldwide, with an estimated 95 million foodborne illnesses yearly (Havelaar et al., 2015). In the US, *Campylobacter* has had the highest rate of yearly incidence since 2013 (Tack et al., 2019), with an estimated 1.5 million foodborne *Campylobacter* infections occurring every year (Centers for Disease Control and Prevention, 2019a) and an economic impact of about $1.9 billion (in 2014 dollars) (Economic Research Service, 2014; Hoffmann et al., 2012; Michael et al., 2014). *Campylobacter* is found in contaminated water and is frequently associated with animals, including farm animals such as poultry and cows.

Several *Campylobacter* bacteriophages have been isolated, generally from poultry sources (e.g., faecal matter, surfaces, and internal tissues such as livers and ceca), and some of them have been examined for their ability to reduce contamination of various foods by *Campylobacter* (Firlieyanti et al., 2016; Hammerl et al., 2014; Kittler et al., 2013; Zampara et al., 2017). Interestingly, isolation of *Campylobacter*-specific phages has almost exclusively been performed utilizing just two *C. jejuni* isolates as a host strains for phage isolation. Most of those were with *C. jejuni* NCTC 12662 and the isolated phages generally specifically targeted the capsular polysaccharide, termed Group III phages (Sorensen et al., 2015). In contrast, a few studies used *C. jejuni* RM1221 as the target strain and these phages, termed Group II phages, typically utilize the flagella as a route of entry (Sorensen et al., 2015). Most reports of using phages for biocontrol are for pre-harvest interventions and studies examining the ability to reduce post-harvest contamination of various foods by *Campylobacter* are limited. In a study on chicken neck skin (Zampara et al., 2017), two Group III phages were able to reduce contamination by ca. 0.4 logs each when applied as single phages, but their efficacy increased to 0.7 log reduction of *Campylobacter* when used in combination with one another. The application of a Group II phage showed no effect on the *Campylobacter* levels. Another group of investigators reported 1-3 logs reduction in *Campylobacter* levels on artificially contaminated cooked and raw beef slices (Bigwood et al., 2008). The effects were more pronounced when higher levels of bacteria were present and when the phage was applied at a higher MOI. Reductions were maintained for up to 8 days at 5°C storage. Currently, there are no commercial phage biocontrol products available in the US but, given the importance of *Campylobacter* from the standpoint of food safety, it seems likely that such products would be made commercially available in the not too distant future.

## Considerations for Phage Biocontrol

As interest in bacteriophage biocontrol for food safety purposes increases, several considerations need to be taken into account for the optimal implementation of this novel approach. Some of these aspects are briefly discussed below.

*Regulation of Bacteriophage Preparations*

Despite the more than 100-year history of using phages therapeutically in humans, their use for food safety applications is a relatively novel concept pioneered by a US-based company Intralytix, Inc., which was the first company in the world to obtain FDA-approval for bacteriophage product for the phage biocontrol applications (note: the FDA does not endorse or approve any interventions; however, the term "approval" is commonly used – including throughout this review - to indicate that the agency allows the product to be used commercially). That first approval was for the *L. monocytogenes* specific phage cocktail called ListShield™ (formerly LMP-102; Figure 1) and it came in the form of an amendment of the FDA's "food additive regulations" in 2006 (Food and Drug Administration, 2006). The same product was also listed by FSIS as a suitable antimicrobial intervention for ready-to-eat meat and poultry products (FSIS Directive 7120.1). The success of this original petition paved the way for other companies in the US and abroad to invent and develop new phage products for food safety applications – and other approvals indeed soon followed. For example, later that same year, the FDA issued a no objection letter for the *Listeria*-specific preparation Listex™ (currently PhageGuard Listex™) as a Generally Recognized as Safe (GRAS) substance. As of the day of this writing (December 31, 2019), fourteen phage biocontrol products have been approved for food safety applications under various regulatory frameworks in the United States (Table 2). Of these fourteen products, eleven products were approved as GRAS by the Center for Food Safety and Applied Nutrition (CFSAN) of the FDA, and application for GRAS designation now is the most commonly used route for regulating phage biocontrol products for post-harvest food applications. The approval process includes engaging the USDA, if necessary, to determine if the phage preparation should be included in their guidelines for safe and suitable ingredients used in the production of meat, poultry, and egg products. The FSIS Directive 7120.1 includes both pre- and post-harvest applications, such as the application of phage to the hides of cattle and targeted phage application onto poultry or meat. While the FSIS guidelines are developed based upon specific phage preparation efficacy data, in general, any phage product that meets the description in the directive may be considered to be compliant. The use of phage preparations for food safety purposes has also received approval in several other countries, such as Israel, Canada, Switzerland, Australia, and New Zealand (Table 2); these approvals are frequently issued based upon US regulatory approvals. One notable exception is the European Food Safety Authority (EFSA) which has been slow in adopting this new intervention and has been delaying the introduction of phage biocontrol in the European Union.

*Efficacy*

Phage biocontrol typically reduces the levels of targeted bacteria by 1-3 logs, which is lower than the up to 5 logs reduction claimed for some other, harsher interventions, e.g., irradiation. However, several of the studies reporting such a high log reduction have also used a very high inoculum dose for the challenge. For example, Nagel et al. (Nagel et al., 2013) used an 8 log inoculum of *Salmonella* and *Campylobacter* to achieve a 5.5 log reduction in poultry carcasses treated with various antimicrobials (including chlorine and peracetic acid) in a post-chill immersion tank; whereas the majority of phage biocontrol studies have used much lower challenge inoculum (e.g. Soffer et al. (2017) used a 3 log inoculum). Thus, higher reductions may be possible if a very high dose of bacterial inoculum is used to experimentally contaminate foods during phage efficacy studies; e.g., a reduction of up to 5 logs was reported as a result of phage treatment by several authors (Guenther et al., 2009; Leverentz et al., 2004)). Notably, when a lower inoculum dose is used in challenge studies, many chemical antimicrobials (e.g. PAA) also demonstrate lower reduction (Moore et al., 2017) – similar to reductions achieved by phage biocontrol. These potentially lower log reductions by phage biocontrol may also be more of a perception problem than a real technical issue. For instance, ListShield™ reduced the levels of *Listeria monocytogenes* by approximately 1 log in artificially contaminated smoked salmon fillets, but when the same cocktail was applied to naturally contaminated salmon fillets in a commercial processing facility, the *L. monocytogenes* was completely eliminated (no detectable *L. monocytogenes*) (Perera et al., 2015). Few, if any, foods are contaminated with 5 logs of foodborne bacteria; nevertheless, a 5-log reduction sounds much more effective than a 1-3 log reduction and the companies marketing phage preparations for the food industry may have to overcome this perception challenge. Thus, future phage biocontrol studies may need to (i) utilize comparably high initial bacterial challenge levels to demonstrate that phage biocontrol can (or cannot) also result in 4-5 log reduction similar to chemical sanitizers, and/or (ii) provide compelling evidence that lower reductions (1-3 log) still provide a strong, real life-pertinent protection of foods (e.g., result in a significant reduction in the incidence/ occurrence of foodborne bacteria in foods contaminated with the levels of foodborne bacteria commonly found in real-life settings). In support of the second point, a 2003 risk-assessment study included a model predicting how reductions in *L. monocytogenes* contamination of pre-retail deli meat would affect the mortality rate associated with that pathogen. The predictions indicated that even when only reduced by 1 or 2 logs, the mortality rate of elderly people due to *L. monocytogenes* would decrease by ca. 50% - 74% (Food and Drug Administration, 2003), suggesting even small reductions in contamination may yield significant improvements in food safety and public health.

### *Additional Technical Challenges*

Phage biocontrol often significantly reduces the levels of the targeted pathogen but does not always eliminate it completely from foods. This issue is not phage biocontrol-specific and many other antimicrobial treatments of food products exhibit similar shortcomings (Kalchayanand et al., 2016; Penney et al., 2007), but the possible explanations for the incomplete inactivation by phages differ from standard interventions. Three major causes are discussed below.

### *Physical Impedance*

A number of factors related to application methods on food and certain inherent properties of bacteriophages may hinder the efficacy of phage biocontrol. One of the key factors restricting phage efficacy is the food matrix. Phages are required to physically adsorb onto the bacteria to exert their lytic activity. This interaction can be impeded by the food matrix. Unlike some bacteria, phages are immobile and depend on Brownian motion to reach their target (Kasman and Porter, 2019). In an event where phages are not evenly sprayed or mixed with the food to ensure direct contact of phages with their targeted bacterial cells, the efficacy of phage treatment may be significantly reduced. Efficacy may also be lower in drier foods where not enough liquid is available to enable Brownian motion. Even in the case of foods with adequate moisture, such as meats and cut fruits and vegetables, the phages may get trapped in microscopic surface structures and not reach the target bacteria, rendering them ineffective. Thus, ensuring thorough coverage of the food surface area with phages is perhaps the most critical prerequisite for phage biocontrol efficacy. Therefore, carefully designed application of the phages for the particular food processing facility is necessary. Various approaches, such as adjusting the phage concentration and/or the spray volumes, using fine (mist-like) sprays, thoroughly mixing foods during phage application, and otherwise ensuring thorough phage application should enhance the effectiveness of phage biocontrol.

*Resistance*

Another important challenge is the potential emergence of phage-resistant bacterial isolates. The emergence of phage-resistant isolates has not yet been reported to be an efficacy-hindering problem during industrial phage biocontrol applications, but it remains a valid, and perhaps guaranteed, an eventuality which has been documented under laboratory conditions (Hong et al., 2016; Sillankorva et al., 2012). For example, in a study of *Salmonella* in raw eggs and ground pork, the frequency of phage resistant isolates was higher in the raw eggs treated with phage than in the untreated eggs; however, the ground pork did not show a similar trend (Hong et al., 2016). The food matrices may have played a role, as the liquid nature of the raw egg could have allowed more opportunities for the phage and *Salmonella* to interact, which was likely limited on the more solid ground pork. Additionally, this study used a single bacteriophage preparation, as opposed to a phage cocktail, which might also have contributed to the higher phage resistance rates observed. The use of phage cocktail is reported to reduce the frequency of emergence of phage resistant mutants (Örmälä and Jalasvuori, 2013; Woolston et al., 2013; Yuan et al., 2019), as phages can use a variety of bacterial surface structures to adsorb (Lindberg and Holme, 1969; Marti et al., 2013; Zhang et al., 2016). Thus, the smart design of phage preparations, ones that contain lytic bacteriophages capable of targeting multiple strains of the same species and/or distinct bacterial receptors, can strengthen the potency and effectiveness of phage biocontrol. To this end, a PhageSelector™ program has been recently reported to help design an optimally effective phage preparation with a broad target range against *E.*

*coli* (Cieplak et al., 2018). Additionally, applying phages as close to the end of food processing as possible could help reduce the emergence of phage-resistant mutants by limiting the exposure of the bacteria to the phages.

Even when efficacy-hindering resistance evolves, it should be possible to quickly update phage preparations by replacing old phages that are no longer effective against their targeted strains with new phages that have lytic potency against those strains. The technical feasibility of such a "product update" has been demonstrated for a *Salmonella* phage preparation (Woolston et al., 2013). The study examined the ability of a six-phage *Salmonella* phage cocktail to reduce the levels of *Salmonella* on glass and stainless steel surfaces which are commonly used as food contact surfaces in the food processing industry (Woolston et al., 2013). Initial studies demonstrated that the bacteriophage cocktail significantly reduced the population of *S.* Kentucky and *S.* Brandenburg by ca. 2-4 logs, on all surfaces examined. But the phage preparation was ineffective in reducing *S.* Paratyphi B S661 (which was resistant to this phage preparation) levels on the hard surfaces contaminated with that strain. Updating the phage preparation by replacing two of its original component phages by two new phages with lytic potency against the *S.* Paratyphi B S661 strain, instantly restored the efficacy of the preparation. This new phage preparation reduced the *S.* Paratyphi levels by ca. 2 logs on glass and stainless steel surfaces, while also maintaining efficacy against the Kentucky and Brandenburg serotypes (Woolston et al., 2013). The study provided compelling evidence that phage cocktails could be easily modified when needed, e.g., if and when efficacy-hindering phage-resistant mutants emerge against a given commercial phage preparation. Implementation of this approach in a real-life setting, however, is still not fully delineated, including regulatory approvals. Encouragingly, the FDA and USDA have started to allow such updates of some commercial phage preparations without the need to go through the entire regulatory approval process with every update; e.g., GRN 834 for an *E. coli*-targeted phage cocktail EcoShield PX™ (Table 2) allows the product manufacturer to modify the phage preparation to include three to eight lytic phages in order to achieve optimal efficacy, including in the event of bacterial resistance emerging against the original preparation.

Temporal resistance to phages on the food matrix may also contribute to incomplete elimination of the target bacteria (Hoskisson and Smith, 2007; Tokman et al., 2016). Several studies tested this conjecture by evaluating randomly selected isolates recovered following phage treatment, however, a clear answer is still elusive. The observance of transient phage resistance in studies using pure bacterial culture led to the speculation that temporal resistance may contribute to the survival of bacteria on the food matrix (Orquera et al., 2015), but the evidence from food application of commercial phages is lacking. For example, two studies found that the recovered *L. monocytogenes* isolates following phage treatment were susceptible to the phage preparations suggesting that temporal resistance was either a very short-termed and an unstable phenomenon or not a major contributor to incomplete eradication of the target bacteria on food (Carlton et al., 2005; Chibeu et al., 2013).

*Implementation*

Phage biocontrol is envisioned to be a part of a multi-hurdle approach for improving food safety. Therefore, it requires planning to achieve optimal efficacy when combining bacteriophages with other food safety interventions. As almost all commonly used chemical sanitizers are capable of inactivating phages, therefore phages must be applied separately to ensure that they remain effective (Sukumaran et al., 2015). This was illustrated by the decreased efficacy of a simultaneous application of bacteriophages and chemical preservatives when compared to either treatment individually (Rodríguez et al., 2004). But, if phage preparations are carefully incorporated with other approaches, the efficacy of each could be – and often is – improved. For instance, in the presence of high organic loads, pretreatment of fruits and vegetables with a bacteriophage preparation boosted the efficacy of a produce wash by up to 2 logs (Magnone et al., 2013). Similarly, studies suggest an additive effect also occurs when phages are applied after chemical treatments, as higher reductions were observed in the combined treatments than when the interventions were used alone on apples, cantaloupes, lettuce, and chicken breast (Moye et al., 2020; Sukumaran et al., 2015). Phage biocontrol has also been shown to be effective when combined with modified atmospheric conditions, having better reductions in bacterial counts on chicken breast compared to their storage under aerobic conditions (Sukumaran et al., 2016). As modified atmosphere packaging is widely used by the various food industries, this observation has direct implication to improve product safety. In summary, proper integration of phage biocontrol in the existing Hazard Analysis and Critical Control Points (HACCP) protocols is key to it becoming an integral part of an effective multi-hurdle approach for improving food safety.

## Acceptance of phage biocontrol

Today's consumer is increasingly looking for "green appeal" in their food products (Atchley, 2019). There is a growing trend towards the purchase of food products that are not treated with chemical sanitizers or antibiotics and that are sustainably and naturally grown and processed, including not "genetically modified" (Lewis and Hill, 2020). Demand for health and wellness foods, including organic and locally produced food products, such as available at local farmer's markets and community-supported agriculture, has been on the rise (Reganold and Wachter, 2016; Woods et al.). While this is a welcome trend from the perspective of improving consumer health, one of the major challenges for the food industry is to keep this minimally treated food safe from microbial contamination. Phages have multiple properties that make them the perfect microbial control tool to improve the safety of minimally treated food including (i) organic nature of wild-type lytic phages, (ii) specific target pathogen control as opposed to generalized killing by antibiotics and chemical sanitizers, (iii) minimal or no residual effect on the natural microflora of food products, and (iv) no effect on food sensory or health-promoting qualities. However, the use of phages is not without challenges. Among these hindrances, besides the technical challenges described above, are (1) consumer perception and (2) willingness by food producers to adopt this new green technology. Consumer perception is perhaps the biggest challenge, as there is a potential risk that the phage technology can be misconstrued as "viruses on my food," where viruses are being inherently identified as "bad" by the general populace. Food producers take their cue from consumer trends and may be reluctant to use phages for microbial control if phage technology is miscomprehended. On the other hand, a recognition of the true benefits of phage mediated biocontrol of foodborne pathogens, including reduced chemical sanitizer use while controlling the pathogens, should encourage consumers and food producers to adopt the phage technology.

Although the science behind the phage technology is unambiguous, overcoming the problem of consumer perception and adoption by industry will require a fair amount of education of consumers and food processors on basic facts about phages and their use. While a limited amount of research has been conducted regarding the perception of phages, current evidence suggests that the phage biocontrol technology is viewed favourably (Naanwaab et al., 2014). Generally, consumers were supportive and even willing to pay a higher price for phage treated fresh produce when they were properly educated on the nature of lytic bacteriophages and the phage biocontrol approach (Naanwaab et al., 2014). Thus, for the phage technology to be embraced by consumers and food producers, the basic essential facts, such as the ubiquitous nature of phages, their natural presence in all the foods ingested, and their contribution to human health (Dalmasso et al., 2014; Guglielmi, 2017; Hatfull, 2008; Sulakvelidze and Barrow, 2005), need to be conveyed clearly. Ideally, phages will gain the recognition that the now well-received bacterial probiotics have earned, that phages are already a part of our ecosystem and can be beneficial to our health, leading to wider acceptance of phage biocontrol and improving the safety of our foods.

## Concluding Remarks

Despite some lingering challenges, phage biocontrol is increasingly being recognized for its effectiveness in controlling foodborne pathogens in the food processing industry. Food producers are adopting phage biocontrol as part of a multiple hurdle approach to achieve a greater reduction in target pathogens and improve food safety. Moreover, a supportive consumer perception will likely further propel the adoption of phage biocontrol. A number of commercial phage preparations have been approved or are under consideration by the regulatory bodies in the US and other countries. The flexibility of commercial phage preparations from an application standpoint is helpful for food producers, as phages can be employed at either pre-harvest or post-harvest stages, at a variety of processing steps, and through various mechanisms such as spraying, dipping, etc. The biggest advantage of using phage biocontrol is that wild type lytic phages are natural antimicrobials that allow targeted elimination of problem foodborne pathogens in foods without deleteriously impacting the natural microflora of foods and other nutritional or organoleptic qualities of foods.

## Acknowledgements / Conflict of Interest Statement

All the authors are employees of Intralytix, Inc., a Maryland-based corporation engaged in the development and commercialization of bacteriophage preparations for various applications, including phage biocontrol. AS and JW also hold an equity stake in Intralytix, Inc.

## References

Abuladze, T., Li, M., Menetrez, M.Y., Dean, T., Senecal, A., and Sulakvelidze, A. (2008). Bacteriophages reduce experimental contamination of hard surfaces, tomato, spinach, broccoli, and ground beef by *Escherichia coli* O157:H7. Appl Environ Microbiol *74*, 6230-6238.10.1128/AEM.01465-08

Allerberger, F., and Wagner, M. (2010). Listeriosis: a resurgent foodborne infection. Clin Microbiol Infec *16*, 16-23.https://doi.org/10.1111/j.1469-0691.2009.03109.x

Ashelford, K.E., Day, M.J., and Fry, J.C. (2003). Elevated abundance of bacteriophage infecting bacteria in soil. Appl Environ Microb *69*, 285-289.10.1128/aem.69.1.285-289.2003

Atchley, C. (2019). Top trends driving the global food industry. 10th Jan 2019. https://www.foodbusinessnews.net/articles/13554-top-trends-driving-the-global-food-industry

Atterbury, R.J., Connerton, P.L., Dodd, C.E., Rees, C.E., and Connerton, I.F. (2003). Application of host-specific bacteriophages to the surface of chicken skin leads to a reduction in recovery of *Campylobacter jejuni*. Appl Environ Microbiol *69*, 6302-6306.10.1128/aem.69.10.6302-6306.2003

Aw, T.G., Wengert, S., and Rose, J.B. (2016). Metagenomic analysis of viruses associated with field-grown and retail lettuce identifies human and animal viruses. Int J Food Microbiol *223*, 50-56.https://doi.org/10.1016/j.ijfoodmicro.2016.02.008

Bai, J., Jeon, B., and Ryu, S. (2019). Effective inhibition of Salmonella Typhimurium in fresh produce by a phage cocktail targeting multiple host receptors. Food Microbiol *77*, 52-60.https://doi.org/10.1016/j.fm.2018.08.011

Bajovic, B., Bolumar, T., and Heinz, V. (2012). Quality considerations with high pressure processing of fresh and value added meat products. Meat Sci *92*, 280-289.10.1016/j.meatsci.2012.04.024

Bandara, N., Jo, J., Ryu, S., and Kim, K.P. (2012). Bacteriophages BCP1-1 and BCP8-2 require divalent cations for efficient control of *Bacillus cereus* in fermented foods. Food Microbiol *31*, 9-16.10.1016/j.fm.2012.02.003

Bergh, Ø., BØrsheim, K.Y., Bratbak, G., and Heldal, M. (1989). High abundance of viruses found in aquatic environments. Nature *340*, 467-468.10.1038/340467a0

Besser, R.E., Griffin, P.M., and Slutsker, L. (1999). *Escherichia coli* O157:H7 gastroenteritis and the Hemolytic Uremic Syndrome: An emerging infectious disease. Ann Rev Medicine *50*, 355-367.10.1146/annurev.med.50.1.355

Beuchat, L.R., and Ryu, J.H. (1997). Produce handling and processing practices. Emerg Infect Dis *3*, 459-465.10.3201/eid0304.970407

Bigot, B., Lee, W.J., McIntyre, L., Wilson, T., Hudson, J.A., Billington, C., and Heinemann, J.A. (2011). Control of *Listeria monocytogenes* growth in a ready-to-eat poultry product using a bacteriophage. Food Microbiol *28*, 1448-1452.10.1016/j.fm.2011.07.001

Bigwood, T., Hudson, J.A., Billington, C., Carey-Smith, G.V., and Heinemann, J.A. (2008). Phage inactivation of foodborne pathogens on cooked and raw meat. Food Microbiol *25*, 400-406.10.1016/j.fm.2007.11.003

Boyacioglu, O., Sharma, M., Sulakvelidze, A., and Goktepe, I. (2013). Biocontrol of *Escherichia coli* O157:H7 on fresh-cut leafy greens. Bacteriophage *3*, e24620.10.4161/bact.24620

Brüssow, H. (2007). Bacteria between protists and phages: from antipredation strategies to the evolution of pathogenicity. Mol Microbiol *65*, 583-589.10.1111/j.1365-2958.2007.05826.x

Bueno, E., García, P., Martínez, B., and Rodríguez, A. (2012). Phage inactivation of *Staphylococcus aureus* in fresh and hard-type cheeses. Int J Food Microbiol *158*, 23-27.10.1016/j.ijfoodmicro.2012.06.012

Cadwell, K. (2015). The Virome in Host Health and Disease. Immunity *42*, 805-813.https://doi.org/10.1016/j.immuni.2015.05.003

Carlton, R.M., Noordman, W.H., Biswas, B., de Meester, E.D., and Loessner, M.J. (2005). Bacteriophage P100 for control of *Listeria monocytogenes* in foods: genome sequence, bioinformatic analyses, oral toxicity study, and application. Regul Toxicol Pharmacol *43*, 301-312.10.1016/j.yrtph.2005.08.005

Carter, C.D., Parks, A., Abuladze, T., Li, M., Woolston, J., Magnone, J., Senecal, A., Kropinski, A.M., and Sulakvelidze, A. (2012). Bacteriophage cocktail significantly reduces *Escherichia coli* O157:H7 contamination of lettuce and beef, but does not protect against recontamination. Bacteriophage *2*, 178-185.10.4161/bact.22825

Centers for Disease Control and Prevention (2019a). Campylobacter (Campylobacteriosis). 9th Jan, 2020. https://www.cdc.gov/campylobacter/faq.html

Centers for Disease Control and Prevention (2019b). Outbreak of *Salmonella* infections linked to pre-cut melons. 7th Jan, 2020. https://www.cdc.gov/salmonella/carrau-04-19/index.html

Centers for Disease Control and Prevention (2019c). Salmonella. 18th Nov 2019. https://www.cdc.gov/salmonella/

Centers for Disease Control and Prevention (2020). Outbreak of *Salmonella* infections linked to cut fruit. 7th Jan, 2020. https://www.cdc.gov/salmonella/javiana-12-19/index.html

Centers for Disease Control and Preventions (2018). National Shiga toxin-producing *Escherichia coli* (STEC) surveillance annual report, 2016 (Atlanta, Georgia: US Department of Health and Human Services, CDC)

Centers for Disease Control and Preventions (2019). Foodborne outbreaks. 17th July 2019. https://www.cdc.gov/foodsafety/outbreaks/multistate-outbreaks/outbreaks-list.html

Chen, F.-C., Godwin, S., Green, A., Chowdhury, S., and Stone, R. (2018). Prevalence of *Salmonella*, *Campylobacter*, and Shiga Toxin–Producing *Escherichia coli* on the surfaces of raw poultry packages. J Food Prot *81*, 1707-1712.10.4315/0362-028x.Jfp-18-149

Chen, L.-K., Kuo, S.-C., Chang, K.-C., Cheng, C.-C., Yu, P.-Y., Chang, C.-H., Chen, T.-Y., and Tseng, C.-C. (2017). Clinical antibiotic-resistant *Acinetobacter baumannii* strains with higher susceptibility to environmental phages than antibiotic-sensitive strains. Scientific Rep *7*, 6319.10.1038/s41598-017-06688-w

Chibeu, A., Agius, L., Gao, A., Sabour, P.M., Kropinski, A.M., and Balamurugan, S. (2013). Efficacy of bacteriophage LISTEX$^{TM}$ P100 combined with chemical antimicrobials in reducing *Listeria monocytogenes* in cooked turkey and roast beef. Int J Food Microbiol *167*, 208-214.10.1016/j.ijfoodmicro.2013.08.018

Cieplak, T., Soffer, N., Sulakvelidze, A., and Nielsen, D.S. (2018). A bacteriophage cocktail targeting *Escherichia coli* reduces *E. coli* in simulated gut conditions, while preserving a non-targeted representative commensal normal microbiota. Gut Microbes *9*, 391-399.10.1080/19490976.2018.1447291

Dalmasso, M., Hill, C., and Ross, R.P. (2014). Exploiting gut bacteriophages for human health. Trends Microbiol *22*, 399-405.10.1016/j.tim.2014.02.010

Delwart, E. (2012). Animal virus discovery: improving animal health, understanding zoonoses, and opportunities for vaccine development. Curr Opin Virol 2, 344-352.https://doi.org/10.1016/j.coviro.2012.02.012

Dietert, R.R., and Dietert, J.M. (2015). The microbiome and sustainable healthcare. Healthcare *3*, 100-129.10.3390/healthcare3010100

Economic Research Service (2014). Cost Estimates of Foodborne Illnesses. http://ersusdagov/data-products/cost-estimates-of-foodborne-illnessesaspx

Endersen, L., Coffey, A., Neve, H., McAuliffe, O., Ross, R.P., and O'Mahony, J.M. (2013). Isolation and characterisation of six novel mycobacteriophages and investigation of their antimicrobial potential in milk. Int Dairy J *28*, 8-14.10.1016/j.idairyj.2012.07.010

Endersen, L., O'Mahony, J., Hill, C., Ross, R.P., McAuliffe, O., and Coffey, A. (2014). Phage therapy in the food industry. Annu Rev Food Sci Technol *5*, 327-349.10.1146/annurev-food-030713-092415

Enteric Disease Laboratory Branch (2016). National Enteric Disease Surveillance: Shigella Annual Report, 2016. 9$^{th}$ Dec 2019. https://www.cdc.gov/shigella/surveillance.html

Fatica, M.K., and Schneider, K.R. (2009). The use of chlorination and alternative sanitizers in the produce industry. CAB Reviews: Perspectives in Agriculture,

Veterinary Science, Nutrition and Natural Resources *4*, 1-10.10.1079/ PAVSNNR20094052

Ferens, W.A., and Hovde, C.J. (2011). *Escherichia coli* O157:H7: Animal reservoir and sources of human infection. Foodborne Pathogens Dis *8*, 465-487.10.1089/fpd.2010.0673

Ferguson, S., Roberts, C., Handy, E., and Sharma, M. (2013). Lytic bacteriophages reduce *Escherichia coli* O157:H7 on fresh cut lettuce introduced through cross-contamination. Bacteriophage *3*, e24323.10.4161/ bact.24323

Figueiredo, A.C.L., and Almeida, R.C.C. (2017). Antibacterial efficacy of nisin, bacteriophage P100 and sodium lactate against *Listeria monocytogenes* in ready-to-eat sliced pork ham. Braz J Microbiol *48*, 724-729.10.1016/j.bjm. 2017.02.010

Firlieyanti, A.S., Connerton, P.L., and Connerton, I.F. (2016). *Campylobacters* and their bacteriophages from chicken liver: The prospect for phage biocontrol. Int J Food Microbiol *237*, 121-127.10.1016/j.ijfoodmicro. 2016.08.026

Flynn, D. (2019). Oregon wedding and reception last August was a big Shigella event. Food Safety News. https://www.foodsafetynews.com/2019/04/oregon-wedding-and-reception-last-august-was-a-big-shigella-event/

Food and Drug Administration (1997). Irradiation in the Production, Processing and Handling of Food. 21 CFR 179, *62 FR 64107* 64107-64121

Food and Drug Administration (2003). Quantitative assessment of relative risk to public health from foodborne *Listeria monocytogenes* among selected categories of ready-to-eat foods. March 19. https://www.fda.gov/downloads/ food/scienceresearch/researchareas/riskassessmentsafetyassessment/ ucm197330.pdf

Food and Drug Administration (2006). Food Additives Permitted for Direct Addition to Food for Human Consumption; Bacteriophage Preparation. 21 CFR 172, *71 FR 47729.*

Food and Drug Administration (2015). Standards for the growing, harvesting, packing, and holding of produce for human consumption, Food and Drug Administration, ed.

Food and Drug Administration (2019). FDA investigation of *Shigella* illnesses linked to imported raw oysters. 9th Jan, 2020. https://www.fda.gov/food/alerts-advisories-safety-information/fda-investigation-shigella-illnesses-linked-imported-raw-oysters

Food Safety and Inspection Service (2012). Shiga Toxin-Producing *Escherichia coli* in certain raw beef products, US Department of Agriculture, ed.

Galarce, N.E., Bravo, J.L., Robeson, J.P., and Borie, C.F. (2014). Bacteriophage cocktail reduces *Salmonella enterica* serovar Enteritidis counts in raw and smoked salmon tissues. Revista Argentina de Microbiología *46*, 333-337.https://doi.org/10.1016/S0325-7541(14)70092-6

Garcia, P., Madera, C., Martinez, B., and Rodriguez, A. (2007). Biocontrol of *Staphylococcus aureus* in curd manufacturing processes using bacteriophages. Int Dairy J *17*, 1232-1239.10.1016/j.idairyj.2007.03.014

Gomez-Lopez, V.M. (2012). Decontamination of Fresh and Minimally Processed Produce (John Wiley & Sons)

Goode, D., Allen, V.M., and Barrow, P.A. (2003). Reduction of experimental *Salmonella* and *Campylobacter* contamination of chicken skin by application of lytic bacteriophages. Appl Environ Microbiol *69*, 5032-5036.10.1128/Aem.69.8.5032-5036.2003

Grant, A., Parveen, S., Schwarz, J., Hashem, F., and Vimini, B. (2017). Reduction of *Salmonella* in ground chicken using a bacteriophage. Poult Sci *96*, 2845-2852.10.3382/ps/pex062

Greer, G.G. (2005). Bacteriophage control of foodborne bacteria. J Food Prot *68*, 1102-1111.10.4315/0362-028x-68.5.1102

Guenther, S., Herzig, O., Fieseler, L., Klumpp, J., and Loessner, M.J. (2012). Biocontrol of Salmonella Typhimurium in RTE foods with the virulent bacteriophage FO1-E2. Int J Food Microbiol *154*, 66-72.10.1016/j.ijfoodmicro.2011.12.023

Guenther, S., Huwyler, D., Richard, S., and Loessner, M.J. (2009). Virulent bacteriophage for efficient biocontrol of *Listeria monocytogenes* in ready-to-eat foods. Appl Environ Microbiol *75*, 93-100.10.1128/AEM.01711-08

Guenther, S., and Loessner, M.J. (2011). Bacteriophage biocontrol of Listeria monocytogenes on soft ripened white mold and red-smear cheeses. Bacteriophage *1*, 94-100.10.4161/bact.1.2.15662

Guglielmi, G. (2017). Do bacteriophage guests protect human health? Science *358*, 982-983.10.1126/science.358.6366.982

Hagens, S., de Vegt, B., and Peterson, R. (2018). Efficacy of a Commercial Phage Cocktail in Reducing Salmonella Contamination on Poultry Products-Laboratory Data and Industrial Trial Data. Meat Muscle Biol *2*, 156-156.10.221751/rmc2018.136

Hammerl, J.A., Jäckel, C., Alter, T., Janzcyk, P., Stingl, K., Knüver, M.T., and Hertwig, S. (2014). Reduction of *Campylobacter jejuni* in broiler chicken by successive application of group II and group III phages. Plos One *9*, e114785. https://doi.org/10.1371/journal.pone.0114785

Hannigan, G.D., Meisel, J.S., Tyldsley, A.S., Zheng, Q., Hodkinson, B.P., SanMiguel, A.J., Minot, S., Bushman, F.D., and Grice, E.A. (2015). The human skin double-stranded DNA virome: Topographical and temporal diversity, genetic enrichment, and dynamic associations with the host microbiome. mBio *6*, e01578-01515.10.1128/mBio.01578-15

Hatfull, G.F. (2008). Bacteriophage genomics. Curr Opin Microbiol *11*, 447-453.10.1016/j.mib.2008.09.004

Havelaar, A.H., Kirk, M.D., Torgerson, P.R., Gibb, H.J., Hald, T., Lake, R.J., Praet, N., Bellinger, D.C., de Silva, N.R., Gargouri, N., *et al.* (2015). World Health Organization Global Estimates and Regional Comparisons of the Burden of Foodborne Disease in 2010. PLoS Med *12*, e1001923.10.1371/journal.pmed.1001923

Hesse, S., and Adhya, S. (2019). Phage therapy in the twenty-first century: Facing the decline of the antibiotic era; is it finally time for the age of the phage? Ann Rev Microbiol *73*, 155-174.10.1146/annurev-micro-090817-062535

Higgins, J.P., Higgins, S.E., Guenther, K.L., Huff, W., Donoghue, A.M., Donoghue, D.J., and Hargis, B.M. (2005). Use of a specific bacteriophage treatment to reduce *Salmonella* in poultry products. Poult Sci *84*, 1141-1145.10.1093/ps/84.7.1141

Hoffmann, S., Batz, M.B., and J. Glenn Morris, J. (2012). Annual cost of illness and quality-adjusted life year losses in the United States due to 14 foodborne pathogens. J Food Prot *75*, 1292-1302.10.4315/0362-028x.Jfp-11-417

Hong, Y., Schmidt, K., Marks, D., Hatter, S., Marshall, A., Albino, L., and Ebner, P. (2016). Treatment of *Salmonella*-contaminated eggs and pork with a broad-spectrum, single bacteriophage: Assessment of efficacy and resistance development. Foodborne Pathog Dis *13*, 679-688.10.1089/fpd.2016.2172

Hoskisson, P.A., and Smith, M.C. (2007). Hypervariation and phase variation in the bacteriophage 'resistome'. Curr Opin Microbiol *10*, 396-400.10.1016/j.mib.2007.04.003

Hudson, J.A., Billington, C., Cornelius, A.J., Wilson, T., On, S.L.W., Premaratne, A., and King, N.J. (2013a). Use of a bacteriophage to inactivate *Escherichia coli* O157:H7 on beef. Food Microbiol *36*, 14-21.https://doi.org/10.1016/j.fm.2013.03.006

Hudson, J.A., Billington, C., Wilson, T., and On, S.L. (2013b). Effect of phage and host concentration on the inactivation of *Escherichia coli* O157:H7 on cooked and raw beef. Food Sci Technol Int *21*, 104-109.10.1177/1082013213513031

Hui, Y.H. (2003). Food Plant Sanitation (New York, NY: Marcel Dekker, Inc)

Hungaro, H.M., Mendonça, R.C.S., Gouvêa, D.M., Vanetti, M.C.D., and Pinto, C.L.D. (2013). Use of bacteriophages to reduce *Salmonella* in chicken skin in comparison with chemical agents. Food Res Int *52*, 75-81.10.1016/j.foodres.2013.02.032

Interagency Food Safety Analytics Collaboration (2018). Foodborne illness source attribution estimates for 2016 for *Salmonella*, *Escherichia coli* O157, *Listeria monocytogenes*, and *Campylobacter* using multi-year outbreak surveillance data, United States, C. US Department of Health and Human Services, ed. (Atlanta, GA: https://www.cdc.gov/foodsafety/ifsac/pdf/P19-2016-report-TriAgency-508.pdf)

Islam, M.S., Zhou, Y., Liang, L., Nime, I., Liu, K., Yan, T., Wang, X., and Li, J. (2019). Application of a phage cocktail for control of *Salmonella* in foods and reducing biofilms. Viruses *11*, 841.10.3390/v11090841

Jun, J.W., Giri, S.S., Kim, H.J., Yun, S.K., Chi, C., Chai, J.Y., Lee, B.C., and Park, S.C. (2016). Bacteriophage application to control the contaminated water with *Shigella*. Scientific Rep *6*, 22636.10.1038/srep22636

Kalchayanand, N., Koohmaraie, M., and Wheeler, T.L. (2016). Effect of exposure time and organic matter on efficacy of antimicrobial compounds against shiga toxin–producing *Escherichia coli* and *Salmonella*. J Food Prot *79*, 561-568.10.4315/0362-028x.Jfp-15-204

Kang, H.W., Kim, J.W., Jung, T.S., and Woo, G.J. (2013). Wksl3, a new biocontrol agent for *Salmonella enterica* serovars enteritidis and typhimurium in foods: Characterization, application, sequence analysis, and oral acute toxicity study. Appl Environ Microbiol *79*, 1956-1968.10.1128/Aem.02793-12

Kasman, L.M., and Porter, L.D. (2019). Bacteriophages. 10th Dec 2019. https://www.ncbi.nlm.nih.gov/books/NBK493185

Kim, K.P., Klumpp, J., and Loessner, M.J. (2007). *Enterobacter sakazakii* bacteriophages can prevent bacterial growth in reconstituted infant formula. Int J Food Microbiol *115*, 195-203.10.1016/j.ijfoodmicro.2006.10.029

Kirk, M.D., Pires, S.M., Black, R.E., Caipo, M., Crump, J.A., Devleesschauwer, B., Döpfer, D., Fazil, A., Fischer-Walker, C.L., Hald, T., *et al.* (2015). World Health Organization estimates of the global and regional disease burden of 22 foodborne bacterial, protozoal, and viral diseases, 2010: A data synthesis. PLOS Med *12*, e1001921.10.1371/journal.pmed.1001921

Kittler, S., Fischer, S., Abdulmawjood, A., Glunder, G., and Klein, G. (2013). Effect of bacteriophage application on *Campylobacter jejuni* loads in commercial broiler flocks. Appl Environ Microbiol *79*, 7525-7533.10.1128/AEM.02703-13

Kortright, K.E., Chan, B.K., Koff, J.L., and Turner, P.E. (2019). Phage therapy: A renewed approach to combat antibiotic-resistant bacteria. Cell Host Microbe *25*, 219-232.https://doi.org/10.1016/j.chom.2019.01.014

Kotloff, K.L., Winickoff, J.P., Ivanoff, B., Clemens, J.D., Swerdlow, D.L., Sansonetti, P.J., Adak, G.K., and Levine, M.M. (1999). Global burden of *Shigella* infections: implications for vaccine development and implementation of control strategies. Bull World Health Organ *77*, 651-666.https://apps.who.int/iris/handle/10665/267903

Kutter, E., and Sulakvelidze, A. (2004). Bacteriophages: biology and applications (Boca Raton, FL: CRC Press).https://doi.org/10.1201/9780203491751

Leverentz, B., Conway, W.S., Alavidze, Z., Janisiewicz, W.J., Fuchs, Y., Camp, M.J., Chighladze, E., and Sulakvelidze, A. (2001). Examination of bacteriophage as a biocontrol method for *Salmonella* on fresh-cut fruit: a model study. J Food Prot *64*, 1116-1121.https://doi.org/10.4315/0362-028X-64.8.1116

Leverentz, B., Conway, W.S., Camp, M.J., Janisiewicz, W.J., Abuladze, T., Yang, M., Saftner, R., and Sulakvelidze, A. (2003). Biocontrol of *Listeria monocytogenes* on fresh-cut produce by treatment with lytic bacteriophages and a bacteriocin. Appl Environ Microbiol *69*, 4519-4526.10.1128/aem.69.8.4519-4526.2003

Leverentz, B., Conway, W.S., Janisiewicz, W., and Camp, M.J. (2004). Optimizing concentration and timing of a phage spray application to reduce *Listeria monocytogenes* on honeydew melon tissue. J Food Prot *67*, 1682-1686.10.4315/0362-028x-67.8.1682

Lewis, R., and Hill, C. (2020). Overcoming barriers to phage application in food and feed. Curr Opin Biotechnol *61*, 38-44.https://doi.org/10.1016/j.copbio.2019.09.018

Lindberg, A.A., and Holme, T. (1969). Influence of O side chains on the attachment of the Felix O-1 bacteriophage to *Salmonella* bacteria. J Bacteriol *99*, 513-519

Lone, A., Anany, H., Hakeem, M., Aguis, L., Avdjian, A.C., Bouget, M., Atashi, A., Brovko, L., Rochefort, D., and Griffiths, M.W. (2016). Development of prototypes of bioactive packaging materials based on immobilized bacteriophages for control of growth of bacterial pathogens in foods. Int J Food Microbiol *217*, 49-58.10.1016/j.ijfoodmicro.2015.10.011

Magnone, J.P., Marek, P.J., Sulakvelidze, A., and Senecal, A.G. (2013). Additive approach for inactivation of *Escherichia coli* O157:H7, *Salmonella*, and *Shigella* spp. on contaminated fresh fruits and vegetables using bacteriophage cocktail and produce wash. J Food Prot *76*, 1336-1341.10.4315/0362-028X.JFP-12-517

Mai, V., Ukhanova, M., Reinhard, M.K., Li, M., and Sulakvelidze, A. (2015). Bacteriophage administration significantly reduces *Shigella* colonization and shedding by *Shigella*-challenged mice without deleterious side effects and distortions in the gut microbiota. Bacteriophage *5*, e1088124.10.1080/21597081.2015.1088124

Marti, R., Zurfluh, K., Hagens, S., Pianezzi, J., Klumpp, J., and Loessner, M.J. (2013). Long tail fibres of the novel broad-host-range T-even bacteriophage

S16 specifically recognize *Salmonella* OmpC. Mol Microbiol *87*, 818-834.10.1111/mmi.12134

Mathusa, E.C., Chen, Y., Enache, E., and Hontz, L. (2010). Non-O157 Shiga Toxin–Producing *Escherichia coli* in foods. J Food Prot *73*, 1721-1736.10.4315/0362-028x-73.9.1721

Maukonen, J., Matto, J., Wirtanen, G., Raaska, L., Mattila-Sandholm, T., and Saarela, M. (2003). Methodologies for the characterization of microbes in industrial environments: a review. J Ind Microbiol Biotechnol *30*, 327 - 356.10.1007/s10295-003-0056-y

McLean, S.K., Dunn, L.A., and Palombo, E.A. (2013). Phage inhibition of *Escherichia coli* in ultrahigh-temperature-treated and raw milk. Foodborne Pathog Dis *10*, 956-962.10.1089/fpd.2012.1473

Michael, B., Sandra, H., and Jr., M.J.G. (2014). Disease-outcome trees, eq-5d scores, and estimated annual losses of quality-adjusted life years (qalys) for 14 foodborne pathogens in the united states. Foodborne Pathogens Dis *11*, 395-402.10.1089/fpd.2013.1658

Modi, R., Hirvi, Y., Hill, A., and Griffiths, M.W. (2001). Effect of phage on survival of *Salmonella* Enteritidis during manufacture and storage of cheddar cheese made from raw and pasteurized milk. J Food Prot *64*, 927-933.10.4315/0362-028x-64.7.927

Mohiuddin, M., and Schellhorn, H. (2015). Spatial and temporal dynamics of virus occurrence in two freshwater lakes captured through metagenomic analysis. Front Microbiol *6*, 960.https://doi.org/10.3389/fmicb.2015.00960

Moore, A., Nannapaneni, R., Kiess, A., and Sharma, C.S. (2017). Evaluation of USDA approved antimicrobials on the reduction of *Salmonella* and *Campylobacter* in ground chicken frames and their effect on meat quality. Poultry Sci *96*, 2385-2392.https://doi.org/10.3382/ps/pew497

Moye, Z., Rajanna, C., Tokman, J., Fanelli, B., Hasan, N., Marek, P., Senecal, A., and Sulakvelidze, A. (2020). Treatment of fresh produce with a *Salmonella*-targeted bacteriophage cocktail is compatible with chlorine or peracetic acid and more consistently preserves the microbial community on produce. J Food Safety *In Press*.10.1111/jfs.12763

Moye, Z.D., Woolston, J., and Sulakvelidze, A. (2018). Bacteriophage applications for food production and processing. Viruses *10*.10.3390/v10040205

Naanwaab, C., Yeboah, O.A., Ofori Kyei, F., Sulakvelidze, A., and Goktepe, I. (2014). Evaluation of consumers' perception and willingness to pay for bacteriophage treated fresh produce. Bacteriophage *4*, e979662.10.4161/21597081.2014.979662

Nagel, G.M., Bauermeister, L.J., Bratcher, C.L., Singh, M., and McKee, S.R. (2013). *Salmonella* and *Campylobacter* reduction and quality characteristics of poultry carcasses treated with various antimicrobials in a post-chill immersion tank. Int J Food Microbiol *165*, 281-286.https://doi.org/10.1016/j.ijfoodmicro.2013.05.016

NAMI (2020). The United States Meat Industry at a Glance. 01/10/2020. https://www.meatinstitute.org/index.php?ht=d/sp/i/47465/pid/47465

O'Flynn, G., Ross, R.P., Fitzgerald, G.F., and Coffey, A. (2004). Evaluation of a cocktail of three bacteriophages for biocontrol of *Escherichia coli* O157:H7. Appl Environ Microbiol *70*, 3417-3424.10.1128/AEM.70.6.3417-3424.2004

Oliveira, M., Viñas, I., Colàs, P., Anguera, M., Usall, J., and Abadias, M. (2014). Effectiveness of a bacteriophage in reducing *Listeria monocytogenes* on fresh-cut fruits and fruit juices. Food Microbiol *38*, 137-142.10.1016/j.fm.2013.08.018

Örmälä, A.-M., and Jalasvuori, M. (2013). Phage therapy. Bacteriophage *3*, e24219.10.4161/bact.24219

Orquera, S., Hertwig, S., Alter, T., Hammerl, J.A., Jirova, A., and Golz, G. (2015). Development of transient phage resistance in Campylobacter coli against the group II phage CP84. Berl Munch Tierarztl Wochenschr *128*, 141-147

Park, E.-J., Kim, K.-H., Abell, G.C.J., Kim, M.-S., Roh, S.W., and Bae, J.-W. (2011). Metagenomic Analysis of the Viral Communities in Fermented Foods. Appl Environ Microbiol *77*, 1284-1291.10.1128/aem.01859-10

Penney, N., Bigwood, T., Barea, H., Pulford, D., Leroux, G., Cook, R., Jarvis, G., and Brightwell, G. (2007). Efficacy of a peroxyacetic acid formulation as an antimicrobial intervention to reduce levels of inoculated *Escherichia coli*

O157:H7 on external carcass surfaces of hot-boned beef and veal. J Food Prot *70*, 200-203.10.4315/0362-028x-70.1.200

Perera, M.N., Abuladze, T., Li, M., Woolston, J., and Sulakvelidze, A. (2015). Bacteriophage cocktail significantly reduces or eliminates *Listeria monocytogenes* contamination on lettuce, apples, cheese, smoked salmon and frozen foods. Food Microbiol *52*, 42-48.10.1016/j.fm.2015.06.006

Reganold, J.P., and Wachter, J.M. (2016). Organic agriculture in the twenty-first century. Nat Plants 2, 15221.10.1038/nplants.2015.221

Rodríguez, E., Seguer, J., Rocabayera, X., and Manresa, A. (2004). Cellular effects of monohydrochloride of L-arginine, $N^{\alpha}$-lauroyl ethylester (LAE) on exposure to *Salmonella typhimurium* and *Staphylococcus aureus*. J Appl Microbiol *96*, 903-912.10.1111/j.1365-2672.2004.02207.x

Salmond, G.P.C., and Fineran, P.C. (2015). A century of the phage: past, present and future. Nature Rev Microbiol *13*, 777.10.1038/nrmicro3564

Sanath Kumar, H., Otta, S.K., Karunasagar, I., and Karunasagar, I. (2001). Detection of Shiga-toxigenic *Escherichia* coli (STEC) in fresh seafood and meat marketed in Mangalore, India by PCR. Lett Applied Microbiol *33*, 334-338.10.1046/j.1472-765X.2001.01007.x

Scharff, R.L. (2012). Economic burden from health losses due to foodborne illness in the United States. J Food Prot *75*, 123-131.10.4315/0362-028X.JFP-11-058

Schmelcher, M., and Loessner, M.J. (2016). Bacteriophage endolysins: applications for food safety. Curr Opin Biotechnol *37*, 76-87.10.1016/j.copbio.2015.10.005

Shahin, K., and Bouzari, M. (2018). Bacteriophage application for biocontrolling *Shigella flexneri* in contaminated foods. J Food Sci Technol *55*, 550-559.10.1007/s13197-017-2964-2

Shahin, K., Bouzari, M., Wang, R., and Yazdi, M. (2019). Prevalence and molecular characterization of multidrug-resistant *Shigella* species of food origins and their inactivation by specific lytic bacteriophages. Int J Food Microbiol *305*, 108252.https://doi.org/10.1016/j.ijfoodmicro.2019.108252

Sharma, C.S., Dhakal, J., and Nannapaneni, R. (2015). Efficacy of lytic bacteriophage preparation in reducing *Salmonella* in vitro, on turkey breast cutlets, and on ground turkey. J Food Prot *78*, 1357-1362.10.4315/0362-028x.Jfp-14-585

Sharma, M., Dashiell, G., Handy, E.T., East, C., Reynnells, R., White, C., Nyarko, E., Micallef, S., Hashem, F., and Millner, P.D. (2017). Survival of *Salmonella* Newport on whole and fresh-cut cucumbers treated with lytic bacteriophages. J Food Prot *80*, 668-673.10.4315/0362-028x.Jfp-16-449

Sharma, M., Patel, J.R., Conway, W.S., Ferguson, S., and Sulakvelidze, A. (2009). Effectiveness of bacteriophages in reducing *Escherichia coli* O157:H7 on fresh-cut cantaloupes and lettuce. J Food Prot *72*, 1481-1485.10.4315/0362-028x-72.7.1481

Shkoporov, A.N., Clooney, A.G., Sutton, T.D.S., Ryan, F.J., Daly, K.M., Nolan, J.A., McDonnell, S.A., Khokhlova, E.V., Draper, L.A., Forde, A., *et al.* (2019). The human gut virome is highly diverse, stable and individual-specific. bioRxiv, 657528.10.1101/657528

Sillankorva, S.M., Oliveira, H., and Azeredo, J. (2012). Bacteriophages and their role in food safety. Int J Microbiol *2012*, 13.10.1155/2012/863945

Silva, E.N., Figueiredo, A.C., Miranda, F.A., and de Castro Almeida, R.C. (2014). Control of *Listeria monocytogenes* growth in soft cheeses by bacteriophage P100. Braz J Microbiol *45*, 11-16.10.1590/s1517-83822014000100003

Snyder, A.B., Perry, J.J., and Yousef, A.E. (2016). Developing and optimizing bacteriophage treatment to control enterohemorrhagic *Escherichia coli* on fresh produce. Int J Food Microbiol *236*, 90-97.10.1016/j.ijfoodmicro.2016.07.023

Soffer, N., Abuladze, T., Woolston, J., Li, M., Hanna, L.F., Heyse, S., Charbonneau, D., and Sulakvelidze, A. (2016). Bacteriophages safely reduce *Salmonella* contamination in pet food and raw pet food ingredients. Bacteriophage *6*, e1220347.10.1080/21597081.2016.1220347

Soffer, N., Woolston, J., Li, M., Das, C., and Sulakvelidze, A. (2017). Bacteriophage preparation lytic for *Shigella* significantly reduces *Shigella sonnei* contamination in various foods. PLoS One *12*, e0175256.10.1371/journal.pone.0175256

Sohaib, M., Anjum, F.M., Arshad, M.S., and Rahman, U.U. (2016). Postharvest intervention technologies for safety enhancement of meat and meat based products; a critical review. J Food Sci Technol *53*, 19-30.10.1007/s13197-015-1985-y

Soni, K.A., Desai, M., Oladunjoye, A., Skrobot, F., and Nannapaneni, R. (2012). Reduction of *Listeria monocytogenes* in queso fresco cheese by a combination of listericidal and listeriostatic GRAS antimicrobials. Int J Food Microbiol *155*, 82-88.10.1016/j.ijfoodmicro.2012.01.010

Soni, K.A., and Nannapaneni, R. (2010). Bacteriophage significantly reduces *Listeria monocytogenes* on raw salmon fillet tissue. J Food Prot *73*, 32-38.10.4315/0362-028x-73.1.32

Soni, K.A., Nannapaneni, R., and Hagens, S. (2010). Reduction of *Listeria monocytogenes* on the surface of fresh channel catfish fillets by bacteriophage Listex P100. Foodborne Pathog Dis *7*, 427-434.10.1089/fpd.2009.0432

Sorensen, M.C., Gencay, Y.E., Birk, T., Baldvinsson, S.B., Jackel, C., Hammerl, J.A., Vegge, C.S., Neve, H., and Brondsted, L. (2015). Primary isolation strain determines both phage type and receptors recognised by *Campylobacter jejuni* bacteriophages. PLoS One *10*, e0116287.10.1371/journal.pone.0116287

Spricigo, D.A., Bardina, C., Cortes, P., and Llagostera, M. (2013). Use of a bacteriophage cocktail to control *Salmonella* in food and the food industry. Int J Food Microbiol *165*, 169-174.10.1016/j.ijfoodmicro.2013.05.009

Stanaway, J.D., Parisi, A., Sarkar, K., Blacker, B.F., Reiner, R.C., Hay, S.I., Nixon, M.R., Dolecek, C., James, S.L., Mokdad, A.H., *et al.* (2019). The global burden of non-typhoidal salmonella invasive disease: a systematic analysis for the Global Burden of Disease Study 2017. Lancet Infec Dis *19*, 1312-1324.10.1016/S1473-3099(19)30418-9

Suklim, K., Flick, G.J., and Vichitphan, K. (2014). Effects of gamma irradiation on the physical and sensory quality and inactivation of *Listeria monocytogenes* in blue swimming crab meat (*Portunas pelagicus*). Radiat Phys Chem *103*, 22-26.10.1016/j.radphyschem.2014.05.009

Sukumaran, A.T., Nannapaneni, R., Kiess, A., and Sharma, C.S. (2015). Reduction of *Salmonella* on chicken meat and chicken skin by combined or sequential application of lytic bacteriophage with chemical antimicrobials. Int J Food Microbiol *207*, 8-15.10.1016/j.ijfoodmicro.2015.04.025

Sukumaran, A.T., Nannapaneni, R., Kiess, A., and Sharma, C.S. (2016). Reduction of *Salmonella* on chicken breast fillets stored under aerobic or modified atmosphere packaging by the application of lytic bacteriophage preparation SalmoFresh™. Poult Sci *95*, 668-675.10.3382/ps/pev332

Sulakvelidze, A. (2013). Using lytic bacteriophages to eliminate or significantly reduce contamination of food by foodborne bacterial pathogens. J Sci Food Agric *93*, 3137-3146.10.1002/jsfa.6222

Sulakvelidze, A., Alavidze, Z., and Morris, J.G., Jr. (2001). Bacteriophage therapy. Antimicrob Agents Chemother *45*, 649-659.10.1128/AAC.45.3.649-659.2001

Sulakvelidze, A., and Barrow, P.A. (2005). Phage Therapy in Animals and Agribusiness. In Bacteriophages: Biology and Applications, E. Kutter, and A. Sulakvelidze, eds. (Boca Raton, FL: CRC Press), pp. 335-380

Sulakvelidze, A., and Kutter, E. (2005). Bacteriophage therapy in humans. In Bacteriophages: Biology and Application, E. Kutter, and A. Sulakvelidze, eds. (Boca Raton, FL: CRC Press), pp. 381-436.https://doi.org/10.1201/9780203491751

Sulakvelidze, A., and Morris, J.G. (2015). Bacteriophage therapy: a promising approach for managing antibiotic resistant infections. APUA Newsletter *33*, 5-10

Summers, W.C. (2001). Bacteriophage therapy. Annu Rev Microbiol *55*, 437-451

Summers, W.C. (2012). The strange history of phage therapy. Bacteriophage *2*, 130-133.10.4161/bact.20757

Tack, D.M., Marder, E.P., Griffin, P.M., Cieslak, P.R., Dunn, J., Hurd, S., Scallan, E., Lathrop, S., Alison Muse, M., Ryan, P., *et al.* (2019). Preliminary incidence and trends of infections with pathogens transmitted commonly through food — Foodborne Diseases Active Aurveillance Network, 10 U.S. sites, 2015–2018. MMWR Morb Mortal Wkly Rep *68*, 369-372. http://dx.doi.org/10.15585/mmwr.mm6816a2

Talbot, G.H., Jezek, A., Murray, B.E., Jones, R.N., Ebright, R.H., Nau, G.J., Rodvold, K.A., Newland, J.G., Boucher, H.W., and America, T.I.D.S.o. (2019). The Infectious Diseases Society of America's 10 × '20 Initiative (10 New Systemic Antibacterial Agents US Food and Drug Administration Approved by 2020): Is 20 × '20 a Possibility? Clin Infec Dis *69*, 1-11.10.1093/cid/ciz089

Tokman, J.I., Kent, D.J., Wiedmann, M., and Denes, T. (2016). Temperature significantly affects the plaguing and adsorption efficiencies of *Listeria* phages. Front Microbiol *7*, 631.10.3389/fmicb.2016.00631

U. S. Department of Agriculture (2012). USDA targeting six additional strains of *E. coli* in raw beef trim. USDA News Release 2012; Release No. 0171.12

Viator, C.L., Muth, M.K., Brophy, J.E., and Noyes, G. (2017). Costs of food safety investments in the meat and poultry slaughter industries. J Food Sci *82*, 260-269.10.1111/1750-3841.13597

Viazis, S., Akhtar, M., Feirtag, J., and Diez-Gonzalez, F. (2011). Reduction of *Escherichia coli* O157:H7 viability on leafy green vegetables by treatment with a bacteriophage mixture and trans-cinnamaldehyde. Food Microbiol *28*, 149-157.10.1016/j.fm.2010.09.009

Vikram, A., Tokman, J., Woolston, J., and Sulakvelidze, A. (2020). Phage biocontrol improves food safety by significantly reducing both the concentration and occurrence of *Escherichia coli* O157:H7 in various foods. J Food Prot *In Press*

Wheeler, T.L., Shackelford, S.D., and Koohmaraie, M. (1999). Trained sensory panel and consumer evaluation of the effects of gamma irradiation on palatability of vacuum-packaged frozen ground beef patties. J Animal Sci *77*, 3219-3224.10.2527/1999.77123219x

Whichard, J.M., Sriranganathan, N., and Pierson, F.W. (2003). Suppression of *Salmonella* growth by wild-type and large-plaque variants of bacteriophage Felix O1 in liquid culture and on chicken frankfurters. J Food Prot *66*, 220-225.10.4315/0362-028x-66.2.220

Williamson, K.E., Wommack, K.E., and Radosevich, M. (2003). Sampling natural viral communities from soil for culture-independent analyses. Appl Environ Microbiol *69*, 6628-6633.10.1128/aem.69.11.6628-6633.2003

Wolbang, C.M., Fitos, J.L., and Treeby, M.T. (2008). The effect of high pressure processing on nutritional value and quality attributes of *Cucumis melo* L. Innov Food Sci Emerg *9*, 196-200.https://doi.org/10.1016/j.ifset.2007.08.001

Woods, T., Ernst, M., and Tropp, D. (2017). Community Supported Agriculture – New Models for Changing Markets. . 10th Jan, 2020. https://www.ams.usda.gov/publications/content/community-supported-agriculture-new-models-changing-markets

Woolston, J., Parks, A.R., Abuladze, T., Anderson, B., Li, M., Carter, C., Hanna, L.F., Heyse, S., Charbonneau, D., and Sulakvelidze, A. (2013). Bacteriophages lytic for *Salmonella* rapidly reduce *Salmonella* contamination on glass and stainless steel surfaces. Bacteriophage *3*, e25697.10.4161/bact. 25697

Woolston, J., and Sulakvelidze, A. (2015). Bacteriophages and Food Safety. In eLS, John Wiley & Sons Ltd, ed. https://doi.org/ 10.1002/9780470015902.a0025962

Ye, J., Kostrzynska, M., Dunfield, K., and Warriner, K. (2010). Control of *Salmonella* on sprouting mung bean and alfalfa seeds by using a biocontrol preparation based on antagonistic bacteria and lytic bacteriophages. J Food Prot *73*, 9-17.10.4315/0362-028x-73.1.9

Yeh, Y., de Moura, F.H., Van Den Broek, K., and de Mello, A.S. (2018). Effect of ultraviolet light, organic acids, and bacteriophage on *Salmonella* populations in ground beef. Meat Sci *139*, 44-48.10.1016/j.meatsci.2018.01.007

Yuan, Y., Wang, L., Li, X., Tan, D., Cong, C., and Xu, Y. (2019). Efficacy of a phage cocktail in controlling phage resistance development in multidrug resistant *Acinetobacter baumannii.* Virus Res *272*, 197734.https://doi.org/ 10.1016/j.virusres.2019.197734

Zampara, A., Sørensen, M.C.H., Elsser-Gravesen, A., and Brøndsted, L. (2017). Significance of phage-host interactions for biocontrol of *Campylobacter jejuni* in food. Food Control *73*, 1169-1175.10.1016/j.foodcont.2016.10.033

Zhang, H., Li, L., Zhao, Z., Peng, D., and Zhou, X. (2016). Polar flagella rotation in *Vibrio parahaemolyticus* confers resistance to bacteriophage infection. Sci Rep *6*, 26147.10.1038/srep26147

Zhang, H., Wang, R., and Bao, H.D. (2013). Phage inactivation of foodborne *Shigella* on ready-to-eat spiced chicken. Poultry Sci *92*, 211-217.10.3382/ps. 2011-02037

Zhang, W., Li, L., Deng, X., Kapusinszky, B., and Delwart, E. (2014). What is for dinner? Viral metagenomics of US store bought beef, pork, and chicken. Virology *468-470*, 303-310.https://doi.org/10.1016/j.virol.2014.08.025

Zhang, X., Niu, Y.D., Nan, Y., Stanford, K., Holley, R., McAllister, T., and Narváez-Bravo, C. (2019). SalmoFresh™ effectiveness in controlling

*Salmonella* on romaine lettuce, mung bean sprouts and seeds. Int J Food Microbiol *305*, 108250.https://doi.org/10.1016/j.ijfoodmicro.2019.108250

Zinno, P., Devirgiliis, C., Ercolini, D., Ongeng, D., and Mauriello, G. (2014). Bacteriophage P22 to challenge *Salmonella* in foods. Int J Food Microbiol *191*, 69-74.10.1016/j.ijfoodmicro.2014.08.037

Zuber, S., Boissin-Delaporte, C., Michot, L., Iversen, C., Diep, B., Brussow, H., and Breeuwer, P. (2008). Decreasing *Enterobacter sakazakii* (*Cronobacter* spp.) food contamination level with bacteriophages: prospects and problems. Microb Biotechnol *1*, 532-543.10.1111/j.1751-7915.2008.00058.x

# Chapter 9

# Bacteriophage Therapy in Food Animals

**Yongping Xu[1,3]*, Huijun Geng[1], Xiaoyu Li[1], Bingdong Wei[2], Cong Cong[1], Xiaowen Sun[1] and Jibin Li[3]**

[1]School of Bioengineering, Dalian University of Technology, No. 2, Linggong Road, Ganjingzi District, Dalian City, 116024, Liaoning Province, China
[2]Institute of Animal Science, Jilin Academy of Agriculture Sciences, No. 186, Dongxinghua Road, Gongzhuling City, 136100, Jilin Province, China
[3]Postdoctoral Workstation of Dalian SEM Bio-Engineering Technology Co. Ltd, No. 58, Life 1st Road, double D Port, Development Zone, Dalian City, 116620, Liaoning Province, China

*Corresponding author: xyping@dlut.edu.cn

DOI: https://doi.org/10.21775/9781913652517.09

**Abstract**

In animals destined for food use, bacteriophages have considerable potential, both for biocontrol of zoonotic pathogens such as *Campylobacter*, *E. coli* and *Salmonella* and also for therapeutic uses in animal diseases in much the same way as phages applied to treat human infections. Nevertheless, animal phage applications theoretically face less stringent regulations than those faced in the human context. The success of phage application for controlling target bacteria in animals generally has similar biological hurdles to those in other application areas. These include the phage(s) having a broad host range covering many strains within a bacterial species associated with a particular disease and the ability to physically survive long enough to reach the target area in the animal. Also, successful phage selection for biocontrol and therapeutic applications requires reliable monitoring of the epidemiology of strains or serotypes of the problematic bacterial species in the field. The explorations of phage therapy in the major food animals are discussed in this article and the findings outlined show these biological agents have considerable promise as part of an integrated disease management strategy in food animals.

## Introduction

Veterinary applications of bacteriophages (phages) date back to the same time period when phages were first applied to treat human infections (Abedon et al., 2017). As with many early human phage therapy experiments, results were quite variable with some investigations showing considerable success, and many others not (Taylor, 2014). This is understandable in hindsight when one considers the poor understanding of the biological characteristics of phages at the time (Chanishvili, 2012). The commercialisation of antibiotics in the 1940s solved very many problems in veterinary science, essentially paralleling the spectacular success of these simple mass-produced chemical drugs in human medicine. Over several decades, the widespread use of antibiotics has led to many target bacteria becoming resistant to several antibiotics leading ultimately to the same problems in veterinary sciences as those encountered in human medicine (Tang et al., 2017). Contributing to this global problem is the fact that antibiotics have also been used as animal feed supplements because of their growth-promoting properties in food animals (Kahn et al., 2019). The high prevalence of antibiotic resistance in pathogenic bacteria has led to phage therapy being reevaluated in recent decades with the important benefit of the far better understanding of phage biology that exists today. The use of phages has several documented advantages over antibiotics (Sulakvelidze, 2005; Harper and Kutter, 2009; Abedon, 2014; Abedon, 2018; Lehman, 2018; Wienhold et al., 2019) as discussed in several other chapters in this book. Although it is well documented that bacteria can develop resistance to phages, mixtures of phages can and have been compiled where phage therapy is practised and bacteria which show resistance to one phage in a cocktail and be susceptible to others. Phages are abundant in the environment and the normal animal and human microflora, generally sharing a common ecology with their bacterial hosts (Calendar and Abedon, 2006; Brussow and Kutter, 2005). Scientists involved in the isolation of phages for therapeutic applications agree that they must be strictly lytic (virulent), i.e., they kill their target bacteria on infection without integrating their genome into that of the bacterial host where they may persist in a dormant state in bacterial cells, potentially changing the characteristics of the bacteria without killing them. The control of zoonotic pathogenic bacteria in food animals is a persistent human health concern globally. This challenge is best achieved with judicious flock and herd management and various biologically active agents including antibiotics where justified (Tang et al., 2017), vaccines (Clift and Salisbury, 2017), and in recent decades with phages being considered by many as a valid complementary approach as reviewed in depth by Johnson et al. (2008), Kahn et al. (2019) and Gigante and Atterbury (2019). As discussed elsewhere in this book, phages have already been approved for multiple applications in food safety on many post-harvest animal materials. Research on the use of phages in live animal production has intensified, particularly in the last twenty years. Although showing considerable potential, the work has only been experimental to date.

## Control of *Escherichia coli*

Considerable pioneering phage therapy work was conducted in the UK and published in the 1980s by the group of Smith for the effective treatment of *E. coli* induced diarrhoea in calves, piglets and lambs - and their results have been much reviewed over the years (Smith and Huggins, 1982; Smith and Huggins, 1983, Smith et al., 1987). In the last of their three papers on the topic, experimentally induced *E. coli* diarrhoea in the calves could be cured by a single dose of phage and it could be prevented by spraying the litter in the calf rooms with aqueous phage suspensions or simply by keeping the calves in uncleaned rooms previously occupied by calves whose *E. coli* infections had been treated with phage. Microbiological examinations of calves used in these experiments revealed that the phages multiplied rapidly and profusely after gaining entry to the *E. coli*-infected small intestine, quickly reducing the *E. coli* to numbers that were virtually harmless. The only phage-resistant *E. coli* that emerged in the studies were much less virulent than the pathogenic organisms from which they were derived. (Smith et al., 1987). Also in the UK, Barrow et al. (1998) described a phage, designated phage R, which attaches to the K1 capsule of bacteremic strains of *E. coli* (strain O18:K1:H7). Interestingly, the phage was evaluated for the treatment of experimental *E. coli* septicemia in colostrum-deprived Aberdeen

Angus calves. Two of the calves inoculated orally with the *E. coli* did not survive, but three out of four animals given a high titre of phage R intramuscularly 8 hours after the same *E. coli* challenge remained healthy. While this calf study was quite limited statistically, the phage treatment was concluded to have delayed onset of signs of disease in the animals. Also in the context of cattle, other workers have looked at the potential of phages to reduce haemorrhagic *E. coli* shedding in animals (Sheng et al., 2006; Niu et al., 2009), but further work is warranted in this area.

*E. coli* infections were also researched by Jamalludeen et al. (2009) who showed a beneficial effect of administration of high titre coliphages on weaned pigs infected with an enterotoxigenic *E. coli* (O149:H10). The pigs were colonised with the *E. coli* by oral inoculation and this was followed by treatment with six phages (GJ1-GJ7) either individually or combined. The phages were administered both 15 min after bacterial challenge and 24 hours after. The latter was considered a therapeutic dose while the former was considered a prophylactic treatment. The prophylactic use of the six phages individually significantly reduced the duration and severity of the diarrhoea. The therapeutic administration of a two phage cocktail significantly reduced the symptoms, the development of diarrhoea and also the shedding of the pathogenic *E. coli* without changes to the commensal *E. coli* numbers (Jamalludeen et al. 2009). Other studies, such as Lee et al. (2016) have indicated that there are no adverse effects from the feeding of phages to pigs.

In poultry, pathogenic *E. coli* strains are also responsible for colibacillosis leading to considerable mortality rates in flocks where the colonisation of the respiratory tract and progression to the air sacs often leads to septicaemia and death (Huff et al. 2002a; 2002b; 2003; 2004). Several phage challenge studies by the above authors have looked at different modes of administration of high-titre phage preparations to broiler chickens to prevent the development of the disease in creatures that had the pathogen injected into their thoracic air sacs. The different phage treatments included spraying as an aerosol, oral administration in drinking water and intramuscular injection (thigh). Oral administration was the least effective while the other routes lead to significant reductions in mortality compared with untreated birds, where mortality was high. The timing of phage administration was also important. Combination of phage with the antibiotic enrofloxacin administered orally also gave favourable results and suggested a synergistic effect of the antibiotic with the phage. These above studies by the group of Huff and the results they achieved are very significant for the poultry sector. The group followed the above findings with an extensive range of studies by El-Gohary et al. (2014). Their aim was to establish an experimental model of colibacillosis induced by indirect exposure to *E. coli* and to determine if bacteriophage could protect the birds from developing colibacillosis. In study 1 there were six treatments with two replicate pens of 25 birds. The treatments were control warm brooded; control cold stressed; litter inoculated with *E. coli*, warm brooded; litter inoculated with *E. coli*, cold stressed; seeder birds (5 per pen) challenged with *E. coli*, warm brooded; and seeder birds (5 per pen), cold stressed. The study concluded when the birds were 3 wk of age. Body weights at 1, 2, and 3 wk of age were significantly decreased by cold stress, decreased at 1 and 2 wk of age by both the litter and seeder bird treatments compared with the control treatment and by the seeder bird treatment at 3 wk of age. Study 2 consisted of eight treatments with two replicate pens of 20 birds per treatment. The treatments were control, warm brooded; control, cold stressed; litter inoculated with *E. coli*, cold stressed; and seeder birds (5/pen) challenged with *E. coli*, cold stressed with and without bacteriophage treatment. In the bacteriophage treatments the bacteriophages were sprayed on the litter. The study was concluded at 3 wk of age. Body weights at 1 wk of age were significantly decreased from the control treatment by the seeder bird treatment and were significantly higher in all the bacteriophage treatments compared with their matched untreated treatments, except in the control cold stressed treatment. Mortality was significantly decreased by bacteriophage in the litter challenged treatment. Their data suggest that adding bacteriophage to the environment is a practical and effective way to prevent colibacillosis in broiler chickens. Studies on the same topic by Oliveira et al., (2010), who used a fine-drop scatter spray to deliver phages to Rhode Island Red chickens supported many of the above findings. In a different study with Rhode Island Red chickens, Barrow et al. (1998) used the previously-mentioned phage R targeting bacteremic strains of *E. coli* (strain

O18:K1:H7) to explore prevention and treatment septicemia and cerebritis or meningitis in the birds. The pathogen and phage were both administered intramuscularly and the phage was found to be very effective in preventing disease in the chickens.

## Control of *Salmonella*

Salmonellosis is a serious form of gastroenteritis in humans, resulting from the ingestion of the bacterium from contaminated food. Its main reservoirs include poultry and pigs (Suzuki, 1994) Although, in addition to animals being a source of the pathogen for humans, several serotypes of the bacterium are also associated with disease in the animals themselves (Henzler et al. 1994). The first reports on the control of *Salmonella* with phages came from Georgia in the late 1930s from the work of Kveitadze (Kvesitadze and Mikhailova, 1957). He prepared a poly-phage targeting *B. garthner* (now identified as *Salmonella* Enteriditis), *B. breslau* (now identified as *Salmonella* Typhimurium). The phage mixes were approved in 1940 and were used on several farms and resulted in a marked decreased incidence in calf mortality due to paratyphoid and also colibacillosis. The therapeutic was reported to be five times more effective that the vaccinations, which were available at the time and were also recommended for pigs and poultry (Kvesitadze and Mikhailova, 1957). In more recent years, phage therapy has been studied to control *Salmonella* in many food animals. Sklar Joerger (2001) demonstrated that *Salmonella* colonisation of the cecum could be significantly reduced by administering a cocktail of four high titre phages in the feed, and in addition, noted that the phage-treated birds had a lower incidence of airsacculitis. In another similar study, Fiorentin et al. (2005) demonstrated that an oral dose of three high-titre phages could reduce colonisation of broiler chickens by over 3 logs. Significant reductions in chicken colonisation levels were also reported by Atterbury et al. (2007), who used a mix of three broad-host range phages serotypes Enteritidis, Hadar and Typhimurium. More recently, Lim et al. (2012) reported that phages administered in feed could be used to prevent the horizontal transfer of Enteritidis in a commercial layer chick seeder model. Phages were also administered by coarse spray or in drinking water by Borie et al. (2008) who found that the three phages incorporated successfully reduced levels of intestinal Enteriditis by 1.6

logs, on average. Applying phages for the control of serovar Enteriditis was also reported to be successful in a study by Ahmadi et al. (2016), in which case the target poultry was quail. They compared phage administration in both prophylactic and therapeutic contexts, via either oral gavage or vent lip administration to reduce *Salmonella* Enteritidis colonisation of quail cecal tonsils. Their results demonstrated that administration of phages as a preventive agent could reduce colonisation more effectively than post-challenge administration. Furthermore, they reported that oral administration of phage is a more effective prophylactic tool for reduction of *Salmonella* Enteritidis shedding in poultry than is vent lip administration.

Pigs have been another important target for anti-*Salmonella* phage therapy studies as they readily become colonised with the bacterium during transport in contaminated trailers and holding pens, resulting in increased pathogen shedding just prior to processing. Wall et al. (2009) adopted microencapsulation of phages with alginate beads and administered them to 3 to 4 week old pigs by gavage. This resulted in a 2 to 3 log10 CFU/g reduction in the serovar Typhimurium counts in the ileum, caecum and tonsils. The same authors also used microencapsulated phages with market-weight pigs inoculated also with Typhimurium but in this case followed by comingling of the animals with *Salmonella*-infected seeder pigs. Caecal counts of *Salmonella* in this latter case, revealed a 1.4 log reduction compared with non-phage-treated animals. Microencapsulation was also used to administer phages to pigs by Saez et al. (2011), who also reported good reductions in *Salmonella* Typhimurium counts compared with untreated animals and also reporting lower faecal shedding of the bacterium. Faecal shedding was also analysed by Seo et al. (2018) to demonstrate elimination of orally-administered Typhimurium (at $10^8$ CFU/mL) in phage treated animals, compared with non-phage-treated animals, who were found to continue to shed the bacterium. This study demonstrated that the phages had successfully retained activity in the potentially hostile conditions of the gastrointestinal tract despite not being encapsulated in this case. It is worth mentioning that many approaches for encapsulation have been reported to mitigate against phage damage in the porcine intestine as reviewed by Gigante and Atterbury (2019).

## Control of *Campylobacter*

*Campylobacter,* and in particular *C. jejuni*, is the most common causative agent of bacterial gastroenteritis in humans where it has a relatively low infectious dose (Black et al., 1988). The bacterium readily colonises the chicken caecum and this results in chicken carcass contamination (Lynch et al., 2020). Antibiotic resistance is also becoming increasingly prevalent in the bacterium (Lynch et al., 2020). Phage interventions have been considered worthy of investigation by several scientists in recent decades to restrict the spread of *Campylobacter* to humans. Several studies have been undertaken to date. Wagenaar et al. (2005) tested the effect of phage therapy in the control of *Campylobacter jejuni* colonisation in young broilers, either as a preventive or a therapeutic measure. In the prevention study, the birds were infected with *C. jejuni* at day 4 of a 10 day phage treatment. With the therapeutic group, a phage treatment of 6 days, starting 5 days after *C. jejuni* colonisation of the broilers had been established. Treatment was monitored by enumerating *Campylobacter* colony forming units (CFU) and phage plaque forming units (PFU) from caecal contents and counts were compared with control birds not receiving phage therapy. A clear 3 log decline in *C. jejuni* counts was initially observed in the therapeutic group. However, after five days bacterial counts stabilised at a level ten-fold lower than that of the control group. The colonisation of *C. jejuni* in the prevention group was delayed by the phage treatment but colonisation stabilised within a week at levels comparable to the phage-treated group. There were no adverse health effects from the phage treatment. Two different phages were combined as therapeutic treatment of *Campylobacter* positive chickens challenged at the age approaching broiler harvest. This again resulted in a significant decrease in *Campylobacter* colonisation. These authors concluded that phage treatment is a promising measure for reducing *C. jejuni* colonisation in broilers (Wagenaar et al., 2005).

In another similar study, Loc-Carrillo et al. (2005) used two therapeutic phages from a selection of 53 isolated from chicken faeces in a study with Ross broiler chickens that had been experimentally infected with *C. jejuni.* The birds had been infected by oral gavage at 18 to 20 days of age and within a week were administered with the phages at high titre. A single phage treatment was administered after which *C. jejuni* counts in the upper intestine and ceca were analysed and found to be significantly reduced (by between 0.5 and 5 log10 CFU/g). While some phage resistant mutants were isolated in the study, their mutations appeared to be unstable. The same researchers also showed that in 90 broiler flocks, counts of *C. jejuni* in the presence of naturally occurring bacteriophages were almost 2 log cycles lower when compared to levels in samples where phage were not detected (Atterbury et al., 2005).

Other studies on *Campylobacter* lytic phages have led to their division into three groups (I to III) based on genome size, morphological structure and binding receptor (Jäckel, et al., 2019) with phages from group II and III apparently using multiple host cell receptors for binding (Sørensen et al., 2012). Group II phages specifically were shown to give rise to a 2-log reduction in caecal counts of *Campylobacter jejuni* and *coli* in a study by El-Shibiny et al. (2009). In another study, phages of groups II and III were administered at high titre in different combinations to chicks that had previously been infected with *C. jejuni*. On subsequent analysis, reductions in *Campylobacter* counts of up to 3 logs (depending on the phage combination used) were observed (Hammerl et al., 2014). Similar reductions in birds were reported in other studies on *Campylobacter jejuni* and *coli* by Carvalho et al. (2010).

**Bovine mastitis and clostridiosis in poultry**

Bovine mastitis presents a major economic problem in the dairy sector (Aghamohammadi et al., 2018). Conventional treatment with antibiotics, often administered in combination with teat seals has shown limited success as a result of antimicrobial resistance and subclinical persistence of the infection in dairy herds with relatively low cure rates, (Luby and Middleton, 2005). An early study in France in 1980 evaluated the use of polyvalent phage K (Lerondelle and Poutrel, 1980), which inspired further studies. Nevertheless, despite reported successes in treating artificially-induced mastitis in murine models (Breyne et al, 2017; Geng et al., 2019), phage treatment of bovine mastitis has presented challenges to date. A study by Gill et al. (2006) used five daily intramammary infusions of high titre phage K in udder quarters with sub-clinical

*S. aureus* mastitis, but it resulted in the cure of only 3 of 18 quarters. The observed *S. aureus* counts in milk fluctuated after treatment and recovered *S. aureus* isolates were still susceptible to lysis by phage K. Suggested reasons for the lack of efficacy included enzymatic inactivation of phages in the mammary gland, inhibition of the binding of active phages to the target bacteria by milk proteins, and phage aggregation by milk proteins (Gill et al., 2006). This agreed with studies around the same time by O'Flaherty et al. (2005) who reported that raw milk possessed inhibitory agents that were inactivated in pasteurised milk: in raw milk, bacterial agglutination had occurred, which was not observed in pasteurised milk. Phage adsorption was also significantly reduced in the raw milk compared with the heat-treated milk. Similar observations were made by Titze et al. (2020). Such studies have implications for the successful use of phages for control of staphylococcal mastitis and it can be concluded that the mammary gland environment is not favourable for efficient staphylococcal phage replication. These findings can be extrapolated to mastitis-causing streptococci. It is worth bearing in mind that clinical mastitis involves the internalisation of staphylococcal or streptococcal cells into the mammalian cells of the mammary gland (Almeida, 2007), making them inaccessible to the phages. Nevertheless, the above findings with staphylococcal mastitis may not necessarily be extrapolated to other common bacterial causative agents of bovine mastitis such as *E. coli* (Porter et al., 2016), where phages could still potentially play a role as a prophylactic. Also, the above findings certainly should not discourage efforts to challenge Gram positive mastitis, whether staphylococcal or streptococcal with purified phage endolysins (Fischetti, 2010; Angelopoulou et al., 2019). And it is worth mentioning that technology does now exist for internalisation of Gram-positive phage endolysins into mammalian cells (Shen et al., 2016). Further investigations in mastitis are thus worthwhile. In poultry flocks, *Clostridium perfringens* causes necrotic enteritis, which, while it is not an issue in the context of zoonosis, like mastitis it does cause economic losses in farming (Prescott et al., 2016). The application of *Clostridium* phages were studied in a flock of 900 birds, which were included in a variety of experimental designs by Miller et al. (2010). These authors reported that oral administration of a five phage cocktail by oral gavage or drinking water to experimentally infected broiler chickens with *C. perfringens* resulted in a 92%

reduction in mortality compared with the untreated control group. Thus, this disease problem would appear to be another promising candidate for successful phage therapy in the future.

## Conclusions

Alternative strategies for the control of antibiotic resistant pathogens in the animal population are urgently needed. Various measures have been proposed including flock and herd management, vaccination and the use of phages. While the efficacy of the latter varies depending on the type and location of the infection, several studies outlined herein flocks in livestock show considerable promise in that levels of many pathogens can be considerably reduced. This is significant not only from the point of view of animal health but also the health of human consumers. A lot of the experiments performed shows that there are a wide variety of convenient phage application modes ranging from sprays and aerosols to microencapsulated phages in feed. Nevertheless, regulatory frameworks, which would govern the deliberate introduction of phages to animals still need to be adapted/developed and formalised because as of yet, they represent a significant hurdle. It is well known that phages naturally exist in and on animals, in their feed, their water and their general environment. Thus, on the basis of the extensive experimental phage therapy work done with the strictly lytic phages, it is difficult to envisage health or environmental issues that might arise from their deliberate application to animals. It should be stated that the emergence of phage resistance is not considered with the same perspective or apprehension as antibiotic resistance. This is because phages naturally adapt and evolve as quickly as do the various defence systems in their bacterial targets and the perceived issue of resistance is readily circumvented by the updating of phage therapeutics through the addition of new phage components. Once regulatory frameworks for phage therapy are formalised, considerable opportunities should present themselves for existing and new commercial phage companies.

## References

Abedon, S.T. (2014). Phage therapy: eco-physiological pharmacology. Scientifica *2014*, 581639.

Abedon S.T., García P., Mullany P.and Aminov R. (2017). Phage Therapy: Past, Present and Future. Front Microbiol 8. 981 . DOI: 10.3389/fmicb.2017.00981

Abedon, S.T. (2018). Bacteriophage-mediated biocontrol of wound infections, and ecological exploitation of biofilms by phages. In Recent Clinical Techniques, Results, and Research in Wounds, M.Shiffman, ed. Springer International Publishing AG).

Aghamohammadi, M., Haine, D., and Kelton, D.F. (2018). Herd-level mastitis-associated costs on Canadian dairy farms. Front Vet Sci. **5:** 100.

Ahmadi, M., Karimi, Torshizi, M.A., Rahimi, S., Dennehy, J.J. (2016). Prophylactic bacteriophage administration more effective than post-infection administration in reducing *Salmonella* enterica serovar Enteritidis shedding in quail. Front Microbiol. 7, 1253.

Almeida, R.A., Patel, D., Friton, G.M. and Oliver, S.P. (2007) Intracellular killing of mastitis pathogens by penethamate hydriodide following internalization into mammary epithelial cells. J Vet Pharmacol Ther. 30(2), 151-156.

Angelopoulou, A., Warda, A.K., Hill, C. and Ross, R.P. (2019). Non-antibiotic microbial solutions for bovine mastitis – live biotherapeutics, bacteriophage, and phage lysins. Crit Rev Microbiol 45, 5-6, 564-580.

Atterbury, R.J., Dillon, E., Swift, C., Connerton, P.L., Frost, J.A., Dodd, C.E.R., et al. (2005). Correlation of *Campylobacter* bacteriophage with reduced presence of hosts in broiler chicken ceca. Appl Environ Microbiol. 2005;71(8):4885–4887.

Atterbury, R.J., Van Bergen, M.A.P., Ortiz, F., Lovell, M.A., Harris, J.A., De Boer, A, et al. (2007). Bacteriophage therapy to reduce Salmonella colonization of broiler chickens. Appl Environ Microbiol. 73(14), 4543–4549.

Barrow P, Lovell M, Berchieri A. Use of lytic bacteriophage for control of experimental Escherichia coli septicemia and meningitis in chickens and calves. Clin Diagn Lab Immunol. 5(3):294–8.

Black, R.E., Levine, M.M., Clements, M.L., Hughes, T.P. and Blaser, M.J. (1988). Experimental *Campylobacter jejuni* infection in humans. J Infect Dis. 157(3), 472-479.

Borie, C., Albala, I., Sànchez, P., Sánchez, M.L., Ramírez, S., Navarro, C., et al. (2008). Bacteriophage treatment reduces *Salmonella* colonization of infected chickens. Avian Dis. 52(1), 64–**6**7.

Breyne, K., Honaker, R.W., Hobbs, Z., Richter, M., Żaczek, M., Spangler, T., Steenbrugge, J., Lu, R., Kinkhabwala, A., Marchon, B., Meyer, E. and Mokres, L. (2017). Efficacy and safety of a bovine-associated *Staphylococcus aureus* phage cocktail in a murine model of mastitis. Front Microbiol. 2017; 8: 2348. doi: 10.3389/fmicb.2017.02348

Calendar, R. and Abedon, S.T. (2006). The Bacteriophages. (Oxford: Oxford University Press).

Carvalho, C.M., Gannon, B.W., Halfhide, D.E., Santos, S.B., Hayes, C.M., Roe, J.M. et al. (2010). The in vivo efficacy of two administration routes of a phage cocktail to reduce numbers of *Campylobacter coli* and *Campylobacter jejuni* in chickens. BMC Microbiol. 10(1), 232.

Chanishvili, N. (2012). Phage therapy–history from Twort and d'Herelle through Soviet experience to current approaches. *Adv Virus Res. 83, 3-40. doi: 10.1016/B978-0-12-394438-2.00001-3.*

El-Gohary, F.A., Huff, W.E., Huff, G.R., Rath, N.C., Zhou, Z.Y., Donoghue, A.M. (2014). Environmental augmentation with bacteriophage prevents colibacillosis in broiler chickens1. Poult Sci. 93(11):2788–**27**92.

El-Shibiny, A., Scott, A., Timms, A., Metawea, Y., Connerton, P. and Connerton, I. (2009). Application of a group II Campylobacter bacteriophage to reduce strains of *Campylobacter jejuni* and *Campylobacter coli* colonizing broiler chickens. J Food Prot. 72(4), 733–740.

Fiorentin, L., Vieira, N.D. and Barioni, W. Jr. (2005). Oral treatment with bacteriophages reduces the concentration of *Salmonella* Enteritidis PT4 in caecal contents of broilers. Avian Pathol. 34(3), 258–**2**63.

Fischetti VA. (2010). Bacteriophage endolysins: A novel anti-infective to control Gram-positive pathogens. Int J Med Microbiol. 300, 357–62.

Gigante, A. and Atterbury R.J. (2019). Veterinary use of bacteriophage therapy in intensively-reared livestock. Virol J. 16: 155. DOI:10.1186/s12985-019-1260-3

Hammerl, J.A., Jäckel, C., Alter, T., Janzcyk, P., Stingl, K., Knüver, M.T. et al. (2014). Reduction of Campylobacter jejuni in broiler chicken by successive application of Group II and Group III phages. PLoS One. 9(12):e114785.

Harper, D.R. and Kutter, E. (2009). Bacteriophage: therapeutic use. In Encyclopedia of Life Sciences, John Wiley & Sons, Ltd.), pp. 1-7.

Henzler, D.J., Ebel, E., Sanders, J., Kradel, D. and Mason, J. (1994). *Salmonella* Enteritidis in eggs from commercial chicken layer flocks implicated in human outbreaks. Avian Dis. 38(1), 37–43.

Huff, W.E., Huff, G.R., Rath, N.C., Balog, J.M., Xie, H., Moore, P.A., et al. (2002a). Prevention of *Escherichia coli* respiratory infection in broiler chickens with bacteriophage (SPR02). Poult Sci. 81(4):437–441.

Huff, W., Huff, G., Rath, N., Balog, J. and Donoghue A. (2002b). Prevention of *Escherichia coli* infection in broiler chickens with a bacteriophage aerosol spray. Poult Sci. 81(10),1486–1491.

Huff, W., Huff, G., Rath, N., Balog, J. and Donoghue, A. (2003). Evaluation of aerosol spray and intramuscular injection of bacteriophage to treat an *Escherichia coli* respiratory infection. Poult Sci. 82(7):1108–1112.

Huff W.E., Huff, G.R., Rath, N.C., Balog, J.M. and Donoghue, A.M. (2004). Therapeutic efficacy of bacteriophage and Baytril (enrofloxacin) individually and in combination to treat colibacillosis in broilers. Poult Sci. 2004;83(12): 1944–1947.

Jäckel, C., Hammerl, J., Hertwig, S. (2019). *Campylobacter* phage isolation and characterization: what we have learned so far. Methods Protoc. 2(1),18.

Jamalludeen, N., Johnson, R.P., Shewen, P.E. and Gyles, C.L. (2009). Evaluation of bacteriophages for prevention and treatment of diarrhea due to experimental enterotoxigenic Escherichia coli O149 infection of pigs. Vet Microbiol. 136(1–2):135–41.

Johnson, R.P., Gyles, C.L., Huff, W.E., Ojha, S., Huff, G.R., Rath, N.C. and Donoghue, AM. (2008). Bacteriophages for prophylaxis and therapy in cattle, poultry and pigs[J]. Animal Health Research Reviews. 9(2); 201–215. DOI: 10.1017/S1466252308001576

Kahn, L.H., Bergeron, G., Bourassa, M.W., De Vegt, B., Gill, J., Gomes, F., Malouin, F., Opengart, K., Ritter, G.D., Singer, R.S., Storrs, C. and Topp, E. (2019). From farm management to bacteriophage therapy: strategies to reduce antibiotic use in animal agriculture. Ann N Y Acad Sci. 1441(1), 31-39. doi: 10.1111/nyas.14034.

Lee, S., Hosseindoust, A., Goel, A., Choi, Y., Kwon, I.K. and Chae, B. (2016). Effects of dietary supplementation of bacteriophage with or without zinc oxide on the performance and gut development of weanling pigs. Ital J Anim Sci.

15(3), 412–8.

Lehman, S.M. (2018). Bacteriophage diversity. In Viruses of Microorganisms, P.Hyman and S.T.Abedon, eds. (Caister Academic Press: Norwich, UK), pp. 145-165.

Lerondelle, C. and Poutrel, B. (1980). Essais de traitment par les bacteriophages de l'infection mammaire staphylococcique chez la vache en lactation. *Ann. Rech. Vet.* 11:421-426.

Lim, T.-H., Kim, M.-S., Lee, D.-H., Lee, Y.-N., Park, J.-K., Youn, H.-N., et al. (2012). Use of bacteriophage for biological control of *Salmonella* Enteritidis infection in chicken. Res Vet Sci. 93(3), 1173–**117**8.

Loc-Carrillo C., Atterbury R.J., El-Shibiny, A., Connerton, P.L., Dillon, E., Scott A, et al. (2005). Bacteriophage therapy to reduce Campylobacter jejuni colonization of broiler chickens. Appl Environ Microbiol. 71(11), 6554–6563.

Luby, C. D. and Middleton, J.R. (2005). Efficacy of vaccination and antibiotic therapy against *Staphylococcus aureus* mastitis in dairy cattle. *Vet. Rec.* 157:89-90.

Lynch, C.T., Lynch, H., Egan, J., Whyte, P., Bolton, D., Coffey, A and Lucey, B. (2020) Antimicrobial resistance of *Campylobacter* isolates recovered from broilers in the Republic of Ireland in 2017 and 2018: an update, Brit Poultry Sci, DOI: 10.1080/00071668.2020.1758300

Miller, R.W., Skinner, J., Sulakvelidze, A., Mathis, G.F. and Hofacre, C.L. (2010). Bacteriophage therapy for control of necrotic enteritis of broiler chickens experimentally infected with *Clostridium perfringens*. Avian Dis. 54(1), 33-40.

Niu, Y.D., McAllister, T.A., Xu, Y.P., et al. (2009). Prevalence and impact of bacteriophages on the presence of *E. coli* O157:H7 in feedlot cattle and their environment . Appl Environ Microbiol. 75(5), 1271-1278.

O'Flaherty, S., Coffey, A., Meaney, W.J., Fitzgerald, G.F. and Ross, R.P. (2005). Inhibition of bacteriophage K proliferation on Staphylococcus aureus in raw bovine milk. Lett Appl Microbiol. 41(3), 274-279.

Oliveira, A., Sereno, R. and Azeredo, J. (2010). In vivo efficiency evaluation of a phage cocktail in controlling severe colibacillosis in confined conditions and experimental poultry houses. Vet Microbiol. 146(3–4), 303-308.

Porter, J., Anderson, J., Carter, L., Donjacour, E. and Paros, M. (2016). In vitro evaluation of a novel bacteriophage cocktail as a preventative for bovine coliform mastitis. J Dairy Sci 99(3), 2053-2062. doi: 10.3168/jds.2015-9748.

Prescott, J.F., Parreira, V.R., Mehdizadeh Gohari, I., Lepp, D. and Gong, J. (2016). The pathogenesis of necrotic enteritis in chickens: what we know and what weneed to know: a review, Avian Pathol. 45(3), 288-294.

Saez, A.C., Zhang, J., Rostagno, M.H., Ebner, P.D. (2011). Direct feeding of microencapsulated bacteriophages to reduce *Salmonella* colonization in pigs. Foodborne Pathog Dis. 8(12), 1269–1274. https://doi.org/10.1089/fpd.2011.0905.

Seo, B.-J., Song, E.-T., Lee, K., Kim, J.-W., Jeong, C.-G., Moon, S.-H., et al. (2018) Evaluation of the broad-spectrum lytic capability of bacteriophage cocktails against various *Salmonella* serovars and their effects on weaned pigs infected with *Salmonella* Typhimurium. J Vet Med Sci. 80(6), 851–60.

Shen, Y., Barros, M., Vennemann, T., Gallagher, D. T., Yin, Y., Linden, S. B., Heselpoth, R. D., Spencer, D. J., Donovan, D. M., Moult, J., Fischetti, V. A., Heinrich, F., Lösche, M., & Nelson, D. C. (2016). A bacteriophage endolysin that eliminates intracellular streptococci. *eLife*, *5*, e13152.

Sheng, H., Knecht, H.J., Kudva, I.T, et al. (2006). Application of bacteriophages to control intestinal Escherichia coli O157:H7 levels in ruminants. Appl Environ Microbiol. 72(8), 5359-5366. DOI: 10.1128/AEM.00099-06

Sklar, I.B. and Joerger, R.D. (2001). Attempts to utilize bacteriophage to combat Salmonella enterica serovar Enteritidis infection in chickens. J Food Safety. 21(1), 15–29.

Smith, H.W., Huggins, M.B., (1982), Successful treatment of experimental Escherichia coli infections in mice using phage: its general superiority over antibiotics. J Gen Microbiol 128, 307-318. DOI: 10.1099/00221287-128-2-307

Smith, H.W. and Huggins, M.B. (1983). Effectiveness of phages in treating experimental Escherichia coli Diarrhoea in calves, piglets and lambs. J Gen Microbiol 129(8), 2659–2675.

Smith, H.W., Huggins, M.B. and Shaw, K.M. (1987). The control of experimental *Escherichia coli* diarrhoea in calves by means of bacteriophages. J Gen Microbiol 133, 1111-1126.

Sørensen, M.C.H., van Alphen, L.B., Fodor, C., Crowley, S.M., Christensen, B.B., Szymanski, C.M. et al. (2012). Phase variable expression of capsular polysaccharide modifications allows *Campylobacter jejuni* to avoid bacteriophage infection in chickens. Front Cell Infect Microbiol. 2, 11.

Sulakvelidze, A. (2005). Phage therapy: an attractive option for dealing with antibiotic-resistant bacterial infections. Drug Discov. Today 10, 807-809.

Suzuki, S. (1994). Pathogenicity of *Salmonella* Enteritidis in poultry. Int J Foot Microbiol. 21(1-2), 89–105.

Tang, K.L., Caffrey, N.P., Nóbrega D. B., Ronksley, P.E. et al., (2017). Restricting the use of antibiotics in food-producing animals and its associations with antibiotic resistance in food-producing animals and human beings: a systematic review and meta-analysis. Lancet Planet Health. 1(8), e316-e327. doi: 10.1016/S2542-5196(17)30141-9.

Taylor, M.W., (2014). The Discovery of bacteriophage and the d'Herelle Controversy. In: Viruses and Man: A History of Interactions. Springer, Cham. DOI: https://doi.org/10.1007/978-3-319-07758-1_4

Titze, I., T. Lehnherr, H. Lehnherr & V. Krmker. (2020). Efficacy of bacteriophages against *Staphylococcus aureus* isolates from bovine mastitis. Pharmaceuticals. 13: pii E35 . doi 10.3390/ph13030035

Wagenaar JA, Van Bergen MAP, Mueller MA, Wassenaar TM, Carlton RM. Phage therapy reduces Campylobacter jejuni colonization in broilers. Vet Microbiol. 2005;109(3–4):275–83.

Wall, S.K., Zhang, J., Rostagno, M.H., Ebner, P.D. (2010). Phage therapy to reduce preprocessing *Salmonella* infections in market-weight swine. Appl Environ Microbiol. 76(1), 48–53.

Wienhold, S.M., Lienau, J., and Witzenrath, M. (2019). Towards inhaled phage therapy in Western Europe. Viruses. 11.

# Chapter 10

# Bacteriophages in Aquaculture

**Yongping Xu[1,4*], Hongyu Ren[1], Yongsheng Ma[2], Zhen Li[3,4], Xiaoyu Li[1], Lili Wang[1] and Shuying Li[4]**

[1]School of Bioengineering, Dalian University of Technology, Dalian, 116024, People's Republic of China
[2]College of Food Science and Technology, Dalian Ocean University, Dalian, 116000, People's Republic of China
[3]School of Marine and Biological Engineering, Yancheng Institute of Technology, Yancheng, 224051, People's Republic of China
[4]Postdoctoral Workstation of Dalian SEM Bio-Engineering Technology Co. Ltd, Dalian, 116620, People's Republic of China

*Corresponding author: xyping@dlut.edu.cn

DOI: https://doi.org/10.21775/9781913652517.10

**Abstract**

Aquaculture is one of the fastest-growing industries in the world and an essential part of economic development for numerous countries worldwide. However, large-scale intensive farming in this sector has led to increased occurrence of aquatic diseases leading to substantial financial losses. Additionally, the application of antibiotics and chemical drugs has become increasingly limited, which brings additional challenges to the prevention and treatment of diseases. Bacteriophages (phages) are bacterial viruses that naturally occur in nature and have the potential to be viable alternatives to chemical-based antibacterial agents. In this chapter, we will examine the progress and challenges towards the development of phage-based disease control for this industry.

## Introduction

Aquaculture (or aquafarming) is the breeding, rearing and harvesting of fish, shellfish, algae and other organisms in all types of water environments. It is one of the fastest-growing industries in the world, being a significant contributor to economic development in numerous countries worldwide, such as those in Europe and Asia (Bostock et al., 2010). However, the development of large-scale and intensive aquaculture production systems has led to frequent losses by the outbreaks of aquatic animal diseases (Sweet and Bateman, 2015). Further complicating this issue is the fact that the application of antibiotics in marine systems is problematic and limited, making disease control challenging (Liu et al., 2017; Santos and Ramos, 2018).

In Western countries, there has been increased research activity towards the utilisation of bacterial viruses (phages) for the control of various bacterial diseases in humans, animals and plants (Balogh et al., 2010; Kutter et al., 2010; Nobrega et al., 2015; other chapters in this book). This concept is known as phage therapy, with the driving factor for its development being the emerging threat posed by antibiotic-resistant bacteria, as well as the increasing restrictions for the use of antibiotics in medicine and agriculture, and the lack of the development of new antibiotics with new modes of action by pharmaceutical companies (Laxminarayan et al., 2013). The concept is not new, during the Cold War, the embargo on antibiotics from Western countries forced the former Soviet Union to use phage therapy for human bacterial infections (Kortright et al., 2019; Myelnikov, 2018). With phage therapy still being practised in former Soviet Union states, such as Georgia, where it is still widely used today as described in other chapters of this book.

As phages are natural agents, they can be considered as an environmentally friendly alternative to antibiotics. For example, phage preparations for the treatment of plants diseases have been given organic status by The Organic Materials Review Institute (OMRI) in the United States (Monk et al., 2010). Thus, making phages a promising agent for the control of problematic bacteria in agriculture as well as aquaculture. Based on the biological characteristics of phages and their high abundance in water, researchers in various countries are gradually exploring the effects of phages on the prevention and control of bacterial diseases in aquaculture. Phage research in aquaculture is on the rise year by year, which may lead to the development of phage therapy as a valid antibacterial method for the aquaculture industry throughout the world, with the application of phage in the aquaculture industry seeming to be quite practical and feasible.

## Phages in water

Phages are the most abundant biological entities in nature, with their numbers estimated to be as high as $10^{31}$ in the biosphere (Bergh et al., 1989; Middelboe and Brussaard, 2017). They are widely distributed throughout the environment, present in the oceans, rivers, soil, human and animal intestines, etc. There is a total of $10^{30}$ phages in marine environments, making this one of the most abundant settings for phages on earth. Additionally, the density of phages in these environments differ, with surface seawater estimated to possess $10^{7}$ phages per mL and in deep water only reaching $10^{6}$ phages per mL (Wommack and Colwell, 2000).

In 1955, Spencer isolated the first phage from a marine environment (Spencer, 1955). However, it was not until the 1980s, when more scientists started to show interest in isolating these viruses and studying their impact on the ecological system of marine environments. Thus, revealing their importance with regards to the role they play in the carbon cycle of aquatic ecosystems, it is estimated that are $10^{28}$ cases of phage infections daily in the earth's oceans, contributing to the release of 10 billion tons of carbon due to the lysis of their bacterial hosts (Breitbart et al., 2018). Furthermore, about 25% of the global carbon is fixed by the photosynthesis cycle due to the viral shunt (Breitbart et al., 2018).

## Status of bacterial disease prevention and control in aquaculture

### *Bacterial diseases in aquaculture*

The Food and Agriculture Organization of the United Nations (FAO) estimates that half the world's seafood supply will come from aquaculture by 2020 while overfishing in offshore fisheries will lead to a decline in production (Frankic and Hershner, 2003). However, the occurrence of aquatic diseases in aquaculture has become a considerable bottleneck towards the development of this industry. The primary disease onset times are between April and November, with the peak period being between May and September (Lafferty et al., 2015; Mohamad et al., 2019). The principal diseases are bacterial and parasitic in origin, especially among fish. Shellfish diseases are also being increasingly reported (Sweet and Bateman, 2015). The widely known bacterial pathogens of aquatic animals include *Aeromonas, Vibrio, Pseudomonas, Streptococcus* and *Actinomycetes* (Toranzo et al., 2005).

*Traditional prevention and control measures for common bacterial diseases in aquaculture*

Antibiotics and chemical therapy are conventional methods for the rapid and effective control of bacterial infections in aquaculture. However, their frequent use has led to the continuous development of drug-resistant strains. Vaccines are another possible way to prevent infectious diseases. Although, the availability of commercially available vaccines remains limited in aquaculture, mainly because of their low market demand, long development cycles and the poor performance of candidate vaccines (Dadar et al., 2017; Ringo et al., 2014). Biocontrol technologies, using alternative methods to control aquatic diseases, have been extensively studied, such as probiotics and natural active substances but their ability to combat bacterial disease outbreaks is limited. Additionally, they interact with the micro-ecosystems of targeted animals (Pérez-Sánchez et al., 2018).

Antibiotics employed as either growth promoters or disease prophylactics are often mixed with the feed used to raise marine animals in aquaculture. However, fish treated in this manner do not usually metabolize these antibiotics effectively. This application can result in large quantities of unmetabolized antibiotics (as much as 75%) being released into the environment through faeces (Cabello et al., 2016, Burridge et al., 2010). Additionally, it is expected that the use of veterinary antimicrobials will be highly restricted in the near future in an attempt to limit veterinary drug residues in the ecosystem, thereby helping to prevent the development of antibacterial resistance in the food chain.

### *Advantages of phage in the prevention and control of bacterial diseases in aquaculture*

Compared with antibiotics, phages possess several features that make them useful for aquaculture, as summarised in Table 1, such as having high specificity, strong lytic potential, rapid self-replication, green environmental potential and can be used in combination with antibiotics under required circumstances. Additionally, they possess small genomes which make them relatively straightforward for scientific study. This makes them highly amenable for genetic engineering, allowing the improvement of their anti-bacterial virulence, such as improved host range and the delivery of payloads (Pires et al., 2016).

**Table 1.** Advantages of Phage therapy compared to antibiotics*

| **Advantages** |
| --- |
| Natural agents. |
| Self-replicating, easy to obtain and reproduce |
| Can be applied to both Gram-positive and Gram-negative bacteria |
| Highly specific, can effectively kill pathogen bacteria without affecting a healthy microflora. |
| Easy application, they can be sprayed or mixed directly into the water. |
| They may have synergistic effects when used in mixtures (multi-component phage preparations). |
| They can act synergistically with antibiotics, preservatives and disinfectants. |
| Phages are compatible with food and feed. |
| They can be used as therapeutic or disinfectant agents. |
| To date, no phage related antibodies have not been found in aquatic animals, such as fish, implying there is a low immune response in such organisms |
| Phages are ubiquitous and considered safe, and because no adverse effects have been reported with their use in phage therapy, thus considered as Generally Recognized as Safe (GRAS). |

*(Kortright et al., 2019; Madhusudana Rao and Lalitha, 2015)

## Fish

Fish are often threatened by microbial infections, creating complex interactions between host, pathogen and the environment. They are hypothermic organisms that are highly sensitive to changes in environmental conditions (water temperature, dissolved oxygen, metabolites, chemical additives, etc.). As a result, their environment and environmental changes can significantly affect the sensitivity of fish to infectious factors (Plumb and Hanson, 2010). This complexity makes it difficult to develop methods for controlling infectious diseases in aquatic systems.

In the early 1980s, a team of researchers in Taiwan reported in a pilot study on phage therapy against *Aeromonas hydrophila*. Phages were found to be effective in treating eels against redfin disease caused by *A. hydrophila,* improving survival by at least 60% (Wu et al., 1981). Since the 1990s, more attention has been given to the application of phage for the control of pathogenic bacteria. In the following sections, we will discuss fish infections species among the genera of *Pseudomonas, Aeromonas* and *Flavobacterium* and challenges associated with them and the status of *in-vitro* assessment of phage therapy among them to date. This work and phage research on other fish species and diseases are summarised in Table 2.

Table 2 Phage therapy in fish bacterial diseases

| Host fish | Pathogenic bacteria | Disease | Results | Reference |
|---|---|---|---|---|
| Ayu (*Plecoglossus altivelis*) | *Pseudomonas plecoglossicida* | Bacterial hemorrhagic ascites | Mixed phage can reduce mortality by more than 60% | (Park and Nakai, 2003) |
| | *Flavobacterium psychrophilum* | Bacterial coldwater disease (BCWD) | phage PFpW-3 increased survival rates and mean times to death during net-scratch challenge | (Kim and Park, 2018) |
| Yellowtail (*Seriola quinqueradiata*) | *Lactococcus garvieae* | Streptococcicosis | Injection of PLgY-16 can increase the survival rate by nearly 90% | (Nakai et al., 1999) |

| | | | | |
|---|---|---|---|---|
| Japanese Flounder (*Paralichthys olivaceus*) | *Streptococcus iniae* | Septicemia | Mortalities of fish receiving phages PSiJ41 were significantly lower than those of control | (Matsuoka et al., 2007) |
| | *Shewanella putrefaciens* | Food spoilage | Bacterial growth of the phage-treated groups was significantly inhibited | (Li et al., 2014) |
| Atlantic salmon (*Salmo salar L.*) | *Aeromonas salmonicida* | Furunculosis | Phage cocktail can delay the death of fish, but it cannot provide protection | (Verner–Jeffreys et al., 2007) |
| Brook trout (*Salvelinus fontinalis*) | *Aeromonas salmonicida* | Furunculosis | Phage HER110 delayed the onset of furunculosis in brook trout by 7 days | (Imbeault et al., 2006) |
| Rainbow trout (*Oncorhynchus mykiss*) | *Flavobacterium psychrophilum* | Rainbow trout fry syndrome (RTFS) | Some phages could decrease mortality | (Castillo et al., 2012) |
| | *Aeromonas salmonicida* | Furunculosis | Phage PAS-1 can reduce mortality by more than 26% | (Kim et al., 2015) |
| Senegalese sole (*Solea senegalensis*) | *Aeromonas salmonicida* | Furunculosis | Fish juveniles showed no mortality with phage AS-A | (Silva et al., 2016) |
| African catfish (*Clarias gariepinus*) | *Pseudomonas aeruginosa* | Ulcerative Lesions | Phage therapy effectively cured the ulcerative lesions of the infected fish after 8–10 days of treatment | (Khairnar et al., 2013) |
| Striped catfish (*Pangasianodon hypophthalmus*) | *Aeromonas hydrophila* | Motile Aeromonas Septicemia (MAS) | Phage treatments resulted in a significant improvement in the survival rates with up to 100% survival, compared to 18.3% survival observed in control experiments | (Le et al., 2018) |
| Cyprinid loach (*Misgurnus anguillicaudatus*) | *Aeromonas hydrophila* | Hemorrhagic Sepsis | Each phage had a noticeable protective effect, with fish survival rates up 60%. | (Jun et al., 2013) |

| | | | | |
|---|---|---|---|---|
| Indian catfish (*Clarias batrachus*) | *Flavobacterium columnare* | Columnaris Disease | Phage treatment led to the disappearance of gross symptoms and 100% survival in experimentally infected *C. batrachus* | (Prasad et al., 2011) |
| Atlantic cod (*Gadus morhua L.*) | *Vibrio anguillarum* | Bacterial Infection | Phage KVP40 was able to reduce and/or delay the mortality of cod and turbot larvae challenged with *V. anguillarum*. | (Rørbo et al., 2018) |
| Turbot (*Scophthalmus maximus L.*) | *Vibrio anguillarum* | Bacterial Infection | | (Rørbo et al., 2018) |

### *Application of phages against Pseudomonas spp. infection of fish*

Ayu (*Plecoglossus altivelis*) is a small fish species, with the reputation of being the "king of the freshwater fish". The breeding grounds of ayu have been profoundly affected by increased industrialization and environmental pollution. The fish is quite selective about its living environment, and it is challenging to breed. Japan is a major consumer of ayu, where the artificial breeding of this fish first began. However, the fish species is also cultivated in the Zhejiang and Hebei provinces of China and Taiwan (Li et al., 2009; Zhou et al., 2018). Bacterial hemorrhagic ascites caused by *Pseudomonas plecoglossicida* (Nishimori et al., 2000) has been detected in aquaculture of the ayu since 1990. There are no officially approved drugs for this *Pseudomonas* infection.

Park et al. (2000), obtained two different lytic phages specific for *Pseudomonas plecoglossicida* from diseased fish tissues and cultured pond water. One isolate, phage PPpW-3, was morphologically identified as *Myoviridae*, forming small plaques on plaque assays while the other isolate, phage PPpW-4, was morphologically identified as *Podoviridae*, forming larger plaques than that of PPpW-3. *In-vitro* analysis showed that PPpW-4 could inhibit the growth of *P. plecoglossicida* more effectively than PPpW-3, with the bacteriostatic activity of a two-phage cocktail being higher than either used alone. At the same time, the study found that *P. plecoglossicida*, phage resistant mutants were less infective to fish that the wild-type strain.

In the context of *in-vivo* testing of the above phages, ayu were given a feed containing the *Pseudomonas* and were then immediately administered feeds containing either a single phage or a phage mixture. After two weeks, the cumulative mortality of the phage-free control group reached 93.3%, with that of the phage cocktail (20.0%) being significantly reduced compared to either phage PPpW-3 (53.3%) and PPpW-4 (40.0%) used alone. This study shows the synergistic effect of using a combination of phages. Subsequently, the phage cocktail was applied under conditions of cross-infection in water; the results showed that the phage cocktail could reduce the mortality rate by more than 60% improving the ability of ayu to resist *Pseudomonas* infection (Park and Nakai, 2003).

Trials were then conducted at a commercial ayu farm heavily infected with *Pseudomonas*, which involved the testing of 120,000 fish. Two weeks before phage treatment, roughly 1,000 ayu died each day. Feeds containing PPpW-3 and PPpW-4 ($10^7$ PFU/fish) were fed to ayu by mechanized feeders on days 0, 1 and 8. After three hours of phage administration on day 0, 90% of the fish kidneys were found to have phages present. After two weeks of treatment, the death rate also dropped steadily, eventually to about a third of what it had been prior to phage administration. Notably, no phage-resistant strains and phage-neutralizing antibodies were found in sick or healthy fish during the observation period. Active immunity against phages in ayu was found to be low, as no phage-neutralizing antibodies were identified in fish, even after seven days of oral administration of $10^7$ PFU/fish per day or indeed, four weeks of intramuscular injection of $10^8$ PFU/fish per week. However, this study showed the emergence of *Flavobacterium psychrophilum* infection, resulting in fish death, suggesting an interweaving ecology of these pathogenic organisms.

*Application of phages against Aeromonas salmonicida infection of fish*

*Aeromonas salmonicida* is the primary pathogenic agent of furunculosis among several fish species. During infection, bacteria can gather in heart, kidney, spleen, muscle, gills and other tissues causing tissue lesions. Fish may also become drowsy, blackened in appearance, as well as losing appetite. Bleeding of the ventral fin is widespread, especially near the base of the chest, the ventral fin and the gluteal fin, and the lateral skeletal muscle can develop into liquefying necrosis and "bloody furuncle.

Verner-Jeffreys et al. (2007) were the first to investigate the therapeutic effect of phages against furunculosis caused by *A. salmonicida.* Their findings showed that phage therapy had a limited impact on furunculosis of Atlantic salmon (*Salmo salar L.*) and rainbow trout (*Oncorhynchus mykiss*). As *A. salmonicida* is an intracellular pathogen, it likely makes it difficult for phages to reach this bacterium at the later stages of disease development. However, other work has indicated that the prophylactic application of phages against this pathogen has the potential for the prevention by blocking initial infection (Matsuoka and Nakai, 2004; Yamamoto and Maegawa, 2008).

Silva et al. (2016) isolated phage AS-A against *A. salmonicida* for the treatment of juvenile Senegalese sole. An *in-vivo* experiment showed that cumulative mortality after 72 h from phage treatment was 36% among the control group and 0% among the experimental group. The result indicated that the resistance of juvenile Senegalese sole towards *A. salmonicida* was significantly improved by phage immersion before pathogen exposure. Additionally, the application of phage had no significant effect on the natural flora of water, but some differences were identified among the intestinal flora of phage treated fish. Interestingly, no significant difference was found between the treatment group and control group when phages were added in the presence of the host bacterium.

To control the occurrence of phage resistance of *A. salmonicida*, Duarte et al. (2018) isolated two phages designated AS-D and AS–E, which were used alongside another phage designated AS-A. All three phages were identified to belong to the *Myoviridae* posessing a similar morphology. However, compared with the phages AS-D and AS-E, phage AS-A was found to have a larger burst size and a shorter latent period. However, the *in-vitro* analysis showed phage AS-D had the best antibacterial effect by inducing less bacterial resistance. Additionally, the best phage cocktail was composed of phage AS-A with one or both of the newly isolated phages, with phage mixtures resulting in the generation of less resistant bacteria than a single phage used alone. Furthermore, the effects of pH, salinity, temperature, solar and UV radiation on the stability of phages were also evaluated to determine their possible impact on phage therapy in an aquaculture environment. Phages were found to be mainly affected by UV (up to 3 logs PFU/mL reduction after 12 h exposure) and high temperature. This result would indicate that the best time for phage treatment would be at the end of the day or at night.

*Application of phage against Flavobacterium psychrophilum infection of fish*

*Flavobacterium psychrophilum* is a Gram-negative bacterium that causes rainbow trout fry syndrome (RTFS) or bacterial coldwater disease (BCWD). The bacteria mainly infect rainbow trout fry and fingerlings resulting in high mortalities. Additionally, diseased fish can have necrosis near their caudal fin, accompanied by a large amount of mucus. An autopsy of infected fish can show ascites, intestinal inflammation, splenomegaly and liver discolouration. Infected fish spread the bacteria in water, as do dead fish leading to infection of other fish. There is no commercial vaccine available and antibiotics exhibit limited efficacy (Gomez et al., 2014). In recent years, the application of phages of *F. psychrophilum* infections has been given increased attention, especially among salmonid fishes.

In 2008, twenty-two *F. psychrophilum* phages were isolated from a rainbow trout farm in Denmark (Stenholm et al., 2008). The phage genome sizes of these differed considerably, falling into three major size classes (8.5 to 12 kb, 48 kb, and 90 kb). Eighteen phages were found to have unique host spectrums, with each phage having different lysis activity, burst size and latent period among various bacteria, being able to infect up 24 different strains of *F. psychrophilum*. This work shows that it is possible to isolate a large variety of phages against this pathogen with diverse host infectivity.

Madsen et al. (2013) administered juvenile rainbow trout (2 to 5 g) with *F. psychrophilum* and its phage FpV-9. After ten days of treatment, the phage was found to stay longer in the fish body when the bacteria and phage were inoculated simultaneously than when the phage was injected alone. The bacteria and phage were mainly found in the kidney and spleen, and occasionally in the brain. The phage was found to remain infective between pH 4.5 ~ 7.5, but was inactivated at pH 3. In the absence of a host bacterium, phage titres decreased by a factor of $10^4$ after storage in untreated pond water for 55 days. After storage of the phage for approximately two months, phage titres were found to be unaffected at -80 °C in glycerol but decreased by a factor of 10 when stored in a buffer at 20 °C. Additionally, when phages were mixed with fish-feed, only minimum phage titre losses were detected.

Christiansen et al. (2014) then explored the distribution and survival of phages in rainbow trout after oral administration against *F. psychrophilum*. Three different administration methods: bath, oral intubation into the stomach and phage-coated feed, were designed and studied with regards to phage FpV-9 diffusion and survival in juvenile rainbow trout, with emphasis on oral administration. After half an hour after administration, phages could be detected in all organs (intestine, spleen, brain, kidney), with the concentration of phage being highest in fish intestines. Phages persisted in organs for up to 24 hours, then decreased exponentially. Phages given via feed resulted in their detection in intestines, spleen and kidneys after one hour of administration. These experiments demonstrated the ability of phages to survive in the stomach of fish and to penetrate the intestinal barrier into the circulatory system after oral administration. Furthermore, their results showed that phages could tolerate long-term drying on feed pellets, with a 60% survival rate at 80 °C and a 10% survival rate at 5 °C after eight months of storage. This result demonstrates that the delivery of phage via coated feed pellets could be a viable route for phage application to fish in aquaculture.

*F. psychrophilum* is also pathogenic to ayu in Korea and Japan, with it presenting different clinical symptoms than those observed in rainbow trout (Iida and Mizokam, 1996). Kim and Park (2018) showed that phage PFpW-3 could be used as a biological control agent for the prevention of *F. psychrophilum* infection in the farming of ayu. Preliminary tests showed that muscle stimulation did not induce sufficient levels of disease. Therefore, they designed a new net scratch method to cause bacterial infection. The application of phage PFpW-3 had significant protective effects on ayu infected with *F. psychrophilum*, with 50% higher survival rate and a longer meantime for the occurrence of death than the control group. Also, the growth and decline of bacteria and phages in fish were studied. The results showed that phages could permeate into the tissues of a treated fish reaching the kidneys, regardless of whether the skin was damaged or not. Interestingly, changes in phage concentrations within 72 to 120 hours after administration were not consistent with changes in bacterial concentrations. Phage levels decreased significantly and could not be detected, despite the presence of *F. psychrophilum* for 120h in the kidneys. Therefore, it was speculated that the sudden decrease of phage concentration in the kidney might have resulted from the fish's immune system targeting phage.

**Sea cucumber**

Sea cucumber (*Apostichopus japonicus*) is one of the most significant economic mariculture animals in the northern coastal areas of China. Its cultivation is significantly impacted by a disease called skin ulceration syndrome, causing substantial financial losses, being as high as 2 billion RMB between 2003 to 2005 in China (Deng et al., 2009). Species belonging to the genera of *Vibrio* are the primary agents of this disease. However, other genera (*Aeromonas, Pseudomonas and Pseudoalteromonas*) have also been identified to play a role.

*Phage therapy for skin ulceration syndrome of sea cucumber*

Skin ulceration syndrome is commonly known as "ulcer disease", the initial symptoms are manifested as swelling of the mouth, shaking of the head, spitting of the intestine and their decreased absorption capacity of the intestines. The peak period for disease occurrence is between January and April, where the average water temperature can fall below 8°C. Rapid changes in the ambient room temperature and slow changes in water temperature cause the stratification of dissolved oxygen of sea cucumber cultivation pools. This creates a shortage of dissolved oxygen at the bottom of these pools, leading to increased ammonia and nitrogen content, as well as the acceleration of bacterial reproduction. These bacteria can then colonise damaged parts of sea cucumbers, inducing the autolysis mechanisms of their bodies, resulting in death. Additionally, such colonisation can cause the inhibition of growth, reduce intake of food and harm their breeding (Deng et al., 2009).

The main strategies to control the disease involve prevention and supplementary treatment. It is reported that treatment with antibiotics after the onset of the disease is not ideal and that bacterial resistance and drug residues can occur (Zhu et al., 2018). Therefore, effectively preventing the deterioration of water quality and the inhibition of the proliferation of microorganisms remains the primary focus for the prevention and control of this disease.

*Application of phage to treat Vibrio alginolyticus infection in sea cucumber*

At Dalian University of Technology, our research team applied phages for the prevention and control of infection of *Vibrio alginolyticus* (a causative agent of skin ulceration syndrome) of sea cucumber. In order to assess phage treatments, they replicated the aquaculture environment used for the cultivation of sea cucumber in which *V. alginolyticus* induces disease. Our results showed that a mixture of phages PVA1 and PVA2 could significantly improve the rate of survival (71.1%) of *A. japonicus*. However, the therapeutic effect of phages was considerably lower than treatment with 5 mg/L doxycycline. Furthermore, we found that MOI was an essential factor in treatment against *V. alginolyticus*, although with higher MOI values not necessarily giving the best protective effect (Zhang et al., 2015; Zhang et al., 2014).

*Application of phages to treat Vibrio splendidus and Vibrio cyclitrophicus infection of sea cucumbers*

Li et al (2016a) isolated a phage (Vc1) against *Vibrio cyclitrophicus.* A prevention experiment using this phage was shown to effectively improve the survival rate of juvenile sea cucumbers (18±2 g) from 18% to 81% when fed a lyophilized phage powder as an ingredient in the basic feed (at 5%). This treatment was found to be more effective than a treatment involving phage injection or immersion. Phages could be detected in the water of the testing environment, as well as the intestines of sea cumbers but at lower levels. Notably, feeding phages provided nearly the same protection to sea cucumbers as antibiotic treatment. Additionally, it was found that a lyophilized phage powder using skim milk provided the best protective effect, maintaining a phage titre ≥ $10^7$ PFU/mL at 4°C over 14 months.

Li et al. (2016b) conducted *in vitro* and *in vivo* analysis of three phages (PVS-1, PVS-2, PVS-3) of *Vibrio splendidus* when used as a cocktail and separately. All phage test groups yielded encouraging results, especially the protection afforded by the phage cocktail, which was similar to that obtained with an antibiotic. The effect of injecting phages on the immune system of sea cucumbers was examined. A phage cocktail was injected immediately after challenge with the pathogen. Forty-eight hours after phage administration, elevated levels of lysozyme, phosphatase and superoxide dismutase were found among the untreated phage group. In contrast, these parameters among the test group were similar to that of the blank control. Phage treatment had no adverse effect on the normal growth of sea cucumbers, while the titre level of intestinal phage was lower than that of the test water environment. Additionally, this study found that feeding phage to sea cucumbers did not affect the main growth indicators, namely, weight gain, feeding rate, food conversion efficiency and wall-to-wall ratio.

*Phage therapy pilot study on a sea cucumber farm*

A phage cocktail of PVA1, PVc1 and PVS3 (used in equal ratio) was used to assess the effectiveness of a phage cocktail on a sea farm against *V. splendidus* (Li et al., 2020). Over a 48-day cultivation period, *Vibrio* counts among the phage treated group decreased, with phage particles being detected in the coelomic fluids of sea cucumbers over the cultivation time. Furthermore, incubation with phages showed no significant loss of cell viability compared to the control group, while the level of superoxide dismutase and lysozyme in the phage immersion group remained lower than that of the control group. At the end of the 48-day cultivation period, the total weight gain of the sea cucumbers in the phage cocktail immersion group increased significantly compared to the control group, although remaining lower than the antibiotic treatment group. No significant differences were detected in the viscera body wall ratio and feed conversion of the three groups. Examination of the intestinal flora of sea cucumbers showed that phage treatment resulted in reduce *Vibrio* abundance, while that of other microorganisms was unaffected, indicating that the application of phage could selectively target the bacterial species of interest.

## Shrimp

Intensive farming of shrimp makes them prone to disease outbreak by *Vibrio* infection, especially by the species *Vibrio harveyi, Vibrio parahaemolyticus, Vibrio anguillarum, Vibrio vulnificus* and *V. splendidus*. These strains have been reported to cause disease outbreaks in shrimp, with a mortality rate as high as 100%, causing huge losses within the aquaculture industry. Antibiotics are often used to prevent bacterial infection in shrimp hatcheries, which leads to the emergence of drug resistance, as often seen with *V. harveyi* (Karunasagar et al., 1994).

*Effect of phages on V. parahaemolyticus infection in shrimp*

Acute Hepatopancreatic Necrosis Disease (AHPND) caused by *V. parahaemolyticus* is one of the most problematic diseases in shrimp farming in Southeast Asia and Latin America. Alagappan et al. (2010) isolated five phages (Vp1, Vp3, Vp5, Vp7 and Vp9) from healthy shrimp (*Penaeus monodon*) and shrimp culture ponds using *V. parahaemolyticus* strain MTCC-451 as a host. Among them, Vp1 phage could lyse all isolated *V. parahaemolyticus* strains, while the other four phages had a narrow host spectrum, with all five phages being unable to lyse *V. splendidus*, *V. harveyi*, *E. coli*, *Pseudomonas* and *Bacillus* isolated from shrimp ponds. Based on their initial analysis, the team took three of the five isolated phages (VP1, VP7 and VP9) to evaluate their effectiveness at controlling *V. parahaemolyticus* infection within shrimp ponds (Alagappan et al., 2016). Besides examining phage antibacterial potential, related immune indexes, intestinal phage count and bacterial count of shrimp were monitored in the phage feeding experiment. Shrimp in the phage cocktail group showed better blood cell counts, higher total phenolic oxidase and respiratory burst activity (superoxide anions) than those in the control group. Lysozyme activity increased gradually in all phage treatment groups and decreased significantly in control groups. Notably, the bacterial fragments produced by phage lysis stimulated the shrimp's immune response, making the phage an immunostimulant against any nonspecific infection.

Martinez-Diaz and Hipólito-Morales (2013) used phage Vpms1 to treat brine shrimp (*Artemia franciscana*) infected with *V. parahaemolyticus* strain ATCC-17802. They evaluated the effect of reducing phage dose and delay in treatment. The efficacy of the lytic phage was not affected by a dose reduction (1000, 100, 50, or 10mL of Vpms1 at a titre of $2.7\times10^{9}$ PFU/mL) with a decrease in mortality by approximately 9%, compared with 19.4% in the control group. However, the beneficial effects of phage treatment were diminished with a delay in treatment. At the late stage of infection (5 to 30 hours after challenge), phage had the limited ability to control the pathogen, with mortality rates similar to the control group. Therefore, it was concluded that phage therapy must be combined with effective tools for the early diagnosis and prediction of infection risk, as part of an integrated disease management plan for shrimp farming.

Lomelí-Ortega and Martínez-Díaz (2014) used juvenile white leg shrimp (*Penaeus vannamei*) as subjects to assess the effect of different doses and delayed treatment of lytic phage A3S to control *V. parahaemolyticus*. The results showed that the therapeutic effect could be dependent on MOI. As the survival rate using phage A3S at an MOI of 0.1 was highest of all the experimental groups, and significantly different to observations with an MOI of 10. Furthermore, they showed that phages were effective in reducing the juvenile mortality (survival rate of 75% by phage A3S vs 53% in the control group), regardless of an early or delayed application of phage therapy.

*Effect of phages on V. harveyi infection in shrimp*

The species *V. harveyi*, a rod-shaped, motile and halophilic luminous bacterium is a well-known shrimp pathogen world-wide. It causes exceptionally high death levels at the larval stage of giant tiger prawn *Penaeus monodon* (Lavilla-Pitogo et al., 1990). Antibiotics are often used to prevent bacterial infection in shrimp hatcheries, but this leads to the emergence of antibiotic resistance *V. harveyi* (Karunasagar et al., 1994).

Vinod et al. (2006) isolated a phage of *V. harveyi* from a screening of samples obtained from shrimp farms, hatcheries and seawater. This phage could infect

50 different strains of *V. harveyi* obtained from various sources. The ability of the phage to control the infection of *V. harveyi* in a laboratory setting was evaluated. *V. harveyi* was inoculated at a concentration of $10^5$ CFU/mL into 18-day old shrimp larvae, after which the phage was applied. In the phage-treated group, the survival rate of larval shrimp was 70% after 48 hours, with the luminescent bacteria count decreased by 2 logs - while in the control group, the survival rate was 25%, with bacterial counts increasing by 1 log. Under the condition of natural infection with *V. harveyi*, the therapeutic or preventive effect of phage was demonstrated. Nauplii (*n* = 35000), i.e. shrimp egg hatchlings, the most vulnerable stage of the copepod life cycle were temporarily raised in a 500 L water tank. In separate tanks, phage $10^5$ PFU/mL or antibiotics (5 ppm oxytetracycline, 10 ppm kanamycin) were applied. Over 17 days, the shrimp moved through the stages of zoea, mysis and post-larvae. By the end of the experiment, the survival rate of antibiotic- and phage-treated tanks were 40% and 86%, respectively.

## Abalone

Abalone (*Haliotis*) are commercially important marine molluscs in many countries, including China, the United States, South Africa, Australia, and Japan (Dang et al., 2012). Vibriosis is responsible for the deaths of large numbers of farmed abalone with *harveyi*, *V. alginolyticus* and *V. parahaemolyticus* identified as the primary pathogens (Cai et al., 2006; Jiang et al., 2013; Liu et al., 2001; Liu et al., 2000).

Abalone is highly sensitive to changes in water temperature. Some studies showed that abalone begins to die on the fourth day after a challenge at 19°C. However, death starts within 24 hours after a challenge at 24°C (Travers et al., 2008). When abalone is exposed to high water temperatures of 26°C, damage can occur to the epithelium of their gills providing a pathway for pathogens such as *V. harveyi* to infiltrate (Hooper et al., 2014; Lange et al., 2014). Since controlling water temperature is not financially feasible in the farming of abalone, effective vibriosis treatment is necessary. Under simulated industrial conditions of high temperatures in abalone farming, researchers have evaluated phage therapy for the control of *V. harveyi* infection of *Haliotis laevigata*.

Wang et al. (2017) demonstrated that the cumulative survival rate of abalone could be significantly improved by phage treatment shortly after infection with *V. harveyi.* Two *Siphoviridae* phages, vB_VhaS-a and vB_VhaS-tm, were isolated with both possessing different antibacterial characteristics against different *V. harveyi* strains. An *In vivo* experiment showed that the survival of abalone after a five day exposure to a temperature stress of 24°C and infected with *V. harveyi,* could be improved by phage therapy (70% death with phage treatment compared to complete death of abalone in the non-phage-treated group).

In recent years, new pathogens of abalone have been discovered, and among them *Vibrio owensii*, a newly discovered species of the *Harveyi* branch, has been identified as a pathogen of marine crustaceans (Cano-Gomez et al., 2010). A strain of this *Vibrio* species called B8D, which is most closely related to *V. owensii*, was isolated from abalone farms. Under laboratory conditions, the strain was found to cause death of abalone (data not shown). Five lytic phages were isolated and identified (using B8D as the host bacterium), all of which were long-tailed phages. One-step growth curve experiments showed that the phages differed in their incubation period, ascending period, and burst size. All phages are thermally stable and can tolerate a wide range of pH values (Yu et al., 2013). The results indicate that these phages may be potential candidates for the biological control of *V. owensii* infection in aquaculture systems.

### Phage applications in oyster safety

*Vibrio* species are often the primary pathogens found to contaminate oysters, causing major food safety problems, with the species *V. parahaemolyticus* and *V. vulnificus* being the main culprits. Due to the high risk of such contamination with oysters, it is essential to develop effective strategies for the control of human vibriosis.

Zhang et al. (2008) designed an experimental model to replicate a seafood market environment to evaluate the biosafety effect of phages in a mass food poisoning event, involving *V. parahaemolyticus* with shellfish. A *V. parahaemolyticus* specific phage (vB_VpaS_OMN) was isolated from oysters, which possessed good pH (5-9) and temperature resistance (<50°C), as well as having a broad host range against strains of *V. parahaemolyticus* (20/31). When the phage was applied to oyster samples for 48 h and 72 h, the inactivation rate of *V. parahaemolyticus* on the surface of oyster meat was 90% and 99%, respectively.

## Conclusion

Phages have been reported as green biological agents that can be used for the prevention and treatment of bacterial diseases in aquaculture, mariculture as well as other applications in agriculture. They are highly specific and versatile antibacterial agents, and it's possible to combine mixtures of phages to create preparations (phage cocktails) with desired host ranges against bacterial pathogens of interest. Additionally, they can be combined with other antimicrobial agents, such as antibiotics, with synergic effects. Furthermore, unlike other antibacterial agents, the development of bacterial resistance against them isn't necessarily an adverse outcome, as it can lead to a host bacterium becoming less pathogenic. However, there is no one-size-fits-all approach for their application as antibacterial agents in aquaculture, as with different marine animals and their bacterial pathogens, various factors have to be taken into account. Such considerations include the route of administration, application phage during the night, MOI, etc and these have a critical influence on the success of phage therapy in aquaculture.

Although phage therapy has been demonstrated to possess good antibacterial ability in a number of different fish models, its large-scale use for the prevention and control of bacterial infection has not been approved by any regulatory agent to date. However, phage therapy for other applications has received such approval. As early as 2006, the U.S. Food and Drug Administration (FDA) approved the first phage-based product called ListShield™ for use as a food disinfectant to control *Listeria monocytogenes* in ready to eat food and poultry meat products. In 2007, the FDA approved another phage-based product called Listeria phage LISTEX™ P100 as a safety additive (GRAS). Its anti-Listeria activity is effective on salmon and fresh catfish, improving food preservation as well as preventing food-borne bacteria from exceeding the required standards.

In 2011, the FDA approved the phage-based product EcoShield$^{TM}$ for the control of *E. coli* O157: H7 contamination in minced red meat

Phage therapy has been demonstrated to be capable of controlling several diseases that profoundly affect the aquaculture industry. Furthermore, with an ever-increasing number of commercial phage-based products in the market place, it is expected that it is only a matter of time when phage therapy becomes a common practice in aquaculture.

## References

Alagappan, K., Deivasigamani, B., Somasundaram, S., and Kumaran, S. (2010). Occurrence of Vibrio parahaemolyticus and Its Specific Phages from Shrimp Ponds in East Coast of India. Current microbiology *61*, 235-240.https://xs.scihub.ltd/https://doi.org/10.1007/s00284-010-9599-0

Alagappan, K., Karuppiah, V., and Deivasigamani, B. (2016). Protective effect of phages on experimental V. parahaemolyticus infection and immune response in shrimp (Fabricius, 1798). Aquaculture *453*, 86-92.https://doi.org/10.1016/j.aquaculture.2015.11.037

Balogh, B., Jones, J.B., Iriarte, F.B., and Momol, M.T. (2010). Phage Therapy for Plant Disease Control. Current Pharmaceutical Biotechnology *11*, 48-57. DOI: 10.2174/138920110790725302

Bergh, Ø., BØrsheim, K.Y., Bratbak, G., and Heldal, M. (1989). High abundance of viruses found in aquatic environments. Nature *340*, 467-468. DOI: 10.1038/340467a0

Bostock, J., McAndrew, B., Richards, R., Jauncey, K., Telfer, T., Lorenzen, K., Little, D., Ross, L., Handisyde, N., Gatward, I., *et al.* (2010). Aquaculture: global status and trends. Philosophical Transactions of the Royal Society B-Biological Sciences *365*, 2897-2912. DOI:10.1098/rstb.2010.0170

Breitbart, M., Bonnain, C., Malki, K., and Sawaya, N.A. (2018). Phage puppet masters of the marine microbial realm. Nature Microbiology *3*, 754-766. DOI: 10.1038/s41564-018-0166-y

Burridge, L., Weis, J.S., Cabello, F., Pizarro, J., and Bostick, K. (2010). Chemical use in salmon aquaculture: a review of current practices and possible environmental effects. Aquaculture *306*, 7-23.https://doi.org/10.1016/

j.aquaculture.2010.05.020

Cabello, F.C., Godfrey, H.P., Buschmann, A.H., and Dölz, H.J. (2016). Aquaculture as yet another environmental gateway to the development and globalisation of antimicrobial resistance. The Lancet Infectious Diseases *16*, e127-e133.https://doi.org/10.1016/S1473-3099(16)00100-6

Cai, J., Han, Y., and Wang, Z. (2006). Isolation of *Vibrio parahaemolyticus* from abalone (Haliotis diversicolor supertexta L.) postlarvae associated with mass mortalities. Aquaculture *257*, 161-166.https://doi.org/10.1016/j.aquaculture.2006.03.007

Cano-Gomez, A., Goulden, E.F., Owens, L., and Høj, L. (2010). Vibrio owensii sp. nov., isolated from cultured crustaceans in Australia. FEMS microbiology letters *302*, 175-181.https://doi.org/10.1111/j.1574-6968.2009.01850.x

Castillo, D., Higuera, G., Villa, M., Middelboe, M., Dalsgaard, I., Madsen, L., and Espejo, R. (2012). Diversity of Flavobacterium psychrophilum and the potential use of its phages for protection against bacterial cold water disease in salmonids. Journal of fish diseases *35*, 193-201. DOI: 10.1111/j.1365-2761.2011.01336.x

Christiansen, R.H., Dalsgaard, I., Middelboe, M., Lauritsen, A.H., and Madsen, L. (2014). Detection and Quantification of *Flavobacterium psychrophilum*-Specific Bacteriophages in vivo in rainbow trout upon oral administration: implications for disease control in aquaculture. Applied and Environmental Microbiology *80*, 7683. DOI: 10.1128/AEM.02386-14

Dadar, M., Dhama, K., Vakharia, V.N., Hoseinifar, S.H., Karthik, K., Tiwari, R., Khandia, R., Munjal, A., Salgado-Miranda, C., and Joshi, S.K. (2017). Advances in aquaculture vaccines against fish pathogens: global status and current trends. Reviews in Fisheries Science & Aquaculture *25*, 184-217. DOI: 10.1080/23308249.2016.1261277

Dang, V.T., Speck, P., and Benkendorff, K. (2012). Influence of elevated temperatures on the immune response of abalone, Haliotis rubra. Fish & Shellfish Immunology *32*, 732-740. DOI:10.1016/j.fsi.2012.01.022

Deng, H., He, C., Zhou, Z., Liu, C., Tan, K., Wang, N., Jiang, B., Gao, X., and Liu, W. (2009). Isolation and pathogenicity of pathogens from skin ulceration disease and viscera ejection syndrome of the sea cucumber *Apostichopus japonicuss*. Aquaculture *287*, 18-27. DOI:10.1016/j.aquaculture.2008.10.015

Duarte, J., Pereira, C., Moreirinha, C., Salvio, R., Lopes, A., Wang, D.P., and Almeida, A. (2018). New insights on phage efficacy to control *Aeromonas salmonicida* in aquaculture systems: An in vitro preliminary study. Aquaculture *495*, 970-982. DOI:10.1016/j.aquaculture.2018.07.002

Frankic, A., and Hershner, C. (2003). Sustainable aquaculture: developing the promise of aquaculture. Aquaculture International *11*, 517-530.https://xs.scihub.ltd/https://doi.org/10.1023/B:AQUI.0000013264.38692.91

Gomez, E., Mendez, J., Cascales, D., and Guijarro, J.A. (2014). *Flavobacterium psychrophilum* vaccine development: a difficult task. Microbial Biotechnology *7*, 414-423. DOI:10.1111/1751-7915.12099

Hooper, C., Day, R., Slocombe, R., Benkendorff, K., Handlinger, J., and Goulias, J. (2014). Effects of severe heat stress on immune function, biochemistry and histopathology in farmed Australian abalone (hybrid Haliotis laevigata× Haliotis rubra). Aquaculture *432*, 26-37.https://doi.org/10.1016/j.aquaculture.2014.03.032

Iida, Y., and Mizokam, A. (1996). Outbreaks of Coldwater Disease in Wild Ayu and Pale Chub. 魚病研究 *31*, 157-164. https://doi.org/10.3147/jsfp.31.157

Imbeault, S., Parent, S., Lagace, M., Uhland, C.F., and Blais, J.F. (2006). Using Bacteriophages to prevent furunculosis caused by *Aeromonas salmonicida* in farmed brook trout Journal of Aquatic Animal Health *18*, 203-214. DOI: 10.1577/h06-019.1

Jiang, Q., Shi, L., Ke, C., You, W., and Zhao, J. (2013). Identification and characterization of Vibrio harveyi associated with diseased abalone Haliotis diversicolor. Diseases of aquatic organisms *103*, 133-139.https://doi.org/10.3354/dao02572

Jun, J.W., Kim, J.H., Shin, S.P., Han, J.E., Chai, J.Y., and Park, S.C. (2013). Protective effects of the *Aeromonas* phages pAh1-C and pAh6-C against mass mortality of the cyprinid loach (Misgurnus anguillicaudatus) caused by Aeromonas hydrophila. Aquaculture *416*, 289-295. DOI: 10.1016/j.aquaculture.2013.09.045

Karunasagar, I., Pai, R., Malathi, G., and Karunasagar, I. (1994). Mass mortality of Penaeus monodon larvae due to antibiotic-resistant *Vibrio harveyi* infection. Aquaculture *128*, 203-209.https://doi.org/10.1016/0044-8486(94)90309-3

Khairnar, K., Raut, M.P., Chandekar, R.H., Sanmukh, S.G., and Paunikar, W.N.

(2013). Novel bacteriophage therapy for controlling metallo-beta-lactamase producing *Pseudomonas aeruginosa* infection in Catfish. Bmc Veterinary Research *9*. DOI:10.1186/1746-6148-9-264

Kim, J.H., Choresca, C.H., Shin, S.P., Han, J.E., Jun, J.W., and Park, S. (2015). Biological Control of *Aeromonas salmonicida* subsp. *salmonicida* Infection in Rainbow Trout (Oncorhynchus mykiss) Using *Aeromonas* phage PAS-1. Transboundary and emerging diseases *62*, 81-86. https://doi.org/10.1111/tbed.12088

Kim, J.H., and Park, S.C. (2018). Biological control of *Flavobacterium psychrophilum* infection in ayu (*Plecoglossus altivelis*) using a bacteriophage PFpW-3. Korean Journal of Veterinary Research *58*, 39-43.https://doi.org/10.14405/kjvr.2018.58.1.39

Kortright, K.E., Chan, B.K., Koff, J.L., and Turner, P.E. (2019). Phage therapy: a renewed approach to combat antibiotic-resistant bacteria. Cell host & microbe *25*, 219-232. DOI: 10.1016/j.chom.2019.01.014

Kutter, E., De Vos, D., Gvasalia, G., Alavidze, Z., Gogokhia, L., Kuhl, S., and Abedon, S.T. (2010). Phage Therapy in Clinical Practice: Treatment of Human Infections. Current Pharmaceutical Biotechnology *11*, 69-86. DOI: 10.2174/138920110790725401

Lafferty, K.D., Harvell, C.D., Conrad, J.M., Friedman, C.S., Kent, M.L., Kuris, A.M., Powell, E.N., Rondeau, D., and Saksida, S.M. (2015). Infectious Diseases Affect Marine Fisheries and Aquaculture Economics. In Annual Review of Marine Science, Vol 7, C.A. Carlson, and S.J. Giovannoni, eds., pp. 471-496. DOI:10.1146/annurev-marine-010814-015646

Lange, B., Currie, K.-L., Howarth, G.S., and Stone, D.A. (2014). Grape seed extract and dried macroalgae, Ulva lactuca Linnaeus, improve survival of greenlip abalone, Haliotis laevigata Donovan, at high water temperature. Aquaculture *433*, 348-360.https://doi.org/10.1016/j.aquaculture.2014.06.028

Lavilla-Pitogo, C.R., Baticados, M.C.L., Cruz-Lacierda, E.R., and Leobert, D. (1990). Occurrence of luminous bacterial disease of *Penaeus monodon* larvae in the Philippines. Aquaculture *91*, 1-13.https://doi.org/10.1016/0044-8486(90)90173-K

Laxminarayan, R., Duse, A., Wattal, C., Zaidi, A.K.M., Wertheim, H.F.L., Sumpradit, N., Vlieghe, E., Levy Hara, G., Gould, I.M., Goossens, H., *et al.*

(2013). Antibiotic resistance-the need for global solutions. Lancet Infectious Diseases *13*, 1057-1098. DOI:10.1016/s1473-3099(13)70318-9

Le, T.S., Nguyen, T.H., Vo, H.P., Doan, V.C., Nguyen, H.L., Tran, M.T., Tran, T.T., Southgate, P.C., and Kurtboke, D.I. (2018). Protective effects of bacteriophages against *Aeromonas hydrophila* causing motile *Aeromonas* septicemia (MAS) in Striped Catfish Antibiotics-Basel *7*, 11. DOI:10.3390/antibiotics7010016

Li, C., Chen, J., Shi, Y., and Li, M. (2009). Characterization of *Listonella anguillarum* as the aetiological agent of vibriosis occurred in cultured ayu (*Plecoglossus altivelis*) in Ninghai country, China. Wei sheng wu xue bao = Acta microbiologica Sinica *49*, 931-937

Li, M., Lin, H., Khan, M.N., Wang, J., and Kong, L. (2014). Effects of bacteriophage on the quality and shelf life of Paralichthys olivaceus during chilled storage Journal of the Science of Food and Agriculture *94*, 1657-1662. DOI: 10.1002/jsfa.6475

Li, Z., Zhang, J., Li, X., Wang, X., Cao, Z., Wang, L., and Xu, Y. (2016a). Efficiency of a bacteriophage in controlling vibrio infection in the juvenile sea cucumber *Apostichopus japonicus*. Aquaculture *451*, 345-352.https://doi.org/10.1016/j.aquaculture.2015.09.024

Liu, P., Chen, Y., and Lee, K. (2001). Pathogenicity of *Vibrio alginolyticus* isolated from diseased small abalone Haliotis diversicolor supertexta. Microbios *104*, 71-77

Li, Z., Ren, H., Li, Q., Murtaza, B., Li, X., Zhang, J., and Xu, Y. (2020). Exploring the effects of phage cocktails in preventing *Vibrio* infections in juvenile sea cucumber (*Apostichopus japonicus*) farming. Aquaculture *515*, 734599.https://doi.org/10.1016/j.aquaculture.2019.734599

Li, Z., Li, X., Zhang, J., Wang, X., Wang, L., Cao, Z., and Xu, Y. (2016b). Use of phages to control Vibrio splendidus infection in the juvenile sea cucumber *Apostichopus japonicus*. Fish & shellfish immunology *54*, 302-311.https://doi.org/10.1016/j.fsi.2016.04.026

Liu, P.C., Chen, Y.C., Huang, C.Y., and Lee, K.K. (2000). Virulence of Vibrio parahaemolyticus isolated from cultured small abalone, *Haliotis diversicolor* , with withering syndrome. Letters in applied microbiology *31*, 433-437.https://doi.org/10.1046/j.1365-2672.2000.00843.x

Liu, X., Steele, J.C., and Meng, X.-Z. (2017). Usage, residue, and human health risk of antibiotics in Chinese aquaculture: A review. Environmental Pollution *223*, 161-169.https://doi.org/10.1016/j.envpol.2017.01.003

Lomelí-Ortega, C.O., and Martínez-Díaz, S.F. (2014). Phage therapy against *Vibrio parahaemolyticus* infection in the whiteleg shrimp (*Litopenaeus vannamei*) larvae. Aquaculture *434*, 208-211.https://doi.org/10.1016/j.aquaculture.2014.08.018

Madhusudana Rao, B., and Lalitha, K.V. (2015). Bacteriophages for aquaculture: Are they beneficial or inimical. Aquaculture *437*, 146-154.https://doi.org/10.1016/j.aquaculture.2014.11.039

Madsen, L., Bertelsen, S.K., Dalsgaard, I., and Middelboe, M. (2013). Dispersal and survival of *Flavobacterium psychrophilum* phages in vivo in rainbow trout and in vitro under laboratory conditions: implications for their use in phage therapy. Appl Environ Microbiol *79*, 4853-4861. DOI: 10.1128/AEM.00509-13

Martinez-Diaz, S.F., and Hipólito-Morales, A. (2013). Efficacy of phage therapy to prevent mortality during the vibriosis of brine shrimp. Aquaculture *400*, 120-124.https://doi.org/10.1016/j.aquaculture.2013.03.007

Matsuoka, S., Hashizume, T., Kanzaki, H., Iwamoto, E., Park, S.C., Yoshida, T., and Nakai, T. (2007). Phage Therapy against β-hemolytic streptococcicosis of japanese flounder *Paralichthys olivaceus*. DOI:10.3147/jsfp.42.181

Matsuoka, S., and Nakai, T. (2004). Seasonal appearance of *Edwardsiella tarda* and its bacteriophages in the culture farms of Japanese flounder. Fish Pathology *39*, 145-152. https://doi.org/10.3147/jsfp.39.145

Middelboe, M., and Brussaard, C.P.D. (2017). Marine Viruses: Key Players in Marine Ecosystems. Viruses *9*, 302. DOI: 10.3390/v9100302

Mohamad, N., Amal, M.N.A., Yasin, I.S.M., Saad, M.Z., Nasruddin, N.S., Al-saari, N., Mino, S., and Sawabe, T. (2019). Vibriosis in cultured marine fishes: a review. Aquaculture *512*, 17. DOI:10.1016/j.aquaculture.2019.734289

Monk, A.B., Rees, C.D., Barrow, P., Hagens, S., and Harper, D.R. (2010). Bacteriophage applications: where are we now? Letters in Applied Microbiology *51*, 363-369. DOI:10.1111/j.1472-765X.2010.02916.x

Myelnikov, D. (2018). An Alternative Cure: The Adoption and Survival of Bacteriophage Therapy in the USSR, 1922-1955. Journal of the History of Medicine and Allied Sciences *73*, 385-411. DOI:10.1093/jhmas/jry024

Nakai, T., Sugimoto, R., Park, K.-H., Matsuoka, S., Mori, K.-i., Nishioka, T., and Maruyama, K. (1999). Protective effects of bacteriophage on experimental *Lactococcus garvieae* infection in yellowtail. Diseases of aquatic organisms *37*, 33-41. DOI:10.3354/dao037033

Nishimori, E., Kita-Tsukamoto, K., and Wakabayashi, H. (2000). *Pseudomonas plecoglossicida* sp. nov., the causative agent of bacterial haemorrhagic ascites of ayu, Plecoglossus altivelis. International Journal of Systematic and Evolutionary Microbiology *50*, 83-89.https://doi.org/10.1099/00207713-50-1-83

Nobrega, F.L., Costa, A.R., Kluskens, L.D., and Azeredo, J. (2015). Revisiting phage therapy: new applications for old resources. Trends in Microbiology *23*, 185-191. DOI:10.1016/j.tim.2015.01.006

Park, S.C., and Nakai, T. (2003). Bacteriophage control of *Pseudomonas plecoglossicida* infection in ayu *Plecoglossus altivelis*. Diseases of aquatic organisms *53*, 33-39. DOI:10.3354/dao053033

Park, S.C., Shimamura, I., Fukunaga, M., Mori, K.-I., and Nakai, T. (2000). Isolation of bacteriophages specific to a fish pathogen, *Pseudomonas plecoglossicida*, as a candidate for disease control. Appl Environ Microbiol *66*, 1416-1422. DOI: 10.1128/AEM.66.4.1416-1422.2000

Pérez-Sánchez, T., Mora-Sánchez, B., and Balcázar, J.L. (2018). Biological approaches for disease control in aquaculture: advantages, limitations and challenges. Trends in microbiology *26*, 896-903.https://doi.org/10.1016/j.tim.2018.05.002

Pires, D.P., Cleto, S., Sillankorva, S., Azeredo, J., and Lu, T.K. (2016). Genetically engineered phages: a review of advances over the last decade. Microbiology and Molecular Biology Reviews *80*, 523-543. DOI:10.1128/mmbr.00069-15

Plumb, J.A., and Hanson, L.A. (2010). Health maintenance and principal microbial diseases of cultured fishes (John Wiley & Sons)

Prasad, Y., Arpana, Kumar, D., and Sharma, A.K. (2011). Lytic bacteriophages specific to *Flavobacterium columnare* rescue catfish, *Clarias batrachus* (Linn.) from columnaris disease. Journal of Environmental Biology *32*, 161-168

Ringo, E., Olsen, R.E., Jensen, I., Romero, J., and Lauzon, H.L. (2014). Application of vaccines and dietary supplements in aquaculture: possibilities and challenges. Reviews in Fish Biology and Fisheries *24*, 1005-1032. DOI:

10.1007/s11160-014-9361-y

Rong, R., Lin, H., Wang, J., Khan, M.N., and Li, M. (2014). Reductions *of Vibrio parahaemolyticus* in oysters after bacteriophage application during depuration. Aquaculture *418*, 171-176.https://doi.org/10.1016/j.aquaculture.2013.09.028

Rørbo, N., Rønneseth, A., Kalatzis, P.G., Rasmussen, B.B., Engell-Sørensen, K., Kleppen, H.P., Wergeland, H.I., Gram, L., and Middelboe, M. (2018). Exploring the effect of phage therapy in preventing *Vibrio anguillarum* infections in cod and turbot larvae. Antibiotics *7*, 42.https://doi.org/10.3390/antibiotics7020042

Santos, L., and Ramos, F. (2018). Antimicrobial resistance in aquaculture: Current knowledge and alternatives to tackle the problem. International Journal of Antimicrobial Agents *52*, 135-143.https://doi.org/10.1016/j.ijantimicag.2018.03.010

Silva, Y.J., Moreirinha, C., Pereira, C., Costa, L., Rocha, R.J., Cunha, Â., Gomes, N.C., Calado, R., and Almeida, A. (2016). Biological control of *Aeromonas salmonicida* infection in juvenile Senegalese sole (*Solea senegalensis*) with Phage AS-A. Aquaculture *450*, 225-233.https://doi.org/10.1016/j.aquaculture.2015.07.025

Spencer, R. (1955). A marine bacteriophage. Nature *175*, 690-691. DOI: 10.1038/175690a0

Stenholm, A.R., Dalsgaard, I., and Middelboe, M. (2008). Isolation and characterization of bacteriophages infecting the fish pathogen *Flavobacterium psychrophilum*. Appl Environ Microbiol *74*, 4070-4078. DOI: 10.1128/AEM.00428-08

Sweet, M.J., and Bateman, K.S. (2015). Diseases in marine invertebrates associated with mariculture and commercial fisheries. J Sea Res *104*, 16-32. DOI:10.1016/j.seares.2015.06.016

Toranzo, A., Magarinos, B., and Romalde, J. (2005). A review of the main bacterial fish disease in mariculture systems, Vol 246. DOI:10.1016/j.aquaculture.2005.01.002

Travers, M.-A., Le Goïc, N., Huchette, S., Koken, M., and Paillard, C. (2008). Summer immune depression associated with increased susceptibility of the European abalone, Haliotis tuberculata to Vibrio harveyi infection. Fish & shellfish immunology *25*, 800-808.https://doi.org/10.1016/j.fsi.2008.08.003

Verner–Jeffreys, D.W., Algoet, M., Pond, M.J., Virdee, H.K., Bagwell, N.J., and Roberts, E.G. (2007). Furunculosis in Atlantic salmon (*Salmo salar* L.) is not readily controllable by bacteriophage therapy. Aquaculture *270*, 475-484.https://doi.org/10.1016/j.aquaculture.2007.05.023

Vinod, M.G., Shivu, M.M., Umesha, K., Rajeeva, B., Krohne, G., Karunasagar, I., and Karunasagar, I. (2006). Isolation of *Vibrio harveyi* bacteriophage with a potential for biocontrol of luminous vibriosis in hatchery environments. Aquaculture *255*, 117-124.https://doi.org/10.1016/j.aquaculture.2005.12.003

Wang, Y.H., Barton, M., Elliott, L., Li, X.X., Abraham, S., O'Dea, M., and Munro, J. (2017). Bacteriophage therapy for the control of *Vibrio harveyi* in greenlip abalone (Haliotis laevigata) Aquaculture *473*, 251-258.https://doi.org/10.1016/j.aquaculture.2017.01.003

Wommack, K.E., and Colwell, R.R. (2000). Virioplankton: viruses in aquatic ecosystems. Microbiol Mol Biol Rev *64*, 69-114. DOI: 10.1128/MMBR.64.1.69-114.2000

Wu, J.-L., LIN, H.-M., JAN, L., Hsu, Y.-L., and CHANG, L.-H. (1981). Biological control of fish bacterial pathogen, *Aeromonas hydrophila*, by bacteriophage AH 1. Fish Pathology *15*, 271-276.https://doi.org/10.3147/jsfp.15.271

Yamamoto, A., and Maegawa, T. (2008). Phage typing of *Edwardsiella tarda* from eel farm and diseased eel. Aquaculture Science *56*, 611-612. https://doi.org/10.11233/aquaculturesci.56.611

Yu, Y.P., Gong, T., Jost, G., Liu, W.H., Ye, D.Z., and Luo, Z.H. (2013). Isolation and characterization of five lytic bacteriophages infecting a Vibrio strain closely related to Vibrio owensii. FEMS Microbiology Letters *348*, 112-119.https://doi.org/10.1111/1574-6968.12277

Zhang, H., Yang, Z., Zhou, Y., Bao, H., Wang, R., Li, T., Pang, M., Sun, L., and Zhou, X. (2018). Application of a phage in decontaminating Vibrio parahaemolyticus in oysters. International journal of food microbiology *275*, 24-31.https://doi.org/10.1016/j.ijfoodmicro.2018.03.027

Zhang, J., Cao, Z., Li, Z., Wang, L., Li, H., Wu, F., Jin, L., Li, X., Li, S., and Xu, Y. (2015). Effect of bacteriophages on *Vibrio alginolyticus* infection in the sea cucumber, *Apostichopus japonicus* (Selenka). Journal of the World Aquaculture Society *46*, 149-158.https://doi.org/10.1111/jwas.12177

Zhang, J., Cao, Z., Xu, Y., Li, X., Li, H., Wu, F., Wang, L., Cao, F., Li, Z., and Li,

S. (2014). Complete genomic sequence of the *Vibrio alginolyticus* lytic bacteriophage PVA1. Archives of virology *159*, 3447-3451. DOI:10.1007/s00705-014-2207-z

Zhou, Q.J., Chai, F.C., and Chen, J. (2018). First record of *Glugea plecoglossi* (Takahashi & Egusa, 1977), a microsporidian parasite of ayu (Plecoglossus altivelis altivelis Temminck & Schlegel, 1846) in Mainland China. Journal of Fish Diseases *41*, 165-169. DOI:10.1111/jfd.12674

Zhu, M.H., Zhao, H.X., Chen, J.W., Xie, H.J., and Du, J. (2018). Investigation of antibiotics in sea cucumbers: occurrence, pollution characteristics, and human risk assessment. Environ Sci Pollut Res *25*, 32081-32087. DOI:10.1007/s11356-018-3131-7

## Chapter 11

# *Mycobacterium avium* subspecies *paratuberculosis*: Are Mycobacteriophages the Answer?

**Gillian Crowley[1], Colin Buttimer[2], Lorraine Endersen[1], Aidan Coffey[1,2] and Jim O'Mahony[1*]**

[1]Department of Biological Sciences, Cork Institute of Technology, Cork, Ireland
[2]APC Microbiome Institute, University College, Cork, Ireland

*Corresponding author: Jim.OMahony@cit.ie

DOI: https://doi.org/10.21775/9781913652517.11

**Abstract**

Johne's disease is a chronic wasting disease of ruminants and other animals. It impacts significantly on animal welfare and the livelihood of farmers. Due to the complexity of the causative organism (*Mycobacterium avium* ssp *paratuberculosis*), effective research into the detection and treatment of Johne's disease is problematic. Mycobacteriophages offer much hope on both fronts given their ubiquity in nature and their potential to target specific mycobacterial pathogens. This chapter traces the evolution of Johne's disease and catalogues the importance of mycobacteriophages. It also explores potential avenues of future research that may raise the hope of effectively treating the global problem of Johne's disease.

### Biological Properties and Taxonomy

The etiological agent of many animal based diseases, in particular, Johne's disease is the pathogen *Mycobacterium avium* ssp *paratuberculosis* (MAP). This bacterium is extremely problematic as a result of its slow growth rate, acid fast nature and thick-waxy, Gram-positive cell wall (Harris and Barletta, 2001). MAP typically resides in the environment and has the ability to survive pasteurisation of milk and certain dairy processing steps (Gerrard et al., 2018). Human populations are therefore widely exposed to MAP. Multiple reviews have suggested that MAP is also pathogenic to humans (Pierce, 2010), particularly in its association with Crohn's disease, an inflammatory condition of the gastrointestinal tract, similar to that observed in animals affected with Johne's disease (Tasara and Stephan, 2005).

MAP belongs to the *M. avium complex* (MAC), however, various phenotypic differences exist between itself and other *M. avium* subspecies (Naser et al., 2004). Slow growing mycobacterial species are characteristic of the MAC complex which are ubiquitous in the environment (soil and water) and can cause a wide range of disease in domestic and wild animal's as well as humans. MAP is pathogenic in ruminants and primates who are non-immunocompromised, however, the *M. avium complex* is thought only to affect immunocompromised hosts (Pierce, 2010). This would suggest that if MAP can cause Crohn's disease and infect immunocompromised patients, then there should be sufficient levels of MAP found in the blood and tissues of patients; however, no studies have found sufficient levels of MAP to date.

The main distinguishing features between MAP and other mycobacterial species include its slow growth rate, its inability to produce mycobactin, the presence of the insertion sequence IS900, and it's extremely lipid rich cell wall (Castellanos et al., 2012). The slow growth rate of MAP is attributed to the presence of its cell wall, which is made up of approximately 60% covalently attached lipids. Destruction of MAP by chemicals and heat is difficult, allowing MAP to be one of the most persistent pathogens of the *Mycobacterium avium complex* (MAC). Certain antibiotic regimens have proven successful in treating MAP, however, due to the rise in antibiotic resistance; the number of effective antibiotics against MAP have started to decrease in recent times.

The IS900 insertion sequence is unique to MAP being the first insertion sequence to be identified in mycobacteria. Based on a study in 2001, southern blot analysis estimated that MAP contains roughly 18 copies of these insertion sequences in its genome (Harris and Barletta, 2001). This insertion sequence is a member of the IS116 family with each member sharing 60 to 80% identity. It also has been speculated that IS900 plays a role in genomic plasticity. A variety of other insertion elements are thought to migrate between mycobacterial genomes. An example of such is the insertion element IS110 (Green et al., 1989).

**The process of invasion in MAP based infections**

MAP upon infection has the ability to invade tight junctions and evade antimicrobial peptides present in the intestinal epithelium, thereby causing chronic infection in the host (Koets et al., 2015). Upon entering the gastrointestinal tract, MAP reaches the mucosa which lines the small intestine. Here MAP can proceed into two cell types (microfold cells and enterocytes). A particular study indicates that the uptake of MAP in calves by microfold cells is larger than enterocytes when they enter a calf's intestine (Ponnusamy et al., 2013). The larger number of microfold cells (M-cells) compared to enterocytes is associated with the early life of calves, in which MAP susceptibility is high due to the presence of large numbers of M-cells, which are the major portal of entry for MAP combined with the developing immune system. In the small intestine, MAP particularly affects the M-cells of Peyer's patches due to their antigen sampling role, which makes them susceptible to other invasive pathogens such as *Escherichia coli* and *Vibrio cholera*, as well as *Salmonella* and *Shigella* species (Momotani et al., 1988). The potential reasoning is that M cells display β1 integrins containing a fibronectin bridge which serve to target MAP to M cells (Secott et al., 2004). This has been demonstrated through the use of electron microscopy. The entry of MAP into M cells, however, does not appear to be the only method for evading the host's immune response. The under-development of the Peyer's patches and M-cells in a neonatal calf due to antigen exposure is thought to be another reason (Secott et al., 2004). Additionally, it has been demonstrated that when M-cells are absent, MAP still can invade (Secott et al., 2004). Furthermore, the uptake of MAP can also be affected by goblet cells of the small intestine (Sigurdardóttir et al., 2005). The ability of MAP to grow in a diverse range of hosts is crucial to its survival as it does not grow once outside a host (it only persists in the environment).

## The diverse host range of MAP infection

MAP primarily affects ruminants such as cows, sheep and goats, but in recent years the organism has infected a much larger host range including monogastric species (Morand et al., 2014). The existence of wildlife reservoirs has also provided evidence that they play a role in the epidemiology of the disease, for example, rabbits infected with MAP in Scotland have been shown to cause problems in the spread of the disease. However, this data still has to be clarified (Carta et al., 2013). Clinical incidences of MAP in non-domesticated animals are extremely variable in their prevalence and timing but are comparable to those of domesticated origin. It has been noted that little research has been specifically published in relation to MAP infection in non-ruminant wildlife.

MAP has been reported in every continent of the world in domesticated animals, although no systematic survey has been completed. Numerous MAP prevalence studies have been undertaken on domesticated animals in various countries (Claudia et al., 2013). Differences in sampling strategies make the evaluation of the studies difficult; nonetheless, they imply significant variation in the numbers of infected herds within different countries and geographical areas.

## The environmental distribution of MAP

MAP is an environmental pathogen that is commonly found in soil and water, which are associated with low pH conditions and high iron content (Norby et al., 2007). MAP can persist in the environment for extended periods of time and remain dormant outside the host until transmission to a host occurs via the environment (Emery and Whittington, 2004). The survival of MAP does, however, decrease over time and is finite. Plants or grass that grow in infected soil eventually become contaminated and have the ability to infect grazing animals (Pribylova et al., 2011). Once a ruminant ingests MAP from the environment and infection occurs, eventually shedding follows, and the infection cycle commences again. MAP is shed in several ways from animals, but faecal shedding followed by oral ingestion is a primary means of transmission.

A variety of physical, chemical and biological factors of the soil may have an impact on the movement of MAP through soil, as was established for other bacteria. However, the ability of MAP to survive in soil has been attributed to many factors such as soil substrates and surfaces, the temperature of the soil and pH (Pavlik and Falkinham, 2009). In a study conducted in 2011, the survival of MAP in soil environments was evaluated (Salgado et al., 2015). From this study, the presence and survival of MAP in soil has been associated with moisture, organic material and clay, temperature, pH with warmer soils reducing the rate of survival. The study also demonstrated the ability of MAP to survive in a shaded environment for up to 48 weeks. However, it should be noted that this was an artificial laboratory study and moisture levels made no difference in other relevant field trials.

The survival of MAP in tap, pond and sealed water bottles is estimated to be 6 to 18 months (Amin et al., 2015). In a neutral pH, MAP survives optimally (17 months). However, an increase or decrease in pH from neutral allows MAP to survive for 14 months. Other studies have demonstrated that in lake water environments MAP remains culturable for up to 841 days (Pickup et al., 2005). The survival of MAP in water is based on the presence of fatty acids, lipids and waxes in its cell wall which creates hydrophobicity. The presence of the cell wall in MAP promotes its absorption to surfaces and allows phagocytosis by macrophages and protozoa (Bolster et al., 2009; Strahl et al., 2001). Consequently, as MAP has the ability to survive in water, this creates the potential for disease transmission.

**Antibiotic regimens for the treatment of MAP based infections**

For reasons covered earlier, the treatment of MAP with standard antibiotics is difficult. Additionally, the overuse of antibiotics in agricultural settings has also contributed to the increase in MAP resistance (Garvey, 2018). Antibiotic therapy is a regime which has been utilised continuously for the treatment of MAP infection. To date, there are no antibiotics specifically designed for the treatment of MAP infections. Instead, drugs are utilised on a "trial and error" basis. Alternative mycobacterial diseases such as TB, leprosy and *M. avium* are treated with multidrug therapy with different types of antibiotics; thereby this may be necessary for treating MAP based infections and help avoid the development of MAP resistance. The general type of antibiotics used in combination for MAP treatment are macrolide protein synthesis inhibitors such as clarithromycin and azithromycin which are typically combined with ethambutol and rifamycin which together inhibit cell metabolism and RNA synthesis (Savarino et al., 2019). Anti-inflammatory agents such as monoclonal antibodies may also have the ability to directly affect MAP based infections by inhibiting the immune response (Bannantine et al., 2007).

In the infection of humans with MAP, the cell wall is absent thereby attempts to treat the bacterial cell wall with an antibiotic is ineffective. Efforts which involve targeting peptidoglycan biosynthesis (penicillin's, cephalosporin's and vancomycin) are also ineffective against MAP (Beran et al., 2006). In fact, the utilisation of such antibiotics may lead to the development of multi-drug resistant bacteria. As mentioned the slow growth rate of the microorganism and an inactive, dormant stage outside of the host may make the treatment of MAP with antimicrobials difficult. At present, there is no long term *in vitro* infection model for testing the efficacy of antimicrobials against MAP infections.

Due to lack of effective diagnostic assays and the increase in antibiotic resistance in MAP, the development of effective diagnostic assays, which have the ability to incorporate both MAP viability and drug susceptibilities, would be highly beneficial to this field of research.

## The relationship between Johne's disease and MAP

### *Transmission, monitoring and control of MAP*

Transmission of the pathogen is most commonly caused by contaminated milk passing from mothers to offspring, through the environment or contaminated faeces in farm environments. The most important step to prevent and control the spread of Johne's is to monitor animals and prevent the spread of contamination. To monitor animals, a classification system is in place based on the bacterial level detected in their faeces, using the following scoring system, low (<10 CFU/tube), moderate (10 to 50 CFU/tube) or high (>50 CFU/tube) faecal shedders (Koets et al., 2015). Farmers typically lack the knowledge in Johne's disease research to identify clinical symptoms before treatment is too late. There are four stages of MAP infection based on the severity of clinical symptoms (Table 1). Diagnosis of a cow with Johne's disease is very difficult at Stages 1 and 2 as animals are asymptomatic. Common blood tests and shedding detect no traces of MAP. Detection of MAP typically occurs at Stages 3 and 4 as clinical symptoms begin to develop, in which treatment is not effective.

**Table 1.** The clinical stages of Johne's disease in cattle caused by MAP determined by blood tests and faecal shedding

| Stage | Clinical signs | Positive with blood tests | Shedding in faeces | References |
|---|---|---|---|---|
| Stage 1: Silent/ Early infection | None | No | No | Jesse et al., 2016; Whittington et al., 2017 |
| Stage 2: Subclinical infection | None | Yes/No | Yes (low numbers) | |
| Stage 3: Clinical infection | Decreased appetite, weight loss, persistent or intermittent diarrhoea | Yes | Yes (high numbers) | |
| Stage 4: Advanced infection | Weakness, anaemia, dehydration | Yes | Yes (extremely high numbers) | |

Diagnosis and treatment of Johne's disease can be extremely subjective. Therefore, the most effective way to prevent the spread of Johne's disease is using proper control strategies. Much of the pioneering work on Johne's disease control programmes can be attributed to Mike Collins with subsequent authors developing his research on farm practices (Collins, 1996; Krishnan et al., 2009; Shin et al., 2010). From the literature, it is evident that proper control programs are present in only 22 countries worldwide (Whittington et al., 2019). These are more developed among countries with advanced veterinary facilities. Countries lacking a formal control program were more apparent in South America, and

Asia (76%) while the remainder were in Europe (20%) (Whittington et al., 2019). Reducing the prevalence of Johne's disease is the most important consideration in most countries; however, Norway and Sweden are concerned with surveillance in order to eradicate the disease entirely.

The majority of European countries are concerned with proper manure disposal on their farms as cross contamination may occur (Over et al., 2011). Removal of young calves from herds after birth if the diagnosis of Johne's disease occurs is another major control strategy. Segregating infected ruminants from non-infected ruminants and the manure of each is another major consideration in preventing the spread of infection. In reading various control programs, particularly in Ireland, cleaning of water tanks, milking parlours, and cowsheds are the most extensive measures in controlling the disease (Garcia and Shalloo, 2015). The main advice for farmers with Johne's disease on their farms is to follow the exact procedures listed above, although certain countries have different levels of standards. Control programs were reported to be successful in 16 (73%) of the 22 countries (Whittington et al., 2017).

*The prevalence and impact of Johne's disease on dairy industries globally*

One of the largest profitability losses in the dairy industry is caused by Johne's disease, which in turn has a direct impact on the global economy. Many studies have documented the impact of profitability on the US dairy industry. From a period of 2004 to 2005 in the US, the number of cattle with Johne's disease increased from 7,879 to 10,280 (Garcia and Shalloo, 2015). In the US from 2009 to 2010, Johne's disease had caused losses of $200 million to 1.5 billion due to infection (Johnston et al., 2010; Pillars et al., 2009). Increased culling and decreased milk production have also been associated with these costs. Assessing profitability and productivity is difficult at the herd level. Therefore, it is difficult to estimate the financial impact of Johne's disease worldwide. From a period of 1999 to 2018, the total annual economic losses per cow in infected US dairy herds were estimated at US$21 to US$79 while dairy herds in Canada had losses of CDN$49, French herds with losses of €234 and UK herds with losses of £27 (Chiu et al., 2018; Ott et al., 1999).

The US had the highest number of Johne's disease in ruminants from a period of 1999 to 2018, and Norway and Sweden had the lowest levels, clearly illustrating that levels of Johne's disease differ between each country and region (Whittington et al., 2019). Agriculture in Ireland has contributed significantly to the Irish economy. The presence of Johne's disease, particularly among dairy herds, has resulted in substantial economic losses, with clinical cases of MAP from 1992 to 2002 having increased from 92 to 232 per year (Cashman et al., 2008). The importation of animals from continental Europe between 1992 and 2004 was suggested as an increase in the levels of Johne's disease during this period (Good et al., 2009). In 2009, 376 clinical cases of bovine Johne's disease was detected in Ireland according to the animal disease surveillance report in 2010. No data has yet been released about the number of cases of bovine Johne's disease in Ireland from the period of 2011 to 2018. From Table 2 it can be observed that the number of cases of bovine Johne's disease has dramatically increased from the period of 1932 to 2009, indicating that a biocontrol mechanism needs to be developed.

**Table 2.** The number of clinical cases of bovine Johne's disease reported in Ireland during specified time periods

| Year | Number of cases of bovine Johnes disease reported in Ireland | References |
|---|---|---|
| 1932 to 1992 | 92 | Cashman et al., 2008; Good et al., 2009 |
| 1993 to 2002 | 232 | |
| 2009 | 376 | |
| 2011 to 2018 | Unknown but expected to increase | |

## Understanding the viral entities termed "bacteriophages"

*Bacteriophages and their evolving taxonomy*

Bacteriophages (phages) are viruses which specifically infect and lyse bacterial hosts and have been described as the most abundant biological entities on Earth, playing a significant role in maintaining the microbial balance of the planet, with an estimated number being in the order of $10^{31}$ (Mushegian, 2020; Wittebole et al., 2014). With host lysis following an infection once they have managed to subvert the metabolism of their hosts to generate progeny phage. The majority have been identified as members of the order of *Caudovirales* (the tailed phages) and harbour proteinaceous virions consisting of icosahedral capsids containing double-stranded DNA with tails of varying morphology (Ackermann, 2007). The International Committee on Taxonomy of Viruses (ICTV) has recently introduced several changes to this taxonomic ranking. This has resulted in the addition of six new families (*Ackermannviridae*, *Autographiviridae*, *Chaseviridae*, *Demerecviridae*, *Drexlerviridae* & *Herelleviridae*) along with the families of *Podoviridea*, *Sipovoridea* and *Myoviridea* (Adriaenssens et al., 2020). The latter families were initially constructed around the criteria of virion morphology (Ackermann, 2007). For example, those that possessed short, non-contractile tails were placed in the family *Podoviridea*, those with long, flexible tails were placed in the family *Siphoviridea*, and those with rigid contractile tails were placed in the family *Myoviridea*. However, with the ever-increasing availability of genome sequence data, at the time of this writing (May 2020), a total of 2,488 phage genomes sequences had their genomes placed onto Genbank (RefSeq). It has become alarmingly apparent that the original taxonomy arrangement of the past is insufficient to realise the full extent of phage genetic diversity. With the original three families of *Podoviridea*, *Siphoviridea* and *Myoviridea* now having become like taxonomic bins, awaiting further revision of their members (subfamilies and genera) to form new families that better reflect their genetic properties. For example, members of the newly formed *Herelleviridae* were extracted from the family *Myovirdea*, as they form a monophyletic clade amongst the tailed phages when using single and concatenated marker genes in phylogenetic analysis (Barylski et al., 2019) Phages that constitute the family *Autographiviridea* were extracted from the family *Podoviridea*; these phages possess the defining feature of a genome that encodes an RNA polymerase, as well as a conserved gene arrangement along with other properties (Adriaenssens et al., 2020). Phages outside the order *Caudovirales*, possess virions of varying morphology and genomes of variable nucleic acid composition. Like phages of *Caudovirales*, these too have undergone reconfiguration of their taxonomic rankings (Adriaenssens et al., 2020).

## Insights into the mycobacteriophage and their genomes

### *Mycobacteriophage and their diversity*

Mycobacteriophages are phages that infect mycobacterial hosts. Such phages have been identified to infect *Mycobacterium* species of *M. smegmatis*, *M. tuberculosis*, *M. ulcerans*, *M. bovis*, *M avium* subspecies *and M. abscessus* subspecies, *M. aichiense M. aurum* and *M. phlei* (Broxmeyer et al., 2002; Fullner and Hatfull, 1997; Kraiss et al.; Sassi et al., 2013; Sula et al., 1981; Trigo et al., 2013, Actobacteriphage database). The vast majority of such phages having been isolated and characterised through the Science Education Alliance Phage Hunters Advancing Genomics and Evolutionary Science (SEA-PHAGES) program (https://seaphages.org/). An undergraduate research course ran jointly by the University of Pittsburgh and the Howard Hughes Medical Insititute's Science Education division in the USA. To date (May 2020) their Actobacteriphage database lists the isolation of 11,268 phages, with 1,899 having undergone genome sequencing. These phages are divided into 29 Clusters based on overall nucleotide sequence similarity (A, AA, AB, AC, B, C, D, E, F, G, H, I, J, K, L, M, N, O, P, Q, R, S, T, U, V, W, X, Y & Z) with at least nine phages defined as singletons that do not fall into these major classifications (look to Figure 1 for a phylogram showing the relationship of these clusters to each other). The majority of these have been isolated using *M. smegmatis* $mc^2155$ as a host (11,240 isolates), but phages infecting another six species are also listed (24 isolates). Genome sizes range from 38,341 bp to 164,602 bp with GC content varying from 50.3% to 70.1%. The diversity of these phages likely reflects the different evolutionary pathways they have taken, possibly centring on infecting hosts of a particular genetic composition and resulting in their genomes possessing a GC content reflecting the hosts that they originally evolved infecting. Most of these isolates have been obtained from environmental samples such as soil and water (Hatfull, 2012). However, it's also worth noting that mycobacterium phages have also been isolated from faeces (Basra et al., 2014, Cater and Redmond, 1963). Host range analysis of the *M. smegmatis* phages show they can possess highly limited infectivity, only targeting a limited number of strains, with others showing higher infection capacity, targeting multiple, with others again being able to target strains of other species like *M. tuberculosis* (Clusters K and A) (Jacobs-Sera et al., 2012). Mycobacteriophages are often described to possess a virion morphology of *Siphoviridea* (the different Clusters can vary with virions with differing tail length and capsids with either an isometric or prolate structure) and uncommonly *Myovirdea* (Cluster C), there are highly limited descriptions of those with a *Podoviridea* morphology (look to Figure 2) (Hatfull, 2014; Stella et al., 2013). Possible reasoning for the absence for the *Podoviridae* morphotype among mycobacteriophages is due to their short stubby short tails, which are unable to penetrate the thick-waxy, Gram-positive cell wall of *Mycobacterium*.

*Genome Organisation of mycobacteriophage genomes*

Although there exists great diversity with gene composition among mycobacteriophage genomes, there exists a common trend to the organisation of their genes. First of all, those phages with a *Siphoviridea*-like morphology (those not belonging to Cluster C) typically possess a common arrangement with virion structural and assembly genes positioned to the left end of their genomes. These are often arranged in the order of terminase (small and large subunit genes), portal, capsid maturation protease, scaffold, capsid, tail capping and head-tail connectors, major tail subunit, tail assembly chaperones, tape measure protein, and minor tail genes (of which there can be 4–10). In phages with shorter genomes (40-45 kbp), these genes can make up as much as 50% of a genome. In other phage genomes, the gene order is conserved, but it is interspersed with non-structural proteins (Hatfull, 2018).

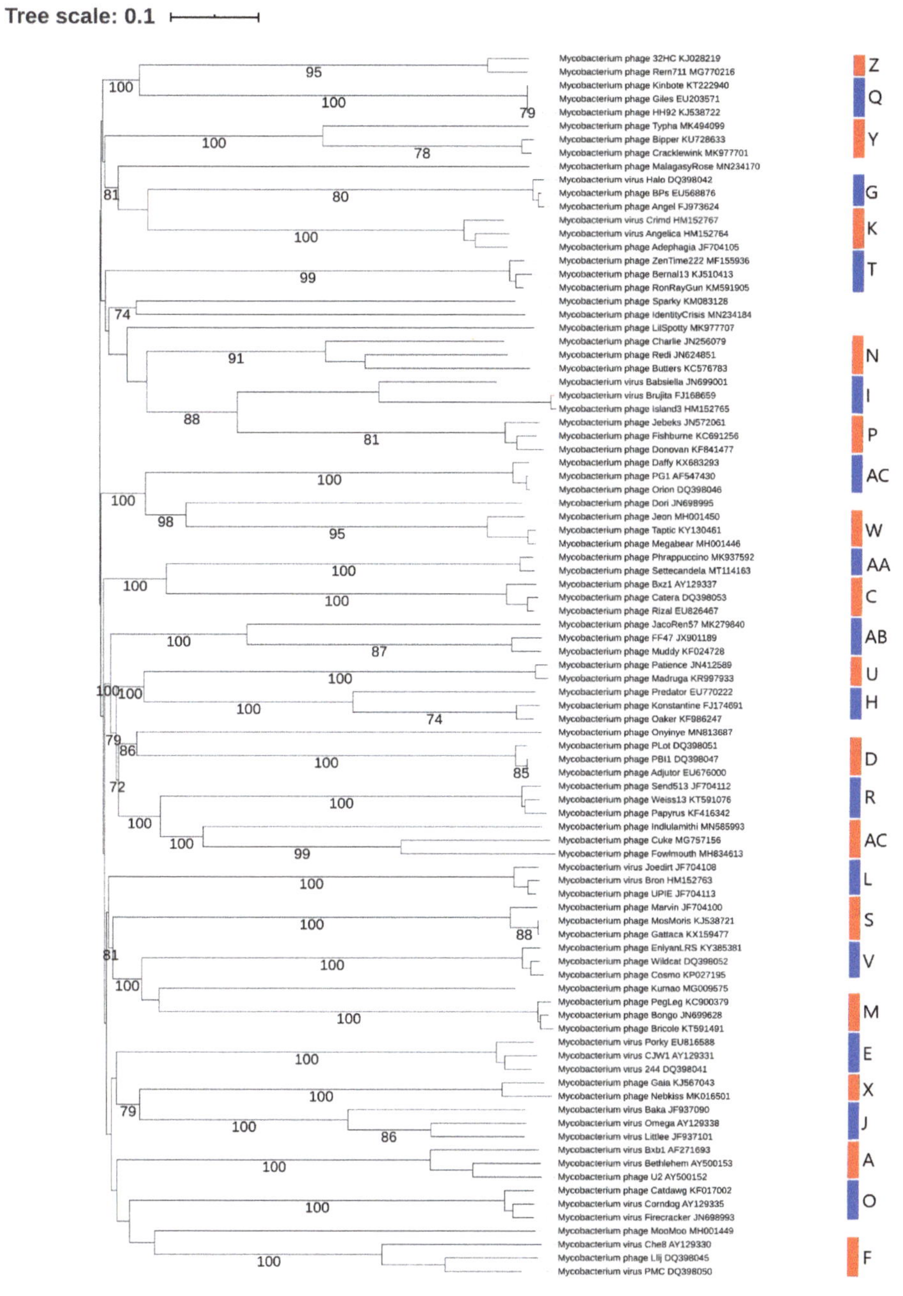

**Figure 1.** Nucleotide VICTOR (Meier-Kolthoff and Göker, 2017) generated phylogenomic Genome-BLAST Distance Phylogeny (GBDP) tree of 91 mycobacterium phages that represent the different Clusters (max of 3 phages per Cluster) and singletons found among these phages inferred using the formula D6 and yielding average support of 72%. The numbers above branches are GBDP pseudo-bootstrap support values from 100 replications. Cluster are highlighted in red and blue

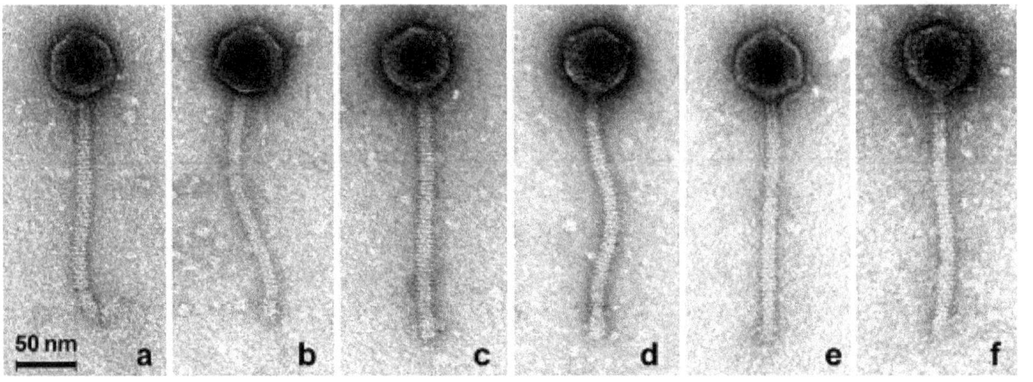


**Figure 2.** Electron micrographs (2% uranyl acetate) of six mycobacteriophages isolated on *M. smegmatis* mc²155 from environmental soil samples (Endersen et al., 2013).

All mycobacterium phages possess genes for a lysis system, made up of a holin, an endolysin (lysin A) and a serine esterase (lysin B). These genes are usually located downstream of the virion structural proteins, or in the case of those of Cluster A, are positioned between the left end of the genome (cos site) and the structural genes (Hatfull, 2018). During cell lysis at the end of phage infection of a host bacterium, holins are responsible for allowing endolysins to pass through the inner cell membrane allowing degradation of cell wall peptidoglycan. As species of the genus *Mycobacterium* possess a mycolic acid-containing outer membrane, serine esterase's are utilised to degrade this layer, allowing lysis to be achieved (Catalão and Pimentel, 2018).

Of the mycobacterium phages characterised, over half appear to be temperate possessing an integration cassette. This is typically located towards the centre of their genome containing an integrase gene and an *attP* site. Both families of tyrosine and serine integrases have been observed. In addition to the mentioned categories of genes, these phages can also have additional genes, the vast majority being of unknown functions with their roles having yet to be explained. Nevertheless, some have been identified to possess functions associated with DNA metabolism, such as recombination and replication. (Hatfull, 2018). Furthermore, among phages of other bacterial families, such genes can also encode for functions contributing to host virulence. It is not well established if such genes exist among phages/prophages of *Mycobacterium*. However, it has been observed that mycobacteriophages within Cluster D can possess vegetative insecticidal protein (VIP2)-like insect toxin genes (Hatfull, 2014). It remains to be elucidated if such genes actual contribute to *Mycobacterium* virulence.

## Phage therapy considerations

### *Phage therapy as a possible alternative to antibiotics against MAP?*

About 75% of the antibiotics used in the EU and USA are in agriculture, with their administration to livestock (cattle, swine and poultry) for the prevention of disease and the promotion of growth (Marston et al., 2016). Of the 27 classes of antibiotics currently available, only seven classes are exclusive to agriculture, with the remaining used in both medicine and animal production (Marston et al., 2016). Several jurisdictions have banned the use of antibiotics for the promotion of growth in livestock, so as to limit the development of antibiotic resistance arising from this activity (Marston et al., 2016). However, agriculture remains a concern for the emergence of resistant strains among medically essential antibiotics. As described earlier in this chapter, even the treatment of Johne's disease among cattle can lead to the development of antibiotic resistance among MAP. Therefore, there is a further need to reduce this industry's dependence to treat animal infections with the development of alternative strategies.

Phage therapy has the potential to be such an alternative treatment strategy. The concept involves the application of phage to specifically infect a bacterial pathogen of interest, resulting in its subsequent death and clearance of infection. This concept can also be employed in prophylactic treatment. This concept is not new, being as old as the discovery of phage themselves. As d'Hérelle, one of the co-discoverers of phages, employed their use for the treatment of cholera (D'Herelle, 1946). To this day, phage therapy is widely practised in eastern European countries and former Soviet states for human infections (Borysowski et al., 2014). Furthermore, Western companies, such as OmniLytics in the US, have begun to develop phage therapy for horticultural applications (https://www.omnilytics.com/). There are several advantages and disadvantages to the approach, see Table 3. Furthermore, they can also be used to combat antibiotic resistance, having been proven effective in the treatment of antibiotic resistant infections as outlined in various studies (Górski et al., 2016; Lin et al., 2017; Weber-Dabrowska et al., 2016a). Phage therapy can also be used in conjunction with conventional antibiotic therapy where enhanced therapeutic effects have been demonstrated (Kamal and Dennis, 2015; Oechslin et al., 2017). However, care should be taken with the types of antibiotic and phage used together as antagonistic effects have also been observed (Abedon, 2019). An additional property that makes their use in animal treatment desirable is that they can be deemed as organic agents, having obtained an OMRI listing, making phage therapy suitable for organic farming (Buttimer et al., 2017). The application of phages in the treatment of human and animal diseases would also be highly beneficial due to their cheap cost of propagation and their fast generation time.

**Table 4.** Advantages and disadvantages of phage therapy

| Advantages of Phage Therapy | Disadvantages of Phage Therapy |
|---|---|
| Highly specific for target bacteria | Antibody-mediated response *in vivo* |
| Replication occurs at the site of infection | Narrow host range |
| Little or no side effects | Safety profile not fully established |
| Ability to overcome bacterial resistance | Pharmacodynamics not fully understood |
| Selecting new phages is a rapid process that can be accomplished in days or weeks | |

References (Drulis-Kawa et al., 2015; Górski et al., 2016; Lin et al., 2017; Sulakvelidze et al., 2001; Weber-Dabrowska et al., 2016b)

### *Characteristics required of therapeutic phage*

Not any phage can be utilised for phage therapy. There are several desirable characteristics for such phages to possess. The first being the infection strategy used by a phage of a host bacterium, as it can differ among phages. The lysogenic and lytic cycle of infection have been described among the majority of characterised phages to date. When such phages have inserted their genome in a sensitive host, they can enter either cycle, depending on their own genetic composition (Lin et al., 2017). The lytic cycle is exclusively followed among the strictly virulent phages, as infection ultimately leads to the lysis of a host cell with the release of progeny phage upon expression of their genomes. This is an essential characteristic of phages used in phage therapy. Undesirable are those that can also follow the lysogenic cycle of infection (such phage can be termed temperate). During the lysogenic pathway, a phage incorporates its own genome into that of its host bacterium (termed a prophage). With this strategy, its own genome replicates as part of its host until a switch (UV, chemical agent) triggers it to switch to the lytic route of infection (Brunner and Pootjes, 1969; Müller et al., 2012; Olszak et al., 2017). These phages can usually be identified with genomes possessing genes for repressor, integrase or excisionase proteins. Other infection strategies, such as pseudolysogeny and chronic infection also exist but are less commonly described (Łoś and Wegrzyn, 2012; Roux et al., 2019)

Therapeutic phage genomes must also not possess genes for virulent factors, as identified among prophages of *Escherichia coli* and *Vibrio cholerae* possessing Shiga and Vibrio toxin genes, respectively (Aertsen et al., 2005; Fan and Kan, 2015). They also must not possess genes for antibiotic resistance. However, limited examples of such genes have actually been detected among phage genomes (Enault et al., 2017). Yet, cases do exist such as *Bacillus* phages Gamma and Cherry encoding fosfomycin resistance genes, as well as such genes being found among several *S. aureus* satellite phages (Novick et al., 2010; Schuch and Fischetti, 2006)

These phages must also possess a limited capacity for horizontal gene transfer. There are currently three known forms for such an event to occur among phages: 1) General transduction: where a lytic phage accidentally packages host DNA into a phage capsid instead of its own DNA, known to occur with the *pac*-type headful mechanism (Schmieger, 1982); 2) Specialised transduction: where bacterial genes adjacent to a cos-site are excised with prophage DNA due to an aberrant excision event (Morse et al., 1956); 3) Lateral transduction: the most recently described, where prophages genomes do not excise from the host chromosome; instead they amplified while integrated which are then processively packed into the phage capsid (Chen et al., 2018). For lytic phage, only generalised transduction is an issue. This is best avoided by the selection of phage that do not employ the *pac*-like headful mechanism. Those most ideal are phage who pack their genomes with exact direct terminal repeats as seen with those related to *Bacillus* phage SPO1, as they have been demonstrated to show limited generalised transduction capability (Klumpp et al., 2008)

Characteristics that result in more efficient infection of a host by a phage at the site of infection (i.e. fast absorption, big burst size, short latent period) leading to higher phage titres. As well as greater virion stability (temperature, pH) enabling a phage to persist in harsh conditions (i.e. gastric acids) are also likely to be important factors that contribute to the therapeutic success of an applied phage

**Mycobacteriophages that infect MAP and their application towards the control of the bacterium**

Recently, the application of phage therapy in humans employing mycophages against antibiotic resistant *Mycobacterium abcessus* has been successfully demonstrated (Dedrick et al., 2019). However, the description of similar work with the treatment of *Mycobacterium* related disease, such as Johne's disease, among animals is limited. Much of the work described with the application of phages towards MAP have been for its detection (discussed in the following section). As MAP is a fastidious pathogen to cultivate, often requiring an incubation duration of one to six months in optimal conditions to see colony formation. This has likely contributed to the scarcity of literature describing phages that infect it and descriptions of phage therapy to control it. However, protocols are described to enable the isolation of phages on it, such as that of Basra et al. (2014). Using this protocol the authors demonstrated the isolation of phage directly on MAP from bovine faeces from cattle suspected to have Johne's disease. However, the majority of phages known to infect MAP to date were initially isolated the fast growing species *M. smegmatis* (Swift et al., 2014). It is interesting to see that phages currently known to infect MAP can be placed among different clusters, indicating a diversity of phage types capable of infecting it. (Table 5).

The possible control of MAP with phages has been indicated in a study by Endersen et al. (2013). As previously mentioned, an essential source of transmission of MAP between a cow and its calf is the mother's own milk. If levels of MAP can be controlled in contaminated milk, it would be a useful strategy to limit its transmission among cattle. In the mentioned study, the authors demonstrated the concept of using phages against *Mycobacterium* in milk to be feasible. Here they applied a phage cocktail of six phages ($1x10^9$ PFU/mL) which could effectively remove *M. smegmatis* ($1x10^3$ CFU/mL) in reconstituted skim milk to non-detectable levels after 96 h of incubation. Effort should now be taken to demonstrate the concept with MAP and additionally look to other agricultural environments (fields, animal housing, etc.) where mycobacteriophages could be used as sanitising (biocontrol) agents.

**Table 5.** Mycobacteriophages that are known to infect MAP

| Name | GenBank accession | Original isolation host species | Genome size (bp) | GC content (%) | Morphology | Cluster |
|---|---|---|---|---|---|---|
| D29 | NC_001900 | Mycobacterium smegmetis | 49136 | 63.5 | *Siphoviridae* | A |
| TM4 | NC_003387 | Mycobacterium smegmetis | 52.8 | 68.1 | *Siphoviridae* | K |
| FF47 | NC_021063.1 | MAP | 47.72 | 58.6 | *Siphoviridea* | AB |

## Exploration of current mycobacteriophage based diagnostic assays for Johne's disease and alternative detection approaches using phages

### *Current detection methods for* MAP

As mentioned previously, proper management of the spread and control of MAP is directly correlated with its rapid diagnosis. Current methods of disease diagnosis for Johne's disease caused by (MAP) are slow and subject to bias. As Johne's disease is extremely difficult to detect at the early stages of infection, the development of a novel detection method for MAP would be highly desirable to prevent further economic losses. The diagnosis and detection of MAP from various sources through traditional (culturing) and molecular methods have proven challenging over the last century.

From the literature relating to the diagnosis of MAP, it has been noted that five distinct test types are frequently available and used. These tests utilise antibodies and antigens against MAP, detection of specific genes in MAP, bacterial cultivation of faecal samples and visual inspection of symptoms (Clark et al., 2008). There can be significant variation with the sensitivity and accuracy for each test which can vary depending at the stage of when the test procedure was performed. The most well-known test for diagnosing MAP infection is the use of the enzyme-linked immunoassay (ELISA) to detect MAP. However, these have limitations, in particular, reduced sensitivity which may be as a result of the slow growth rate of MAP. Animals in an early stage of infections with low faecal shedding may obtain bias results in utilising an ELISA test due to an inadequate detection capacity (Diéguez et al., 2009). Immunohistochemistry (IHC) is a method similar to ELISA based on detecting the immune reaction caused by MAP. The advantage of using IHC is the ability to identify MAP in tissues. The sensitivity of the method in animals with subclinical MAP infection is relatively good, but interaction can occur with other mycobacterial species resulting in cross reactivity (Moravkova et al., 2008).

Although extremely slow, the use of culturing methods from faecal and tissue samples is the most traditional method used for detecting MAP infected animals. In conjunction with other detection methods such as PCR of a region other than IS900, the utilisation of culture methods can be as accurate as 100% (Ellingson et al., 2005). Determining the infection status of MAP infected animals' faecal culture is typically used to measure the sensitivity of other diagnostic methods, such as ELISA and PCR. Faecal cultures are also used as a confirmation of MAP positive animals with immunological tests. Thus culturing is considered to be the "gold standard" for MAP detection. However, the method possesses limitations such as long turnaround time, high cost and high levels of contamination (Donat et al., 2014). The accuracy of faecal culture may also be affected by the stage of MAP infection and disease dynamics.

The interest in IS900 PCR for the detection of MAP soon became popular in 1989, although the utilisation of this specific region in MAP has proven problematic over recent years. The insertion element is considered to be "unique" to MAP, although IS900 like sequences have been found in other mycobacterial species. Considering this statement, various studies have supported this claim. False positive and false negative results appear when reading studies documented in the literature which utilise the IS900 gene (Debroy et al., 2012). Subsequently, more established "traditional" methods such as faecal culture were thought to be more reliable, and their utilisation became more common. This statement does not imply that the application of PCR in detecting MAP is unreliable as many methods have limitations. PCR of the IS900 gene in the detection of MAP genes has proven useful over many years for its speed and specificity and low contamination levels, in which no additional tests are required for confirmation. As mentioned previously, the IS900 gene is typically targeted for MAP detection but is subject to false positive results. A minority of mycobacteria (*M. cookii*, *M. marinum*, *M. paraffinicum*, and *M. scrofulaceum* isolates) are found to contain IS900 like elements with 94% identity to the MAP IS900 gene (Plain et al., 2013). There is a slight possibility that DNA from MAP and various other microorganisms may also have the potential to produce false positive results. Additionally, the extraction of DNA from complex origins such as milk, faeces and blood has led to be problematic, thereby decreasing the sensitivity of PCR (Acharya et al., 2017).

The final method for the detection of MAP is the Johnin Skin Tuberculin Test, which is commonly used in addition to other tests. The test is based on the visual inspection of the skin of the neck after a certain time period when injected with Johnin PPD or *M. avium* PPD. The interpretation of test results may prove difficult based on observation (Casal et al., 2014). The test can be utilised by farmers once trained based on test observations.

*The development of a novel mycobacteriophage based diagnostic assays for MAP*

There is a lack of papers in the literature in relation to the development of new phage based diagnostic assays. The development of a novel phage based diagnostic assay, for the detection of MAP that incorporates specificity, sensitivity, speed and robustness would be of high importance. Mycobacteriophage-based assays have been reported as potentially useful tools for the detection of viable bacilli (mycobacterial species) as well as for antimicrobial susceptibility testing.

In applying a phage-based detection system, no expensive instruments are required, and a large majority of the mycobacterial strains utilised are avirulent. The use of mycobacteriophage in a MAP based diagnostic assay would distinguish viable bacterial cells from non-viable bacterial cells quickly via a surrogate fast growing mycobacterial strain and a turnaround time of 2 – 4 days compared to 2 months for culturing methods.

In applying mycobacteriophages to this type of assay, data may be obtained within a shorter time, with increased accuracy as mycobacteriophages are specific to their host strain but also a reduced cost as propagation is relatively cheap. In 2003, rifampicin resistance was assessed against Mycobacterium tuberculosis using mycobacteriophage as an indicator of drug resistance and drug sensitivity (Gali et al., 2003). No specific information in relation to the phage type used or TB strains used were disclosed; therefore, it is difficult to translate this work to other studies. Alternative areas of mycobacteriophage research in mycobacterial detection have been explored.

Various other attempts were made at developing an assay which has the potential to assess TB viability. The FASTPlaqueTB assay utilises a similar principle. The assay is mentioned repeatedly throughout various papers from 2003 to 2005, indicating its wide applicability in identifying TB from samples (Albay et al., 2003). It was developed in 2000 by Biotec Laboratories Ltd. The principle of the FASTPlaque TB assay is the application of mycobacteriophages (Actiphage™) to detect drug resistant or sensitive TB isolates and viability of TB in a clinical sample (Phulpoto et al., 2005). The addition of a virucidal solution (Virusol™) has the ability to destroy phages that have not infected the TB cells and the remaining phage infect a surrogate mycobacterial strain (*M. smegmatis* mc2 155), with results are interpreted based on plaque numbers. Since its development in 2000, various other researchers have combined this novel phage amplification assay with other reliable diagnostic technologies.

*The establishment of a bioluminescence system to detect viable MAP from infected animals*

The investigation of bioluminescence systems can be traced back to 1918 (Harvey, 1918). The application of bioluminescence systems in microbiology and molecular biology have increased over recent years. Luciferase, which is one type of bioluminescence system, has played an important role in disease prevention, detection and treatment. Over the past decade, luciferase reporter mycobacteriophages (LRPs) have been developed that show great promise for diagnostic microbiology in research environments (Fu et al., 2012; van der Merwe et al., 2014). These mycobacteriophage systems have been trialled using *Mycobacterium paratuberculosis* (Mtb) as a host. As mentioned, previous studies have shown that under research conditions, LRP-based assays can rapidly identify mycobacteria and perform susceptibility tests. It is also possible to perform a luciferase reporter mycobacteriophage assay in a clinical setting as outlined elsewhere (Banaiee et al., 2001) and results were comparable to a well-established drug susceptibility method (BACTEC method). A luciferase reporter phage assay may also have the ability to increase the speed of disease detection at a lower cost.

*The utilisation of fluoromycobacteriophages as a novel diagnostic tool in the detection of MAP infected animals*

The creation of fluoromycobacteriophages in the last 10 years has led to an increase in the research of *Mycobacterium paratuberculosis* detection. Fluoromycobacteriophages differ from luciferase reporter mycobacteriophages in that they are tagged with a green fluorescent protein as opposed to a lux gene (Piuri et al., 2009). A small number of bacterial cells can be identified with no substrate required (<100/ml). In applying fluoromycobacteriophages to a

MAP based detection system, it will be feasible to detect phage propagation if the mycobacterial host is present in Real-time. This type of assay will address problems associated with the slow growth rate of MAP and eliminate the need for plaque assays. The sensitivity of mycobacteriophages appears to vary between assays. When comparing studies, it emerges that as the fluoromycobacteriophage attaches to the bacterial host, the fluorescence may suddenly stop (Piuri et al., 2013; Rondon et al., 2011; Rondón et al., 2018).

*The application of qPCR in conjunction with phage detection methods can improve detection levels of MAP infected animals*

The introduction of qPCR in commercial and research labs over the last two decades has led to the publication and development of thousands of high impact studies in diagnostics laboratories worldwide. By combining qPCR and mycobacteriophage technology, the detection accuracy of MAP from various sources such as blood, milk etc. has the potential for improved accuracy and sensitivity. It is quite evident from the literature that many phage-based studies integrate qPCR with more advanced detection techniques to improve MAP infection. Determining drug susceptibility of MAP to various drugs utilising qPCR and mycobacteriophages in combination, has gained extensive research (Marinelli et al., 2012; Pholwat et al., 2011). The utilisation of qPCR in combination with mycobacteriophage to determine drug susceptibilities, has been studied extensively by Pholwat et al. (2011). Applying this type of assay relies on the change of phage DNA levels, which is correlated to a bacterium's susceptibility to a particular drug under investigation.

## Conclusion

The diagnosis and treatment of MAP, the causative agent of Johne's disease, has proven challenging over recent years. The complexity of its diagnosis is mainly caused by the slow growth rate of the organism, making detection difficult. The diagnostic assays that exist for MAP are immune, visual and molecular based detection; however, these diverse methods can suffer from low sensitivity, low specificity and slow turnaround time. Novel diagnostic assays that can improve on these parameters are in need of development. Furthermore, treatment of Johne's disease with antibiotics is limited and can result in antibiotic resistance; alternative methods are also in need of development for its treatment and control.

The alternative methods could possibly be answered with the mycophages of MAP. However, a greater effort is needed to understand the full diversity of phages infecting this bacterium. This will not only allow a diverse collection of phages from which to build phage-based preparations for the treatment and control of the bacterium, but also a greater range of phage-based resources to develop biotechnology applications towards MAP. Such resources being a greater pool of mycobacteriophage genes (i.e. endolysins, serine esterase's, receptor binding proteins) from which to develop tools for its control and detection. But additionally, the availability of diverse mycobacterium phages that could be engineered towards luminescent/fluorescent-based detection systems targeting the bacterium.

**Databases**

Actobacteriophage database: https://phagesdb.org/

Genbank: https://www.ncbi.nlm.nih.gov/genbank/

**References**

Abedon, S.T. (2019). Phage-antibiotic combination treatments: Antagonistic impacts of antibiotics on the pharmacodynamics of phage therapy? Antibiotics *8*.

Acharya, K.R., Dhand, N.K., Whittington, R.J., and Plain, K.M. (2017). PCR Inhibition of a Quantitative PCR for Detection of *Mycobacterium avium* subspecies *paratuberculosis* DNA in feces: diagnostic implications and potential solutions. Front. Microbiol. *08*.

Ackermann, H.-W. (2007). 5500 Phages examined in the electron microscope. Arch. Virol. *152*, 227–243.

Adriaenssens, E.M., Sullivan, M.B., Knezevic, P., van Zyl, L.J., Sarkar, B.L., Dutilh, B.E., Alfenas-Zerbini, P., Łobocka, M., Tong, Y., Brister, J.R., et al. (2020). Taxonomy of prokaryotic viruses: 2018-2019 update from the ICTV Bacterial and Archaeal Viruses Subcommittee. Arch. Virol. *165*, 1253–1260.

Aertsen, A., Faster, D., and Michiels, C.W. (2005). Induction of Shiga toxin-converting prophage in *Escherichia coli* by high hydrostatic pressure. Appl. Environ. Microbiol. *71*, 1155–1162.

Albay, A., Kisa, O., Baylan, O., and Doganci, L. (2003). The evaluation of FASTPlaqueTBTM test for the rapid diagnosis of tuberculosis. Diagn. Microbiol. Infect. Dis. *46*, 211–215.

Amin, A.S., Behour, T.S., Talaat, A.M., Ghosh, P., Hsu, C.-Y., AbdEl-Fatah, E.M., and Darwish, S.F. (2015). Ecology and genomic features of infection with *Mycobacterium avium* subspecies *paratuberculosis* in Egypt. Microbiology *161*, 807–818.

Banaiee, N., Bobadilla-del-Valle, M., Bardarov S., J., Riska, P.F., Small, P.M., Ponce-de-Leon, A., Jacobs, J., Hatfull, G.F., and Sifuentes-Osornio, J. (2001). Luciferase reporter mycobacteriophages for detection, identification, and antibiotic susceptibility testing of *Mycobacterium tuberculosis* in Mexico. J. Clin. Microbiol. *39*, 3883–3888.

Barylski, J., Enault, F., Dutilh, B.E., Schuller, M.B., Edwards, R.A., Gillis, A., Klumpp, J., Knezevic, P., Krupovic, M., Kuhn, J.H., et al. (2019). Analysis of Spounaviruses as a case study for the overdue reclassification of tailed phages. Syst. Biol. 69(1):110–123

Basra, S., Anany, H., Brovko, L., Kropinski, A.M., and Griffiths, M.W. (2014). Isolation and characterization of a novel bacteriophage against *Mycobacterium avium* subspecies *paratuberculosis*. Arch. Virol. *159*, 2659–2674.

Beran, V., Havelkova, M., Kaustova, J., Dvorska, L., and Pavlik, I. (2006). Cell wall deficient forms of mycobacteria: a review.

Bolster, C.H., Cook, K.L., Haznedaroglu, B.Z., and Walker, S.L. (2009). The transport of *Mycobacterium avium* subsp. *paratuberculosis* through saturated aquifer materials. Lett. Appl. Microbiol. *48*, 307–312.

Borysowski, J., Miedzybrodzki, R., and Górski, A. (2014). Phage therapy : current research and applications. (Caister Academic Press)

Broxmeyer, L., Sosnowska, D., Miltner, E., Chacón, O., Wagner, D., McGarvey, J., Barletta, R.G., and Bermudez, L.E. (2002). Killing of *Mycobacterium avium* and *Mycobacterium tuberculosis* by a Mycobacteriophage Delivered by a Nonvirulent Mycobacterium: A Model for Phage Therapy of Intracellular Bacterial Pathogens . J. Infect. Dis. *186*, 1155–1160.

Buttimer, C., McAuliffe, O., Ross, R.P.P., Hill, C., O'Mahony, J., Coffey, A., O'Mahony, J., and Coffey, A. (2017). Bacteriophages and bacterial plant diseases. Front. Microbiol. *8*, 34.

Carta, T., Álvarez, J., de la Lastra, J.M.P., and Gortázar, C. (2013). Wildlife and paratuberculosis: A review. Res. Vet. Sci. *94*, 191–197.

Casal, C., Díez-Guerrier, A., Álvarez, J., Rodriguez-Campos, S., Mateos, A., Linscott, R., Martel, E., Lawrence, J.C., Whelan, C., Clarke, J., et al. (2014). Strategic use of serology for the diagnosis of bovine tuberculosis after intradermal skin testing. Vet. Microbiol. *170*, 342–351.

Cashman, W., Buckley, J., Quigley, T., Fanning, S., More, S., Egan, J., Berry, D., Grant, I., and O\textquotesingleFarrell, K. (2008). Risk factors for the introduction and within-herd transmission of *Mycobacterium avium* subspecies *paratuberculosis* (MAP) infection on 59 Irish dairy herds. Ir. Vet. J. *61*.

Castellanos, E., de Juan, L., Domínguez, L., and Aranaz, A. (2012). Progress in molecular typing of *Mycobacterium avium* subspecies *paratuberculosis*. Res. Vet. Sci. *92*, 169–179.

Catalão, M.J., and Pimentel, M. (2018). Mycobacteriophage lysis enzymes: Targeting the mycobacterial cell envelope. Viruses *10*.

Cater, J.C., and Redmond, W.B. (1963). Mycobacterial phages isolated from stool specimens of patients with pulmonary disease. Am. Rev. Respir. Dis. *87*, 726–729.

Chen, J., Quiles-Puchalt, N., Chiang, Y.N., Bacigalupe, R., Fillol-Salom, A., Chee, M.S.J., Fitzgerald, J.R., and Penadés, J.R. (2018). Genome hypermobility by lateral transduction. Science (80-. ). *362*, 207–212.

Chiu, L.J.V., Tauer, L.W., Al-Mamun, M.A., Kaniyamattam, K., Smith, R.L., and Grohn, Y.T. (2018). An agent-based model evaluation of economic control strategies for paratuberculosis in a dairy herd. J. Dairy Sci. *101*, 6443–6454.

Clark, D.L., Koziczkowski, J.J., Radcliff, R.P., Carlson, R.A., and Ellingson, J.L.E. (2008). Detection of *Mycobacterium avium* subspecies *paratuberculosis*: Comparing Fecal Culture Versus Serum Enzyme-Linked Immunosorbent Assay and Direct Fecal Polymerase Chain Reaction. J. Dairy Sci. *91*, 2620–2627.

Claudia, A., de Lurdes Pinto, M., Matos, A., Matos, M., and dos Anjos Pires, M. (2013). *Mycobacterium avium* complex in domestic and wild animals. In

Insights from Veterinary Medicine (InTech)

Collins, M.T. (1996). Diagnosis of paratuberculosis. Vet. Clin. North Am. Food Anim. Pract. *12*, 357–371.

D'Herelle, F. (1946). L'étude d'une maladie : le choléra, maladie à paradoxes. (Lausanne: F. Rouge & Cie, S.A).

Debroy, B., Tripathi, B.N., Sonawane, G.G., and Bind, R.B. (2012). Detection of *Mycobacterium avium* subspecies *paratuberculosis* (MAP) in alcohol-fixed tissues of sheep by ISMav2 gene PCR and its comparison with histopathology, bacterial culture and IS900 PCR. Small Rumin. Res. *105*, 329–334.

Dedrick, R.M., Guerrero-Bustamante, C.A., Garlena, R.A., Russell, D.A., Ford, K., Harris, K., Gilmour, K.C., Soothill, J., Jacobs-Sera, D., Schooley, R.T., et al. (2019). Engineered bacteriophages for treatment of a patient with a disseminated drug-resistant *Mycobacterium abscessus*. Nat. Med. *25*, 730–733.

Diéguez, F.J., González, A.M., Menéndez, S., Vilar, M.J., Sanjuán, M.L., Yus, E., and Arnaiz, I. (2009). Evaluation of four commercial serum ELISAs for detection of *Mycobacterium* avium subsp. *paratuberculosis* infection in dairy cows. Vet. J. *180*, 231–235.

Donat, K., Kube, J., Dressel, J., Einax, E., Pfeffer, M., and Failing, K. (2014). Detection of *Mycobacterium avium* subspecies *paratuberculosis* in environmental samples by faecal culture and real-time PCR in relation to apparent within-herd prevalence as determined by individual faecal culture. Epidemiol. Infect. *143*, 975–985.

Drulis-Kawa, Z., Majkowska-Skrobek, G., and Maciejewska, B. (2015). Bacteriophages and Phage-Derived Proteins – Application Approaches. Curr. Med. Chem. *22*, 1757–1773.

Ellingson, J., Anderson, J.L., Kozixzkowski, J.J., Radcliff, R., Sloan, S.J., Allen, S.E., and Sullivan, N.M. (2005). Detection of Viable *Mycobacterium avium* subsp. *paratuberculosis* in Retail Pasteurized Whole Milk by Two Culture Methods and PCR. J. Food Prot. *68*, 966–972.

Emery, D., and Whittington, R. (2004). An evaluation of mycophage therapy, chemotherapy and vaccination for control of subsp. infection. Vet. Microbiol. *104*, 143–155.

Enault, F., Briet, A., Bouteille, L., Roux, S., Sullivan, M.B., and Petit, M.A.

(2017). Phages rarely encode antibiotic resistance genes: A cautionary tale for virome analyses. ISME J. *11*, 237–247.

Endersen, L., Coffey, A., Neve, H., McAuliffe, O., Ross, R.P., and O'Mahony, J.M. (2013). Isolation and characterisation of six novel mycobacteriophages and investigation of their antimicrobial potential in milk. Int. Dairy J. *28*, 8–14.

Fan, F., and Kan, B. (2015). Survival and proliferation of the lysogenic bacteriophage CTXΦ in Vibrio cholerae. Virol. Sin. *30*, 19–25.

Fu, X., Ding, M., Zhang, N., and Li, J. (2012). Mycobacteriophages: An important tool for the diagnosis of *Mycobacterium tuberculosis* (Review). Mol. Med. Rep. *12*, 13–19.

Fullner, K.J., and Hatfull, G.F. (1997). Mycobacteriophage L5 infection of *Mycobacterium bovis* BCG: implications for phage genetics in the slow-growing mycobacteria . Mol. Microbiol. *26*, 755–766.

Gali, N., Dominguez, J., Blanco, S., Prat, C., Quesada, M.D., Matas, L., and Ausina, V. (2003). Utility of an In-house mycobacteriophage-based assay for rapid detection of rifampin resistance in Mycobacterium tuberculosis clinical Isolates. J. Clin. Microbiol. *41*, 2647–2649.

Garcia, A.B., and Shalloo, L. (2015). Invited review: The economic impact and control of paratuberculosis in cattle. J. Dairy Sci. *98*, 5019–5039.

Garvey, M. (2018). *Mycobacterium avium* subspecies *paratuberculosis*: A possible causative agent in human morbidity and risk to public health safety. Open Vet. J. *8*, 172.

Gerrard, Z.E., Swift, B.M.C., Botsaris, G., Davidson, R.S., Hutchings, M.R., Huxley, J.N., and Rees, C.E.D. (2018). Survival of *Mycobacterium avium* subspecies *paratuberculosis* in retail pasteurised milk. Food Microbiol. *74*, 57–63.

Good, M., Clegg, T., Sheridan, H., Yearsely, D., O'Brien, T., Egan, J., and Mullowney, P. (2009). Prevalence and distribution of paratuberculosis (Johne's disease) in cattle herds in Ireland. Ir. Vet. J. *62*.

Górski, A., Mi\kedzybrodzki, R., Weber-D\kabrowska, B., Fortuna, W., Letkiewicz, S., PawełRogóż, Jończyk-Matysiak, E., D\kabrowska, K., Majewska, J., and Borysowski, J. (2016). Phage therapy: combating infections with potential for evolving from merely a treatment for complications to targeting diseases. Front. Microbiol. *7*.

Green, E.P., Tizard, M.L. V, Moss, M.T., Thompson, J., Winterbourne, D.J., McFadden, J.J., and Hermon-Taylor, J. (1989). Sequence and characteristics or IS900, an insertion element identified in a human Crohn's disease isolate or *Mycobacterium paratuberculosis*. Nucleic Acids Res. *17*, 9063–9073.

Harris, N.B., and Barletta, R.G. (2001). *Mycobacterium avium* subsp. *paratuberculosis* in veterinary medicine. Clin. Microbiol. Rev. *14*, 489–512.

Harvey, E.N. (1918). Studies on bioluminescence. J. Gen. Physiol. *1*, 133–145.

Hatfull, G.F. (2012). The Secret Lives of Mycobacteriophages. In Bacteriophages, Part A, (Elsevier), pp. 179–288.

Hatfull, G.F. (2014). Mycobacteriophages: Windows into Tuberculosis. PLoS Pathog. *10*, e1003953.

Hatfull, G.F. (2018). Mycobacteriophages. Microbiol. Spectr. *6*.

Jacobs-Sera, D., Marinelli, L.J., Bowman, C., Broussard, G.W., Guerrero Bustamante, C., Boyle, M.M., Petrova, Z.O., Dedrick, R.M., Pope, W.H., Modlin, R.L., et al. (2012). On the nature of mycobacteriophage diversity and host preference. Virology *434*, 187–201.

Jesse, F., Bitrus, A., Abba, Y., Sadiq, M., Chung, E., Hambali, I., Lila, M., and Haron, A. (2016). Clinical and gross pathological findings of Johne's disease in a calf: A case report. J. Adv. Vet. Anim. Res. *3*, 292.

Johnston, C., Coffey, A., and Mahony, J.O. (2010). Development of a novel oral vaccine against *Mycobacterium avium paratuberculosis* and Johne disease. Bioeng. Bugs *1*, 155–163.

Kamal, F., and Dennis, J.J. (2015). *Burkholderia cepacia* complex phage-antibiotic synergy (PAS): Antibiotics stimulate lytic phage activity. Appl. Environ. Microbiol. *81*, 1132–1138.

Klumpp, J., Dorscht, J., Lurz, R., Bielmann, R., Wieland, M., Zimmer, M., Calendar, R., and Loessner, M.J. (2008). The terminally redundant, nonpermuted genome of *Listeria* bacteriophage A511: A model for the SPO1-like myoviruses of gram-positive bacteria. J. Bacteriol. *190*, 5753–5765.

Koets, A.P., Eda, S., and Sreevatsan, S. (2015). The within host dynamics of *Mycobacterium avium* ssp. *paratuberculosis* infection in cattle: where time and place matter. Vet. Res. *46*.

Kraiss, J.P., Gelbart, S.M., and Juhasz, S.E. A Comparison of Three Mycobacteriophages.

Krishnan, M.Y., Manning, E.J.B., and Collins, M.T. (2009). Comparison of three methods for susceptibility testing of *Mycobacterium avium* subsp. *paratuberculosis* to 11 antimicrobial drugs. J. Antimicrob. Chemother. *64*, 310–316.

Lin, D.M., Koskella, B., and Lin, H.C. (2017). Phage therapy: An alternative to antibiotics in the age of multi-drug resistance. World J. Gastrointest. Pharmacol. Ther. *8*, 162.

Marinelli, L.J., Hatfull, G.F., and Piuri, M. (2012). Recombineering. Bacteriophage *2*, 5–14.

Meier-Kolthoff, J.P., and Göker, M. (2017). VICTOR: genome-based phylogeny and classification of prokaryotic viruses. Bioinformatics *33*, 3396–3404.

van der Merwe, R.G., van Helden, P.D., Warren, R.M., Sampson, S.L., and Gey van Pittius, N.C. (2014). Phage-based detection of bacterial pathogens. Analyst *139*, 2617–2626.

Momotani, E., Whipple, D.L., Thiermann, A.B., and Cheville, N.F. (1988). Role of m cells and macrophages in the entrance of *Mycobacterium paratuberculosis* into domes of ileal Peyer's patches in calves. Vet. Pathol. *25*, 131–137.

Morand, S., McIntyre, K.M., and Baylis, M. (2014). Domesticated animals and human infectious diseases of zoonotic origins: Domestication time matters. Infect. Genet. Evol. *24*, 76–81.

Moravkova, M., Hlozek, P., Beran, V., Pavlik, I., Preziuso, S., Cuteri, V., and Bartos, M. (2008). Strategy for the detection and differentiation of *Mycobacterium avium* species in isolates and heavily infected tissues. Res. Vet. Sci. *85*, 257–264.

Morse, M.L., Lederberg, E.M., and Lederberg, J. (1956). Transduction in *Escherichia coli* K-12. Genetics *41*.

Mushegian, A.R. (2020). Are There 1031 Virus Particles on Earth, or More, or Fewer? J. Bacteriol. *202*.

Naser, S.A., Ghobrial, G., Romero, C., and Valentine, J.F. (2004). Culture of *Mycobacterium avium* subspecies paratuberculosis from the blood of patients with Crohn's disease. Lancet *364*, 1039–1044.

Norby, B., Fosgate, G.T., Manning, E.J.B., Collins, M.T., and Roussel, A.J. (2007). Environmental mycobacteria in soil and water on beef ranches: Association between presence of cultivable mycobacteria and soil and water physicochemical characteristics. Vet. Microbiol. *124*, 153–159.

Novick, R.P., Christie, G.E., and Penadés, J.R. (2010). The phage-related chromosomal islands of Gram-positive bacteria. Nat. Rev. Microbiol. *8*, 541–551.

Oechslin, F., Piccardi, P., Mancini, S., Gabard, J., Moreillon, P., Entenza, J.M., Resch, G., and Que, Y.A. (2017). Synergistic interaction between phage therapy and antibiotics clears *Pseudomonas aeruginosa* infection in endocarditis and reduces virulence. J. Infect. Dis. *215*, 703–712.

Ott, S.L., Wells, S.J., and Wagner, B.A. (1999). Herd-level economic losses associated with Johne's disease on US dairy operations. Prev. Vet. Med. *40*, 179–192.

Over, K., Crandall, P.G., O'Bryan, C.A., and Ricke, S.C. (2011). Current perspectives on *Mycobacterium avium* subsp. *paratuberculosis*, Johne's disease, and Crohn's disease: a Review. Crit. Rev. Microbiol. *37*, 141–156.

Pavlik, I., and Falkinham, J.O. (2009). The occurrence of pathogenic and potentially pathogenic mycobacteria in animals and the role of the environment in the spread of infection. In The Ecology of Mycobacteria: Impact on Animals and Humans Health, (Springer Netherlands), pp. 199–281.

Pholwat, S., Ehdaie, B., Foongladda, S., Kelly, K., and Houpt, E. (2011). Real-Time PCR Using Mycobacteriophage DNA for Rapid Phenotypic Drug Susceptibility results for *Mycobacterium tuberculosis*. J. Clin. Microbiol. *50*, 754–761.

Phulpoto, M.A., Qayyum, S., Rizvi, N., and Khuhawar, S.M. (2005). Diagnostic yield of fast plaque TB test for rapid detection of *Mycobacterium* in tuberculosis suspects. J Pak Med Assoc *55*, 57–60.

Pickup, R.W., Rhodes, G., Arnott, S., Sidi-Boumedine, K., Bull, T.J., Weightman, A., Hurley, M., and Hermon-Taylor, J. (2005*). Mycobacterium avium* subsp. *paratuberculosis* in the catchment area and water of the river Taff in South Wales, United Kingdom, and its potential relationship to clustering of Crohn's Disease Cases in the City of Cardiff. Appl. Environ. Microbiol. *71*, 2130–2139.

Pierce, E.S. (2010). Ulcerative colitis and Crohn's disease: is *Mycobacterium avium* subspecies *paratuberculosis* the common villain? Gut Pathog. *2*, 21.

Pillars, R.B., Grooms, D.L., Wolf, C.A., and Kaneene, J.B. (2009). Economic evaluation of Johne's disease control programs implemented on six Michigan dairy farms. Prev. Vet. Med. *90*, 223–232.

Piuri, M., Jacobs, W.R., and Hatfull, G.F. (2009). Fluoromycobacteriophages for rapid, specific, and sensitive antibiotic susceptibility testing of *Mycobacterium tuberculosis*. PLoS One *4*, e4870.

Piuri, M., Rondón, L., Urdániz, E., and Hatfull, G.F. (2013). generation of affinity-tagged fluoromycobacteriophages by mixed assembly of phage capsids. Appl. Environ. Microbiol. *79*, 5608–5615.

Plain, K.M., Marsh, I.B., Waldron, A.M., Galea, F., Whittington, A.-M., Saunders, V.F., Begg, D.J., de Silva, K., Purdie, A.C., and Whittington, R.J. (2013). High-throughput direct fecal pcr assay for detection of *Mycobacterium avium* subsp. *paratuberculosis* in sheep and cattle. J. Clin. Microbiol. *52*, 745–757.

Ponnusamy, D., Periasamy, S., Tripathi, B.N., and Pal, A. (2013). *Mycobacterium avium* subsp. *paratuberculosis* invades through M cells and enterocytes across ileal and jejunal mucosa of lambs. Res. Vet. Sci. *94*, 306–312.

Pribylova, R., Slana, I., Kralik, P., Kralova, A., Babak, V., and Pavlik, I. (2011). Correlation of *Mycobacterium avium* subsp. *paratuberculosis* counts in gastrointestinal tract, muscles of the diaphragm and the masseter of dairy cattle and potential risk for consumers. Int. J. Food Microbiol. *151*, 314–318.

Rondon, L., Piuri, M., Jacobs, W.R., de Waard, J., Hatfull, G.F., and Takiff, H.E. (2011). evaluation of fluoromycobacteriophages for detecting drug resistance in *Mycobacterium tuberculosis*. J. Clin. Microbiol. *49*, 1838–1842.

Rondón, L., Urdániz, E., Latini, C., Payaslian, F., Matteo, M., Sosa, E.J., Porto, D.F. Do, Turjanski, A.G., Nemirovsky, S., Hatfull, G.F., et al. (2018). Fluoromycobacteriophages can detect viable Mycobacterium tuberculosis and Determine phenotypic rifampicin resistance in 3-5 days from sputum collection. Front. Microbiol. 9:1471.

doi: 10.3389/fmicb.2018.01471 Salgado, M., Alfaro, M., Salazar, F., Badilla, X., Troncoso, E., Zambrano, A., González, M., Mitchell, R.M., and Collins, M.T. (2015). Application of cattle slurry containing *Mycobacterium avium* subsp. *paratuberculosis* (MAP) to grassland soil and its effect on the relationship between MAP and free-living amoeba. Vet. Microbiol. *175*, 26–34.

Sassi, M., Bebeacua, C., Drancourt, M., and Cambillau, C. (2013). The first structure of a mycobacteriophage, the *Mycobacterium abscessus* subsp.

*bolletii* phage Araucaria. J. Virol. *87*, 8099–8109.

Savarino, E., Bertani, L., Ceccarelli, L., Bodini, G., Zingone, F., Buda, A., Facchin, S., Lorenzon, G., Marchi, S., Marabotto, E., et al. (2019). Antimicrobial treatment with the fixed-dose antibiotic combination RHB-104 for *Mycobacterium avium* subspecies *paratuberculosis* in Crohn's disease: pharmacological and clinical implications. Expert Opin. Biol. Ther. *19*, 79–88.

Schmieger, H. (1982). Packaging signals for phage P22 on the chromosome of *Salmonella* typhimurium. MGG Mol. Gen. Genet. *187*, 516–518.

Schuch, R., and Fischetti, V.A. (2006). Detailed genomic analysis of the Wβ and γ phages infecting *Bacillus anthracis*: Implications for evolution of environmental fitness and antibiotic resistance. J. Bacteriol. *188*, 3037–3051.

Secott, T.E., Lin, T.L., and Wu, C.C. (2004). *Mycobacterium avium* subsp. *paratuberculosis* fibronectin attachment protein facilitates M-Cell targeting and invasion through a fibronectin bridge with host integrins. Infect. Immun. *72*, 3724–3732.

Shin, A.-R., Kim, H.-J., Cho, S.N., Collins, M.T., Manning, E.J.B., Naser, S.A., and Shin, S.J. (2010). Identification of seroreactive proteins in the culture filtrate antigen of *Mycobacterium avium* ssp.*paratuberculosis* human isolates to sera from Crohn's disease patients. FEMS Immunol. Med. Microbiol. *58*, 128–137.

Sigurdardóttir, Ó.G., Bakke-McKellep, A.M., Djønne, B., and Evensen, Ø. (2005). *Mycobacterium avium* subsp. *paratuberculosis* enters the small intestinal mucosa of goat kids in areas with and without Peyer's patches as demonstrated with the everted sleeve method. Comp. Immunol. Microbiol. Infect. Dis. *28*, 223–230.

Stella, E.J., Franceschelli, J.J., Tasselli, S.E., and Morbidoni, H.R. (2013). Analysis of Novel mycobacteriophages indicates the existence of different strategies for phage inheritance in mycobacteria. PLoS One *8*, e56384.

Strahl, E.D., Gillaspy, G.E., and Falkinham, J.O. (2001). fluorescent acid-fast microscopy for measuring phagocytosis of *Mycobacterium avium*, *Mycobacterium intracellulare*, and *Mycobacterium scrofulaceum* by tetrahymena pyriformis and their intracellular growth. Appl. Environ. Microbiol. *67*, 4432–4439.

Sula, L., Sulova, J., and Stolcpartova, M. (1981). Therapy of experimental tuberculosis in guinea pigs with mycobacterial phages DS-6A, GR-21 T, MY-327. Czech. Med. *4*, 209–214.

Sulakvelidze, A., Alavidze, Z., and Morris, J. (2001). Bacteriophage therapy. Antimicrob. Agents Chemother. *45*, 649–659.

Swift, B.M.C.C., Gerrard, Z.E., Huxley, J.N., and Rees, C.E.D.D. (2014). Factors affecting phage d29 infection: a tool to investigate different growth states of mycobacteria. PLoS One *9*, e106690.

Tasara, T., and Stephan, R. (2005). development of an f57 sequence-based real-time pcr assay for detection of *Mycobacterium avium* subsp. *paratuberculosis* in milk. Appl. Environ. Microbiol. *71*, 5957–5968.

Trigo, G., Martins, T.G., Fraga, A.G., Longatto-Filho, A., Castro, A.G., Azeredo, J., and Pedrosa, J. (2013). phage therapy is effective against infection by *Mycobacterium ulcerans* in a murine footpad model. PLoS Negl. Trop. Dis. *7*.

Weber-Dabrowska, B., Jończyk-Matysiak, E., Zaczek, M., Łobocka, M., Łusiak-Szelachowska, M., Górski, A., Weber-D\kabrowska, B., Jończyk-Matysiak, E., Żaczek, M., Łobocka, M., et al. (2016b). Bacteriophage procurement for therapeutic purposes. Front. Microbiol. *7*,1177

Whittington, R., Donat, K., Weber, M.F., Kelton, D., Nielsen, S.S., Eisenberg, S., Arrigoni, N., Juste, R., Sáez, J.L., Dhand, N., et al. (2019). Control of paratuberculosis: Who, why and how. A review of 48 countries. BMC Vet. Res. *15*, 1–29.

Whittington, R.J., Begg, D.J., de Silva, K., Purdie, A.C., Dhand, N.K., and Plain, K.M. (2017). Case definition terminology for paratuberculosis (Johne's disease). BMC Vet. Res. *13*, 328.

Wittebole, X., De Roock, S., and Opal, S.M. (2014). A historical overview of bacteriophage therapy as an alternative to antibiotics for the treatment of bacterial pathogens. Virulence *5*, 226–235.

## Chapter 12

# Phage Structural Antimicrobial Proteins

**Sílvio B. Santos, Luís D.R. Melo and Hugo Oliveira***

Centre of Biological Engineering (CEB), University of Minho, Braga, Portugal

*Corresponding author: hugooliveira@deb.uminho.pt

DOI: https://doi.org/10.21775/9781913652517.12

### Abstract

Bacterial polysaccharides (capsules, lipopolysaccharides, exopolysaccharides and peptidoglycan) play an important role in protection, by blocking the entry of antimicrobials, evading microbial defences and avoiding phage predation. To be able to enter the cell and replicate, phages rely mostly on two structural proteins to break down several host carbohydrate barriers. First, polysaccharide depolymerases are used by phages to bind and degrade the outermost polymers (capsules, lipopolysaccharides and exopolysaccharides) to reach the final host receptor. Second, the virion-associated lysins locally puncture small holes in the rigid cell wall (peptidoglycan) to enable the injection of the phage DNA into the cytoplasm, leading to the subsequent infection and replication of progeny phage. These polysaccharide depolymerases and virion-associated lysins are emerging as alternative antimicrobial agents to control pathogens by either reducing bacterial virulence or lysing bacteria, respectively. Herein we provide a comprehensive critical overview on the diversity of the structure, functions and biotechnological applications of these enzymes. This overview underlines the great potential of using phage structural proteins as anti-virulence and antimicrobial agents to control multidrug resistant pathogens in food, veterinary and human medicine.

## Bacterial carbohydrate barriers against phage infection

Bacteriophages (phages) are viruses that infect bacteria making use of their cellular machinery to reproduce. Under the lytic cycle, phages take over the host's metabolic mechanisms in order to replicate and produce their progeny, leading ultimately to the death of the host cell (Hatfull and Hendrix, 2011). Phage invasion is initiated through their adsorption to the host cell. This process dictates the host range of each phage, as it is the initial phage-host contact. After an irreversible recognition, phage particles position themselves to inject their nucleic acid into the bacterial cells, after which the replication begins (Olszak et al., 2017a).

Phages are described as the most abundant entities on planet earth and are supposed to be present in all environments where bacteria prosper. Due to their obligate bacterial parasitism, phages play a role in ecosystems shaping and in the evolution of host cells. Indeed, the co-evolution processes occurring between prey (bacteria) and predator (phage) are recognized as the most important process modulation in microbial communities, consequently affecting the ecological cycles (Koskella and Brockhurst, 2014). For instance, the emergence of phage-resistant cell mutants with altered fitness or phenotypes proves the impact of phage predation (Hosseinidoust et al., 2013). Actually, bacteria have an innate and/or adaptive immune mechanism to protect themselves from phage predation. These strategies used by bacteria to inhibit phage infections were recently reviewed (Safari et al., 2020) and the first-line defence is the surface modification, a process that consists in masking, mutating or losing the phage receptor proteins that will therefore inhibit phage adsorption. However, phages may co-evolved with the bacterial cells to become infective again, in an endless phage-host arms race.

During phage infection, phage particles must overcome several physical barriers of bacteria that can affect phage infection (Figure 1). Due to their size, phage particles cannot get into the bacterial cell envelope, which is usually composed by an inner cytoplasmic membrane, a cell wall and an outer cytoplasmic membrane (only in Gram-negative) (Safari et al., 2020). Also, some bacteria produce an S-layer (a surface layer), which consists of proteins or glycoproteins located at the surface of the cell envelope. These proteins are poorly conserved, possessing large variability even within related species (Fagan and Fairweather, 2014). Besides the cell envelope, some bacteria produce capsular polysaccharides (CPS), a layer outside the cell envelope. While capsules are well-defined outer layers of bacterial cells, slime layers are extended outwards from the bacterial surface in a non-organized manner (Ruqeya, 2019). This slime layer, in biofilms, is also known as the biofilm matrix. In a simple way, biofilms can be defined by a community of microbes adhered to a surface and embedded in an organic polymer matrix (Carpentier and Cerf, 1993). The composition and key biological functions of these cellular components that can have a physical impact on phage infection will described herein.

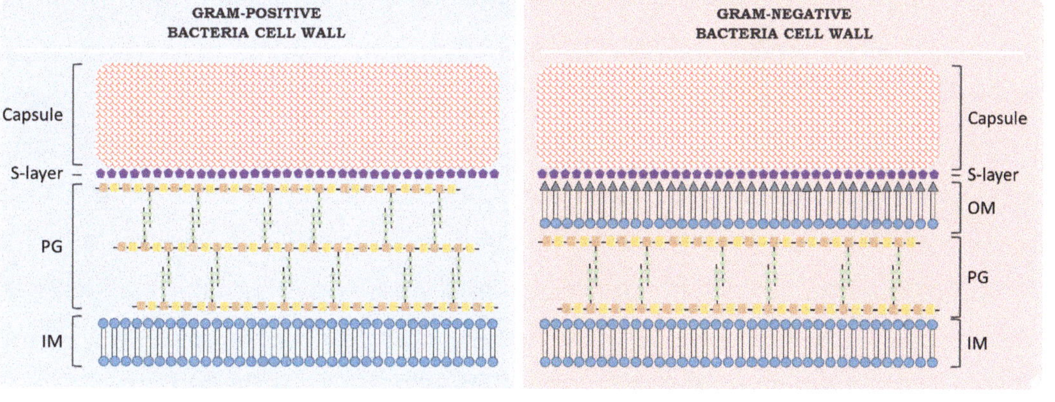


**Figure 1.** Schematic representation of the structure and composition of the bacterial cell envelope in Gram-positive bacteria (left) and Gram-negative bacteria (right). OM, outer membrane; PG, peptidoglycan; IM, inner membrane. The capsule and S-layer are extracellular structures.

In nature, most bacterial cells live under a biofilm phenotype. In these communities, extracellular polymeric substances (EPS) halt biofilm cells in close proximity allowing intense cell-cell communications and horizontal gene transfer events. The EPS that confer the self-produced biofilm matrix are mainly polysaccharides, proteins, nucleic acids and lipids. This structure protects biofilm cells from physical (desiccation, ultraviolet radiation), chemical (biocides, antibiotics, heavy metals), immune defences and phage and protozoan predation (Flemming and Wingender, 2010). The biofilm matrix composition is species or even strain-dependent (Oliveira et al., 2015). Nevertheless, in several cases, the main polysaccharide that forms the capsule is the main constituent of the biofilm matrix. For example, alginate and poly-N-acetylglucasamine are the main components of *Pseudomonas aeruginosa* and *Staphylococcus aureus* capsule and biofilm matrix, respectively (Arciola et al., 2015; Boyd and Chakrabarty, 1995).

A capsule is an outer layer involving the bacterial cell. In Gram-negative bacteria, CPS are connected to the outer membrane via a lipid anchor (Liston et al., 2016). CPS have different designations according to the bacterial species in which they are present. As such, in *Escherichia coli, Klebsiella pneumoniae* and *Acinetobacter* spp. they are termed K-antigens, while in *Salmonella* spp. and *Citrobacter* spp. CPS are named Vi antigens (Knecht et al., 2019). CPS are usually much larger than other polysaccharides, namely lipopolysaccharide (LPS) (Wen, 2015). There is a wide range of chemical types of CPS. The monosaccharide units, glycosidic linkages and substitutions with non-carbohydrates vary with different CPS. For example, for *P. aeruginosa* only one capsular type has been described, while in *E. coli* and *K. pneumoniae* 80 or more different antigens have been described so far (Sachdeva et al., 2017). In addition to Gram-negative bacteria, also some Gram-positive species produce CPS. As in Gram-negative microorganisms, there is also a huge diversity among them. While *Bacillus anthracis* only has one capsular type, *Streptococcus pneumoniae* has a total of 98 different capsule serotypes (Paton and Trappetti, 2019). The molecules composing the capsules are also different, while in *S. aureus* CPS are mainly composed by poly-N-acetylglucosamine, *Streptococcus pyogenes* produces a hyaluronic acid capsule (Knecht et al., 2019). CPS production is also considered a bacterial virulence determinant, while non-capsulated bacteria (CPS mutants included) have attenuated virulence. In pathogenic bacteria, CPS can impair host immune recognition and consequent phagocytic killing (Cress et al., 2014).

The outer membrane is only present on Gram-negative bacteria. This structure is an asymmetric bilayer composed of phospholipids from the inside and glycolipids from the outside. On the outer membrane, two types of proteins are commonly found: lipoproteins (anchored to the peptidoglycan) and integral outer membrane proteins (usually working as transporters). LPS is the major component of the outer membrane and has the function of providing stability and protection to the cell against external stresses. Although some bacteria can survive without them (*e.g. Acinetobacter baumannii* and *Neisseria meningitidis*), the LPS is essential to the majority of Gram-negative bacteria (Zhang et al., 2013). LPS molecules are composed of three portions: the lipid A (lipid moiety), a core oligosaccharide and the hypervariable O-antigen (polysaccharide) (Ruiz et al., 2006). When the O-antigen is attached to the core lipid A it results in a smooth LPS type (S-type), while LPS without O-antigen are named rough type (R-type). If bacteria exhibit one repeating unit of the O-antigen to core lipid A, it results in a mixed LPS named SR-LPS (Paramonov et al., 2009). Along with the capsule, the O-antigen is one of the main surface antigens having an impact on serotype differentiation on Gram-negative bacteria. LPS is associated with surface adhesion, phage resistance and interaction with predators (Abu-Lail and Camesano, 2003; Pflaum et al., 2012; Wright et al., 2019). Moreover, these molecules are seen as bacterial virulence factors, as they are potent cellular activators and also stimulate cells from the animal immune system.

Surrounding the cell membrane is the cell wall, which is responsible for maintaining the cell shape and integrity. This structure can counter the high cytoplasmic osmotic pressure. Peptidoglycan, which is a polymeric network surrounding the inner membrane is the major structural element of the cell wall. The peptidoglycan is usually framed with diverse proteins (in all bacteria) and anionic wall teichoic acids (only in Gram-positive bacteria). This multi-layered structure is composed by repeating layers of N-acetylglucosamine and N-acetylmuramic acid that are linked by glycosidic bonds β (1 → 4). The glycan strands are cross-linked by peptides that are connected to N-acetylmuramic acid residues, usually involving -Lys or m-DAP or D-Ala amino acid residues of complementary peptide chains. On most Gram-negative bacteria (and a few Gram-positive) the peptide stem is linked by an inter-peptide bond, while on most Gram-positive species the linkage is made by an inter-peptide bridge (Vollmer et al., 2008). In nature, there are several peptidoglycan-degrading enzymes from exogenous or endogenous origin. The latter are involved in several functions, namely peptidoglycan maturation, cell division and defence against other bacteria (Alcorlo et al., 2017). From the exogenous point of view, almost all biological entities produce peptidoglycan-degrading enzymes, namely phages (endolysins and structural virion-associated lysins) and animals (lysozymes).

The cytoplasmic membrane is another structure present in all bacterial species. This semipermeable barrier protects the intracellular content from the external environments. In opposition to the outer membrane, the inner membrane is a symmetric phospholipid bilayer containing proteins, lipoproteins and glycolipids (Paradis-Bleau, 2015). In Gram-positive bacteria cell membranes, it is common to observe lipoteichoic acids anchored to the cell wall (Percy and Grundling, 2014). Electron carriers and transporters on the cell membrane surface are responsible for the creation of ion gradients throughout the membrane which generate differences in charges and concentration of ions that lead to the proton motive force (Fernandes and Sao-Jose, 2018).

All these referred layers can act as barriers against phage predation. Consequently, phages have developed strategies to overcome them. In this chapter, the focus will be given to two phage structural enzymes, namely polysaccharide depolymerases and virion-associated lysins, that degrade important carbohydrate barriers. Particular attention is given to their location in the phage, structure and mode of action, as well as their biotechnology application to control bacterial pathogens.

## Phage structural proteins that degrade the cell carbohydrate barriers

During the lytic cycle, tailed phages adsorb onto the host cells through receptors, inject their DNA, replicate and lyse the cells to release their progeny. As mentioned above, phages are faced with a number of diversified carbohydrate structures (CPS, LPS and EPS) that need to be overcome before injecting their DNA and establish infection. Tailed phages have therefore evolved to encode polysaccharide depolymerases to degrade external carbohydrate barriers (CPS- and LPS O-antigens). This allows the phage to irreversible bind to the bacterial host cells, remaining one last barrier for DNA ejection: the peptidoglycan. This barrier is overcome through enzymatic degradation by another specific phage protein called virion-associated lysins. The biological, structural insights as well as the antimicrobial applications of these proteins are discussed herein.

*Polysaccharide depolymerases*

Phage-encoded polysaccharide depolymerases are located at the virion particle to degrade polymers that shield inner cell receptors needed for phage binding and further cell infection. The presence of phage plaques surrounded by hazy rings that usually grow over time has been the hallmark for the detection of phages carrying polysaccharide depolymerases, and was first described in 1956 (Adams and Park, 1956). Since then, and after thousands of phages isolated, it seems that only a small fraction of phages encode such enzymes. The majority of recombinantly produced depolymerases degrade CPS of *E. coli (Lin et al., 2017)*, *K. pneumoniae* (Hsu et al., 2013; Lin et al., 2014; Solovieva et al., 2018) and *A. baumannii* hosts (Oliveira et al., 2019a; Oliveira et al., 2017; Oliveira et al., 2019b; Popova et al., 2017), and LPS antigens of *P. aeruginosa* (Olszak et al., 2017b) and *Salmonella enterica* (Steinbacher et al., 1996) hosts, among others (Table 1). Moreover, some depolymerases are reported to have activity against other surface-associated carbohydrates (*e.g.* teichoic acids) (Myers et al., 2015).

Several pieces of evidence suggest both CPS and LPS antigens to be not only protecting barriers from phage predation but also to function as primary receptors for those phages encoding depolymerases. Depolymerases have been described to drill through the capsule layers, possibly to reach the secondary or irreversible receptor and trigger the subsequent DNA injection process (Leiman et al., 2007). Moreover, phage adsorption onto cells previously treated with recombinant CPS depolymerases are drastically reduced, demonstrating the vital role of the CPS to enable phage infection (Oliveira et al., 2017; Oliveira et al., 2019b). Furthermore, there are also several LPS O-antigen dependent phages with depolymerases that become completely inactivated in rough variants, which lack the O-antigen portion of the LPS structure (Broeker and Barbirz, 2017; Prokhorov et al., 2017). As all these bacterial carbohydrate barriers play major virulence factors in bacteria, blocking antibiotics and host immune defences mechanisms (phagocytosis or complement-mediated killing), the ability of phage depolymerases to strip or modify cells from these surface polysaccharides coats have become an attractive anti-virulence strategy to control pathogenic bacteria.

Location, structure and enzymatic activity

As mentioned, phage plaques with surrounding haloes is the typical phenotype associated with phages carrying polysaccharide depolymerases. In light of such data, it is tempting to speculate that the halo expansion of the phage plaque could be related to the diffusion of free depolymerases, which are not assembled into the phage particle during its replication, thus being released during cells lysis. Indeed, a few reports have associated this characteristic with the free diffusion of soluble depolymerases but all characterised depolymerases described to date are associated with the virion particle. In particular, they seem to be predominantly located at the phage tail spikes (*e.g. Acinetobacter* phage vB_ApiP_P1, *Pseudomonas* phage LKA1, *Escherichia* phage K1F) (Lin et al., 2017; Oliveira et al., 2017; Olszak et al., 2017b) (Table 1), with some examples also found at baseplate (*Lactobacillus* phage Ld17) (Cornelissen et al., 2016) and pre-neck appendage proteins (*Bacillus* phage phi29 and *Staphylococcus* phage vB_SepiS-IPLA7) (Gutierrez et al., 2015; Myers et al., 2015). It is important to state that several reports have also pinpointed polysaccharide depolymerases at the phage tail fibres, however, the growing body of structural data available now suggests that they are likely located at the tail spikes. If true, the presence of enzymatic activity would then be a major difference between tail spikes (recognition with enzymatic activity) and tail fibres (recognise without enzymatic activity), which has already been recognized by others (Nobrega et al., 2018). The acquisition of depolymerases seem to be independent of tail morphology, being identified in viruses of the *Podoviridae*, *Siphoviridae*, *Myoviridae* and *Ackermannviridae* families of the *Caudovirales* order (Table 1). Interestingly, while most phages encode a single depolymerase-encoding gene, there are others that encode more, in order to increase their host range. The *Escherichia* phage K1-5 for instance, encodes both a K1 and a K5 depolymerase allowing the phage to infect K1 and K5 *E. coli* strains, respectively (Leiman et al., 2007). The *Klebsiella* phage K64-1 is an extreme example found to encode nine experimental validated depolymerases to target specific *Klebsiella* CPS antigens (K1, K11, K21, K25, K30/K69, K35, K64, KN4, and KN5) (Pan et al., 2017).

To our knowledge, phage depolymerases typically fold as trimers (*e.g.* *Pseudomonas* phage LKA1, *Acinetobacter* phage phiAB6) (Lee et al., 2017; Olszak et al., 2017b), being only one case reported as a tetramer (*Escherichia* phage K1-dep(4), previously named as 63D) (Machida et al., 2000) (Table 1). They contain three elongated and right-handed monomers with β-helical topology, which associate side-by-side into trimers. Several pieces of evidence suggest that this elongated trimeric β-helix architecture favour 1) scanning of polysaccharide polymers, 2) stability and 3) enzymatic degrading activity. First, the ability of the *Salmonella* phage P22 depolymerase to bind to the entire LPS O-antigen layer and at levels similar to antibodies, agrees with a possible optimized scanning of the polymer till intra or inter-subunit substrate binding residues are reached (Baxa et al., 1996; Waseh et al., 2010). Second, depolymerases have been deemed resistant to high temperatures, protease and sodium dodecyl sulfate denaturation. Several *Acinetobacter*, *Klebsiella* and *Escherichia* phage depolymerases have shown to melt at temperatures above 65 °C (Guo et al., 2017; Majkowska-Skrobek et al., 2016; Oliveira et al., 2019b). In particular, the *Salmonella* phage P22 depolymerase has been used as a model to relate structure and stability, as native trimers are resistant to temperature and sodium dodecyl sulfate denaturation, but their monomeric, dimeric, and protrimeric folding intermediates lose stability and become thermolabile (Kreisberg et al., 2002). Third, the inefficient ability of depolymerases to assemble into trimers has been directly linked to a diminished enzymatic activity. There is one study that reported the therapeutic application of phage depolymerases against *E. coli* in mice, being highly active when in the protein trimer form (K1F, K1H, K5, and K30) and less active when as a 18-mer (K1E), instead of the expected trimer, an observation that was correlated to possible limited *in vivo* distribution of such proteins (Lin et al., 2017). Overall, the parsimonious explanation for the extraordinary depolymerase robustness is that they are structural virion proteins evolved to endure harsh external conditions to maintain the phage infectivity.

**Table 1**. Polysaccharide depolymerases described in current literature (experimental validated or non-validated). Only CPS- and LPS-depolymerases with known antigen substrates (an example for each) are listed. All the depolymerases described are located at the tail spike (given through structure availability) and the corresponding phage name and family (without the suffix *viridae*) is indicated. Monomer mass in kDa (Mass), oligomerization (oligom), NCBI Reference Sequence accession number (Accession n.), and specificity (antigen) of the protein is also given. ND, not determined.

| Host | Phage (*family*) | Accession n. | Mass (oligom)* | Antigen | Ref. |
|---|---|---|---|---|---|
| **CPS depolymerases** | | | | | |
| *E. coli* | K1-dep (4) (*Sipho*) | ADA82273 | 108 (tetramer) | K1 | Machida et al., 2000 |
| | K1F (*Podo*) | YP_338127 | 119 (trimer) - 3JU4 | K1 | Lin et al., 2017 |
| | K1-5 (*Podo*) | YP_654147 | 67 (trimer) - 2X3H | K5 | Lin et al., 2017 |
| | K30 (*Podo*) | YP_004678762 | 97 (trimer) | K30 | Lin et al., 2017 |
| | phi92 (*Myo*) | CBY99579 | 87 (trimer) - 6E0V | K1, K92 | Schwarzer et al., 2015 |
| *K. pneumoniae* | NTUH-K2044-K1–1 (*Podo*) | YP_009098385 | 70 (ND) | K1 | Lin et al., 2014 |
| | KpV74 (*Podo*) | APZ82768 | 61 (ND) | K2 | Solovieva et al., 2018 |
| | KP32 (*Podo*) | YP_003347555 | 95 (ND) | K3 | Majkowska-Skrobek et al., 2018 |
| | K5-2 (*Podo*) | APZ82804 | 85 (ND) | K5 | Hsieh et al., 2017 |
| | K5-4 (*Podo*) | APZ82848 | 71 (ND) | K8 | Hsieh et al., 2017 |
| | K64–1 (*Myo*) | BAW85694 | 76 (ND) | K11 | Pan et al., 2017 |
| | KP32 (*Podo*) | YP_003347556 | 62 (ND) | K21 | Majkowska-Skrobek et al., 2018 |
| | K64–1 (*Myo*) | BAW85696 | 62 (ND) | K25 | Pan et al., 2017 |
| | K64–1 (*Myo*) | BAW85697 | 123 (ND) | K35 | Pan et al., 2017 |
| | K5-2 (*Podo*) | APZ82805 | 71 (ND) | K30, K69 | Hsieh et al., 2017 |
| | KpV767 (*Podo*) | AOZ65519 | 92 (ND) | K57 | Solovieva et al., 2018 |
| | KP36 (*Sipho*) | YP_009226011 | 94 (trimer) | K63 | Majkowska-Skrobek et al., 2016 |
| | K64–1 (*Myo*) | BAQ02780 | 110 (ND) | K64 | Pan et al., 2017 |
| | KN1-1 (*Podo*) | BBF66844 | 90 (ND) | KN1 | Pan et al., 2019 |
| | 0507–KN2–1 (*Myo*) | YP_008532047 | 134 (ND) | KN2 | Hsu et al., 2013 |
| | K64–1 (*Myo*) | BAW85692 | 81 (ND) | KN4 | Pan et al., 2017 |
| | K64–1 (*Myo*) | BAW85695 | 130 (ND) | KN5 | Pan et al., 2017 |

| | | | | | |
|---|---|---|---|---|---|
| ***A. baumannii*** | P1 (*Podo*) | ASN73504 | 80 (trimer) - 6E1R | K1 | Oliveira et al., 2017 |
| | phiAB6 (*Podo*) | YP_009288671 | 76 (trimer) - 5JS4 | K2 | Lee et al., 2017 |
| | B1 (*Podo*) | ASN73353 | 85 (ND) | K9 | Oliveira et al., 2017 |
| | Fri1 (*Podo*) | YP_009203055 | 86 (trimer) - 6C72 | K19 | Popova et al., 2017 |
| | AS12 (*Podo*) | APW79830 | 99 (trimer) - 6EU4 | K27 | Popova et al., 2017 |
| | B9 (*Myo*) | AWD93192 | 94 (ND) | K30, K45 | Oliveira et al., 2019a |
| **LPS depolymerases** | | | | | |
| ***E. coli*** | HK620 (*Podo*) | NP_112090 | 77 (trimer) - 2VJI | O18 | Barbirz et al., 2008 |
| | phiV10 (*Podo*) | YP_512279 | 94 (trimer) | O157 | Scholl et al., 2009 |
| ***S. enterica*** | P22 (*Podo*) | NP_059644 | 72 (trimer) - 1TYU | A, B, D1 antigens | Steinbacher et al., 1996 |
| | SP6 (*Podo*) | NP_853609 | 59 (trimer) | B, D-antigens | Gebhart et al., 2017 |
| | SP6 (*Podo*) | NP_853610 | 53 (ND) | C2, C3-antigens | Gebhart et al., 2017 |
| | 9NA (*Sipho*) | AIB07058 | 73 (trimer) - 3RIQ | O2, O4, O9 | Schmidt et al., 2016 |
| | Det7 (*Acker*) | YP_009140380 | 85 (trimer) - 6F7D | E | Broeker et al., 2019 |
| ***Salmonella anatum*** | epsilon15 (*Podo*) | NP_848228 | 116 (trimer) | O10 | Guichard et al., 2013; Kropinski et al., 2007 |
| ***Shigella flexneri*** | Sf6 (*Podo*) | AAQ12204 | 67 (trimer) - 2VBE | Y, 2a | Kunstmann et al., 2018 |
| ***P. aeruginosa*** | LKA1 (*Podo*) | YP_001522890 | 80 (trimer) - 4RU4 | O5 | Olszak et al., 2017b |
| ***Vibrio cholerae*** | JA-1 (*Podo*) | YP_008126828 | 87 (ND) | O139 | Linnerborg et al., 2001 |

* The Protein Data Bank ID may be indicated on the Mass (oligom) column after (oligom)

As a reflection of the many bacterial antigen possibilities to bind and degrade, phage depolymerase-encoding genes have high genetic plasticity, enabling constant adaptation of phages to target new hosts barrier/receptors. Depolymerase are therefore amongst the most diverse proteins in the phage proteome, of the limited fraction studied so far. To our knowledge, depolymerases can be generally divided through their cleavage mode of action, as hydrolases or lyases. Hydrolases cleave glycosidic bonds via consumption of a water molecule. They include sialidases, rhamnosidases and phosphoesterases. The sialidases are mostly found in phages that degrade *E. coli* polysialic acid-rich CPS (*e.g. Escherichia* phage K1E, K1F, 63D and phi92) (Machida et al., 2000; Schwarzer et al., 2015; Stummeyer et al., 2006). The rhamnosidases seem to be specialized depolymerases that target the Gram-negative LPS O-antigens, being frequently found in phages infecting *Salmonella* (*e.g. Salmonella* phages epsilon15, KB1, P27, 9NA ,P22) (Baxa et al., 1996; Guichard et al., 2013; Wollin et al., 1981) and *Shigella* (*e.g. Shigella* phage Sf6) species (Chua et al., 1999). Phosphoesterases compose another less known group of hydrolases that interact with teichoic acids or other surface-associated carbohydrates of distant host taxa. Examples have been described either with phosphodiesterase and phosphomonoesterase activity (*e.g. Lactobacillus* phage Ld17, *Bacillus* phage phi29 and *Escherichia* phage G7C) (Cornelissen et al., 2016; Myers et al., 2015; Prokhorov et al., 2017). Lyases also degrade glycosidic bonds but through a β-elimination mechanism. The pectin/pectate and alginate lyases and hyaluronate are included in this group. Pectin/pectate lyases are by far the most represented depolymerases found to degrade CPS of *Klebsiella* (*e.g. Klebsiella* phage KP36 and K64-1) (Majkowska-Skrobek et al., 2016; Pan et al., 2017) and *Acinetobacter* (*e.g. Acinetobacter* phages P1, B3 and B9) (Oliveira et al., 2019a; Oliveira et al., 2017; Oliveira et al., 2019b), but enzymes that degrade the *Pseudomonas* LPS O-antigen (*Pseudomonas* phage LKA1) (Olszak et al., 2017b) have been also found (Gutierrez et al., 2015). Finally, the hyaluronate and alginate lyases are a less explored class of enzymes predicted to cleave bacterial hyaluronate-capsules (Hynes et al., 1995) and alginate-based polymers (Davidson et al., 1977; Glonti et al., 2010).

### Activity screening assays

There are several methods described to measure the polysaccharide depolymerase activity. The choice is dependent if one aims to obtain a qualitative or quantitative measurement. The study of a particular polysaccharide degradation is, however, challenging due to the difficulties of efficiently purifying CPS- and LPS O-antigens from the EPS mesh.

The drop test is the most popular one but only allows qualitative measurement. The drop test is an inexpensive and quick assay to observe the presence/ absence of depolymerase activity, as a depolymerase solution only needs to be spotted into bacterial lawns prepared on agar plates using the overlay technique. After a few hours of incubation, the formation of a halo indicates depolymerase activity (Oliveira et al., 2017).

There are also methods that can quantitatively measure the depolymerase activity. The most commonly applied methods can be sub-categorized in turbidimetric (cetylpyridinium chloride assay) and colourimetric tests (3,5-Dinitrosalicylic acid, bicinchoninic acid and crystal violet assays). In the cetylpyridinium chloride assay, digested polysaccharide products are first precipitated with the cetylpyridinium chloride and then the residual optical density of the solution is measured (Majkowska-Skrobek et al., 2016). As for the 3,5-Dinitrosalicylic acid (DNS) assay, measures the sugar reducing ends in a solution that results after the incubation of extracted polysaccharides with depolymerases (Oliveira et al., 2019a). The results can be correlated into glucose reducing ends equivalents if glucose is used as a standard. Similarly, the bicinchoninic acid which is an original method to quantify proteins can also be adapted to determine the reducing ends of polysaccharides in solution (Hernandez-Morales et al., 2018). Regarding crystal violet, this assay has been mostly applied to measure the degradation of the biofilm matrix, which are heavily composed of EPS. After biofilm formation and subsequence depolymerase treatment, the biofilm matrix degradation can be stained with crystal violet dye and visualized colourimetrically (Cornelissen et al., 2012; Cornelissen et al., 2011).

There are also other methods assisted by powerful instruments that mostly focus on qualitative measurement. However, some techniques enable the distinction of the depolymerase enzymatic activity (lyase or hydrolase) as well as the determination of the polysaccharide cleavage sites. The simpler methods only allow the visualization of the polysaccharide content (*e.g.* CPS), and do not take into account the digested products (Hsieh et al., 2017). In these methods, the polysaccharides can be visualized either by positively staining (*e.g.* alcian blue/silver or methylene blue in polyacrylamide electrophoresis gels) (Hsieh et al., 2017), negatively staining (*e.g.* congo red and nigrosine under the optical microscope) or unstained and visualized through electronic microscopy techniques (*e.g.* Atomic Force Microscopy, Confocal Laser Scanning Microscopy, Transmission Electron Microscopy and Scanning Electron Microscopy) (Geisinger and Isberg, 2015; Oliveira et al., 2017; Stukalov et al., 2008). The most powerful methods are the biochemistry-assisted techniques such as Fast-atom bombardment mass spectrometry (FABMS), Matrix Assisted Laser Desorption/Ionization mass spectrometry (MALDI MS), Gas chromatography–mass spectrometry (GC–MS) and NMR spectroscopy (Linnerborg et al., 2001; Olszak et al., 2017b), which enable the examination of the enzymatic activity and polysaccharide cleavage sites.

### Polysaccharide depolymerases *in vivo* applications

As stated, bacterial polysaccharides (*e.g.* CPS- and LPS-O-antigens) are regarded as major virulence factors. These polysaccharides are sometimes identical to mammalian glycans, which help bacteria to hide from the host defences. As phage depolymerases are specialized enzymes able to degrade and strip bacterial cells from their polysaccharides coats, they have the potential of turning pathogenic bacteria into avirulent forms that can be easily managed by the immune system. Therefore, in an era of multidrug-resistance, depolymerases have been gaining visibility as alternative anti-virulence agents with several pre-clinical studies established already (Table 2).

Pathogenic *E. coli* is able to cause broad and potentially lethal human infection, being the capsules frequently implicated with increased virulence (Korhonen et

**Table 2.** Depolymerase *in vivo* applications. All the depolymerases described are located at the tail spike and the corresponding phage name and family (without the suffix *viridae*) and the NCBI Reference Sequence accession number (Accession n.) is indicated.

| Host | Phage (*family*) | Accession n. | Model (infection) | Main outcome | Ref. |
|---|---|---|---|---|---|
| ***E. coli*** | K1E (*Podo*) | YP_425027 | Rat (bacteraemia) | ≤60% buffer-treated and ≥70% of K1 depolymerase treated rats survived. Enzymes become less effective if added days after infection. | Mushtaq et al., 2004 |
| | K1E (*Podo*) | YP_425027 | Rat (bacteraemia) | 10% buffer-treated and 80% K1 depolymerase treated rats survived. Dose low as 0.25 µg are equally effective as larger doses (up to 20 µg). | Mushtaq et al., 2005 |
| | K1E (*Podo*) | YP_425027 | Rat (bacteraemia) | 0% buffer-treated and 100% K1-depolymerase treated rats survived. | Zelmer et al., 2010 |
| | K1E (*Podo*)<br>K1F (*Podo*)<br>K1H (*Podo*)<br>K1-5 (*Podo*)<br>K30 (*Podo*) | YP_425027<br>YP_338127<br>ADA82322<br>YP_654147<br>YP_004678762 | Mice (bacteraemia) | 0% buffer-treated and ≥70% of K1F, K1H, K5, and K30 depolymerases treated mice survived. K1E-depolymerase only rescued 10% of mice and was the only one that did not form a trimer. | Lin et al., 2017 |
| | K1E (*Podo*)<br>K1F (*Podo*)<br>K1H (*Podo*)<br>K1-5 (*Podo*)<br>K30 (*Podo*) | YP_425027<br>YP_338127<br>ADA82322<br>YP_654147<br>YP_004678762 | Mice (bacteraemia) | 0% buffer-treated and ≥70% of K1F, K1H, K5, and K30 depolymerases treated mice survived. K1E-depolymerase did not rescue mice. All enzymes become less effective if added days after infection. Intramuscular or intraperitoneal injections influence success. | Lin et al., 2018 |
| ***K. pneumoniae*** | NTUH-K2044-K1-1 (*Podo*) | YP_009098385 | Mice (bacteraemia) | 12.5% buffer-treated and 100% K1 depolymerase treated mice survived. | Lin et al., 2014 |
| | KP36 (*Podo*) | YP_009226011 | *G. mellonella* (bacteraemia) | 0% buffer-treated and 43% K63 depolymerase treated larvae survived. | Majkowska-Skrobek et al., 2016 |
| | K64-1 (*Myo*) | AB897513 | Mice (bacteraemia) | 0% buffer-treated and 100% K64 depolymerase treated mice survived. Enzyme becomes ineffective if added several hours post infection. | Pan et al., 2015 |
| | KP32 (*Podo*) | YP_003347555<br>YP_003347556 | *G. mellonella* (bacteraemia) | 0% buffer-treated and <10% K3 depolymerase treated larvae survived. <0% buffer-treated and <80% K21 depolymerase treated larvae survived. K21 depolymerase treatments were low or completely ineffective, depending on the K21 strains used. | Majkowska-Skrobek et al., 2018 |
| ***A. baumannii*** | B3 (*Podo*) | ASN73401 | Mice (bacteraemia) | 0% buffer-treated and 70% K2 depolymerase treated mice survived. | Oliveira et al., 2019b |
| | IME200 (*Podo*) | YP_009216489 | Mice (bacteraemia) | 0% buffer-treated and 100% capsule depolymerase treated mice survived. The knowledge of the specific capsular type is unknown. | Liu et al., 2019 |
| ***S. enterica*** | P22 (*Podo*) | NP_059644 | Chicken (intestine infection) | LPS O depolymerase that recognises serotypes A, B and D1, significantly reduced (at least 10-fold) bacteria in the cecum and internal organs. | Waseh et al., 2010 |
| ***P. aeruginosa*** | LKA1 (*Podo*) | YP_001522890 | *G. mellonella* (bacteremia) | 0% buffer-treated and 20% LPS O5 depolymerase-treated larvae survived. | Olszak et al., 2017b |

al., 1985; Orskov et al., 1977; Whitfield, 2006). The *E. coli* K1 type, for instance, is made of polysialic acid, a carbohydrate also presents in human epithelial cells, which cannot be recognized by the immune system, causing potentially lethal bacteraemia and meningitis in new-born infants. A K1 depolymerase was first discovered in 1985 (Tomlinson and Taylor, 1985) and several studies since showed that intraperitoneally injections (20-25 μg) could markedly reduce the mortality of neonatal rats infected by systemic infections of K1 *E. coli* (Mushtaq et al., 2004, 2005; Zelmer et al., 2010). The therapeutic effect was attributed not to the enzyme killing effect, but instead its ability to strip cells from their protective capsule, thus sensitizing the pathogen to the complement system. These encouraging early results have motivated other researchers to assess if similar outcomes could be achieved with more depolymerases targeting similar and different capsular types. In one study, K1, K5 and K30 depolymerase treatments via intramuscular injections 0.5 h after bacterial challenge protected mice from deadly *E. coli* systemic infections. A set of different doses were tested (2 to 20 μg), being 2 μg to 20 μg considered critical depending on the enzyme used, to achieve effective treatments. Interestingly, different K1 depolymerases had different therapeutic successes, being the results directly related to their structures. While both K1E and K1F depolymerases were successful and formed trimers, K1E depolymerase was not successful and produced 18-mers, instead of the expected trimer, a fact speculated to have limited the enzyme *in vivo* distribution. More recently, the same authors tested if different time treatments (0.5 h *vs* 8 h after challenge), route of administrations (intramuscular *vs* intraperitoneal) and mice models (healthy *vs* leukopenic) would affect the effectiveness of the same depolymerases (Lin et al., 2018). Results showed that delayed treatments were much less efficient for all tested enzymes. The different route of administrations also affected the treatment outcomes in some but not all depolymerases. Finally, K1 depolymerases also proved to be successful in protecting immune-suppressed, leukopenic mice infected with K1 *E. coli*, even with delayed treatments.

*K. pneumoniae* is an important human pathogen able to cause nosocomial and community-acquired infections. *K. pneumoniae* forms thick and highly diverse capsules. Currently, >130 capsular types exist, being some certain types associated with specific diseases. There are several studies that have validated the depolymerase anti-virulence effect *in vivo*. A K1 depolymerase sensitized the *Klebsiella* K1 strain to the serum complement activity and single intraperitoneal injections (25 μg) added 0.5 h after challenge protected mice infected with K1 *K. pneumoniae* (Lin et al., 2014). A K64 depolymerase enhanced both the serum and the neutrophil killing effect, and intraperitoneal injections (18.75 μg) added 1 h after challenge increased the survival rate of mice infected with K1 *K. pneumoniae* (Pan et al., 2015). However, the enzyme becomes ineffective when added 8 h after challenge and at higher concentrations. A K63 depolymerase demonstrated not to affect the antibiotic susceptibility of cells *in vitro* and to increase the survival rate of *Galleria mellonella* larvae infected with K63 *K. pneumoniae* in a time-dependent manner (Majkowska-Skrobek et al., 2016). Finally, K3 and K21 depolymerases sensitized cells to the complement-mediated killing and decreased the survival of phagocytized intracellular cells in macrophages. Similarly, enzyme treatments increased the lifespan of *G. mellonella* larvae infected with *K. pneumoniae* in a time- and strain-dependent manner (Majkowska-Skrobek et al., 2018).

*A. baumannii* has been recently listed in the priority pathogen list of the World Health Organization for the development of novel effective antimicrobials. Strains of this species are already resistant to most antibiotics and cause important nosocomial infections primarily in immunocompromised patients (Wong et al., 2017). In contrast with *E. coli* and *K. pneumoniae*, typing schemes of *A. baumannii* capsules are not available. Earlier studies showed that *Acinetobacter* phages encode depolymerases specific towards capsular types and that these structures are sometimes shared among species of the *Acinetobacter calcoaceticus-Acinetobacter baumannii* complex (Oliveira et al., 2017). Although some CPS depolymerases targeting different structures have been cloned (Oliveira et al., 2019a; Oliveira et al., 2017; Popova et al., 2017), only two were explored *in vivo* (Liu et al., 2019; Oliveira et al., 2017). A K2 depolymerase protected *G. mellonella* larvae from K2 *A. baumannii* infections, independent whether if added simultaneously or 0.5 h after bacterial challenge. The K2 depolymerase intraperitoneal injections (50 μg) also protected mice with septicaemia caused by K2 *A. baumannii*. Interestingly, and for the first time, the

K2 depolymerase demonstrated to be highly refractory to resistance upon selective pressure. Another CPS depolymerase also demonstrated to be therapeutically effective in *G. mellonella* and murine models of infection, although the authors did not describe the capsular type that was being degraded by the depolymerase (Liu et al., 2019). In both cases, depolymerases made cells susceptible to the serum killing, even in immunocompromised mice.

There are also successful treatments with LPS depolymerases that target other important bacterial pathogens. Among the best studied examples is that of the *Salmonella* phage P22 depolymerase, which is specific against *S. enterica* LPS O antigens (A, B, D1 serotypes) (Steinbacher et al., 1996). When the enzyme was orally administrated to chickens, it could reduce *Salmonella* colonization in the gut as well as their further penetration into internal organs (Waseh et al., 2010). The results were explained by the LPS hydrolytic activity of the depolymerase which negatively affects *Salmonella* mobility. Recently, an LPS O5 depolymerase has also been shown to sensitized O5 *P. aeruginosa* cells to serum complement activity and reduced their virulence in *G. mellonella* larvae, although with an inferior effect when compared with other tested depolymerases. Moreover, the enzyme did not affect the ciprofloxacin and gentamicin activity *in vitro* (Olszak et al., 2017b).

*Virion-associated lysins (VALs)*

Phages, during their adsorption step, either with or without previous digestion of bacterial polysaccharides by depolymerases, will eventually bind irreversibly to a specific bacterial cell receptor (Leiman et al., 2007). Such binding will induce changes in the phage particle structure ultimately triggering the ejection of its DNA (Fokine and Rossmann, 2014) that must enter the cell for efficient phage infection, but in-between there is still an obstacle: the peptidoglycan, another rigid carbohydrate barrier that comprises up to 40 layers in case of Gram-positive bacteria (up to 3 in case of Gram-negative bacteria) (Schleifer and Kandler, 1972). To surpass this obstacle, phages are armoured with specific enzymes, associated with their virion, which degrades peptidoglycan, called virion-associated lysins (VALs).

Although very similar to phage endolysins, VALs degrade the peptidoglycan only locally in order to produce a hole in the peptidoglycan that is large enough to allow translocation of the phage genome into the cytoplasm, but without compromising the cell (Kanamaru et al., 2002). Phage T4 was shown to drill the outer membrane of *E. coli* through its structural baseplate protein and further degrade the peptidoglycan layer enlarging the initial hole by its integrated lysozyme domain, allowing the phage genome to be injected (Kanamaru et al., 2002; Nishima et al., 2011). This ability to degrade the peptidoglycan locally maintaining the cell integrity is crucial for phage survival in order to allow the bacterial machinery to produce the phage progeny. Although, when multiple phages adsorb to a single bacterial cell, the local degradation of peptidoglycan at multiple sites will cause cell lysis in a phenomenon known as “lysis from without” that does not result in phage replication (Abedon, 2011). The phage T4 baseplate was shown to be required for “lysis from without” and mutations on this baseplate protein with peptidoglycan enzymatic activity reduced the phage ability to produce this phenomenon (Kao and Mcclain, 1980). The presence of a peptidoglycan barrier in bacteria should turn these enzymes into a prerequisite in phages. Even so, the activity of such an enzyme is not required for some phages when infecting bacteria under laboratory conditions but phages seem to benefit from this activity when infecting cells under physiological conditions that promote murein thickening or increased cross-linking (consequently leading to a decreased porosity of the peptidoglycan) (Moak and Molineux, 2000; Piuri and Hatfull, 2006; Rodriguez-Rubio et al., 2013c) .

Due to their association to the phage particle, these enzymes are frequently known as virion-associated peptidoglycan hydrolases (VAPGHs) (Rodriguez-Rubio et al., 2013b), tail-associated muralytic enzymes (TAMEs) (Paul et al., 2011), and structural lysins (Oliveira et al., 2013), but all these terms may mislead. Not all are hydrolases, not all are associated with the tail, and not all are a component of the phage structure. Some are transglycosylases (Moak and Molineux, 2000), some are located at the phage neck (Tang et al., 2011) or baseplate (Kanamaru et al., 2002), and some may be enclosed in the phage capsid (Hu et al., 2013) free or linked to the phage genomic DNA (Moak and Molineux, 2004). Exolysins is another term used but which is commonly associated with enzymes secreted by bacteria (Schmelcher et al., 2012) and thus also discouraged. The most correct and consensual term for these enzymes is "Virion-Associated Lysins", or simply VALs.

Location, structure and enzymatic activity

Being enzymes associated to the virion, VALs are frequently structural components, or domains thereof, of the phage particle (or associated to it), mainly tail proteins (tail fibres, tail measure protein, tail tip) as in the *Lactococcus lactis* phage TP901-1 (Mahony et al., 2016) and *E. coli* phage T5 (Boulanger et al., 2008; Zivanovic et al., 2014). Other structural components include the baseplate, as in the case of the *E. coli* phage T4 (Kanamaru et al., 2002) and *Pseudomonas* phage phiKZ (Briers et al., 2008), the capsid in the coliphages PRD1 (Rydman and Bamford, 2000, 2002), phi6 (Caldentey and Bamford, 1992) and T7 (Moak and Molineux, 2000), and the phage neck being the only example described the *Salmonella* phage P22 (Tang et al., 2011). Cases exist where the VAL is not part of the virus structure but enclosed inside the phage capsid, associated or not with the phage DNA genome. VALs of phages T7, T3, phiYeO3–12 and SP6 are examples of internal head proteins (Table 3) (Moak and Molineux, 2004; Pajunen et al., 2001; Serwer et al., 2014; Struthers-Schlinke et al., 2000). The VAL of phage phi29 is also enclosed on the phage capsid covalently bound to the genomic DNA which is ejected prior to the DNA in order to digest the peptidoglycan and allowing the entry of the phage genome into the cell (Moak and Molineux, 2004). Interestingly, a small portion of this VAL is found free inside the capsid (not linked to the phage DNA), which aids the linked portion on the peptidoglycan digestion.

They seem to be present in all phage families and although being in their majority phage structural proteins, recombinant expression of these VALs has been successful in *E. coli* allowing their study. Even so, when compared to their close related endolysins, knowledge on their structure, muralytic and antibacterial activity is much scarcer. They present a high diversity, with molecular weights ranging from 37 kDa (P7, the internal membrane of the *Pseudoalteromonas* phage PM2) (Kivela et al., 2004) to 252 kDa (CwlP, from the *Bacillus* prophage SP-β) (Sudiarta et al., 2010). Also, VALs tend to be

**Table 3.** VALs described in the literature (experimentally validated or not). The VALs location, corresponding phage name and family (without the suffix *viridae*) is indicated. Monomer mass in kDa (Mass), oligomerization (oligom), the NCBI Reference Sequence accession number (Accession n.), and enzymatic activity of the protein is also given. ND, not determined.

| Host | Phage (*family*) | | Location | Accession n. | Mass (oligom)* | Enzymatic activity | |
|---|---|---|---|---|---|---|---|
| ***Bacillus subtilis*** | GA-1 | (*Podo*) | associated with DNA | (NP_073686) | 30 | lysozyme | Moak and Molineux, 2004 |
| | phi29 | (*Podo*) | associated with DNA in free form and packed within capsid | (YP_002004530) | 31 (dimeric) - 2EX3 | lysozyme | Moak and Molineux, 2004 |
| | M2 | (*Podo*) | associated with DNA | | 31 | lysozyme | Moak and Molineux, 2004 |
| | SP-β | (*Sipho*) | tail | (NP_046584) | 252 | lysozyme and endopeptidase | Sudiarta et al., 2010 |
| ***E. coli*** | T3 | (*Podo*) | internal head protein | (NP_523341) | 143 (trimer) | lytic transglycosylase | Serwer et al., 2014 |
| | T7 | (*Podo*) | internal head protein | (NP_042004) | 143 (trimer) | lytic transglycosylase | Molineux, 2001 |
| | K1–5 | (*Podo*) | tail | (YP_654136) | 109 | lysozyme | Moak and Molineux, 2004 |
| | T5 | (*Sipho*) | tail tip | (YP_006968) | 124 (multimeric) | unknown | Boulanger et al., 2008 |
| | T4 | (*Myo*) | baseplate | (NP_049757) | 63 (homotrimer) - 1PDL | lysozyme | Arisaka et al., 2003 |
| | PRD1 | (*Tecti*) | membrane-associated protein | (YP_009639979) | 27 (multimeric) | lytic transglycosylase | Rydman and Bamford, 2000 |
| | | | | (YP_009639980) | 17 (multimeric) | lytic transglycosylase | |
| ***Lactococcus lactis*** | Tuc2009 | (*Sipho*) | tail tip | (NP_108727) | 101 (trimer) | endopeptidase | Kenny et al., 2004 |
| | TP901–1 | (*Sipho*) | tail tip | (NP_112710) | 102 (trimer) | endopeptidase | Stockdale et al., 2013 |
| | PM2 | (*Cortico*) | lipid core-associated, integral membrane protein | (NP_049902) | 37 | unknown | Kivela et al., 2004 |

| | | | | | | | |
|---|---|---|---|---|---|---|---|
| ***P. aeruginosa*** | phiKMV | (*Podo*) | internal head protein | (NP_877475) | 115 | lysozyme | Briers et al., 2006 |
| | phiKZ | (*Myo*) | needle-like cell-puncturing device | (NP_803747) | 246 (trimer) | glycosidase (lysozyme or lytic transglycosylase) | Briers et al., 2008 |
| ***Pseudomonas savastanoi*** | phi6 | (*Cysto*) | nucleocapsid surface (outside) | (NP_620343) | 24 (monomer) | endopeptidase | Caldentey and Bamford, 1992 |
| ***Ralstonia solanacearum*** | phiAP1 | (*Podo*) | internal virion protein | (APU03186) | 99 | lysozyme | da Silva Xavier et al., 2018 |
| | | | internal virion protein | (APU03187) | 138 | lytic transglycosylase | |
| ***Salmonella typhimurium*** | SP6 | (*Podo*) | internal head protein | (NP_853596) | 107 | lysozyme | Moak and Molineux, 2004 |
| | P22 | (*Podo*) | neck (head-tail connector) | (NP_059632) | 18 (27-meric) 5GAI | lysozyme | Moak and Molineux, 2004; Tang et al., 2011 |
| ***S. aureus*** | K | (*Myo*) | tail | (AAO47505) | 91 | amidase/peptidase | Paul et al., 2011 |
| | phiP68 | (*Podo*) | ND | (NP_817333) | 75 (11-meric) 6IAC | unknown | Takac and Blasi, 2005 |
| | phiMR11 | (*Sipho*) | tail tip | (YP_001604152) | 71 | amidase (N-term) and lysozyme (C-term) | Rashel et al., 2008 |
| | phiIPLA88 | (*Sipho*) | baseplate | (YP_002332533) | 72 | amidase (N-term) and lysozyme (C-term) | Rodriguez-Rubio et al., 2012; Rodriguez et al., 2011 |
| | 187 | (*Sipho*) | | (YP_239513) | 71 | amidase | Yang et al., 2014 |
| | SA2 | (*Sipho*) | tail | (AWY02947) | 58 | endopeptidase | Wang et al., 2019 |
| | UPMK_1 | (*Sipho*) | tail | (ATW69216) | 71 | endopeptidase | Dakheel et al., 2019 |
| | UPMK_2 | (*Sipho*) | tail | (AUM57559) | 71 | endopeptidase | Dakheel et al., 2019 |
| | phi11 | (*Sipho*) | baseplate | (NP_803302) | 72 | lysozyme and amidase | Li et al., 2016; Moak and Molineux, 2004; Rodriguez-Rubio et al., 2013c |
| | | | baseplate | (NP_803297) | 71 | endopeptidase (N-term) and lipase or esterase (C-term) | |
| ***Staphylococcus pseudintermedius*** | QT1 | (*Sipho*) | tape-measure protein | (QBJ05125) | 222 | M23 peptidase | Zeman et al., 2019 |
| ***Yersinia enterocolitica*** | phiYeO3–12 | (*Podo*) | internal head protein | (NP_052116) | 143 | lytic transglycosylase | Moak and Molineux, 2004 |

* The Protein Data Bank ID may be indicated on the Mass (oligom) column after (oligom).

multifunctional and large proteins with some forming oligomers (usually trimeric) and sometimes playing key roles in the morphogenesis, stability and infectivity of the phage virus particles (Kanamaru et al., 2002; Rodriguez-Rubio et al., 2013c; Stockdale et al., 2013).

VALs can present specificity and activity onto the different bonds composing the peptidoglycan allowing for their classification into three classes: (i) glycosidases (which include lysozymes, glucosaminidases and lytic transglycosylases), that cleave one of the two glycosidic bonds in the glycan chain; (ii) amidases, specific for the amide bond between the lactyl group of N-acetylmuramic acid and L-alanine of the stem peptide; and (iii) endopeptidases, that cleave within the stem peptide or the cross-link (Latka et al., 2017).

Looking into the VALs reported to date, they belong in their majority to the glycosidase class, which includes the well-known lysozyme VAL from phage T4 (Yap and Rossmann, 2014) and the lytic transglycosylase VAL from phage T7 (Moak and Molineux, 2000). Of note is to mention that the lytic transglycosylases do not compose hydrolases since a water molecule is not used during the enzymatic cleavage of these enzymes (Scheurwater et al., 2008). The less common amidases include VALs from the *Staphylococcus* phages phiMR11, phiIPLA88 and phi11 (Rashel et al., 2008; Rodriguez et al., 2011). Endopeptidases are also frequent and almost exclusively found in phages infecting Gram-positive bacteria as those from phages K, phi11, Tuc2009, TP901–1, and SP-β (Paul et al., 2011; Stockdale et al., 2013; Sudiarta et al., 2010).

The diversity of VALs is also reflected in their structure, specifically on the number of domains and their organization. Their modular structure is usually composed of a domain that makes the linkage to the phage particle and a domain with peptidoglycan enzymatic activity without a fixed order. A few VALs have been identified with two lytic activity domains combining different enzymatic activities (lysozyme and endopeptidase, endopeptidase and hydrolase, lysozyme and amidase), an architecture that seems to be restricted to phages infecting Gram-positive bacteria (Li et al., 2016; Olszak et al., 2017b;

Rashel et al., 2008; Rodriguez-Rubio et al., 2012; Rodriguez et al., 2011; Sudiarta et al., 2010). The reason for the presence of double lytic domains might rely on the enhanced enzymatic degradation that they offer for the thicker peptidoglycan found in Gram-positive cells. A different architecture was found on the VAL of the staphylococcal phage P68, which possesses an additional domain responsible for binding to the cell surface in what seems to be an isolated case (Takac and Blasi, 2005). Exceptionally, phages may also possess two different VALs on their virion particle, being *E. coli* phage PDR1, *Staphylococcus* phage phi11 and the *Ralstonia* phage phiAP1 the only reported cases (Moak and Molineux, 2004; Rydman and Bamford, 2000, 2002).

Although closely related to endolysins, there are many features that discriminate VALs. At first glance, the obvious difference is the fact that VALs, as suggested by their name, are peptidoglycan degrading enzymes that are associated with the viral particle. They are larger proteins (37–252 kDa) when comparing with the typical sizes of endolysins (15–40 kDa), and usually form oligomers (Kivela et al., 2004; Sudiarta et al., 2010; Tang et al., 2011). This is a consequence of their function in the phage replication cycle. VALs need to digest the peptidoglycan from the outside of bacterial cells that are actively growing in order to allow phage genome ejection. Consequently, they need to be present in the phage particle (associated to), often composing specific phage structures that require larger proteins and their oligomerization. On the other hand, endolysins degrade the peptidoglycan from the inside to release the phage progeny from a cell that is already dying, aided by other phage proteins that produce holes in the inner membrane to allow access of endolysins to the peptidoglycan (Young, 2013). By sharing the same substrate, the peptidoglycan, VALs and endolysins also share the same enzymatic domains (ED), but the M23 endopeptidase family is usually found only in VALs, as well in other bacterial enzymes that digest the peptidoglycan from the outside (*e.g.* lysostaphin and enterolysin A), something that might explain this difference. The presence of a cell-wall binding domain (CBD) is something that is attributed to endolysins (characteristic from phages infecting Gram-positive bacteria) but which is not found in VALs (with the mentioned exception of the staphylococcal phage P68 VAL) constituting another difference between these enzymes (Takac and Blasi,

2005). The association of VALs to the phage particle requires endurance under the harsh conditions found in phage habitats otherwise the phage infectivity, and consequently their survival, could be compromised. Consequently, VALs have evolved to be highly thermostable enzymes, able to resist temperatures up to 100 °C and even conditions of autoclaving (Lavigne et al., 2004).

### Activity screening assays

Assessment of the enzymatic activity of VALs is not as challenging as that of depolymerases since there is less variability in the peptidoglycan composition than in the substrates of depolymerases (EPS, CPS, O-polysaccharide). Even so, there is still no standard procedure. As for their cognate endolysins, a number of protocols can be found in the literature and rely on the effect of VALs on purified peptidoglycan or whole cells (either dead or alive). The method of choice will depend on the aim, laboratory logistics, easiness of implementation and state of enzyme characterization, leading to results that can be qualitative or quantitative.

The initial characterization of a VAL, in which its activity needs to be validated, often involves the drop test or spot-on-lawn test. This is a simple, inexpensive and fast method, which allows for a qualitative measure resulting in the determination of the lytic spectrum of the enzyme when testing against different strains (Donovan et al., 2006). Similar to what was described for depolymerases, in this method a small drop of a VAL solution is spotted onto a bacterial lawn, incorporated or not in the agar. The drop may be added right after overlaying the bacteria or after the bacterial lawn has been formed and the appearance of inhibition halos is a result of the enzyme lytic activity through growth inhibition or bacteriolysis in a few hours or minutes after incubation, respectively (Proenca et al., 2015). When testing against Gram-negative bacteria an additional treatment with chloroform has to be performed to permeabilize the outer membrane so that the VAL may reach the peptidoglycan layer (Oliveira et al., 2014). Worth mentioning, with Gram-positive bacteria, the use of heavily dense bacterial lawns may lead to false-negatives.

Spectrophotometry is probably the most used method, better known as turbidimetry. Here, the lytic activity of the VAL is measured by the rate of reduction of the optical density overtime of the enzyme substrate, which might be live cells, dead cells or extracted peptidoglycan (Becker et al., 2009; Donovan et al., 2006; Donovan and Foster-Frey, 2008; Rodriguez-Rubio et al., 2012). This quantitative method calculates the steepest slope of the killing curve or the necessary amount of enzyme to drop a certain percentage of the optical density when compared with the initial value. As in the case of the spot-on-lawn test, when using Gram-negative cells, a pre-treatment with chloroform is mandatory to permeabilize the outer membrane and to allow the VAL to surpass this barrier (Lavigne et al., 2004). An important observation on this method when using cells as a substrate is the fact that often they are prepared in a buffer and thus the results cannot reflect the potential of VALs *in vivo* as such conditions do not allow bacterial growth. Moreover, the lack of a standard calculation of the enzyme activity coupled with the day-to-day variability of the determined values and consequent inconsistency, makes it difficult or even impossible to compare results from different studies (Donovan and Foster-Frey, 2008). To avoid this, Briers and colleagues proposed a standardized approach to determine the enzymatic activity of peptidoglycan degrading enzymes against Gram-negative strains that uses the linear region of the turbidity reduction to determine the enzymatic activity (Briers et al., 2007).

Enzymatic activity of VALs is also commonly demonstrated through zymogram analysis, an electrophoretic technique that detects the presence or absence of enzymatic degradation of the substrate, cells or peptidoglycan, by the appearance or not of transparent bands in zymogram gels, indicative of the presence of lysis zones (Moak and Molineux, 2004; Rodriguez-Rubio et al., 2012). This methodology directly detects the presence of enzymatic degradation but the result is only qualitative.

The antibacterial potential of VALs is calculated through cell viability, usually determined by measuring the reduction of the number of bacterial Colony Forming Units (CFUs) as a result of the enzymatic activity of the VAL, when compared with the control (Donovan et al., 2006). Briefly, a known concentration of live bacterial suspension is challenged with a known concentration of the VAL and incubated for some minutes or hours. The antibacterial activity is then expressed as reductions of orders of magnitude or a percentage of killing allowing an understanding of the bacterial load that the VAL is able to treat. Even so, and because this assay is usually carried with bacteria suspended in a buffer, as referred for the turbidimetry assay, the results cannot reflect the potential of VALs *in vivo*. Also here, the activity measurement against Gram-negative bacteria requires a previous step of membrane permeabilization (Oliveira et al., 2016; Plotka et al., 2015).

To allow the comparison with antibiotics and understand the potential of VALs as alternatives to those drugs, or even their synergistic effect in combined therapy, the minimum inhibitory concentration (MIC) assay, routinely used with antibiotics, is the best and a common choice (Briers et al., 2014; Daniel et al., 2010; Diez-Martinez et al., 2015). Besides, this assay can be used to set the initial dosage for *in vivo* trials (Gilmer et al., 2017). By testing serial dilutions of a VAL solution against a fixed bacterial load suspended in growth media, the MIC test allows determination of the minimum enzyme concentration that eradicates the bacteria into an undetectable number of CFUs (Ceri et al., 1999; Shen et al., 2013). This assay can also be modified to determine the minimum biofilm eradication concentration (MBEC), allowing to understand and determine the effect of VALs in biolfilms.

### Virion-associated lysins *in vivo* applications

The ability of VALs to kill bacteria when added externally and their resemblance to endolysins has hypothesized their therapeutic use as a novel antibacterial. As endolysins, VALs application at this time point seems to be restricted to treat infections caused by Gram-positive bacteria due to the presence of the outer membrane in Gram-negative bacteria that precludes VALs to achieve the peptidoglycan when externally added. Despite the potential application of these enzymes being the same as endolysins, the number of studies reporting the application of VALs is far lower. In fact, when we look for those that report *in vivo* studies, the number is quite scarce and, to our knowledge, none exists assessing a natural VAL *in vivo*. The existing ones (Table 4) refer all to chimeric proteins derived from only two VALs encoded on the genomes of phages infecting *S. aureus* and thus the therapeutic potential was assessed on this problematic pathogen.

**Table 4.** Viron-associated lysin *in vivo* applications. The VALs corresponding phage name and family (without the suffix *viridae*) and the NCBI Reference Sequence accession number (Accession n.) is indicated.

| Host | Phage (*family*) | Protein accession n. | Model (infection) | Main outcome* | Ref. |
|---|---|---|---|---|---|
| ***S. aureus*** | K (*Myo*) | tail VAL (AAO47505)+ SH3b of lysostaphin | Rat (nasal colonization) | 44.4% rats were completely decolonized with P128, while all rats in the control remained colonized. Reduction of two orders of magnitude in the CFUs of MRSA. | Paul et al., 2011 |
| | K (*Myo*) | tail VAL (AAO47505)+ SH3b of lysostaphin | Rat (bacteraemia) | Control with 80-100% mortality and abscesses in the kidneys. P128 administration resulted in up to 100% survival, with very few or no abscesses in kidneys. | Channabasappa et al., 2018 |
| | 187 (*Sipho*) | Ply187 (CAA69022)+ SH3b from LysK, the endolysin of phage K | mice (endophthalmitis) | Intravitreally administration resulted in 90-95% retention of eyes, significant reduction of the bacterial burden, reduced cell death as well as reduced inflammation, and maintained the visual function | Singh et al., 2014 |
| | 187 (*Sipho*) | Ply187 (CAA69022)+ CBD of phage phiNM3 VAL | Mice (bacteraemia) | The intraperitoneal administration of ClyH 3 h after infection could rescue 100% of the mice from the deadly infection. | Yang et al., 2014 |
| | 187 (*Sipho*) | Ply187 (CAA69022)+ CBD of phage phiNM3 VAL | Mice (bacteraemia) | ClyF was able to rescue 100% of the infected mice. A reduced bioburden of up to 3.3 orders of magnitude as a result of biofilm degradation. | Yang et al., 2017 |

*S. aureus* is a ubiquitous pathogen that colonizes the nasopharynx, skin, and gastrointestinal tract, but also an opportunistic that causes severe infections such as bacteraemia, infective endocarditis, osteomyelitis, skin and soft tissue infections, pneumonia, and device-related infections that can progress to life-threatening diseases (Boucher et al., 2010; Chuang and Huang, 2013; Daum, 2007; Holland et al., 2014; Lowy, 1998). It is also a leading cause of both community and health care acquired infections (Thwaites et al., 2011). In addition to the severity of these infections, *S. aureus* is able to rapidly develop antibiotic resistance with methicillin-resistant *S. aureus* (MRSA) being a serious problem worldwide (Miller et al., 2005; Mylotte et al., 1987).

The VAL of the *Staphylococcus* phage K was found to be a tail protein with muralytic activity at its C-terminus. By combining an N-terminal truncated version of the VAL (Lys16) with the *Staphylococcus*-specific CBD SH3b of lysostaphin, the chimeric protein P128 was obtained which possessed an improved bactericidal activity against *Staphylococcus* (Paul et al., 2011). Intranasal application of a 50 mg hydrogel, containing 100 µg of P128, twice a day (and for two days), after 24 h of a rat nares MRSA infection, was able to completely decolonize four of nine (44.4%) rats, while all rats in the control remained colonized. Also, on average, a two order of magnitude reduction with the CFUs of MRSA was observed. These results demonstrate the high potential of VALs to control pathogenic bacteria *in vivo*. A combined Phase I and Phase II (ClinicalTrials.gov Identifier NCT01746654) are being carried by GangaGen to assess the safety and efficacy of P128 as a topical treatment.

The efficacy of P128 was also evaluated in a newly developed rat model of *S. aureus* bacteremia (Channabasappa et al., 2018). In this model, 80-100% mortality is observed by day 14 and abscesses in the kidneys were detected. The treatment of the infected animals with P128 either by single intravenous bolus administration of 2.5 mg/kg via a tail vein or by 1 h infusion of 10 mg/kg via the jugular vein at 2 h post-infection resulted in 100% and 84% survival, respectively, with very few or no abscesses in the kidneys. This study shows that P128 can protect from lethal bacteremia caused by *S. aureus*, demonstrating its therapeutic potential and that this chimeric VAL deserves further experiments to assess treatment efficacy in bacteremia in humans.

P128 seems to be the only VAL reported for which *in vivo* therapeutic potential was assessed. Although, Ply187 which was presented as an *S. aureus* endolysin is likely a VAL due to its genetic organization and high similarity to other VALs (Sao-Jose, 2018) and for which *in vivo* studies were conducted. The chimaera Ply187A-NKSH3b was constructed by combining the ED of the Ply187

VAL (endopeptidase) with the CBD (SH3b-type) of LysK, the *S. aureus* phage K endolysin (Singh et al., 2014). The chimeric protein presented an improved antimicrobial activity, when compared to the wild-type full-length ply187 VAL, the truncated versions, and to the LysK itself and it was able to disrupt biofilms formed by antibiotic-resistant *S. aureus* strains. Importantly, no resistance development was observed. This new enzyme was evaluated for its therapeutic efficacy on an *S. aureus* endophthalmitis mouse model. The chimeric Ply187 was administered at 1 µg/eye intravitreally at 6 h (group I) or 12 h (group II) post-infection and results revealed 90-95% retention of eyes, a significant reduction of the bacterial burden, reduced cell death as well reduced inflammation, and preservation of the retina integrity maintaining the visual function. Taken together, the results demonstrate the potential of Ply187 treatment in improving the endophthalmitis disease outcome.

A novel chimeric VAL, ClyH, was constructed by fusing the Ply187 N-terminal with the C-terminal of the *Staphylococcus* phage phiNM3 VAL and its antimicrobial potential assessed on a mouse model (Yang et al., 2014). In this model, the intraperitoneal injection of MRSA leads to 100% mortality within 24-48 h. The chimeric VAL was found to present a MIC against clinical MRSA as low as 50 µg/L to 1610 µg/L and the intraperitoneal administration of 900 µg of ClyH 3 h after infection could rescue 100% of the mice while the infected control group resulted in 100% death at 24 h. The ClyH was able to protect mice from death caused by MRSA, without obvious toxicity or other harmful effects, suggesting ClyH as an alternative therapeutic agent for the treatment of infections caused by MRSA.

This same VAL, PlyP187, was used to construct a new chimaera formed by the N-terminal part of the PlyP187 (the same used for ClyH) and the CBD from the PlySs2 endolysin, giving rise to ClyF (Yang et al., 2017). Using the same mouse model as ClyH, all mice infected with MRSA of the untreated group died within 24 h but a single intraperitoneal administration of a high dose (50 mg/kg) of ClyF or repeated intraperitoneal administration of a lower dose (25 mg/kg/day, for five continuous days) was able to rescue 100% of mice. This same chimaera was tested against a mouse model of burn wounds infected by MRSA, through a single or double (in a 6 h interval) application of 100 µg of ClyF. Burn wounds infected with MRSA may progress into septicaemia and eventually death. The administration of ClyF resulted in a reduced bioburden of up to 3.3 orders of magnitude as a result of biofilm degradation. Dosing ClyF twice demonstrated much better antibacterial activity showing that repeated treatment may be needed when applying ClyF topically.

## Synthetic proteins: Improving activity through engineering

Phages have co-evolved with their hosts for millions of years to efficiently predate and parasite bacteria in an efficient manner. To accomplish this, they have collected and ameliorated their protein armament. Even so, phages continue to evolve beside their host constituting an almost perfect machine programmed to kill bacteria, However it is still possible to improve the characteristics of both proteins and phage particles. The high diversity of phage proteins or even their enzymatic and binding domains open the possibility to genetically engineer not only the phage proteins but also the phage itself by domain and protein shuffling respectively. Domains can be switched to generate chimeric proteins, while proteins can be switched to generate chimeric phages. Nature has provided scientists with the raw material (phages) and protein engineering offers the techniques to improve both the proteins and phages. While engineering of VALs has been focused on improving their properties, depolymerase engineering focused on the improvement of phages. This *a la carte* design can lead to the creation of proteins and phages with enhanced antimicrobial activity against specific pathogenic bacteria.

### *Depolymerase engineering*

Depolymerases have been exploited through genetic engineering to increase the anti-bacterial activity of fruits, phages and bacterial bacteriocins. In fruits, there is one elegant study where transgenic pears were constructed to express depolymerases, which led to a higher immunity against fire blight *in vitro* and greenhouse experiments (Malnoy et al., 2005). In phages, some examples are described in which swapping depolymerase-encoding genes granted new host ranges, offering the possibility to design tailor-made phages for specific applications. This was demonstrated with the *Acinetobacter* phages phiAB1 an phiAB6, offering the possibility of the synthetic phage to infect a new *Acinetobacter* capsular type (Malnoy et al., 2005). There are also studies aimed at increasing the phage anti-biofilm activity through expression of depolymerases during their lytic cycle. This was first demonstrated with the *E. coli* phage T7 to express dispersin B, a depolymerase of bacterial origin. The synthetic phage reduced biofilms by two orders of magnitude higher when compared with the non-enzymatic wild-type phage (Lu and Collins, 2007). Later, a similar concept was applied to an *Erwinia* phage *Y2::dpoL1-C*, a non-enzymatic phage equipped to express a depolymerase (Born et al., 2017). The synthetic phage greatly enhanced the bacterial killing and significantly reduced the ability of *E. amylovora* to colonize the surface of detached flowers. Finally, there is one innovative study that constructed a chimeric polysaccharide depolymerase by genetically fusing it with an R-type pyocin (*Pseudomonas* bacteriocin) (Scholl et al., 2009). This experiment demonstrated that it is also possible to add lytic activity to a phage depolymerase, as the chimeric protein efficiently and specifically kill 100% of the *E. coli* O157:H7 isolates, but no other serotypes tested.

*VAL engineering*

The described intrinsic properties of VALs turns them into efficient and powerful weapons in the fight against bacterial pathogens, including multi-resistant strains. They have a limited host range avoiding to kill beneficial bacteria, resistance development and spread is not likely, since they target highly conserved bonds in the peptidoglycan and they are efficient in killing antibiotic resistant bacteria and with the same efficiency as antibiotic sensitive bacteria due to a different mode of action to that of antibiotics. Although sharing many characteristics with endolysins, the native mode of action of VALs poses an advantage over those since they were designed to degrade the peptidoglycan from the outside. Moreover, VALs are known for their thermostability, ionic strength tolerance and can act synergistically with antibiotics and endolysins (Latka et al., 2017). Even so, there is still a wide room available for improving their efficiency in killing pathogenic bacteria. The common modular structure of VALs allows to swap and combine different domains to create novel chimeric proteins with better antimicrobial properties by taking advantage of the domain functions.

One limitation of VALs is their often lack of a CBD which may compromise their performance and thus, the majority of engineered VALs have been designed to accommodate a CBD to the full length of the VAL or to its ED. Engineered VALs have shown to present improved results over the natural ones in such a way that the only *in vivo* studies reported for these enzymes (see above) include only chimeric proteins obtained through the combinations of VALs with other specific proteins or peptides. The chimeric protein P128 (a combination of Lys16, the truncated VAL from the *Staphylococcus* phage K, with the staphylococcal CBD of lysostaphin) was able to reduce the CFU of an MRSA suspension by three orders of magnitude when using 750 ng/ml, while the parental Lys16 could not achieve such reduction even when using a 100-fold higher concentration (Paul et al., 2011). The application of P128 reduced significantly the turbidity of bacterial cultures through cell lysis in just 15 minutes and importantly, in a specific way, targeting only *Staphylococcus* and including MRSA strains. Reductions on cell counts of up to 5 orders of magnitude were reported and the application of a hydrogel containing P128 *in vivo* in a rat *S. aureus* nasal colonization model completely or significantly reduced cell numbers. This chimeric VAL is currently under testing in a clinical trial.

A similar chimaera was constructed by combining the highly thermostable HydH5 VAL of *S. aureus* phage philPLA88 with the SH3b CBD of lysostaphin (Rodriguez-Rubio et al., 2013a). The obtained HydH5SH3b presented an almost 2-fold enhanced and specific staphylolytic activity when compared with the parental VAL but using only the endopeptidase domain of HydH5 with the same SH3b domain of lysostaphin (CHAPSH3b) resulted in an even higher activity (more than 3-fold) against *Staphylococcus* in zymograms, plate lysis and turbidity reduction assays. Moreover, an antimicrobial synergy with the endolysin LysH5 was observed. Interestingly, the CHAPSH3b chimaera retained some lytic activity after pasteurization and after storage at 4°C for 3 days showing its high thermostability. Its addition to raw milk immediately reduced *S. aureus* CFUs to undetectable levels without an observable re-growth during the assay. CHAPSH3b was able to inhibit biofilm development as well to reduce CFUs of total adhered biomass of preformed biofilms, demonstrating its potential to control *S. aureus* biofilms, even considering that the assays were carried under conditions that did not allow bacterial growth (Fernandez et al., 2017).

A chimeric protein was constructed by fusing the catalytic N-terminus CHAP domain of the staphylococcal phage 187 (Ply187AN) with the SH3b CBD of the well-known and strong staphylococcal endolysin from phage K (LysK) (Mao et al., 2013). In the plate lysis assay, the chimaera could produce a zone of clearing with just 0.003 µmol of protein while the parent Ply187AN and LysK needed 0.5 µmol and 0.02 µmol respectively. Also, the lytic spectra, which included staphylococcal strains isolated from bovine mastitis, MRSA strains and coagulase negative staphylococci, the chimaera exhibited higher activity than Ply187 and LysK and with specificity to the *Staphylococcus* genus (inactive to other genera). Although presenting a higher MIC than LysK (but lower when compared to Ply187AN), *in vitro* activity assays in milk inoculated with *S. aureus*, Ply187AN-KSH3b immediately eradicated the bacterial population (and without observed growth for at least 3 h), while Ply187-AN and LysK showed no significant difference compared to buffer alone. These results showed that the chimeric Ply187AN-KSH3b is a more effective antimicrobial agent than the full-length Ply187 or the truncated Ply187 (Ply187AN), and it also outperformed the known high-activity of the LysK endolysin, with a noticeable superior activity in complex matrices (milk).

Another chimeric protein based on the Ply187 ED and the non-SH3b CBD of the *Staphylococcus* phage phiNM3 endolysin, ClyH, also produced a rapid decrease in turbidity and a cell wall damage of *Staphylococcus* cells in suspension, including methicillin-resistant strains (Yang et al., 2014). The lytic activity of ClyH was found to be higher than lysostaphin, lysing all tested clinical MRSA isolates, including 2 strains insensitive to lysostaphin. Besides this broader range when compared to lysostaphin it also presented a higher activity than the parental Ply187 and even higher than the Ply187 ED (up to 13.6-fold), comparable to that of Ply187AN-KSH3b. This higher lytic activity coupled with the observable low MIC validates the potential use of ClyS as an antimicrobial agent for the treatment of infections caused by MRSA *in vivo*.

A random combination of seven EDs and three CBDs gave rise to a library of 126 chimaeras from which, one was distinguished for its lytic ability against *S. aureus*, being active against all staphylococcal isolates tested in a buffer (Yang et al., 2017). This clone was found to be a result of the combination between the ED of the Ply187 VAL and the CBD from the PlySs2 endolysin (from a *Streptococcus suis* prophage) and was named ClyF. This chimeric protein at a concentration of 25 µg/ml was able to decrease the turbidity of an *S. aureus* suspension from 1.0 to 0.1 in just 5 min, resulting in a CFU reduction of 3 orders of magnitude. The fast decrease in turbidity produced by ClyF contrasted with the 15 min lag phase observed for the parental PlySs2 endolysin and an improved lytic rate up to 8-fold higher was determined for ClyF. Although a comparable lytic activity was determined for ClyF and the ED of Ply187, with much higher activity was observed for ClyF in milk, serum and on pigskin. Besides the absence of neutralization, the remarkable *S. aureus* CFUs reduction in these bio-matrices ascended to seven orders of magnitude, an activity not observable with many other staphylococcal VALs or endolysins.

While endolysins usually present a high lytic activity when tested on conditions that do not support robust bacterial growth, actively growing bacteria usually pose a certain level of resistance to endolysins. This fact seems to be related with the nature of endolysins that are programmed to lyse cells at the end of the lytic cycle when they are already dying (Catalao et al., 2013). On the other hand, VALs exert their natural enzymatic activity on actively growing bacteria, presenting thus a marked advantage against bacteria in active growing conditions. This knowledge was the rationale behind the design of a new chimeric protein, the EC300 (Proenca et al., 2015), composed of a fusion between the M23 endopeptidase domain of the *Enterococcus* phage F170/08 VAL and the endolysin (Lys170) from the same phage. While the endolysin Lys170 presented a broad spectrum and high lytic activity in buffer, a very poor or no lytic activity was observed when testing the enzyme in rich media containing actively growing cells. The chimeric EC300 provoked faster and more extensive lysis than the endolysin in a buffered suspension of *E. faecalis*, and more important, EC300 was still able to elicit lysis in cell suspensions prepared in a fresh TSB culture medium. Although actively growing *E. faecalis* cells are intrinsically resistant to Lys170 attack from the outside, an improved lytic and killing performance was observed when testing the chimeric EC300 on these cells. Growth of almost all tested *E. faecalis* clinical strains was inhibited, including strains with a high-level resistance to gentamicin and vancomycin.

The few existing examples of engineered VALs are reported in the literature and presented here. This works seem to be restricted to work involving the creation of chimaeras resulting from the fusion of the ED of a VAL and the CBD from an endolysin. Such an approach has enabled the creation of chimaeras with improved features as extended lytic spectra, higher and faster lytic activity, ability to control pathogenic bacteria in complex bio-matrices and harsh conditions, and activity on actively growing cells. These features allow the application of engineered VALs in a number of situations including disinfection of surfaces, food industry, agriculture, animal and human health. Consequently, VALs engineering is a promising strategy for developing novel antibacterials with improved properties for a wide range of applications.

## Conclusions and Perspectives

Over the last few years, phage depolymerases and VALs have been used as alternative agents to control bacterial pathogens. Polysaccharide depolymerases have demonstrated to be powerful anti-virulence weapons. Their effect can thwart the innate immunity of bacterial pathogens, but their treatments may need to be individually optimized. Given the specificity of these enzymes to a highly diverse group of CPS and LPS antigens, fast typing schemes and epidemiological studies are necessary to narrow down the most important bacterial virulent antigens, and subsequently, the useful cognate depolymerases, so as to make innovative depolymerase-based treatments a reality in the future. Although the use of VALs is less explored, they seem to have advantages in comparison with endolysins, mainly for their higher thermostability and likely better ability to lyse the cell wall of actively growing cells, crucial for *in vivo* applications. Through well implemented protein engineering techniques, several enzymes are now expected to be quickly generated with altered traits for optimal performance, surpassing limitations of their native enzymes. Thus, we expect that biotechnological applications of both depolymerases and VALs to control pathogenic bacteria will be widely explored and play a relevant role in the near future as antimicrobial agents.

## Acknowledgements

This study was supported by the Portuguese Foundation for Science and Technology (FCT) under the scope of the strategic funding of UID/BIO/04469/2013 unit, COMPETE 2020 (POCI-01-0145-FEDER-006684) and Projects PTDC/SAU-PUB/29182/2017 (POCI-01-0145-FEDER-029182) and PTDC/CVT-CVT/29628/2017 (POCI-01-0145-FEDER-029628). This work was also supported by BioTecNorte operation (NORTE-01-0145-FEDER-000004) funded by the European Regional Development Fund under the scope of Norte2020 - Programa Operacional Regional do Norte.

## References

Abedon, S.T. (2011). Lysis from without. Bacteriophage 1, 46-49. https://dx.doi.org/ 10.4161/bact.1.1.13980.

Abu-Lail, N.I., and Camesano, T.A. (2003). Role of lipopolysaccharides in the adhesion, retention, and transport of *Escherichia coli* JM109. Environ Sci Technol 37, 2173-2183. https://dx.doi.org/ 10.1021/es026159o.

Adams, M.H., and Park, B.H. (1956). An enzyme produced by a phage-host cell system. II. The properties of the polysaccharide depolymerase. Virology 2, 719-736. https://dx.doi.org/ 10.1016/0042-6822(56)90054-x.

Alcorlo, M., Martinez-Caballero, S., Molina, R., and Hermoso, J.A. (2017). Carbohydrate recognition and lysis by bacterial peptidoglycan hydrolases. Curr Opin Struc Biol 44, 87-100. https://dx.doi.org/ 10.1016/j.sbi.2017.01.001.

Arciola, C.R., Campoccia, D., Ravaioli, S., and Montanaro, L. (2015). Polysaccharide intercellular adhesin in biofilm: structural and regulatory aspects. Frontiers in cellular and infection microbiology 5. https://dx.doi.org/ 10.3389/fcimb.2015.00007.

Arisaka, F., Kanamaru, S., Leiman, P., and Rossmann, M.G. (2003). The tail lysozyme complex of bacteriophage T4. The international journal of biochemistry & cell biology 35, 16-21. https://dx.doi.org/ 10.1016/ s1357-2725(02)00098-5.

Barbirz, S., Muller, J.J., Uetrecht, C., Clark, A.J., Heinemann, U., and Seckler, R. (2008). Crystal structure of *Escherichia coli* phage HK620 tailspike: podoviral tailspike endoglycosidase modules are evolutionarily related. Mol Microbiol 69, 303-316. https://dx.doi.org/ 10.1111/j.1365-2958.2008.06311.x.

Baxa, U., Steinbacher, S., Miller, S., Weintraub, A., Huber, R., and Seckler, R. (1996). Interactions of phage P22 tails with their cellular receptor, *Salmonella* O-antigen polysaccharide. Biophys J 71, 2040-2048. https://dx.doi.org/ Doi 10.1016/S0006-3495(96)79402-X.

Becker, S.C., Dong, S.L., Baker, J.R., Foster-Frey, J., Pritchard, D.G., and Donovan, D.M. (2009). LysK CHAP endopeptidase domain is required for lysis of live staphylococcal cells. FEMS microbiology letters 294, 52-60. https:// dx.doi.org/ 10.1111/j.1574-6968.2009.01541.x.

Born, Y., Fieseler, L., Thony, V., Leimer, N., Duffy, B., and Loessner, M.J. (2017). Engineering of Bacteriophages Y2::dpoL1-C and Y2::luxAB for Efficient Control and Rapid Detection of the Fire Blight Pathogen, *Erwinia amylovora*. Applied and environmental microbiology 83. https://dx.doi.org/ 10.1128/AEM. 00341-17.

Boucher, H., Miller, L.G., and Razonable, R.R. (2010). Serious infections caused by methicillin-resistant *Staphylococcus aureus*. Clinical infectious diseases : an official publication of the Infectious Diseases Society of America 51 Suppl 2, S183-197. https://dx.doi.org/ 10.1086/653519.

Boulanger, P., Jacquot, P., Plancon, L., Chami, M., Engel, A., Parquet, C., Herbeuval, C., and Letellier, L. (2008). Phage T5 straight tail fiber is a multifunctional protein acting as a tape measure and carrying fusogenic and muralytic activities. Journal of Biological Chemistry 283, 13556-13564. https:// dx.doi.org/ 10.1074/jbc.M800052200.

Boyd, A., and Chakrabarty, A.M. (1995). *Pseudomonas aeruginosa* Biofilms - Role of the Alginate Exopolysaccharide. J Ind Microbiol 15, 162-168. https:// dx.doi.org/ Doi 10.1007/Bf01569821.

Briers, Y., Lavigne, R., Plessers, P., Hertveldt, K., Hanssens, I., Engelborghs, Y., and Volckaert, G. (2006). Stability analysis of the bacteriophage phiKMV lysin gp36C and its putative role during infection. Cellular and molecular life sciences : CMLS 63, 1899-1905. https://dx.doi.org/ 10.1007/ s00018-006-6183-7.

Briers, Y., Lavigne, R., Volckaert, G., and Hertveldt, K. (2007). A standardized approach for accurate quantification of murein hydrolase activity in high-throughput assays. Journal of biochemical and biophysical methods 70, 531-533. https://dx.doi.org/ 10.1016/j.jbbm.2006.10.009.

Briers, Y., Miroshnikov, K., Chertkov, O., Nekrasov, A., Mesyanzhinov, V., Volckaert, G., and Lavigne, R. (2008). The structural peptidoglycan hydrolase gp181 of bacteriophage phiKZ. Biochem Biophys Res Commun 374, 747-751. https://dx.doi.org/ 10.1016/j.bbrc.2008.07.102.

Briers, Y., Walmagh, M., Grymonprez, B., Biebl, M., Pirnay, J.P., Defraine, V., Michiels, J., Cenens, W., Aertsen, A., Miller, S., *et al.* (2014). Art-175 is a highly efficient antibacterial against multidrug-resistant strains and persisters of *Pseudomonas aeruginosa*. Antimicrob Agents Chemother 58, 3774-3784. https://dx.doi.org/ 10.1128/AAC.02668-14.

Broeker, N.K., and Barbirz, S. (2017). Not a barrier but a key: How bacteriophages exploit host's O-antigen as an essential receptor to initiate infection. Mol Microbiol 105, 353-357. https://dx.doi.org/ 10.1111/mmi.13729.

Broeker, N.K., Roske, Y., Valleriani, A., Stephan, M.S., Andres, D., Koetz, J., Heinemann, U., and Barbirz, S. (2019). Time-resolved DNA release from an O-antigen-specific *Salmonella* bacteriophage with a contractile tail. The Journal of biological chemistry 294, 11751-11761. https://dx.doi.org/ 10.1074/ jbc.RA119.008133.

Caldentey, J., and Bamford, D.H. (1992). The lytic enzyme of the *Pseudomonas* phage phi 6. Purification and biochemical characterization. Biochimica et biophysica acta 1159, 44-50. https://dx.doi.org/ 10.1016/0167-4838(92)90073-m.

Carpentier, B., and Cerf, O. (1993). Biofilms and Their Consequences, with Particular Reference to Hygiene in the Food-Industry. J Appl Bacteriol 75, 499-511. https://dx.doi.org/ DOI 10.1111/j.1365-2672.1993.tb01587.x.

Catalao, M.J., Gil, F., Moniz-Pereira, J., Sao-Jose, C., and Pimentel, M. (2013). Diversity in bacterial lysis systems: bacteriophages show the way. FEMS microbiology reviews 37, 554-571. https://dx.doi.org/ 10.1111/1574-6976.12006.

Ceri, H., Olson, M.E., Stremick, C., Read, R.R., Morck, D., and Buret, A. (1999). The Calgary Biofilm Device: new technology for rapid determination of antibiotic susceptibilities of bacterial biofilms. Journal of clinical microbiology 37, 1771-1776.

Channabasappa, S., Durgaiah, M., Chikkamadaiah, R., Kumar, S., Joshi, A., and Sriram, B. (2018). Efficacy of Novel Antistaphylococcal Ectolysin P128 in a Rat Model of Methicillin-Resistant *Staphylococcus aureus* Bacteremia. Antimicrob Agents Chemother 62. https://dx.doi.org/ 10.1128/AAC.01358-17.

Chua, J.E., Manning, P.A., and Morona, R. (1999). The *Shigella flexneri* bacteriophage Sf6 tailspike protein (TSP)/endorhamnosidase is related to the bacteriophage P22 TSP and has a motif common to exo- and endoglycanases, and C-5 epimerases. Microbiology 145 ( Pt 7), 1649-1659. https://dx.doi.org/ 10.1099/13500872-145-7-1649.

Chuang, Y.Y., and Huang, Y.C. (2013). Molecular epidemiology of community-associated meticillin-resistant *Staphylococcus aureus* in Asia. The Lancet Infectious diseases 13, 698-708. https://dx.doi.org/ 10.1016/S1473-3099(13)70136-1.

Cornelissen, A., Ceyssens, P.J., Krylov, V.N., Noben, J.P., Volckaert, G., and Lavigne, R. (2012). Identification of EPS-degrading activity within the tail spikes of the novel *Pseudomonas putida* phage AF. Virology 434, 251-256. https://dx.doi.org/ 10.1016/j.virol.2012.09.030.

Cornelissen, A., Ceyssens, P.J., T'Syen, J., Van Praet, H., Noben, J.P., Shaburova, O.V., Krylov, V.N., Volckaert, G., and Lavigne, R. (2011). The T7-Related *Pseudomonas putida* Phage phi 15 Displays Virion-Associated Biofilm Degradation Properties. PloS one 6. https://dx.doi.org/ 10.1371/journal.pone.0018597.

Cornelissen, A., Sadovskaya, I., Vinogradov, E., Blangy, S., Spinelli, S., Casey, E., Mahony, J., Noben, J.P., Dal Bello, F., Cambillau, C., *et al.* (2016). The Baseplate of *Lactobacillus delbrueckii* Bacteriophage Ld17 Harbors a Glycerophosphodiesterase. Journal of Biological Chemistry 291, 16816-16827. https://dx.doi.org/ 10.1074/jbc.M116.728279.

Cress, B.F., Englaender, J.A., He, W., Kasper, D., Linhardt, R.J., and Koffas, M.A. (2014). Masquerading microbial pathogens: capsular polysaccharides mimic host-tissue molecules. FEMS microbiology reviews 38, 660-697. https://dx.doi.org/ 10.1111/1574-6976.12056.

da Silva Xavier, A., da Silva, F.P., Vidigal, P.M.P., Lima, T.T.M., de Souza, F.O., and Alfenas-Zerbini, P. (2018). Genomic and biological characterization of a new member of the genus Phikmvvirus infecting phytopathogenic *Ralstonia* bacteria. Arch Virol 163, 3275-3290. https://dx.doi.org/ 10.1007/s00705-018-4006-4.

Dakheel, K.H., Rahim, R.A., Neela, V.K., Al-Obaidi, J.R., Hun, T.G., Isa, M.N.M., and Yusoff, K. (2019). Genomic analyses of two novel biofilm-degrading methicillin-resistant *Staphylococcus aureus* phages. Bmc Microbiol 19, 114. https://dx.doi.org/ 10.1186/s12866-019-1484-9.

Daniel, A., Euler, C., Collin, M., Chahales, P., Gorelick, K.J., and Fischetti, V.A. (2010). Synergism between a novel chimeric lysin and oxacillin protects against infection by methicillin-resistant *Staphylococcus aureus*. Antimicrob Agents Chemother 54, 1603-1612. https://dx.doi.org/ 10.1128/AAC.01625-09.

Daum, R.S. (2007). Clinical practice. Skin and soft-tissue infections caused by methicillin-resistant *Staphylococcus aureus*. The New England journal of medicine 357, 380-390. https://dx.doi.org/ 10.1056/NEJMcp070747.

Davidson, I.W., Lawson, C.J., and Sutherland, I.W. (1977). An alginate lysate from *Azotobacter vinelandii* phage. Journal of general microbiology 98, 223-229. https://dx.doi.org/ 10.1099/00221287-98-1-223.

Diez-Martinez, R., De Paz, H.D., Garcia-Fernandez, E., Bustamante, N., Euler, C.W., Fischetti, V.A., Menendez, M., and Garcia, P. (2015). A novel chimeric phage lysin with high *in vitro* and *in vivo* bactericidal activity against *Streptococcus pneumoniae*. The Journal of antimicrobial chemotherapy 70, 1763-1773. https://dx.doi.org/ 10.1093/jac/dkv038.

Donovan, D.M., Dong, S.L., Garrett, W., Rousseau, G.M., Moineau, S., and Pritchard, D.G. (2006). Peptidoglycan hydrolase fusions maintain their parental specificities. Applied and environmental microbiology 72, 2988-2996. https://dx.doi.org/ 10.1128/Aem.72.4.2988-2996.2006.

Donovan, D.M., and Foster-Frey, J. (2008). LambdaSa2 prophage endolysin requires Cpl-7-binding domains and amidase-5 domain for antimicrobial lysis of streptococci. FEMS microbiology letters 287, 22-33. https://dx.doi.org/ 10.1111/j.1574-6968.2008.01287.x.

Fagan, R.P., and Fairweather, N.F. (2014). Biogenesis and functions of bacterial S-layers. Nature reviews Microbiology 12, 211-222. https://dx.doi.org/ 10.1038/ nrmicro3213.

Fernandes, S., and Sao-Jose, C. (2018). Enzymes and Mechanisms Employed by Tailed Bacteriophages to Breach the Bacterial Cell Barriers. Viruses-Basel 10. https://dx.doi.org/ 10.3390/V10080396.

Fernandez, L., Gonzalez, S., Campelo, A.B., Martinez, B., Rodriguez, A., and Garcia, P. (2017). Downregulation of Autolysin-Encoding Genes by Phage-Derived Lytic Proteins Inhibits Biofilm Formation in *Staphylococcus aureus*. Antimicrob Agents Chemother 61. https://dx.doi.org/ 10.1128/AAC.02724-16.

Flemming, H.C., and Wingender, J. (2010). The biofilm matrix. Nature reviews Microbiology 8, 623-633. https://dx.doi.org/ 10.1038/nrmicro2415.

Fokine, A., and Rossmann, M.G. (2014). Molecular architecture of tailed double-stranded DNA phages. Bacteriophage 4, e28281. https://dx.doi.org/ 10.4161/ bact.28281.

Gebhart, D., Williams, S.R., and Scholl, D. (2017). Bacteriophage SP6 encodes a second tailspike protein that recognizes *Salmonella enterica* serogroups C-2 and C-3. Virology 507, 263-266. https://dx.doi.org/ 10.1016/j.virol.2017.02.025.

Geisinger, E., and Isberg, R.R. (2015). Antibiotic modulation of capsular exopolysaccharide and virulence in *Acinetobacter baumannii*. PLoS pathogens 11, e1004691. https://dx.doi.org/ 10.1371/journal.ppat.1004691.

Gilmer, D.B., Schmitz, J.E., Thandar, M., Euler, C.W., and Fischetti, V.A. (2017). The Phage Lysin PlySs2 Decolonizes *Streptococcus suis* from Murine Intranasal Mucosa. PloS one 12, e0169180. https://dx.doi.org/ 10.1371/ journal.pone.0169180.

Glonti, T., Chanishvili, N., and Taylor, P.W. (2010). Bacteriophage-derived enzyme that depolymerizes the alginic acid capsule associated with cystic fibrosis isolates of *Pseudomonas aeruginosa*. J Appl Microbiol 108, 695-702. https://dx.doi.org/ 10.1111/j.1365-2672.2009.04469.x.

Guichard, J.A., Middleton, P.C., and McConnell, M.R. (2013). Genetic analysis of structural proteins in the adsorption apparatus of bacteriophage epsilon 15. World journal of virology 2, 152-159. https://dx.doi.org/ 10.5501/wjv.v2.i4.152.

Guo, Z., Huang, J., Yan, G., Lei, L., Wang, S., Yu, L., Zhou, L., Gao, A., Feng, X., Han, W., *et al.* (2017). Identification and Characterization of Dpo42, a

Novel Depolymerase Derived from the *Escherichia coli* Phage vB_EcoM_ECOO78. Frontiers in microbiology 8, 1460. https://dx.doi.org/ 10.3389/fmicb.2017.01460.

Gutierrez, D., Briers, Y., Rodriguez-Rubio, L., Martinez, B., Rodriguez, A., Lavigne, R., and Garcia, P. (2015). Role of the Pre-neck Appendage Protein (Dpo7) from Phage vB_SepiS-philPLA7 as an Anti-biofilm Agent in Staphylococcal Species. Frontiers in microbiology 6. https://dx.doi.org/ 10.3389/Fmicb.2015.01315.

Hatfull, G.F., and Hendrix, R.W. (2011). Bacteriophages and their genomes. Curr Opin Virol 1, 298-303. https://dx.doi.org/ 10.1016/j.coviro.2011.06.009.

Hernandez-Morales, A.C., Lessor, L.L., Wood, T.L., Migl, D., Mijalis, E.M., Cahill, J., Russell, W.K., Young, R.F., and Gill, J.J. (2018). Genomic and Biochemical Characterization of Acinetobacter Podophage Petty Reveals a Novel Lysis Mechanism and Tail-Associated Depolymerase Activity. Journal of virology 92. https://dx.doi.org/10.1128/JVI.01064-17.

Holland, T.L., Arnold, C., and Fowler, V.G., Jr. (2014). Clinical management of *Staphylococcus aureus* bacteremia: a review. Jama 312, 1330-1341. https:// dx.doi.org/ 10.1001/jama.2014.9743.

Hosseinidoust, Z., van de Ven, T.G.M., and Tufenkji, N. (2013). Evolution of *Pseudomonas aeruginosa* Virulence as a Result of Phage Predation. Applied and environmental microbiology 79, 6110-6116. https://dx.doi.org/ 10.1128/ Aem.01421-13.

Hsieh, P.F., Lin, H.H., Lin, T.L., Chen, Y.Y., and Wang, J.T. (2017). Two T7-like Bacteriophages, K5-2 and K5-4, Each Encodes Two Capsule Depolymerases: Isolation and Functional Characterization. Scientific reports 7. https:// dx.doi.org/ 10.1038/S41598-017-04644-2.

Hsu, C.R., Lin, T.L., Pan, Y.J., Hsieh, P.F., and Wang, J.T. (2013). Isolation of a Bacteriophage Specific for a New Capsular Type of *Klebsiella pneumoniae* and Characterization of Its Polysaccharide Depolymerase. PloS one 8. https:// dx.doi.org/10.1371/journal.pone.0070092.

Hu, B., Margolin, W., Molineux, I.J., and Liu, J. (2013). The bacteriophage t7 virion undergoes extensive structural remodeling during infection. Science 339, 576-579. https://dx.doi.org/ 10.1126/science.1231887.

Hynes, W.L., Hancock, L., and Ferretti, J.J. (1995). Analysis of a second bacteriophage hyaluronidase gene from *Streptococcus pyogenes*: evidence for a third hyaluronidase involved in extracellular enzymatic activity. Infection and immunity 63, 3015-3020.

Kanamaru, S., Leiman, P.G., Kostyuchenko, V.A., Chipman, P.R., Mesyanzhinov, V.V., Arisaka, F., and Rossmann, M.G. (2002). Structure of the cell-puncturing device of bacteriophage T4. Nature 415, 553-557. https://dx.doi.org/ Doi 10.1038/415553a.

Kao, S.H., and Mcclain, W.H. (1980). Baseplate Protein of Bacteriophage-T4 with Both Structural and Lytic Functions. Journal of virology 34, 95-103. https://dx.doi.org/ Doi 10.1128/Jvi.34.1.95-103.1980.

Kenny, J.G., McGrath, S., Fitzgerald, G.F., and van Sinderen, D. (2004). Bacteriophage Tuc2009 encodes a tail-associated cell wall-degrading activity. J Bacteriol 186, 3480-3491. https://dx.doi.org/ 10.1128/JB.186.11.3480-3491.2004.

Kivela, H.M., Daugelavicius, R., Hankkio, R.H., Bamford, J.K., and Bamford, D.H. (2004). Penetration of membrane-containing double-stranded-DNA bacteriophage PM2 into *Pseudoalteromonas* hosts. J Bacteriol 186, 5342-5354. https://dx.doi.org/ 10.1128/JB.186.16.5342-5354.2004.

Knecht, L.E., Veljkovic, M., and Fieseler, L. (2019). Diversity and Function of Phage Encoded Depolymerases. Frontiers in microbiology 10, 2949. https://dx.doi.org/ 10.3389/fmicb.2019.02949.

Korhonen, T.K., Valtonen, M.V., Parkkinen, J., Vaisanen-Rhen, V., Finne, J., Orskov, F., Orskov, I., Svenson, S.B., and Makela, P.H. (1985). Serotypes, hemolysin production, and receptor recognition of *Escherichia coli* strains associated with neonatal sepsis and meningitis. Infection and immunity 48, 486-491.

Koskella, B., and Brockhurst, M.A. (2014). Bacteria-phage coevolution as a driver of ecological and evolutionary processes in microbial communities. FEMS microbiology reviews 38, 916-931. https://dx.doi.org/ 10.1111/1574-6976.12072.

Kreisberg, J.F., Betts, S.D., Haase-Pettingell, C., and King, J. (2002). The interdigitated beta-helix domain of the P22 tailspike protein acts as a molecular clamp in trimer stabilization. Protein science : a publication of the Protein Society 11, 820-830. https://dx.doi.org/ 10.1110/ps.3440102.

Kropinski, A.M., Kovalyova, I.V., Billington, S.J., Patrick, A.N., Butts, B.D., Guichard, J.A., Pitcher, T.J., Guthrie, C.C., Sydlaske, A.D., Barnhill, L.M., *et al.* (2007). The genome of epsilon15, a serotype-converting, Group E1 *Salmonella enterica*-specific bacteriophage. Virology 369, 234-244. https:// dx.doi.org/ 10.1016/j.virol.2007.07.027.

Kunstmann, S., Scheidt, T., Buchwald, S., Helm, A., Mulard, L.A., Fruth, A., and Barbirz, S. (2018). Bacteriophage Sf6 Tailspike Protein for Detection of *Shigella flexneri* Pathogens. Viruses-Basel 10. https://dx.doi.org/ 10.3390/ V10080431.

Latka, A., Maciejewska, B., Majkowska-Skrobek, G., Briers, Y., and Drulis-Kawa, Z. (2017). Bacteriophage-encoded virion-associated enzymes to overcome the carbohydrate barriers during the infection process. Appl Microbiol Biotechnol 101, 3103-3119. https://dx.doi.org/ 10.1007/s00253-017-8224-6.

Lavigne, R., Briers, Y., Hertveldt, K., Robben, J., and Volckaert, G. (2004). Identification and characterization of a highly thermostable bacteriophage lysozyme. Cellular and molecular life sciences : CMLS 61, 2753-2759. https:// dx.doi.org/ 10.1007/s00018-004-4301-y.

Lee, I.M., Tu, I.F., Yang, F.L., Ko, T.P., Liao, J.H., Lin, N.T., Wu, C.Y., Ren, C.T., Wang, A.H., Chang, C.M., *et al.* (2017). Structural basis for fragmenting the exopolysaccharide of *Acinetobacter baumannii* by bacteriophage PhiAB6 tailspike protein. Scientific reports 7, 42711. https://dx.doi.org/ 10.1038/ srep42711.

Leiman, P.G., Battisti, A.J., Bowman, V.D., Stummeyer, K., Muhlenhoff, M., Gerardy-Schahn, R., Scholl, D., and Molineux, I.J. (2007). The structures of bacteriophages K1E and K1-5 explain processive degradation of polysaccharide capsules and evolution of new host specificities. Journal of molecular biology 371, 836-849. https://dx.doi.org/ 10.1016/j.jmb.2007.05.083.

Li, X., Koc, C., Kuhner, P., Stierhof, Y.D., Krismer, B., Enright, M.C., Penades, J.R., Wolz, C., Stehle, T., Cambillau, C., *et al.* (2016). An essential role for the baseplate protein Gp45 in phage adsorption to *Staphylococcus aureus*. Scientific reports 6, 26455. https://dx.doi.org/ 10.1038/srep26455.

Lin, H., Paff, M.L., Molineux, I.J., and Bull, J.J. (2017). Therapeutic Application of Phage Capsule Depolymerases against K1, K5, and K30 Capsulated *E. coli* in Mice. Frontiers in microbiology 8. https://dx.doi.org/ 10.3389/Fmicb. 2017.02257.

Lin, H., Paff, M.L., Molineux, I.J., and Bull, J.J. (2018). Antibiotic Therapy Using Phage Depolymerases: Robustness Across a Range of Conditions. Viruses-Basel 10. https://dx.doi.org/ 10.3390/V10110622.

Lin, T.L., Hsieh, P.F., Huang, Y.T., Lee, W.C., Tsai, Y.T., Su, P.A., Pan, Y.J., Hsu, C.R., Wu, M.C., and Wang, J.T. (2014). Isolation of a bacteriophage and its depolymerase specific for K1 capsule of *Klebsiella pneumoniae*: implication in typing and treatment. The Journal of infectious diseases 210, 1734-1744. https://dx.doi.org/ 10.1093/infdis/jiu332.

Linnerborg, M., Weintraub, A., Albert, M.J., and Widmalm, G. (2001). Depolymerization of the capsular polysaccharide from *Vibrio cholerae* O139 by a lyase associated with the bacteriophage JA1. Carbohyd Res 333, 263-269. https://dx.doi.org/ Doi 10.1016/S0008-6215(01)00159-8.

Liston, S.D., Ovchinnikova, O.G., and Whitfield, C. (2016). Unique lipid anchor attaches Vi antigen capsule to the surface of *Salmonella enterica* serovar Typhi. Proceedings of the National Academy of Sciences of the United States of America 113, 6719-6724. https://dx.doi.org/ 10.1073/pnas.1524665113.

Liu, Y., Leung, S.S.Y., Guo, Y., Zhao, L., Jiang, N., Mi, L., Li, P., Wang, C., Qin, Y., Mi, Z., *et al.* (2019). The Capsule Depolymerase Dpo48 Rescues *Galleria mellonella* and Mice From *Acinetobacter baumannii* Systemic Infections. Frontiers in microbiology 10, 545. https://dx.doi.org/ 10.3389/fmicb. 2019.00545.

Lowy, F.D. (1998). *Staphylococcus aureus* infections. The New England journal of medicine 339, 520-532. https://dx.doi.org/ 10.1056/ NEJM199808203390806.

Lu, T.K., and Collins, J.J. (2007). Dispersing biofilms with engineered enzymatic bacteriophage. Proceedings of the National Academy of Sciences of the United States of America 104, 11197-11202. https://dx.doi.org/ 10.1073/pnas. 0704624104.

Machida, Y., Miyake, K., Hattori, K., Yamamoto, S., Kawase, M., and Iijima, S. (2000). Structure and function of a novel coliphage-associated sialidase.

FEMS microbiology letters 182, 333-337. https://dx.doi.org/ 10.1111/j.1574-6968.2000.tb08917.x.

Mahony, J., Stockdale, S.R., Collins, B., Spinelli, S., Douillard, F.P., Cambillau, C., and van Sinderen, D. (2016). *Lactococcus lactis* phage TP901-1 as a model for *Siphoviridae* virion assembly. Bacteriophage 6, e1123795. https://dx.doi.org/ 10.1080/21597081.2015.1123795.

Majkowska-Skrobek, G., Latka, A., Berisio, R., Maciejewska, B., Squeglia, F., Romano, M., Lavigne, R., Struve, C., and Drulis-Kawa, Z. (2016). Capsule-Targeting Depolymerase, Derived from *Klebsiella* KP36 Phage, as a Tool for the Development of Anti-Virulent Strategy. Viruses-Basel 8. https://dx.doi.org/ 10.3390/V8120324.

Majkowska-Skrobek, G., Latka, A., Berisio, R., Squeglia, F., Maciejewska, B., Briers, Y., and Drulis-Kawa, Z. (2018). Phage-Borne Depolymerases Decrease *Klebsiella pneumoniae* Resistance to Innate Defense Mechanisms. Frontiers in microbiology 9. https://dx.doi.org/ 10.3389/Fmicb.2018.02517.

Malnoy, M., Faize, M., Venisse, J.S., Geider, K., and Chevreau, E. (2005). Expression of viral EPS-depolymerase reduces fire blight susceptibility in transgenic pear. Plant cell reports 23, 632-638. https://dx.doi.org/ 10.1007/s00299-004-0855-2.

Mao, J., Schmelcher, M., Harty, W.J., Foster-Frey, J., and Donovan, D.M. (2013). Chimeric Ply187 endolysin kills *Staphylococcus aureus* more effectively than the parental enzyme. FEMS microbiology letters 342, 30-36. https://dx.doi.org/ 10.1111/1574-6968.12104.

Miller, L.G., Perdreau-Remington, F., Rieg, G., Mehdi, S., Perlroth, J., Bayer, A.S., Tang, A.W., Phung, T.O., and Spellberg, B. (2005). Necrotizing fasciitis caused by community-associated methicillin-resistant *Staphylococcus aureus* in Los Angeles. The New England journal of medicine 352, 1445-1453. https://dx.doi.org/ 10.1056/NEJMoa042683.

Moak, M., and Molineux, I.J. (2000). Role of the Gp16 lytic transglycosylase motif in bacteriophage T7 virions at the initiation of infection. Mol Microbiol 37, 345-355. https://dx.doi.org/ DOI 10.1046/j.1365-2958.2000.01995.x.

Moak, M., and Molineux, I.J. (2004). Peptidoglycan hydrolytic activities associated with bacteriophage virions. Mol Microbiol 51, 1169-1183. https://dx.doi.org/ 10.1046/j.1365-2958.2003.03894.x.

Molineux, I.J. (2001). No syringes please, ejection of phage T7 DNA from the virion is enzyme driven. Mol Microbiol 40, 1-8. https://dx.doi.org/ DOI 10.1046/ j.1365-2958.2001.02357.x.

Mushtaq, N., Redpath, M.B., Luzio, J.P., and Taylor, P.W. (2004). Prevention and cure of systemic *Escherichia coli* K1 infection by modification of the bacterial phenotype. Antimicrob Agents Chemother 48, 1503-1508. https://dx.doi.org/ 10.1128/aac.48.5.1503-1508.2004.

Mushtaq, N., Redpath, M.B., Luzio, J.P., and Taylor, P.W. (2005). Treatment of experimental *Escherichia coli* infection with recombinant bacteriophage-derived capsule depolymerase. The Journal of antimicrobial chemotherapy 56, 160-165. https://dx.doi.org/ 10.1093/jac/dki177.

Myers, C.L., Ireland, R.G., Garrett, T.A., and Brown, E.D. (2015). Characterization of Wall Teichoic Acid Degradation by the Bacteriophage phi 29 Appendage Protein GP12 Using Synthetic Substrate Analogs. Journal of Biological Chemistry 290, 19133-19145. https://dx.doi.org/ 10.1074/ jbc.M115.662866.

Mylotte, J.M., McDermott, C., and Spooner, J.A. (1987). Prospective study of 114 consecutive episodes of *Staphylococcus aureus* bacteremia. Reviews of infectious diseases 9, 891-907. https://dx.doi.org/ 10.1093/clinids/9.5.891.

Nishima, W., Kanamaru, S., Arisaka, F., and Kitao, A. (2011). Screw Motion Regulates Multiple Functions of T4 Phage Protein Gene Product 5 during Cell Puncturing. Journal of the American Chemical Society 133, 13571-13576. https://dx.doi.org/ 10.1021/ja204451g.

Nobrega, F.L., Vlot, M., de Jonge, P.A., Dreesens, L.L., Beaumont, H.J.E., Lavigne, R., Dutilh, B.E., and Brouns, S.J.J. (2018). Targeting mechanisms of tailed bacteriophages. Nature reviews Microbiology 16, 760-773. https:// dx.doi.org/ 10.1038/s41579-018-0070-8.

Oliveira, F., Lima, C.A., Bras, S., Franca, A., and Cerca, N. (2015). Evidence for inter- and intraspecies biofilm formation variability among a small group of coagulase-negative staphylococci. FEMS microbiology letters 362. https:// dx.doi.org/ 10.1093/femsle/fnv175.

Oliveira, H., Costa, A.R., Ferreira, A., Konstantinides, N., Santos, S.B., Boon, M., Noben, J.P., Lavigne, R., and Azeredo, J. (2019a). Functional Analysis and Antivirulence Properties of a New Depolymerase from a Myovirus That Infects

*Acinetobacter baumannii* Capsule K45. Journal of virology 93. https://dx.doi.org/ 10.1128/JVI.01163-18.

Oliveira, H., Costa, A.R., Konstantinides, N., Ferreira, A., Akturk, E., Sillankorva, S., Nemec, A., Shneider, M., Dotsch, A., and Azeredo, J. (2017). Ability of phages to infect *Acinetobacter calcoaceticus-Acinetobacter baumannii* complex species through acquisition of different pectate lyase depolymerase domains. Environmental microbiology 19, 5060-5077. https://dx.doi.org/ 10.1111/1462-2920.13970.

Oliveira, H., Melo, L.D., Santos, S.B., Nobrega, F.L., Ferreira, E.C., Cerca, N., Azeredo, J., and Kluskens, L.D. (2013). Molecular aspects and comparative genomics of bacteriophage endolysins. Journal of virology 87, 4558-4570. https://dx.doi.org/ 10.1128/JVI.03277-12.

Oliveira, H., Mendes, A., Fraga, A.G., Ferreira, A., Pimenta, A.I., Mil-Homens, D., Fialho, A.M., Pedrosa, J., and Azeredo, J. (2019b). K2 Capsule Depolymerase Is Highly Stable, Is Refractory to Resistance, and Protects Larvae and Mice from *Acinetobacter baumannii* Sepsis. Applied and environmental microbiology 85. https://dx.doi.org/ 10.1128/AEM.00934-19.

Oliveira, H., Thiagarajan, V., Walmagh, M., Sillankorva, S., Lavigne, R., Neves-Petersen, M.T., Kluskens, L.D., and Azeredo, J. (2014). A thermostable *Salmonella* phage endolysin, Lys68, with broad bactericidal properties against gram-negative pathogens in presence of weak acids. PloS one 9, e108376. https://dx.doi.org/ 10.1371/journal.pone.0108376.

Oliveira, H., Vilas Boas, D., Mesnage, S., Kluskens, L.D., Lavigne, R., Sillankorva, S., Secundo, F., and Azeredo, J. (2016). Structural and Enzymatic Characterization of ABgp46, a Novel Phage Endolysin with Broad Anti-Gram-Negative Bacterial Activity. Frontiers in microbiology 7, 208. https://dx.doi.org/ 10.3389/fmicb.2016.00208.

Olszak, T., Latka, A., Roszniowski, B., Valvano, M.A., and Drulis-Kawa, Z. (2017a). Phage Life Cycles Behind Bacterial Biodiversity. Curr Med Chem 24, 3987-4001. https://dx.doi.org/ 10.2174/0929867324666170413100136.

Olszak, T., Shneider, M.M., Latka, A., Maciejewska, B., Browning, C., Sycheva, L.V., Cornelissen, A., Danis-Wlodarczyk, K., Senchenkova, S.N., Shashkov, A.S., *et al.* (2017b). The O-specific polysaccharide lyase from the phage LKA1 tailspike reduces *Pseudomonas* virulence. Scientific reports 7, 16302. https://dx.doi.org/ 10.1038/s41598-017-16411-4.

Orskov, I., Orskov, F., Jann, B., and Jann, K. (1977). Serology, chemistry, and genetics of O and K antigens of *Escherichia coli*. Bacteriological reviews 41, 667-710.

Pajunen, M.I., Kiljunen, S.J., Soderholm, M.E., and Skurnik, M. (2001). Complete genomic sequence of the lytic bacteriophage phiYeO3-12 of *Yersinia enterocolitica* serotype O:3. J Bacteriol 183, 1928-1937. https://dx.doi.org/ 10.1128/JB.183.6.1928-1937.2001.

Pan, Y.J., Lin, T.L., Chen, C.C., Tsai, Y.T., Cheng, Y.H., Chen, Y.Y., Hsieh, P.F., Lin, Y.T., and Wang, J.T. (2017). Klebsiella Phage PhiK64-1 Encodes Multiple Depolymerases for Multiple Host Capsular Types. Journal of virology 91. https://dx.doi.org/ 10.1128/JVI.02457-16.

Pan, Y.J., Lin, T.L., Chen, Y.Y., Lai, P.H., Tsai, Y.T., Hsu, C.R., Hsieh, P.F., Lin, Y.T., and Wang, J.T. (2019). Identification of three podoviruses infecting *Klebsiella* encoding capsule depolymerases that digest specific capsular types. Microbial biotechnology 12, 472-486. https://dx.doi.org/ 10.1111/1751-7915.13370.

Pan, Y.J., Lin, T.L., Lin, Y.T., Su, P.A., Chen, C.T., Hsieh, P.F., Hsu, C.R., Chen, C.C., Hsieh, Y.C., and Wang, J.T. (2015). Identification of capsular types in carbapenem-resistant *Klebsiella pneumoniae* strains by wzc sequencing and implications for capsule depolymerase treatment. Antimicrob Agents Chemother 59, 1038-1047. https://dx.doi.org/ 10.1128/AAC.03560-14.

Paradis-Bleau, K.D. (2015). Biology and Assembly of the Bacterial Envelope. In Prokaryotic Systems Biology, M.B. Nevan J. Krogan, ed. (Springer, Cham), pp. 41-76. https://dx.doi.org/ doi.org/10.1007/978-3-319-23603-2.

Paramonov, N.A., Aduse-Opoku, J., Hashim, A., Rangarajan, M., and Curtis, M.A. (2009). Structural Analysis of the Core Region of O-Lipopolysaccharide of *Porphyromonas gingivalis* from Mutants Defective in O-Antigen Ligase and O-Antigen Polymerase. J Bacteriol 191, 5272-5282. https://dx.doi.org/ 10.1128/Jb.00019-09.

Paton, J.C., and Trappetti, C. (2019). *Streptococcus pneumoniae* Capsular Polysaccharide. Microbiol Spectr 7. https://dx.doi.org/ 10.1128/Microbiolspec.

Paul, V.D., Rajagopalan, S.S., Sundarrajan, S., George, S.E., Asrani, J.Y., Pillai, R., Chikkamadaiah, R., Durgaiah, M., Sriram, B., and Padmanabhan, S. (2011). A novel bacteriophage Tail-Associated Muralytic Enzyme (TAME) from Phage K and its development into a potent antistaphylococcal protein. Bmc Microbiol 11, 226. https://dx.doi.org/ 10.1186/1471-2180-11-226.

Percy, M.G., and Grundling, A. (2014). Lipoteichoic Acid Synthesis and Function in Gram-Positive Bacteria. Annual Review of Microbiology, Vol 68 68, 81-100. https://dx.doi.org/ 10.1146/annurev-micro-091213-112949.

Pflaum, K., Gerdes, K., Yovo, K., Callahan, J., and Snyder, M.L.D. (2012). Lipopolysaccharide induction of autophagy is associated with enhanced bactericidal activity in *Dictyostelium discoideum*. Biochem Bioph Res Co 422, 417-422. https://dx.doi.org/ 10.1016/j.bbrc.2012.05.006.

Piuri, M., and Hatfull, G.F. (2006). A peptidoglycan hydrolase motif within the mycobacteriophage TM4 tape measure protein promotes efficient infection of stationary phase cells. Mol Microbiol 62, 1569-1585. https://dx.doi.org/ 10.1111/j.1365-2958.2006.05473.x.

Plotka, M., Kaczorowska, A.K., Morzywolek, A., Makowska, J., Kozlowski, L.P., Thorisdottir, A., Skirnisdottir, S., Hjorleifsdottir, S., Fridjonsson, O.H., Hreggvidsson, G.O., *et al.* (2015). Biochemical Characterization and Validation of a Catalytic Site of a Highly Thermostable Ts2631 Endolysin from the *Thermus scotoductus* Phage vB_Tsc2631. PloS one 10, e0137374. https:// dx.doi.org/ 10.1371/journal.pone.0137374.

Popova, A.V., Lavysh, D.G., Klimuk, E.I., Edelstein, M.V., Bogun, A.G., Shneider, M.M., Goncharov, A.E., Leonov, S.V., and Severinov, K.V. (2017). Novel Fri1-like Viruses Infecting *Acinetobacter baumannii*-vB_AbaP_AS11 and vB_AbaP_AS12-Characterization, Comparative Genomic Analysis, and Host-Recognition Strategy. Viruses 9. https://dx.doi.org/ 10.3390/v9070188.

Proenca, D., Leandro, C., Garcia, M., Pimentel, M., and Sao-Jose, C. (2015). EC300: a phage-based, bacteriolysin-like protein with enhanced antibacterial activity against *Enterococcus faecalis*. Appl Microbiol Biotechnol 99, 5137-5149. https://dx.doi.org/ 10.1007/s00253-015-6483-7.

Prokhorov, N.S., Riccio, C., Zdorovenko, E.L., Shneider, M.M., Browning, C., Knirel, Y.A., Leiman, P.G., and Letarov, A.V. (2017). Function of bacteriophage

G7C esterase tailspike in host cell adsorption. Mol Microbiol 105, 385-398. https://dx.doi.org/ 10.1111/mmi.13710.

Rashel, M., Uchiyama, J., Takemura, I., Hoshiba, H., Ujihara, T., Takatsuji, H., Honke, K., and Matsuzaki, S. (2008). Tail-associated structural protein gp61 of *Staphylococcus aureus* phage phi MR11 has bifunctional lytic activity. FEMS microbiology letters 284, 9-16. https://dx.doi.org/ 10.1111/j. 1574-6968.2008.01152.x.

Rodriguez-Rubio, L., Martinez, B., Donovan, D.M., Garcia, P., and Rodriguez, A. (2013a). Potential of the virion-associated peptidoglycan hydrolase HydH5 and its derivative fusion proteins in milk biopreservation. PloS one 8, e54828. https://dx.doi.org/ 10.1371/journal.pone.0054828.

Rodriguez-Rubio, L., Martinez, B., Donovan, D.M., Rodriguez, A., and Garcia, P. (2013b). Bacteriophage virion-associated peptidoglycan hydrolases: potential new enzybiotics. Critical reviews in microbiology 39, 427-434. https:// dx.doi.org/ 10.3109/1040841X.2012.723675.

Rodriguez-Rubio, L., Martinez, B., Rodriguez, A., Donovan, D.M., and Garcia, P. (2012). Enhanced staphylolytic activity of the *Staphylococcus aureus* bacteriophage vB_SauS-philPLA88 HydH5 virion-associated peptidoglycan hydrolase: fusions, deletions, and synergy with LysH5. Applied and environmental microbiology 78, 2241-2248. https://dx.doi.org/ 10.1128/AEM. 07621-11.

Rodriguez-Rubio, L., Quiles-Puchalt, N., Martinez, B., Rodriguez, A., Penades, J.R., and Garcia, P. (2013c). The Peptidoglycan Hydrolase of *Staphylococcus aureus* Bacteriophage phi 11 Plays a Structural Role in the Viral Particle. Applied and environmental microbiology 79, 6187-6190. https://dx.doi.org/ 10.1128/Aem.01388-13.

Rodriguez, L., Martinez, B., Zhou, Y., Rodriguez, A., Donovan, D.M., and Garcia, P. (2011). Lytic activity of the virion-associated peptidoglycan hydrolase HydH5 of *Staphylococcus aureus* bacteriophage vB_SauS-philPLA88. Bmc Microbiol 11. https://dx.doi.org/ 10.1186/1471-2180-11-138.

Ruiz, N., Kahne, D., and Silhavy, T.J. (2006). Advances in understanding bacterial outer-membrane biogenesis. Nature Reviews Microbiology 4, 57-66. https://dx.doi.org/ 10.1038/nrmicro1322.

Ruqeya, S.R., Marofull Nisa, Uqab ali Baba (2019). Freshwater Microbiology: Perspectives of Bacterial Dynamics in Lake Ecosystems. In Exploring bacterial diversity: from cell to sequence, S.S. Syhaib A. Bandh, Nowsheen Shameem, ed., pp. 263-306. https://dx.doi.org/ doi.org/10.1016/ B978-0-12-817495-1.00007-4.

Rydman, P.S., and Bamford, D.H. (2000). Bacteriophage PRD1 DNA entry uses a viral membrane-associated transglycosylase activity. Mol Microbiol 37, 356-363. https://dx.doi.org/ 10.1046/j.1365-2958.2000.01996.x.

Rydman, P.S., and Bamford, D.H. (2002). The lytic enzyme of bacteriophage PRD1 is associated with the viral membrane. J Bacteriol 184, 1502-1502. https://dx.doi.org/ 10.1128/jb.184.1.104-110.2002.

Sachdeva, S., Palur, R.V., Sudhakar, K.U., and Rathinavelan, T. (2017). *E. coli* Group 1 Capsular Polysaccharide Exportation Nanomachinary as a Plausible Antivirulence Targetin the Perspective of Emerging Antimicrobial Resistance. Frontiers in microbiology 8. https://dx.doi.org/ 10.3389/Fmicb.2017.00070.

Safari, F., Sharifi, M., Farajnia, S., Akbari, B., Ahmadi, M.K.B., Negahdaripour, M., and Ghasemi, Y. (2020). The interaction of phages and bacteria: the co-evolutionary arms race. Crit Rev Biotechnol 40, 119-137. https://dx.doi.org/ 10.1080/07388551.2019.1674774.

Sao-Jose, C. (2018). Engineering of Phage-Derived Lytic Enzymes: Improving Their Potential as Antimicrobials. Antibiotics (Basel) 7. https://dx.doi.org/ 10.3390/antibiotics7020029.

Scheurwater, E., Reid, C.W., and Clarke, A.J. (2008). Lytic transglycosylases: Bacterial space-making autolysins. Int J Biochem Cell B 40, 586-591. https:// dx.doi.org/ 10.1016/j.biocel.2007.03.018.

Schleifer, K.H., and Kandler, O. (1972). Peptidoglycan Types of Bacterial Cell-Walls and Their Taxonomic Implications. Bacteriological reviews 36, 407-477. https://dx.doi.org/ Doi 10.1128/Mmbr.36.4.407-477.1972.

Schmelcher, M., Donovan, D.M., and Loessner, M.J. (2012). Bacteriophage endolysins as novel antimicrobials. Future Microbiol 7, 1147-1171. https:// dx.doi.org/ 10.2217/Fmb.12.97.

Schmidt, A., Rabsch, W., Broeker, N.K., and Barbirz, S. (2016). Bacteriophage tailspike protein based assay to monitor phase variable glucosylations in

*Salmonella* O-antigens. Bmc Microbiol 16. https://dx.doi.org/ 10.1186/ S12866-016-0826-0.

Scholl, D., Cooley, M., Williams, S.R., Gebhart, D., Martin, D., Bates, A., and Mandrell, R. (2009). An Engineered R-Type Pyocin Is a Highly Specific and Sensitive Bactericidal Agent for the Food-Borne Pathogen *Escherichia coli* O157:H7. Antimicrob Agents Ch 53, 3074-3080. https://dx.doi.org/ 10.1128/ Aac.01660-08.

Schwarzer, D., Browning, C., Stummeyer, K., Oberbeck, A., Muhlenhoff, M., Gerardy-Schahn, R., and Leiman, P.G. (2015). Structure and biochemical characterization of bacteriophage phi92 endosialidase. Virology 477, 133-143. https://dx.doi.org/ 10.1016/j.virol.2014.11.002.

Serwer, P., Wright, E.T., Liu, Z., and Jiang, W. (2014). Length quantization of DNA partially expelled from heads of a bacteriophage T3 mutant. Virology 456-457, 157-170. https://dx.doi.org/ 10.1016/j.virol.2014.03.016.

Shen, Y., Koller, T., Kreikemeyer, B., and Nelson, D.C. (2013). Rapid degradation of *Streptococcus pyogenes* biofilms by PlyC, a bacteriophage-encoded endolysin. The Journal of antimicrobial chemotherapy 68, 1818-1824. https://dx.doi.org/ 10.1093/jac/dkt104.

Singh, P.K., Donovan, D.M., and Kumar, A. (2014). Intravitreal injection of the chimeric phage endolysin Ply187 protects mice from *Staphylococcus aureus* endophthalmitis. Antimicrob Agents Chemother 58, 4621-4629. https:// dx.doi.org/ 10.1128/AAC.00126-14.

Solovieva, E.V., Myakinina, V.P., Kislichkina, A.A., Krasilnikova, V.M., Verevkin, V.V., Mochalov, V.V., Lev, A.I., Fursova, N.K., and Volozhantsev, N.V. (2018). Comparative genome analysis of novel Podoviruses lytic for hypermucoviscous *Klebsiella pneumoniae* of K1, K2, and K57 capsular types. Virus Res 243, 10-18. https://dx.doi.org/ 10.1016/j.virusres.2017.09.026.

Steinbacher, S., Baxa, U., Miller, S., Weintraub, A., Seckler, R., and Huber, R. (1996). Crystal structure of phage P22 tailspike protein complexed with *Salmonella* sp O-antigen receptors. Proceedings of the National Academy of Sciences of the United States of America 93, 10584-10588. https://dx.doi.org/ DOI 10.1073/pnas.93.20.10584.

Stockdale, S.R., Mahony, J., Courtin, P., Chapot-Chartier, M.P., van Pijkeren, J.P., Britton, R.A., Neve, H., Heller, K.J., Aideh, B., Vogensen, F.K., *et al.*

(2013). The lactococcal phages Tuc2009 and TP901-1 incorporate two alternate forms of their tail fiber into their virions for infection specialization. The Journal of biological chemistry 288, 5581-5590. https://dx.doi.org/ 10.1074/jbc.M112.444901.

Struthers-Schlinke, J.S., Robins, W.P., Kemp, P., and Molineux, I.J. (2000). The internal head protein Gp16 controls DNA ejection from the bacteriophage T7 virion. Journal of molecular biology 301, 35-45. https://dx.doi.org/ 10.1006/ jmbi.2000.3940.

Stukalov, O., Korenevsky, A., Beveridge, T.J., and Dutcher, J.R. (2008). Use of atomic force microscopy and transmission electron microscopy for correlative studies of bacterial capsules. Applied and environmental microbiology 74, 5457-5465. https://dx.doi.org/ 10.1128/Aem.02075-07.

Stummeyer, K., Schwarzer, D., Claus, H., Vogel, U., Gerardy-Schahn, R., and Muhlenhoff, M. (2006). Evolution of bacteriophages infecting encapsulated bacteria: lessons from *Escherichia coli* K1-specific phages. Mol Microbiol 60, 1123-1135. https://dx.doi.org/ 10.1111/j.1365-2958.2006.05173.x.

Sudiarta, I.P., Fukushima, T., and Sekiguchi, J. (2010). *Bacillus subtilis* CwlP of the SP-{beta} prophage has two novel peptidoglycan hydrolase domains, muramidase and cross-linkage digesting DD-endopeptidase. The Journal of biological chemistry 285, 41232-41243. https://dx.doi.org/ 10.1074/ jbc.M110.156273.

Takac, M., and Blasi, U. (2005). Phage P68 virion-associated protein 17 displays activity against clinical isolates of *Staphylococcus aureus*. Antimicrob Agents Chemother 49, 2934-2940. https://dx.doi.org/ 10.1128/AAC. 49.7.2934-2940.2005.

Tang, J., Lander, G.C., Olia, A.S., Li, R., Casjens, S., Prevelige, P., Jr., Cingolani, G., Baker, T.S., and Johnson, J.E. (2011). Peering down the barrel of a bacteriophage portal: the genome packaging and release valve in p22. Structure 19, 496-502. https://dx.doi.org/ 10.1016/j.str.2011.02.010.

Thwaites, G.E., Edgeworth, J.D., Gkrania-Klotsas, E., Kirby, A., Tilley, R., Torok, M.E., Walker, S., Wertheim, H.F., Wilson, P., and Llewelyn, M.J. (2011). Clinical management of *Staphylococcus aureus* bacteraemia. The Lancet Infectious diseases 11, 208-222. https://dx.doi.org/ 10.1016/ S1473-3099(10)70285-1.

Tomlinson, S., and Taylor, P.W. (1985). Neuraminidase Associated with Coliphage-E That Specifically Depolymerizes the *Escherichia Coli* K1 Capsular Polysaccharide. Journal of virology 55, 374-378.

Vollmer, W., Blanot, D., and de Pedro, M.A. (2008). Peptidoglycan structure and architecture. FEMS microbiology reviews 32, 149-167. https://dx.doi.org/ 10.1111/j.1574-6976.2007.00094.x.

Wang, J., Zhao, F., Sun, H., Wang, Q., Zhang, C., Liu, W., Zou, L., Pan, Q., and Ren, H. (2019). Isolation and characterization of the *Staphylococcus aureus* bacteriophage vB_SauS_SA2. AIMS microbiology 5, 285-307. https:// dx.doi.org/ 10.3934/microbiol.2019.3.285.

Waseh, S., Hanifi-Moghaddam, P., Coleman, R., Masotti, M., Ryan, S., Foss, M., MacKenzie, R., Henry, M., Szymanski, C.M., and Tanha, J. (2010). Orally administered P22 phage tailspike protein reduces *salmonella* colonization in chickens: prospects of a novel therapy against bacterial infections. PloS one 5, e13904. https://dx.doi.org/ 10.1371/journal.pone.0013904.

Wen, Z., and Zhang, J.-R. (2015). Chapter 3 - bacterial capsules. In Molecular Medical Microbiology, M.S. Y.-W. Tang, D. Liu, I. Poxton, and J. Schwartzman, ed. (Boston: Academic Press), pp. 33–53. https://dx.doi.org/ 10.1016/ b978-0-12-397169-2.00003-2.

Whitfield, C. (2006). Biosynthesis and assembly of capsular polysaccharides in *Escherichia coli*. Annual review of biochemistry 75, 39-68. https://dx.doi.org/ 10.1146/annurev.biochem.75.103004.142545.

Wollin, R., Eriksson, U., and Lindberg, A.A. (1981). *Salmonella* bacteriophage glycanases: endorhamnosidase activity of bacteriophages P27, 9NA, and KB1. Journal of virology 38, 1025-1033.

Wong, D., Nielsen, T.B., Bonomo, R.A., Pantapalangkoor, P., Luna, B., and Spellberg, B. (2017). Clinical and Pathophysiological Overview of *Acinetobacter* Infections: a Century of Challenges. Clinical microbiology reviews 30, 409-447. https://dx.doi.org/ 10.1128/CMR.00058-16.

Wright, R.C.T., Friman, V.P., Smith, M.C.M., and Brockhurst, M.A. (2019). Resistance Evolution against Phage Combinations Depends on the Timing and Order of Exposure. Mbio 10. https://dx.doi.org/ 10.1128/mBio.01652-19.

Yang, H., Zhang, H., Wang, J., Yu, J., and Wei, H. (2017). A novel chimeric lysin with robust antibacterial activity against planktonic and biofilm methicillin-resistant *Staphylococcus aureus*. Scientific reports 7, 40182. https://dx.doi.org/ 10.1038/srep40182.

Yang, H., Zhang, Y., Yu, J., Huang, Y., Zhang, X.E., and Wei, H. (2014). Novel chimeric lysin with high-level antimicrobial activity against methicillin-resistant *Staphylococcus aureus in vitro* and *in vivo*. Antimicrob Agents Chemother 58, 536-542. https://dx.doi.org/ 10.1128/AAC.01793-13.

Yap, M.L., and Rossmann, M.G. (2014). Structure and function of bacteriophage T4. Future Microbiol 9, 1319-1337. https://dx.doi.org/ 10.2217/fmb.14.91.

Young, R. (2013). Phage lysis: do we have the hole story yet? Curr Opin Microbiol 16, 790-797. https://dx.doi.org/ 10.1016/j.mib.2013.08.008.

Zelmer, A., Martin, M.J., Gundogdu, O., Birchenough, G., Lever, R., Wren, B.W., Luzio, J.P., and Taylor, P.W. (2010). Administration of capsule-selective endosialidase E minimizes upregulation of organ gene expression induced by experimental systemic infection with *Escherichia coli* K1. Microbiology 156, 2205-2215. https://dx.doi.org/ 10.1099/mic.0.036145-0.

Zeman, M., Bardy, P., Vrbovska, V., Roudnicky, P., Zdrahal, Z., Ruzickova, V., Doskar, J., and Pantucek, R. (2019). New Genus Fibralongavirus in *Siphoviridae* Phages of *Staphylococcus pseudintermedius*. Viruses 11. https:// dx.doi.org/ 10.3390/v11121143.

Zhang, G., Meredith, T.C., and Kahne, D. (2013). On the essentiality of lipopolysaccharide to Gram-negative bacteria. Curr Opin Microbiol 16, 779-785. https://dx.doi.org/ 10.1016/j.mib.2013.09.007.

Zivanovic, Y., Confalonieri, F., Ponchon, L., Lurz, R., Chami, M., Flayhan, A., Renouard, M., Huet, A., Decottignies, P., Davidson, A.R., *et al.* (2014). Insights into Bacteriophage T5 Structure from Analysis of Its Morphogenesis Genes and Protein Components. Journal of virology 88, 1162-1174. https://dx.doi.org/ 10.1128/Jvi.02262-13.

Chapter 13

# Crystallographic Structure Determination of Bacteriophage Endolysins

**Marta Sanz-Gaitero[1,2,3] and Mark J. van Raaij[1*]**

[1]Centro Nacional de Biotecnologia, CSIC, Calle Darwin 3, E-28049 Madrid, Spain.
[2]Cork Institute of Technology, Bishopstown, Cork T12 P928, Ireland.
[3]Current address: Centre for Molecular Medicine Norway, University of Oslo, Gaustadalléen 21, 0349 Oslo, Norway.

*Corresponding author: mjvanraaij@cnb.csic.es

DOI: https://doi.org/10.21775/9781913652517.13

## Abstract

Bacteriophages produce endolysins that target and cleave the hosts peptidoglycan to release their progeny at the end of the infection cycle. These proteins can be used for the eradication of pathogenic bacteria, but also for their detection. Endolysins may contain a single catalytic domain or several domains, including a cell wall binding domain. To understand their function in detail and design mutated or chimeric molecules with novel properties, knowledge of their structures and detailed mechanisms is necessary. X-ray protein crystallography is an excellent method to obtain high-resolution structures of biological macromolecules, and here we describe the method and the folds of known endolysin domains.

## Introduction

Bacteriophages (phages) are a promising treatment for antibiotic-resistant pathogens (see chapter by Hoyle et al., this book). However, they also present some drawbacks. A safety concern is their possible impact on non-target microbiota. Phages can modify their bacterial targets causing them to express virulence genes or transduce DNA between bacteria, although these risks can be reduced by avoiding the use of temperate phages. As they are unknown entities to the host, they may induce both adaptive and innate immunological reactions (Majewska et al., 2019). Phage formulations may also contain harmful products, such as endotoxins, due to incomplete purification. Specific self-replication of phages at the infection site is a major advantage. However, a self-replicating entity may change its properties by spontaneous mutation or recombination with host sequences such as prophages. Self-replication also means that the pharmacological dose is difficult to calculate (see chapter by Danis-Wlodarczyk et al., this book). To avoid these drawbacks, in some cases, it may be better to use isolated phage proteins.

At the end of the infection cycle, once the phage progeny has been generated, the new virions need to be released from the bacteria. To this end, phages generate proteins called endolysins, able to break peptidoglycan links. This causes the bacteria to burst, due to the osmotic pressure. However, to reach the peptidoglycan layer, endolysin molecules first have to cross the cytoplasmic membrane. To allow this passage, phages produce pore-generating proteins called holins (Young, 2013), which insert themselves in the membrane and generate pores, through which the endolysins pass. Endolysins may have a single domain, with a catalytic activity. Other endolysins are multi-modular, combining cell wall binding domains and catalytic domains. Different catalytic domains cleave different chemical bonds of the peptidoglycan: muramidases, glycosidases, peptidases or amidases (Loessner, 2005).

Endolysins can cause the lysis of Gram-positive bacteria when applied exogenously (Fischetti, 2008; See chapter by Arroyo-Moreno et al., this book). In the case of Gram-negative bacteria, endolysins can be genetically engineered to allow them to pass through the outer membrane and lyse their host, for example by adding an outer-membrane-permeabilizing peptide (O'Sullivan et al., 2016; see chapter by Gutiérrez and Briers, this book).

Endolysin applications are varied. They can be used as antimicrobial agents, for bacterial identification or for surface sterilization. Each of these applications requires the use of endolysins with different characteristics. For some applications, a wide lytic range me be necessary, for others a high specificity may be desired. Resistance to different temperature, pH, salinity or humidity ranges may also be required.

To obtain custom-designed endolysins, several techniques can be applied. Mutagenesis and protein truncation can be used to modify the lytic and antibacterial activity, cell-wall binding domain dependence, specificity, half-life and thermostability of endolysins. Recombining domains or domain swapping has been used successfully to create endolysins with altered properties (chimeolysins), such as activity against a certain type of bacteria or increased solubility (Yang et al., 2014; Gerstmans et al., 2018). Sequences can be added to these domains to increase their solubility or hydrophobicity, as in the case of artilysins, where an outer membrane permeabilizing region is added to allow them to pass through the bacterial outer membrane, allowing them to be effective even when applied exogenously. Cell wall binding domains of endolysins have been fused to fluorescent proteins or antibodies for bacterial detection purposes.

To reliably design tailor-made endolysins, knowledge about their characteristics, catalytic mechanisms, regulation is very advantageous. High-resolution structures of endolysins help to determine these properties. Determining the structure not only helps to elucidate the mechanism of action of that specific endolysin, but also allows to predict the structure of closely related enzymes.

### Structure determination methods

Three main techniques can be used for structure determination: NMR spectroscopy, cryo-electron microscopy and X-ray crystallography (Mateu, 2013; Gauto et al., 2019). These techniques can be used alone or in combination (Everett et al., 2015; Wang et al., 2017).

*Nuclear Magnetic Resonance (NMR) spectroscopy*

Atomic nuclei are charged spinning particles. When exposed to a strong magnetic field and radiofrequency radiation, they will produce an electromagnetic signal with a frequency characteristic of the magnetic field at the nucleus. This signal will be different depending on the chemical environment, and the magnetic properties of every atom, making it useful for structure determination (Krishnan, 2019). The main advantages of NMR spectroscopy are that structures are determined in solution and the possibility of investigating protein dynamics and ligand binding properties. The main disadvantages are the requirement of high amounts of isotopically labelled sample and the fact that for large proteins and protein complexes, signals start to overlap, limiting the technique to smaller proteins (generally up to 30-50 kDa).

*Cryo-Electron Microscopy (Cryo-EM)*

In cryo-electron microscopy, small samples, in which the macromolecular particles are randomly oriented, are vitrified and kept at very low temperature. In the microscope, they are exposed to electrons to produce images of individual particles. These images are averaged to produce several two-dimensional classes that will be used to reconstruct the three-dimensional structure of the macromolecule. The main advantages of cryo-EM are the small amount of sample required and the fact that it is not necessary to obtain protein crystals. It is also suitable for large complexes, such as phage particles. On the other hand, cryo-EM does not work so well for smaller proteins (less than around 150 kDa). It is also not routinely possible to achieve atomic resolution structures.

*X-ray crystallography*

Crystallography uses diffraction to determine the position and arrangement of the atoms in a crystal. Using X-rays, atomic or near-atomic resolution can be achieved for small molecules, proteins and, in favourable cases, large complexes such as whole viruses. The main drawback of this technique is that crystals are necessary, and crystallisation of proteins requires high amounts of very pure and homogeneous sample and time. Often, a well-diffracting crystal can simply not be obtained. In addition to that, a static crystalline structure is obtained, although molecular dynamic techniques may provide clues about the dynamics of the protein. Detailed protein-ligand interactions can be determined in favourable cases by co-crystallisation with a ligand or by soaking of ligands into the crystal. The latter is possible because protein crystals contain large solvent channels. Full-length endolysins vary in size. Their molecular weight may be too large to use NMR spectroscopy to determine their structure but too small to use cryo-EM. Due to this and to the nearly-atomic resolution obtained, X-ray crystallography is often the technique of choice.

*Main steps in X-ray crystallography*

The first step of X-ray crystallography is obtaining a protein sample with enough quality to be crystallised (Figure 1). This sample should be highly pure, stable, homogeneous, reproducible and abundant (Bergfors, 2009). If the protein contains several domains connected by flexible regions, the chances of getting protein crystals decrease. In these cases, truncated versions of the protein, i.e. removing several of the domains or limited proteolysis techniques to get rid of the flexible loops, are often useful.

After purification and concentration, the protein will remain in a soluble state as long as its concentration remains below its saturation point. In a supersaturated solution, protein molecules will tend to aggregate and, often, precipitate. To crystallise a protein, it is necessary to achieve protein supersaturation in a controlled way, causing the protein to aggregate in an ordered fashion, forming protein crystals. If supersaturation is reached slowly, precipitation will not occur immediately, and the protein will form nuclei. These nuclei are formed by a small number of protein molecules ordered in a lattice. Once these nuclei are formed, additional protein molecules will be incorporated in an ordered way, contributing to the crystal growth. The crystal will continue growing until the protein concentration decreases and gets below the supersaturation level.

Crystallisation is influenced by variables such as protein concentration, pH, temperature, and the use of different precipitants, salts and additives that can promote or delay protein aggregation (McPherson et al., 2014; Kirkwood et al., 2015). To crystallise a protein, different setups can be used. The most common is sitting-drop vapour diffusion. A drop containing a mix of the protein and the crystallisation condition in different proportions sits on a pedestal next to a well or reservoir that contains a larger volume of the crystallisation condition. The system is covered by a non-permeable film. Due to the difference in concentration between the drop and the well, the solute of the drop will evaporate and will be transferred to the reservoir, causing the protein to slowly concentrate. If the conditions are optimal, supersaturation and controlled aggregation occur, leading to nucleation and crystal growth (Rupp, 2009).

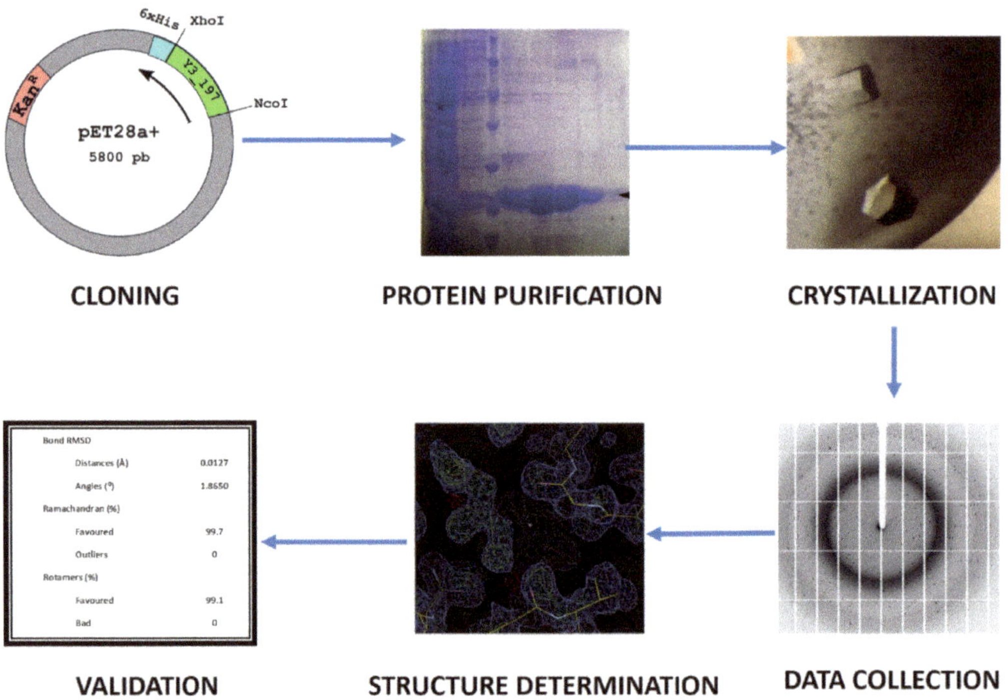


**Figure 1**: Main steps for protein structure determination using X-ray crystallography. See text for more details.

Once a good quality crystal is obtained, the three-dimensional structure of the protein forming it can be determined. This can be achieved by analysing the diffraction pattern generated by a highly parallel X-ray beam aimed at the crystal. After cryo-protecting and flash-cooling the crystal using liquid nitrogen, it is exposed to a source of X-rays. X-rays are electromagnetic radiation with wavelengths ranging between 0.1 and 100 Å. Diffraction will occur if the wavelength of the radiation is similar to the spacing between the atoms of the lattice. In protein crystallography, the wavelengths most commonly used range from 0.6 Å to 2.5 Å, similar to the length of the chemical bonds in these molecules (Als-Nielsen et al., 2011). Nowadays, the most commonly used X-ray sources are synchrotron radiation sources. Diffracted beams form a diffraction pattern, collected on a detector. The crystal is mounted on a goniometer head that rotates during X-ray data collection. That way, diffraction data can be collected at different angles, allowing for a complete diffraction data set.

X-rays are scattered from every point of the crystal, with a strength proportional to the concentration of electrons at that point. X-rays scattered along any particular direction interfere with each other and originate features in the diffraction pattern through constructive or destructive interference phenomena. For a large enough and well-ordered crystal, diffraction only occurs when Braggs' law is satisfied (Bragg, 1913). The spacing of the diffraction spots depends on the crystal symmetry and the size and shape of the crystal asymmetric unit (the smallest identifiable unique feature in the crystal that, when repeated, give rise to the complete, regular crystal). The intensity of the diffraction spots, however, depends on the internal structure of the asymmetric unit, which may contain one or more protein molecules.

When performing an X-ray crystallography experiment on a protein crystal, we obtain, for each spot in the diffraction pattern, the exact position and its intensity. Using this information, the symmetry of the crystal can be determined. Afterwards, the intensities of all the diffraction spots are determined (Leslie, 2006). The electron density in each asymmetric unit is related to the intensities of the diffraction spots by a mathematical formula known as the Fourier transform. Using this formula, if we know the exact content of the asymmetric unit (location and nature of each atom), we can predict the diffraction pattern. However, the inverse, and the goal of our experiment, i.e. determining the asymmetric unit contents from the diffraction pattern, is not that simple. To calculate the electron density within the asymmetric unit, we need not only the intensities, but also the phase of the scattered X-rays (Taylor, 2010). Unfortunately, only the intensity of the diffraction spots can be measured experimentally, as the phase is lost. This is known as the phase problem (Cowtan, 2003).

To solve the phase problem, several approaches can be used (Taylor, 2010). In molecular replacement, we use the phases calculated from a protein model derived from a structure with high sequence similarity to ours (greater than 30% sequence identity as a rule of thumb), assuming their structures will be similar. For molecular replacement, first the model has to be placed in the correct location and orientation in the unit cell, using sophisticated software. In experimental phasing, a limited number of heavy atoms incorporated at specific places into the protein structure (for example mercury ions bound to accessible cysteine residues). This is usually done by trial and error soaking of heavy atom compounds into pre-formed crystals. A commonly used alternative method is to incorporate seleno-methionine residues into the protein by expressing the protein in special growth media and crystallising this seleno-methionine derivative version of the protein. The substructure of the heavy atoms is simpler and easier to calculate than the complete protein structure and can be used for phase determination.

Once the phases have been estimated, an electron density map of the protein can be obtained. A model that fits this data is then built, and if built correctly, this model then provides improved phases for calculation of an improved electron density map. The model is refined against the data, incorporating restraints limiting excessive variation of bond lengths, bond angles and other known protein structural parameters. As a final step, the structural model of the protein will be validated. The geometry of the final model will be analysed to detect potential outliers in rotamer prevalence, Ramachandran configurations, close contacts and other parameters that might have been introduced by the building or refinement process (Kleywegt, 2000). If the geometry of the model is correct, and the model to data statistics are reasonable, the model is deposited in the PDB, the Protein Data Bank (www.pdbe.org).

## Endolysin structure

Catalytic domains of endolysins belong to one of the following enzyme families (Loessner, 2005; Fenton et al., 2010): N-acetylmuramidases (lysozymes), N-acetyl-β-D-glucosaminidases (glycosidases), N-acetylmuramoyl-L-alanine amidases, L-alanoyl-D-glutamate endopeptidases and interpeptide bridge-specific endopeptidases (Figure 2). Single-domain endolysins are exclusive to phages infecting Gram-negative bacteria and consist of just one catalytic domain. However, most endolysins exhibit a modular configuration with one or several catalytic domains and a cell wall binding domain. These domains are connected by flexible linkers. The lytic spectrum of an endolysin depends on its cell wall binding domain and on the conservation among bacteria of the cell wall motif that the cell wall binding domain recognises (Nelson et al., 2012, Navarre et al., 1999; O'Flaherty et al., 2005; Fernandes et al., 2018).

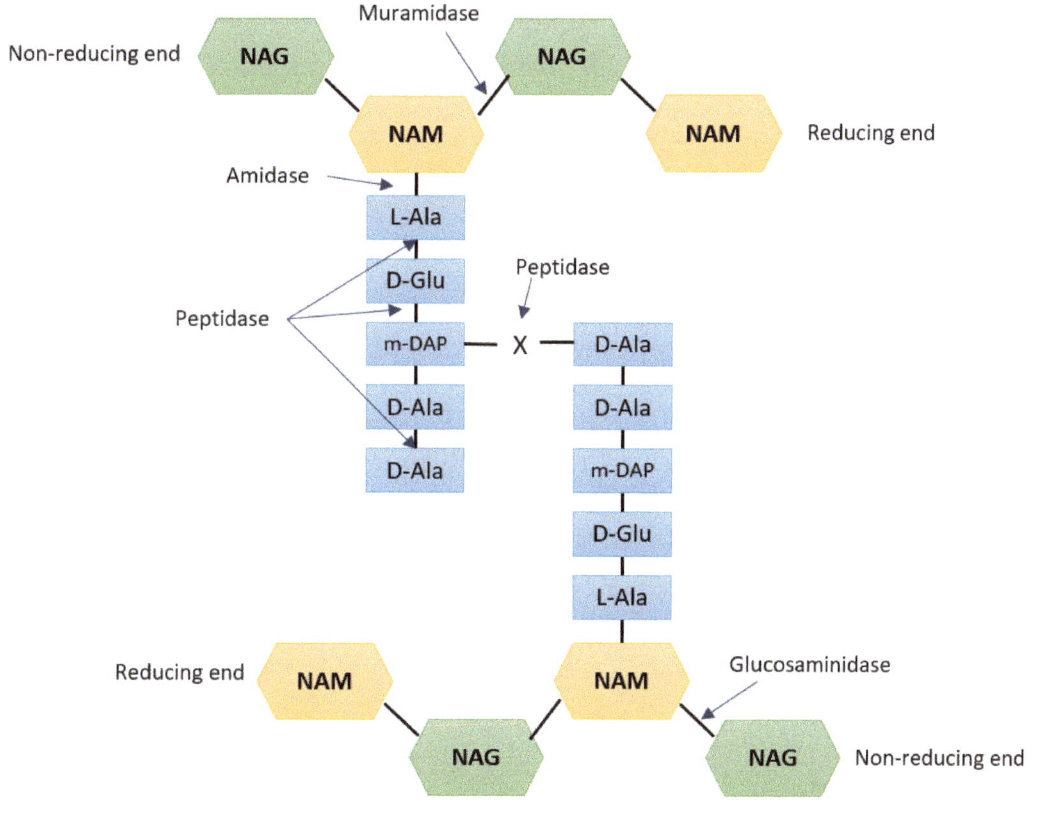


**Figure 2**: Enzymatic activities of peptidoglycan degrading enzymes. The basic structure of Gram-negative cell wall peptidoglycan, indicating the possible enzymatic activities of peptidoglycan-degrading enzymes and the bonds they cleave.

*N-acetylglucosaminidases or glycosidases*

N-acetylglucosaminidases hydrolyse the β-(1→4) glycosidic bond between N-acetylmuramic acid (MurNAc) and N-acetylglucosamine (GlcNAc). The only structure of a phage endolysin showing a N-acetylglycosidase domain is PlyC, the lysin from *Streptococcus* phage C1 (McGowan et al., 2012; Broendum et al., 2018). PlyC is a powerful multimeric endolysin that consists of two separate proteins: a single copy of the 50 kDa PlyCA subunit and at least eight copies of the 8 kDa PlyCB subunit. The eight PLyCB subunits form the cell wall binding domain of the protein and are arranged in a planar octameric ring 80 Å in diameter and 20 Å high. The PlyCA subunit contains two different domains, a carboxy-terminal cysteine, histidine-dependent amidohydrolase/peptidase (CHAP) domain and an amino-terminal N-acetylglycosidase domain. The N-acetylglycosidase domain is structurally related to N-acetylglucosaminidase domains from the glycosyl hydrolase 73 (GH73) family, with a highly conserved core formed by six α-helices. This core contains a conserved glutamic acid residue (glutamate 78) essential for catalytic activity and a substrate-binding groove (Figure 3A).

*N-acetylmuramidases (lysozymes or muramidases)*

N-acetylmuramidase domains hydrolyse the β-(1→4) glycosidic bond between N-acetylglucosamine (GlcNAc) and N-acetylmuramic acid (MurNAc) in the sugar moiety of the cell wall. Phage muramidase domains belong to at least three different families of glycosyl hydrolases called GH19, GH24 and GH25. Most belong to the GH25 family, including Cpl-1, the endolysin from the *Streptococcus pneumoniae* phage Cp-1 (Perez-Dorado et al., 2007). The Cpl-1 structure is formed by two domains connected to each other with a linker. The cell wall binding domain presents six choline-binding repeats. The muramidase domain forms an irregular $(\beta/\alpha)_5\beta_3$ barrel, in which the catalytic residues are located. All members of the GH25 family share this barrel fold. Hydrolysis in this group of muramidases takes place via a general acid/base mechanism that requires two acidic amino acids. One acidic residue behaves as the proton donor and the other acts as the nucleophile, promoting hydrolytic attack by a water molecule. These residues are located in a catalytic groove and are strongly conserved. In the case of Cpl-1, the residues are glutamate 94 and aspartate 10 (Figure 3B, left).

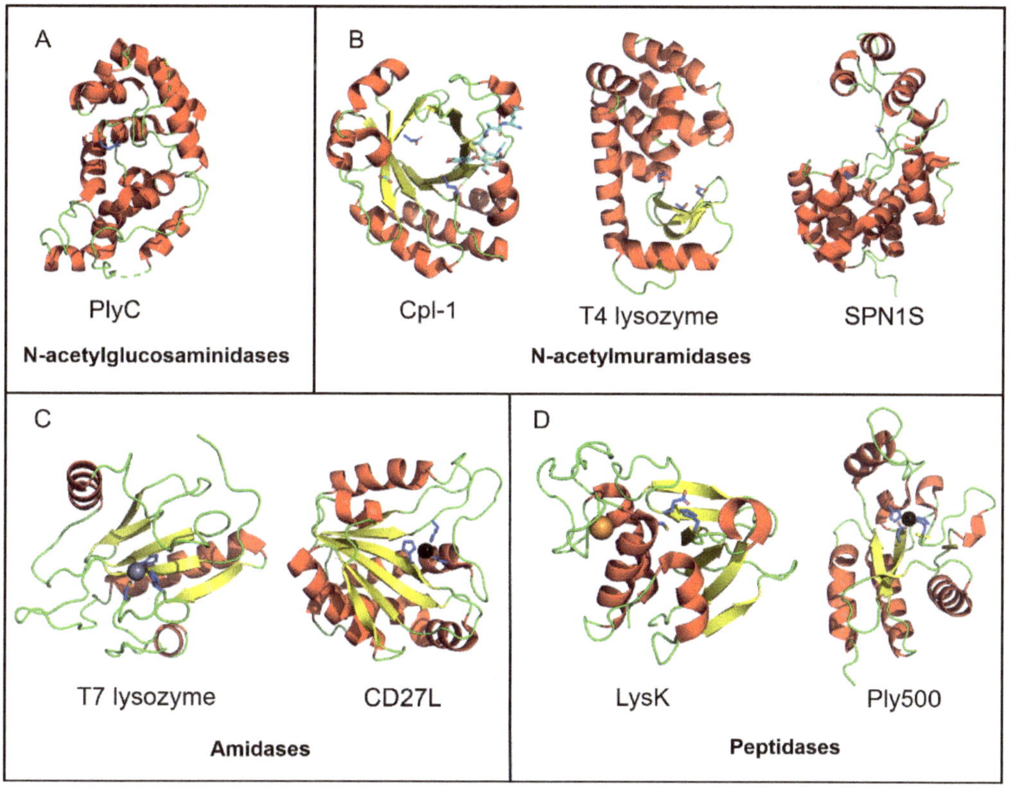


**Figure 3**. Representation of the variety of enzymatically active domains in phage endolysins. The proteins are shown in cartoon representation, coloured according to their secondary structure. α-helices are represented in red, β-sheets in yellow and loops in green. Carbon atoms of catalytic residues are depicted in dark blue, carbon atoms of ligands in cyan, zinc ions as dark grey or black spheres and calcium ions as orange spheres. **A.** N-acetylglucosaminidase domain from the streptococcal phage C1 lysin PlyC (PDB entry 4F88). **B.** Different families of N-acetylmuraminidases. Left: Cpl-1, endolysin from the *Streptococcus pneumoniae* phage Cp-1, representative of the GH25 family, (PDB entry 2IXU). Centre: T4 lysozyme, representative of the GH24 family (PDB entry 4LZM). Right: Endolysin from *Salmonella* phage SPN1S, representative of the GH19 family (PDB entry 4OK7). **C.** Different families of N-acetylmuramoyl-L-alanine amidases. Left: T7 lysozyme, representative of the Pfam amidase_2 family (PDB entry 1LBA). Right: CD27L, *Clostridium difficile* φCD27 phage endolysin, representative of the VanY d-Ala-d-Ala carboxypeptidase subfamily (PDB entry 3QAY). **D.** Different families of peptidases. Left: $CHAP_K$, CHAP domain of the staphylococcal phage K endolysin LysK (PDB entry 4CSH). Right: pLy500, L-alanoyl-D-glutamate peptidase from *Listeria* phage A500 (PDB entry 2VO9), representative of the M15C family of endopeptidases.

One of the most studied phage muramidases is T4 lysozyme (Bell et al., 1991). T4 lysozyme has been studied in depth and exhaustive mutational analyses have been performed to infer the function of every residue (Rennell et al., 1991). This enzyme is an inverting glycoside hydrolase that belongs to the GH24 family. T4 lysozyme is formed by two domains, that can move with respect to each other. The active-site cleft is located in the interface between the domains. Without the peptidoglycan substrate, T4 lysozyme remains in its open conformation. After peptidoglycan is added, the lysozyme transitions to a closed processive conformation. The change of conformation is caused by an 8 Å hinge motion that brings the two lobes of the enzyme closer to each other (Akhterov et al., 2015). Once the enzyme is in a closed conformation, a nucleophilic water molecule, activated by aspartate 20, attacks the β-(1→4) linked anomeric C1 carbon of the peptidic N-acetyl muramic acid, while the general acid glutamate 11 donates a proton to the leaving aglycone, N-acetyl glucosamine (Kuroki et al., 1995). Mediated by hydrogen bonds, threonine 26 helps to position the water molecule that performs the nucleophilic attack (Figure 3B, centre).

The endolysin from *Salmonella* phage SPN1S (Park et al., 2014) belongs to the GH19 chitinase family, although it has muramidase activity. It is a helical monomer with a Pacman-like shape, formed by two domains: a large domain made up of the amino-terminal and carboxy-terminal regions of the protein and a smaller domain made up of the central part of the protein. The large domain is formed by a total of eight α-helices. Helix α3 is located in the central part of the domain, surrounded by helices α1, α2 and α10-α13. The small domain is formed by the five central α-helices (α4-α8) and a $3_{10}$-helix. Helices α4, α6 and α7 form an antiparallel three-helix bundle whose axis is perpendicular to one of the bundles formed by the helices in the large domain. These two regions are connected by two long loops. As a result, a deep wide groove is formed between both domains, where the catalytic cleft is located. The catalytic residues from family 19 chitinases are two glutamates. A glutamate located at the end of the third α-helix (glutamate 49 in SPN1) acts as the catalytic general acid, whereas another glutamate located in a more variable loop-like structure (glutamate 58 in SPN1S) acts as the catalytic general base (Figure 3B, right).*N*-

### *Acetylmuramoyl-L-alanine amidases*

Amidases cleave the amide bond connecting the sugar and peptide moieties of the bacterial cell wall. According to the classification performed by Broendum et al. (2018), two different zinc-dependent N-acetylmuramoyl-l-alanine amidase folds exist. The most common amidase domain belongs to the Pfam amidase_2 family (Firczuk et al., 2007). Although several phage endolysins contain this kind of fold, the more studied one has been T7 lysozyme (Cheng et al., 1994). These enzymes contain a five-stranded, mostly parallel β-sheet with conserved topology, flanked by a varying number of α-helices. The active site zinc ion is located on one of the sides of the β-sheet, at the bottom of the peptidoglycan binding groove. The zinc ion is tetrahedrally coordinated to a water molecule, two histidine residues and a cysteine residue. These residues are highly conserved (Figure 3C, left).

Other amidases, like the *Clostridium difficile* φCD27 phage endolysin CD27L (Mayer et al., 2011), belong to the VanY d-Ala-d-Ala carboxypeptidase subfamily. Despite belonging to this subfamily, they have N-acetylmuramoyl-L-alanine amidase activity. They exhibit an α/β fold, with a core of six beta strands surrounded by five alpha helices. A zinc ion is present in the catalytic groove, coordinated by two histidine residues and a glutamic acid. In the case of CD27L, these residues are histidine 9, histidine 84 and glutamate 26 (Figure 3C, right).

### *Peptidases*

Peptidases cleave the peptide moiety of the peptidoglycan. Depending on the links in the peptidoglycan they target, there are two types, L-alanoyl-D-glutamate endopeptidases and interpeptide bridge-specific peptidases.

Most of the domains with peptidase activity found in phage endolysins are histidine-dependent amidohydrolase/peptidase (CHAP) domains. CHAP domains show a great versatility and can have N-acetylmuramoyl-L-alanine amidase or D-alanyl-glycyl endopeptidase activities (Bateman et al., 2003). CHAP domains are formed by two α-helices clustered against a six-stranded antiparallel β-sheet, accommodating a long and deep hydrophobic catalytic groove between them. The catalytic triad is formed by a cysteine, a histidine and a glutamic acid residue, which are strongly conserved. Some CHAP domains, such as $CHAP_K$, the CHAP domain of the staphylococcal phage K endolysin LysK (Sanz-Gaitero et al., 2014), contain a calcium ion bound to an EF-hand-like motif located in the amino-terminal part of the protein, involving residues from the long loop connecting the first and second α-helices (Figure 3D, left). This ion is necessary for the enzyme to be catalytically active. However, some other CHAP domains, like the one in pLyC, the lysin from the streptococcal phage C1, do not contain a calcium binding motif (McGowan et al., 2012).

The only phage endolysin with endopeptidase activity that does not contain a CHAP domain is pLy500, the L-alanoyl-D-glutamate peptidase from *Listeria* phage A500 (Korndoerfer et al., 2008). Its catalytic domain belongs to the M15C family of endopeptidases. The overall shape of Ply500 catalytic domain resembles a sofa, with an α-helix and the three antiparallel β-strands forming the 'seat'. Another α-helix and a loop give rise to the 'backrest', and an additional loop conforms to the 'left armrest' of the sofa. A catalytically active zinc ion is tetrahedrally coordinated by a highly conserved set of residues (histidine 80, aspartate 87 and histidine 133) and a water molecule is located in a groove formed between the 'seat' and the 'backrest' (Figure 3D, right).

*Cell wall binding domains*

Cell wall binding domains are combined with enzymatically active domains to form modular endolysins. Cell wall binding domains bind non-covalently to a specific substrate (usually carbohydrate) found in the cell wall of the host bacterium, offering some degree of specificity to the endolysins. In general, while the catalytic domains of endolysins show a high sequence similarity among them, sequence similarity between cell wall binding domains is much lower (Fischetti, 2008). Cell wall binding domains show high affinity to their substrate, controlling diffusion of the enzymes and preventing the killing of neighbouring cells that have not yet produced phage particles (Loessner et al., 2002).

According to their structure, there are four main cell wall binding domain classes (Broendum et al., 2018):

1. SH3b domains, with a β-barrel fold, which can contain between five and seven β-strands connected by linkers. PlyPSA, the endolysin from the *Listeria* phage PSA (Korndoerfer et al., 2006) contains two SH3b repeats in its cell wall binding domain. These copies are held together by means of swapped β-strands (Figure 4A).
2. Choline binding modules (ChBr), often found in pneumococcal endolysins such as Cpl-1, which recognise the teichoic acids of the pneumococcal cell wall (Perez-Dorado et al., 2007). They are formed by multiple tandem copies of a short amino acid sequence. The Cpl-1 cell wall binding domain contains an amino-terminal super-helical subdomain formed by the first four ChBr repeats, where the choline-binding pockets are located, and a carboxy-terminal β-sheet that interacts with the catalytic domain (Figure 4B).
3. Three-helix bundles. In endolysins targeting Gram-positive bacteria, such as the *Streptococcus* phage CP-7 endolysin Cpl-7 (Bustamante et al., 2017), several copies of this fold compose the cell wall binding domain. Endolysins targeting Gram-negative bacteria only contain one copy of the tree-helix bundle (Figure 4C).
4. α/β multimers. Several endolysin cell wall binding domains show a similar fold consisting of a central parallel or antiparallel four-stranded β-sheet flanked by two helices. One example of this kind of fold is PlyC, the lysin from the streptococcal phage C1 (McGowan et al., 2012). These cell wall binding domains oligomerize to form a functional assembly (Figure 4D).

*Endolysin variability*

Phage-encoded endolysins show a lot of diversity. They can contain a single globular catalytically active domain or combine different cell wall binding domains and enzymatically active domains to form a multi-modular endolysin. This flexibility in their design makes phage endolysins highly versatile proteins, with different catalytic activities, host ranges and stabilities. A comparative genomic analysis of 723 endolysins identified 24 different enzymatically active domains and 13 cell wall binding domain types arranged in 89 different architectures, highlighting the high variety among phage endolysins (Oliveira et al., 2013).

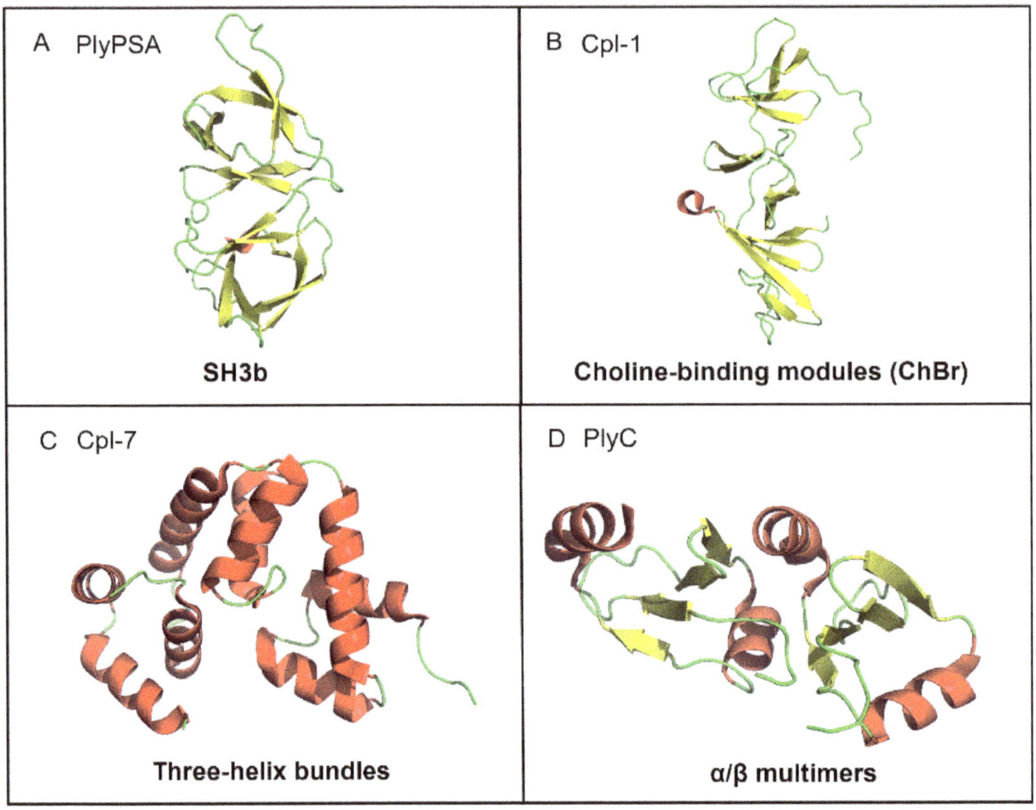


**Figure 4.** Representation of the variability of cell wall binding domains in phage endolysins. The proteins are shown in cartoon representation, coloured according to their secondary structure. α-helices are represented in red, β-sheets in yellow and loops in green. **A.** PlyPSA, endolysin from the *Listeria* phage PSA (PDB entry 1XOV) contains two SH3b copies in its cell wall binding domain. **B.** Choline binding module (ChBr) in the cell wall binding domain of Cpl-1, the endolysin from the *Streptococcus pneumoniae* phage Cp-1 (PDB entry 2IXU). **C.** *Streptococcus* phage CP-7 endolysin Cpl-7 cell wall binding domain showing three copies of a three-helix bundle (PDB entry 5I8L). **D.** Dimer of PlyCB, the cell wall binding domain of the endolysin from the *Streptococcus* phage C1 (PDB entry 4F87), forming α/β multimers.

The multi-modular nature of many endolysins offers the opportunity to design tailor-made proteins with properties adapted for different applications, by combining, truncating or mutating these domains. Due to the flexibility between modules, not many full-length multi-modular endolysins structures have been published. A common approach in structure determination of multi-modular endolysins is to study truncated versions of the endolysin, containing just one of the stable domains.

Some endolysins are formed by more than one peptide chain. Staphylococcal phage C1 endolysin, PlyC, is formed by single copy of the 50 kDa PlyCA subunit and at least eight copies of the 8 kDa PlyCB subunit (see the section *N-acetylglucosaminidases*). *Enterococcus* phage IMEEF1 endolysin LysIME-EF1 (Zhou et al., 1994) is also a two-component endolysin. The active protein is a hetero-tetramer formed by one full-length LysIME-EF1 copy and three additional carboxy-terminal cell wall binding domains. The cell wall binding domains are truncated versions of the full-length protein translated from an alternative ribosomal binding site within the gene.

*Clostridium difficile* phage φCD27 endolysin CD27L (Dunne et al., 2014) contains an amino-terminal amidase domain and a carboxy-terminal cell wall binding domain. A self-cleavage site is located in the linker between both domains. CD27L exists in two different dimeric states. The tensed inactive state is a head-on dimer, where the two autocleavage sites are far away from each other. The relaxed active state is the side-by-side dimer, which promotes autocleavage and the release of the amino-terminal catalytic domain from the carboxy-terminal cell wall binding domain. Autocleavage increases the action radius of the catalytic module and may allow it to penetrate the bacterial cell wall independently of the cell wall binding domain.

The structures from two different 'SAR' endolysins have been solved using X-ray crystallography, Lyz of coliphage P1 (Xu et al., 2005) and R21 of the lambdoid phage 21 (Sun et al., 2009). These enzymes contain an amino-terminal SAR (Signal-Anchor-Release) domain that allows the translocation to the periplasm through the Sec system, in a holin-independent manner. This SAR domain is not cleaved by the signal peptidase after the translocation, and the SAR endolysin remains attached to the membrane in its inactive conformation during phage maturation thanks to the amino-terminal transmembrane domain embedded in the inner membrane. To become active, the endolysin must be released from the inner membrane and refolded into the catalytically active form. Dissipation of the proton motive force causes the transmembrane domain to be released from the cell membrane. After leaving the bilayer, the protein refolds into a periplasmic active form (Fernandes et al., 2018). The holins facilitate this process by mediating the collapse of the proton motive force (Xu et al., 2004; Xu et al., 2005; Sun et al., 2009).

## Conclusion

X-ray crystallography has contributed significantly towards the knowledge of endolysin structure and mechanism. Structures of catalytic domains have provided clues as to their enzymatic mechanisms, while structures of cell wall binding domains have informed us as to how they specifically recognise the peptidoglycan of their bacterial hosts. However, structures of endolysin domains complexed with peptidoglycan substrates are still scarce. Future improvements in the organic synthesis of peptidoglycan analogues will hopefully fill this void. In any case, structural biology in general, and X-ray crystallography in particular, will continue to play a role in determining structures of endolysins, endolysin domains and their complexes with substrate analogues. In turn, this structural information will guide the design of more stable and more efficient enzybiotics to eradicate pathogenic bacteria and also positively influence the use of endolysin-based bacterial detection methods.

## Acknowledgements

This work was supported by grant BFU2017-82207-P from the Spanish Ministry of Science and Innovation to MJvR and by a RISAM PhD fellowship from the Cork Institute of Technology to MS-G.

## References

Akhterov, M.V., Choi, Y., Olsen, T.J., Sims, P.C., Iftikhar, M., Gul, O.T., Corso, B.L., Weiss, G.A. and Collins, P.G. (2015) Observing lysozyme's closing and opening motions by high-resolution single-molecule enzymology. ACS Chem. Biol. 10, 1495–1501. https://doi.org/10.1021/cb500750v.

Alcorlo, M., Martinez-Caballero, S., Molina, R. and Hermoso, J.A. (2017) Carbohydrate recognition and lysis by bacterial peptidoglycan hydrolases. Curr. Opin. Struct. Biol. 44, 87–100. https://doi.org/10.1016/j.sbi.2017.01.001.

Als-Nielsen, J. and McMorrow, D. (2011) Elements of modern X-ray physics. (New York, USA: John Wiley & Sons). https://doi.org/10.1002/9781119998365.

Bateman A., Rawlings N.D. (2003) The CHAP domain: a large family of amidases including GSP amidase and peptidoglycan hydrolases. Trends. Biochem. Sci. 28, 234–7. https://doi.org/10.1016/S0968-0004(03)00061-6.

Bell, J. A., Wilson, K. P., Zhang, X.J., Faber, H. R., Nicholson, H. and Matthews, B. W. (1991) Comparison of the crystal structure of bacteriophage T4 lysozyme at low, medium, and high ionic strengths. Proteins 10, 10–21. https://doi.org/10.1002/prot.340100103.

Bergfors, T. (2009) Protein crystallization. (La Jolla, California, USA: International University Line).

Bragg, W.L. (1913) The diffraction of short electromagnetic waves by a crystal, Proceedings of the Cambridge Philosophical Society, 17, 43–57.

Broendum, S.S., Buckle, A. M. and McGowan, S. (2018) Catalytic diversity and cell wall binding repeats in the phage-encoded endolysins. Mol. Microbiol. 110, 879–896. https://doi.org/10.1111/mmi.14134.

Bustamante, N., Iglesias-Bexiga, M., Bernardo-García, N., Silva-Martín, N., García, G., Campanero-Rhodes, M.A., et al. (2017) Deciphering how Cpl-7 cell wall-binding repeats recognize the bacterial peptidoglycan. Sci. Rep., 7, 16494. https://doi.org/10.1038/s41598-017-16392-4.

Cheng X., Zhang X., Pflugrath J.W., Studier F.W. (1994) The structure of bacteriophage T7 lysozyme, a zinc amidase and an inhibitor of T7 RNA polymerase. Proc. Natl. Acad. Sci. U. S. A. 91 ,4034-4038. https://doi.org/10.1073/pnas.91.9.4034.

Cowtan K (2003) Phase problem in X-ray crystallography, and its solution. In Encyclopedia of life sciences. (Chichester, UK: Wiley). https://doi.org/10.1038/npg.els.0002722.

Dunne, M., Mertens, H. D., Garefalaki, V., Jeffries, C. M., Thompson, A., Lemke, E. A., Svergun, D. I., Mayer, M. J., Narbad, A., and Meijers, R. (2014). The CD27L and CTP1L endolysins targeting Clostridia contain a built-in trigger and release factor. PLoS Pathog. 10, e1004228. https://doi.org/10.1371/journal.ppat.1004228

Everett, J.K., Tejero, R., Murthy, S.B., Acton, T.B., Aramini, J.M., Baran, M.C., Benach, J., Cort, J.R., Eletsky, A., Forouhar, F., Guan, R., Kuzin, A.P., Lee,

H.W., Liu, G., Mani, R., Mao, B., Mills, J.L., Montelione, A.F., Pederson, K., Powers, R., et. al. (2016) A community resource of experimental data for NMR / X-ray crystal structure pairs. Protein Sci. 25, 30–45. https://doi.org/10.1002/pro.2774.

Fenton, M., Ross, P., Mcauliffe, O., O'Mahony, J. and Coffey, A. (2010) Recombinant bacteriophage lysins as antibacterials. Bioeng. Bugs. 1, 9-16. https://doi.org/10.4161/bbug.1.1.9818.

Firczuk, M. and Bochtler, M. (2007) Folds and activities of peptidoglycan amidases. FEMS Microbiol. Revi. 31, 676–69. https://doi.org/10.1111/j.1574-6976.2007.00084.x.

Fischetti, V. A. (2008) Bacteriophage lysins as effective antibacterials. Curr. Opin. Microbiol. 11, 393-400. https://doi.org/10.1016/j.tim.2005.08.007.

Gauto, D.F., Estrozi, L.F., Schwieters, C.D., Effantin, G., Macek, P., Sounier, R., Sivertsen, A.C., Schmidt, E., Kerfah, R., Mas, G., Colletier, J.P., Güntert, P., Favier, A., Schoehn, G., Schanda, P and Boisbouvier, J. (2019). Integrated NMR and cryo-EM atomic-resolution structure determination of a half-megadalton enzyme complex. Nat. Comm. 10, 2697. https://doi.org/10.1038/s41467-019-10490-9.

Gerstmans, H., Criel, B. and Briers, Y. (2018) Synthetic biology of modular endolysins. Biotechnol. Adv. 36, 624–640. https://doi.org/10.1016/j.biotechadv.2017.12.009.

Kakasis, A. and Panitsa G. (2019) Bacteriophage therapy as an alternative treatment for human infections. A comprehensive review. Int. J. Antimicrob. Agents. 53, 16-21. https://doi.org/10.1016/j.ijantimicag.2018.09.004.

Kirkwood, J., Hargreaves, D., O'Keefe, S. and Wilson, J. (2015) Analysis of crystallization data in the Protein Data Bank. Acta. Cryst. F71, 1228–1234. https://doi.org/10.1107/S2053230X15014892.

Kleywegt, G.J. (2000) Validation of protein crystal structures. Acta. Cryst. D56, 249–265. https://doi.org/10.1107/S0907444999016364.

Korndoerfer, I.P., Danzer, J., Schmelcher, M., Zimmer, M., Skerra, A. and Loessner, M.J. (2006) The crystal structure of the bacteriophage PSA endolysin reveals a unique fold responsible for specific recognition of *Listeria* cell walls. J. Mol. Biol. 364, 678–689. https://doi.org/10.1016/j.jmb.2006.08.069.

Korndoerfer, I.P, Kanitz, A., Danzer, J., Zimmer, M, Loessner, M.J., Skerra, A. (2008) Structural analysis of the l-alanoyl-d-glutamate endopeptidase domain of Listeria bacteriophage endolysin Ply500 reveals a new member of the LAS peptidase family. Acta Cryst. D64, 644–650. https://doi.org/10.1107/S0907444908007890.

Krishnan V.V. (2019) Molecular thermodynamics using nuclear magnetic resonance (NMR) spectroscopy. Inventions *4*, 13. https://doi.org/10.3390/inventions4010013.

Kuroki, R., Weaver, L.H. and Matthews, B.W. (1995) Structure-based design of a lysozyme with altered catalytic activity. Nat. Struct. Biol. 2 ,1007-1011. https://doi.org/10.1038/nsb1195-1007.

Leslie, A.G. (2006) The integration of macromolecular diffraction data. Acta Cryst. D62, 48–57. https://doi.org/10.1107/S0907444905039107.

Loessner, M.J. (2005) Bacteriophage endolysins-current state of research and applications. Curr. Opin. Microbiol. 8, 480-487. https://doi.org/10.1016/j.mib.2005.06.002.

Loessner, M.J., Kramer, K., Ebel, F. and Scherer, S. (2002) C-terminal domains of Listeria monocytogenes bacteriophage murein hydrolases determine specific recognition and high-affinity binding to bacterial cell wall carbohydrates. Mol. Microbiol. 335-349. https://doi.org/10.1046/j.1365-2958.2002.02889.x.

Mateu M.G. (2013) Introduction: The Structural Basis of Virus Function. In Structure and Physics of Viruses. Subcellular Biochemistry. M.G Mateu. ed. (Dordrecht, Netherlands: Springer) vol 68, pp. 3-51. https://doi.org/10.1007/978-94-007-6552-8_1.

Majewska, J., Kaźmierczak, Z., Lahutta, K., Lecion, D., Szymczak, A., Miernikiewicz, P., Drapała, J., Harhala, M., Marek-Bukowiec, K., Jędruchniewicz, N., Owczarek, B., Górski, A., and Dąbrowska, K. (2019). Induction of phage-specific antibodies by two therapeutic staphylococcal bacteriophages administered *per os*. Front. Immunol., 14 November 2019. doi: 10.3389/fimmu.2019.02607

Mayer, M.J., Garefalaki, V., Spoerl, R., Narbad, A., and Meijers, R. (2011) Structure-based modification of a Clostridium difficile-targeting endolysin affects activity and host range. J. Bacteriol. 193, 5477–5486. https://doi.org/10.1128/JB.00439-11.

McGowan, S., Buckle, A.M., Mitchell, M.S., Hoopes, J.T., Gallagher, D.T., Heselpoth, R.D., et al. (2012) X-ray crystal structure of the streptococcal specific phage lysin PlyC. Proc. Natl. Acad. Sci. U. S. A. 109, 12752–12757. https://doi.org/10.1073/pnas.1208424109.

McPherson, A. and Cudney, B. (2014) Optimization of crystallization conditions for biological macromolecules. Acta. Cryst. F70, 1445–1467. https://doi.org/10.1107/S2053230X14019670.

Navarre, W.W., Ton-That, H., Faull, K.F. and Schneewind, O. (1999) Multiple enzymatic activities of the murein hydrolase from staphylococcal phage φ11. Identification of a D-alanyl-glycine endopeptidase activity. J. Biol. Chem., 274, 15847-15856. https://doi.org/10.1074/jbc.274.22.15847.

Nelson, D.C., Schmelcher, M., Rodriguez-Rubio, L., Klumpp, J., Pritchard, D.G., Dong, S. and Donovan, D.M. (2012) Endolysins as antimicrobials. Adv. Virus. Res. 83, 299–365. https://doi.org/10.1016/B978-0-12-394438-2.00007-4.

O'Flaherty, S., Coffey, A., Meaney, W., Fitzgerald, G.F. and Ross, R.P. (2005) The recombinant phage lysin LysK has a broad spectrum of lytic activity against clinically relevant staphylococci, including methicillin-resistant *Staphylococcus aureus*. J. Bacteriol. 187, 7161-1764. http://doi.org/ 10.1128/JB.187.20.7161-7164.2005.

O'Sullivan, L., Buttimer, C., McAuliffe, O., Bolton, D. and Coffey, A. (2016) Bacteriophage-based tools: recent advances and novel applications. F1000Research 5, 2782. https://doi.org/10.12688/f1000research.9705.1.

Oliveira H., Melo L.D., Santos S.B., Nóbrega F.L., Ferreira E.C., Cerca N., Azeredo J. and Kluskens L.D. (2013) Molecular aspects and comparative genomics of bacteriophage endolysins. J. Virol. 87, 4558–4570. https://doi.org/10.1128/JVI.03277-12.

Park, Y., Lim, J.A, Kong, M., Ryu, S. and Rhee, S. (2014) Structure of bacteriophage SPN1S endolysin reveals an unusual two-module fold for the peptidoglycan lytic and binding activity. Mol. Microbiol. 92, 316–325. https://doi.org/10.1111/mmi.12555.

Pérez-Dorado, I., Campillo, N.E., Monterroso, B., Hesek, D., Lee, M., Páez, J.A., et al. (2007) Elucidation of the molecular recognition of bacterial cell wall by modular pneumococcal phage endolysin CPL-1. J. Biol. Chem., 282, 24990–24999. https://doi.org/10.1074/jbc.M704317200.

Rennell, D., Bouvier, S.E., Hardy, L.W. and Poteete, A.R. (1991) Systematic mutation of bacteriophage T4 lysozyme. J. Mo.l Biol. 222, 67–88. https://doi.org/10.1016/0022-2836(91)90738-R.

Rupp, B. (2009) Biomolecular crystallography: principles, practice and application to structural biology (New York, USA:Garland Science).

Sanz-Gaitero, M., Keary, R., Garcia-Doval, C., Coffey, A., and van Raaij, M.J. (2014) Crystal structure of the lytic CHAPk domain of the endolysin LysK from *Staphylococcus aureus* bacteriophage K. Virol. J. 11, 133. https://doi.org/10.1186/1743-422X-11-133.

Sun, Q., Kuty, G.F., Arockiasamy, A., Xu, M., Young, R. and Sacchettini, J.C. (2009) Regulation of a muralytic enzyme by dynamic membrane topology. Nat Struct. Mol. Biol. 16, 1192–1194. https://doi.org/10.1038/nsmb.1681.

Taylor, G.L. (2010) Introduction to phasing. Acta Cryst. D66, 325–328. https://doi.org/10.1107/S0907444910006694.

Wang, H.W. and Wang, J.W. (2017). How cryo-electron microscopy and X-ray crystallography complement each other. Protein Sci. 26, 32–39. https://doi.org/10.1002/pro.3022.

Xu, M., Arulandu, A., Struck, D. K., Swanson, S., Sacchettini, J. C. and Young, R. (2005) Disulfide isomerization after membrane release of its SAR domain activates P1 lysozyme. Science 307, 113–117. https://doi.org/10.1126/science.1105143.

Xu, M., Struck, D.K., Deaton, J., Wang, I.N. and Young, R. (2004) The signal arrest-release (SAR) sequence mediates export and control of the phage P1 endolysin. Proc. Natl. Acad. Sci. U. S. A. 101, 6415–6420. https://doi.org/10.1073/pnas.0400957101.

Yang, H., Yu, J. and Wei, H. (2014) Engineered bacteriophage lysins as novel anti-infectives. Front. Microbiol. 5, 1–6. https://doi.org/10.3389/fmicb.2014.00542.

Young, R. (2013) Phage lysis: Do we have the hole story yet? Curr. Opin. Microbiol. 16, 790–797. https://doi.org/10.1016/j.mib.2013.08.008.

Zhou, B., Zhen, X., Zhou, H., Zhao, F., Fan, C., Perčulija, V., Tong, Y., Mi, Z., and Ouyang, S. (2020). Structural and functional insights into a novel two-component endolysin encoded by a single gene in *Enterococcus faecalis* phage. PLoS pathogens 16, e1008394. https://doi.org/10.1371/journal.ppat.1008394.

## Chapter 14

# Peptidoglycan Hydrolases from Phages of Gram-positive Bacteria

**Sara Arroyo-Moreno[1], Colin Buttimer[2] and Aidan Coffey*[1,2]**

[1]Department of Biological Sciences, Cork Institute of Technology, Cork, Ireland
[2]APC Microbiome Institute, University College, Cork, Ireland

*Corresponding author: aidan.coffey@cit.ie

DOI: https://doi.org/10.21775/9781913652517.14

### Abstract

Bacterial resistance to conventional antibiotics has become an urgent healthcare issue worldwide. In this light, a renewed interest has emerged for therapy with bacteriophages, which are viruses that can specifically infect bacteria. These viruses encode different proteins that degrade specific bonds within the bacterial cell wall, both to inject their genomic material into the bacterial cells and to release the viral progeny: virion-associated hydrolases and endolysins, respectively. Elimination of Gram-positive pathogens with recombinant versions of phage enzymes represents a promising alternative to antibiotics. In some studies, these enzymes have been used successfully to eliminate/control bacterial pathogens in various anatomical locations in mice and other animal models. This review discusses the strategies developed to improve efficacy in terms of lytic activity and therapeutic delivery of these enzymes.

### Introduction

Bacteriophages (phages) are viruses that can infect bacterial cells. They specifically target bacteria, sometimes only recognising limited strains within a bacterial species without causing damage to other bacteria. After recognition, the virus injects its DNA/RNA into the bacterial cell, with the help of virion-associated peptidoglycan hydrolases that can break bonds within the cell wall. These viruses can generally undergo two different types of infection cycles: lytic if progeny phages are released from the host, or lysogenic if phage DNA gets integrated into the bacterial chromosome (prophage). In some cases, prophages can also exist as plasmids. Other phages can undergo pseudolysogeny, a scenario where the genome of a phage lies dormant within a host cell without either undergoing multiplication or having its replication being synchronised with the host cell (Cenens et al., 2013). During the phage lytic replication cycle, progeny phages are released from their host by the action of an enzyme known as an endolysin (lysin) that degrades the host's peptidoglycan cell wall layer at the end of the lytic cycle leading to cell lysis and death. These lysins are associated with other phage-encoded proteins called holins, responsible for pore formation in the bacterial cell membrane permitting the endolysin to access the peptidoglycan (Young et al., 2000). In the case of Gram-negative phages, additional proteins called spanins are responsible for the disruption of the lipopolysaccharide outer membrane (Kongari et al., 2018).

As described elsewhere in this book, lytic phages were employed in the treatment of bacterial infections as far back as the 1920s. However, the commercialisation of antibiotics from the 1940s caused phage therapy to decline in the West. Recent decades have seen increased research aimed at the development of phage therapy as a viable alternative to antibiotics to eliminate Gram-positive and Gram-negative multidrug-resistant pathogens by using either whole phages or recombinant versions of their lytic enzymes (Nelson et al., 2012). This research is fuelled by an ever-growing concern caused by increasing reports of antibiotic resistance among major human pathogens and also the pharmaceutical industry not developing antibiotics with new modes of action (Fair and Tor, 2014).

Multi-drug resistant Gram-positive pathogens that have a significant detrimental impact on healthcare environments worldwide, as well as in the food industry and agriculture. Among the Gram-positive pathogens, those that are especially problematic include members of the genera *Staphylococcus*, *Enterococcus, Listeria* and *Streptococcus*. These can develop diverse mechanisms to become resistant to different types of antibiotics (Munita et al., 2015) including drug inactivation, target replacement, target modification and changes in cell surface characteristics. Factors contributing to the development of resistance include failure to use narrow-spectrum drugs, and overuse of a range of antibiotics in medical and veterinary practise and also in animal husbandry.

Phage therapy involves the use of whole phages in mixtures (using a combination of different phages can minimize the development of phage resistance) or the production recombinant versions of phage lytic proteins. This latter option avoids the limitations of using whole phages. Their application is more similar to conventional molecular therapeutics, having a generally broader host range and being less likely to give rise to bacterial resistance. Bacteria can resist phage attack through a variety of mechanisms, including spontaneous mutations, restriction-modification systems, abortive infection systems and adaptive immunity via CRISPR-Cas systems (Oliveira, 2018). Despite attempts to identify bacterial resistance to phage lytic proteins, none has been reported to date; and this is understandable given that resistance to a peptidoglycan hydrolase would entail a structural change in the bacterial peptidoglycan (Love et al., 2018). Phages encode two main enzymes capable of degrading peptidoglycan: endolysins (to release viral progeny at the end of the infection cycle) and virion-associated lysins (VALs) (to allow the phage to inject its nucleic acid inside the bacterial cell at the beginning of a phage infection). Recombinant versions of these enzymes can be effective against Gram-positive bacteria when they are exogenously applied as they have direct access to the cell wall target, given the absence of an outer membrane. This fact contrasts with Gram-negative phage endolysins, where penetration of the outer membrane is an additional hurdle (Lim et al., 2014). Indeed, some VALs also include permeabilization or penetration domains (Oliveira, 2018 and elswhere in this book). Endolysins and virion-associated lysins in general, can have multiple uses ranging from treatment of human and animal infections to food and beverage preservation, surface disinfection in industry and hospital environments, as well as to control bacterial populations in horticulture. As such, this chapter will discuss the current state of Gram-positive phage lysin research. Gram-negative endolysins are covered in detail in another chapter in this book.

## Classical structures of Gram-positive phage lytic proteins

Recombinant versions of endolysins and VALs have been proven to be effective in killing Gram-positive bacteria *in vitro* and *in vivo* (Rodríguez-rubio et al., 2016), as these bacteria have their peptidoglycan cell wall exposed, having no outer membrane to block the enzymatic activity of these proteins. Both enzymes are highly specific and thermostable, making them the optimum candidates not only for clinical uses but also for food safety applications and surface disinfection within industry and hospital environments.

### *Endolysins*

A feature of Gram-positive phage lysins is a two-domain modular structure, containing a peptidoglycan-lytic domain, a flexible linker region and a cell-binding domain (CBD), as shown in Figure 1. This organization contrasts with the globular structure typical of Gram-negative endolysins, containing only a single domain. However, it has been reported that some endolysins from Gram-negative phages also have a modular structure such as endolysins OBPgp279 (from *Pseudomonas fluorescens* phage OBP), PVP-SE1gp146 (*Salmonella enterica* serovar Enteritidis phage PVP-SE1) and 201φ2-1gp229 (*Pseudomonas chlororaphis* phage 201φ2-1). All three modular endolysins showed strong muralytic activity on the peptidoglycan of a broad range of Gram-negativebacteria, partly due to the presence of the cell wall binding domain (Walmagh et al., 2012).

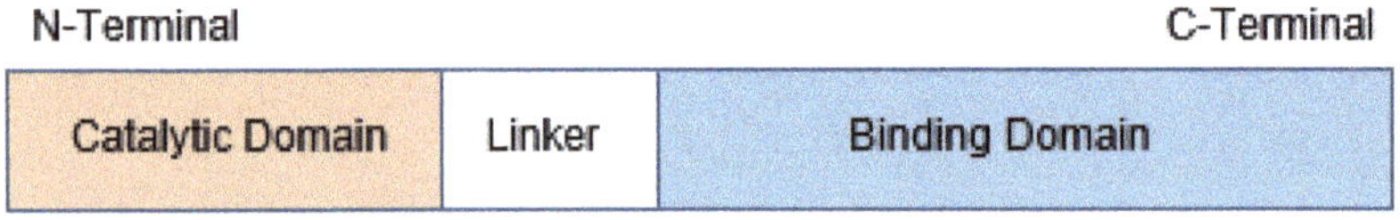


**Figure 1**. Typical modular organization of endolysins encoded by phages that target Gram-positive bacteria. A lytic domain is present in the amino-terminus of the protein and cell-binding domain (CBD) can be spotted in the carboxy-terminus, and both domains are linked by the interdomain or linker region.

Generally, all endolysins are the product of a single gene with their sizes ranging between 25 and 40 kDa. The endolysin PlyC of *Streptococcus phage* $C_1$ is an exception, made up of PlyCA and PlyCB genes and having a size of 114 kDa. PlyCA encodes the enzymatic domain and PlyCB contains eight subunits that harbour the CBD (Wang et al., 2015). Both parts of the enzyme are interconnected by an intergenic locus, which is essential for the enzymatic activity, as shown in Figure 2 (Nelson et al., 2006). Different *in silico* analyses have illustrated the large diversity and complexity of endolysins. It has been observed that up to four domains can be comprised in the same coding sequence, having 24 different types of catalytic and 15 binding domains, and 89 possible architectural organizations (Oliveira et al., 2016).

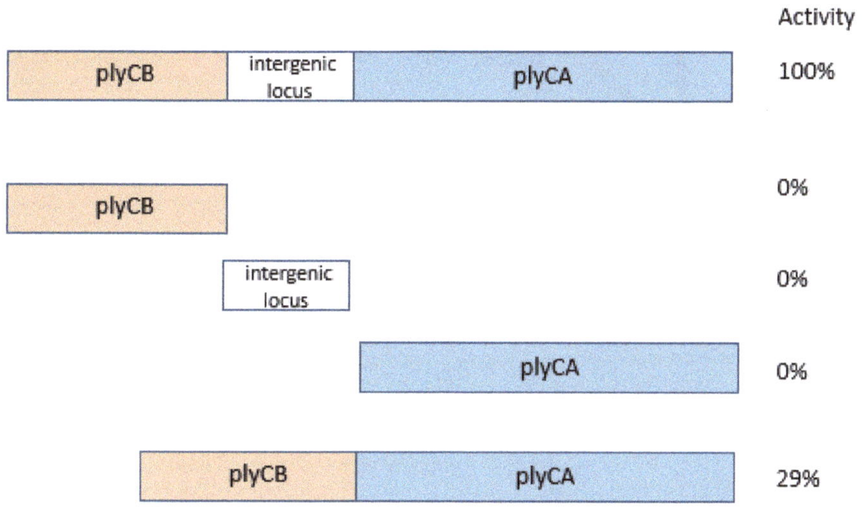


**Figure 2**. Gene structure of the multimeric endolysin PlyC (modification from Nelson et al., 2006). ORF1 contains the sequence for the PlyC light-chain, PlyCB. ORF2 is positioned between the genes for the two chains of PlyC and it is called intergenic locus. ORF3 contains the sequence for the PlyC heavy-chain, PlyCA. Expression of any of these ORFs by themselves results in no activity. If the intergenic locus is removed, only 29% of the enzymatic activity can be observed.

## N-Terminal region

The N-terminal of an endolysin typically contains the catalytic domain (Fischetti, 2011). This domain cleaves the peptidoglycan bacterial cell wall, producing holes leading to extrusion of the cytoplasmic membrane and ultimate cell collapse. In some cases, two or even three different catalytic domains can be associated with an endolysin. For example, two catalytic domains were observed in several staphylococcal lysins such as LysK (Harhala et al., 2018) and the phage phi11 endolysin that has both an endopeptidase and an amidase. These catalytic domains can be of different types depending on the bonds they break within the peptidoglycan (Figure 3), such as glucosamidase, lytic transglycolase, lysozyme, amidase, endopeptidase and CHAP (cysteine, histidine-dependent amidohydrolases/peptidases).

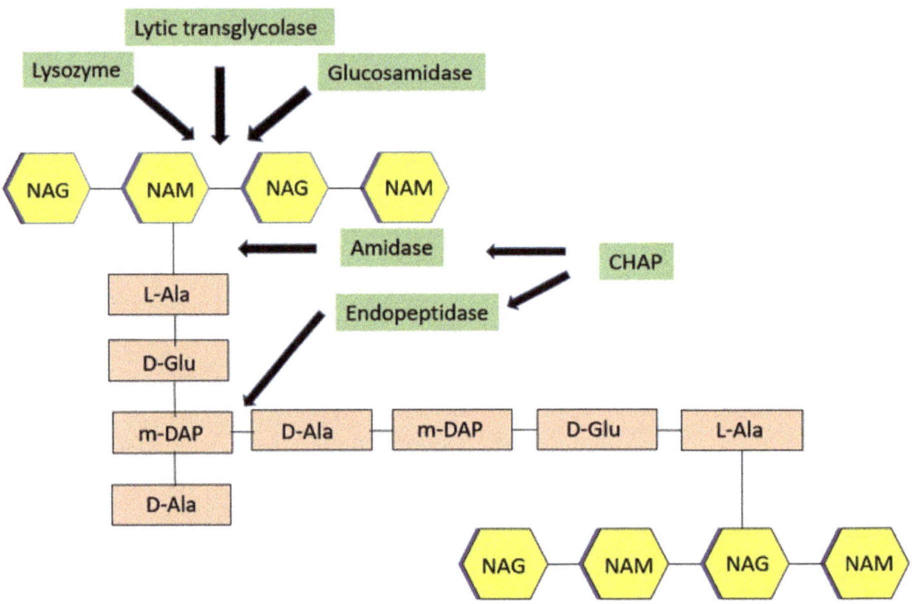


**Figure 3**. Diagram of the different endolysin catalytic domains cleavage sites in the bacterial peptidoglycan. The peptidoglycan is composed of repeating reducing sugar units, N-Acetylglucosamine (NAG) and N-acetylmuramic acid (NAM) linked by β (1-4) glycosidic bonds, which are cross-linked via an interpeptide bridge between the meso-diaminopimelic acid (m-DAP) and D-Alanine (D-Ala) residues of adjacent tetrapeptide chains. The chains also contain L-Alanine (L-Ala) and D-glutamic acid (D-Glu). The different types of endolysins catalytic domains include glucosamidase, lytic transglycolase, lysozyme (these target the bond between NAM and NAG), amidase (target the bond between m-DAP and NAM), endopeptidase (target the bond between m-DAP and L-Ala) and CHAP (can have the function of both amidase and endopeptidase).

It has also been described that in some Gram-positive endolysins (as well as Gram-negative endolysins) the amino-terminal can have a signalling function, enabling the enzymes to cross the bacterial cytoplasmatic independently from the action of holins. In this light, endolysins can use two different mechanisms: Signal arrest release (SAR) or peptide signal sequences. In the case of SAR endolysins, they possess an N-terminal transmembrane domain that engages the host sec translocon. These endolysins rely on a type of proteins called pinholins that are able to induce depolarization in the host cytoplasmic membrane. A strong example of SAR endolysin is Lyz, from the *E. coli* phage P1 (Xu et al., 2004). On the other hand, some phages encode endolysins with a typical Sec-type signal peptide, being dependant on the host's general secretion pathway (the Sec system). A Sec-dependant endolysin has been identified in the *Lactobacillus plantarum* phage Øg1e. As an alternative pathway, endolysins might bind precursors of cell-wall secondary polymers in the cytoplasm and be co-transported to the cell-wall. This is the case with the endolysin of *Streptococcus pneumoniae* phage SV1 (Escobedo et al., 2019).

### C-terminal region

A cell-binding domain (CBD) can be found in the C-terminal region of Gram-positive endolysins, allowing specific binding to the cell wall substrate. This binding specificity is usually very narrow, usually at the bacterial strain level, making the CBDs useful candidates to be alternative tools for the detection of microorganisms. The different types of CBD can recognize different ligands in the cell wall of the host organism, and these can be carbohydrates, proteins, choline, teichoic acids and lipoteichoic acids (Oliveira et al., 2013). The binding affinity of these domains to their ligands is equal to or higher than that of antibody-antigen interaction, as well as having greater stability. The main types of cell-binding domains found among phages endolysins share some common domain motifs, such as LysM (Lysin Motif), PG binding (Peptidoglycan Binding Domain), SH3 (Domain Binds to Proline-rich proteins) and Cpl-7 (Choline-Binding Modules), among others (Jarábková et al., 2015). Modular endolysins of phages infecting *Lactobacillus*, *Lactococcus* and *Bacillus* strains possess the broadest diversity of CBDs, with some having three different CBDs, for example, the endolysins of the phages A2 and phiAt3 infecting *Lactobacillus* (Mercanti et al., 2016). Although CBDs are usually located at the C-terminus, they can also be located at N-terminal end, and in some cases, they can be located as central domains (Santos et al., 2019) such as the streptococcal lysin λSA2, which presents a centrally located CBD with two flanking catalytic domains. The most accepted hypothesis for the presence of CBD on endolysins of phages infecting Gram-positive hosts is that these domains cause endolysins to bind to the cell wall debris after cell lysis, preventing their release leading to the destruction of neighbouring potential host cells for progeny phages to infect (Oliveira, 2018).

*Virion associated lysins (VALs)*

These enzymes are typically associated with the phage DNA injection machinery, and they can be part of the phage tail, baseplates structural proteins, or can be embedded in viral membranes (Wu et al., 2017). The function of VALs among phages of *Caudovirales* (tailed phages) allows the creation of a hole in the peptidoglycan cell wall, through which the phage tail tube can penetrate allowing the ejection of phage genetic material to initiate infection (Fernandes and São-José, 2018). These enzymes are also called virion-associated peptidoglycan hydrolases (VAPGH). However, some of these enzymes can be lytic transglycolases and not hydrolases, so for that reason, some authors established that it was better to call them VALs (Latka et al., 2017). VALs also play a role in the morphogenesis and stability of phage virions (Wittekind and Schuch, 2016). However, some VALs do not serve as structural elements but are either associated with the phage DNA or occur in a free form enclosed in the capsid. These enzymes were discovered due to the phenomenon of "lysis from without" when the attachment of different phages to the bacterial cell surface caused a detectable degree of lysis. Recombinant versions of these enzymes have been shown to be effective in *in vitro* and in *in vivo* bacterial lysis (Paul et al., 2011).

The VALs that have been described to date have a similar modular organization to endolysins possessing one or two catalytic domains for the digestion of the peptidoglycan cell wall and a domain bound to the virion (Hermoso et al., 2007). The catalytic domains are similar to those present in endolysins (Figure 2). In terms of structure, they are highly diverse enzymes, differing in the number of domains, their organization, molecular weight and oligomeric state. Endopeptidase activity is generally attributed to VALs of Gram-positive-specific phages such as ORF56 encoded by the *Staphylococcus* phage K (Paul et al., 2011). The endopeptidases from the M23 family are rarely found in endolysins, being very characteristic of VALs from Gram-positive phages. The majority of VALs only have one catalytic domain, and interestingly, all the VALs with two catalytic domains are encoded by phages specific for Gram-positive bacteria, which can be related to the thick cell wall of Gram-positive cells (Latka et al., 2017). Some lytic domains can be located in the N-terminus and others can be located in the C-terminus, making the domain order in VALs variable. The vast majority of known VALs do not exhibit any obvious CBD under *in silico* analysis, and this constitutes a significant difference to endolysins. The presence of this domain is considered to be dispensable in VALs because their contact with the cell wall is guaranteed by the tight interaction between phage tails and host cell surface receptors (São-José, 2018). Despite this, the presence of a cell wall binding domain was reported for P17 VAL of *Staphylococcus* phage P68 (Takáč and Bläsi, 2005).

VALs are much less studied than endolysins regarding their biochemical properties, and represent an emerging group of phage lytic enzymes. Typically, they are much larger than endolysins (37-342 kDa) and possess interchangeable catalytic domains. Although all types of endolysin enzymatic activities have been observed in VALs, many of these enzymes exhibit extraordinary thermostability, which may be related to their apparent fast refolding kinetics, making them very promising for commercial applications and incorporation into stable nanocomposites (Wu et al., 2017).

*Other potential antimicrobial phage proteins against Gram-positive bacteria*

Phages encode other proteins that do not degrade peptidoglycan *per se*, but still exhibit other properties that would make them useful as potential antibacterial agents or as a complement for phage lytic proteins and other antimicrobial compounds.

### Holins

At the end of the phage lytic cycle, lysins need to move across the cytoplasmatic membrane to degrade their targets in the peptidoglycan. This is usually aided by proteins designated as holins (except for SAR endolysins and endolysins that encode signal peptides). Holins produce lesions in the cytoplasmic membrane of the host bacteria, with the formation of pores through which endolysins can pass. They are classified depending on the number of transmembrane domains they possess (Wu et al., 2017). It is believed that their production during the phage cycle is involved in the activation of cellular stress pathways. This involves the release of gene transfer agents, facilitation of programmes of differentiation such as those that allow sporulation and spore germination and contribution to biofilm formation (Saier and Reddy, 2015). Homologues of holins exist in eukaryotic, bacterial and archaeal organisms. Cell lysis cannot be achieved by applying recombinant versions of holins exogenously; however, different studies suggest that they can show some bacteriostatic activity (Roach and Donovan, 2015). In this light, it was observed that the holin-like gene designated *tmp*1, which was isolated from a goatskin surface metagenome analysis, could inhibit the growth of several Gram-positive food contaminants such as *Kurthia gibsonii*, *Staphylococcus epidermidis*, and *Listeria innocula* (Rajesh et al., 2011). It was also observed that the holin from the *Streptococcus suis* phage SMP could cause weak lysis on *S. suis*, *S. aureus* and *Bacillus subtilis* when applied alone. Additionally, it could expand the lytic spectrum of the endolysin from SMP phage (Wang et al., 2009). Therefore, the combination of holins and endolysins might be a potential strategy to improve lytic potency.

### Polysaccharide depolymerases

Other proteins derived from phage virions should be taken also into consideration in terms of their use as therapeutics and diagnosis tools, as well as in combination with phage lytic proteins. One class is the carbohydrate-active enzymes termed polysaccharide depolymerases, and the role and potential of these are discussed in detail in another chapter.

### *In vitro* and *in vivo studies*

In the past few years, several studies have been carried out to assess the ability of phage lytic enzymes to lyse Gram-positive bacterial cells *in vitro* and *in vivo*, to disrupt and prevent biofilm formation, and also to guarantee safety for use in clinical and non-clinical applications.

#### *In vitro studies*

The first report of *in vitro* activity of a recombinant endolysin dates back to 1959 (Reiter and Oram, 1963; Freimer et al., 1959). Since then, nanogram quantities of certain endolysins have shown to be efficient in eliminating bacterial suspensions in seconds, and numerous studies have proved their antibacterial potential *in vitro* (Oliveira, 2018). In Gram-positive hosts, VALs from *Lactococcus* phages Tuc2009 and TP901-1 and *Bacillus subtilis* phages SP and phi29 were tested and proven to be able to degrade the bacterial peptidoglycan cell wall. Interestingly, the VALs of all these phages carry at least one catalytic domain of the M23 family (endopeptidase). In the lactococcal phages, this domain was shown to facilitate cell-wall penetration and phage genome delivery (Stockdale et al., 2013).

The majority of *in vitro* studies to assess bactericidal activity consist of turbidity reduction assays (monitorization of absorbance reduction in the bacterial cultures), reduction in CFU/mL counts, and zymogram assays - all accompanied by MIC (minimum inhibitory concentration) determination. Each have individual advantages and limitations. High lytic activity has been observed when using endolysins and VALs against log-phase cultures of different Gram-positive bacteria.

In general, endolysins whether truncated or not are easier to apply in medical applications than whole phages given that the latter are self replicating and frequently need to have individual therapeutic phage compositions updated from time to time, with each variation requiring new clinical trials (Zhang et al., 2013). Recombinant versions of endolysins and VALs can easily be obtained for all the different types of phages and prophages present on the genome of a Gram-positive bacteria. However, the translation of *in vitro* activity observations to *in vivo* applications is not always straightforward given the complexity of different bacterial infections and possible challenges with endolysin access to the site of infection.

Multiple endolysins have been identified within the genomes of isolated phages, but their diversity in uncultured prophages has been largely overlooked until recently. A metagenomics study on uncultured phages has resulted in the discovery of 2,628 putative endolysins. A number of them displayed novel domain architectures and were predicted to be active against many pathogenic bacteria, greatly enhancing the diversity of known endolysins (Fernández-Ruiz et al., 2018).

Phage lytic enzymes have been shown to be able to disrupt biofilms, therefore being able to lyse even sessile antibiotic-resistant cells of problematic Gram-positive pathogens such as *Staphylococcus* and *Streptococcus* (Fenton et al., 2013; Meng et al., 2011). Biofilms play an essential role in bacterial infections of both humans and animals and as a source of contamination in food production and processing (Kashani et al., 2017), and usually, bacteria in this state are highly resistant to antimicrobials. The ability of phage lytic enzymes to disrupt biofilms makes them excellent candidates for therapeutics and disinfectants. In 2017, it was proved that the recombinant staphylococcal endolysin HY-133 was not only effective in eliminating and preventing biofilms in vascular grafts but it also exhibited no significant cytotoxicity against vascular cells, unlike antibiotics such as rifampin, vancomycin and daptomycin, which exhibited high levels of cytotoxicity and reduction of vascular cells viability (Herten et al., 2017). Therefore, phage lytic proteins are good candidates for the treatment and prevention of infections in vascular grafts, increasing their chances of success.

Lytic activity of endolysins and VALs have also been tested in mammalian cell cultures. The native endolysin, PlyC, has been demonstrated to target intracellular *Streptococcus pyogenes*, (Shen et al., 2016). However, the vast majority of these enzymes cannot traverse the epithelial cell membrane, making it necessary for the development of delivery strategies such as liposomes (Bai et al., 2019).

*Studies in vivo*

Both phages and their lytic enzymes have been demonstrated to be efficient and safe in extensive preclinical studies. In addition, the safety of specific phages and lytic enzymes has been proven in human case studies having completed phase I in clinical trials as reviewed recently (Abdelkader et al., 2019).

Studies in animal models (mainly mice, and in some occasions, monkeys and cows) have demonstrated the antibacterial efficacy of phage lysins, when applied intravenously, topically or subcutaneously, against relevant Gram-positive pathogens such as *Streptococcus spp*, *S.aureus*, *B.anthracis* and *E. faecalis*. Some of these studies have proved the enzyme's ability to protect mice against some infections such as mastitis and endophthalmitis (Kashani et al., 2017; Park et al., 2018; Singh and Donovan, 2014). Bacterial elimination was also observed when examining *S. aureus* in mice nares by CHAPk, the truncated derivate of the LysK staphylococcal endolysin (Fenton et al., 2010). In addition, the staphylococcal endolysin SAL200 was proved to be efficient against severe pneumonia caused by *S. aureus* in mice model with intranasal administration (Bae et al., 2019). Time is an important factor in guaranteeing the efficacy of enzymatic treatments in murine models. Much higher rates of animal survival have been seen when an early application of the endolysin was carried out (usually 1 hour after the infection challenge in the animal) compared to later additions (Oliveira, 2018). The first clinical trial of a staphylococcal lysin was carried out with 36 healthy men from South Korea, to prove its safety (Jun et al., 2017). No allergic reactions or toxicity of any kind was observed when the recombinant endolysin SAL200-1A was administrated intravenously in concentrations of 18 mg/mL. These observations suggest the possibility of carrying out further clinical trials to study the efficacy of endolysins and VALs in humans with infections caused by Gram-positive pathogens. Currently, there are two commercially available staphylococcal endolysins, produced by Micreos Human Health (StaphefektTM) and ContraFect (CF-301), respectively (Cooper et al., 2018). Both of them have been proven to be capable of disrupting biofilms, and they can be used for treating staphylococcal bacteremia. After the application of StaphefektTM for one week, *S. aureus* was eradicated from the lesions of *S. aureus* positive rosacea patients, while other skin inhabitants remained present (Totté et al., 2017). Relevant clinical trials with endolysins are compiled in Table 1.

**Table 1**. Most relevant clinical trials of phage endolysins to date (modification of Abdelkader et al., 2019). All studies are targeting *S. aureus* ranging from topical and nasal to systemic infections.

| Study | Phase | Results | Reference |
|---|---|---|---|
| A study to evaluate the safety, pharmacokinetics and pharmacodynamics of the staphylococcal endolysin SAL200 in healthy male volunteers. | I | No serious adverse events were reported. A dosing regimen of more than 1 mg/kg was recommended as a treatment option. | (Jun et al., 2017) |
| A placebo-controlled, dose-escalating study to examine the safety and tolerability of single intravenous doses of the staphylococcal endolysin CF-301 in healthy subjects | I | A single dose of CF-301 has a low propensity to induce an inflammatory response. Long term immunological monitoring revealed no relation between specific antibody production and hypersensitivity factors. | (Cooper et al., 2018) |
| The effect of the staphylococcal endolysin StaphefektTM on disease severity and the skin microbiome. | I/II | 3 cases where patients with chronic and recurrent *S. aureus*-related dermatoses were successfully treated with StaphefektTM | (Totté et al., 2017) |

In 2018, another endolysin safety study was carried out to examine two pneumococcal lysins, Cpl-1 and Pal (Harhala et al., 2018) to monitor their safety and toxicity profiles using human macrophages and pharyngeal cells, as well as mice models. The findings demonstrated that no gene expression was observed in the cells mentioned above after adding 0.5 μM of each endolysin for 6 hours and additionally, the mice did not experience an inflammatory response and no changes in their intestinal flora were observed with doses of 15mg/kg of each of the endolysins.

These findings support further preclinical and clinical studies. Moreover, they are unlikely to produce bacterial resistances makes them even optimal candidates for prophylactics treatments, in clinical and non-clinical applications.

## Protein engineering

Employing endolysins and VALs as antibacterial agents imply that their lytic activity will be exerted in conditions substantially different from those found in their native context of action, which may negatively impact in their activity (São-José, 2018). For this reason, it is worth considering performing some improvements in phage lytic enzymes to ensure their maximum efficacy when used as antibacterial agents.

Lytic activity of VALs and endolysins can be improved by domain swapping, removal, shuffle or combinations of domains from multiple enzymes (Vázquez et al., 2018). It is expected that the presence of multiple lytic domains on VALs and endolysins may further reduce the likelihood of resistance development as well improve lytic activity (Hancock, 2005) Another possibility to alter these proteins is by the mutation of amino acids in their active sites or other locations resulting in changes in conformation or net charge, with the possibility of achieving improved lytic and/or binding properties.

Moreover, a possible improvement for endolysins could be attaching small cationic peptides to increase permeability. This attachment does not only enable the protein to cross the lipopolysaccharide membrane for endolysins targeting Gram-negative bacteria but also to increase affinity to the bacterial cell wall for Gram-positive endolysins and VALS, as well as allowing access across eukaryotic cytoplasmic membrane and killing intracellular pathogens. A compilation of strategies to perform protein engineering is shown in in Table 2.

**Table 2**. Strategies for molecular and protein engineering of phage lytic proteins to enhance their activities against Gram-positive pathogens.

| **Modification** | | **Properties** | **Example** | **Reference** |
|---|---|---|---|---|
| Domain engineering | Truncation of the full-length enzyme | Increased catalytic ability and solubility | CHAPk | (Fenton et al., 2010b) |
| | The fusion of catalytic domains of endolysins or VALs with different CBDs | Increased catalytic ability, solubility, lytic spectrum and thermostability | P128 | (Paul et al., 2011a) |
| Changes in protein net charge | Mutagenesis | Increased efficacy of catalytic domains without a CBD | PlyGBS | (Shang and Nelson, 2019) |
| Combined treatments | Combination of phage lytic proteins with classical antibiotics | *in vitro* synergies and enhanced *in vivo* activities | Cpl-711/ cefotaxime | (Letrado et al., 2018) |
| | Combination of different phage lytic proteins | | Lysostaphin/ LysK | (Becker et al., 2008) |
| | Combination of phage lytic proteins with bacteriocins | | Cpl-1 and Pal | (Wittekind and Schuch, 2016) |
| Other improvements | Cell-penetrating peptides (cationic/ amphipathic) | Enhanced structural order, permeability and stability | Art-240 | (Rodríguez-Rubio et al., 2016) |
| | Combination of CBD with fluorescent molecules | Bacterial diagnosis | CTP1L | (Gómez-Torres et al., 2018) |

*Domain engineering*

Combinations of multiple enzymatic domains can result in higher specificity, efficacy. In this light, a chimeric protein has been constructed by fusing the staphylococcal endolysin LysK and the bacteriocin lysostaphin (proceeding from *Staphylococcus simulans*), having triple the anti-*Staphylococci* activity (amidase, alanyl-glycyl endopeptidase and glycyl-glycyl endopeptidase) (Filatova et al., 2016). As well, it is known that to date that the most potent lysins tested against *S. pneumoniae* are chimeric proteins, which combine several catalytic domains (Vázquez et al., 2018). Treatment with the chimeric lysin Cpl-711 sharply reduced the attachment of *S. pneumoniae* to human epithelial cells. A single intranasal dose of Cpl-711 significantly reduced nasopharyngeal colonization in a mouse model (Corsini et al., 2018).

A variety of results can be observed when domains are removed or shuffled, denoting their importance in the lytic activities of the different enzymes. In some cases, a truncated derivate containing only a single catalytic domain can have a higher activity than the whole enzyme. For example, in the case of CHAPk, a truncated derivate from the staphylococcal endolysin LysK was observed to have a higher lytic activity than the full-length endolysin in vitro (Fenton et al., 2011). A similar observation was made with the truncated derivate of the *Streptococcus* phage endolysin PlyGBS, having a 20-fold increase in specific activity when the CBD was removed (Shang and Nelson, 2019). It was also attempted to express a truncated derivate of LysK consisting of just the central amidase domain, although no activity was observed (Sanz-gaitero et al., 2014), suggesting that domains that are located downstream of the N-terminal enzymatic domain can be challenging to clone and show expression on their own. Another approach was to optimize primary, secondary and tertiary structures of the CHAP and amidase domain of LysK to improve the stability and solubility of the protein, designing an *in silico* model (Kashani et al., 2017).

Sometimes, the removal of a cell-binding domain can make the enzyme have a broader lytic spectrum (Wu et al., 2017), although host range can also be determined by the net charge of the lytic domain. However, in other cases, the CBD can have a vital role in the enzyme activity, participating as well in the lytic activity, such as in the case of *Clostridium* phage phi3626 endolysin in which the absence of the cell-binding domain results in a significant decrease of lytic activity (Zimmer et al., 2002).

VALs have also been used to generate chimeric lytic enzymes by functioning as an alternative source of catalytic domains, which may increase thermostability compared to those derived from endolysins. Moreover, by following their native mode of action, it has been suggested that VALs may be better prepared to promote lysis from without than endolysins (São-José, 2018). Since VALs typically do not possess cell-binding domains, their activity can be improved by adding those to full-length enzymes or single catalytic domains. Impressive results have been observed when combining CHAP catalytic domains from staphylococcal VALs and lysostaphin cell-binding domain. For example, the CHAP domains proceeding from the VALs of phage phiIPLA88 and phage K. The first one, designated as CHAPSH3b, was proven to be highly thermostable and capable of eradicating *S. aureus* present in raw and pasteurized milk and of disrupting biofilms (Rodríguez-Rubio et al., 2012). The second chimaera, P128 exhibited a 100-fold higher bactericidal activity when compared to the isolated CHAP domain (Paul et al., 2011b).

It has also been observed that the combination of a cell-binding domain and a single catalytic domain can result in improved thermostability. In this light, a gene that fused the N-acetylmuramyl-L-alanine amidase domain from the endolysin of the thermophilic phage ΦGVE2 to the cell-wall binding domain (CWB) from the endolysin of the *C. perfringens*-specific phage ΦCP26F. The resulting protein, PlyGVE2CpCWB, lysed *C. perfringens* in liquid and solid cultures and was able to withstand temperatures up to 50ºC and still lyse *C. perfringens* (Swift et al., 2015).

In some cases, a single catalytic domain can have more activity by itself. If a single catalytic domain has a positive net charge, it might be potentially less dependent on a CDB as cell walls surfaces of Gram-positives are generally negatively charged (Low et al., 2011). Therefore, the dependence on a CBD could be determined by the catalytic domains net charge. Additionally, reduced molecular size may also be an explanation for improved activity as seen against bacteria from the genus *Bacillus*. Members of this genus often possess S-layers or capsules, and it has been observed that these can act as molecular sieves that are size-dependent, and thus favouring the use of single catalytic domains instead of full-length endolysin proteins to treat such groups of bacteria.

Therefore, it is advantageous to try different combinations of domain removal and shuffling to establish optimal activities. In this light, the use of *in silico* modelling can facilitate discovering what protein characteristics produce optimal bactericidal activity.

*Changes in protein net charge*

Mutagenesis can be performed in the DNA sequences of the phage lytic proteins to bring about amino acid substitutions and/or deletions causing changes in protein net charge. As stated above, net charge in a catalytic domain of a phage lytic protein can play a significant role in its efficiency and host range against Gram-positive pathogens (Low et al., 2011), as these have a negatively charged surface. Then, if the catalytic domains or other parts of the catalytic domain have a net positive charge, this allows for stronger attraction between the lytic domain and the bacterial cell surface, which could potentially replace the interactions between the CBDs and their ligands in the bacterial cell wall. It has been observed that inversion of net charge could eliminate or bring about dependence on cell binding domains (Shang and Nelson, 2019).

Other experiments demonstrated that after creating mutants of the CHAP domain of the PlyC streptococcal endolysin with an increased net positive charge, they did not have higher bactericidal activity than the wild type protein (Shang and Nelson, 2019). These mutants were created by inducing point mutations in five surface and unstructured residues, incorporating neutral and positive charges into the CHAP domain. It was also observed that unlike other CHAP domains, that the PlyC endolysin needs a CBD for full activity, and that making its net charge more positive does not remove this dependence. Therefore, the mechanisms used by a catalytic domain to establish ionic interactions with the cell surface of Gram-positive bacteria are complicated, and they must be studied in more detail to improve bactericidal activity and host range. Additionally, it is important to investigate engineering different domain combinations as well as engineering changes in net charge.

*Combined treatments*

Another possibility to achieve maximum bactericidal activity from phage lytic proteins against Gram-positive pathogens can be their administration in combination, by either using combinations of different endolysins/VALs together or using in conjunction with other antimicrobial agents/methods. Synergic properties have been observed when antibiotics are combined with whole phages (Torres-Barceló et al., 2014), suggesting that the same effect could be achieved using recombinant versions of their lytic proteins. Several studies suggest that the combination of endolysins, and possibly VALs, with conventional antibiotics, can increase antibacterial activity to Gram-positive bacteria. An excellent example of this is the combination of the pneumococcal endolysin Cpl-711 with cefotaxime, and other antibiotics like amoxicillin, which shows a synergic effect and this can be helpful for the treatment of multiresistant pneumococcal strains (Letrado et al., 2018). MV-L lysin from the *Staphylococcus* phage MR11 was observed to efficiently kill vancomycin-resistant strains in combination with vancomycin (Corsini et al., 2018). In addition, the staphylococcal lysin SAL200 acts synergistically when combined with standard-of-care (SOC) antibiotics and it also enhances the antistaphylococcal activities of SOC antibiotics in a murine bacteremia model as well as an invertebrate model using larvae of *Galleria mellonella* (Kim et al., 2018). Some more examples of combinations between phage derived lytic proteins and antibiotics are compiled in Table 3.

**Table 3**. Other examples of combinations of phage-derived lytic proteins with antibiotics or other disinfectant agents targeting Gram-positive pathogens (Modification from Cooper et al.,2018).

| **Bacterial Target** | **Combination tested** | **Results** | **Reference** |
|---|---|---|---|
| *S. aureus* | CF-301/ daptomycin and vancomycin | The synergy between CF-301 and DAP/VAN resulted in increased murine survival compared to when just treated with antibiotics alone ($p < 0.0001$) | (Schuch et al., 2014) |
| | Endolysin LysSA97 (376 nM)/carvacrol (3.33 mM) | Individual components showed a reduction in bacterial content of approx one log10. The combination reduced bacterial content >4log10. | (Chang et al., 2017) |
| *Streptococcus pneumoniae* | Endolysin Cpl-1/ Penicillin | Synergy with penicillin was dependent on the level of penicillin resistance. | (Djurkovic et al., 2005) |
| *Listeria monocytogenes* | Endolysin PlyP100/ nisin | Endolysin stable for up to 28 days under cold storage PlyP100/nisin combination maintained activity over four weeks. | (Ibarra-Sánchez et al., 2018) |
| | Endolysin PlyP825/ High hydrostatic pressure (HHP) | Synergistic inactivation of *L. monocytogenes* in milk, cheese and smoked fish. Allows for lower pressure level to be used with the same antimicrobial efficacy when treated in combination with phage. | (Misiou et al., 2018) |

In addition, synergistic effects can also be spotted when combining different endolysins or in combinations of endolysins and bacteriocins like lysostaphin. For example, the combination of lysostaphin with LysK resulted in a synergic effect against staphylococci (Becker et al., 2008). Two anti-pneumococcal lysins Cpl-1 and Pal (with muraminidase and amidase activities, respectively) exhibited *in vitro* synergy, and the activity of the combination was shown to be superior to either individual lysin in a murine pneumococcal infection model (Wittekind and Schuch, 2016). Therefore, as evidenced by many studies, combinations of different cell wall hydrolases proteins can exhibit *in vitro* synergies and enhanced *in vivo* activities against relevant Gram-positive pathogens.

**Other improvements**

As mentioned above, attaching a cationic peptide to a phage lytic protein can make it able to cross eukaryotic cell membranes, allowing them to target intracellular pathogens. So far, many studies have disclosed that cell-penetrating peptides (CPPs) are short peptides that are able to penetrate biological membranes and drive the bioactive cargo into the cells. A good example of this is the fusion of a lysin from the *Staphylococcus* phage JD007 (JDlys) to a cell-penetrating peptide to facilitate JDlys entry into keratinocytes, enhancing its ability to inhibit the proliferation of *S. aureus* in murine skin (Wang et al., 2018). The permeabilization achieved by attaching cationic or amphipathic peptides into phage lytic proteins (a process known as "Artilysation"), is not only beneficial to improve the lysis from without for Gram-negative pathogens. It can also suppose an improvement in the activity and stability of lysins that target Gram-positive pathogens because this allows for enhanced structural ordering and breaking of aggregates due to electrostatic interactions (Haddad Kashani et al., 2018). In this light, a streptococcal "Artilysin" (Art-240) was developed by making use of the polycationic nonapeptide (PCNP) which was previously successfully used in the construction of Artilysins against Gram-negative *P. aeruginosa*. Art-240 outperforms its parental endolysin in an increased enzymatic activity under a broad range of pH and NaCl conditions (Rodríguez-Rubio et al., 2016). Furthermore, the stability of the staphylococcal endolysin LysK in complexes with cationic polymers and their block copolymers with polyethene glycol (PEG) was improved by both hydrophobic and electrostatic mechanisms.

In the case of Gram-negative endolysins, an alternative engineering strategy to "Artilysation" can be the development of "Innolysins". Innolysins combine the enzymatic activity of endolysins with the binding capacity of phage receptor binding proteins (RBPs). This allows the enzymes to pass through the outer membrane, hydrolyse the peptidoglycan and kill the target bacterium. An example of "Innolysin" is the combination of *E. coli* phage T5 endolysin and receptor binding protein Pb5 (Zampara et al., 2018).

Another possible application for phage endolysins could be for bacterial detection (very relevant for food-borne pathogens). In this light, recombinant versions of the endolysin CBDs can be used, as their affinity for the target bacterium is equal to or higher than the antibody-antigen interaction. Moreover, CBDs can circumvent the usual drawbacks of antibodies: being easier and less time-consuming to produce, less expensive as well as having greater stability at different temperatures and pH conditions. Thus, a possibility for enhanced detection can be by attaching fluorescent peptides such as green fluorescent protein (GFP) to cell-binding domains, full-size endolysins or VALs to allow bacterial detection in industrial, food or clinical environments. It has already been done for detection *in situ* of *Clostridium* strains implicated in cheese spoilage, for which CTP1L, a phage endolysin active against *C. tyrobutyricum* was investigated, and its enzymatically active and cell wall-binding domains were attached to GFP to detect the clostridia by fluorescence microscopy (Dunne et al., 2018). As such, a novel CBD targeting *Paenibacillus larvae* was also attached to a GFP molecule, a breakthrough for the detection of this Gram-positive spore-forming bacteria that causes the American Foulbrood disease of honey bees (Santos et al., 2019).

### Delivery systems

As discussed in other sections of this review, phage lytic proteins are promising alternatives for classical antibiotics in terms of drug development. The mode of delivery of any new potential antimicrobial candidate must ensure activity, stability and correct dosage conditions so that the therapeutic benefit can be maximized, and the potential side effects can be minimized. Additionally, the delivery system has to ensure that the active molecule reaches its right destination within the body. A proper delivery can be achieved by using microorganisms as probiotics (delivery in the intestinal system), or through triggered release using encapsulation in nanoparticles. The triggered release of a therapeutic agent (small molecule, protein or virus) may rely on a variety of external stimuli in order to release the active cargo, including pH, temperature, ultrasound, magnetism or biomarker signals (Hathaway et al., 2017a). Another possibility to make sure that these enzymes reach their destination within the body is to encapsulate these enzymes into liposomes (to combat intracellular pathogens) or make aerosol formulations (to treat respiratory infections).

## Probiotics

Lactic acid bacteria (LAB) are food-grade organisms with the potential of delivering proteins of interest to sites in the digestive tract. Lactic acid bacteria have been utilized for years in different food preservation and fermentation processes. Both lactococci and lactobacilli possess GRAS (Generally Regarded As Safe) status and are involved in the production of particular metabolite and toxins with inhibitory effect on bacterial pathogens. Some of these organisms are well characterized, and they are promising candidates for delivering heterologous proteins to the digestive tract, as genetic tools for bioengineering these organisms are widely available. In this light, they can be used to deliver phage lytic proteins to combat Gram-positive pathogens present in the intestinal system, especially from the genus *Clostridium*, acting at the same time as exclusion agents against them. Moreover, these phage lytic bacteria, unlike other antimicrobials, will not cause any collateral damage to commensal bacteria in the colon are beneficial. These approaches can be applied for human and animal therapy, as well as for disinfection in food industry environments.

The *Listeria* phage endolysins Ply118 and Ply511 were the first endolysins successfully produced and secreted by *Lactococcus lactis* (Gaeng et al., 2000), in order to improve biopreservation and food safety properties using dairy starter strains. For that, the nisRK two-component regulatory system was used, allowing constitutive expression of the endolysin under the control of the PnisA promoter while the use of a signal peptide led to successful secretion of the active endolysin to the surrounding media (Figure 4).

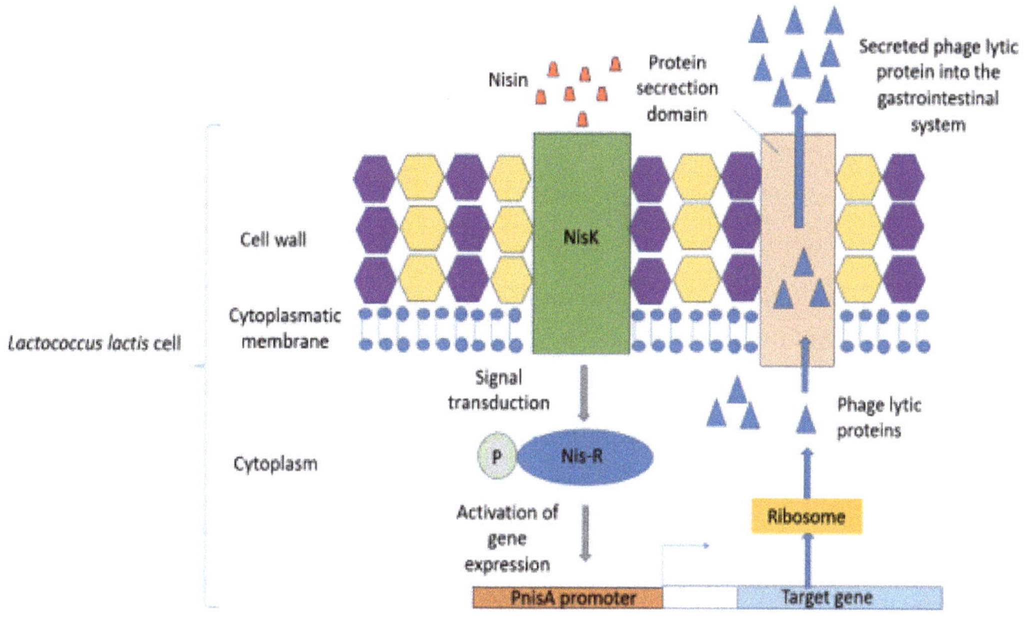


**Figure 4**. Protein secretion pathway in the nisin-controlled gene expression system in *L. lactis*. The regulator NisR is activated by phosphorylation in the presence of nisin, and this activates the PnisA promoter, and then follows the production of the recombinant phage lytic enzyme. The recombinant target gene must include a signal peptide, to allow for protein secretion outside the cell.

This gene expression regulation system from *L. lactis* has been implemented in other lactic acid bacteria to produce other endolysins against foodborne and intestinal Gram-positive pathogens. The *Listeria* endolysin Ply511 and lysostaphin (anti-staphylococcal bacteriocin) were produced and secreted in various lactic acid bacteria, including different strains of *Lactobacillus plantarum*, and in some cases, a 7-log reduction in staphylococcal cells was achieved (Turner et al., 2007). An endolysin against *Clostridium perfringens*, designated as CP25L, was expressed in a probiotic strain of *Lactobacillus johnsonii* to be expressed and explicitly delivered in the intestinal tract (Gervasi et al., 2014).

## Nanoparticles

Nanoparticles are now being used extensively for the delivery of therapeutic agents such as anticancer drugs to eukaryotic cells, and they also offer possibilities to target bacterial cells effectively. Since surfaces of various nanoparticles can be modified to allow conjugation with biological molecules, these particles can be utilized as efficient vehicles that are able to deliver a predetermined number of endolysins and VALs to the target the cell walls of Gram-positive pathogens (Hathaway et al., 2017b). Similarly, there are possibilities for engineering surfaces or interiors of different nanocomposites with fluorophores in order to achieve quantitative detection, due to the high specificity of phage lytic proteins. A good example of encapsulation of a Gram-positive endolysin is the truncated form of the staphylococcal endolysin LysK, denoted CHAPk, in combination with the bacteriocin lysostaphin, which was encapsulated in thermo-sensitive PNIPAM nanoparticles, producing controlled release of this synergic enzybiotic cocktail at high temperatures (37ºC) that associated with *Staphylococcus* infection in skin (Hathaway et al., 2017a). Nanoparticles can also allow higher stability in phage lytic enzymes via immobilization. A good example is the covalent conjugation of the anti-Listeria endolysin Ply500 with silica nanoparticles, which produced a much higher lytic activity retention at 4ºC compared to the single endolysin. However, although immobilization usually results in improved enzymatical stability, it can produce some loss in catalytical activity (Solanki et al., 2013). The role of encapsulation in phage therapy is discussed in detail in another chapter of this book.

## Conclusion and future perspective

Phage therapy is a promising alternative to classic antibiotics. Moreover, the inherent biological limitations of using whole phages can be overcome with the use of phage proteins instead of whole phages. In the case of Gram-positive bacteria, since their peptidoglycan cell wall is exposed, phage lytic proteins can be highly effective against them when applied exogenously. Moreover, they have many advantages compared to classic antibiotics, such as high specificity and very low likeness to develop bacterial resistance. In this light, it is highly essential to continue their development as novel antimicrobials, testing their activity and safety in more clinical trials, and at the same time, continue investigating ways to improve their activity and delivery for different clinical and non-clinical purposes.

## References

Abdelkader, K., Gerstmans, H., Saafan, A., Dishisha, T., and Briers, Y. (2019). The preclinical and clinical progress of bacteriophages and their lytic enzymes: The parts are easier than the whole. Viruses *11*, 1–16.10.3390/v11020096.

Bae, JY., Jun, KI., Kang, CK., Song KH. ,Choe, PG., Bang, JH., Kim, ES., Park, SW., Kim, HB., Kim, NJ., and Park, WB. (2019) .Efficacy of intranasal administration of the recombinant endolysin sal200 in a lethal murine *Staphylococcus aureus* pneumonia model. Antimicrob. Agents Chemother *63*. 10.1128/AAC.02009-18.

Becker, S.C., Foster-Frey, J., and Donovan, D.M. (2008). The phage K lytic enzyme LysK and lysostaphin act synergistically to kill MRSA. FEMS Microbiol. Lett. *287*, 185–191. 10.1111/j.1574-6968.2008.01308.

Cenens, W., Makumi, A., Mebrhatu, M.T., Lavigne, R., and Aertsen, A. (2013). Phage–host interactions during pseudolysogeny. Bacteriophage *3*, e25029. 10.4161/bact.25029

Chang, Y., Yoon, H., Kang, D.H., Chang, P.S., and Ryu, S. (2017). Endolysin LysSA97 is synergistic with carvacrol in controlling *Staphylococcus aureus* in foods. Int. J. Food Microbiol. *244*, 19–26. 10.1016/j.ijfoodmicro.2016.12.007

Cooper, C.J., Koonjan, S., and Nilsson, A.S. (2018). Enhancing whole phage therapy and their derived antimicrobial enzymes through complex formulation. Pharmaceuticals *11*. 10.3390/ph11020034.

Corsini, B., Díez-Martínez, R., Aguinagalde, L., González-Camacho, F., García-Fernández, E., Letrado, P., García, P., and Yustea, J. (2018). Chemotherapy with phage lysins reduces pneumococcal colonization of the respiratory tract. Antimicrob. Agents Chemother. *62*, 1–13. 10.1128/AAC.02212-17

Djurkovic, S., Loeffler, J.M., and Fischetti, V.A. (2005). Synergistic killing of *Streptococcus pneumoniae* with the bacteriophage lytic enzyme Cpl-1 and penicillin or gentamicin depends on the level of penicillin resistance. Antimicrob. Agents Chemother. *49*, 1225–1228. 10.1128/AAC.49.3.1225-1228.2005

Escobedo, S., Campelo, A.B., Wegmann, U., García, P., Rodríguez, A., and Martínez, B. (2019). Insight into the lytic functions of the lactococcal prophage TP712. Viruses *11*, 1–16. 10.3390/v11100881.

Fair, R. J., & Tor, Y. (2014). Antibiotics and bacterial resistance in the 21st century. *Perspectives in medicinal chemistry*, *6*, 25–64.10.4137/PMC.S14459

Fenton, M., Casey, P.G., Hill, C., Gahan, C.G.M., Ross, R.P., Mcauliffe, O., O'Mahony, J., Maher, F., and Coffey, A. (2010). The truncated phage lysin CHAPk eliminates *Staphylococcus aureus* in the nares of mice. Bioeng. Bugs *1*, 404–407. 10.4161/bbug.1.6.13422.

Fenton, M., Cooney, J.C., Ross, R.P., Sleator, R.D., Mcauliffe, O., and Coffey, A. (2011). In silico modeling of the staphylococcal bacteriophage-derived peptidase CHAPk. Bacteriophage 1:4, 198–206. 10.4161/bact.1.4.18245

Fenton, M., Keary, R., McAuliffe, O., Ross, R.P., O'Mahony, J., and Coffey, A. (2013). Bacteriophage-derived peptidase CHAPk eliminates and prevents staphylococcal biofilms. Int J Microbiol. *2013*:625341. 10.1155/2013/625341.

Fernandes, S., and São-José, C. (2018). Enzymes and mechanisms employed by tailed bacteriophages to breach the bacterial cell barriers. Viruses *10*, 1–22. 10.3390/v10080396.

Fernández-Ruiz, I., Coutinho, F.H., and Rodriguez-Valera, F. (2018). Thousands of novel endolysins discovered in uncultured phage genomes. Front Microbiol. *9*, 1–8. 10.3389/fmicb.2018.01033.

Filatova, L.Y., Donovan, D.M., Ishnazarova, N.T., Foster-Frey, J.A., Becker, S.C., Pugachev, V.G., Balabushevich, N.G., Dmitrieva, N.F., and Klyachko, N.L. (2016). A chimeric LysK-lysostaphin fusion enzyme lysing Cells: *Staphylococcus aureus* a study of both kinetics of inactivation and specifics of interaction with anionic polymers. Appl Biochem Biotechnol. *180*, 544–557. 10.1007/s12010-016-2115-7.

Fischetti, V.A. (2011). Exploting what phage have evolved to control Gram-positive pathogens. Bacteriophage *1*, 188–194. 10.4161/bact.1.4.17747.

Freimer, E.H.; Krause, R.M.; Mc, C.M. (1959) Studies of l forms and protoplasts of group A streptococci. I. Isolation, growth, and bacteriologic characteristics.

J. Exp. Med. *110*, 853–874.

Gaeng, S., Scherer, S., Neve, H., and Loessner, M.J. (2000). Gene cloning and expression and secretion of *Listeria monocytogenes* bacteriophage-lyric enzymes in Lactococcus lactis. Appl. Environ. Microbiol. *66*, 2951–2958. 10.1128/aem.66.7.2951-2958.2000.

Gervasi, T., Horn, N., Wegmann, U., Dugo, G., Narbad, A., and Mayer, M.J. (2014). Expression and delivery of an endolysin to combat *Clostridium perfringens*. App Microbiol Biotechnol. *98*, 2495–2505. 10.1007/s00253-013-5128-y.

Gómez-Torres, N., Dunne, M., Garde, S., Meijers, R., Narbad, A., Avila, M., and Mayer, M.J. (2018). Development of a specific fluorescent phage endolysin for in situ detection of Clostridium species associated with cheese spoilage. Microb. Biotechnol. *11*, 332-345. 10.1111/1751-7915.12883.

Haddad Kashani, H., Schmelcher, M., Sabzalipoor, H., Seyed Hosseini, E., and Moniri, R. (2018). Recombinant endolysins as potential therapeutics against antibiotic-resistant *Staphylococcus aureus*: Current status of research and novel delivery strategies. Clin. Microbiol. Rev. *31*, e00071-17. 10.1128/CMR.00071-17.

Hancock, R.E.W. (2005). Mechanisms of action of newer antibiotics for Gram-positive pathogens major targets. Lancet Infect Dis. *5*, 209-218. 10.1016/S1473-3099(05)70051-7.

Harhala, M., Nelson, D.C., Miernikiewicz, P., Heselpoth, R.D., Brzezicka, B., Majewska, J., Linden, S.B., Shang, X., Szymczak, A., Lecion, D., Marek-Bukowiec, K., Kłak M., Wojciechowicz, B., Lahutta, K., Konieczny, A., and Dąbrowska K (2018). Safety Studies of pneumococcal endolysins Cpl-1 and Pal. Viruses. *11*., e638.10.3390/v10110638.

Hathaway, H., Ajuebor, J., Stephens, L., Coffey, A., Potter, U., Sutton, J.M., and Jenkins, A.T.A. (2017a). Thermally triggered release of the bacteriophage endolysin CHAP K and the bacteriocin lysostaphin for the control of methicillin resistant *Staphylococcus aureus* (MRSA). J. Control. Release *245*, 108–115. 10.1016/j.jconrel.2016.11.030.

Hathaway, H., Milo, S., Sutton, J.M., and Jenkins, T.A. (2017b). Recent advances in therapeutic delivery systems of bacteriophage and bacteriophage-encoded endolysins. Ther. Deliv. *8*, 543–556. 10.4155/

tde-2017-0040.

Hermoso, J.A., García, J.L., and García, P. (2007). Taking aim on bacterial pathogens: from phage therapy to enzybiotics. Curr. Opin. Microbiol. *10*, 461–472. 10.1016/j.mib.2007.08.002.

Herten, M., Idelevich, E.A., Sielker, S., Becker, K., Scherzinger, A.S., Osada, N., Torsello, G.B., and Bisdas, T. (2017). Vascular Graft Impregnation with Antibiotics : the influence of high concentrations of rifampin , vancomycin , daptomycin , and bacteriophage endolysin HY-133 on viability of vascular cells. Med Sci Monit Basic Res. *3,* 250–257. 10.12659/msmbr.902879.

Ibarra-Sánchez, L.A., Van Tassell, M.L., and Miller, M.J. (2018). Antimicrobial behavior of phage endolysin PlyP100 and its synergy with nisin to control *Listeria monocytogenes* in Queso Fresco. Food Microbiol. *72*, 128–134. 10.1016/j.fm.2017.11.013.

Jarábková, V., Tišáková, L., and Godány, A. (2015). Phage endolysin : a way to understand a binding function of c-terminal domains a mini review. Nova Biotechnologica et Chimica *14*, 117–134. 10.1515/nbec-2015-0021.

Jun, S.Y., Jang, I.J., Yoon, S., Jang, K., Yu, K., Youn, J., Seong, M., Jung, G.M., Yoon, S.J., and Kang, S.H. (2017). Pharmacokinetics and tolerance of the phage endolysin-based candidate drug SAL200 after a single intravenous administration among healthy volunteers. Antimicrob Agents Chemother. *61*. e02629-16. 10.1128/AAC.02629-16.

Kashani, H.H., Fahimi, H., Goli, Y.D., and Moniri, R. (2017). A Novel Chimeric Endolysin with Antibacterial Activity against Methicillin-Resistant *Staphylococcus aureus*. Front Cell Infect Microbiol. 7, *7*, 1–12. 10.3389/fcimb. 2017.00290.

Kim, N., Park, W.B., Cho, J.E., Choi, Y.J., Choi, S.J., Jun, S.Y., Kang, C.K., Song, K., Choe, P.G., Bang, J., Kim, E., Park, S., Kim. N., Oh, M., and Kima, B. (2018). Effect of Phage Endolysin SAL200 Combined with antibiotics on *Staphylococcus aureus* Infection. Antimicrob Agents Chemother. *62*, 1–10. 10.1128/AAC.00731-18.

Kongari, R., Rajaure, M., Cahill, J., Rasche, E., Mijalis, E., Berry, J., and Young, R. (2018). Phage spanins: Diversity, topological dynamics and gene convergence. BMC Bioinformatics *19*, 1–26. 10.1186/s12859-018-2342-8.

Latka, A., Maciejewska, B., and Majkowska-skrobek, G. (2017). Bacteriophage-encoded virion-associated enzymes to overcome the carbohydrate barriers during the infection process. Appl Microbiol Biotechnol. 101, 3103–3119. 10.1007/s00253-017-8224-6.

Letrado, P., Corsini, B., Díez-Martínez, R., Bustamante, N., Yuste, J.E., and García, P. (2018). Bactericidal synergism between antibiotics and phage endolysin Cpl-711 to kill multidrug-resistant pneumococcus. Future Microbiol. *13*, 1215–1223. 10.2217/fmb-*2018*-0077.

Lim, J.A., Shin, H., Heu, S., and Ryu, S. (2014). Exogenous lytic activity of SPN9CC endolysin against gram-negative Bacteria. J. Microbiol. Biotechnol. *24*, 803–811. 10.4014/jmb.1403.03035.

Love, M.J., Bhandari, D., Dobson, R.C.J., and Billington, C. (2018). Potential for bacteriophage endolysins to supplement or replace antibiotics in food production and clinical care. Antibiotics *7*, 1–25._10.3390/antibiotics7010017.

Low, L.Y., Yang, C., Perego, M., Osterman, A., and Liddington, R. (2011). Role of net charge on catalytic domain and influence of cell wall binding domain on bactericidal activity, specificity, and host range of phage lysins. J. Biol. Chem. *286*, 34391–34403. 10.1074/jbc.M111.244160.

Meng, X., Shi, Y., Ji, W., Meng, X., Zhang, J., Wang, H., Lu, C., Sun, J., and Yan, Y. (2011). Application of a bacteriophage lysin to disrupt biofilms formed by the animal pathogen *Streptococcus suis*. Appl. Environ. Microbiol. *77*, 8272–8279. 10.1128/AEM.05151-11.

Mercanti, D.J., Rousseau, G.M., Capra, M.L., Quiberoni, A., Tremblay, D.M., Labrie, S.J., and Moineau, S. (2016). Genomic diversity of phages infecting probiotic strains of *Lactobacillus paracasei*. Appl. Environ. Microbiol. *82*, 95–105. 10.1128/AEM.02723-15.

Misiou, O., van Nassau, T.J., Lenz, C.A., and Vogel, R.F. (2018). The preservation of Listeria-critical foods by a combination of endolysin and high hydrostatic pressure. Int. J. Food Microbiol. *266*, 355–362. 10.1016/j.ijfoodmicro.2017.10.004.

Munita, J.M., Bayer, A.S., and Arias, C.A. (2015). Evolving Resistance Among Gram-positive Pathogens. Clin Infect Dis. *61*, 48–57. 10.1093/cid/civ523.

Nelson, D., Schuch, R., Chahales, P., Zhu, S., and Fischetti, V.A. (2006). PlyC: A multimeric bacteriophage lysin. Proc. Natl. Acad. Sci. U. S. A. *103*, 10765–10770. 10.1073/pnas.0604521103.

Nelson, D.C., Schmelcher, M., Rodriguez-rubio, L., Klumpp, J., Pritchard, D.G., Dong, S., and Donovan, D.M. (2012). Endolysins as Antimicrobials. Adv Virus Res. *83*, 299-365. 10.1016/B978-0-12-394438-2.00007-4.

Oliveira, H.,São-José, C., and Azeredo, J. (2018). Phage-derived peptidoglycan degrading enzymes : Challenges and Future Prospects for In Vivo Therapy. Viruses. *10*, e292.10.3390/v10060292.

Oliveira, H., Melo, L.D.R., Santos, S.B., Nobrega, F.L., Ferreira, E.C., Cerca, N., Azeredo, J., and Kluskens, L.D. (2013). Molecular aspects and comparative genomics of bacteriophage endolysins. J. Virol. *87*, 4558–4570. 10.1128/JVI. 03277-12.

Oliveira, H., Boas, D.V., Mesnage, S., Kluskens, L.D., Lavigne, R., Sillankorva, S., Secundo, F., and Azeredo, J. (2016). Structural and enzymatic characterization of ABgp46, a novel phage endolysin with broad anti-Gram-negative bacterial activity. Front. Microbiol. *7*, 1–9. 10.3389/fmicb.2016.00208.

Park, S., Jun, S.Y., Kim, C., Jung, G.M., Son, J.S., and Jeong, S.T. (2018). Characterisation of the antibacterial properties of the recombinant phage endolysins AP50-31 and LysB4 as potent bactericidal agents against *Bacillus anthracis*. Sci. Rep. *8*, 1–11. 10.1038/s41598-017-18535-z.

Paul, V., Rajagopalan, S., Sundarrajan, S., George, S.E., Asrani, J.Y., Pillai, R., Chikkamadaiah, R., Durgaiah, M., Sriram, B., and Padmanabhan, S. (2011a). A novel bacteriophage Tail-Associated Muralytic Enzyme (TAME) from Phage K and its development into a potent antistaphylococcal protein. BMC Microbiol. *11*, 226.

Rajesh, T., Anthony, T., Saranya, S., Pushpam, P.L., and Gunasekaran, P. (2011). Functional characterization of a new holin-like antibacterial protein coding gene tmp1 from goat skin surface metagenome. Appl. Microbiol. Biotechnol. *89*, 1061–1073. 10.1007/s00253-010-2907-6.

Reiner B., and Oram, J.D. (1963). Group N Streptococcal phage lysin. J. gen. Microbiol. *32*, 90-32

Roach, D.R., and Donovan, D.M. (2015). Antimicrobial bacteriophage-derived proteins and therapeutic applications. Bacteriophage *5*, e1062590. 10.1080/21597081.2015.1062590.

Rodríguez-rubio, L., Gerstmans, H., Thorpe, S., Mesnage, S., and Lavigne, R. (2016). DUF3380 domain from a *Salmonella* phage endolysin shows potent n

-acetylmuramidase activity. Appl Environ Microbiol. *82*, 4975–4981. 10.1128/AEM.00446-16.

Rodríguez-Rubio, L., Martínez, B., Rodríguez, A., Donovan, D.M., and García, P. (2012). Enhanced staphylolytic activity of the *Staphylococcus aureus* bacteriophage vB_SauS-phiiPla88 HydH5 Virion-associated peptidoglycan hydrolase: Fusions, deletions, and synergy with LysH5. Appl. Environ. Microbiol. *78*, 2241–2248. 10.1128/AEM.07621-11.

Rodríguez-Rubio, L., Chang, W.L., Gutiérrez, D., Lavigne, R., Martínez, B., Rodríguez, A., Govers, S.K., Aertsen, A., Hirl, C., Biebl, M., et al. (2016). "Artilysation" of endolysin λSa2lys strongly improves its enzymatic and antibacterial activity against streptococci. Sci. Rep. *6*, 1–11. 10.1038/srep35382.

Saier, M.H., and Reddy, B.L. (2015). Holins in bacteria, eukaryotes, and archaea: Multifunctional xenologues with potential biotechnological and biomedical applications. J. Bacteriol. *197*, 7–17. 10.1128/JB.02046-14.

Santos, S.B., Oliveira, A., Melo, L.D.R., and Azeredo, J. (2019). Identification of the first endolysin Cell Binding Domain (CBD) targeting *Paenibacillus larvae*. Sci. Rep. *9*, 1–9. 10.1038/s41598-019-39097-2.

Sanz-gaitero, M., Keary, R., Garcia-doval, C., Coffey, A., and Raaij, M.J. Van (2014). Crystal structure of the lytic CHAP K domain of the endolysin LysK from *Staphylococcus aureus* bacteriophage K Crystal structure of the lytic CHAP K domain of the endolysin LysK from *Staphylococcus aureus* bacteriophage K. Virol J. *11*, 1–10. 10.1186/1743-422X-11-133.

São-José, C. (2018). Engineering of phage-derived lytic enzymes: Improving their potential as antimicrobials. Antibiotics *7*, e29. 10.3390/antibiotics7020029.

Schuch, R., Lee, H.M., Schneider, B.C., Sauve, K.L., Law, C., Khan, B.K., Rotolo, J.A., Horiuchi, Y., Couto, D.E., Raz, A., Fischetti, V.A., Huang, D.B., Nowinski, R.C., and Wittekind, M. (2014). Combination therapy with lysin CF-301 and antibiotic is superior to antibiotic alone for treating methicillin-resistant staphylococcus aureus-induced murine bacteremia. J. Infect. Dis. *209*, 1469–1478. doi: 10.1093/infdis/jit637.

Shang, X., and Nelson, D.C. (2019). Contributions of net charge on the PlyC endolysin chap domain. Antibiotics *8*, 1–11. 10.3390/antibiotics8020070.

Shen, Y., Barros, M., Vennemann, T., Gallagher, D.T., Yin, Y., Linden, S.B., Heselpoth, R.D., Spencer, D.J., Donovan, D.M., Moult, J., Fischetti, V.A., Heinrich, F., Lösche, M., and Nelson, D.C. (2016). A bacteriophage endolysin that eliminates intracellular streptococci. eLife *5*, 1–26. 10.7554/eLife.13152.

Singh, P.K., and Donovan, D.M. (2014). Intravitreal injection of the chimeric phage endolysin Ply187 protects mice from *Staphylococcus aureus* endophthalmitis. Antimicrob Agents Chemother. *58*, 4621–4629. 10.1128/AAC.00126-14.

Solanki, K., Grover, N., Downs, P., Paskaleva, E.E., Mehta, K.K., Lee, L., Schadler, L.S., Kane, R.S., and Dordick, J.S. (2013). Enzyme-based listericidal nanocomposites. Sci. Rep. *3*, 1–6. 10.1038/srep01584.

Stockdale, S.R., Mahony, J., Courtin, P., Chapot-Chartier, M.P., Van Pijkeren, J.P., Britton, R.A., Neve, H., Heller, K.J., Aideh, B., Vogensen, F.K., and Van Sinderen, D. (2013). The lactococcal phages Tuc2009 and TP901-1 incorporate two alternate forms of their tail fiber into their virions for infection specialization. J. Biol. Chem. *288*, 5581–5590. 10.1074/jbc.M112.444901.

Swift, S.M., Seal, B.S., Garrish, J.K., Oakley, B.B., Hiett, K., Yeh, H.Y., Woolsey, R., Schegg, K.M., Line, J.E., and Donovan, D.M. (2015). A thermophilic phage endolysin fusion to a *Clostridium perfringens*-specific cell wall binding domain creates an anti-clostridium antimicrobial with improved thermostability. Viruses *7*, 3019–3034. 10.3390/v7062758.

Takáč, M., and Bläsi, U. (2005). Phage P68 virion-associated protein 17 displays activity against clinical isolates of *Staphylococcus aureus*. Antimicrob. Agents Chemother. *49*, 2934–2940. 10.1128/AAC.49.7.2934-2940.2005.

Torres-Barceló, C., Arias-Sánchez, F.I., Vasse, M., Ramsayer, J., Kaltz, O., and Hochberg, M.E. (2014). A window of opportunity to control the bacterial pathogen *Pseudomonas aeruginosa* combining antibiotics and phages. PLoS One *9*, 1–7. 10.1371/journal.pone.0106628.

Totté, J.E.E., van Doorn, M.B., and Pasmans, S.G.M.A. (2017). Successful Treatment of Chronic *Staphylococcus aureus*-Related Dermatoses with the Topical Endolysin Staphefekt SA.100: A Report of 3 Cases. Case Rep. Dermatol. *9*, 19–25. 10.1159/000473872.

Turner, M.S., Waldherr, F., Loessner, M.J., and Giffard, P.M. (2007). Antimicrobial activity of lysostaphin and a *Listeria monocytogenes* bacteriophage endolysin produced and secreted by lactic acid bacteria. Syst. Appl. Microbiol. *30*, 58–67. 10.1016/j.syapm.2006.01.013.

Vázquez, R., García, E., and García, P. (2018). Phage lysins for fighting bacterial respiratory infections: A new generation of antimicrobials. Front. Immunol. *9*, 2252. 10.3389/fimmu.2018.02252.

Walmagh, M., Briers, Y., dos Santos, S.B., Azeredo, J., and Lavigne, R. (2012). Characterization of modular bacteriophage endolysins from myoviridae phages OBP, 201Q2-1 and PVP-SE1. PLoS One *7*, 1–10.10.1371/ journal.pone.0036991.

Wang, Q., Euler, C.W., Delaune, A., and Fischetti, V.A. (2015). Using a novel lysin to help control *Clostridium difficile* infections. Antimicrob. Agents Chemother. *59*, 7447-7457. 10.1128/AAC.01357-15.

Wang, Y., Sun, J.H., and Lu, C.P. (2009). Purified recombinant phage lysin LySMP: An extensive spectrum of lytic activity for swine streptococci. Curr. Microbiol. *58*, 609–615. 10.1007/s00284-009-9379-x.

Wang, Z.F., Kong, L.C., Liu, Y., Fu, Q., Cui, Z.L., Wang, J., Ma, J.J., Wang, H.A., Yan, Y.X., and Sun, J.H. (2018). A phage lysin fused to a cell-penetrating peptide kills intracellular methicillin-resistant *Staphylococcus aureus* in keratinocytes and has potential as a treatment for skin infections in mice. Appl. Environ. Microbiol. *84*, 1–13. 10.1128/AEM.00380-18.

Wittekind, M., and Schuch, R. (2016). Cell wall hydrolases and antibiotics: Exploiting synergy to create efficacious new antimicrobial treatments. Curr. Opin. Microbiol. *33*, 18–24. 10.1016/j.mib.2016.05.006.

Wu, X., Kwon, S., Kim, J., Kane, R.S., and Dordick, J.S. (2017). Biocatalytic Nanocomposites for Combating Bacterial Pathogens. Annu. Rev. Chem. Biomol. Eng. *8*, 87–113. 10.1146/annurev-chembioeng-060816-101612.

Xu, M., Struck, D.K., Deaton, J., Wang, I.N., and Young, R. (2004). A signal-arrest-release sequence mediates export and control of the phage P1 endolysin. Proc. Natl. Acad. Sci. U. S. A. *101*, 6415–6420. 10.1073/pnas. 0400957101.

Young, R., Wang, I.N., and Roof, W.D. (2000). Phages will out: strategies of host cell lysis. Trends Microbiol. *8*, 120–128. 10.1016/s0966-842x(00)01705-4.

Zhang, W., Mi, Z., Yin, X., Fan, H., An, X., Zhang, Z., and Chen, J. (2013). Characterization of *Enterococcus faecalis* phage IME-EF1 and its endolysin.

PloS One *8*, e80435. 10.1371/journal.pone.0080435.

Zimmer, M., Vukov, N., Scherer, S., and Loessner, M.J. (2002). The murein hydrolase of the bacteriophage φ3626 dual lysis system is active against all tested *Clostridium perfringens* strains. Appl. Environ. Microbiol. *68*, 5311–5317. 10.1128/AEM.68.11.5311-5317.2002.

Chapter 15

# Developments and Opportunities of Bacteriophage Lytic Proteins for Therapeutics Against Gram-negative Pathogens

**Diana Gutiérrez[1] and Yves Briers[1*]**

[1]Department of Biotechnology. Ghent University, Valentin Vaerwyckweg 1 - building C, 9000 Gent, Belgium.

*Corresponding author: yves.briers@ugent.be

DOI: https://doi.org/10.21775/9781913652517.15

**Abstract**

The globally increasing antimicrobial resistance levels urge for the introduction of novel classes of antimicrobials. In this regard, enzyme-based antibiotics (or enzybiotics) derived from phage lytic proteins represent a promising strategy. These enzymes kill bacteria through peptidoglycan degradation followed by osmotic lysis and come along with significant advantages such as a rapid bactericidal effect, a novel mode of action and a low probability of resistance development. They have been progressed significantly on the route for clinical application to treat infections caused by Gram-positive bacteria. The latter have a thick but easily accessible peptidoglycan layer. Gram-negative bacteria have been for long a no go zone for enzybiotics due to their protective outer membrane, principally excluding access to the thin peptidoglycan layer. This view has shifted drastically, particularly driven by the lack of remaining therapeutic options for infections caused by Gram-negatives. Diverse strategies grouped in three classes (phage lytic proteins with a natural intrinsic antibacterial activity, combination strategies with physical or chemical means and protein engineering) have been proposed and developed in the last decade, ranging from the conceptual to the preclinical level, and will be discussed in this chapter.

## Introduction

The rapid rise and dissemination of multidrug-resistant (MDR) pathogenic bacteria represents a major global threat to human health, sustainable food production and development. The antimicrobial resistance (AMR) crisis is nowadays one of the greatest challenges that the global community faces since morbidity and mortality associated with AMR has increased exponentially. In this context, the World Health Organization (WHO), the Food and Agriculture Organization of the United Nations (FAO) and the World Organization for Animal Health are working together in the 'One Health' approach (WHO, 2017a). This approach is a call for global action encouraging several strategies, including: 1) optimization of the use of antibiotics by implementing 'antimicrobial stewardship' in programs and interventions to avoid the widespread of antimicrobial resistance (Dyar et al., 2017); 2) establishment of preventive and control intervention in healthcare facilities (Pires et al., 2017); 3) reduction of the use of antimicrobials in livestock production and along the food chain, including control of the environmental pollution generated by these production systems (Collignon et al., 2016); 4) search for new alternatives to the currently used antibiotics (Tacconelli et al., 2018). For the latter action point, WHO has created a priority list of antibiotic-resistant bacteria for which research, discovery and development on new antibacterials are urgently needed (WHO, 2017b). This list includes MDR bacteria with a global impact on human health, including *Mycobacterium tuberculosis*, some members of the ESKAPE group such as *Enterococcus faecium, Staphylococcus aureus, Acinetobacter baumannii, Pseudomonas aeruginosa* and Enterobacteriaceae (WHO, 2017c). Notably, the majority of the MDR bacteria included in the priority list are Gram-negative pathogens. These species represent a major health risk due to the particular structure of their envelope that consists of three layers, i.e., the inner cytoplasmic membrane, the peptidoglycan and the outer membrane. The outer membrane, which is only present in Gram-negative bacteria, acts as a barrier that prevents the entrance of many antimicrobials (Exner et al., 2017). Specifically, this barrier does not only exclude many antibiotics currently used for the treatment of infections, but also makes the development of new antibacterials for this group of pathogens a riskier undertaking. This is particularly troublesome for risk-averse industry and accordingly, the lack of development of new antibiotics by the pharmaceutical industry is even more outspoken in the case of Gram-negative bacteria. Therefore, Gram-negative infections being the most worrisome due to the lack of therapeutic options in combination with the least efforts by the pharmaceutical industry take global health systems in a chokehold that needs to be urgently addressed.

## Therapeutic potential of phage lytic proteins

In the search for new antimicrobials, the renewed interest in the use of bacteriophages (phages) and their proteins has intensified during the last years. These viruses infect and kill bacteria in a highly specific way, and they have been considered and used for the treatment of bacterial infections since their discovery in 1915. Phage therapy has, however, only been actively continued in some Eastern Europe countries with promising results. Nowadays the use of phages is leading around the world to new regulations that speed up the availability of phage therapy as an alternative to conventional antibiotics (Cisek et al., 2017; Moelling et al., 2018). Lytic bacteriophages are able to lyse antibiotic-resistant bacteria at the end of their replication cycle. Most phages utilize therefore a two-component lysis system (holin-endolysin) to release the new viral progeny (Drulis-Kawa et al., 2015; Loessner, 2005). Holins perforate the bacterial cytoplasmic membrane at a strictly programmed time point, forming pores through which the endolysin gains access to the peptidoglycan, followed by immediate peptidoglycan degradation (Schmelcher et al., 2012a). Finally, due to the high internal osmotic pressure, the bacterial cell lyses and the viral particles are released in a process known as lysis from within (Cisek et al., 2017). Many phages infecting Gram-negative bacteria carry additional proteins called spanins that destabilize the outer membrane and in combination with the holin-endolysin system, facilitate the release of the viral progeny (Catalao et al., 2013). Recently, a new mechanism of outer membrane disruption, unrelated to spanins, has been revealed, and is mediated by a disruptin. The disruptin is a cationic peptide with a predicted amphipathic helical structure that is proposed to destabilize the outer membrane in a similar way as many amphipathic antimicrobial peptides (Holt et al., 2019).

Besides an endolysin, phages also encode virion-associated peptidoglycan hydrolases (VAPGHs) that locally degrades peptidoglycan during the initial steps of infection. Therefore, a VAPGH is a structural component of the phage particle and its role is to facilitate the entrance of the phage genetic material into the bacterial cytoplasm (Gutiérrez et al., 2018; Rodríguez-Rubio et al., 2013). Endolysins and VAPGHs are grouped as phage lytic proteins.

Over the last few years, the study of phage lytic proteins has attracted great interest for use as antimicrobial due to their unique features: 1) rapid bacterial death immediately upon application; 2) low probability to provoke bacterial resistance development since they act on conserved and immutable sites of the peptidoglycan, that cannot be altered without significant loss in fitness (Gilmer et al., 2013); 3) high specificity against the target pathogen, not affecting beneficial microbiota; 4) an activity generally independent of the physiological state of the bacteria (Vázquez et al., 2017); and 5) able to remove bacteria forming a biofilm on animal tissues or inert surfaces (Becker et al., 2016; Gutiérrez et al., 2016; Schmelcher et al., 2012b); 6) bactericidal activity against MDR bacteria and persisters (Briers et al., 2014a; Gutiérrez et al., 2018); and 7) synergy with other antimicrobials increasing the therapeutic effect (Schmelcher et al., 2012b).

Therefore, the use of phage lytic proteins as antibiotics (also called enzyme-based antibiotics or enzybiotics) represents a promising research field to combat MDR bacteria (Dams and Briers, 2019; Fischetti, 2018). In fact, a recent pipeline portfolio review about alternatives to antibiotics in The Lancet Infectious Diseases highlighted enzybiotics as alternatives with great potential to be used as therapeutics (Theuretzbacher et al., 2019). There is abundant evidence that many recombinant phage lytic proteins are effective in several animal models of Gram-positive bacterial infection, rescuing the animals from an otherwise deadly infection (Gerstmans et al., 2018; Gutiérrez et al., 2018; Nelson et al., 2001; Schmelcher et al., 2015, 2012b; Schuch et al., 2014). Recently, two phase 1 human clinical trials have been completed with iNtRON-N-Rephasin® SAL200 (Tonabacase) (iNtRON Biotecnology, 2020a) and CF-301 (Exebacase) (ContraFect, 2019a) against *S. aureus* infections with promising results and without observing any adverse effects (reviewed by Abdelkader et al., 2019; Jun et al., 2017). Tonabacase is currently being evaluated in phase 2 clinical trial (iNtRON Biotecnology, 2020b) to treat bacteremia caused by *S. aureus*, whereas Exebacase has finished a phase 2 trial and has entered into phase 3 to test its activity against bacteremia and endocarditis caused by methicillin-resistant *S. aureus* (ContraFect, 2019b).

Besides their clinical application, many phage lytic proteins have demonstrated efficacy to control and eliminate pathogenic bacteria along the food chain. In veterinary therapy, their usefulness has been demonstrated in the treatment of staphylococcal mastitis in a murine model of infection (Schmelcher et al., 2015). Recently, their potential has been also shown for the biocontrol of *Vibrio parahaemolyticus* in aquaculture production systems (Srinivasan et al., 2019). In addition, phage lytic proteins have been proposed as disinfectants for the food industry (Gutiérrez et al., 2016) as well as biopreservatives (Rodríguez-Rubio et al., 2013).

A major initial shortcoming in the therapeutic potential of enzybiotics was their inactivity against Gram-negative bacteria, as peptidoglycan is generally not accessible due to the presence of the protective outer membrane in the bacterial cell envelope. Given the above-cited outspoken lack of new antibiotics for Gram-negative pathogens, there is a logical and urgent interest to expand the therapeutic potential of this novel class of antibacterials to Gram-negative bacteria as well. This chapter focuses on the recent advances realized in this field.

## The barrier function of the Gram-negative envelope

The envelope of Gram-negative bacteria has a three-layered structure (Figure 1), including the inner membrane (IM), the peptidoglycan layer and the outer membrane (OM). This structure contrasts to the two-layered structure of the Gram-positive cell wall, which lacks the outer membrane.

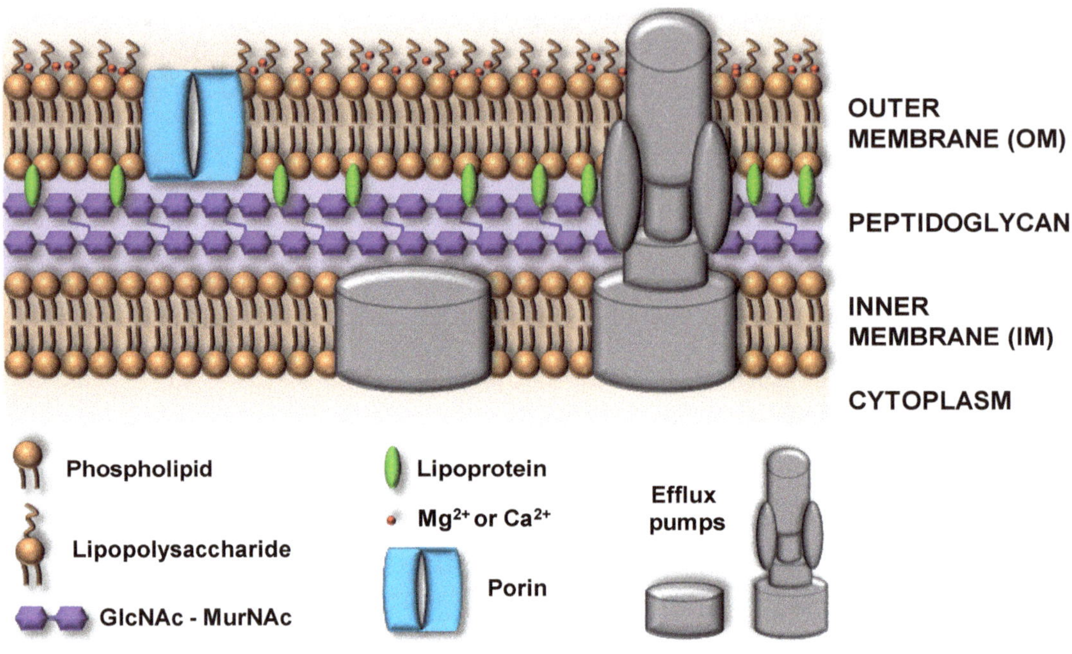


**Figure 1.** Schematic representation of the cell envelope of Gram-negative bacteria.

The IM is a phospholipid bilayer that forms a semi-permeable barrier between the cytoplasm and the periplasm. The IM creates an osmotic barrier with tight control of the energy-dependent transport of ions, polymers, protein secretion channels and lipid and protein biosynthesis (De Geyter et al., 2019).

Peptidoglycan is a large meshwork composed of chains of repeated units of alternating, β1-4-linked N-acetylglucosamine (GlcNAc) and N-acetylmuramic acid (MurNAc) residues. Each chain is terminated by a 1,6-anhydroMurNAc residue. These chains are cross-linked via the stem peptide attached to each MurNAc residue. The main difference with Gram-positive bacteria is that the cell wall of Gram-negative bacteria possesses a thin peptidoglycan layer (1-3 layers), which is only partially cross-linked (Vollmer et al., 2008). In addition, all Gram-negative bacteria share a common peptidoglycan chemotype (A1$\gamma$), whereas a large variation in chemotypes is observed among Gram-positive bacteria. Peptidoglycan is the largest molecule of a bacterium and forms a rigid structure that determines the bacterial shape. Antimicrobial agents such as enzybiotics that damage the peptidoglycan cause cell lysis due to the turgor pressure of the cytoplasm. Lysis can be prevented in media which are isotone to the cytoplasmic osmolarity, although without peptidoglycan, bacteria lose their particular shape. Under such suitable conditions, cell wall-deficient bacteria or L-forms can be created by (partial) removal of the peptidoglycan layer. L-forms represent one of the most extreme examples of bacterial plasticity, since L-forms can propagate and multiply indefinitely on high osmolarity media in the absence of an intact cell wall (Errington, 2013).

The periplasm is a multipurpose compartment separated from the cytoplasm, located in between the IM and OM. Its distinct reducing environment allows more efficient and diverse mechanisms of protein oxidation, folding, and quality control. The periplasm also contains structural elements and important environmental sensing modules, and it allows complex nanomachines to span the cell envelope (Miller and Salama, 2018).

The most external layer of the bacterial envelope is the OM. The OM acts as a semi-permeable barrier separating the extracellular environment from the periplasm (Silhavy et al., 2010). The OM is basically composed of an asymmetric bilayer with an inner stratum of phospholipids and an outer stratum consisting of lipopolysaccharides (LPS). LPS, also known as endotoxins, are large molecules consisting of a lipid anchor and a polysaccharide composed of the core and O-antigen. They are the most prominent indicators of Gram-negative bacteria during infection. Consequently, the human innate immunity is highly responsive to the presence of this molecule that can potentially lead to a fatal septic shock when LPS is present in a too high amount (Opal, 2010). LPS is essential for bacterial survival in hazardous environments because the lack of LPS results in extreme sensitivity of Gram negative bacteria to antibiotics. Individual, adjacent LPS molecules mutually interact through divalent cations ($Mg^{2+}$ and $Ca^{2+}$) to form a barrier that protects from the entrance of hydrophobic and hydrophilic compounds larger than 600 Da such as antibiotics, bile salts and detergents (Tamaki et al., 1971; Zhang et al., 2013). The LPS layer is one of the main factors contributing to the intrinsic antibiotic resistances observed in Gram-negative bacteria (Exner et al., 2017).

Two main types of protein can be found in the OM: lipoproteins and β-barrel proteins (Figure 1). The lipoproteins are located in the inner part of the OM. Although the function of most of them is unknown, it is recognized that lipoprotein Lpp links the peptidoglycan to the OM. As the most abundant protein in *E. coli* ($10^6$ molecules per cell), Lpp plays also an important role in the integrity of the cell envelope and is involved in protein secretion, conjugation, signal transduction, cell wall metabolism, and antimicrobial resistance. The human innate immunity recognizes these lipoproteins leading to inflammation and the adaptive immune response (Braun and Hantke, 2019). On the other hand, the β-barrel proteins are transmembrane proteins wrapped into cylinders. Their main function is to allow diffusion of small molecules such as monosaccharides, disaccharides and amino acids across the OM. Well-known examples of these molecules are porins such as OmpF, OmpC, LamB or PhoE to name a few (Vergalli et al., 2019). Recent studies have also demonstrated that porins can interact with a variety of host tissues for invasion and adhesion, but their role in pathogenesis is still under investigation (Galdiero et al., 2012). Porins play also a key role in the acquisition of antimicrobial resistance. In a normal scenario, small molecules such as β-lactams, tetracycline, chloramphenicol and fluoroquinolones diffuse through the porins into the periplasm, but in the case of antibiotic-resistant bacteria, two major porin-based mechanisms of antibiotic resistance may counteract this mechanism by 1) modification of the OM by losing/reducing the number of porins or replacing the two major porins by another; 2) reduction of the permeability of porins by specific mutations (Galdiero et al., 2012).

Besides those OM-specific proteins, there is a third group of transenvelope proteins, spanning the whole distance from the inner to the outer membrane, thereby crossing the peptidoglycan layer (Figure 1). These transenvelope proteins include appendages needed for bacterial motility such as flagella, Type III secretion systems, which inject toxins in eukaryotic cells during the bacterial infection cycle and efflux pumps, which pump toxic molecules like antibiotics out of the bacterial cell (Silhavy et al., 2010). In particular, the efflux pumps are of major concern in the field of antibiotic discovery since they can lead to antimicrobial resistance by lowering the amount of antibiotic inside the bacteria (Li et al., 2015).

In summary, the cell envelope of Gram-negative bacteria is a complex and dynamic structure with a variety of protective and adaptive roles. Since the peptidoglycan, the target of enzybiotics, is protected by the OM, strategies should be established to facilitate the passage of phage lytic proteins through the OM barrier.

## Structural and biochemical diversity of phage lytic proteins active against Gram-negative bacteria

All phage lytic proteins share the biological function of peptidoglycan degradation, yet the constant co-evolution between bacteriophages and their hosts has led to a huge variety of phage lytic proteins, both at the structural and biochemical level (Nelson et al., 2012). Bioinformatic analyses show that endolysins vary enormously in terms of length, size and domain composition (Oliveira et al., 2013). In this work, a total of 727 endolysins obtained from previously sequenced phage genomes available in databases, were analyzed. Nowadays the number of phage lytic proteins available in databases has exponentially increased to more than 12707 phage lytic proteins, which can be extensively studied using the recently created database of phage lytic proteins (PhaLP) (http://www.phalp.org/; accessed January, 2020). This database provides information of conserved domains, enzymatic specificity, gene ontologies and experimental evidence of all phage lytic proteins described to date, with a convenient Biomart-based search engine. The very large majority of endolysins have been identified from cultured phage or prophage genomes, which have been previously sequenced. Preliminary analyses of uncultured viral metagenomes have revealed hundreds of phage lytic proteins with an unprecedented diversity (Fernandez-Ruiz et al., 2018), indicating much insights in structural and biochemical diversity remain to be gained. Theoretically, the high diversity of already published phage endolysins and the ones in viral metagenomes represent an almost infinite source of new potential antimicrobials to be explored against any MDR pathogen.

Phage lytic proteins that are encoded by phages infecting Gram-negative bacteria possess a different structure compared to those encoded by phages infecting Gram-positives (Oliveira et al., 2013). This feature is most likely due to the significant differences in cell wall composition. On the one hand, endolysins directed to Gram-positives display a modular structure composed of one or more catalytic domains (EADs) and one or more cell wall binding domains (CBDs) (Gutiérrez et al., 2018; Oliveira et al., 2013). The CBD recognizes and binds to the cell wall to facilitate peptidoglycan degradation by the EAD. The domains are usually connected by a linker, which can vary in size and is the ultimate responsible for the autonomous function of the different domains (Roach and Donovan, 2015; Schmelcher et al., 2012a). On the other hand, most endolysins active against Gram-negatives possess a globular structure with just one EAD, although different exceptions to this rule have been reported as well (Oliveira et al., 2013). Globular endolysins encoded by Gram-negative infecting phages may not need a CBD to direct the EAD to the peptidoglycan due to the thin peptidoglycan layer and the presence of the OM that facilitates the accumulation of the endolysins within the periplasm (Briers et al., 2007). A highly-cited hypothesis that explains the presence of a CBD in endolysins targeting Gram-positives is the need to retain the 'used' endolysin molecules bound to degraded cell wall debris, preventing endolysins from killing neighboring bacteria that may become new host cells of the released bacteriophages (Loessner et al., 2002). The presence of the OM prevents this possible collateral damage in the case of Gram-negative bacteria, also excluding the need of a CBD for this reason.

Besides this structural diversity, endolysins also show a high biochemical diversity. A classification is based on the specific chemical bond that is targeted in the peptidoglycan (Figure 2). Four different specificity groups can be observed: 1) *N*-acetyl-β-D-muramidases (E.C. 3.2.1.17) or lysozymes, which cleave the MurNAc-N-GlcNAc bond; 2) *N*-acetyl-β-D-glucosaminidases (E.C. 3.2.1.52), which cleave the GlcNAc-N-MurNAc bond; 3) *N*-acetylmuramoyl-L-alanine amidases (E.C. 3.5.1.28), which hydrolyse the amide bond between N-acetylmuramoyl and L-alanine; 4) endopeptidases (E.C. 3.4.X.X), which cleave the peptide bond between two amino acids of the interpeptide bridge of the peptidoglycan (Figure 2) (Oliveira et al., 2013). In addition, lytic transglycosylases cleave the same bond as *N*-acetyl-β-D-muramidases but with the concomitant formation of a 1,6-anhydro bond in MurNAc (E.C. 4.2.2.X). The latter enzymes do not use a water molecule in their reaction and are therefore not hydrolases by lyases.

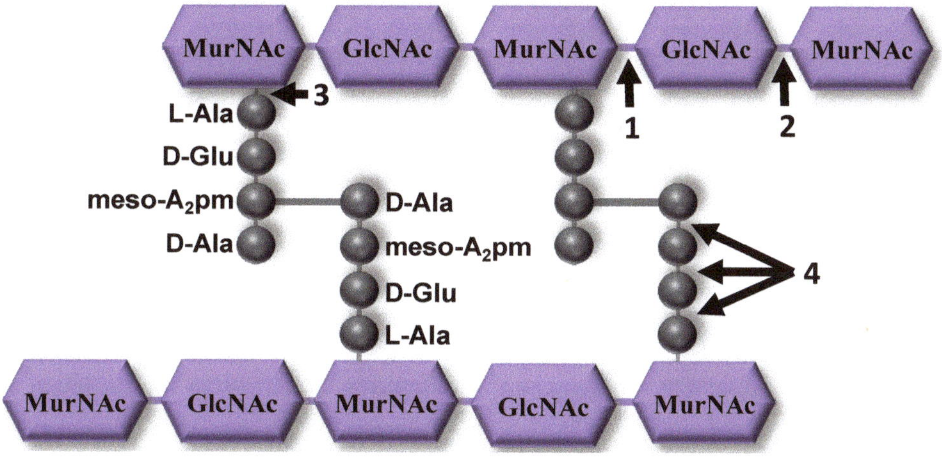


**Figure 2.** Schematic representation of the peptidoglycan of Gram-negative bacteria (chemotype A1γ) indicating the four main specificities of phage encoded endolysins: 1) *N*-acetyl-β-D-muramidases (lysozymes) and transglycosylases; 2) *N*-acetyl-β-D-glucosaminidases; 3) *N*-acetylmuramoyl-L-alanine amidases; and 4) endopeptidases.

In contrast to Gram-positive acting endolysins, the majority of Gram-negative acting endolysins carry a lysozyme-like (LYSO) catalytic domain. This LYSO domain possesses *N*-acetyl-β-D-muramidase activity, which is shared with other domains like GH19, GH108 and MURA. The latter domains are also found among endolysins from Gram-negative infecting phages but with lower frequencies. The amidase-2 (Ami-2) domain is also present in a high percentage of endolysins. Ami-2 possesses a *N*-acetylmuramoyl-L-alanine amidase activity, similar to the less frequent Ami-3 domain. Other enzymatic activity domains that appear in lower frequencies are GLUCO, VANY, PET-M15-3, PET-M15-4 and PET-M23 (Oliveira et al., 2013).

Besides the prevailing globular structure with a single EAD, there are few exceptions of modular endolysins that carry two EADs such as endolysins encoded by phages infecting *Prochlorococcus marinus* and *Synechococcus*. These endolysins carry a muramidase domain and a peptidase domain (Oliveira et al., 2013). Moreover, different endolysins carry a N-terminal CBD and a C-terminal EAD. Most of the endolysins that also have a CBD are encoded by jumbo phages. All these endolysins carry conserved CBD domains PG-1 or PG-3 that were demonstrated to bind to the peptidoglycan chemotype A1γ, which is conserved among all Gram-negative species (Briers et al., 2007; Antonova et al., 2019; Oliveira et al., 2013; Srinivasan et al., 2019; Walmagh et al., 2012; Wang et al., 2016). Moreover, fusion of the CBD of endolysin KZ144 to the VAPGH domain KMV36C resulted in a chimeric protein with a three-fold higher specific activity than the native catalytic domain (Briers et al., 2009). This illustrates that creating chimeric proteins combining different functional domains is a feasible strategy to improve the specific activity of phage lytic proteins acting against Gram-negative bacteria.

**Development of enzybiotics against Gram-negative bacteria**

Since the first publication describing the use of a phage lytic protein to prevent and to eliminate upper respiratory colonization of mice by group A streptococci (Nelson et al., 2001), numerous studies from laboratories all around the world have been reported on the identification and use of phage lytic proteins as antimicrobials (Dams and Briers, 2019; Drulis-Kawa et al., 2015; Fischetti, 2018; Gutiérrez et al., 2018; Vázquez et al., 2018). Obviously, the majority are endolysins encoded by phages infecting Gram-positive bacteria, with a particular focus on *S. aureus* and streptococci. This is not surprising since the peptidoglycan is easily accessible in the case of Gram-positive bacteria. Whereas the peptidoglycan layer of Gram-negative bacteria is much thinner and can thus theoretically be degraded by a much lower number of enzymatic cleavages, the challenge is to get the phage lytic proteins in the periplasm. Driven by the particularly high need for novel antimicrobials killing Gram-negative bacteria, several different approaches have been developed to expand this novel class of antibacterials to Gram-negatives. These approaches can be categorized in three groups: (1) phage lytic proteins that can naturally permeate the OM and show intrinsic antibacterial activity; (2) the combination of phage lytic proteins with physical forces or chemicals to induce bacterial lysis; and (3) the application of protein engineering to equip phage lytic proteins with the capacity to pass the outer membrane, either by disrupting the OM or by a receptor-mediated uptake mechanism. While a surprisingly large variation of effective approaches has now been proposed, the efficacy of killing, the dose-dependency and the activity spectrum is highly diverse (Table 1).

**Table 1.** Major findings related to the use of phage lytic proteins in combination with different methods to enhance antibacterial activity against Gram-negative bacteria.

| Methodology | Protein | Susceptible bacteria | Treatment conditions | Reference |
|---|---|---|---|---|
| **Natural intrinsic activity** | | | | |
| Wild type enzyme | T4 lysozyme | *E. coli* | Exposure to 100 µg/ml of protein caused bacterial killing >99.85% (~3 logs) after 1 h of incubation. A C-terminal amphipathic peptide is responsible for the bactericidal effect. | Düring et al., 1999 |
| Wild type enzyme | Lys1521 | *P. aeruginosa* and *E. coli* | Dose-dependent effect after 1 h of incubation. Maximum decrease of bacterial viability with >99.9% (~3 logs) using >100 µg/ml of protein. | Muyombwe et al., 1999 |
| Wild type enzyme | LysAB2 | Gram-negative and Gram-positive bacteria | Treatment with 500 µg/ml for 1 h altered the cell shape of *A. baumannii* and *S. aureus* and killed up to 99 % (2 logs) of *A. baumannii, E. coli* and *Streptococcus sanguis* cells; up to 82 % of *S. aureus* and up to 70% of *Bacillus subtilis, C. freundii* and *Salmonella enterica.* | Lai et al., 2011 |
| Wild type enzyme | OBPgp279 | *P. aeruginosa* and *E. coli* | Treatment with 1.5 µM (56 µg/ml) of protein for 30 min resulted in ~1 log reduction of *P. aeruginosa* and 0.3 logs of the *E. coli* cell number. | Walmagh et al., 2012 |

| | | | | |
|---|---|---|---|---|
| Wild type enzyme | LysAB3 and LysAB4 | *A. baumannii* | Treatment with 10 µg/ml of each protein reduced 99% of *A. baumannii* cells (~2 logs). | Lai et al., 2013 |
| Wild type enzyme | SPN9CC | *E. coli* | Treatment with 300 µg/ml for 1 h resulted in a reduction of 2 logs. | Lim et al., 2014 |
| Wild type enzyme | PlyAB1 | *A. baumannii* | Treatment with 2.4 mg/ml of protein resulted in a reduction of the turbidity (OD600nm) of multi-drug (n=206) and pan-drug resistant (n=48) *A. baumannii* strains. | Huang et al., 2014 |
| Wild type enzyme | PlyF307 | *A. baumannii* | Treatment with 100 µg/ml for 2 h resulted in a decrease of more than 5 logs of all tested strains. Treatment with two doses of 1 mg of protein reduced bacteria with 2 logs of in biofilms formed on a catheter implanted in a mouse model. The injection of 1 mg of protein to cure a mouse model of systemic infection resulted in the rescue of 50% of the mice. | Lood et al., 2015 |
| Wild type enzyme | EndoT5 | *E. coli* | Treatment with 40 µg/ml for 24 h resulted in a reduction of 0.6 log units. | Shavrina et al., 2016 |
| Wild type enzyme | LysABP-01 | *A. baumannii* | Dose-dependent effect of the protein after 24 h of incubation. Maximum bacterial growth inhibition (>80 %) with protein concentrations over 5 µM. | Thummeepak et al., 2016 |
| Wild type enzyme | LysCfP1 | *Citrobacter* spp. | Treatment with 3 µM for 2 h resulted in a reduction of *Citrobacter freundii* (>2 logs) and *Citrobacter koseri* (>4 logs). | Oliveira et al., 2016a |
| Wild type enzyme | ABgp46 | *A. baumannii* | Treatment with 2 µM for 2 h resulted in a decrease up to 2 logs. | Oliveira et al., 2016b |

| | | | | |
|---|---|---|---|---|
| Wild type enzyme | LysPA26 | Gram-negatives and *S. aureus* | Dose-dependent effect of the treatment after 24 h of incubation. Maximum bacterial growth inhibition (>80 %) with protein concentrations over 200 µg/ml. Treatment with 500 µg/ml reduced >4 logs of *P. aeruginosa*, *E. coli*, *Klebsiella pneumoniae*, *E. coli* and *S. aureus* in 30 min. | Guo et al., 2017 |
| Wild type enzyme | PlyE146 | *P. aeruginosa*, *E. coli* and *A. baumannii* | Treatment with 400 µg/ml for 2 h resulted in a reduction of 3.6 logs, ~3 logs and ~4 logs of the number of *P. aeruginosa*, *E. coli* and *A. baumannii*, respectively. | Larpin et al., 2018 |
| Wild type enzyme | LysAm24, LysECD7, and LysSi3 | Gram-negatives | Treatment with concentrations ranging from 5 to 15 µg/ml of each protein for 30 min reduced >5 logs of *P. aeruginosa, A. baumannii, K. pneumoniae, E. coli,* and *S.* Typhimurium. LysSi3 partially inhibited the growth of *Staphylococcus haemolyticus* up to 45%. | Antonova et al., 2019 |
| **OM permeabilization by physical and chemical methods** | | | | |
| High hydrostatic pressure | T4 lysozyme and Lambda lysozyme | *Yersinia enterocolitica, Shigella flexneri*, *E. coli* O157:H7 and *P. aeruginosa* | Sensitization of the bacteria (*Yersinia enterocolitica*, *Shigella flexneri*, *E. coli* O157:H7, *P. aeruginosa*) with a HHP treatment of 130 -250 MPa sensitized the bacteria and increased bacterial death when combined with T4 lysozyme (2.8 × $10^2$ U/ml) and Lambda lysozyme (3.5 × $10^2$ U/ml). The maximum increase of activity was observed for *P. aeruginosa*, specifically a 29-fold and 20-fold increase for T4 lysozyme and Lambda lysozyme, respectively. | Nakimbugwe et al., 2006 |
| High hydrostatic pressure | KZ144 and EL188 | *P. aeruginosa* | Treatment for 15 min with 25 µg of the proteins and pressure levels of 175 MPa, reduced the bacteria with up to 4 logs. | Briers et al., 2008 |

| | | | | |
|---|---|---|---|---|
| EDTA | EL188 | *P. aeruginosa* | Treatment with 50 µg/ml and 10 mM of EDTA resulted in a bacterial decrease between 3 and 4 logs after 30 min. Increasing the EL188 concentration to 250 µg/ml further reduced the most antibiotic-resistant strain (4 logs). | Briers et al., 2011 |
| EDTA | OBPgp279 | *P. aeruginosa* and *E. coli* | Treatment with 1.5 µM (56 µg/ml) of protein and 0.5 mM of EDTA for 30 min increased the antibacterial activity up to a ~4 logs reduction of *P. aeruginosa* and 1 log reduction of *E. coli*. | Walmagh et al., 2012 |
| EDTA | LoGT-001 and LoGT-008 | *P. aeruginosa* | Treatment with 1.5 µM LoGT-001 or 5 µM LoGT-008 in presence of 0.5 mM of EDTA for 30 min, resulted in a reduction up to 5 logs of the bacteria. | Briers et al., 2014b |
| EDTA | SPN9CC | *E. coli*, *P. putida* and *Salmonella* Typhimurium | Treatment with 300 µg/ml for 1 h resulted in a reduction of 2 logs. Treatment for 5 min with 0.5 µg of protein and 1 mM of EDTA increased the bacterial spectrum of activity to Gram-negative bacteria. Reaching a reduction of turbidity up to 70% against *E. coli* and *P. putida* and up to 40% of *Salmonella* Typhimurium. | Lim et al., 2014 |
| EDTA | L-KPP10 | *P. aeruginosa* | Treatment for 30 min with protein doses higher than 12 µg/ml and 0.5 mM of EDTA reduced the bacteria number up to 5 logs. | Antonova et al., 2018 |
| EDTA | LysAm24, LysECD7, and LysSi3 | *E. coli* and *A. baumannii*. | Treatment with 10 µg/ml of each protein in the presence of 0.5 mM of EDTA for 30 min, resulted in a complete inhibition (100%) of exponentially growing *E. coli* and *A. baumannii*. | Antonova et al., 2019 |

| | | | | |
|---|---|---|---|---|
| EDTA or weak organic acids | Lys68 | Gram-negatives | Treatment with 2 μM of protein in the presence of 0.5 mM of EDTA resulted in 2 log reduction of *P. aeruginosa* after 2 h. The presence of 2 mM citric acid or 5 mM malic acid broadened the activity of the protein to a panel of 13 different Gram-negative bacteria with a range of reduction between 1 and 5 logs. | Oliveira et al., 2014 |
| Weak organic acids | ABgp46 | Gram-negatives | Treatment with 2 μM of protein in the presence of citric or malic acid reduced *A. baumannii* below the detection limit (>5 logs) and more than 4 logs of *P. aeruginosa* and *S.* Typhimurium strains | Oliveira et al., 2016b |
| Essential oils (carvacrol) | lysozyme Cpl-7S | *E. coli* and *P. putida* | Pre-incubation of bacteria with 0.01 % of carvacrol followed by treatment with 5 μg/ml of Cpl-7S resulted in a bacterial reduction up to 3 log units after 1 h. | Díez-Martínez et al., 2013 |
| Polymyxin B, gramicidin D, chlorhexidine, miramistin or ε–poly–L–lysine | EndoT5 | *E. coli* | Treatment of 24 h with 40 μg/ml of protein in combination with polymyxin B (0.4 μg/ml); gramicidin D (100 μmol/l); (chlorhexidine (0.5 μg/ml); miramistin (0.5 μg/ml); or ε–poly–L–lysine (80 μg/ml), resulted in a bacterial reduction of ~5 logs for the different chemical agents except for chlorhexidine were the reduction was of 6 logs. | Shavrina et al., 2016 |
| Colistin | LysABP-01 | *A. baumannii* | The combination of colistin with LysABP-01 resulted in a reduction of MIC values of the endolysin from 20 μM to 1.25 μM, and of colistin from 2 μg/ml to 0.25 μg/ml. | Thummeepak et al., 2016 |
| Cationic peptides (PGLa and poly-L-Arg) | Lys394 | *E. coli* | Treatment of 1 h with 250 μg/ml in combination with 20 μg/ml of PGLa or 25 μg/ml of poly-L-arginine resulted in an increase (>4 times) of lytic activity (DOD/min). | Legotsky et al., 2014 |

| | | | | |
|---|---|---|---|---|
| Silver nanoparticles (AgNPs) | KP27 endolysin | *P. aeruginosa* | Treatment with 5 µM of KP27 and AgNPs (5–50 µg/mL) resulted in a reduction of turbidity under 20% of the bacteria in a dose-dependent manner, with the highest AgNPs concentrations showing better results. | Ciepluch et al., 2019 |
| Encapsulation (liposomes) | BSP16Lys | *S.* Typhimurium and *E. coli* | Treatment for 1 h with BSP16Lys-encapsulated liposomes reduced the number of *S.* Typhimurium and *E. coli* with 2.2 and 1.6 logs, respectively. | Bai et al., 2019 |
| **Protein engineering** | | | | |
| Fusion of outer membrane permeabilizing peptides to endolysins | LoGT-001 and LoGT-008 | *P. aeruginosa* | Treatment with 1.5 µM LoGT-001 or 5 µM LoGT-008 for 30 min resulted in a reduction of 2.6 and 1.5 logs, respectively.<br>A treatment with 2 µM of LoGT-008 protected 71% of keratinocytes infected with the bacteria.<br>LogT-008 also rescued 63% of nematodes in an *in vivo* model of gut infection with *P. aeruginosa.* | Briers et al., 2014b |
| Fusion of an antimicrobial peptide to an endolysin | Artilysin® Art-175 | *P. aeruginosa*, *A. baumannii* | Treatment with 3 µM Art-175 for 1 h reduced the cell number of *P. aeruginosa* with more than 4 logs. Also persisters selected after ofloxacin exposure were reduced with 4 logs, and up to 5 logs (total eradication) in the presence of 0.5 mM of EDTA.<br>Stationary-phase cultures of different *A. baumannii* strains were completely eliminated (approximately 8- to 9-log reduction) after a single treatment with 30×MIC of Art-175.<br>*In vivo* efficacy of Art-085 has been reported in two case studies of dogs presenting ear infections, achieving complete healing of the infection. | Briers et al., 2014a; Defraine et al., 2016; Briers and Lavigne, 2015 |

| Fusion of an antimicrobial peptide to an endolysin | PlyA | *P. aeruginosa*, *A. baumannii* | Treatment for 1 h with 50 or 100 μg/ml reduced the bacterial cell number of *A. baumannii* (2-5 logs) and *P. aeruginosa* (1.5-2 logs), respectively. | Yang et al., 2015 |
|---|---|---|---|---|
| Fusion of an antimicrobial peptide to an endolysin | AL-KPP10 | *P. aeruginosa* | Treatment with 25 μg/ml for 30 min reduced the bacterial cell number up to 5 logs | Antonova et al., 2018 |
| Fusion of polycationic peptides to an endolysin | Lysep3 + polycationic peptides at the C-terminus, Lysep3-5 and Lysep3-D8 | *E. coli*, *Salmonella, Shigella* and *P. aeruginosa* and *A. baumannii* | Lysep3 carrying 15 extra amino acids showed an antibacterial activity (9 - 42 % reduction) against *E. coli*, *Salmonella, Shigella* and *P. aeruginosa* strains when treated with 0.75-1.5 mg/ml protein.<br>Lysep3-5 showed antibacterial activity against *A. baumannii* when adding 1.75 mg/ml of the protein at pH 5.<br>Lysep3-D8 is bacteriostatic against *E. coli, P. aeruginosa* and *A. baumannii* and *Streptococcus* and bactericidal (2 logs, 100 μg/ml, 3 h) against *E. coli* in a strain-dependent manner. | Ma et al., 2017; Yang et al., 2019; Wang et al., 2017 |
| Fusion of an OM translocating domain to an endolysin | Pesticin translocation domain-T4 lysozyme | *Yersinia pestis* and *E. coli* | Treatment with 100 μg/ml of the protein reduced the bacterial cell number under 10%.<br>Treatment of *E. coli* with the chimeric protein resulted in ruptured cells and cells presenting vesiculated membranes. | Lukacik et al., 2012 |
| Fusion of an OM translocating domain to an endolysin | Colicin translocation domain-Lysep3 | *E. coli* | Treatment with 8 μg/ml of the chimeric protein for 15 min reduced bacteria up to 3 logs.<br>In a mouse intestinal infection model with *E. coli*, treatment with 100 μg Colicin-Lysep3, reduced the bacteria from 15 to 4% at the forepart of the intestine, and from 15 to 12% in the cecum. | Yan et al., 2017 |

| | | | | |
|---|---|---|---|---|
| Fusion of an OM translocating domain to an endolysin | Lysocin PyS2-GN4 | *P. aeruginosa* | Treatment with 0.1-100 µg/ml of PyS2-GN4 in combination with EDDHA resulted in a 4 logs reduction of the bacteria after 12 h. Complete elimination was obtained in 50% human serum supplemented with the chimeric protein (≥10 µg/ml) and EDDHA after 24 h. A dose of 25 mg/kg of PyS2-GN4 fully rescued mice from bacteremia. | Heselpoth et al., 2019 |
| Fusion of phage receptor binding protein (RBP) to an endolysin | Innolysin Innolysin#6 | *P. aeruginosa* and *Shigella sonnei* | Treatment with 200 µg/ml for 30 min resulted in a 1 log bacterial reduction. | Zampara et al., 2018 |

*Natural intrinsic activity of endolysins active against Gram-negatives*

Some endolysins have the surprising capacity to permeate the OM of Gram-negative bacteria, to degrade the peptidoglycan and to kill cells when added exogenously. A large diversity of pathogens has been shown to be susceptible to such endolysins. One of the first reported examples of a endolysin with natural intrinsic activity was T4 lysozyme, which has been demonstrated to be active against both Gram-negative and Gram-positive bacteria (Düring et al., 1999b). Later, the *Bacillus amyloliquefaciens* phage endolysin Lys1521 was shown to lyse *P. aeruginosa* and *Escherichia coli* (Morita et al., 2001a). OBPgp279 (Walmagh et al., 2012) and PlyE146 (Larpin et al., 2018) are further examples of endolysins active against these two pathogens. On the other hand, EndoT5 (Shavrina et al., 2016) and SPN9CC are active against *E. coli*, with SPN9CC also being active against *Salmonella* Typhimurium (Lim et al., 2014). CfP1 has intrinsic activity against *Citrobacter* sp. (Oliveira et al., 2016a).

Besides these examples, *A. baumannii* appears a particularly susceptible bacterium for endolysins with intrinsic activity, with many reported examples, including LysAB2 (Lai et al., 2011), LysAB3, LysAB4 (Lai et al., 2013), PlyAB1 (Huang et al., 2014), PlyF307 (Lood et al., 2015), LysABP-01 (Thummeepak et al., 2016), ABgp46 (Oliveira et al., 2016b), LysPA26 (Guo et al., 2017),

LysAm24, LysECD7 and LysSi3 (Antonova et al., 2019). It is important to highlight that some of these endolysins only show a narrow activity restricted to *A. baumannii* with no activity against other Gram-negative or Gram-positive bacteria. This is the case for LysAB3 and LysAB4 that reduce 99% of *A. baumannii* cells (2 logs) after one hour of incubation when treated with 10 µg/ml of protein (Lai et al., 2013). PlyAB1 is another example with a narrow activity spectrum. A high dose of 2.4 mg/ml PlyAB1 completely eliminated multi-drug (n=206) and pan-drug resistant (n=48) *A. baumannii* strains after 30 min of exposure (Huang et al., 2014). PlyF307 was only tested against *A. baumannii* strains and has a high bactericidal activity reducing the bacterial number with more than 5 logs when treated with 100 µg/ml of protein. Moreover, in a mouse *in vivo* catheter model of infection with *A. baumannii*, a 2 log reduction of biofilm-associated cells was observed after treatment with two 1 mg doses. Finally, the injection of 1 mg of protein to cure a mouse model of systemic infection resulted in the rescue of 50% of the mice (Lood et al., 2015). The remainder of endolysins with intrinsic activity were proven to be active against a variety of Gram-negative bacteria and interestingly, LysAB2, LysAm24 and LysSi3 are also active against staphylococci (Antonova et al., 2019; Lai et al., 2011).

In general, the majority of these studies reports relatively high concentrations of endolysins (up to 2.4 mg/ml), indicating that the OM is still hindering with only a minority of proteins getting effectively through (Antonova et al., 2019; Lim et al., 2014; Shavrina et al., 2016; Thummeepak et al., 2016). In this perspective, it is interesting that a clear synergistic effect has been reported for LysABP-01 and colistin, and for EndoT5 and polymyxin B, respectively (Thummeepak et al., 2016; Shavrina et al., 2016). Both antibiotics disturb the OM. Moreover, the activity of ABgp46 was improved by the addition of a weak organic acid such as citric or malic acid (Oliveira et al., 2016b), which may act as weak chelating agents. These results reinforce the idea that the OM is still acting as a barrier for the action of these endolysins and that removal, permeabilization or damaging of the OM can further enhance the antimicrobial activity of endolysins with intrinsic activity. However, on the other hand, EDTA could not increase the antimicrobial activity of endolysins with intrinsic activity (Larpin et al., 2018; Raz et al., 2019).

The antibacterial mode of action of the *B. amyloliquefaciens* endolysin Lys1521 against *P. aeruginosa* PAO1 has been investigated in more detail. In this work, the authors created C-terminal truncations that were unable to kill *P. aeruginosa*, although the enzymatic activity remained intact (Morita et al., 2001b). Two homologous regions rich in cationic and hydrophobic amino acids were predicted in its C-terminus. These regions may fold as amphipathic helices and may allow the endolysin to interact with the OM (Orito et al., 2004). Similar observations were done for LysAB2 and LysAB3 since truncated variants lacking the C-terminal region completely lost their antibacterial activity against the bacteria tested (Lai et al., 2013, 2011). Based on these observations the following mechanism of intrinsic activity has been proposed: first, the C-terminal region of the endolysin interacts with the OM, facilitating OM penetration; second, the N-terminal part carrying the EAD domain degrades the peptidoglycan, causing bacterial lysis (Orito et al., 2004). Interestingly, the presence of a positively charged, amphipathic C-terminal region (amino acids 143-155) was already demonstrated for T4 lysozyme. However, here the authors found that antibacterial activity is independent of its muralytic activity, as shown with heat-denatured T4 lysozyme, but is instead caused by a C-terminal amphipathic helix alone. A synthetic peptide corresponding to amino acids 143-155 retained the full bactericidal activity (Düring et al., 1999b). In line with these findings, two other studies demonstrated the bactericidal effect of a C-terminal peptide isolated from two different endolysins (Thandar, et al., 2016; Peng et al., 2017). Thandar et al. (2016) showed that C-terminal peptide P307 from the *A. baumannii* phage endolysin PlyF307 has a strong bactericidal effect (3.8 logs). Further engineering of this peptide resulted in a modified peptide that can fully eliminate a $10^6$ CFU/ml *A. baumannii* culture and is not cytotoxic or hemolytic. Although the peptide is inactive in 100% plasma, it reduces the bacterial load in a murine skin infection model (~2 logs). Conform to these findings, also the C-terminus of endolysin LysAB2 was found to possess a peptide sequence (aa 113-145) with antibacterial activity against *A. baumannii* strains (MIC = 64 μM). Modifications of the peptide sequence to increase the net charge from +4 to +6 resulted in a peptide with enhanced antibacterial activity with MIC values ranging from 4 to 8 µM (8-16 times lower that the parental peptide). This modified peptide is also not cytotoxic or hemolytic and could rescue 60% of mice in a bacteremia model of infection (Peng, et al., 2017).

The cationic P307 peptide was also successfully used as a query to screen *E. coli* proteomes, resulting in the identification of a new prophage endolysin PlyE146, which has 57.7% similarity to PlyF307. The C-terminal cationic peptide has a predicted hairpin-like di-alpha helical structure linked by a flexible unstructured region. PlyE146 has a pH, salt and growth phase-dependent, and strain-specific bactericidal activity against *E. coli* and *P. aeruginosa*, whereas all three tested *A. baumannii* strains were susceptible. No activity was detected against *K. pneumoniae* and *S. enterica* isolates. Raz et al. (2019) used a phylogenetic approach (BLAST analysis) starting from PlyF307 to identify lytic proteins with natural intrinsic activity in *P. aeruginosa* phages. Two proteins with high bactericidal activity against *P. aeruginosa* were withheld and extensively characterized. These proteins show differences in killing rate, salt dependency, efficacy against biofilms and inactivation rate by serum. PlyPa03 was demonstrated to be effective in a mouse model of skin infection, whereas PlyPa91 protected mice in a lung infection model (Raz et al., 2019). Altogether, these data underscore the relevance of a polycationic C-terminus for natural intrinsic activity.

Notably, most endolysins with intrinsic antimicrobial activities are globular endolysins with a lysozyme-like domain. There are three reported exceptions to this rule, i.e., the modular endolysin LysAm24 with a PG-1 cell wall binding domain at the N-terminus and a lysozyme-like domain at the C-terminus (Antonova et al., 2019), and LysECD7 and CfP1, which are globular endolysins with a peptidase domain (Antonova et al., 2019; Oliveira et al., 2016a). In the particular case of LysECD7 and CfP1, there is no positively charged C-terminal region. The authors hypothesize that the purification tag comprising eight histidine residues in combination with some preceding amino acids form a positively charged cluster (KLEHHHHHHHH) might act similar to the above described cationic helix. Nevertheless, further analysis of the protein will be needed to confirm this hypothesis (Antonova et al., 2019). Since the pKa value of histidine is ~6-7, the tag is expected to be only positively charged at acidic conditions.

*OM permeabilization by physical and chemical methods*

In the search for strategies to overcome the barrier of the OM, physical or chemical mechanisms that disrupt, weaken or permeabilize the OM have been broadly investigated.

High hydrostatic pressure (HHP) can be used to transiently permeabilize the OM for phage lytic proteins (Briers et al., 2008a; Nakimbugwe et al., 2006). In this regard, T4 lysozyme (2.8 × $10^2$ U/ml) and Lambda lysozyme (3.5 × $10^2$ U/ml) were applied against a panel of Gram-negative bacteria, including *Yersinia enterocolitica*, *Shigella flexneri*, *E. coli* O157:H7, *P. aeruginosa* and *S.* Typhimurium, in combination with HHP ranging from 130-250 MPa. The results showed that under high pressure, bacteria are significantly sensitized for the lysozymes. The highest increase of antimicrobial activity was observed for *P. aeruginosa*, specifically a 29- and 20-fold for T4 lysozyme and Lambda lysozyme, respectively. On the contrary, no sensitization to any of the lysozymes has been observed for *S.* Typhimurium, and the lysozymes even protected the cells against pressure inactivation (Nakimbugwe et al., 2006). In another study, it was demonstrated that pressure levels around 175 MPa, applied to stationary phase *P. aeruginosa* bacteria in the presence of 25 µg of either KZ144 or EL188, for 15 min, reduced the number of bacteria with a range between 2 and 4 logs. About two-thirds of the total inactivation was caused by the lytic enzymes. Experiments with a recombinant CBD-GFP fusion protein showed that OM permeabilization by HHP is transient (Briers et al., 2008a). Although these results show a clear synergistic effect between HHP and endolysins, this strategy cannot be applied for therapeutic purposes, but rather in some particular niches of the food industry. HHP is a food preservation technique that can substitute for pasteurization in cases where pasteurization cannot be applied (e.g., vegetables, meats, seafood, dairy, and egg products) (Wang et al., 2016). Yet, the implementation of this technology is hampered due its high costs associated with instruments to apply pressures of 200-500 MPa or even higher.

Including endolysins can reduce the required pressure and thus the associated costs.

Chemical permeabilization in combination with endolysins may be a more convenient approach. Several options have been evaluated in this regard (Figure 3). The studied permeabilizers are very broad in nature, including chelating compounds (mostly EDTA) (Briers et al., 2011, 2007), weak organic acids (Oliveira et al., 2014), essential oils (Díez-Martínez et al., 2013), antibiotics (Shavrina et al., 2016; Thummeepak et al., 2016), cationic detergents (Shavrina et al., 2016), cationic peptides (Legotsky et al., 2014; Shavrina et al., 2016), cationic dendrimers (Ciepluch et al., 2019a), dendronized silver nanoparticles (Ciepluch et al., 2019b) and cationic liposomes (Bai et al., 2019).

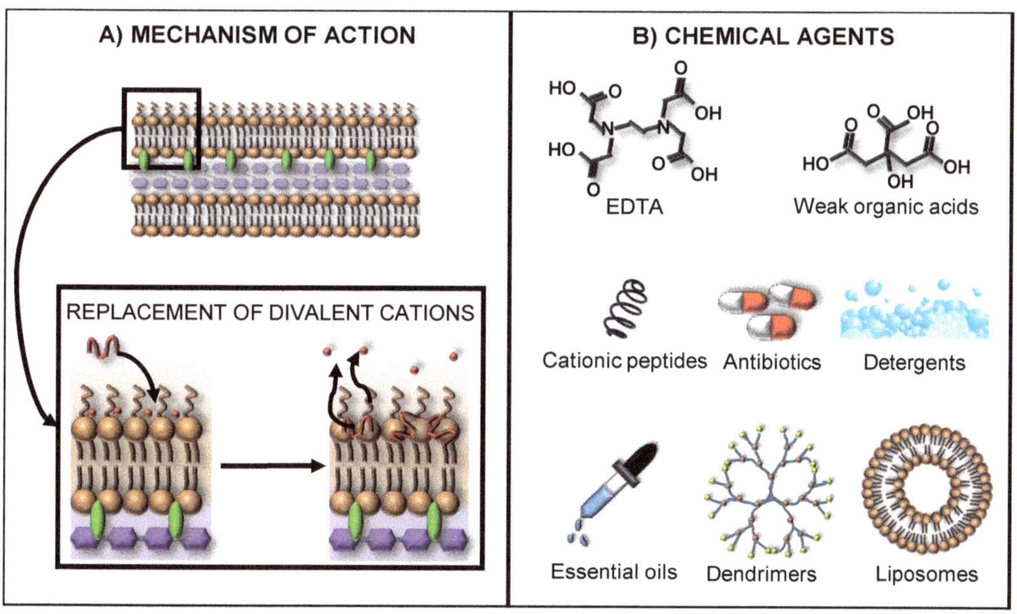


**Figure 3.** Mechanism of OM permeabilization mediated by the displacement of divalent cations ($Mg^{2+}$ and $Ca^{2+}$) carried out by many cationic chemical agents. The chemical compound displaces the divalent cations that are interacting with the charged phosphate groups from the LPS molecules, resulting in destabilization of the OM. B) Principal chemical agents that have proven to enhance endolysin activity against Gram-negative bacteria.

EDTA is a widely used compound to permeabilize the OM because it chelates $Mg^{2+}$ and $Ca^{2+}$ divalent ions that interact with negatively charged phosphate groups from adjacent LPS molecules. Subsequently, phage lytic proteins can more easily pass the damaged OM and reach the peptidoglycan (Finnegan and Percival, 2015). Several publications have shown that endolysins become active or obtain an increased antibacterial activity when Gram-negative bacteria are subjected to a combined treatment of endolysin and EDTA (Antonova et al., 2018; Briers et al., 2011, 2008b, 2007; Walmagh et al., 2012; Yang et al., 2015). However, the potential future application of an endolysin-EDTA combination in therapeutics is limited to topical applications since EDTA possess a strong anticoagulant effect even at low doses (Banfi et al., 2007). The enhancing effect of EDTA is most outspoken when targeting *P. aeruginosa* and *A. baumannii*. To broaden the activity spectrum to Enterobacteriaceae (such as *E. coli* and *S.* Typhimurium) the addition of weak organic acids (citric and malic acid) have been proven to be successful. This is illustrated for endolysin Lys68, which is in the presence of EDTA only active against *P. aeruginosa*. Using citric or malic acid instead broadens its activity to nine and eleven Gram-negative species, respectively. Particularly, the combination of citric or malic acid, and Lys68 led to a reduction of 3-5 logs of *S.* Typhimurium after a 2 h treatment and a 1 log reduction when the bacteria were in the stationary growth phase or forming a biofilm (Oliveira et al., 2014). A similar effect was observed for ABgp46. Its activity was increased up to complete eradication of *A. baumannii* and the cell number of *P. aeruginosa* and *S.* Typhimurium was reduced with more than 4 logs (Oliveira et al., 2016b). Weak organic acids such as citric and malic acid can act as weak chelators, yet, it is proposed that also the acidity may contribute to OM disruption and intracellular damage (Oliveira et al., 2016b). Also carvacrol, an essential oil extracted from oregano and thyme, has been demonstrated as an effective OM permeabilizer for the engineered pneumococcal lysozyme Cpl-7S. This lysozyme is highly active against Gram-positive bacteria (*Streptococcus pneumoniae* and *Enterococcus faecalis*), but in combination with 0.01% of carvacrol its activity is extended to the Gram-negative bacteria, *E. coli* and *Pseudomonas putida* (Díez-Martínez et al., 2013).

Three antibiotics have been evaluated for OM permeabilization: colistin (or polymyxin E), and polymyxin B, both disrupting the OM by displacing the divalent cations ($Ca^{2+}$ and $Mg^{2+}$) from the LPS; and gramicidin D, whose effect is based on the formation of membrane ionophores. The use of colistin with LysABP-01 results in a synergistic effect reducing the MIC value of the endolysin from 20 μM to 1.25 μM in the presence of colistin. The MIC of the latter was also reduced from 2 μg/ml to 0.25 μg/ml (Thummeepak et al., 2016). Similarly, the addition of polymyxin B (0.4 μg/ml) with endolysin EndoT5 (40 μg/ml) resulted in a reduction of 5 logs of *E. coli* (Shavrina et al., 2016). The use of colistin has been kept as a last-resort option for therapeutic purposes due to its renal and neural toxic effects (Spapen et al., 2011). Nevertheless, the combination of colistin or polymyxin B with endolysins may reduce the dose needed for treatment, thus also reducing their possible side effects. Unexpectedly, the use of gramicidin D in combination with EndoT5 also results in a reduction of 5 logs of *E. coli* (Shavrina et al., 2016). This synergistic effect might be explained by the mode of action of this antibiotic, i.e. the formation of pores with a diameter ranging from 30 to 90 nm in the membrane of Gram-positive bacteria; the size of these pores exceeds the average diameter of the EndoT5 molecule by an order of magnitude. Therefore, the authors propose that the *E. coli* OM is permeabilized by gramicidin for EndoT5 (Shavrina et al., 2016). Interestingly, EndoT5 also shows a synergistic effect when combined with cationic detergents that are commonly used as antiseptics, such as chlorhexidine and miramistin (at concentrations 1000- and 25-fold lower than the commonly used, respectively), resulting in a reduction with more than 5 logs (Shavrina et al., 2016).

A different class of compounds that has been investigated as permeabilizer are cationic peptides. Two major cationic peptides (PGLa and poly-L-Arg) used in combination with endolysin Lys394 increase its antimicrobial activity against *E. coli* up to 10-fold in a dose-dependent manner (Legotsky et al., 2014). PGLa is an antimicrobial peptide found in frog skin and has an amphipathic, α-helical structure that can bind and permeabilize the OM of Gram-negatives (Soravia et al., 1988). In a similar way, the combined use of EndoT5 with the polycationic peptide ε-poly-L-lysine results in a synergistic effect against *E. coli* with a reduction of 3 logs. Noteworthy, this synergy is lower than the one observed when combining EndoT5 with polymyxin B, gramicidin, chlorhexidine and miramistin (Shavrina et al., 2016).

Recently, a novel strategy to permeabilize the OM based on the use of polypropylene imine (PPI) dendrimers has been reported. PPI dendrimers are cationic nanoparticles that have been shown to be promising three-dimensional polymers for use in pharmaceutical and biomedical applications (Fuentes-Paniagua et al., 2016). Specifically, PPI enhances the antibacterial activity of endolysin KP27 against wild-type *P. aeruginosa* and mutants with truncated O-antigen in the LPS of the OM. Reduction of the positive charge of PPI increased further the antibacterial effect (Ciepluch et al., 2019a). Another class of nanoparticles with antibacterial properties and a synergic effect with endolysin KP27 are PEGylated dendronized silver nanoparticles (AgNPs) (Ciepluch et al., 2019b). The mode of action of the bactericidal effect of AgNPs remains unclear. Yet, previous results indicate that they might interact with the OM of Gram-negative bacteria, resulting in the production of reactive oxygen species (ROS). This is followed by the release of $Ag^+$ ions from the nanoparticle. The released $Ag^+$ ions interact with these ROS products, affecting different metabolic routes and leading to bacterial death (Qing et al., 2018).

Liposomes can penetrate bacterial membranes by membrane fusion. The aqueous phase in the liposome is therefore well suited to deliver a cargo to the bacterial cell. In this regard, BSP16Lys was purified and encapsulated into a cationic liposome composed of dipalmitoylphosphatidylcholine (DPPC), cholesterol, and hexadecylamine. Cationic liposomes were chosen to stimulate liposome-cell interactions due to the anionic cell surface of Gram-negative bacteria. *Salmonella* Typhimurium and *E. coli* cells treated with encapsulated BSP16Lys were reduced with 2.2 and 1.6 logs in the absence of an OM permeabilizer. These results indicate that the cationic liposome is effectively interacting and disrupting the OM and releases the encapsulated endolysin in the periplasm, although further research is needed to fully understand the mechanism (Bai et al., 2019).

*Protein engineering*

Protein engineering is a process that mimics natural evolution on a lab scale and allows to redirect the natural function or properties of proteins. Specific for phage lytic proteins active on peptidoglycan of Gram-negative bacteria, major protein engineering efforts have been undertaken to provide the enzymes with outer membrane permeabilizing activity. In this context, several strategies were adopted based on the construction of fusion proteins comprising a phage lytic protein on the one hand and a variety of fusion partners on the other hand. Reported fusion partners are outer membrane permeabilizing peptides (including antimicrobial peptides), translocation domains and receptor binding domains (Figure 4).

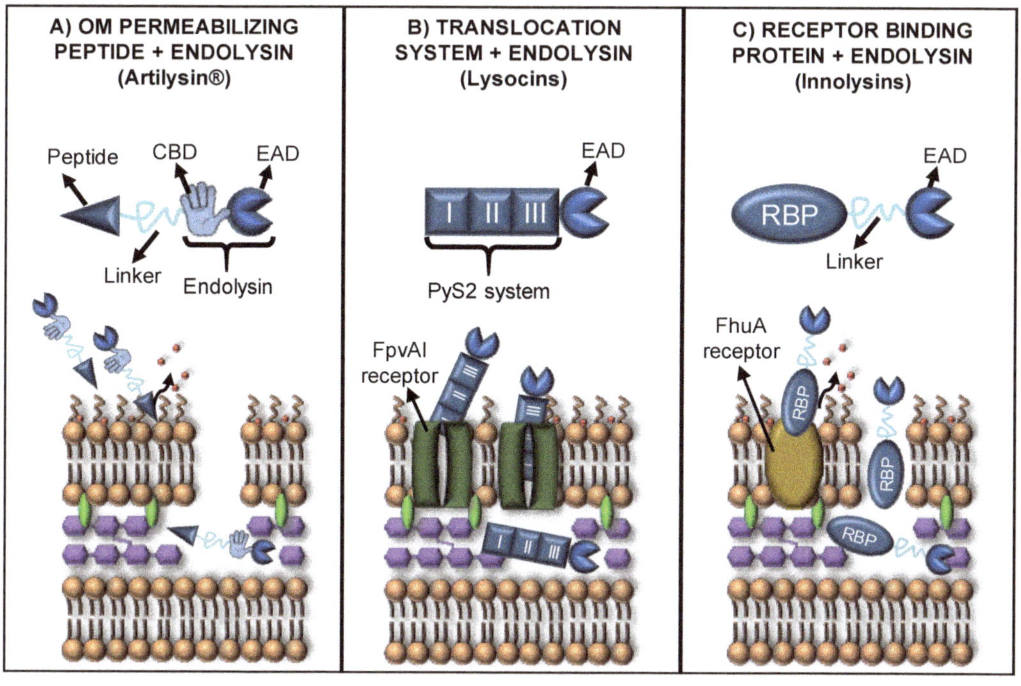


**Figure 4.** Current strategies of protein engineering to improve the antibacterial activity of phage lytic proteins against Gram-negative bacteria. A) Mechanism of action of an Artilysin® carrying and OM permeabilizing peptide and an endolysin. B) Mechanism of translocation of a Lysocin, a chimeric protein carrying a pyocin-based OM translocation system (I: N-terminal receptor-binding domain targeting the ferripyoverdine receptor FpvAI; II: α-helical domain; III: colicin-like domain) and an endolysin. C) Mechanism of translocation of an Innolysin mediated by a fusion of a receptor binding protein, that binds to the receptor protein FhuA involved in ferrichrome transport, and an endolysin.

Phage lytic proteins engineered with outer membrane permeabilizing peptides.

A first approach to create engineered phage lytic proteins with high antimicrobial activity against Gram-negative bacteria relies on the fusion of an outer membrane permeabilizing peptide (OMP) to a phage lytic protein. Such an OMP produces a local OM disruption by competitive displacement of the divalent cations, which otherwise stabilize the LPS layer. The hydrophobic amino acids in an OMP may interfere with the hydrophobic stacking forces in the lipid A. This local disruption is sufficient to get the enzyme moiety through the OM, probably similar to the self-promoted uptake mechanism of cationic peptides. The lytic enzyme (with or without CBD) ends up in the periplasm, where it degrades peptidoglycan, resulting in bacterial killing through osmotic lysis. This principle has been firstly reported with polycationic, hydrophobic and amphipathic peptides in 2014 (Briers et al., 2014b) and is currently commercially developed by Lysando AG (https://www.lysando.com/welcome.html) under the registered trademark Artilysin®. Also the OM permeabilizing activity of an antimicrobial peptide that can intrinsically pass through the OM has been successfully harnessed in a similar way (Briers et al., 2014a). Nevertheless, the mere addition of an OMP to the N- or C-terminus of a phage lytic protein does not automatically result in highly active antibacterial agent, and variable activities can be observed. Instead, the right partner combination and the right configuration has to be identified, aspects that can be further modulated by specific linkers, in order to identify the best variants that perform adequately at low doses. Several polycationic, hydrophobic or amphipathic OMPs (n=7) were initially fused to different endolysins (n=6) (Briers et al., 2014b). The two best initial combinations of OMP and endolysin in terms of antimicrobial activity against *P. aeruginosa* were LoGT-001 and LoGT-008. Both of them carry a polycationic nonapeptide (PCNP) fused to OBPgp279 (LoGT-001) or PVP-SE1gp146 (LoGT-008), respectively. LoGT-001 (1.5 µM) showed the highest antibacterial activity, reducing the *P. aeruginosa* cell number with 2.6 logs in 30 min, while 5 µM LoGT-008 showed a reduction of 1.5 logs. Addition of EDTA has a strong synergistic effect, resulting in complete sterilization of the culture with 1.5 µM protein after maximum 30 min. Time lapse microscopy showed membrane bulging and cell degradation (i.e., lysis) immediately after addition of LoGT-001, which fully aligns with the proposed mechanism relying on osmotic lysis. Furthermore, LoGT-008 is also effective in an *in vitro* skin infection model (human keratinocytes) and *in vivo* in nematodes (*Caenorhabditis elegans*). Increasing the linker length between the PCNP and OBPgp279 (LoGT-023) further increased the antibacterial activity from 2.6 to 3.4 logs. Notably, the fusion of two OMPs (PCNP in combination with a more hydrophobic peptide) enhanced the antibacterial activity against *E. coli* and *S.* Typhimurium, which were less susceptible for a single PCNP fusion protein. These data support the modularity principle stating that the antibacterial activity of this class of engineered proteins is dependent on the modular composition of the engineered protein. Both the OMP, phage lytic protein, linker and specific order affect the eventual outcome (Briers et al., 2014b).

In a further report the development and analysis of Art-175 has been described. Art-175 consists of the fusion of the broad-spectrum sheep myeloid-29 acid (SMAP-29) antimicrobial peptide to the Gram-negative endolysin KZ144. To create a more stable protein, three cysteine residues were exchanged to serines, preventing the formation of oligomers. Art-175 shows a rapid and highly bactericidal activity (>4 logs reduction of *P. aeruginosa* after a 30 min treatment of all environmental, clinical and MDR strains tested. The bactericidal activity is further increased to a full eradication with 0.125 mM EDTA (Briers et al., 2014a). Art-175 also kills MDR *A. baumannii* cells (MIC values between 4 and 20 µg/ml) in the stationary growth phase with a reduction of >8 logs in 2 h. Art-175 can fully eliminate *A. baumannii* persister cells that have been previously selected by antibiotics (Defraine et al., 2016). Persisters represent a small percentage of the bacterial population that can tolerate a high concentration of antibiotics or a prolonged period of treatment by entering a dormant state. When the antibiotic is removed, persisters can regrow normally, explaining their role in chronic, recurrent infections (Keren et al., 2004). No resistance was observed against Art-175 after serial exposure to sub-lethal doses of Art-175 in contrast to the antibiotic controls. Any possible cross-resistance with 22 antibiotics was absent. This includes *mcr-1* driven colistin resistance associated with LPS modifications in *E. coli* (Briers et al., 2014a; Schirmeier et al., 2018). No cytotoxic effect was observed against mouse connective tissue fibroblasts. *In vivo* efficacy of Art-085 (similar to Art-175 but without the three point mutations) has been reported in two case studies of dogs presenting ear infections. Treatment with standard antibiotics for six and three weeks, respectively, failed to heal the infection. However, the treatment with Art-085 was effective when applying seven treatments of Art-085 within one day and three additional doses at day 7 and 8. After two weeks no relapse was observed (Briers and Lavigne, 2015).

Yang et al. (2015) constructed a fusion between the antimicrobial peptide cecropin A and endolysin OBPgp279 (PlyA). PlyA is highly active when applied for 1 h in a concentration of 50 or 100 µg/ml against *A. baumannii* (2-5 logs) and *P. aeruginosa* (1.5-2 logs), respectively. This activity is dependent on the physiological growth state with reduced activities observed in the stationary phase (less than one log for *A. baumannii* and less than 2 logs for *P. aeruginosa*). Yet, this reduction can be reverted by the addition of OM permeabilizers such as EDTA and citric acid. PlyA becomes inactive in the presence of milk or human serum. Recently, the SMAP-29 peptide, which was also used for Art-175, was fused to the C-terminus of endolysin KPP10. The fusion protein named AL-KPP10 reduced the number of *P. aeruginosa* cells with 5 logs after a 30 min treatment with 25 µg/ml (Antonova et al., 2018).

*Escherichia coli* phage lysin Lysep3 has also been intensively subjected to engineering at its C-terminus, reported in three different studies. In the first report, cationic tails between 5 and 15 amino acids, and a 12 amino acids long oligopeptide comprising cationic and hydrophobic amino acids, were fused to Lysep3. Lysep3 carrying 15 extra amino acids was the most active combination causing 9 to 42 % reductions of cell numbers from a panel of *E. coli*, *Salmonella, Shigella* and *P. aeruginosa* strains when treated with 0.75-1.5 mg/ml of the protein for 30 min (Ma et al., 2017). The second report describes the C-terminal addition of hydrophobic tails with a length between 3 and 12 amino acids to Lysep3. A maximum pH-dependent reduction of 60% with 1.75 mg/ml protein is observed (Yan et al., 2019). Overall, these are slight reductions given the high concentrations used. The third study reports on the fusion of the C-terminal region D8 of the above-mentioned *B. amyloliquefaciens* phage with natural intrinsic activity (Morita et al., 2001a; Orito et al., 2004) to the C-terminus of Lysep3. The fusion protein is bacteriostatic against nine out of fourteen *E. coli* strains, all three tested *P. aeruginosa* strains, one out of two tested *A. baumannii* strains but also against *Streptococcus*. A dose of 100 µg/ml has a bactericidal effect of almost 2 logs against five out of fourteen tested *E. coli* strains (Wang et al., 2017).

Endolysins engineered with OM translocating or receptor binding domains.
A different strategy is based on receptor-mediated uptake of lytic domains, relying on natural mechanisms of bacteriocins. Such bacteriocins interact with OM proteins, followed by uptake of the bacteriocins and bacterial killing through a diversity of possible killing domains including muramidases, DNases and RNases. This concept has been firstly demonstrated exploiting pesticin, which is produced by *Yersinia pestis* and has a modular structure with a N-terminal receptor binding domain and a C-terminal domain with muramidase activity. To exert its activity, the N-terminal domain of pesticin binds to the OM protein FyuA, followed by uptake across the OM using the Ton transport system, and peptidoglycan degradation by the C-terminal muramidase domain (Lukacik et al., 2012a; Patzer et al., 2012). For reasons of self-protection, *Y. pestis* produces an immunity protein, called Pim, that binds to the muramidase domain of pesticin and inhibits its activity. To avoid this immunity protein and to create an engineered protein able to eliminate *Y. pestis*, a chimeric protein carrying the N-terminal domain of the pesticin and T4 lysozyme at the C-terminus was constructed. In contrast to the natural muramidase domain of pesticin, Pim does not recognize T4 lysozyme. Using a concentration of 100 µg/ml, the chimeric protein reduces *Y. pestis* and *E. coli* strains expressing the FyuA protein with more than 90%. Moreover, treatment of *E. coli* with the chimeric protein causes morphological defects such as ruptured cells and cells presenting vesiculated membranes (Lukacik et al., 2012b).

A similar approach was applied to colicin A, a bacteriocin produced by *E. coli*. (Yan et al., 2017). Colicin A binds to the BtuB receptor present in *E. coli*, followed by translocation through the Tol-Pal system and peptidoglycan degradation through its muramidase domain. Again, the C-terminal natural muramidase domain was replaced by an endolysin (Lysep3), retaining the N-terminal binding domain of colicin A. The chimeric protein is active against most

*E. coli* strains tested (3 log killing in 15 minutes with 8 µg/ml at pH 5) and inactive against other Gram-negative bacteria like *P. aeruginosa*, *Salmonella* or *Shigella*. The susceptibility of the bacteria to the chimeric protein appears to be correlated with the expression and presence of the BtuB receptor. In a mouse intestinal infection model with mice fed with 100 µg Colicin-Lysep3, the ratio of GFP-labeled susceptible *E. coli* bacteria was reduced from 15 to 4% at the forepart of the intestine, and from 15 to 12% in the cecum (Yan et al., 2017).

S-type pyocins are colicin-like bacteriocins produced by *P. aeruginosa*. Specifically, PyS2 consists of four domains: 1) an N-terminal receptor-binding domain targeting the ferripyoverdine receptor FpvAI; 2) an α-helical domain; 3) a colicin-like domain; and 4) a C-terminal DNase domain. The initial steps in the mechanism of PyS2 are similar to pesticin and colicin A (OM protein interaction and translocation to the periplasm). However, since PyS2 does not bear a muramidase domain, which acts in the periplasm, but a DNase domain that kills the cells by DNA cleavage inside the cytoplasm, the DNase domain is cleaved off in the periplasm by a protease for further translocation to the cytoplasm through an inner membrane protein channel. A 'lysocin' (hybrid lysin-bacteriocin) has been created by the substitution of the DNase domain of PyS2 with the *P. aeruginosa* phage PAJU2 endolysin GN4. Doses of 0.1-100 µg/ml PyS2-GN4 resulted in a 4-log killing of *P. aeruginosa* after 12h in the presence of the iron chelator ethylenediamine hydroxyphenylacetic acid (EDDHA) to mimic free-iron deprivation in human blood. Full sterilization is observed after 24 h in 50 % human serum supplemented with EDDHA using a dose ≥ 10 µg/ml. In addition, PyS2-GN4 is able to degrade biofilms efficiently. A dose of 25 mg/kg PyS2-GN4 fully rescues mice from bacteremia (Heselpoth et al., 2019).

The selection of the bacteriocin chassis for engineering is a highly relevant parameter for at least two reasons: (1) different bacteriocins target different OM proteins and the prevalence of each OM protein is highly diverse among isolates. E.g., Colicin-Lysep3 kills 22 out of 32 strains, and PyS2-GN4 is active against four out of eleven *P. aeruginosa* strains, which is strictly correlated with the presence of the FpvAI receptor. Cocktails of engineered proteins targeting diverse OM proteins may expand the spectrum to cover a whole species; (2)

receptor-binding may be easily inhibited by point mutations giving rise to resistance. Some mutated OM proteins will not affect the strain virulence (as has been suggested for strains resistant to Colicin-Lysep3; Heselpoth et al., 2019), while other mutated OM proteins (e.g. the FpvAI receptor) may turn the isolate strictly avirulent. Loss of virulence would eventually also result in infection control.

A method, which is reminiscent to the bacteriocin-based approach but conceptually different, is the 'innolysin' concept. Innolysins combine the enzymatic activity of an endolysin with the binding capacity of a phage receptor binding protein (RBP) (Zampara et al., 2018). RBPs are structural phage proteins located at the tail spikes or fibers and are responsible for bacterial recognition by binding to specific bacterial receptors (OM proteins, pili, flagella, lipopolysaccharides, capsular polysaccharides or other exopolysaccharides) (Rakhuba et al., 2010). In a proof of concept study, twelve innolysins were constructed based on combinations of the phage T5 endolysin, different linkers and the phage T5 RBP, i.e. Pb5. Pb5 irreversibly binds to the *E. coli* receptor protein FhuA involved in ferrichrome transport and is located in the OM. Innolysin#6 is the variant with the highest antimicrobial activity reducing the cell number with 1 log unit after 30 min using a dose of 200 µg/ml. No reduction was observed in *E. coli* mutants with a deletion in the FhuA receptor, and activity was also observed against other bacterial species carrying homologs of the FhuA receptor such as *Shigella sonnei* and *P. aeruginosa*, underscoring the role of the FhuA receptor in the mode of action. Whereas innolysins share the principle of receptor-binding with bacteriocin-derived chimeric proteins including lysocins, innolysins do not have protein domains for translocation across the OM and are likely too large to pass the narrow FhuA channel (2.5 nm). Although the mode of action of innolysins remains to be elucidated, it appears that mere receptor binding can eventually lead to killing.

## Conclusion

While phage lytic proteins targeting Gram-positive bacteria are rapidly progressing in clinical research and for a multitude of other applications, the scientific community has successfully expanded this novel class of antibacterials to Gram-negative pathogens. Diverse strategies have been developed to assist phage lytic proteins to span the OM and to reach the subjacent peptidoglycan layer. Generally, a high variation in dosing and antibacterial effects has been described among the different strategies (Table 1). While some approaches are in the conceptual stage with significant margin for further development and improvement, other strategies have well evolved to the preclinical stage with Artilysin®s being developed at the commercial level. One conclusion is sure: due to high natural diversity and the large engineering potential of these proteinaceous antibacterials, enzybiotics have a broad potential to target all groups of bacteria.

## References

Abdelkader, K., Gerstmans, H., Saafan, A., Dishisha, T., and Briers, Y. (2019). The preclinical and clinical progress of bacteriophages and their lytic enzymes: The parts are easier than the whole. Viruses 11. (2). pii: E96. https://doi.org/10.3390/v11020096

Antonova, N.P., Balabanyan, V.Y., Tkachuk, A.P., Makarov, V. V., and Gushchin, V.A. (2018). Physical and chemical properties of recombinant KPP10 phage lysins and their antimicrobial activity against *Pseudomonas aeruginosa*. Bull. Russ. State Med. Univ. 7, 22–29. https://doi.org/10.24075/brsmu.2018.010

Antonova, N.P., Vasina, D. V., Lendel, A.M., Usachev, E. V., Makarov, V. V., Gintsburg, A.L., Tkachuk, A.P., and Gushchin, V.A. (2019). Broad bactericidal activity of the myoviridae bacteriophage Lysins LysAm24, LysECD7, and LysSi3 against Gram-Negative ESKAPE Pathogens. Viruses 11. https://doi.org/10.3390/v11030284

Bai, J., Yang, E., Chang, P.S., and Ryu, S. (2019). Preparation and characterization of endolysin-containing liposomes and evaluation of their antimicrobial activities against gram-negative bacteria. Enzyme Microb. Technol. 128, 40–48. https://doi.org/10.1016/j.enzmictec.2019.05.006

Banfi, G., Salvagno, G.L., and Lippi, G. (2007). The role of ethylenediamine tetraacetic acid (EDTA) as in vitro anticoagulant for diagnostic purposes. Clin. Chem. Lab. Med. 45, 565–576. https://doi.org/10.1515/CCLM.2007.110

Becker, S.C., Roach, D.R., Chauhan, V.S., Shen, Y., Foster-Frey, J., Powell, A.M., Bauchan, G., Lease, R.A., Mohammadi, H., Harty, W.J., et al. (2016).

Triple-acting lytic enzyme treatment of drug-resistant and intracellular *Staphylococcus aureus*. Sci Rep 6, 25063. https://doi.org/10.1038/srep25063

Braun, V., and Hantke, K. (2019). Lipoproteins: Structure, Function, Biosynthesis. In Subcellular Biochemistry, (Springer New York), pp. 39–77. https://doi.org/10.1007/978-3-030-18768-2_3

Briers, Y., Volckaert, G., Cornelissen, A., Lagaert, S., Michiels, C.W., Hertveldt, K., and Lavigne, R. (2007). Muralytic activity and modular structure of the endolysins of *Pseudomonas aeruginosa* bacteriophages phiKZ and EL. Mol Microbiol 65, 1334–1344. https://doi.org/10.1111/j.1365-2958.2007.05870.x

Briers, Y., Cornelissen, A., Aertsen, A., Hertveldt, K., Michiels, C.W., Volckaert, G., and Lavigne, R. (2008a). Analysis of outer membrane permeability of *Pseudomonas aeruginosa* and bactericidal activity of endolysins KZ144 and EL188 under high hydrostatic pressure. FEMS Microbiol. Lett. 280, 113–119. https://doi.org/10.1111/j.1574-6968.2007.01051.x

Briers, Y., Miroshnikov, K., Chertkov, O., Nekrasov, A., Mesyanzhinov, V., Volckaert, G., and Lavigne, R. (2008b). The structural peptidoglycan hydrolase gp181 of bacteriophage φKZ. Biochem. Biophys. Res. Commun. 374, 747–751. https://doi.org/10.1016/j.bbrc.2008.07.102

Briers, Y., Schmelcher, M., Loessner, M.J., Hendrix, J., Engelborghs, Y., Volckaert, G., and Lavigne, R. (2009). The high-affinity peptidoglycan binding domain of *Pseudomonas* phage endolysin KZ144. Biochem Biophys Res Commun 383, 187–191. https://doi.org/10.1016/j.bbrc.2009.03.161

Briers, Y., Walmagh, M., and Lavigne, R. (2011). Use of bacteriophage endolysin EL188 and outer membrane permeabilizers against *Pseudomonas aeruginosa*. J. Appl. Microbiol. 110, 778–785. https://doi.org/10.1111/j.1365-2672.2010.04931.x

Briers, Y., Walmagh, M., Grymonprez, B., Biebl, M., Pirnay, J.P., Defraine, V., Michiels, J., Cenens, W., Aertsen, A., Miller, S., et al. (2014a). Art-175 is a highly efficient antibacterial against multidrug-resistant strains and persisters of *Pseudomonas aeruginosa*. Antimicrob. Agents Chemother. 58, 3774–3784. https://doi.org/10.1128/AAC.02668-14

Briers, Y., Walmagh, M., Van Puyenbroeck, V., Cornelissen, A., Cenens, W., Aertsen, A., Oliveira, H., Azeredo, J., Verween, G., Pirnay, J.P., et al. (2014b). Engineered endolysin-based "Artilysins" to combat multidrug-resistant gram-negative pathogens. MBio 5, e01379-14. https://doi.org/10.1128/mBio.01379-14

Briers, Y., and Lavigne, R. (2015). Breaking barriers: expansion of the use of endolysins as novel antibacterials against Gram-negative bacteria. Futur. Microbiol 10, 377–390. https://doi.org/10.2217/fmb.15.8

Catalao, M.J., Gil, F., Moniz-Pereira, J., Sao-Jose, C., and Pimentel, M. (2013). Diversity in bacterial lysis systems: bacteriophages show the way. FEMS Microbiol Rev 37, 554–571. https://doi.org/10.1111/1574-6976.12006

Ciepluch, K., Maciejewska, B., Gałczyńska, K., Kuc-Ciepluch, D., Bryszewska, M., Appelhans, D., Drulis-Kawa, Z., and Arabski, M. (2019a). The influence of cationic dendrimers on antibacterial activity of phage endolysin against *P. aeruginosa* cells. Bioorg. Chem. 91. https://doi.org/10.1016/j.bioorg.2019.103121

Ciepluch, K., Skrzyniarz, K., Barrios-Gumiel, A., Quintana, S., Sánchez-Nieves, J., de la Mata, F.J., Maciejewska, B., Drulis-Kawa, Z., and Arabski, M. (2019b). Dendronized silver nanoparticles as bacterial membrane permeabilizers and their interactions with *P. aeruginosa* lipopolysaccharides, lysozymes, and phage-derived endolysins. Front. Microbiol. 10. https://doi.org/10.3389/fmicb.2019.02771

Cisek, A.A., Dąbrowska, I., Gregorczyk, K.P., and Wyżewski, Z. (2017). Phage therapy in bacterial infections treatment: One hundred years after the discovery of bacteriophages. Curr. Microbiol. 74, 277–283. https://doi.org/10.1007/s00284-016-1166-x

Collignon, P.C., Conly, J.M., Andremont, A., McEwen, S.A., Aidara-Kane, A., Griffin, P.M., Agerso, Y., Dang Ninh, T., Donado-Godoy, P., Fedorka-Cray, P., et al. (2016). World Health Organization Ranking of Antimicrobials According to Their Importance in Human Medicine: A Critical Step for Developing Risk Management Strategies to Control Antimicrobial Resistance from Food Animal Production. In Clinical Infectious Diseases, (Oxford University Press), pp. 1087–1093. https://doi.org/10.1093/cid/ciw475

ContraFect (2019a). Exebacase Pipeline. URL https://www.contrafect.com/pipeline/exebacase (accessed 1.18.20).

ContraFect (2019b). Direct lysis of *Staph aureus* resistant pathogen trial of Exebacase - Full Text View - ClinicalTrials.gov. NIH Clin. URL https://

www.clinicaltrials.gov/ct2/show/NCT04160468?term=CF-301&draw=1&rank=3 (accessed 1.18.20).

Dams, D., and Briers, Y. (2019). Enzybiotics: Enzyme-based antibacterials as therapeutics. In Advances in Experimental Medicine and Biology, (Springer New York LLC), pp. 233–253. https://doi.org/10.1007/978-981-13-7709-9_11

De Geyter, J., Smets, D., Karamanou, S., Economou, A., 2019. Inner membrane translocases and insertases, in: Subcellular biochemistry. Springer New York, pp. 337–366. https://doi.org/10.1007/978-3-030-18768-2_11

Defraine, V., Scheurmans, J., Grymonprez, B., Govers, S.K., Aertsen, A., Fauvart, M., Michiels, J., Lavigne, R., and Briers, Y. (2016). Efficacy of Artilysin(R) Art-175 against resistant and persistent *Acinetobacter baumannii*. Antimicrob. Agents Chemother. https://doi.org/10.1128/AAC.00285-16

Díez-Martínez, R., de Paz, H.D., Bustamante, N., García, E., Menéndez, M., and García, P. (2013). Improving the lethal effect of cpl-7, a pneumococcal phage lysozyme with broad bactericidal activity, by inverting the net charge of its cell wall-binding module. Antimicrob. Agents Chemother. 57, 5355–5365. https://doi.org/10.1128/AAC.01372-13

Drulis-Kawa, Z., Majkowska-Skrobek, G., and Maciejewska, B. (2015). Bacteriophages and Phage-Derived Proteins – Application Approaches. Curr. Med. Chem. 22, 1757–1773. https://doi.org/10.2174/0929867322666150209152851

Düring, K., Porsch, P., Mahn, A., Brinkmann, O., and Gieffers, W. (1999). The non-enzymatic microbicidal activity of lysozymes. FEBS Lett. 449, 93–100. https://doi.org/10.1016/S0014-5793(99)00405-6

Dyar, O.J., Huttner, B., Schouten, J., and Pulcini, C. (2017). What is antimicrobial stewardship? Clin. Microbiol. Infect. 23, 793–798. https://doi.org/10.1098/rsob.120143

Errington, J. (2013). L-form bacteria, cell walls and the origins of life. Open Biol. 3.

Exner, M., Bhattacharya, S., Christiansen, B., Gebel, J., Goroncy-Bermes, P., Hartemann, P., Heeg, P., Ilschner, C., Kramer, A., Larson, E., et al. (2017). Antibiotic resistance: What is so special about multidrug-resistant Gram-negative bacteria? GMS Hyg. Infect. Control 12, Doc05. https://doi.org/10.3205/dgkh000290

Fernandez-Ruiz, I., Coutinho, F.H., Rodriguez-Valera, F., 2018. Thousands of novel endolysins discovered in uncultured phage genomes. Front Microbiol 9, 1033. https://doi.org/10.3389/fmicb.2018.01033

Finnegan, S., and Percival, S.L. (2015). EDTA: An antimicrobial and antibiofilm agent for use in wound care. Adv. Wound Care. 4, 415–421. https://doi.org/10.1089/wound.2014.0577

Fischetti, V.A. (2018). Development of phage lysins as novel therapeutics: A historical perspective. Viruses 10. https://doi.org/10.3390/v10060310

Fuentes-Paniagua, E., Sánchez-Nieves, J., Hernández-Ros, J.M., Fernández-Ezequiel, A., Soliveri, J., Copa-Patiño, J.L., Gómez, R., and Javier De La Mata, F. (2016). Structure-activity relationship study of cationic carbosilane dendritic systems as antibacterial agents. RSC Adv. 6, 7022–7033. https://doi.org/10.1039/c5ra25901k

Galdiero, S., Falanga, A., Cantisani, M., Tarallo, R., Della Pepa, M.E., D'Oriano, V., and Galdiero, M. (2012). Microbe-host interactions: structure and role of Gram-negative bacterial porins. Curr. Protein Pept. Sci. 13, 843–854. https://doi.org/10.2174/138920312804871120

Gerstmans, H., Criel, B., and Briers, Y. (2018). Synthetic biology of modular endolysins. Biotechnol. Adv. 36, 624–640. https://doi.org/10.1016/j.biotechadv.2017.12.009

Gilmer, D.B., Schmitz, J.E., Euler, C.W., and Fischetti, V.A. (2013). Novel bacteriophage lysin with broad lytic activity protects against mixed infection by *Streptococcus pyogenes* and methicillin-resistant *Staphylococcus aureus*. Antimicrob. Agents Chemother. 57, 2743–2750. https://doi.org/10.1128/AAC.02526-12

Guo, M., Feng, C., Ren, J., Zhuang, X., Zhang, Y., Zhu, Y., Dong, K., He, P., Guo, X., and Qin, J. (2017). A Novel Antimicrobial Endolysin, LysPA26, against *Pseudomonas aeruginosa*. Front. Microbiol. 8, 293. https://doi.org/10.3389/fmicb.2017.00293

Gutiérrez, D., Rodríguez-Rubio, L., Martínez, B., Rodríguez, A., and García, P. (2016). Bacteriophages as weapons against bacterial biofilms in the food industry. Front. Microbiol. 7. https://doi.org/10.3389/fmicb.2016.00825

Gutiérrez, D., Fernández, L., Rodríguez, A., and García, P. (2018). Are phage lytic proteins the secret weapon to kill *Staphylococcus aureus*? MBio 9. https://doi.org/10.1128/mBio.01923-17

Heselpoth, R.D., Euler, C.W., Schuch, R., and Fischetti, V.A. (2019). Lysocins: Bioengineered antimicrobials that deliver lysins across the outer membrane of Gram-negative bacteria. Antimicrob. Agents Chemother. 63. https://doi.org/10.1128/AAC.00342-19

Holt, A., Cahill, J., Ramsey, J., O'Leary, C., Moreland, R., Martin, C., Galbadage, D.T., Sharan, R., Sule, P., Bettridge, K., et al. (2019). Phage-encoded cationic antimicrobial peptide used for outer membrane disruption in lysis. BioRxiv 515445. https://doi.org/10.1101/515445

Huang, G., Shen, X., Gong, Y., Dong, Z., Zhao, X., Shen, W., Wang, J., Hu, F., and Peng, Y. (2014). Antibacterial properties of *Acinetobacter baumannii* phage Abp1 endolysin (PlyAB1). BMC Infect. Dis. 14, 681. https://doi.org/10.1186/s12879-014-0681-2

iNtRON Biotecnology (2020a). iNtRON- N-Rephasin® SAL200 (Tonabacase) - AMR Industry Alliance. ClinicalTrials.gov. AMR Industry Alliance [WWW Document]. URL https://www.amrindustryalliance.org/case-study/n-rephasin-sal200-tonabacase/ (accessed 1.19.20).

iNtRON Biotecnology (2020b). Phase IIa Clinical Study of N-Rephasin® SAL200 - Full Text View - ClinicalTrials.gov. URL https://clinicaltrials.gov/ct2/show/NCT03089697 (accessed 1.19.20).

Jun, S.Y., Jang, I.J., Yoon, S., Jang, K., Yu, K.S., Cho, J.Y., Seong, M.W., Jung, G.M., Yoon, S.J., and Kang, S.H. (2017). Pharmacokinetics and tolerance of the phage endolysin-based candidate drug SAL200 after a single intravenous administration among healthy volunteers. Antimicrob Agents Chemother. https://doi.org/10.1128/AAC.02629-16

Keren, I., Kaldalu, N., Spoering, A., Wang, Y., and Lewis, K. (2004). Persister cells and tolerance to antimicrobials. FEMS Microbiol Lett 230, 13–18.

Lai, M.J., Lin, N.T., Hu, A., Soo, P.C., Chen, L.K., Chen, L.H., and Chang, K.C. (2011). Antibacterial activity of *Acinetobacter baumannii* phage ΦaB2 endolysin (LysAB2) against both Gram-positive and Gram-negative bacteria. Appl. Microbiol. Biotechnol. 90, 529–539. https://doi.org/10.1007/s00253-011-3104-y

Lai, M.J., Soo, P.C., Lin, N.T., Hu, A., Chen, Y.J., Chen, L.K., and Chang, K.C. (2013). Identification and characterisation of the putative phage-related endolysins through full genome sequence analysis in *Acinetobacter baumannii* ATCC 17978. Int. J. Antimicrob. Agents 42, 141–148. https://doi.org/10.1016/j.ijantimicag.2013.04.022

Larpin, Y., Oechslin, F., Moreillon, P., Resch, G., Entenza, J.M., and Mancini, S. (2018). In vitro characterization of PlyE146, a novel phage lysin that targets Gram-negative bacteria. PLoS One 13. https://doi.org/10.1371/journal.pone.0192507

Legotsky, S.A., Vlasova, K.Y., Priyma, A.D., Shneider, M.M., Pugachev, V.G., Totmenina, O.D., Kabanov, A. V., Miroshnikov, K.A., and Klyachko, N.L. (2014). Peptidoglycan degrading activity of the broad-range Salmonella bacteriophage S-394 recombinant endolysin. Biochimie 107, 293–299. https://doi.org/10.1016/j.biochi.2014.09.017

Li, X.Z., Plésiat, P., and Nikaido, H. (2015). The challenge of efflux-mediated antibiotic resistance in Gram-negative bacteria. Clin. Microbiol. Rev. 28, 337–418. https://doi.org/10.1128/CMR.00117-14

Lim, J.A., Shin, H., Heu, S., and Ryu, S. (2014). Exogenous lytic activity of SPN9CC endolysin against gram-negative Bacteria. J. Microbiol. Biotechnol. 24, 803–811. https://doi.org/10.4014/jmb.1403.03035

Loessner, M.J., Kramer, K., Ebel, F., and Scherer, S. (2002). C-terminal domains of *Listeria monocytogenes* bacteriophage murein hydrolases determine specific recognition and high-affinity binding to bacterial cell wall carbohydrates. Mol. Microbiol. 44, 335–349. https://doi.org/10.1046/j.1365-2958.2002.02889.x

Loessner, M.J. (2005). Bacteriophage endolysins--current state of research and applications. Curr Opin Microbiol 8, 480–487. https://doi.org/10.1016/j.mib.2005.06.002

Lood, R., Winer, B.Y., Pelzek, A.J., Diez-Martinez, R., Thandar, M., Euler, C.W., Schuch, R., and Fischetti, V.A. (2015). Novel phage lysin capable of killing the multidrug-resistant gram-negative bacterium *Acinetobacter baumannii* in a mouse bacteremia model. Antimicrob. Agents Chemother. 59, 1983–1991. https://doi.org/10.1128/AAC.04641-14

Lukacik, P., Barnard, T.J., and Buchanan, S.K. (2012a). Using a bacteriocin structure to engineer a phage lysin that targets Yersinia pestis. Biochem. Soc. Trans. 40, 1503–1506. https://doi.org/10.1042/BST20120209

Lukacik, P., Barnard, T.J., Keller, P.W., Chaturvedi, K.S., Seddiki, N., Fairman, J.W., Noinaj, N., Kirby, T.L., Henderson, J.P., Steven, A.C., et al. (2012b). Structural engineering of a phage lysin that targets gram-negative pathogens. Proc. Natl. Acad. Sci. U. S. A. 109, 9857–9862. https://doi.org/10.1073/pnas.1203472109

Ma, Q., Guo, Z., Gao, C., Zhu, R., Wang, S., Yu, L., Qin, W., Xia, X., Gu, J., Yan, G., et al. (2017). Enhancement of the direct antimicrobial activity of Lysep3 against Escherichia coli by inserting cationic peptides into its C terminus. Antonie Van Leeuwenhoek 110, 347–355. https://doi.org/10.1007/s10482-016-0806-2

Miller, S.I., and Salama, N.R. (2018). The gram-negative bacterial periplasm: Size matters. PLOS Biol. 16, e2004935. https://doi.org/10.1371/journal.pbio.2004935

Moelling, K., Broecker, F., and Willy, C. (2018). A wake-up call: We need phage therapy now. Viruses 10. https://doi.org/10.3390/v10120688

Morita, M., Tanji, Y., Mizoguchi, K., Soejima, A., Orito, Y., and Unno, H. (2001a). Antibacterial activity of *Bacillus amyloliquefaciens* phage endolysin without holin conjugation. J. Biosci. Bioeng. 91, 469–473. https://doi.org/10.1016/S1389-1723(01)80275-9

Morita, M., Tanji, Y., Orito, Y., Mizoguchi, K., Soejima, A., and Unno, H. (2001b). Functional analysis of antibacterial activity of *Bacillus amyloliquefaciens* phage endolysin against Gram-negative bacteria. FEBS Lett. 500, 56–59. https://doi.org/10.1016/S0014-5793(01)02587-X

Muyombwe, A., Tanji, Y., and Unno, H. (1999). Cloning and expression of a gene encoding the lytic functions of *Bacillus amyloliquefaciens* phage: Evidence of an auxiliary lysis system. J. Biosci. Bioeng. 88, 221–225. https://doi.org/10.1016/S1389-1723(99)80206-0

Nakimbugwe, D., Masschalck, B., Atanassova, M., Zewdie-Bosüner, A., and Michiels, C.W. (2006). Comparison of bactericidal activity of six lysozymes at atmospheric pressure and under high hydrostatic pressure. Int. J. Food Microbiol. 108, 355–363. https://doi.org/10.1016/j.ijfoodmicro.2005.11.021

Nelson, D., Loomis, L., and Fischetti, V.A. (2001). Prevention and elimination of upper respiratory colonization of mice by group A streptococci by using a bacteriophage lytic enzyme. Proc Natl Acad Sci U S A 98, 4107–4112. https://doi.org/10.1073/pnas.061038398

Nelson, D.C., Schmelcher, M., Rodriguez-Rubio, L., Klumpp, J., Pritchard, D.G., Dong, S., and Donovan, D.M. (2012). Endolysins as antimicrobials. Adv Virus Res 83, 299–365. https://doi.org/10.1016/B978-0-12-394438-2.00007-4

Oliveira, H., Melo, L.D., Santos, S.B., Nobrega, F.L., Ferreira, E.C., Cerca, N., Azeredo, J., and Kluskens, L.D. (2013). Molecular aspects and comparative genomics of bacteriophage endolysins. J Virol 87, 4558–4570. https://doi.org/10.1128/JVI.03277-12

Oliveira, H., Thiagarajan, V., Walmagh, M., Sillankorva, S., Lavigne, R., Neves-Petersen, M.T., Kluskens, L.D., and Azeredo, J. (2014). A thermostable *Salmonella* phage endolysin, Lys68, with broad bactericidal properties against gram-negative pathogens in presence of weak acids. PLoS One 9, e108376. https://doi.org/10.1371/journal.pone.0108376

Oliveira, H., Pinto, G., Oliveira, A., Oliveira, C., Faustino, M.A., Briers, Y., Domingues, L., and Azeredo, J. (2016a). Characterization and genome sequencing of a *Citrobacter freundii* phage CfP1 harboring a lysin active against multidrug-resistant isolates. Appl. Microbiol. Biotechnol. 100, 10543–10553. https://doi.org/10.1007/s00253-016-7858-0

Oliveira, H., Vilas Boas, D., Mesnage, S., Kluskens, L.D., Lavigne, R., Sillankorva, S., Secundo, F., and Azeredo, J. (2016b). Structural and Enzymatic Characterization of ABgp46, a Novel Phage Endolysin with Broad Anti-Gram-Negative Bacterial Activity. Front. Microbiol. 7, 208. https://doi.org/10.3389/fmicb.2016.00208

Opal, S.M. (2010). Endotoxins and other sepsis triggers. Contrib. Nephrol. 167, 14–24.

Orit0, Y., Morita, M., Hori, K., Unno, H., and Tanji, Y. (2004). *Bacillus amyloliquefaciens* phage endolysin can enhance the permeability of *Pseudomonas aeruginosa* outer membrane and induce lysis. Appl. Microbiol. Biotechnol. 65, 105–109. https://doi.org/10.1007/s00253-003-1522-1.

Patzer, S.I., Albrecht, R., Braun, V., and Zeth, K. (2012). Structural and mechanistic studies of pesticin, a bacterial homolog of phage lysozymes. J. Biol. Chem. 287, 23381–23396. https://doi.org/10.1159/000315915.

Peng, S.-Y., You, R.-I., Lai, M.-J., Lin, N.-T., Chen, L.-K., Chang, K.-C. (2017). Highly potent antimicrobial modified peptides derived from the *Acinetobacter baumannii* phage endolysin LysAB2. Sci. Rep. 7, 11477. https://doi.org/10.1038/s41598-017-11832-7

Pires, D., de Kraker, M.E.A., Tartari, E., Abbas, M., and Pittet, D. (2017). 'Fight antibiotic resistance—It's in your hands': Call from the World Health Organization for 5th May 2017. Clin. Infect. Dis. 64, 1780–1783. https://doi.org/10.1093/cid/cix226

Qing, Y., Cheng, L., Li, R., Liu, G., Zhang, Y., Tang, X., Wang, J., Liu, H., and Qin, Y. (2018). Potential antibacterial mechanism of silver nanoparticles and the optimization of orthopedic implants by advanced modification technologies. Int. J. Nanomedicine 13, 3311–3327. https://doi.org/10.2147/IJN.S165125

Rakhuba, D. V, Kolomiets, E.I., Dey, E.S., and Novik, G.I. (2010). Bacteriophage receptors, mechanisms of phage adsorption and penetration into host cell. Pol J Microbiol 59, 145–155.

Raz, A., Serrano, A., Hernandez, A., Euler, C.W., and Fischetti, V.A. (2019). Isolation of phage lysins that effectively kill *Pseudomonas aeruginosa* in mouse models of lung and skin infection. Antimicrob. Agents Chemother. 63, e00024-19. https://doi.org/10.1128/AAC.00024-19.

Roach, D.R., and Donovan, D.M. (2015). Antimicrobial bacteriophage-derived proteins and therapeutic applications. Bacteriophage 5, e1062590. https://doi.org/10.1080/21597081.2015.1062590

Rodríguez-Rubio, L., Martínez, B., Donovan, D.M., Rodríguez, A., and García, P. (2013). Bacteriophage virion-associated peptidoglycan hydrolases: potential new enzybiotics. Crit Rev Microbiol 39, 427–434. https://doi.org/10.3109/1040841X.2012.723675

Schirmeier, E., Zimmermann, P., Hofmann, V., Biebl, M., Gerstmans, H., Maervoet, V.E.T., and Briers, Y. (2018). Inhibitory and bactericidal effect of Artilysin® Art-175 against colistin-resistant mcr-1-positive *Escherichia coli* isolates. Int. J. Antimicrob. Agents 51, 528–529. https://doi.org/10.1016/j.ijantimicag.2017.08.027

Schmelcher, M., Donovan, D.M., and Loessner, M.J. (2012a). Bacteriophage endolysins as novel antimicrobials. Futur. Microbiol 7, 1147–1171. https://doi.org/10.2217/fmb.12.97

Schmelcher, M., Powell, A.M., Becker, S.C., Camp, M.J., and Donovan, D.M. (2012b). Chimeric phage lysins act synergistically with lysostaphin to kill mastitis-causing *Staphylococcus aureus* in murine mammary glands. Appl Env. Microbiol 78, 2297–2305. https://doi.org/10.1128/AEM.07050-11

Schmelcher, M., Powell, A.M., Camp, M.J., Pohl, C.S., and Donovan, D.M. (2015). Synergistic streptococcal phage lambdaSA2 and B30 endolysins kill streptococci in cow milk and in a mouse model of mastitis. Appl Microbiol Biotechnol 99, 8475–8486. https://doi.org/10.1007/s00253-015-6579-0

Schuch, R., Lee, H.M., Schneider, B.C., Sauve, K.L., Law, C., Khan, B.K., Rotolo, J.A., Horiuchi, Y., Couto, D.E., Raz, A., et al. (2014). Combination therapy with lysin CF-301 and antibiotic is superior to antibiotic alone for treating methicillin-resistant Staphylococcus aureus-induced murine bacteremia. J Infect Dis 209, 1469–1478. https://doi.org/10.1093/infdis/jit637

Shavrina, M.S., Zimin, A.A., Molochkov, N. V, Chernyshov, S. V, Machulin, A. V, and Mikoulinskaia, G. V (2016). In vitro study of the antibacterial effect of the bacteriophage T5 thermostable endolysin on Escherichia coli cells. J. Appl. Microbiol. 121, 1282–1290. https://doi.org/10.1111/jam.13251

Silhavy, T.J., Kahne, D., and Walker, S. (2010). The bacterial cell envelope. Cold Spring Harb. Perspect. Biol. 2. https://doi.org/10.1101/cshperspect.a000414

Soravia, E., Martini, G., and Zasloff, M. (1988). Antimicrobial properties of peptides from Xenopus granular gland secretions. FEBS Lett. 228, 337–340. https://doi.org/10.1016/0014-5793(88)80027-9

Spapen, H., Jacobs, R., Van Gorp, V., Troubleyn, J., and Honoré, P.M. (2011). Renal and neurological side effects of colistin in critically ill patients. Ann. Intensive Care 1. https://doi.org/10.1186/2110-5820-1-14

Srinivasan, R., Chaitanyakumar, A., Subramanian, P., Mageswari, A., Gomathi, A., Aswini, V., Sankar, A.M., Ramya, M., and Gothandam, K.M. (2019). Recombinant engineered phage-derived enzybiotic in *Pichia pastoris* X-33 as whole cell biocatalyst for effective biocontrol of *Vibrio parahaemolyticus* in aquaculture. Int. J. Biol. Macromol. https://doi.org/10.1016/j.ijbiomac.2019.11.042

Tacconelli, E., Sifakis, F., Harbarth, S., Schrijver, R., van Mourik, M., Voss, A., Sharland, M., Rajendran, N.B., Rodríguez-Baño, J., Bielicki, J., et al. (2018). Surveillance for control of antimicrobial resistance. Lancet Infect. Dis. 18, e99–e106. https://doi.org/10.1016/S1473-3099(17)30485-1

Tamaki, S., Sato, T., and Matsuhashi, M. (1971). Role of lipopolysaccharides in antibiotic resistance and bacteriophage adsorption of *Escherichia coli* K-12. J. Bacteriol. 105, 968–975.

Thandar, M., Lood, R., Winer, B.Y., Deutsch, D.R., Euler, C.W., and Fischetti, V.A. (2016). Novel engineered peptides of a phage lysin as effective antimicrobials against multidrug-resistant *Acinetobacter baumannii*. Antimicrob. Agents Chemother. 60, 2971–2979. https://doi.org/10.1128/AAC.02972-15

Theuretzbacher, U., Outterson, K., Engel, A., and Karlén, A. (2019). The global preclinical antibacterial pipeline. Nat. Rev. Microbiol. 2019 1–11. https://doi.org/10.1038/s41579-019-0288-0

Thummeepak, R., Kitti, T., Kunthalert, D., and Sitthisak, S. (2016). Enhanced Antibacterial Activity of *Acinetobacter baumannii* Bacteriophage ØABP-01 Endolysin (LysABP-01) in Combination with Colistin. Front. Microbiol. 7, 1402. https://doi.org/10.3389/fmicb.2016.01402

Vázquez, R., Domenech, M., Iglesias-Bexiga, M., Menéndez, M., and García, P. (2017). Csl2, a novel chimeric bacteriophage lysin to fight infections caused by *Streptococcus suis*, an emerging zoonotic pathogen. Sci. Rep. 7. https://doi.org/10.1038/s41598-017-16736-0

Vázquez, R., García, E., and García, P. (2018). Phage lysins for fighting bacterial respiratory infections: A new generation of antimicrobials. Front. Immunol. 9. https://doi.org/10.3389/fimmu.2018.02252

Vergalli, J., Bodrenko, I. V, Masi, M., Moynié, L., Acosta-Gutiérrez, S., Naismith, J.H., Davin-Regli, A., Ceccarelli, M., van den Berg, B., Winterhalter, M., et al. (2019). Porins and small-molecule translocation across the outer membrane of Gram-negative bacteria. Nat. Rev. Microbiol. https://doi.org/10.1038/s41579-019-0294-2

Vollmer, W., Blanot, D., and De Pedro, M.A. (2008). Peptidoglycan structure and architecture. FEMS Microbiol. Rev. 32, 149–167. https://doi.org/10.1111/j.1574-6976.2007.00094.x

Walmagh, M., Briers, Y., dos Santos, S.B., Azeredo, J., and Lavigne, R. (2012). Characterization of modular bacteriophage endolysins from myoviridae phages OBP, 201Q2-1 and PVP-SE1. PLoS One 7. https://doi.org/10.1371/journal.pone.0036991

Wang, C.Y., Huang, H.W., Hsu, C.P., and Yang, B.B. (2016a). Recent Advances in Food Processing Using High Hydrostatic Pressure Technology. Crit. Rev. Food Sci. Nutr. 56, 527–540. https://doi.org/10.1080/10408398.2012.745479

Wang, W., Li, M., Lin, H., Wang, J., and Mao, X. (2016b). The Vibrio parahaemolyticus-infecting bacteriophage qdvp001: genome sequence and endolysin with a modular structure. Arch. Virol. 161, 2645–2652. https://doi.org/10.1007/s00705-016-2957-x

Wang, S., Gu, J., Lv, M., Guo, Z., Yan, G., Yu, L., Du, C., Feng, X., Han, W., Sun, C., et al. (2017). The antibacterial activity of *E. coli* bacteriophage lysin lysep3 is enhanced by fusing the *Bacillus amyloliquefaciens* bacteriophage endolysin binding domain D8 to the C-terminal region. J. Microbiol. 55, 403–408. https://doi.org/10.1007/s12275-017-6431-6

WHO, 2017a. WHO | One Health, World Health Organization. World Health Organization.

WHO, 2017b. WHO | Global priority list of antibiotic-resistant bacteria to guide research, discovery, and development of new antibiotics. Geneva, 25-27 January 2017.

WHO 2017. Prioritization of pathogens to guide discovery, research and development of new antibiotics for drug-resistant bacterial infections, including tuberculosis. Geneva: World Health Organization; 2017(WHO/EMP/IAU/2017.12). (License: CC BY-NC-SA 3.0 IGO).

Yan, G., Liu, J., Ma, Q., Zhu, R., Guo, Z., Gao, C., Wang, S., Yu, L., Gu, J., Hu, D., et al. (2017). The N-terminal and central domain of colicin A enables phage lysin to lyse *Escherichia coli* extracellularly. Antonie van Leeuwenhoek, Int. J. Gen. Mol. Microbiol. 110, 1627–1635. https://doi.org/10.1007/s10482-017-0912-9

Yan, G., Yang, R., Fan, K., Dong, H., Gao, C., Wang, S., Yu, L., Cheng, Z., and Lei, L. (2019). External lysis of *Escherichia coli* by a bacteriophage endolysin modified with hydrophobic amino acids. AMB Express 9, 106. https://doi.org/10.1186/s13568-019-0838-x

Yang, H., Wang, M., Yu, J., and Wei, H. (2015). Antibacterial activity of a novel peptide-modified lysin against *Acinetobacter baumannii* and *Pseudomonas aeruginosa*. Front. Microbiol. 6. https://doi.org/10.3389/fmicb.2015.01471

Zampara, A., Sørensen, M.C.H., Grimon, D., Antenucci, F., Briers, Y., and Brøndsted, L. (2018). Innolysins: A novel approach to engineer endolysins to kill Gram-negative bacteria. BioRxiv 408948. https://doi.org/10.1101/408948

Zhang, G., Meredith, T.C., and Kahne, D. (2013). On the essentiality of lipopolysaccharide to Gram-negative bacteria. Curr. Opin. Microbiol. 16, 779–785. https://doi.org/10.1016/j.mib.2013.09.007

## Chapter 16

# Genetically Engineered Bacteriophages

**Rajesh Mamkulathil Devasia[1] and Salim Manoharadas*[1]**

[1]Central Laboratory, College of Science 5, King Saud University, Riyadh 11451, Saudi Arabia

*Corresponding author: smanoharadas@ksu.edu.sa

DOI: https://doi.org/10.21775/9781913652517.16

### Abstract

With the advent of antibiotics, the use of bacteriophages (phages) as antimicrobial agents has been abandoned in the western world. However, the increasing prevalence of multi-drug resistant bacterial pathogens has resulted in a resurgence of research efforts to use phages as antimicrobials. A side effect of many antibiotics as well as of phage therapy is the release of cell wall components, e.g. endotoxins of Gram-negative bacteria, which mediate the general pathological aspects of septicaemia. In the last decade, several strategies based on genetically engineered lysis-deficient phages have been devised with the aim to avoid disintegration of the cell envelope but to kill the bacterial target. These studies indicated that killing-proficient but lysis-defective recombinant phages can be exploited as efficient antimicrobials with reduced side effects. Moreover, genetically engineered phages can be used to augment the antimicrobial efficacy of antibiotics and to reduce bacterial biofilms. Apart from these potential medical applications, modified phages have been used to detect bacterial pathogens in foodstuffs. Here, we provide a review of these studies and briefly discuss the prospects of genetically modified phage in medicine and industry.

## Introduction

The therapeutic potential of lytic bacteriophages to control bacterial pathogens is long known, and phage therapy is still in use in some central and eastern European countries. In contrast, the emergence of effective antibiotics with a broad host range together with their facile administration has resulted in widespread antibiotic resistance. This emergence of multi-drug resistant bacterial pathogens has spawned a renewed interest in phage therapy. For instance, approximately two billion people worldwide are colonized with *Staphylococcus aureus*, out of which approximately 53 million (2.7%) are thought to carry MRSA (Methicillin Resistant *S. aureus*) strains. MRSA strains are resistant to all conventional anti-staphylococcal beta-lactam antibiotics. The glycopeptide antibiotics vancomycin and teicoplanin have served for many years as a last resort to treat MRSA infections (Schentag et al., 1998; Segarra-Newnham and Church, 2012). However, several of the newly discovered MRSA strains have shown resistance to both antibiotics. These MRSA strains are known as VISA (Vancomycin Intermediate-Resistant *S. aureus*) (Sieradzki and Tomasz, 1997; Schito, 2006). *S. aureus* isolates with decreased susceptibility to new antibiotics, including linezolid and daptomycin, are also reported (Sader et al., 2013). In addition to *S. aureus,* several other emerging multi drug-resistant pathogens are beginning to cause severe issues within medicine, to name a few: *Acinetobacter baumanii, Clostridium difficile, Enterococcus faecalis, Klebsiella pneumoniae, Pseudomonas aeruginosa* and *Escherichia coli* (Woodford et al., 2011; Chancey et al., 2012).

Phages can offer advantages over antibiotics. They are natural entities, can be produced at a reasonable cost, are highly specific for a bacterial target, do not harm beneficial commensals, and they vanish once the bacterial target is eliminated (Kutateladze and Adamia, 2010; Burrowes et al., 2011). The number of scientific reports on the treatment of experimental animal infections with intact phages particles or phage-derived proteins either solely or in combination with antibiotics has steadily increased in the last decade. For instance, phage therapy was used to treat respiratory infections caused by *E. coli* in broiler chickens. The intramuscular injection of phages protected chickens from infection when administered at 0 h, 24 h or 48 h after challenge with *Escherichia coli* (Huff et al., 2003). In another study, phages were successfully used to treat fish infections caused by *Pseudomonas plecoglossicida* (Park and Nakai, 2003). The efficacy of phage therapy was also assessed in mice with acute lung infections caused by a bioluminescent *P. aeruginosa* strain. The phage treated mice showed a reduced pathogen load in the chest within 6 h, and were completely devoid of the pathogen after 24 h (Debarbieux et al., 2010). In addition, successful curative and preventive treatment of lung infections in mice caused by *P. aeruginosa* strain CHA was also achieved using phage P3-CHA (Morello et al., 2011).

The efficacy and safety of phage therapy in the treatment of a wide range of human infections has been demonstrated in many uncontrolled studies conducted mostly in Poland, Georgia, and Russia (Kutadeladze and Adamia, 2010; Abedon et al., 2011; Chanishvili et al., 2012; Miedzybrodzki et al., 2012, and also other chapters in this book). Recently, the first controlled clinical trials were conducted (Wright et al., 2009; Harper et al., 2011; Sarker et al., 2012). They seem to confirm the results of earlier uncontrolled studies.

Host cell lysis by double-stranded DNA phage is commonly achieved by a phage encoded holin in conjunction with a muralytic enzyme (endolysin). In general, the holin induces a non-specific lesion in the cytoplasmic membrane, which allows the endolysin to gain access to and to degrade the peptidoglycan, leading to disruption of the cell envelope (Young, 1992; Catalão et al., 2013). In contrast to Gram-negative bacteria, endolysins encoded by phage of Gram-positive bacteria can act as exolysins, i.e. by degrading the peptidoglycan from without (Web resource 1 and 2). Purified phage endolysins have been used as antimicrobials against *Streptococci* in mice (Nelson et al., 2001; Loeffler et al., 2001; Gilmer et al., 2013), and have been proven effective against other Gram-positive pathogens including *S. aureus* (Fenton et al., 2013; Gilmer et al., 2013), *Enterococcus faecalis* and *E. faecium* (Yoong et al., 2004), *Clostridium perfringens* (Zimmer et al., 2002; Simmons et al., 2010), group B *Streptococci* (Cheng et al., 2005) and *Bacillus anthracis* (Schuch et al., 2002). The term enzybiotic has been coined to classify these antimicrobials (Nelson et al., 2001;

Hermoso et al., 2007; Fischetti, 2010), which we will discuss in a later section of this chapter.

The US Food and Drug Administration (FDA) has issued GRAS (Generally Recognized As Safe) status for two phage preparations (Listex™ and SalmoFresh™). Moreover, one phage preparation (EcoShield™) has been approved as a food contact substance, and another one (ListShield™) as a food additive (Sulakvelidze et al., 2013). These approvals have provided a significant boost towards the application of phages in the food industry (Web Resource 3; Guenther et al., 2009; Sillankorva et al., 2012; Sulakvelidze, 2013).

To achieve efficient biocontrol virulent phages possessing a broad host range are desired (O'Flaherty et al., 2005; Bielke et al., 2007). Alternatively, a cocktail or mixture of different phages can be used. A cocktail containing three different phages was used to treat beef contaminated with *E. coli* O157:H7. After storage, no viable *E. coli* cells were observed in the majority of the samples (O'Flynn et al., 2004).

**Genetically engineered phages**

As mentioned above, phage infection generally culminates in lysis of the host and consequently mediates the release of cell-wall components, e.g. endotoxin release from Gram-negative bacteria. Orally administered phages have been detected in the blood of patients (Slopek et al., 1983; Weber-Dabrowska et al., 1987), which would induce endotoxin release upon phage-mediated lysis of Gram-negative pathogens, which in turn may lead to septic shock (Matsuda et al., 2005; Paul et al., 2011). Also, as phages are replicating entities, the application of a defined dose is difficult, which could jeopardize the endeavour to obtain certification for phages as pharmaceuticals. Genetically engineered phages (Figure. 1) that effectively kill the host but do not cause lysis may alleviate these concerns without compromising efficacy. Moreover, phages can be genetically modified to extend their host range, to augment the activity of antibiotics or to reduce biofilms (Moradpour and Ghasemian, 2011).

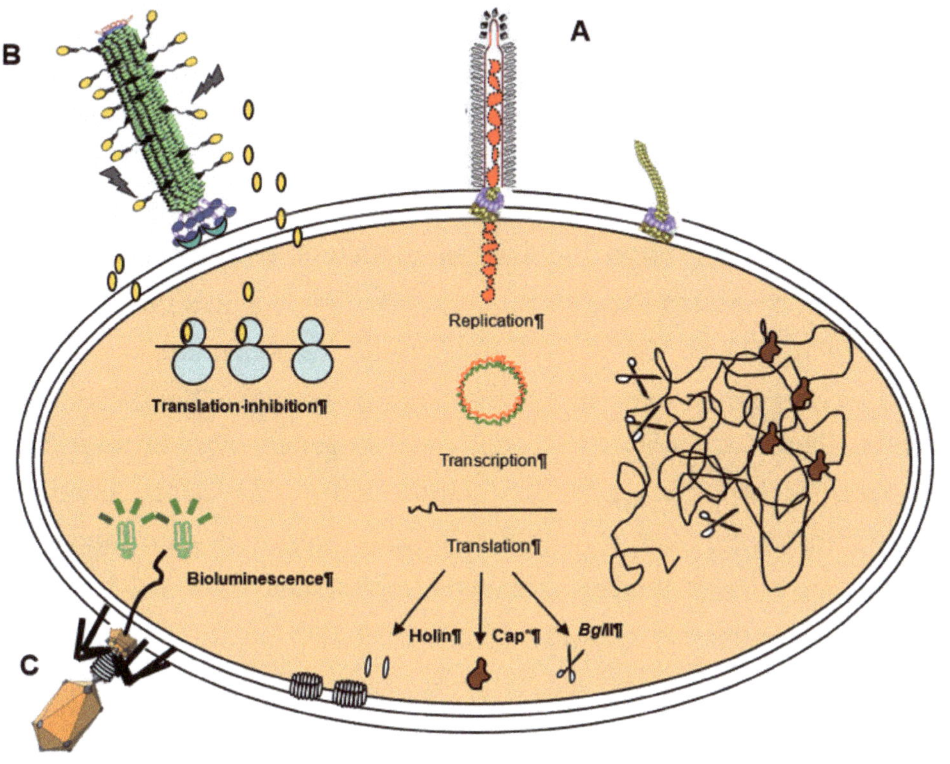


**Figure. 1.** Host cell killing by genetically engineered phages **(A)** Recombinant phages encoding lethal functions. M13 variants encoding a holin (kills the host cell by the generation of lesions in the inner membrane), Cap* (binds chromosomal DNA non-specifically), or *Bgl*II (cleaves host DNA) are used to kill target bacteria without lysis. (B) Recombinant phage with engineered specificity for the bacterial target. M13 phages are used as a drug carrying platform. Upon targeted binding to a specific pathogen, the attached drug (chloramphenicol; yellow oval) is released and causes cell death. (C) Recombinant phage exploited as bio detectors. Phage encoding luciferase or GFP are employed to detect pathogens. For more details and references, see text.

## Lysis-deficient phage

It has been shown that a recombinant filamentous phage encoding lethal but non-lytic proteins can efficiently kill *E. coli* with minimal release of endotoxins when compared to strictly lytic phages (Hagens and Bläsi, 2003). The filamentous M13 phage (Figure. 1) was genetically engineered to encode either the phage lambda holin (λS105) or the *Bgl*II restriction endonuclease (R). The reasoning for these additions being that upon host infection phages M13-λS105 and M13R, with the encoded holin will cause a lethal lesion in the inner membrane and the *Bgl*II enzyme will lead to breaks in the chromosomal DNA, respectively. Two hours following infection with M13-λS105 at an MOI (Multiplicity Of Infection) of 10, 99% of the cells were killed. However, growth of the bacterial cells resumed after two hours, albeit slower than in the control. This result was attributed to the emergence of *E. coli* mutants lacking F-pili, which are required by M13 to adsorb to the cell surface. The infection of *E. coli* cells with the M13R phage led likewise to a rapid decline in viable cell counts. When compared with M13-λS105, the killing efficiency was increased in a manner comparable with the performance of the lytic λcI- phage. Although 99.9% of the cells were killed, the number of *E. coli* cells remained constant over a 6-hour period following M13R infection. In contrast, the number of structurally intact cells mirrored the number of viable cells upon infection with phage *λcI*$^-$. As anticipated, the endotoxin levels released after infection with M13-λS105, M13R and *λcI*$^-$ differed. After 1 h following infection with *λcI*$^-$, an 18-fold increase of the endotoxin levels was noticed as compared to a ~ 2-fold increase with either M13S105 or M13R (Hagens and Bläsi, 2003).

In a follow up study, a non-replicating, non-lytic phage was engineered to combat an experimental *Pseudomonas aeruginosa* infection (Hagens et al., 2004). A gene for an export protein of the filamentous *Pseudomonas* phage Pf3 was replaced with that of the *Bgl*II restriction endonuclease. This replacement rendered the Pf3 variant (Pf3R) non-replicative and concomitantly prevented the release of Pf3R from the target cell. The recombinant phage was propagated on a host strain expressing the *bglII* methylase gene and the Pf3 export function. The Pf3R phage efficiently killed *P. aeruginosa* while the structural integrity of the cells was maintained, and endotoxin release was kept to a minimum. The release of endotoxin after Pf3R infection was approximately 10-fold lower when compared with the lytic *P. aeruginosa* phage Pt1. Pf3R treatment and treatment with the replicating lytic phage Pt1 of *P. aeruginosa* infections of mice resulted in a comparable survival rate upon challenge with a minimal lethal dose of 3. However, when compared with the lytic phage, the survival rate after phage therapy with Pf3R was significantly higher upon challenge with a minimal lethal dose of 5. After 48 h, the Pf3R and Pt1 phage treated mice showed a survival rate of 73% and 20%, respectively. When compared to Pf3R treated mice, increased TNF-α and IL-6 levels were detected in serum samples of mice treated with the lytic phage Pt1. Thus, the increased survival rate of Pf3R treated mice appeared to be a result of reduced endotoxin release (Hagens et al., 2004).

In a conceptually similar study, an M13 phagemid was used to deliver the genes encoding the toxic *E. coli* proteins Gef and ChpBK (Westwater et al., 2003). By forming a pore in the inner membrane, Gef dissipates the membrane potential leading to cell death (Gerdes et al., 1986). ChpBK acts as a sequence-specific endoribonuclease, which preferentially cleaves single-stranded RNAs (Zhang et al., 2005). The *gef* and *chpBK* genes were cloned into a phagemid, and the respective transformants were infected with the helper phage R408. This procedure allowed preferential packaging of the phagemid DNA, resulting in the creation of recombinant M13 phages encoding either the *gef* and *chpBK* genes. Infection of *E. coli* ER2738 with either M13-Gef or M13-ChpBK phages resulted in a substantial reduction of viable cells. The efficacy of these recombinant phages was also studied in a mouse infection model. A single intraperitoneal injection of $10^8$ CFU of *E. coli* ER2738 was used to induce a transient infection in immuno-compromised mice. When compared to the control, a significant reduction (> 90 %) of blood circulating bacteria was observed five hours after treatment with either recombinant phage (Westwater et al., 2003).

Catabolite activator protein (CAP) is a DNA-binding protein involved in carbon utilization (de Crombrugghe et al., 1984). CAP in complex with cyclic-AMP (cAMP) binds to DNA target sites at or near many promoters in the *E. coli* chromosome and stimulates transcription (Musso et al., 1977). A modified CAP variant with a replacement of Glu181 > Gln displays unspecific DNA binding, and over-production of this protein is lethal (Lopata et al., 1997). Using the same strategy as described above, Moradpour *et al.* (2009) cloned the mutant *cap** gene into the pBluescript II SK (+) vector to generate an M13 variant encoding the lethal function (Figure. 1). When compared to the uninfected control, a ~ 650-fold reduction in CFU was observed 60 min post infection of the pathogenic *E. coli* O157:H7 with the M13-*cap** phage. As *E. coli* O157:H7 is sometimes found in dairy products including milk and yoghurt (Kumar et al.,

2013), the efficacy of the M13-*cap** phage in reducing viable *E. coli* O157:H7 was tested in milk. Sterilized milk was inoculated with $10^7$ bacterial cells, and the M13-*cap** phage was added at an MOI of 100. After 120 min, the viable bacterial cell count in milk was reduced by six orders of magnitude (Moradpour *et al.*, 2009).

Bactericidal quinolone antibiotics such as ofloxacin induce hydroxyl radical formation that leads to damage of cellular DNA, proteins and lipids (Kohanski et al., 2007; Cheng et al., 2013). Consequently, the SOS response of the cell is induced, which results in the repair of the damaged DNA (Lewin et al., 1989; Miller et al., 2004; Janion, 2008). The efficacy of such bactericidal antibiotics can be enhanced by disabling the SOS response. An M13 phage was engineered to over-express the *lexA3* gene, encoding a repressor of the SOS response (Lu and Collins, 2009). When compared to ofloxacin treated cells, a 4.5 (log) fold drop in CFU was observed when *E. coli* cells were additionally infected with the M13-*lexA3* phage. The M13-*lexA3* phage was also proficient in killing *E. coli* strains resistant to quinolone antibiotics. *E. coli* RFS289, which harbours a mutation that confers resistance against quinolone antibiotics, was susceptible to a combination therapy with ofloxacin and the modified phage (Lu and Collins, 2009). In addition, the efficacy of the combination therapy was tested in mice. Intraperitoneal injections of 8.8 x $10^7$ CFU per mouse of *E. coli* EMG2 killed the mice within 2 days. In contrast, 80% of the mice that received a combination therapy with M13-*lexA3* phage ($10^9$ phage per mouse) and ofloxacin (0.2 mg / kg) after 1 h of infection survived. In contrast, only 50% and 20% treated with unmodified phage-plus-ofloxacin and ofloxacin alone survived, respectively (Lu and Collins, 2009). Futhermore, the M13-*lexA3* phage augmented the antibacterial efficacy of other classes of antibiotics (especially aminoglycosides and ß-lactams), could reduced the development of antibiotic-resistance upon exposure of bacteria to subinhibitory concentrations and reduce levels of persister cells (Lu and Collins, 2009).

These authors further designed recombinant M13 variants equipped with genes, the products of which target gene networks other than the SOS response. The M13 variants carried the *soxR*, the *csrA* and the *ompF* gene, respectively.

Genes *soxR* and *soxS* form part of the *soxRS* operon, which controls the cellular response to oxidative stress (Hidalgo, 1997; Chiang and Schellhorn, 2012). CsrA regulates glycogen synthesis and acts as a repressor of biofilm formation (Jackson et al., 2002; Romeo et al., 2013), whereas OmpF is a porin through which quinolones enter the bacterial cell (Hirai et al., 1986; Mahendran et al., 2010). In all cases, a combination of the respective recombinant M13 phage and ofloxacin was more efficient in killing *E. coli* than the sole application of either the recombinant phage or the antibiotic.

Although the studies described above proved the efficacy of recombinant phage *in vitro* as well as *in vivo* infection models, genetically modified phages are commonly believed to complicate the endeavour to obtain a certification as pharmaceuticals. An alternative to recombinant phages encoding toxic functions are endolysin-deficient phages. These phages retain the ability to kill through their holin function but do not lyse the host cell (Rietsch and Bläsi, 1993; Paul et al., 2011). *Vice versa*, holin-deficient phages are non-lytic but can kill the host through phage-induced DNases and/or by interference with the bacterial metabolism. In one study, a T4 phage (LyD) carrying an amber mutation in the holin gene *t* was tested for its efficacy in murine peritonitis. The mice were injected intraperitoneally with $1 \times 10^{10}$ CFU of *E. coli*. Thirty minutes after infection, the first and second group was treated with $1 \times 10^{11}$ PFU of T4-LyD and T4 wild-type phages, respectively. The third group was treated with the ß-lactam antibiotic LMOX at a final concentration of 25 mg/kg body weight. The untreated mice died after 20 h following the challenge. However, 81% of the mice treated with a single dose of $1 \times 10^{11}$ PFU of the T4-LyD phage were rescued. Interestingly, the group treated with T4 wild-type phage and LMOX had a survival rate of only 52% and 33%, respectively (Matsuda et al., 2005). The total endotoxin levels in peritoneal lavage fluid were significantly lower in the T4-LyD phage treated group than in the other two groups, again showing that reduced endotoxin release correlated with increased survival.

## Genetically engineered phages with an extended host range

The specificity of a phage is attributed to receptor binding proteins usually present on tail fibres. For instance, the specificity of phage T4 towards *E. coli* B is determined by gene *37,* encoding a 1026 amino-acid distal tail fibre protein (Oliver and Crowther, 1981; Bartual et al., 2010). A mutation resulting in the duplication of a small domain within the P37 tail fibre adhesin can expand the host range of phage T4 (Tétart et al., 1996). There are several studies in which phages have been genetically modified to extend their host range.

In a study by Cao et al., (2000), the filamentous phage M13 was engineered to present the variable fragments of IgG against *Helicobacter pylori* on the surface. The recombinant phage (ScFv) could efficiently bind to spiral and coccoid forms of *H. pylori.* In addition, the ScFv-phage was effective in reducing the viable counts of *H. pylori in vitro* and *in vivo.* The recombinant phage was further able to selectively reduce *H. pylori* cells in a mixed population of bacteria, consisting of *S. aureus*, *Streptococcus* and *E. coli* strains. Moreover, an ScFv-phage/*H. pylori* mixture exhibited a somewhat reduced efficiency to colonize mice when compared with *H. pylori* alone (Cao et al., 2000).

The filamentous phage fd was engineered to infect bacteria bearing both, F and N pili. Phage fd is an F-pilus specific *E. coli* phage (Marvin et al., 2006). The minor coat protein pIII is found on the tip of the virion and accounts for its host specificity. Binding of pIII to the F-pilus of *E. coli* results in pilus retraction and the transfer of phage to host surface. The translocation of phage DNA through the cell envelope requires functions of the *tolQRA* operon (Sun and Webster, 1987). In addition to binding to F-pili, pIII also interacts with the inner membrane protein TolA. Following TolA-pIII interaction, the major coat protein pVIII is inserted into the inner membrane, and the phage DNA is translocated into the cytoplasm of the host cell (Click and Webster, 1998). The specificity for N-pili by phage fd was achieved through the creation of a fusion between genes *3* of phages fd and IKe, the latter of which encodes a receptor specific for N-pili. The recombinant fd phage effectively infected *E. coli* strains bearing either F or N pili (Marzari et al., 1997).

Similarly, phage fd was modified to infect *Vibrio cholerae*. The lysogenic phage CTXΦ infects *V. cholera* by binding to the toxin-co-regulated pilus (TCP) (Davis and Waldor, 2003). The product of CTXΦ OrfU ($pIII^{CTX}$) is homologous to the pIII protein of phage fd. *OrfU* was abutted with fd gene *3* and the fusion protein OrfU-pIII, this was shown to be exposed on the surface of the recombinant fd-$pIII^{CTX}$ phage. Phage fd-$pIII^{CTX}$ was effective in transducing both $TCP^{+}$ (1.1 x $10^{-4}$) and $TCP^{-}$ (3.3 x $10^{-5}$) *V. cholerae* strains without affecting the effectiveness of infection of the host *E. coli* $F^{+}$ TG1. Infection of the TCP- *V. cholerae* strain by phage fd-$pIII^{CTX}$ was attributed to the fact that different segments of $pIII^{CTX}$ mediate distinct steps in phage infection. Phage fd-$pIII^{CTX}$ encompasses aa 15-138 of $pIII^{CTX}$, which was sufficient to support infection of *V. cholerae* in a TolA-dependent, TCP-independent manner (Heilpern and Waldor, 2003).

Genetic recombination has also been used to switch the host range of a lytic phage. The genes 37/38 of *E. coli* phage T2 and IP008 encode long tail fibre proteins that recognize surface receptors on the bacterial cell (Riede et al., 1987). Phage T2 and IP008 were able to infect approximately 7% and 33% of environmental *E. coli* strains (Mahichi et al., 2009), respectively. The host range of phage T2 was successfully extended by exchanging genes *37/38* with the orthologous gene pair of phage IP008 (Mahichi et al., 2009). The recombinant T2 phage acquired specificity towards *E. coli* strains that are not infected by the wild-type T2 (Mahichi et al., 2009).

As a consequence of the co-evolution of bacteria and phage, phage resistant bacteria have emerged and will continue to emerge. One way of overcoming the dilemma of phage-resistant bacterial pathogens is the establishment of phage banks, from which virulent phage can be selected against any pathogen. In a recent pilot study, Pouillot et al. (2010) have created a genetically engineered T4-phage bank by means of nested PCR technology termed TAPE (Targeted Accelerated Protein Evolution) and homologous recombination. Random mutations were created in the variable regions of the T4 genes *37/38*, encoding the tail fibre proteins. Following the selection of the mutant alleles, they were recombined into the T4 *hoc* gene encoding the highly antigenic outer capsid protein. To permit recombination, *E. coli* DK8 was engineered to reversibly interrupt the lytic cycle of phage T4 by over-expressing a mutant Rho protein gene. The recombinant T4 particles were capable of infecting and lysing bacteria that are evolutionarily close (*Yersinia ruckeri* ATCC 29908) and distantly

(*Pseudomonas aeruginosa* ATCC 47053) related to the wild type phage T4 host *E. coli* K12 (Pouillot et al., 2010).

## The exploitation of genetically engineered phages to reduce bacterial biofilms

Bacterial biofilms refer to cellular communities that adhere to an inert or living surface. They can resist treatment with antibiotics or the immune system (Costerton et al., 1999; Costerton et al., 2003; Römling and Balsalobre, 2012). The heterogeneous structure of biofilms is formed by hydrated EPS (Extracellular Polymeric Substance) matrix generally composed of several polysaccharides, proteins, nucleic acids and lipids (Xavier et al., 2005). Given the antibiotic resistance of bacterial biofilms and experimental data that suggests that antibiotics can even induce biofilm formation (Hoffman et al., 2005; Kaplan, 2011), there is an intense need for novel antimicrobials that disperse biofilms and eradicate biofilm communities. Disruption of biofilms can be achieved by enzymatic degradation of EPS components (Kaplan, 2009). For instance, the enzymatic degradation of the cell-bound EPS polysaccharide, adhesin (polymeric β-1,6-N-acetyl-D-glucosamine) by *Actinobacillus actinomycetemcomitans* dispersin B (DspB) has been shown to disrupt biofilms of several bacterial species including *S. aureus* and *E. coli* (Itoh et al., 2005). However, while exogenous enzymes may perturb the biofilm mesh, they do not necessarily kill the embedded bacteria. Another possibility to disrupt biofilms is the use of phages (Donlan, 2009; Brüssow, 2013). It has been reported that phage T4 can infect and replicate within *E. coli* biofilms and disrupt the biofilm by lysing the embedded bacteria (Doolittle et al., 1995; Doolittle et al., 1996). In addition, phages have been shown to diffuse through alginate gels and to penetrate through a *P. aeruginosa* EPS matrix (Hanlon et al., 2001).

With the aim to engineer a phage that could effectively disperse *E. coli* biofilms, phage T7 was genetically modified to encode the *A. actinomycetemcomitans* DspB protein (Lu and Collins, 2007). Upon T7-mediated host cell lysis, the DspB protein was anticipated to be released and to degrade the EPS. The engineered T7 phage was tested on F-plasmid bearing *E. coli* TG1 biofilms (Ghigo, 2001; Da Re et al., 2007). However, phage T7 was unable to infect the F-plasmid bearing *E. coli* strain (García and Molineux, 1995). To counter this problem, the *dspB* harbouring T7 phage was further modified by inserting *gene 1.2* from phage T3 into the T7-*dspB* genome. The resulting phage infected *E. coli* TG1, as well as possessing an increased host range. *In vitro* studies using T7-*dspB-1.2* were promising, as the phage effectively reduced *E. coli* TG1 biofilms by an $A_{600}$ factor of 10.3 (Lu and Collins, 2007). These authors also showed that the observed decrease of biofilms was accompanied by a reduction of viability of the biofilm embedded cells. These studies hold promise that phage harbouring EPS degrading functions can be designed for other biofilm forming pathogens, like *P. aeruginosa,* which causes severe problems in patients afflicted with cystic fibrosis.

**Use of genetically engineered phages as biodetectors**

The available biological methods to identify bacterial pathogens in industries and associated laboratories are time consuming, and they need additional enrichment steps to make exact identification of the species present. Plating methods can detect pathogens, but it fails when the pathogen present at low levels (Almeida et al., 2013). Other methods like antibody-based diagnosis usually fail to detect pathogenic bacteria, especially when the sample is complex and no enrichment steps are availed. The disadvantage of PCR based diagnosis sometimes amplify the DNA even from dead cells and designing of oligonucleotides for the target identification is difficult. These limitations often give false positive and nonspecific results (Lu et al., 2013).

*Salmonella* spp., *Campylobacter* spp., *E. coli* and *L. monocytogenes* are prevalent bacterial pathogens that impose the most significant burden of food-borne illnesses worldwide. The fatality rate in listeriosis reaches 20 to 30% with higher numbers occurring among immuno-compromised individuals (Newell et al., 2010). The rapid detection of food-contaminating pathogens is a primary task for the prevention of food-borne diseases. The currently used detection methods for food-borne pathogens have several drawbacks; they are time-consuming and laborious, costly and sometimes lack sensitivity and specificity. In particular, the detection of a small number of bacteria in clinical, environment or food samples requires time-consuming initial cultivation to increase the cell count. This time might be considerably shortened by using phage-based bioreporters.

Genetically engineered phages have been successfully used to detect food contaminants rapidly. Loessner et al. (1997) used phage A511, which infects 95% of all tested *L. monocytogenes* strains, including serovars 1/2 and 4 that are implicated in human listeriosis. Phage A511 was genetically modified to direct synthesis of luciferase upon infection of the host (Figure. 1). The *Vibrio harveyi* derived chimeric luciferase gene was inserted downstream of the major capsid protein gene (*cps*). Upon addition of the oxidizable aldehyde substrate, luciferase activity, i.e. successful infection of *L. monocytogenes* by the engineered A511 phage, was monitored by a luminometer. As few as 500 - 1000 cells per mL were detected within 2 h. Moreover, the engineered A511 phage permitted detection of *L. monocytogenes* in a variety of foodstuff including meat, cheese, milk and chocolate pudding upon deliberate contamination with the pathogen (Loessner et al., 1997).

In a related study, a T4*e*$^-$ phage, harbouring an amber mutation in the endolysin gene, was engineered to produce a fusion protein composed of GFP (green fluorescent protein) and the phage small outer capsid protein Soc. The GFP-Soc protein did not impair phage adsorption to *E. coli*, and the recombinant endolysin-deficient T4*e*-(*gfp-soc*) phage did not provoke lysis. As monitored by epifluorescence microscopy, the recombinant phage specifically detected *E. coli* K12 in a mixed population. When compared to the lysis-proficient T4 phage harbouring the *gfp-soc* fusion gene, the fluorescence increased over time when the T4*e*-(*gfp-soc*) phage was used as a bioreporter (Tanji et al., 2004).

Furthermore, "fluoromycobacteriophage" were employed to detect *Mycobacterium tuberculosis*. Several diagnostic tools have been developed to detect the presence of slow growing *M. tuberculosis,* including sputum smear microscopy. However, this method requires a minimum of 10.000 bacteria per ml of sputum (Watterson and Drobniewski, 2000). Mycobacteriophage TM4 was genetically engineered to express *gfp* or *Zsyellow* (fluorescent reporter genes) upon infection of the target bacterium. The rapid detection of as few as 100 cells was possible by either fluorescent microscopy or flow cytometry. Another advantage of using GFP as a reporter is its ability to withstand fixation of cells by para-formaldehyde, which in turn confers biosafety while analysing hazardous strains like XDR-TB (Piuri et al., 2009). In addition to specific detection, phage-based assays can also provide a means to assess antibiotic susceptibility. A modified phage expressing the gene encoding an enhanced GFP (EGFP-Phage) was used to detect isoniazid, rifampicin and streptomycin resistance in 155 strains of *M. tuberculosis* with a specificity of 94% (Rondón et al., 2011).

*In vivo* biotinylation of phage T7 followed by conjugation of the T7 phage to streptavidin-coated quantum dots was performed to design a related bioreporter. Quantum dots (QDs) are fluorescent colloidal semiconductor crystals of a few nanometers in diameter. They exhibit broad absorption spectra. The presence of a few atomic layers in the outer shell increases the quantum yield and further increases the photostability of the fluorescent probes. Phage T7 was genetically engineered such that a 15 aa biotinylation peptide was attached to the major capsid protein gp10A (> 400 copies per phage head). Following infection of the host bacterium with the recombinant phage, all progeny virions produced are anticipated to be biotinylated by the host cell biotin-ligase protein machinery (BLP-BirA in *E. coli*). The presence of biotinylated phage was detected by conjugation to streptavidin-linked QDs. The phage biodetector was able to specifically sense the *E. coli* host in quantities as low as 10 CFU/ within a mixed population of *P. aeruginosa*, *Vibrio cholerae*, *Salmonella*, *Yersinia pseudotuberculosis* and *Bacillus subtilis*, all of which are not susceptible to phage T7. This method of bacterial detection appears to be particularly suited for slow growing bacteria or highly infectious agents like *Bacillus anthracis* (Edgar et al., 2006).

**Recombinant phages as vehicles for the delivery of drugs**

Other than DNA molecules, the phages can be used for drug delivery to both eukaryotic and prokaryotic cells, including tumour cells. Phage proteins that specifically bind to the target cells are screened from the library. The phage is modified so that they express functional phage peptides on the phage coat surface (Pande et al., 2010).

*Prokaryotic cell drug delivery*

Filamentous phages have been further engineered to deliver antimicrobials to bacteria other than their natural host. In the study by Yacoby et al. (2006), chloramphenicol was conjugated to M13 phage with the aim to specifically deliver the drug to *S. aureus* SH100 (Figure 1). The authors used a phage display library and selected an M13 variant phage (A12C) that bound to *S. aureus* with high specificity, which resulted from an N-terminal lysine-free peptide displayed on the pVIII protein. In another approach, the sequence encoding the IgG-binding ZZ domain of *S. aureus* protein A was cloned into the phagemid vector fUSE5. This resulted in the polyvalent display of the ZZ domain on all copies of pIII of the recombinant phage (fUSE5-ZZ). Phage fUSE5-ZZ was further equipped with *S. aureus* targeting antibodies that were bound to the P3 linked ZZ domain (Yacoby et al., 2006).

*Eukaryotic cell drug delivery*

Phages were also engineered for the delivery of drugs against Alzheimer's disease (Frenkel and Solomon, 2002). In an experiment conducted in transgenic mice, a filamentous phage was engineered to express amyloid–β-peptide (Aβ) which accumulate in the brain during Alzheimer's disease progression (Murphy and LeVine, 2010). The phage was designed to express Aβ, and it was shown to specifically and effectively bind with the target cells for antibody delivery. Phages engineered in this way reduced the concentration of the drug for a particular treatment, thus reducing side effects and the incurred cost of treatment.

Phage dependent drug delivery of cancer has been reported where genetically modified fUSE5-ZZ filamentous phage is used (Bar et al., 2008). These phages are modified to display the ligand that is specific for the target cancer cells expressing ErbB2. The ligands expressed on the phage coat are further conjugated with the drug hygromycin or doxorubicin. *In vitro* studies show that these phages are targeted towards SKBR3 (Human breast adenocarcinoma cells) that are over-expressing ErbB2. Once these phages are endocytosed and degraded, the attached drug is delivered. It was found in the study that there is a 50% reduction in cancer cells and a thousand-fold increase in the potential of drug delivery. In a similar study by Du et al. (2010), phages were developed to display a peptide molecule conjugated with doxorubicin targeted for delivering to hepatocellular carcinoma. The binding peptide was selected by repeated panning of phage display libraries inside mice. The *in vivo* antitumor test results showed that all treated mice survived and only 40% survival occurred among free doxorubicin treatment.

The screening of a phage library for a specific peptide that targets the metastatic prostate cancer cell line PC-3M was done by Fagbohun et al. (2013). A specific prostate cancer targeting phage was engineered with a phagemid that expresses GFP under the presence of a cytomegalovirus promoter. The development of fluorescence upon binding of phagemid to the cancer cells is observed by fluorescence microscopy showing the potentiality of this method to deliver therapeutic genes to the cancer cells. In a similar study, Wang et al. (2014) screened a phage library to identify the breast cancer MCF-7 specific peptide fused with the phage coat protein. These phages were bound with doxorubicin drug, and the activity was checked. The antitumor activity was found enhanced in mouse models compared to other non-targeted formulations. In yet another study, prostate cancer treatment with refactored M13 phage was performed (Ghosh et al., 2012). In order for refactoring of the M13 genome, initially overlapping genes *VII* and *IX* were separated. This facilitated the manipulation of the N terminal end of protein p9 (encoded by gene *IX*) without affecting protein p7 (encoded by gene *VII*). The refactored M13 phage was displayed with three peptides of different phage components. The SPARC (secreted protein acidic and rich in cysteine) binding peptide (SBP) was displayed on the p3 phage minor coat protein encoded by gene *III*, a DFK amino acid motif attached to p8 major coat protein encoded by gene *VIII*, and a biotin acceptor peptide (BAP) linked to the p9 minor coat protein encoded by gene *IX*. The modified phage was designated M13-983. SPARC is a glycoprotein expressed in cancer cells (Tai and Tang, 2008). DFK peptide sequence is recognized by cathepsin B, a lysosomal cysteine protease over-expressed in prostate cancer (Miyake et al., 2004; Mohamed and Sloane, 2006). The drug doxorubicin was additionally attached to the p8 protein. The p9 protein displaying BAP was enzymatically biotinylated and incubated with streptavidin coated Alexa Flour 488 dye, resulting in phage M13-983-Alexa-Dox. The modified phage was tested *in vitro* for human prostate cancer cell lines which were expressing SPARC at a higher(C42B) or lower levels (DU145). The modified phage was efficient for targeting C42B cells since the normalized fluorescence intensity was ten times higher for C42B cells than DU145 cells. M13-983-Alexa-Dox was 100 times more cytotoxic than that of doxorubicin alone.

## Engineering phage for vaccines development

Phage are effective agents for developing vaccines and such a preliminary study was shown for HIV vaccine, anthrax and foot and mouth disease (FMD) virus. By engineering phage to display antigens on their surface, vaccines can be developed for various disease outbreaks both in human and veterinary needs. In order to develop a multi component vaccine for HIV, different components of the virus need to be displayed on phage surface. The efficiency of the vaccine in terms of immunogenicity developed by a phage engineering is been tested and proven (Sathaliyawala et al., 2006). Similarly, Ren et al., explained the development of vaccine against the foot and mouth disease virus (FMDV) by a T4 phage display system (Ren et al., 2008). The FMDVT4 recombinant vaccine was proved effective in mouse model experiments and its protection against the disease after immunization was also evaluated.

## Mechanisms for the generation of genetically engineered phages

With the recent advancement in the field of genome engineering, the strategies for the application of phages has dramatically increased. Apart from the canonical antibacterial therapeutic applications, phages have been genetically modified to be applicable in synthetic biology (Lemire et al., 2018), biomedical fields (Cao et al., 2018; Tao et al., 2018c) and material science (Cao et al., 2016). The reprogramming of biological systems by synthetic biology often use orthogonal components, which function individually without interfering with their counterparts. Phages offer an excellent source of orthogonal components to be used in synthetic biology applications. For instance, the phage encoded promoter-specific RNA polymerases provide precise control over gene expression since the RNA polymerases selectively recognize promoter sequences that are not recognized by endogenous bacterial polymerases. Furthermore, the specificity of phage encoded RNA polymerases can be altered by mutation without interfering with the enzymatic domain (Temme et al., 2012; Meyer et al., 2015). Filamentous phages have been recently used in biomedical research as supramacromolecules. The phage DNA manipulation facilitated the expression of foreign peptides on the surface. This strategy is beneficial for the identification of target specific peptides and in the construction of organic-inorganic nanohybrids. Similar techniques employing phage in chemical biomedical research has even displayed encouraging results with *in vivo* tissue regeneration, induced by phage (Cao et al., 2018). There is a wide range of genomic techniques employed for the construction of engineered bacteriophages. The routinely used methods are listed below:

## Phage genome engineering

Phage genome engineering is based on the finding that the entire phage genome can be assembled using synthetic oligonucleotides to produce live phage. *In vitro* synthesized and purified oligonucleotides are annealed and assembled by polymerase chain cycling (PCA), followed by phosphorylation. The fully synthesized phage genome is PCR amplified, digested with specific restriction endonucleases and circulated by ligation. The ligated DNA is transformed into *E. coli* by transformation and are plated for plaque development. This technique offers the advantage that desired alterations of the genome can be done by mutation. One potential disadvantage of this technique is that only small phage genomes can be manipulated, specifically, due to the difficulty in synthesizing large genomes *in vitro*. Phage genome engineering using PCR is comparatively easier to introduce alterations into phage genomes as compared to the recombinant based method of phage engineering. The recombinant technique of phage genome engineering is applicable to large phage genomes. The phage genome is engineered and cloned in yeast shuttle vector encompassing overhangs for recombination. The engineered genome is propagated in yeast cells. Following the recombination, the phage genome is isolated and introduced into a host bacterium to generate functional phage particles (Lu et al., 2013). Importantly, the efficiency of this method is dependent on the transformation efficiency of bacterial cells. This approach has been used for genetic modifications in genomes of coliphages T3 (38,208 bp), T7 (39,937) and *Klebsiella* phage K11 (41,181 bp) (Ando et al., 2015).

*Development of novel phages through DNA recombination*

The method of developing mutant phage by recombination of DNA is otherwise known as phage crosses. Recombination is a natural phenomenon where exogenous DNA is recombined with cells own genomic DNA when both the sequences share regions of homology (Friedberg, 2003). The gene to be introduced (genetically engineered) is first cloned into a replicative plasmid and transformed into bacterial cells. The bacterial cells harbouring the engineered plasmid is then infected with the phage that is to be modified. This eventually allows the heterologous DNA to be introduced and packaged into phage particles. This method can be used to develop recombinant phages with desired mutations or phages with gene replacements, deletions or insertions (Sarkis et al., 1995: Rao and Mitchell, 2001: Oda et al., 2004: Tanji et al., 2004: Namura et al., 2008). However, in a scenario where the number of mutations to be incorporated into the recombinant phage is large, the recombination-based method of phage development is not easy since it is time consuming. Bacteriophage recombineering is similar to homologous recombination-based technique. For instance, to generate recombinant phage λ, the phage DNA and the plasmid-based gene is co-electroporated into electro-competent cells that carry the red system of phage lambda and RecE/RecT system of Rac prophage. This system dramatically improves the recombination frequency (Murphy, 2012). This method is mainly used in *E. coli* cells, that are defective of λ prophage and the pL operon. The cells are infected with the phage to be engineered, and the λ red recombination functions are induced by heat shock treatment. This system comprises of the gamma (γ), exo (α) and beta (β) genes. Gamma inhibits the *E. coli* Rec BCD exonuclease complex to prevent degradation of the linear double stranded substrate (Poteete, 2001). Exo degrades the double stranded DNA in a 5' to 3' manner to generate single stranded DNA (ssDNA). Beta is single stranded DNA binding protein that anneals to the single stranded DNA substrate to its recombination target in a phage genome. These proteins promote a high level of recombination. Followed by electroporation of the phage genome and the gene encoding plasmid to the host cells, the cells are plated, and the recombinant phages are selected by plating for plaques. Plaques are then screened by PCR for the identification of the correctly mutated phage genome (Marinelli et al., 2008). The yield of recombination is higher than that of the homologous recombination-based method.

*CRISPR-Cas mediated genome engineering of phages*

Clustered regularly interspaced short palindromic repeats (CRISPR) is an immune protective system in bacteria protecting it from invading DNA or phage DNA (Deveau et al., 2010: Barrangou et al., 2007). Genome engineering by CRISPR-Cas based method has been utilized in phage engineering for specifically inducing a desired mutation or in eliminating a particular mutation (Shen et al., 2018). The two components of the CRISPR-Cas system are the Cas proteins and CRISPR RNA (CrRNA). The Cas protein encompasses the catalytic component of the system, which facilitates the cleavage of the incoming DNA (Shmakov et al., 2017). The CRISPR loci, on the other hand is the genetic locus that functions and directs the catalytic activity against the foreign DNA (Westra et al., 2014). The system undergoes three significant processes for its activity that include CRISPR adaptation, RNA biogenesis and CRISPR-Cas interference. This system specifically binds with the target site of the phage genome and creates a double stranded DNA break on the phage genome during infection. There are various reports on phage genome engineering using CRISPR Cas system for gene deletions and gene replacement (Box et al., 2016; Martel and Moineau, 2014). A method for developing phage genome engineering by using 1-E CRISPR-Cas system was explained by Kiro et al. (2014). In this study, homologous recombination was used to delete a non-essential gene (gene 1.7) of phage T7. The T7 phage was propagated in a bacterial host harbouring a plasmid with a region of homology upstream and downstream of the non-essential gene. The recombination between the DNA will delete the gene 1.7. The resulting phage population would contain both recombinant and non-recombinant phages. The enrichment of recombinant phages was further facilitated by a CRISPR-Cas based selection system. The non-recombinant phages were selectively cleaved by the CRISPR-Cas system, which subsequently leads to the enhancement of the recombinant phages that lacked the target gene 1.7. The advantage of CRISPR-Cas system is that it helps to overcome the difficulty to select the small percentage of recombinants from a large group of wild type phages.

*Development of phage genome in a cell free system*

The development of functional phage particles in the host bacterium is limiting because as it causes toxicity to the organism and reduces transformation efficiency. The development of phage particles in a cell free system avoids these limitations. Shin et al., in 2012, explained the construction and assembly of T7 bacteriophage φX174 in a cell free reaction. Here the phage DNA is mixed with a cell free expression system prepared from *E. coli* BL21 and Rosetta 2. Incubation of the phage DNA with the cells extracts of host bacteria enables transcription, translation, DNA replication and assembly of whole phage particles.

*Refactoring a phage genome*

Manipulation of a phage genome can be done before they are transformed into bacterial cells. This process is known as genome refactoring. Here the individual segment of the whole phage genome is manipulated separately by adding restriction sites to each segment (Chan et al., 2005). The manipulated segments are assembled by molecular cloning and transformed into the host bacterium. However, the main disadvantage of genome refactoring is the difficulty in manipulating large phage genomes. Moreover, it is a time-consuming process. Recombinant phages synthesized by this method develop smaller plaques than wild type phages. This technique was also found to be inefficient for non-domesticated bacteria.

## Phage based antimicrobials and their engineering

As discussed, phages can be used as antimicrobial agents but so too can their proteins, such as endolysins. These proteins can act as exolysins predominantly against Gram-positive bacterial pathogens (Fischetti, 2010: Fischetti, 2011). One of the major disadvantages attributed to endolysins is their inefficiency against Gram-negative bacteria due to their outer membranes. However, endolysins that overcome this barrier which utilise cationic terminals, such as those developed by Briers et al., (2014). These resulting endolysins are known as Artilysins.

Genes encoding antibacterial endolysins were screened from a genomic library based on the prophages induced from *A. baumanni,* with several endolysins identified and isolated which display activity against *A. baumanni* (Lood et al., 2015). Out of the endolysins screened against *A. baumanni*, phage lysin PlyF307 displays the most potent activity. The *in vitro* studies of PlyF307 against *A. baumanni* biofilms have shown a reduction of 1.6 orders of magnitude against biofilm encased cells after 2 hours of treatment. Significantly, 50% of mice were rescued by the activity of PlyF307 from a lethal *A. baumanni* infection, in contrast with the untreated mice, which died within two days.

These proteins can also have synergist effects when used with antibiotics.
For instance, a chimeric enzybiotic consisting of the N-terminal catalytic domain of the P16 endolysin and the C-terminal cell wall binding domain of the virion protein P17 of S. aureus phage P68 was constructed. The recombinant protein was soluble and exerted antimicrobial activity against S. aureus strains, including MRSA when used alone or in combination with antibiotics (Manoharadas et al., 2009). A chimeric enzybiotic (ClyS) encompassing the N-terminal domain of the endolysin of the S. aureus phage Twort and the C-terminal cell wall-targeting domain of the endolysin of S. aureus phage phiNM3, displayed antimicrobial activity against MRSA and VISA strains. ClyS also showed synergistic antimicrobial activity with both vancomycin and oxacillin in vitro, thereby reducing the effective concentration of the drug (Daniel et al., 2010).

In one study reported the construction of four chimeric phage endolysins (Cpl-711, Cpl-771, Cpl-117, Cpl-177) by making combinations of the structural elements of the two pneumococcal phage endolysins Cp1 and Cpl-7s (Diez-Martinez et al., 2015). The phage endolysin, Cpl-711 displayed significant bactericidal activity against *S. pneumoniae*. The chimeric lysin Cpl-711 was composed of the catalytic domain of the Cpl-7s, linker and cell wall binding domain of CPI-1. At a working concentration of 1 µg/ml, CPI-711 reduced the viable cell count of *S. pneumonia* by four orders of magnitude in two hours. *In vivo* experiments with mice infected intraperitoneally with *S. pneumoniae* followed by treatment with CP1-711 was shown to have a 50% greater survival than those treated with CPI-1. This study shows the potential of making endolysins using genetic engineering. Ever since the data emerged about the scope of engineered phage endolysins as antimicrobial agents, similar work has been reported on the activity of phage endolysins against *Listeria* spp. (Scmelcher et al., 2011), *Streptococcus* spp. (Dong et al., 2015) and *Staphylococcus* spp. (Fernandes et al., 2012).

## Conclusion

The described studies with recombinant phage hold promise that phage based antimicrobials can be specifically tailored for a pathogen of interest. An inherent benefit of targeted killing of pathogens by recombinant phage may not only be the reduction of side effects observed with wild-type phage or antibiotics but the preservation of beneficial gut commensals that are vital for immune development and function. A major hurdle in using recombinant phages as antimicrobials in systemic infections will presumably be restrictions imposed by regulatory agencies. To further develop the avenue of "recombinant phage therapy" an increased number of successful studies using experimental infection models will be necessary to reduce concerns. In contrast to the exploitation of recombinant phage, phage derived products such as enzybiotics or biofilm degrading enzymes may encounter fewer obstacles towards certification. Apart from medical applications, we are confident that phage will find acceptance as biocontrol agents and bio detectors in the food industry.

## Future trends

The rapid progress in understanding complex cellular machines, regulatory events and gene networks in bacterial pathogens at the molecular level will continue to deliver many "Achilles heels" in bacterial pathogens. Specific targeting and delivery of chemical or "proteinaceous agents" to distinct pathogens by recombinant phage could provide a means to exploit this knowledge for therapy. Fast high throughput sequencing on the one hand together with the possibility to synthesize whole genomes on the other will open unprecedented opportunities to design phage-based antimicrobials with high selectivity and potency to kill a bacterial target. In addition, *de novo* assembly of phage genomes may entail the creation of phage with an extended host range. Similarly, phage derived enzybiotics can be further developed towards a broader spectrum of pathogens. Moreover, future developments may include optimized synthesis and activity of phage-encoded EPS degrading enzymes as well as the exploitation of different activities that target distinct components of the EPS such as carbohydrates, proteins or lipids.

## Web Resources

1. http://www.rockefeller.edu/vaf/media/bac1a.html
2. http://www.rockefeller.edu/vaf/media/bac2a.html
3. http://www.accessdata.fda.gov/scripts/fcn/gras_notices/701456A.PDF

## References

Abedon, S.T., Kuhl, S.J., Blasdel, B.G., and Kutter, E.M. (2011). Phage treatment of human infections. Bacteriophage. *1*, 66-85.

Almeida, C., Cerqueira, L., Azevedo, N.F., and Vieira, M.J. (2013). Detection of Salmonella enterica serovar Enteritidis using realtime PCR, immune capture assay, PNA FISH and standard culture methods in different types of food samples. Int. J. Food. Microbiol. 161, 16–22.

Ando, H. Lemire, S., Pires, D.P., and Lu, T.K. (2015). Engineering modular viral scaffolds for targeted bacterial population editing. Cell. Syst. 1,187–196.

Bar, H., Yacoby, I., and Benhar, I. (2008). Killing cancer cells by targeted drug carrying phage nano medicines. BMC Biotechnol. 8, 37.

Barrangou, R., Fremaux, C., Deveau, H., Richards, M., Boyaval, P., Moineau, S., Romero, D.A., and Horvath, P. (2007). CRISPR provides acquired resistance against viruses in prokaryotes. Science.315,1709–1712.

Bartual, S.G., Garcia-Doval, C., Alonso, J., Schoehn, G., and van Raaij, M.J. (2010). Two-chaperone assisted soluble expression and purification of the bacteriophage T4 long tail fibre protein gp37. Protein Expr. Purif. *70*, 116-121.

Bielke, L.R., Higgins, S.E., Donoghue, A.M., Donoghue, D.J., Hargis, B.M., and Tellez, G. (2007). Use of wide-host-range bacteriophages to reduce *Salmonella* on poultry products. Int. J. Poult. Sci. *6*, 754-757.

Box, A. M., Mcguffie, M. J., O'hara, B. J., and Seed, K. D. (2016). Functional analysis of bacteriophage immunity through a type I-E CRISPR-Cas system in *Vibrio cholerae* and its application in bacteriophage genome engineering. J. Bacteriol. 198, 578–590.

Briers, Y., Walmagh, M., Van, Puyenbroeck, V., Cornelissen, A., Cenens, W., Aertsen, A., Oliveira, H., Azeredo, J., Verween, G., Pirnay, J-P., Miller, S., Volckaert, G., and Lavigne, R. (2014). Engineered endolysin-based "Artilysins" to combat multidrug-resistant Gram-negative pathogens. mBio 5, e01379-14.

Brüssow, H. (2013). Bacteriophage-host interaction: from splendid isolation into a messy reality. Curr. Opin. Microbiol. *In press.* doi:pii: S1369-5274(13)00064-7. 10.1016/j.mib.2013.04.007.

Burrowes, B., Harper, D.R., Anderson, J., McConville, M., and Enright, M.C. (2011). Bacteriophage therapy: potential uses in the control of antibiotic-resistant pathogens. Expert Rev Anti Infect. Ther. *9*, 775-785.

Cao, J., Sun, Y., Berglindh, T., Mellgård, B., Li, Z., Mårdh, B., and Mårdh, S. (2000). *Helicobacter pylori*-antigen-binding fragments expressed on the filamentous M13 phage prevent bacterial growth. Biochim. Biophys. Acta. *6*, 107-113.

Cao, B., Yang, M., and Mao, C. (2016). Phage as a genetically modifiable supramacromolecule in chemistry, materials and medicine. Acc. Chem. Res. 49, 1111–1120. doi: 10.1021/acs.accounts.5b00557.

Cao, B., Li, Y., Yang, T., Bao, Q., Yang, M., and Mao, C. (2018). Bacteriophage-based biomaterials for tissue regeneration. Adv. Drug Deliv. Rev. pii: S0169-409X(18)30291-6. doi: 10.1016/j.addr.2018.11.004

Catalão, M.J., Gil, F., Moniz-Pereira, J., São-José, C., and Pimentel, M. (2013). Diversity in bacterial lysis systems: bacteriophages show the way. FEMS Microbiol Rev. *37*, 554-71.

Chan, L.Y., Kosuri, S.,and Endy, D. (2005). Refactoring bacteriophage T7. Mol. Syst. Biol. 1, 2005.0018.

Chanishvili, N. (2012). Phage therapy--history from Twort and d'Herelle through Soviet experience to current approaches. Adv Virus Res. 2012; 83:3-40.

Chen, J., Wang, M., Xi, B., Xue, J., He, D., Zhang, J., and Zhao, Y. (2012). SPARC is a key regulator of proliferation, apoptosis and invasion in human ovarian cancer. PLoS One 7:e42413.

Cheng, Q., Nelson, D., Zhu, S., and Fischetti, V.A. (2005). Removal of group B *Streptococci* colonizing the vagina and oropharynx of mice with a bacteriophage lytic enzyme. Antimicrob. Agents Chemother. *49*, 111-117.

Cheng, G., Hao, H., Dai, M., Liu, Z., and Yuan, Z. (2013). Antibacterial action of quinolones: From target to network. Eur J Med Chem. *In press.* doi:pii: S0223-5234(13)00125-6. 10.1016/j.ejmech.2013.01.057.

Chiang, S.M, and Schellhorn, H.E. (2012). Regulators of oxidative stress response genes in *Escherichia coli* and their functional conservation in bacteria. Arch Biochem Biophys. *525*, 161-169.

Click, E.M., and Webster, R.E. (1998). The TolQRA proteins are required for membrane insertion of the major capsid protein of the filamentous phage f1 during infection. J. Bacteriol. *180*, 1723-1728.

Costerton, J.W., Stewart, P.S., and Greenberg, E.P. (1999). Bacterial Biofilms: A common cause of persistent infections. Science. ***284***, 1318.

Costerton, W., Veeh, R., Shirtliff, M., Pasmore, M., Post, C., and Ehrlich, G. (2003). The application of biofilm science to the study and control of chronic bacterial infections. J. Clin. Invest. *112*, 1466-1477.

Da Re, S., Le Quéré, B., Ghigo, J.M., and Beloin, C. (2007). Tight modulation of *Escherichia coli* bacterial biofilm formation through controlled expression of adhesion factors. Appl. Environ. Microbiol. *73*, 3391-3403.

Daniel, A., Euler, C., Collin, M., Chahales, P., Gorelick, K.J., and Fischetti, V.A. (2010). Synergism between a novel chimeric lysin and oxacillin protects against infection by methicillin-resistant *Staphylococcus aureus.* Antimicrob. Agents Chemother. *54,* 1603-1612.

Davis, B.M., and Waldor, M.K. (2003). Filamentous phages linked to virulence of *Vibrio cholerae.* Curr. Opin. Microbiol. 6, 35-42.

de Crombrugghe, B., Busby, S., and Buc, H. (1984). Cyclic AMP receptor protein: role in transcription activation. Science. *224*, 831-838.

Debarbieux, L., Leduc, D., Maura, D., Morello, E., Criscuolo, A., Grossi, O., Balloy, V., and Touqui, L. (2010). Bacteriophages can treat and prevent *Pseudomonas aeruginosa* lung infections. J. Infect. Dis. *201*, 1096-1104.

Deveau, H., Garneau, J.E., and Moineau, S. (2010). CRISPR/Cas system and its role in phage-bacteria interactions. Annu. Rev. Microbiol. 64,475–493.

Díez-Martínez, R., De Paz, H.D., García-Fernández, E., Bustamante, N., Euler, C.W., Fischetti, V.A., Menendez, M., and García, P. (2015). A novel chimeric phage lysin with high in vitro and in vivo bactericidal activity against *Streptococcus pneumonia*. J.Antimicrob.Chemother.70, 1763–1773.

Dong, Q., Wang, J., Yang, H., Wei, C., Yu, J., Zhang, Y., Huang, Y., Zhang, X-E., and Wei, H. (2015). Construction of a chimeric lysin Ply187N-V12C with extended lytic activity against staphylococci and streptococci. Microb. Biotechnol. 8, 210–220.

Donlan, R.M. (2009). Preventing biofilms of clinically relevant organisms using bacteriophage. Trends. Microbiol. *17*, 66-72.

Doolittle, M.M., Cooney, J.J., and Caldwell, D.E. (1995). Lytic infection of *Escherichia coli* biofilms by bacteriophage T4. Can. J. Microbiol. *41*, 12-18.

Doolittle, M.M., Cooney, J.J., and Caldwell, D.E. (1996). Tracing the interaction of bacteriophage with bacterial biofilms using fluorescent and chromogenic probes. J. Ind. Microbiol. *16*, 331-341.

Du, B., Han, H., Wang, Z., Kuang,L., Wang, L., Yu, L., Wu, M., Zhou, Z., and Qian, M. (2010). Targeted drug delivery to hepatocarcinoma in vivo by phage displayed specific binding peptide. Mol.Cancer.Res.8, 135–144.

Edgar, R., McKinstry, M., Hwang, J., Oppenheim, A.B., Fekete, R.A., Giulian, G., Merril, C., Nagashima, K., and Adhya, S. (2006). High-sensitivity bacterial detection using biotin-tagged phage and quantum-dot nanocomplexes. Proc. Natl. Acad. Sci. U. S. A. *103*, 4841-4845.

Edgar, R., Friedman, N., Molshanski-Mor S.,and Qimron, U. (2012). Reversing bacterial resistance to antibiotics by phage-mediated delivery of dominant sensitive genes. Appl. Environ. Microbiol. 78, 744–751.

Fagbohun, O.A., Kazmierczak, R.A., Petrenko, V.A., and Eisenstark, A. (2013). Metastatic prostate cancer cell-specific phage-like particles as a targeted gene-delivery system. J. Nanobiotechnol. 11, 31.

Fenton, M., Casey, P.G., Hill, C., Gahan, C.G., Ross, R.P., McAuliffe, O., O'Mahony, J., Maher, F., and Coffey, A. (2010). The truncated phage lysin CHAP(k) eliminates *Staphylococcus aureus* in the nares of mice. Bioeng. Bugs. 1, 404-407.

Fenton, M., Keary, R., McAuliffe, O., Ross, R.P., O'Mahony, J., and Coffey, A. (2013). Bacteriophage-Derived Peptidase CHAP(K) Eliminates and Prevents Staphylococcal Biofilms. Int. J. Microbiol. 625341. doi: 10.1155/2013/625341.

Fernandes, S., Proença, D., Cantante, C., Silva, F.A., Leandro, C., Lourenço, S., Milheiriço, C., de Lencastre, H., Cavaco-Silva, P., Pimentel, M., and São-José, C. (2012). Novel chimerical endolysins with broad antimicrobial activity against methicillin-resistant *Staphylococcus aureus*. Microb. Drug. Resist. 18, 333–343.

Fischetti, V.A. (2008). Bacteriophage lysins as effective antibacterials. Curr Opin Microbiol. 11,393–400.

Fischetti, V.A. (2010). Bacteriophage endolysins: a novel anti-infective to control Gram-positive pathogens. Int. J. Med. Microbiol. *300*, 357-362.

Fischetti,V.A. (2011).Exploiting what phage have evolved to control gram positive pathogens. Bacteriophage. 1, 188–194.

Frenkel, D., and Solomon, B. (2002). Filamentous phage as vector-mediated antibody delivery to the brain. Proc. Natl. Acad. Sci. USA. 99, 5675–5679.

Friedberg, E.C. (2003). DNA damage and repair. Nature. 421,436–440.

García, L.R., and Molineux, I.J. (1995). Incomplete entry of bacteriophage T7 DNA into F plasmid-containing *Escherichia coli*. J. Bacteriol. *177*, 4077-4083.

Gerdes, K., Bech, F.W., Jørgensen, S.T., Løbner-Olesen, A., Rasmussen, P.B., Atlung, T., Boe, L., Karlstrom, O., Molin, S., and von Meyenburg, K. (1986). Mechanism of postsegregational killing by the hok gene product of the parB system of plasmid R1 and its homology with the relF gene product of the *E. coli* relB operon. EMBO J. *5*, 2023-2039.

Ghigo, J.M. (2001). Natural conjugative plasmids induce bacterial biofilm development. Nature. *412*, 442-445.

Ghosh, D., Kohli., A.G., Moser, F., Endy, D., and Belcher, A.M. (2012). Refactored M13 bacteriophage as a platform for tumor cell imaging and drug delivery. ACS Synth. Biol. 1, 576–582.

Gilmer, D.B., Schmitz, J.E., Euler, C.W., and Fischetti, V.A. (2013). Novel Bacteriophage Lysin with Broad Lytic Activity Protects against Mixed Infection by Streptococcus pyogenes and Methicillin-Resistant *Staphylococcus aureus*. Antimicrob. Agents Chemother. *57*, 2743-50.

Guenther, S., Huwyler, D., Richard, S., and Loessner, M.J. (2009). Virulent bacteriophage for efficient biocontrol of *Listeria monocytogenes* in ready-to-eat foods. Appl. Environ. Microbiol. *75*, 93-100.

Hagens, S., and Bläsi, U. (2003). Genetically modified filamentous phage as bactericidal agents: a pilot study. Lett. Appl. Microbiol. *37,* 318-23.

Hagens, S., Habel, A., von Ahsen, U., von Gabain, A., and Bläsi, U. (2004). Therapy of experimental *Pseudomonas* infections with a nonreplicating genetically modified phage. Antimicrob. Agents Chemother. *48*, 3817-3822.

Hagens, S., Habel, A., and Bläsi, U. (2006). Augmentation of the antimicrobial efficacy of antibiotics by filamentous phage. Microb. Drug Resist. *12*, 164-168.

Hanlon, G.W., Denyer, S.P., Olliff, C.J., and Ibrahim, L.J. (2001). Reduction in exopolysaccharide viscosity as an aid to bacteriophage penetration through *Pseudomonas aeruginosa* biofilms. *67*, 2746-2753.

Harper, D.R., Anderson, J., and Enright, M.C. (2011). Phage therapy: delivering on the promise. Ther. Deliv. *2*, 935-947.

Heilpern, A.J., and Waldor, M.K. (2003). pIIICTX, a predicted CTXphi minor coat protein, can expand the host range of coliphage fd to include *Vibrio cholerae*. J Bacteriol. *185*, 1037-1044.

Hermoso, J.A., García, J.L., and García, P. (2007). Taking aim on bacterial pathogens: from phage therapy to enzybiotics. Curr. Opin. Microbiol. *10*, 461-472.

Hidalgo, E., Ding, H., and Demple, B. (1997). Redox signal transduction: Mutations shifting [2Fe-2S] centers of the SoxR sensor-regulator to the oxidized form. Cell. *88*, 121-129.

Hirai, K., Aoyama, H., Irikura, T., Iyobe, S., and Mitsuhashi, S. (1986). Differences in susceptibility to quinolones of outer membrane mutants of *Salmonella typhimurium* and *Escherichia coli*. Antimicrob. Agents Chemother. *29*, 535-538.

Hoffman, L.R., D'Argenio, D.A., MacCoss, M.J., Zhang, Z., Jones, R.A., and Miller, S.I. (2005). Aminoglycoside antibiotics induce bacterial biofilm formation. Nature. *436*, 1171-1175.

Huff, W.E., Huff, G.R., Rath, N.C., Balog, J.M., and Donoghue, A.M. (2003). Evaluation of aerosol spray and intramuscular injection of bacteriophage to treat an *Escherichia coli* respiratory infection. Poult. Sci. *82*, 1108-1112.

Itoh, Y., Wang, X., Hinnebusch, B.J., Preston, J.F. 3rd, and Romeo, T. (2005). Depolymerization of beta-1,6-N-acetyl-D-glucosamine disrupts the integrity of diverse bacterial biofilms. J. Bacteriol. *187*, 382-387.

Jackson, D.W., Suzuki, K., Oakford, L., Simecka, J.W., Hart, M.E., and Romeo, T. (2002). Biofilm formation and dispersal under the influence of the global regulator CsrA of *Escherichia coli.* J. Bacteriol. *184*, 290-301.

Janion, C. (2008). Inducible SOS response system of DNA repair and mutagenesis in Escherichia coli. Int. J. Biol. Sci. *4*, 338-44.

Kaplan, J.B. (2009). Therapeutic potential of biofilm-dispersing enzymes. Int. J. Artif. Organs. *32*, 545-554.

Kaplan, J.B. (2011). Antibiotic-induced biofilm formation. Int J Artif Organs. *34*, 737-751.

Kiro, R., Shitrit, D., and Qimron, U. (2014). Efficient engineering of a bacteriophage genome using the type I-E CRISPR-Cas system. RNA Biol. 11, 42– 44.

Kohanski, M.A., Dwyer, D.J., Hayete, B., Lawrence, C.A., and Collins, J.J. (2007). A common mechanism of cellular death induced by bactericidal antibiotics. Cell. *130*, 797–810.

Kumar, A., Grover, S., and Kumar Batish, V. (2013). Application of multiplex PCR assay based on uidR and fliCH7 genes for detection of *Escherichia coli* O157:H7 in milk. J. Gen. Appl. Microbiol. *59*, 11-19.

Kutateladze, M., and Adamia, R. (2010). Bacteriophages as potential new therapeutics to replace or supplement antibiotics. Trends Biotechnol. *28*, 591-595.

Lemire, S., Yehl, K. M., and Lu, T. K. (2018). Phage-based applications in synthetic biology. Annu. Rev. Virol. 5, 453–476. doi: 10.1146/annurev-virology-092917-043544.

Lewin, C.S., Howard, B.M., Ratcliffe, N.T., and Smith, J.T. (1989). 4-Quinolones and the SOS response. J. Med. Microbiol. *29*,139-144.

Loeffler, J.M., Nelson, D., and Fischetti, V.A. (2001). Rapid killing of *Streptococcus pneumoniae* with a bacteriophage cell wall hydrolase. Science. *294*, 2170-2172.

Loessner, M., Rees, C., Stewart, G., and Scherer, S. (1996). Construction of luciferase reporter bacteriophage A511: luxAB for rapid and sensitive detection of viable *Listeria* cells. Appl. Environ. Microbiol. 62, 1133–1140.

Loessner, M.J., Rudolf, M., and Scherer, S. (1997). Evaluation of luciferase reporter bacteriophage A511::luxAB for detection of *Listeria monocytogenes* in contaminated foods. Appl. Environ. Microbiol. *63*, 2961-2965.

Loessner, M.J. (2005). Bacteriophage endolysins—current state of research and applications. Curr. Opin. Microbiol. 8,480–487.

Lood, R., Winer, B.Y., Pelzek, A.J., Diez-Martinez, R., Thandar, M., Euler, C.W., Schuch, R., Fischetti, V.A. (2015). Novel phage lysin capable of killing the multidrug-resistant Gram-negative bacterium *Acinetobacter baumannii* in a mouse bacteremia model. Antimicrob. Agents. Chemother. 59, 1983–1991.

Lopata, M., Schlieper, D., von Wilcken-Bergmann, B., and Müller-Hill, B. (1997). A lethal mutant of the catabolite gene activator protein CAP of *Escherichia coli.* Biol. Chem. *378*, 1153-1162.

Lu, T.K., and Collins, J.J. (2007). Dispersing biofilms with engineered enzymatic bacteriophage. Proc. Natl. Acad. Sci. U. S. A. *104*, 11197-11202.

Lu, T.K., and Collins, J.J. (2009). Engineered bacteriophage targeting gene networks as adjuvants for antibiotic therapy. Proc. Natl. Acad. Sci. U. S. A. *106*, 4629-4634.

Lu, T.K, Bowers, J., and Koeris, M.S. (2013). Advancing bacteriophage-based microbial diagnostics with synthetic biology. Trends.Biotechnol.31, 325– 327.

Lu, T.K., Koeris, M.S., Chevalier, B., Holder, J., McKenzie, G., and Brownell, D. (May 2013). Recombinant phage and methods. US patent 13/627,060.

Mahendran, K.R., Hajjar, E., Mach, T., Lovelle, M., Kumar, A., Sousa, I., Spiga, E., Weingart, H., Gameiro, P., Winterhalter, M., and Ceccarelli, M. (2010). Molecular basis of enrofloxacin translocation through OmpF, an outer membrane channel of *Escherichia coli*--when binding does not imply translocation. J Phys Chem B. *114*, 5170-5179.

Mahichi, F., Synnott, A.J., Yamamichi, K., Osada, T., and Tanji, Y. (2009). Site-specific recombination of T2 phage using IP008 long tail fiber genes provides a targeted method for expanding host range while retaining lytic activity. FEMS Microbiol. Lett. *295*, 211-217.

Manoharadas, S., Witte, A., and Bläsi, U. (2009). Antimicrobial activity of a chimeric enzybiotic towards *Staphylococcus aureus*. J. Biotechnol. *139*, 118-23.

Marinelli, L. J., Piuri, M., Swigonova, Z., Balachandran, A., Oldfield, L. M., and Van Kessel, J. C. (2008). BRED: a simple and powerful tool for constructing mutant and recombinant bacteriophage genomes. PLoS One 3:e3957.

Martel, B., and Moineau, S. (2014). CRISPR-Cas: an efficient tool for genome engineering of virulent bacteriophages. Nucleic Acids Res.42, 9504–9513.

Marvin, D.A., Welsh, L.C., Symmons, M.F., Scott, W.R., and Straus, S.K. (2006). Molecular structure of fd (f1, M13) filamentous bacteriophage refined with respect to X-ray fibre diffraction and solid-state NMR data supports specific models of phage assembly at the bacterial membrane. J. Mol. Biol. 355, 294-309.

Marzari, R., Sblattero, D., Righi, M., and Bradbury, A. (1997). Extending filamentous phage host range by the grafting of a heterologous receptor binding domain. Gene. *185*, 27-33.

Matsuda, T., Freeman, T.A., Hilbert, D.W., Duff, M., Fuortes, M., Stapleton, P.P., and Daly, J.M. (2005). Lysis-deficient bacteriophage therapy decreases endotoxin and inflammatory mediator release and improves survival in a murine peritonitis model. Surgery. *137*, 639-646.

Meyer AJ, Ellefson JW, Ellington AD. 2015. Directed evolution of a panel of orthogonal T7 RNA polymerase variants for in vivo or in vitro synthetic circuitry. ACS Synth. Biol 4:1070–76.

Miller, C., Thomsen, L.E., Gaggero, C., Mosseri, R., Ingmer, H., and Cohen, S.N. (2004). SOS response induction by beta-lactams and bacterial defense against antibiotic lethality. Science. *305*, 1629-1631.

Miyake, H., Hara, I., and Eto, H. (2004). Serum level of cathepsin B and its density in men with prostate cancer as novel markers of disease progression. Anticancer Res. 24, 2573–2577.

Mohamed, M.M., and Sloane, B.F. (2006). Cysteine cathepsins: multifunctional enzymes in cancer. Nat.Rev.Cancer.6, 764–775.

Moradpour, Z., Sepehrizadeh, Z., Rahbarizadeh, F., Ghasemian, A., Yazdi, M.T, and Shahverdi, A.R. (2009). Genetically engineered phage harbouring the lethal catabolite gene activator protein gene with an inducer-independent promoter for biocontrol of *Escherichia coli.* FEMS Microbiol. Lett. *296*, 67-71.

Moradpour, Z., and Ghasemian, A. (2011). Modified phages: novel antimicrobial agents to combat infectious diseases. Biotechnol Adv. *29*, 732-738.

Morello, E., Saussereau, E., Maura, D., Huerre, M., Touqui, L., and Debarbieux, L. (2011). Pulmonary bacteriophage therapy on *Pseudomonas aeruginosa* cystic fibrosis strains: first steps towards treatment and prevention. PLoS One. *15*, e16963.

Murphy, M.P., and LeVine, H. (2010). Alzheimer's disease and the amyloid-beta peptide. J Alzheimers. Dis. 19,311–323.

Murphy, K. C. (2012). Phage recombinases and their applications. Adv. Virus Res. 83, 367–414.

Musso, R.E., Di Lauro, R., Adhya, S., and de Crombrugghe, B. (1977). Dual control for transcription of the galactose operon by cyclic AMP and its receptor protein at two interspersed promoters. Cell. *12*, 847-854.

Namura, M., Hijikata, T., Miyanaga, K., and Tanji, Y. (2008). Detection of Escherichia coli with fluorescent labeled phages that have a broad host range to E. coli in sewage water. Biotechnol. Prog. 24, 481–486.

Nelson, D., Loomis, L., and Fischetti, V.A. (2001). Prevention and elimination of upper respiratory colonization of mice by group A *Streptococci* by using a bacteriophage lytic enzyme. Proc. Natl. Acad. Sci. U. S. A. *98*, 4107-4112.

Newell, D.G., Koopmans, M., Verhoef, L., Duizer, E., Aidara-Kane, A., Sprong, H., Opsteegh, M., Langelaar, M., Threfall, J., Scheutz, F., van der Giessen, J., and Kruse, H. (2010). Food-borne diseases - the challenges of 20 years ago still persist while new ones continue to emerge. Int. J. Food Microbiol. *139*, 3-15.

Oda, M., Morita, M., Unno, H., and Tanji, Y. (2004). Rapid detection of Escherichia coli O157:H7 by using green fluorescent protein-labeled PP01 bacteriophage. Appl. Environ. Microbiol. 70, 527–534.

O'Flaherty, S., Ross, R.P., Meaney, W., Fitzgerald, G.F., Elbreki, M.F., and Coffey, A. (2005). Potential of the polyvalent anti-*Staphylococcus* bacteriophage K for control of antibiotic resistant *Staphylococci* from hospitals. Appl. Environ. Microbiol. *71*, 1836-1842.

O'Flynn, G., Ross, R.P., Fitzgerald, G.F., and Coffey, A., (2004). Evaluation of a cocktail of three bacteriophages for biocontrol of *Escherichia coli* O157:H7. Appl. Environ. Microbiol. *70*, 3417-3424.

Oliver, D.B., and Crowther, R.A. (1981). DNA sequence of the tail fibre genes 36 and 37 of bacteriophage T4. J. Mol. Biol. *153*, 545-568.

Pande, J., Szewczyk, M.M., and Grover, A.K. (2010). Phage display: concept, innovations, applications and future. Biotechnol.Adv.28, 849–858.

Park, S.C., and Nakai, T. (2003). Bacteriophage control of *Pseudomonas plecoglossicida* infection in ayu *Plecoglossus altivelis*. Dis. Aquat. Organ. *53,* 33-39.

Paul, V.D., Sundarrajan, S., Rajagopalan, S.S., Hariharan, S., Kempashanaiah, N., Padmanabhan, S., Sriram, B., and Ramachandran, J. (2011). Lysis-deficient phages as novel therapeutic agents for controlling bacterial infection. BMC Microbiol. 11:195. doi: 10.1186/1471-2180-11-195.

Piuri, M., Jacobs, W.R., Jr, and Hatfull, G.F. (2009). Fluoromycobacteriophages for rapid, specific, and sensitive antibiotic susceptibility testing of *Mycobacterium tuberculosis*. PLoS One. *4*, e4870.

Poteete, A. R. (2001). What makes the bacteriophage lambda red system useful for genetic engineering: molecular mechanism and biological function. FEMS Microbiol. Lett. 201, 9–14.

Pouillot, F., Blois, H., and Iris, F. (2010). Genetically engineered virulent phage banks in the detection and control of emergent pathogenic bacteria. Biosecur. Bioterror. *8*, 155-169.

Rao, V. B., and Mitchell, M. S. (2001). The N-terminal ATPase site in the large terminase protein gp17 is critically required for DNA packaging in bacteriophage T4. J. Mol. Biol. 314, 401–411.

Ren, Z.J., Tian, C.J., Zhu, Q.S., Zhao, M.Y., Xin, A.G., Nie, W.X., Ling, S.R., Zhu, M.W., Wu, J.Y., Lan, H.Y., Cao, Y.C., and Bi, Y.Z. (2008). Orally delivered foot-and mouth disease virus capsid protomer vaccine displayed on T4 bacteriophage surface: 100% protection from potency challenge in mice. Vaccine. 26, 1471–1481.

Rhoads, D.D., Wolcott, R.D., Kuskowski, M.A., Wolcott, B.M., Ward, L.S., and Sulakvelidze, A. (2009). Bacteriophage therapy of venous leg ulcers in humans: results of a phase I safety trial. J.Wound Care.18, 237–238,240–243.

Riede, I., Drexler, K., Schwarz, H., and Henning, U. (1987). T-even-type bacteriophages use an adhesin for recognition of cellular receptors. J. Mol. Biol. *194*, 23-30.

Riesenfeld, C., Everett, M., Piddock, L.J.,and Hall, B.G. (1997). Adaptive mutations produce resistance to ciprofloxacin. Antimicrob Agents Chemother. 41, 2059–2060.

Rietsch, A., and Bläsi, U. (1993). Non-specific hole formation in the *Escherichia coli* inner membrane by lambda S proteins in independent of cellular secY and secA functions and of the proportion of membrane acidic phospholipids. FEMS Microbiol. Lett. *107*, 101-105.

Romeo, T., Vakulskas, C.A., and Babitzke, P. (2013). Post-transcriptional regulation on a global scale: form and function of Csr/Rsm systems. Environ Microbiol. 15, 313-324.

Rondón, L., Piuri, M., Jacobs, W.R. Jr, de Waard, J., Hatfull, G.F., and Takiff, H.E. (2011). Evaluation of fluoromycobacteriophages for detecting drug resistance in *Mycobacterium tuberculosis*. J. Clin. Microbiol. *49*, 1838-1842.

Römling, U., and Balsalobre, C. (2012). Biofilm infections, their resilience to therapy and innovative treatment strategies. J. Intern. Med. *272*, 541-561.

Rose, T., Verbeken, G., Vos, D.D., Merabishvili, M., Vaneechoutte, M., Lavigne, R., Jennes, S., Zizi, M., Pirnay, J-P. (2014). Experimental phage therapy of burn wound infection: difficult first steps. Int.J.Burns.Trauma. 4, 66–73.

Sader, H.S., Flamm, R.K., and Jones, R.N. (2013). antimicrobial activity of ceftaroline tested against staphylococci with reduced susceptibility to linezolid, daptomycin, or vancomycin from U.S. hospitals, 2008 to 2011. Antimicrob. Agents Chemother. *57*, 3178-3181.

Sarker, S.A., McCallin, S., Barretto, C., Berger, B., Pittet, A.C., Sultana, S., Krause, L., Huq, S, Bibiloni, R., Bruttin, A., Reuteler, G., and Brüssow, H. (2012). Oral T4-like phage cocktail application to healthy adult volunteers from Bangladesh. Virology *434*, 222-232.

Sarkis, G. J., Jacobs, W. R. Jr., and Hatfull, G. F. (1995). L5 luciferase reporter mycobacteriophages: a sensitive tool for the detection and assay of live mycobacteria. Mol. Microbiol. 15, 1055–1067.

Sathaliyawala, T., Rao, M., Maclean, D.M., Birx, D.L., Alving, C.R., Rao, V.B. (2006). Assembly of human immunodeficiency virus (HIV) antigens on bacteriophage T4: a novel in vitro approach to construct multicomponent HIV vaccines. J. Virol. 80, 7688–7698.

Shen, J., Zhou, J., Chen, G. Q., and Xiu, Z. L. (2018). Efficient genome engineering of a virulent klebsiella bacteriophage using CRISPR-Cas9. J. Virol. 92, pii: e00534-18.

Schentag, J.J., Hyatt, J.M., Carr, J.R., Paladino, J.A., Birmingham, M.C., Zimmer, G.S., and Cumbo, T.J. (1998). Genesis of methicillin-resistant *Staphylococcus aureus* (MRSA), how treatment of MRSA infections has selected for vancomycin-resistant *Enterococcus faecium*, and the importance of antibiotic management and infection control. Clin. Infect. Dis. *26*, 1204-1214.

Shmakov, S., Smargon, A., Scott, D., Cox, D., Pyzocha, N.,and Yan, W. (2017). Diversity and evolution of class 2 CRISPR-Cas systems. Nat. Rev. Microbiol. 15, 169–182.

Schmelcher, M., Tchang, V.S., and Loessner, M.J. (2011). Domain shuffling and module engineering of Listeria phage endolysins for enhanced lytic activity and binding affinity. Microb. Biotechnol. 4,651–662.

Shin, J., Jardine, P., and Noireaux, V. (2012).Genome replication, synthesis, and assembly of the bacteriophage T7 in a single cell-free reaction. ACS. Synth. Biol. 1,408–413.

Schito, G.C. (2006). The importance of the development of antibiotic resistance in *Staphylococcus aureus*. Clin. Microbiol. Infect. *12* Suppl 1, 3-8.

Schmelcher, M., Donovan, D.M., and Loessner, M.J. (2012). Bacteriophage endolysins as novel antimicrobials. Future Microbiol. 7, 1147–1171.

Schuch, R., Nelson, D., and Fischetti, V.A. (2002). A bacteriolytic agent that detects and kills *Bacillus anthracis*. Nature. *418*, 884-889.

Segarra-Newnham, M., and Church, T.J. (2012). Pharmacotherapy for methicillin-resistant Staphylococcus aureus nosocomial pneumonia. Ann. Pharmacother. *46*, 1678-1687.

Sieradzki, K., and Tomasz, A. (1997). Inhibition of cell wall turnover and autolysis by vancomycin in a highly vancomycin-resistant mutant of *Staphylococcus aureus*. J. Bacteriol. *179*, 2557-2566.

Sillankorva, S.M., Oliveira, H., and Azeredo, J. (2012). Bacteriophages and their role in food safety. Int. J. Microbiol. 2012:863945. doi: 10.1155/2012/863945.

Slopek, S., Durlakowa, I., Weber-Dabrowska, B., Kucharewicz-Krukowska, A., Dabrowski, M., and Bisikiewicz, R. (1983). Results of bacteriophage treatment of suppurative bacterial infections. II. Detailed evaluation of the results. Arch. Immunol. Ther. Exp. (Warsz). *31,* 293-327.

Simmons, M., Donovan, D.M., Siragusa, G.R., and Seal, B.S. (2010). Recombinant expression of two bacteriophage proteins that lyse clostridium perfringens and share identical sequences in the C-terminal cell wall binding domain of the molecules but are dissimilar in their N-terminal active domains. J. Agric. Food Chem. 58, 10330-7.

Sulakvelidze, A. (2013). Using lytic bacteriophages to prevent or significantly reduce contamination of food by foodborne bacterial pathogens. J. Sci. Food Agric. *In press.* doi: 10.1002/jsfa.6222.

Sun, T.P., and Webster, R.E. (1987). Nucleotide sequence of a gene cluster involved in entry of E colicins and single-stranded DNA of infecting filamentous bacteriophages into *Escherichia coli.* J. Bacteriol. *169*, 2667-2674.

Tai, I.T., and Tang, M.J. (2008). SPARC in cancer biology: its role in cancer progression and potential for therapy. Drug. Resist. Updat. 11,231–246.

Tanji, Y., Furukawa, C., Na, S.H., Hijikata, T., Miyanaga, K., and Unno, H. (2004). *Escherichia coli* detection by GFP-labeled lysozyme-inactivated T4 bacteriophage. J. Biotechnol. *114*, 11-20.

Tao, P., Zhu, J., Mahalingam, M., Batra, H., and Rao, V. B. (2018c). Bacteriophage T4 nanoparticles for vaccine delivery against infectious diseases. Adv. Drug Deliv. Rev. pii: S0169-409X(18)30164-9. doi: 10.1016/j.addr.2018.06.025.

Temme K, Hill R, Segall-Shapiro TH, Moser F, Voigt CA. 2012. Modular control of multiple pathways using engineered orthogonal T7 polymerases. Nucleic Acids Res. 40, 8773–8781.

Tétart, F., Repoila, F., Monod, C., and Krisch, H.M. (1996). Bacteriophage T4 host range is expanded by duplications of a small domain of the tail fiber adhesin. J. Mol. Biol. 258, 726-731.

Wang, T., Hartner, W.C., Gillespie, J.W., Praveen, K.P., Yang, S., Mei, L.A., Petrenko, V.A., and Torchilin, V.P. (2014). Enhanced tumor delivery and antitumor activity in vivo of liposomal doxorubicin modified with MCF-7specific phage fusion protein. Nanomedicine. 10, 421–430.

Watterson, S.A., and Drobniewski, F.A. (2000). Modern laboratory diagnosis of mycobacterial infections. J. Clin. Pathol. *53*, 727-732.

Weber-Dabrowska, B., Dabrowski, M., and Slopek, S. (1987). Studies on bacteriophage penetration in patients subjected to phage therapy. Archivum Immunologiae Et. Therapiae Experimentalis *35*, 563-568.

Westra, E.R., Buckling, A., and Fineran, P.C. (2014). CRISPR-Cas systems: beyond adaptive immunity. Nat. Rev. Microbiol. 12,317–326.

Woodford, N., Turton, J.F., and Livermore, D.M. (2011). Multiresistant Gram-negative bacteria: the role of high-risk clones in the dissemination of antibiotic resistance. FEMS Microbiol. Rev. *35*, 736-755.

Wright, A., Hawkins, C.H., Anggård, E.E., and Harper, D.R. (2009). A controlled clinical trial of a therapeutic bacteriophage preparation in chronic otitis due to antibiotic resistant *Pseudomonas aeruginosa;* a preliminary report of efficacy. Clin Otolaryngol. 34, 349-357.

Xavier, J.B., Picioreanu, C. Rani, S.A., van Loosdrecht, M.C., and Stewart, P.S. (2005). Biofilm-control strategies based on enzymic disruption of the extracellular polymeric substance matrix-a modelling study. Microbiology. *151*, 3817-3832.

Yacoby, I., Shamis, M., Bar, H., Shabat, D., and Benhar, I. (2006). Targeting antibacterial agents by using drug-carrying filamentous bacteriophages. Antimicrob. Agents Chemother. *50*, 2087-2097.

Yoong, P., Schuch, R., Nelson, D., and Fischetti, V.A. (2004). Identification of a broadly active phage lytic enzyme with lethal activity against antibiotic-resistant *Enterococcus faecalis* and *Enterococcus faecium*. J. Bacteriol. *186*, 4808-4812.

Young, R. (1992). Bacteriophage lysis: mechanism and regulation. Microbiol. Rev. *56*, 430-481.

Zhang, Y., Zhu, L., Zhang, J., and Inouye, M. (2005). Characterization of ChpBK, an mRNA interferase from *Escherichia coli*. J. Biol. Chem. *280*, 26080-26088.

Zimmer, M., Vukov, N., Scherer, S., and Loessner, M.J. (2002). The murein hydrolase of the bacteriophage phi3626 dual lysis system is active against all tested *Clostridium perfringens* strains. Appl. Environ. Microbiol. *68*, 5311-5317.

# Chapter 17

# Bacteriophage Encapsulation Using Spray Drying for Phage Therapy

**Danish J. Malik**

Chemical Engineering Department, Loughborough University, Loughborough LE11 3TU, United Kingdom

Email: d.j.malik@lboro.ac.uk

DOI: https://doi.org/10.21775/9781913652517.17

**Abstract**

Exploiting the potential of bacteriophages for phage therapy is an exciting future prospect. However, in order to be successful, there is a pressing need for the manufacture of safe and efficacious phage drug products to treat patients. Scalable manufacture of phage biologics as a stable solid dry powder form is highly desirable and achievable using the process of spray drying. Spray drying of purified phage suspensions formulated with suitable excipients can be carried out in a single step with high process throughput and at relatively low cost. The resulting phage-containing powders can possess good storage shelf-life. The process allows control over the final phage dose in the powder and production of microparticles suitable for a variety of therapeutic uses. Spray dried powders may include different polymer formulations employing a multitude of different triggers for phage release at the target site including pH, enzymes, virulence factors etc. The activity of the phages in spray dried powders is adversely affected during spray drying due to desiccation and thermal stresses which need to be controlled. The choice of polymers, excipients and moisture content of the dry powders affects the material glass transition temperature and the stability of the phages during storage. The storage temperature and storage humidity are important factors affecting the stability of the phages in the dry powders. A quality by design (QbD) approach for phage drug product development needs to identify drug product characteristics that are critical to quality from the patient's perspective and translates them into the critical quality attributes (CQA) of the drug product. The relationship between the phage drug product CQAs and formulation development and spray drying process conditions are discussed in this chapter.

## Introduction

Phage drug products need careful formulation development and an appreciation of the chemical and physical stresses that bacteriophages (phages) may encounter both during processing as well as during storage once formulated. Phage inactivation and long-term reduction in phage titres upon storage are highly undesirable. Delivery of high titres of phages and their controlled release at the site of treatment affects phage pharmacokinetics (Malik et al., 2017). The physical and chemical properties of the materials used in the formulation need careful consideration when selecting a technique for encapsulation. Many techniques and processes may be used for stabilising, immobilising and encapsulating phages including spray-drying, spray freeze drying, freeze drying, extrusion dripping methods, emulsion and polymerisation techniques. (Malik et al., 2017). Phages are protein structures, and they are therefore susceptible to factors known to denature proteins; and these include exposure to organic solvents (Lee and Belcher, 2004; Puapermpoonsiri et al., 2009), high temperatures (Briers et al., 2008), pH (Briers et al., 2008; Knezevic et al., 2011), ionic strength (Knezevic et al., 2011), and, interfacial effects. Additionally, mechanical stresses during formulation or encapsulation including shear stresses during mixing and agitation, atomisation during spraying (Leung et al., 2016) and, desiccation stresses during drying (Dini and de Urraza, 2013), need careful consideration. Scalable, low cost GMP compliant processes for phage encapsulation are needed.

## Spray drying of phages

Spray drying processes atomise a liquid containing dissolved solids, converting it into a fine mist or aerosol which is contacted with a hot dry gas (typically hot dry air) inside a drying chamber (Figure 1). Due to the high surface area to volume ratio of the very small atomised droplets (droplet size typically 10 – 100 µm), the solvent rapidly evaporates with each droplet forming a particle comprising the non-volatile components (solids fraction in feed containing phage and excipients) along with a small amount of residual moisture. The particles are separated from the airstream exiting the dryer, and cyclones or bag filters are typically used for this. Spray drying is a scalable industrial process technology and is widely used to produce fine powders containing biologicals for pharmaceutical applications including pulmonary delivery via dry powder inhalers (DPI) (Hoe et al., 2014) (Figure 2). Dry powders are increasingly being considered for phage drug products because they show relatively long storage stability without requiring a cold supply chain and refrigeration (Carrigy et al., 2019; Chew and Chan, 2002; Klingler et al., 2009). Spray drying phages using formulations containing synthetic polymers may allow encapsulation, targeted delivery and controlled release aspects such as triggered release in response to pH changes, e.g. for gastrointestinal delivery (Vinner et al., 2019). The powders may also be filled into capsules or compressed into tablets for oral solid dosage (Vinner et al., 2019).

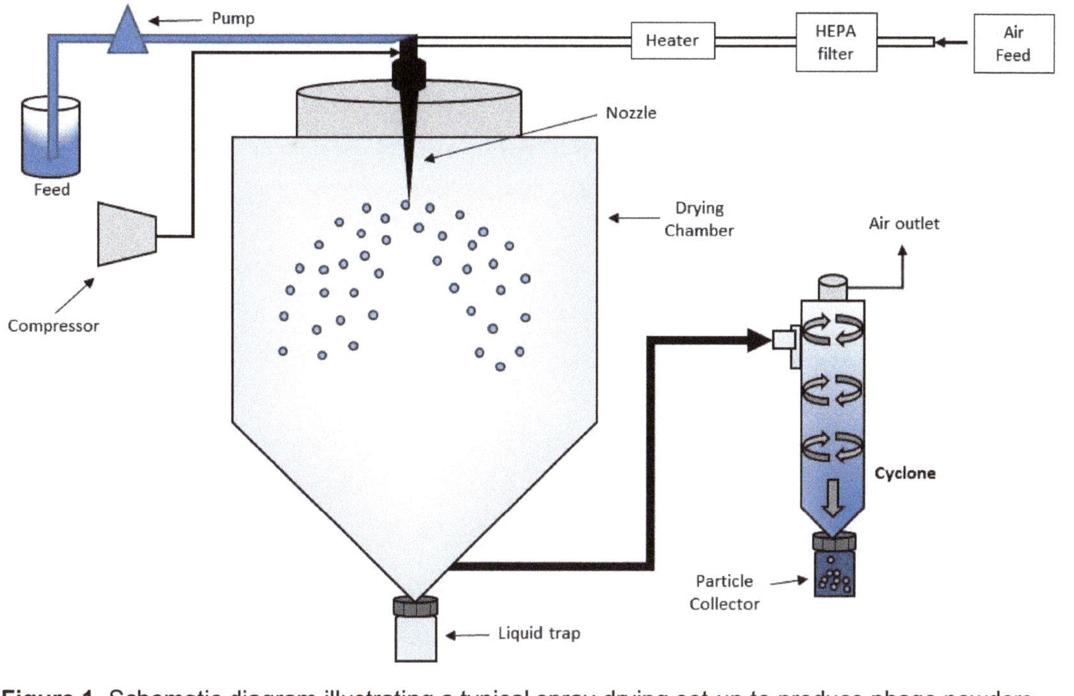


**Figure 1.** Schematic diagram illustrating a typical spray drying set-up to produce phage powders.

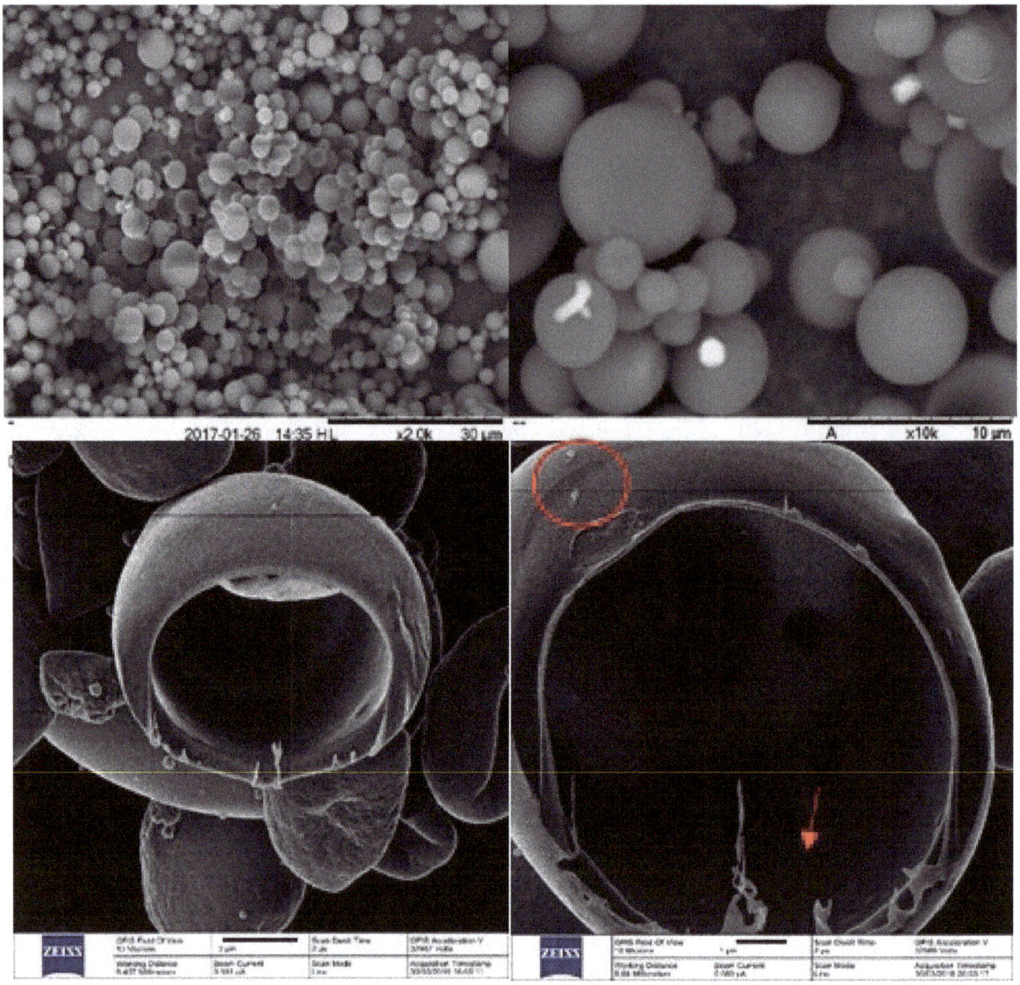

**Figure 2.** Scanning Electron Microscopy (SEM) images of spray dried phage-containing powders dried at an outlet temperature of 80°C using a bench-top spray dryer employing a 2-fluid nozzle for liquid atomisation. The arrow and circle indicate embedded *Myoviridae* phage particles in the core-shell structure of the microparticles.

## *Spray drying of phage biologicals using a Quality-by-Design approach*

Quality by design is a concept pioneered by Joseph Juran and is based around the concept that quality should be designed into a product rather than relying on end-of-pipe testing (Juran, 2008). A high-quality phage drug product needs to be safe and efficacious, delivering the therapeutic benefit promised to the consumer on the product label. Regulators such as the US Food and Drug Administration (US FDA) encourage manufacturers to adopt a risk-based approach using QbD principles for drug product development, manufacturing and regulation (Yu et al., 2014). A QbD-led approach emphasises a deep understanding of the manufacturing process and linking product quality to the desired clinical performance of the drug product. Designing a robust formulation and manufacturing process for phage drug products that consistently deliver the desired phage drug product quality is essential for the future of phage therapy. Performance-based quality specifications are critical in ensuring this is achieved. Improved product and process understanding helps in designing control strategies to ensure the drug product quality remains within acceptable limits without compromising safety and efficacy. Spray drying of phage biologicals requires identification of process and product characteristics that are critical to product quality from the patient's perspective. These then need to be translated into critical quality attributes (CQAs) of the phage drug product. Linking formulation and manufacturing aspects to CQAs is vital in ensuring the phage drug product meets performance quality criteria. This requires defining what the phage biological's quality target product profile (QTPP) looks like and linking it to CQAs. Identification of critical material attributes (CMAs) (explained below) that are relevant to spray drying of phage biologicals needs to be ascertained as well as critical process parameters (CPPs) specific for the spray drying process. Implementation of process control strategies and accounting for variability in process inputs as well as putting specifications on CMAs and the drug substance, excipients and drug products are needed (Yu et al., 2014).

The QTPP of the phage drug product is the basis for the phage drug product development and is a summary of all the quality attributes that the drug product needs to meet to ensure it is safe and efficacious (Yu et al., 2014). Identification of the phage drug product's CQAs is the next step and includes many attributes including phage titre in the powder, the powder moisture content, the material's glass transition temperature etc (see summary Table 1). The QTPP of new phage drug products needs to be adequately defined before drug product development work commences. A focus on the phage product design is just as important as the consideration of the process design here. The stability of the phage drug product throughout its shelf life needs to be confirmed through stability studies. The key objective of product design is to ensure that the phage drug product retains the desired QTPP over the product shelf life (Yu et al., 2014). Failure of a recent phage therapy trial due to a complex liquid phage drug product having stability issues during the trial is a valuable lesson for the future (Jault et al., 2018). Key aspects related to CMAs and CQAs for dry powder phage biologicals relevant to spray drying of phage formulations are discussed below.

**Table 1.** Typical input material attributes, process parameters and quality attributes of spray dried phage powders.

| Input material attributes | Process parameters | Quality attributes |
|---|---|---|
| • Phage type, e.g. *Myoviridae*<br>• Intrinsic susceptibility of a phage drug candidate to spray drying process stresses including atomisation, drying, desiccation, osmotic and re-hydration stresses<br>• Excipient type, e.g. sugars, polymers, proteins<br>• Solution viscosity<br>• Solution solids content w/v%<br>• Phage titre<br>• Lysate purity | • Drying temperature<br>• Drying air gas flow rate<br>• Liquid feed flowrate<br>• Type of atomiser, e.g. two fluid nozzle (and nozzle size), rotary wheel (and rotation speed).<br>• Dryer residence time<br>• Cyclone collection efficiency (affects the yield of material) | • Phage titre in spray dried powders (potency)<br>• Powder glass transition temperature ($T_g$)<br>• Powder moisture content<br>• Phage shelf life (Months stored under storage specific conditions at a given temperature and relative humidity)<br>• Powder re-suspension characteristics<br>• Powder particle size and size distribution<br>• Cohesive/adhesive properties<br>• Flow properties<br>• Bulk/tapped/true density<br>• Powder yield<br>• Amorphous vs crystallinity |

*The role of excipients in stabilising spray dried phage drug products*

Spray drying of phage suspensions typically includes excipients in the formulation to protect the phages from thermal and desiccation stresses. The critical material attributes (CMAs) of excipients used in spray drying needs careful consideration. Trehalose, sucrose and lactose have emerged as the most promising excipients for spray drying phage suspensions with trehalose being the front runner due to its high glass transition temperature and water replacement properties (Grasmeijer et al., 2013). Trehalose is the most frequently reported excipient used in the spray drying of phages and it has been shown to result in spray dried powders with high phage titre and good storage stability (Carrigy et al., 2019; Leung et al., 2016, 2017; Matinkhoo et al., 2011; Vandenheuvel et al., 2014). Upon spray drying, sugar excipients form amorphous structures with high glass transition temperatures (Tg), e.g. in their anhydrous amorphous form Tg of trehalose is 115 °C and Tg for lactose is 108 °C (Chang et al., 2019; Chen et al., 2000; Roe and Labuza, 2005). Below the glass transition temperature, vitrification due to the excipient stabilises the encapsulated bioactive agent and permits storage over long periods at ambient or lower temperatures (Carpenter and Crowe, 1988; Leung et al., 2017). Trehalose has low toxicity and has been shown to protect biological materials including proteins, probiotics and vaccines, against desiccation and thermal stress. Crowe et al. (1996) suggested that the efficacy of trehalose as an excipient in drying is partly due to its high glass transition temperature at all water contents but also because trehalose binds residual water left over from the drying process to form a dihydrate which might otherwise participate in lowering the glass transition temperature to below ambient (Vandenheuvel et al., 2014). In the amorphous state, trehalose may stabilise protein conformation through hydrogen bonding (Allison et al., 1999). Other excipients that have been used to improve the dispersibility of spray dried phage containing powders include dextrans, lactose (common excipient used for DPI powders), glucose, sucrose, mannitol and leucine (Leung et al., 2016; Matinkhoo et al., 2011; Telko and Hickey, 2005; Vandenheuvel et al., 2013b). Reducing sugars such as lactose may be unsuitable excipients for phage (Vandenheuvel et al., 2013b). Alternatives such as mannitol have been suggested in the literature; however, mannitol has a tendency to recrystallize rather than remaining amorphous (Steckel and Bolzen, 2004). Phage spray drying studies using mannitol have reported improved particle dispersion characteristics. However, phage viability results have been mixed (Leung et al., 2016, 2017). Addition of proteins, e.g. casein in combination with trehalose has shown good results (Matinkhoo et al., 2011).

*Effect of spray drying conditions on phage viability and shelf life*

The spray drying conditions need to be carefully optimised to ensure viability of spray dried phages in the final drug product. Two-fluid nozzles typically used for atomisation of the phage suspensions for spray drying have been shown to generate shear induced damage and loss of phage titre (Leung et al., 2016; Vandenheuvel et al., 2013b). Phages are highly sensitive to thermal stresses resulting in loss of phage activity at high spray drying temperatures (for co-current dryers, outlet temperatures exceeding 60°C) (Davies and Kelly, 1969; Jończyk et al., 2011). Typically, spray drying studies with phages have employed co-current dryers operating at low outlet spray drying temperatures (~40-50 °C outlet air temperatures) resulting in higher phage titres in the powders and reduced overall phage losses immediately post spray drying (Leung et al., 2017). High loss of phage titres have been observed at higher spray drying temperatures (Leung et al., 2016; Matinkhoo et al., 2011; Vandenheuvel et al., 2013b; Walbeck, 2013). However, the residual moisture content in the spray dried powders (typically ~5% w/w or more, Figure 3) due to low drying temperatures (40-50 °C dryer outlet temperatures) may result in lowering of the glass transition temperature (Tg) (Roe and Labuza, 2005). The glass transition temperature of an amorphous pharmaceutical solid is a critical physical property (CQA) which can dramatically influence its chemical and physical stability as well as its viscoelastic properties (Figure 4). Water acts as a plasticiser in spray dried powders which may also adsorb water upon storage due to exposure to a humid atmosphere (Hancock and Zografi, 1994). The Tg of amorphous powders such as trehalose, lactose and sucrose fall dramatically when the powders contain low amounts of moisture (Chen et al., 2000; Hancock and Zografi, 1994; Roe and Labuza, 2005). Spray dried phage containing powders tend to be amorphous upon production; however, storage conditions may subject the material to temperatures and humidity conditions lying above the glass transition line. Crystallization rates of glasses depend on the storage temperature in relation to Tg. The recrystallisation of amorphous sugar in spray dried powders negatively affect phage viability, perhaps due to denaturing of phage receptor proteins as sugar molecules can no longer interact through hydrogen bonding with proteins to retain protein conformation upon crystallisation (Carpenter and Crowe, 1988; Roe and Labuza, 2005) (Figures 5 and 6). The use of excipients that are hygroscopic, e.g. lactose and trehalose also require careful handling under controlled atmospheric conditions and the use of a dry atmosphere controlled glove box is advisable during powder handling to prevent exposure of the dry powder to moisture in the atmosphere (Carrigy et al., 2019; Leung et al., 2017; Matinkhoo et al., 2011).

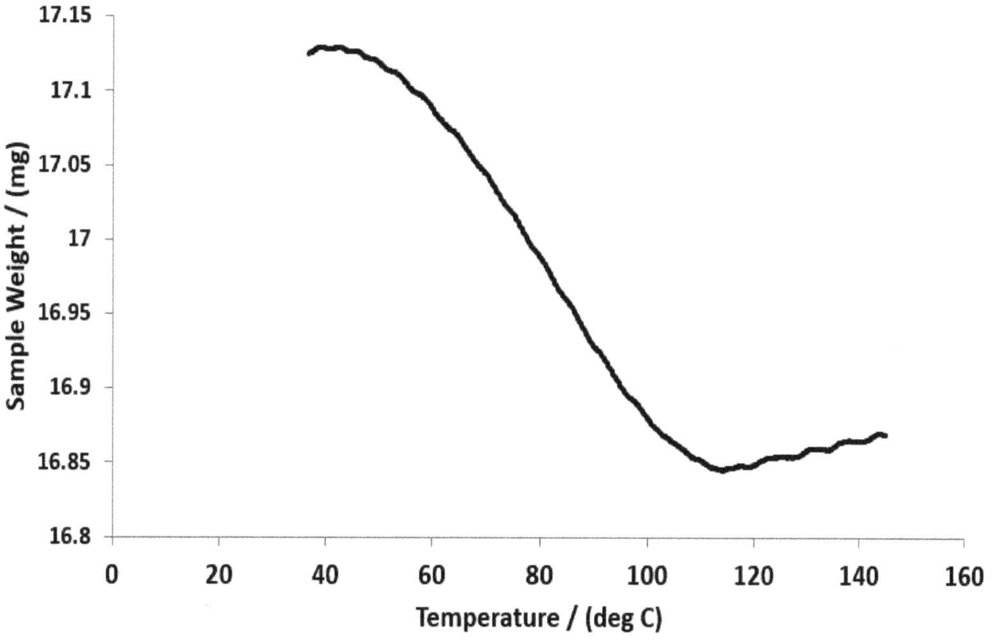


**Figure 3.** Measurement of residual moisture in spray dried powders using thermal gravimetric analysis (TGA). Spray dried powder produced at an outlet temperature of 65 °C.

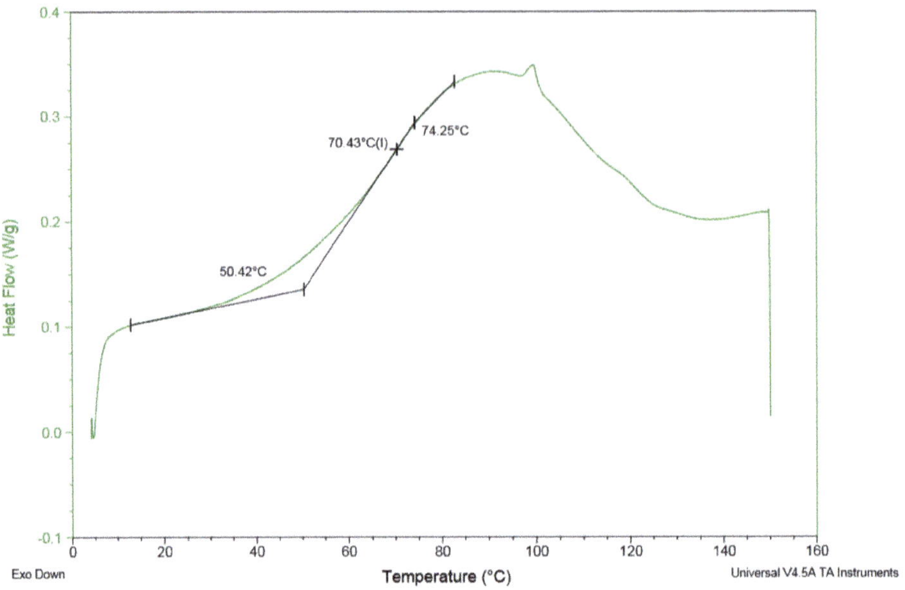


**Figure 4.** Measurement of glass transition temperature using differential scanning calorimetry (DSC) of a spray dried powder (using trehalose as excipient). The powder was produced at an outlet temperature of 65 °C. Onset temperature shown is 50 °C.

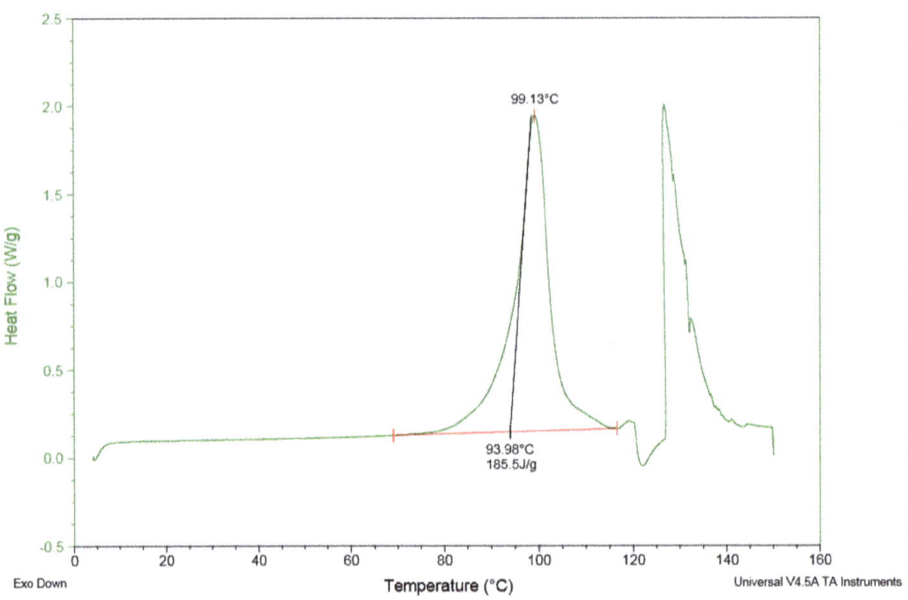


**Figure 5.** Measurement of glass transition temperature using differential scanning calorimetry (DSC) of a spray dried powder (using trehalose as excipient). The powder was produced at an outlet temperature of 65 °C and exposed to storage conditions of 30 °C and 65% relative humidity for 24 hours. Peaks indicate re-crystallisation of the trehalose with endothermic melting peaks for different trehalose polymorphs shown here.

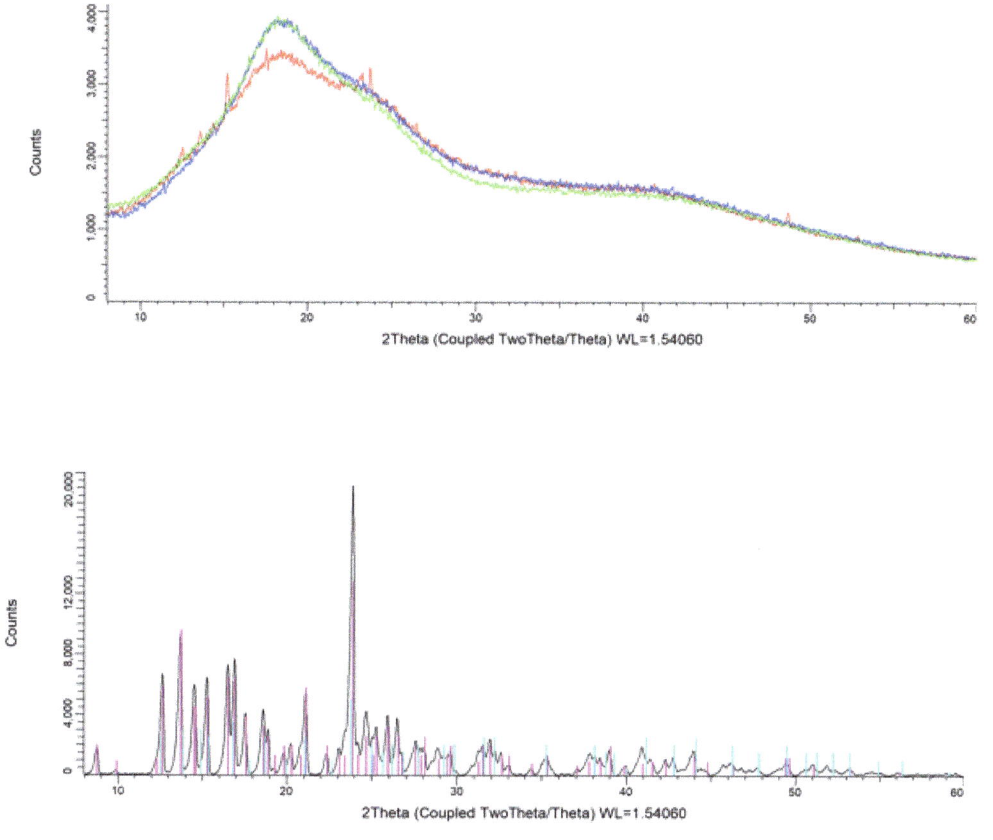


**Figure 6.** Top Panel: Powder X-ray diffraction results showing amorphous trehalose powders immediately, six months and one year following spray drying at 65°C outlet drying temperature and stored at 4°C (blue curve (immediately after spray drying), green curve (six months after spray drying) and red curve (one year after spray drying)). Bottom panel: Same sample exposed to storage conditions of 30 °C and 65 % relative humidity for 24 hours resulting in re-crystallisation of the trehalose.

## *Effect of storage conditions on phage stability in spray dried powders*

Storage temperature and relative humidity are the two main factors shown to affect phage stability in spray dried powders. Amorphous forms of dry powder pharmaceutical biologics are most stable when molecular motion in the disordered state is retarded over a meaningful pharmaceutical timeframe (Hancock and Zografi, 1994). The storage temperature needs to be considerably below the amorphous powder material's Tg to slow down molecular motion such that chemical and physical instability of the drug product does not occur over the lifetime of the pharmaceutical product, which is typically in the order of a few years (Hancock and Zografi, 1994). Storage temperatures 50°C below the glass transition temperature are thought to significantly reduce molecular movement and help in lengthening phage powder storage shelf life (Chang et al., 2020; Hancock et al., 1995). Refrigerated storage of spray dried powders at relative humidity exceeding 20% has been shown to result in loss of phage titre whereas samples stored at low humidity were stable (Leung et al., 2017; Vandenheuvel et al., 2014). Disaccharides such as trehalose, sucrose and lactose preserve the bioactivity of phages in the dry state. Refrigerated storage of spray dried powders was reported to yield higher titres of phage during long term storage compared with storage under ambient conditions (Leung et al., 2017). Different phage strains formulated and spray dried under identical conditions showed significant differences in the resulting phage titre suggesting the need for individually formulating each phage to be used in a phage cocktail (Vandenheuvel et al., 2013b, 2014). The selection of phages as suitable candidates for phage therapy should include consideration of whether they lend themselves to spray drying i.e. a CMA as part of QbD approach to linking drug product design to the QTPP if the final drug product is to be delivered to patients in a stable dry powder form. A number of studies have reported spray dried phage powders to have a suitable mass median aerosolized diameter for pulmonary delivery of phages to treat respiratory infections (Leung et al., 2016; Matinkhoo et al., 2011; Vandenheuvel et al., 2013a, 2014). Here the powder particle size is a CQA linking powder aerosolization to QTPP. Such critical issues need greater consideration early in the product design cycle.

## Conclusion

Formulation and stability of phage drug products need greater consideration during phage drug product design. Spray drying is a scalable, high throughput process that lends itself to the production of stable dry powder phage biologicals. A QbD approach to phage drug product design is advocated in this chapter. CMAs includes the selection of suitable phage candidates at an early stage in drug product design that lend themselves to spray drying and the use of excipients that result in dry powders suitable for phage biologicals. A better understanding of the critical process parameters related to spray drying and consideration of powder handling, packaging and storage conditions is needed. Spray drying of phage biologicals can increase the stability and shelf life of the drug product as well as incorporating targeted delivery and controlled release elements. Spray dried powders can be loaded into capsules or made into tablets using direct compression processes giving added flexibility for ease of use.

## References

Allison, S. D., Chang, B., Randolph, T. W., and Carpenter, J. F. (1999). Hydrogen bonding between sugar and protein is responsible for inhibition of dehydration-induced protein unfolding. Arch. Biochem. Biophys. 365, 289–298. doi:10.1006/abbi.1999.1175.

Briers, Y., Miroshnikov, K., Chertkov, O., Nekrasov, A., Mesyanzhinov, V., Volckaert, G., and Lavigne, R. (2008). The structural peptidoglycan hydrolase gp181 of bacteriophage phiKZ. Biochem. Biophys. Res. Commun. 374, 747–751. doi:10.1016/j.bbrc.2008.07.102.

Carpenter, J. F., and Crowe, J. H. (1988). Modes of stabilization of a protein by organic solutes during desiccation. Cryobiology 25, 459–470. doi: 10.1016/0011-2240(88)90054-5.

Carrigy, N. B., Liang, L., Wang, H., Kariuki, S., Nagel, T. E., Connerton, I. F., and Vehring, R. (2019). Spray-dried anti-Campylobacter bacteriophage CP30A powder suitable for global distribution without cold chain infrastructure. Int. J. Pharm. 569, 118601. doi:10.1016/j.ijpharm.2019.118601.

Chang, R. Y. K., Kwok, P. C. L., Khanal, D., Morales, S., Kutter, E., Li, J., and Chan, H-K. (2020). Inhalable bacteriophage powders: Glass transition temperature and bioactivity stabilization. Bioeng. Transl. Med., e10159. doi: 10.1002/btm2.10159.

Chang, R. Y. K., Wallin, M., Kutter, E., Morales, S., Britton, W., Li, J., and Chan, H-K. (2019). Storage stability of inhalable phage powders containing lactose at ambient conditions. Int. J. Pharm. 560, 11–18. doi:10.1016/j.ijpharm. 2019.01.050.

Chen, T., Fowler, A., and Toner, M. (2000). Literature review: Supplemented phase diagram of the trehalose-water binary mixture. Cryobiology 40, 277–282. doi:10.1006/cryo.2000.2244.

Chew, N. Y. K. ., and Chan, H. K. . (2002). The role of particle properties in pharmaceutical powder inhalation formulations. J. aerosol Med. 15, 325–330.

Crowe, L. M., Reid, D. S., and Crowe, J. H. (1996). Is trehalose special for preserving dry biomaterials? Biophys. J. 71, 2087–93. doi:10.1016/S0006-3495(96)79407-9.

Davies, J. D., and Kelly, M. J. (1969). The preservation of bacteriophage H1 of Corynebacterium ulcerans U103 by freeze-drying. J. Hyg. (Lond). 67, 573–583. doi:10.1017/S0022172400042030.

Dini, C., and de Urraza, P. J. (2013). Effect of buffer systems and disaccharides concentration on Podoviridae coliphage stability during freeze drying and storage. Cryobiology 66, 339–342. doi:10.1016/j.cryobiol.2013.03.007.

Grasmeijer, N., Stankovic, M., De Waard, H., Frijlink, H. W., and Hinrichs, W. L. J. (2013). Unraveling protein stabilization mechanisms: Vitrification and water replacement in a glass transition temperature controlled system. Biochim. Biophys. Acta - Proteins Proteomics 1834, 763–769. doi:10.1016/j.bbapap.2013.01.020.

Hancock, B. C., Shamblin, S. L., and Zografi, G. (1995). Molecular Mobility of Amorphous Pharmaceutical Solids Below Their Glass Transition Temperatures. Pharm. Res. An Off. J. Am. Assoc. Pharm. Sci. 12, 799–806. doi:10.1023/A:1016292416526.

Hancock, B. C., and Zografi, G. (1994). The Relationship Between the Glass Transition Temperature and the Water Content of Amorphous Pharmaceutical Solids. Pharm. Res. An Off. J. Am. Assoc. Pharm. Sci. 11, 471–477. doi: 10.1023/A:1018941810744.

Hoe, S., Ivey, J. W., Boraey, M. A., Shamsaddini-Shahrbabak, A., Javaheri, E., Matinkhoo, S., Finlay, W.H. and Vehring, R. (2014). Use of a fundamental approach to spray-drying formulation design to facilitate the development of multi-component dry powder aerosols for respiratory drug delivery. Pharm. Res. 31, 449–465. doi:10.1007/s11095-013-1174-5.

Jault, P., Leclerc, T., Jennes, S., Pirnay, J. P., Que, Y.-A., Resch, G., Rousseau, A.F., Ravat, F., Carsin, H., Le Floch, R., Schaal, J.V., Soler, C., Fevre, C.,

Arnaud, I., Bretaudeau, L. and Gabard, J. (2018). Efficacy and tolerability of a cocktail of bacteriophages to treat burn wounds infected by *Pseudomonas aeruginosa* (PhagoBurn): a randomised, controlled, double-blind phase 1/2 trial. Lancet Infect. Dis. 0, 1–11. doi:10.1016/S1473-3099(18)30482-1.

Jończyk, E., Kłak, M., Międzybrodzki, R., and Górski, A. (2011). The influence of external factors on bacteriophages--review. Folia Microbiol. (Praha). 56, 191–200. doi:10.1007/s12223-011-0039-8.

Juran, J. M. (2008). Juran on Quality by Design: The New Steps for Planning Quality into Goods and Services. Simon and Schuster.

Klingler, C., Müller, B. W., and Steckel, H. (2009). Insulin-micro- and nanoparticles for pulmonary delivery. Int. J. Pharm. 377, 173–179. doi: 10.1016/j.ijpharm.2009.05.008.

Knezevic, P., Obreht, D., Curcin, S., Petrusic, M., Aleksic, V., Kostanjsek, R., Petrovic, O. (2011). Phages of *Pseudomonas aeruginosa*: Response to environmental factors and in vitro ability to inhibit bacterial growth and biofilm formation. J. Appl. Microbiol. 111, 245–254. doi:10.1111/j. 1365-2672.2011.05043.x.

Lee, S. W., and Belcher, A. M. (2004). Virus-based fabrication of micro- and nanofibers using electrospinning. Nano Lett. 4, 387–390. doi:10.1021/ nl034911t.

Leung, S. S. Y., Parumasivam, T., Gao, F. G., Carrigy, N. B., Vehring, R., Finlay, W. H., Morales, S., Britton, W. J., Kutter, E. and Chan, H-K. (2016). Production of Inhalation Phage Powders Using Spray Freeze Drying and Spray Drying Techniques for Treatment of Respiratory Infections. Pharm. Res. 33, 1486–1496. doi:10.1007/s11095-016-1892-6.

Leung, S. S. Y., Parumasivam, T., Gao, F. G., Carrigy, N. B., Vehring, R., Finlay, W. H., Morales, S., Britton, W. J., Kutter, E. and Chan, H-K.. (2017). Effects of storage conditions on the stability of spray dried, inhalable bacteriophage powders. Int. J. Pharm. doi:10.1016/j.ijpharm.2017.01.060.

Malik, D. J., Sokolov, I. J., Vinner, G. K., Mancuso, F., Cinquerrui, S., Vladisavljevic, G. T., Clokie, M. R. J. Garton, N.J., Stapley, A.G.F. and Kirpichnikova, A. (2017). Formulation , stabilisation and encapsulation of bacteriophage for phage therapy. Adv. Colloid Interface Sci. 249, 100–133. doi:10.1016/j.cis.2017.05.014.

Matinkhoo, S., Lynch, K. H. ., Dennis, J. J. ., Finlay, W. H., and Vehring, R. (2011). Spray-Dried Respirable Powders Containing Bacteriophages for the Treatment of Pulmonary Infections. J. Pharm. Sci. 100, 5197–5205.

Puapermpoonsiri, U., Spencer, J., and Walle, C. F. Van Der (2009). A freeze-dried formulation of bacteriophage encapsulated in biodegradable microspheres. Eur. J. Pharm. Biopharm. 72, 26–33. doi:10.1016/j.ejpb.2008.12.001.

Roe, K. D., and Labuza, T. P. (2005). Glass transition and crystallization of amorphous trehalose-sucrose mixtures. Int. J. Food Prop. 8, 559–574. doi:10.1080/10942910500269824.

Steckel, H., and Bolzen, N. (2004). Alternative sugars as potential carriers for dry powder inhalations. Int. J. Pharm. 270, 297–306. doi:10.1016/j.ijpharm.2003.10.039.

Telko, M. J., and Hickey, A. J. (2005). Dry powder inhaler formulation. Respir. Care 50, 1209–1227. doi:10.2165/00128413-200615470-00016.

Vandenheuvel, D., Meeus, J., Lavigne, R., and Van Den Mooter, G. (2014). Instability of bacteriophages in spray-dried trehalose powders is caused by crystallization of the matrix. Int. J. Pharm. 472, 202–205. doi:10.1016/j.ijpharm.2014.06.026.

Vandenheuvel, D., Singh, A., Vandersteegen, K., Klumpp, J., Lavigne, R., and Mooter, G. Van Den (2013a). European Journal of Pharmaceutics and Biopharmaceutics Feasibility of spray drying bacteriophages into respirable powders to combat pulmonary bacterial infections. Eur. J. Pharm. Biopharm. 84, 578–582. doi:10.1016/j.ejpb.2012.12.022.

Vandenheuvel, D., Singh, A., Vandersteegen, K., Klumpp, J., Lavigne, R., and Van Den Mooter, G. (2013b). Feasibility of spray drying bacteriophages into respirable powders to combat pulmonary bacterial infections. Eur. J. Pharm. Biopharm. 84, 578–582. doi:10.1016/j.ejpb.2012.12.022.

Vinner, G. K., Rezaie-Yazdi, Z., Leppanen, M., Stapley, A. G. F., Leaper, M. C., and Malik, D. J. (2019). Microencapsulation of Salmonella-specific bacteriophage Felix O1 using spray-drying in a pH-responsive formulation and direct compression tableting of powders into a solid oral dosage form. Pharmaceuticals 12, 1–14. doi:10.3390/ph12010043.

Walbeck, A. K. (2013). US Patent, 8501453 B2:Methods for drying bacteriophage and bacteriophage-containing compositions, the resulting dry compositions, and methods of use.

Yu, L. X., Amidon, G., Khan, M. A., Hoag, S. W., Polli, J., Raju, G. K., and Woodcock, J. (2014). Understanding pharmaceutical quality by design. AAPS J. 16, 771–783. doi:10.1208/s12248-014-9598-3.

Chapter 18

# Practical Issues in Setting Up and Maintaining a Collection of Therapeutic Bacteriophages: the Finnish Experience

**Saija Kiljunen[1], Jussi Tervonen[2] and Mikael Skurnik[1,3]***

[1]Department of Bacteriology and Immunology, Human Microbiome Research Program, Faculty of Medicine, University of Helsinki, Helsinki, Finland
[2]HUS Pharmacy, Helsinki University Hospital, Helsinki, Finland
[3]Division of Clinical Microbiology, Helsinki University Hospital, HUSLAB, Helsinki, Finland

*Corresponding author: mikael.skurnik@helsinki.fi

DOI: https://doi.org/10.21775/9781913652517.18

## Abstract

As part of our efforts to make bacteriophage (phage) therapy available to Finnish clinicians to treat patients suffering from infections caused by multi-drug resistant (MDR) bacterial pathogens, we have started a continually growing collection of phages against the most common MDR pathogens. The collection at present contains close to 400 phages that we characterize to select those that are suitable for therapeutic purposes. Roughly half of the phage genomes have been sequenced and analyzed. In this chapter, we describe some of the practical issues related to this project.

## Introduction

Bacteriophage therapy has experienced a renaissance in Western countries during the last decade or so, due to the threat cast by increasing antibiotic resistance among bacterial pathogens causing human and animal infections. In fact, a PubMed search for "phage therapy" reveals that a huge increase in interest occurred around the Millennium shift. Shortly after that, the first discussions on the possibility to set up phage therapy in Finland took place; however, our grant applications in 2002 and 2005 were not successful. By this time, the Skurnik laboratory had already had almost 20-year experience working with bacteriophages, as our first *Yersinia* phages were isolated back in 1988 from the sewage in Turku and were being used as genetic tools in lipopolysaccharide research (Al-Hendy et al., 1992; Skurnik et al., 1999; Skurnik et al., 1995; Zhang and Skurnik, 1994). We carried out genomic sequencing and detailed characterization of the phages as part of a project to generate reporter phages (Kiljunen et al., 2005a; Kiljunen et al., 2005b; Pajunen et al., 2000; Pajunen et al., 2002; Pajunen et al., 2001; Skurnik et al., 2012), and as a demonstration of our continuing interest in phage therapy we also wrote a review article and two book chapters on the topic (Skurnik et al., 2008; Skurnik et al., 2007; Skurnik and Strauch, 2006). Finally, in 2012 we had a successful grant application providing us with enough funding to employ an MSc student who started in March 2013. Thus our small team set up to work towards our goal of setting up a Phage Therapy Unit in Finland. Already, in the beginning, we were aware that this would not be a trivial task as our achievements would need to convince politicians and decision makers that phage therapy is a feasible and economically sound option, especially as phage therapy should be integrated as part of the public health system. Below, we will discuss our progress towards setting up a phage collection to be used as a source for therapeutic phages suitable for patients in need.

### Laboratory space, equipment and human resources

The collection of therapeutic phages necessitates working with pathogenic bacteria; therefore, a biosafety level 2 (BSL-2) laboratory is an absolute requirement. The laboratory should be equipped for microbiology and molecular biology work, and basic devices such as incubators, safety cabinets, laminar flow hoods, shakers, centrifuges, spectrophotometers, and fluorometers should be available. As many phages are highly stable and stay infective even in aerosols and dust, all the precautions to avoid cross-contamination should be taken. Pipetting phage solutions should always be carried out using filter tips with micropipettes. The choice of storage method (see below) dictates the need for storage equipment, but at least the storage of bacterial strains typically requires a deep-freezing (-70°C) capacity. Preferably, to cover possible break-down of the freezers, both bacteria and phages should be stored in more than one aliquot and stored in different freezers. If lyophilisation is chosen as the storage method, access to a lyophiliser is naturally required.

As each phage included in the collection has to be analyzed by whole-genome sequencing, the laboratory needs to have access to Next-Generation Sequencing (NGS). The easiest way for this is to utilize a (commercial) sequencing service. In this case, the phage laboratory only needs to have equipment for phage DNA isolation and analysis, such as devices for running electrophoresis gels and measuring DNA concentration. If one wants to have a possibility for in-house NGS, the laboratory should have separate rooms for pre- and post-PCR steps during the library construction and adequate machinery for library preparation and analysis and sequencing.

Bacteriophage research and maintenance requires specific expertise that is not generally taught in schools or universities. Therefore, any new person entering a phage laboratory has to be trained to master the methods, protocols and tricks related to this field of microbiology. At a minimum, one or two laboratory technicians, a bioinformatician, and a microbiologist are required to accumulate and maintain the phage collection. The number of people in the laboratory doubles or triples when phages are prepared for patient use.

## Isolation of bacteriophages

### *Sources of bacteriophages*

Phages can be found everywhere where bacteria live. As human pathogens are readily released into wastewater, incoming water samples from sewage treatment plants are excellent sources for their phages. Most phages in our collection have been isolated from water samples collected from hospital and municipal water treatment plants (Figure 1).

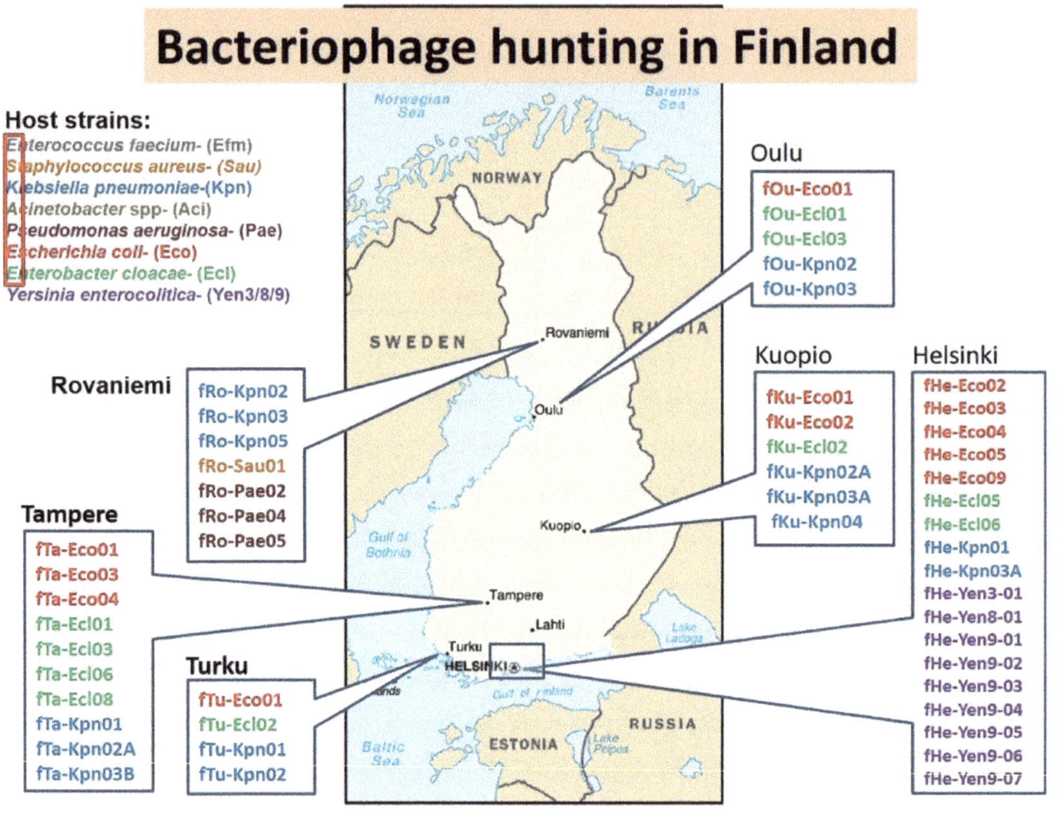


**Figure 1.** Map of Finland showing the locations from which sewage samples were received for phage isolation. As the phages were isolated against the ESKAPEE and *Yersinia* bacteria, we named the phages using the following scheme: **f** followed by a **2-letter** identification of the location, then a **dash** and the **3-letter** identification of the host bacterium and the **number** of the picked plaque from the double-layer plate. Thus, fTu-Kpn02 is a *K.pneumoniae* phage isolated from the sewage sample received from Turku.

In addition to sewage, phages can be found in different environmental or clinical samples such as soil, decomposing plant material, animal and human faeces, saliva, or mucus. For example, we have isolated several phages against *Pseudomonas aeruginosa* from a garden compost, *Yersinia enterocolitica* and *Yersinia pseudotuberculosis* from pig stool samples (Salem and Skurnik, 2018; Salem et al., 2015). Commercial phage products and sharing phages between other phage laboratories are also potential sources to complement a phage collection (see below).

*Isolation methods*

The principle of phage isolation is very simple (Van Twest and Kropinski, 2009): The sample material is first incubated with fresh culture medium containing the bacterial strain(s) against which you want to find phages, in a step called enrichment. Next, bacteria are removed by filtration and/or centrifugation, and the enriched sample is diluted and plated out with the indicator strain(s) to see the possible plaques. To ensure a pure phage culture, a single plaque is picked from the plate and the plaque purification is repeated for three to six times prior to preparing a phage lysate for storage.

When planning the phage isolation process, there are a couple of choices to make that may have a profound effect on the result. The first choice is whether to use chloroform to kill the bacteria after the phage enrichment. Chloroform is often used routinely, as it efficiently lyses bacteria and reduces the regrowth of phage-resistant bacteria in the phage lysate. However, chloroform also inactivates lipid-containing phages, and may thus reduce the number and diversity of obtained phages. The use of 0.2 or 0.45-μm filter to remove bacteria often circumvents the need of using chloroform and may improve the possibility of finding different kinds of phages.

The second choice to make is whether to pool several bacterial strains for the initial enrichment step. This may sound attractive, especially if there are several strains against which one wants to isolate phages. Making one enrichment with a bacterial pool will reduce the labour and material consumption when compared to making several separate enrichments. The drawback of this approach is that the pooling of many strains may increase the likelihood of isolating temperate phages. Prophages present in the genome of one of the strains in the pool may be auto induced at low levels. If another strain in the mixture is sensitive to such a phage, the phage will multiply by lytic cycle and become isolated. When setting up a collection of therapeutic phages, the aim is to isolate strictly lytic phages. Therefore, the temperate phages originating from the strains used in the enrichment will end up causing a lot of extra work and sequencing costs.

## Characterization of bacteriophages

### *Genome sequences*

The nucleotide sequences of phages intended for therapeutic use need to be determined in order to exclude the presence of harmful genes that could be delivered to pathogens by transduction. To this end, during the years we have used the NGS core facilities of our university or our collaborators, or sent the DNA to commercial companies and have now almost 200 phage genomes sequenced and ca. 50 annotated genomes deposited to sequence databases (https://www.helsinki.fi/en/researchgroups/yersinia-and-bacteriophage-research-laboratory/research-data). Due to the growing queues in the core facilities we have lately almost completely relied on companies with whom one can negotiate reasonable deals and will who return the raw sequence data in 3-4 weeks, and in special cases sooner. When selecting the NGS provider we make sure that the DNA libraries are prepared by physical fragmentation of the DNA as such libraries allow the determination of the phage physical ends using the PhageTerm program (Garneau et al., 2017). We mostly ask for paired-end 150 nt Illumina sequences.

The phage nucleic acid (this far only double-stranded DNA) is isolated from phage stocks (>$10^9$ pfu/ml) either using the Sambrook protocol (Sambrook and Russell, 2001) or by the commercial isolation kit. The latter protocol usually has a lower yield that may prevent thorough analysis of the isolated DNA by restriction enzyme digestions to confirm the assembled genomic sequence of the phage.

For the *de novo* assembly of the genomes, we use the A5 pipeline (Coil et al., 2014) installed to the supercomputers at Center for Scientific Computing (CSC) Finland (https://http://www.csc.fi) or the Geneious program package (https://www.geneious.com/) installed to our personal computers. To run the CSC assembly programs the user needs to have a basic knowledge of UNIX and command-line environment, although, these jobs are processed faster on a supercomputer than using Geneious on a personal computer. The A5-pipeline *de novo* assembly produces multiple different files of which the log-file is of immediate interest as from there one gets an overview of the results and quality. We look for the average fragment size of the DNA library, average read coverage, the total length and the number of assembled scaffolds. Rarely do the reads produce a single scaffold with a size suitable for a phage genome. Instead, most times scaffolds will be obtained which represent the total genome of the host bacterium. These can be divided into many scaffolds (>100), of which one or a few may represent the phage genome.

To identify the phage scaffold(s) we use genome browsers such as IGV (Robinson et al., 2011), or Geneious, to search for the scaffold with highest read coverages. The NGS is highly sensitive and will also produce reads for contaminating DNA of the host albeit with marginal read coverage. DNase and RNase are used to degrade any host nucleic acids from the phage lysate before the phage nucleic acid is isolated, however, this might not be complete. Alternatively, the host DNA may accidentally be packaged into phage capsids and this will also be sequenced.

The *de novo* assembly programs usually produce a circular sequence that does not represent correctly the linear sequence isolated from the phage particles. The circular form leaves overlapping flanks to the artificially opened sequence that are usually 1 bp shorter than the sequence reads (*i.e.,* with Illumina PE150, 149 bp). After identification of the phage scaffold, we first remove the overlapping flank and then use the PhageTerm program installed at CSC to determine physical termini and the type of the termini of the genome (Garneau et al., 2017). Using the data we then rearrange the sequence to represent the physical genome and when appropriate copy the identified direct terminal repeat to the end of the genome.

The obtained phage genome is auto-annotated using Rapid Annotation Using Subsystem Technology (RAST, http://rast.nmpdr.org/) and the obtained annotation validated manually using the Artemis tool (Rutherford et al., 2000) or Geneious. The possible presence of harmful genes is checked against the Virulence factor database (http://www.mgc.ac.cn/VFs/) and Antibiotic resistance gene database (http://ardb.cbcb.umd.edu/). Finally, the fully annotated sequence is deposited to the DNA sequence databases.

*Host range determination*

The host range of any phage dictates its usefulness for therapeutic purposes. The larger the host range of a phage the more likely it will infect many more patient isolates than a phage with a narrow host range. To get an overview of the host range we routinely test the phages against a minimum of 50 clinical isolates of the same species that was used to isolate the phage, and if possible, also strains representing closely-related species. To test the host range of a phage against a set of strains we either use the drop-test on bacteria grown in soft agar or the microtiter-plate format where the growth curves of the bacteria are recorded in the presence and absence of the phage. To get reliable results we test several dilutions of the phage stock, preferably followed by determining the efficiency of plating (EOP) on the positive strains.

*Electron microscopy*

We determine the morphologies of the phages by classical transmission electron microscopy (TEM) using either uranyl acetate or phosphotungstate negative staining (Ackermann, 2009). As most of the phages represent *Caudovirales* we conduct measurements of the critical dimensions of a phage, and make comparisons of the capsid dimensions to its genome size.

*Other phage characteristics*

In rare cases, we have the possibility to carry out more thorough studies of our isolated phages. We quite often encounter phages with DNA that cannot be digested with restriction enzymes and/or that cannot be sequenced successfully using NGS. The likely reason is the presence of modified nucleotides in the genome of such phages. This can be analysed by degrading the DNA into deoxyribonucleosides using the New England Biolabs Nucleoside Digestion mix (Cat No 0649) and analyzing the nucleosides in HPLC (Kiljunen et al., 2005a). We have also encountered phages whose heavily modified nucleic acid could not be sequenced using conventional sequencing protocols. In such a case, we have isolated RNA from phage-infected bacteria at different times after infection and determined the sequences of the transcripts. After subtracting the host-genome specific reads, the remaining reads were *de novo* assembled to get phage-specific transcripts (Leskinen et al., 2020).

Another phage characteristic sometimes relevant to therapeutic purposes is the receptor that phage adsorbs to on the bacterial surface. If the receptor is essential for the virulence of the pathogen during an infection such as lipopolysaccharide (LPS), the spontaneous phage-resistant LPS mutants are often less virulent and can be eliminated by the patient's immune system. The easiest approach to identify the phage receptor is to isolate a number of phage-resistant mutants and have them sequenced by NGS along with the wild type strain. Alignment of the sequence reads against the reference genome allows identification of the mutations. In this manner, analyzing the NGS sequence data of 12 phage resistant mutants of *Yersinia pseudotuberculosis* phage, we identified mutations affecting the *ompF* (encoding the outer membrane porin OmpF) and *galU* (involved in the biosynthesis of LPS inner core) genes (Salem et al., 2020). The mutants were successfully complemented *in trans* with wild type genes to phage sensitivity that confirmed their roles.

For selected phages we carry out additional experiments to for detailed characterisation of the phage-host interactions. These include adsorption rate and single step growth curve experiments (Oduor et al., 2020). The latter gives information on the length of the replication cycle of the phage as well as of the average number of phage progeny produced per bacterial cell. Regarding the adsorption rate, in the majority of cases under regular conditions >90% of the phages will adsorb within 5 min onto the bacteria. However, we have encountered phages that are extremely slow with <50% adsorption in 60 min! In such cases, one may consider whether such phages would behave well as therapeutic agents.

**Storage of bacteriophages**

There are several possibilities for the long-term storage of phages. Lyophilisation is a traditional method for phage storage, and it has been shown to retain phage viability for close to 20 years (Zierdt, 1988). Different stabilizers, such as sucrose, trehalose, or gelatin, can further improve the phage stability in the lyophilized state (Manohar and Ramesh, 2019; Merabishvili et al., 2013). If lyophilisation is not a desirable (or possible) option, phages can also be stored deep-frozen using glycerol or dimethyl sulfoxide (DMSO) as cryoprotectant. Different phages may have very different optimal storage conditions; therefore, storing the phages in more than one way (e.g. at both -70°C and +4°C, using different cryoprotectants) may be a good approach to make sure that all the phages will remain viable. We routinely store our phages at -70°C using two different stabilizers, 8% DMSO and 20% glycerol. In addition, phage lysates are stored at +4°C in growth medium supplemented with 8% sucrose, and the final plaque-purified phage sample at +4°C in SM-buffer (100 mM NaCl, 10mM $MgSO_4$, 50 mM Tris-HCl, pH 7.5, 0.01% (w/v) gelatin). Our rule-of-thumb experience is that at -70°C, DMSO is better cryoprotectant than glycerol, and at +4°C, phages are more stable in SM-buffer than in growth medium.

In a phage therapy laboratory, the requirement of performing fast phage typing with incoming clinical bacterial strains sets a need for the constant availability of viable phage lysates. Therefore, the short-term stability of phages in the lysates used for phage typing should be checked regularly. We routinely use sucrose (8%) to improve the phage stability and check the titers every two-three months. We have also made preliminary tests using nanofibrillated cellulose (GrowDex, UPM Biomedicals) as a phage storage matrix. So far, all the tested phages have remained viable without a significant loss of titer for at least six months when stored in 1% or 0.5% GrowDex.

## Collaboration with other phage laboratories

### *Crowdsourcing*

If there is a need for therapeutic phages against a bacterial strain for which there are no phages in one's own collection, the solution is often to ask help from other phage depositories. Phage Directory (https://phage.directory/) maintains a Phage Alerts –service, where the phage depositories and research groups can send phage requests and as well as responses to them.

In 2018, we sent an alert to Phage Directory for phages against an MDR *Klebsiella pneumoniae* strain (Sacher et al., 2019). We received altogether 14 initial replies, 11 of which resulted in further actions. We sent the *Klebsiella* strain for phage typing to six phage research laboratories. Additionally, four academic and one biotech company laboratory sent their phages to us for phage typing. Altogether ~200 phages were screened against the strain, and ten infectious phages were found. Even though this case never led to actual phage therapy, it was an excellent demonstration of the power of crowdsourcing through Phage Directory. Out of one of the contacts, a co-authored publication of two of the phage genomes was published (Yerushalmy et al., 2019).

*Immaterial rights and regulations*

The exchange of phages and their host bacteria between different institutions raises the concern of immaterial rights. There are three aspects that need to be taken into consideration when transporting phages and bacterial strains: Protection of personal data, the rights of the use of phages and strains, and the Nagoya Protocol. In addition, the transportation of pathogenic bacteria to some countries and institutions may require special permits.

In the context of a phage collection, the protection of personal data mainly concerns the clinical bacterial strains used as phage hosts. In Finnish hospital-associated clinical laboratories, the bacterial strains carry an ID code that associates the strains to the patient from whom they were isolated. Therefore, we need to re-code the strains before sharing or publishing them and ensure that the patient ID information is not transferred along the strains. The phage and bacteria ownerships and rights of use form a more complicated issue, as every player has their own policy. After crowdsourcing(s), the collection may contain phages and bacterial strains originating from dozens of institutions, and in the worst case, each institution has their own Material Transfer Agreement (MTA) allowing very different rights of use. A common MTA that could be accepted by all, or at least most organizations would fundamentally aid the collaboration of different phage depositories and research groups.

The Nagoya Protocol on Access to Genetic Resources and the Fair and Equitable Sharing of Benefits Arising from their Utilization (https://www.cbd.int/abs/) also applies to phages and bacteria isolated after 2014. The Nagoya Protocol may cause extra paperwork and set demands for permission from the country of isolation, depending on whether the country has ratified the Protocol and if it has granted unrestricted access to its genetic resources (Sybesma et al., 2018).

**Quality assurance and data management**

Quality assurance (QA) is a system set up to prevent faults in advance by standardizing operations in written instructions, setting specifications for materials and products, and by requiring documentation of performed operations. QA of pharmaceutical production in Europe is controlled by the EU GMP regulations. The most common QA method is a written Standard Operation Procedure (SOP). SOP specifies a certain performance, settings, frequency or any other aspect where standardization reduces the risk of failure.

Setting up quality management for a phage collection can be divided into three "levels" according to the different aspects being controlled: physical storage, digital storage and information storage (Figure 2).

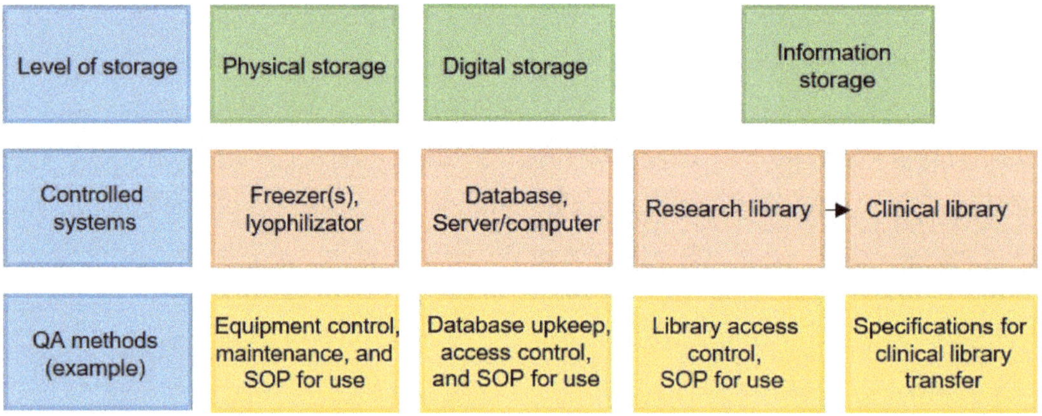


**Figure 2:** Phage collection from three different perspectives and quality assurance methods.

The physical storage of phages requires monitoring and maintenance. An SOP describing the location, access control and maintenance of a physical storage unit(s) is required. This SOP specifies how phages are organized in the storage, how temperature is monitored, who maintains the storage and the maintenance frequency, and how the temperature monitoring and maintenance is documented. Labelling of stored phages should be standardized in the SOP to avoid mixing of samples.

The technical departments of an organization usually maintain deep freezers and other equipment and they typically have their own maintenance programme and instructions. Referring to these programmes and SOPs is acceptable, but the quality management of the phage laboratory must verify the existence and suitability of these.

Digital storage is a database containing the contents of the physical storage. A database is a digital construct that stores data and usually a spreadsheet is sufficient for phage collection, as more complex database systems require more upkeep. The database is treated as a digital piece of equipment, like a physical device. An SOP defines where the database is stored, how/when backups are taken and stored, how the database is protected from unintentional/intentional changes, and who is responsible for the database. Maintenance and granting access to the database should also be defined in the SOP. We currently have a phage database built in Microsoft Access, and the file is stored in the university server where it is automatically backed up on a regular basis. Only the members of our research team have access to the database.

The third level of storage is the actual information in the database, in this case, the phage library. The libraries (or banks) are data collections of the various phages arranged into groups based on chosen preference, for example, basic research and clinical use. If phages are used for clinical purposes, there is a need for quality assurance for traceability and risk management, and this requirement applies to data handling as well as the physical storages. One phage can belong to multiple libraries, but the addition of a phage to a clinical library must have more strict requirements for safety reasons. Arranging phages into different libraries will enable low threshold of collecting new phages while maintaining a controlled system for clinically interesting phages.

In order to add a new phage to a library, there should exist predetermined specifications that must be met at a defined time. Initially, the phage requires a unique identifier code for labelling and information about its origin. On the other hand, implementation of too many requirements may be risky for two reasons: new interesting phages would be difficult to add due to the strict requirements, or the laboratory not abiding by its own quality system by ignoring such requirements.

Moving a phage to a clinical library naturally requires extended specifications. The genetic sequence must be determined and host range evaluated (see above). Temperate phages or phages containing toxin genes are not included in the clinical phage library.

When transferring a phage from one laboratory to another, the associated metadata and collected information of the phage should be transferred. This will reduce the workload at the receiving end. The receiving laboratory is, however, responsible for the accuracy and validity of the data. Quality agreements and audits are tools for assuring the validity of data between different units. In the long term, standardized methods will emerge and national and international authorities will eventually create more guidelines that will become the rules. For data such as genomic sequence and host range breadth, such guidelines currently exist in the form of scientific articles (Hyman, 2019; Philipson et al., 2018).

## Concluding remarks

During the operation of our "phage therapy lab" we have collected altogether close to 400 individual phages, most of which are specific for *Escherichia coli* (>90), *Klebsiella pneumonia* (>75), and *Pseudomonas aeruginosa* (>35). The collection also includes >75 *Yersinia* phages, mostly against enteropathogenic strains (*Y. enterocolitica* and *Y. pseudotuberculosis*) but also a few *Yersinia pestis* phages. We have faced difficulties in isolating *Acinetobacter* and

*Staphylococcus* phages from the Finnish sewage samples, we therefore have tried other sources such as the commercial Russian and Georgian phage cocktails, or sewage and water samples collected from other countries. One of the Finnish *Acinetobacter* phages was published recently (Pulkkinen et al., 2019).

We have also learned that it would have been beneficial to have implemented from the very beginning a standard protocol on how to handle and record all the tens or even hundreds of samples that accumulate for each phage. For example, how phage stocks have been prepared, which were discarded, which have been used to isolate DNA, which have been checked for stability, adsorption rates, growth curves, best conditions for propagation. This is easy when you have less than 10 phages in the collection. However, with increasing phage numbers within a collection, a need for a logistics system arises. Retrospective implementation of the system is tedious but can be carried out.

As as we have accumulated isolates to our phage collection, we also have established small scale methods for the preparation of phages and phage cocktails for therapeutic purposes (Hietala et al., 2019). The protocols we have developed have enabled us to treat patients with personalized phage cocktails. Indeed, we have prepared cocktails against MDR *K. pneumonia, P. aeruginosa*, and *S. aureus* which have been used in compassionate treatments of chronic sinusitis.

## Acknowledgements

The phage therapy project has been supported by grants from the Jane and Aatos Erkko Foundation and the Finnish Special State Subsidy for Health Science Research (projects TYH2013237, TYH2015411, TYH2016249, TYH2018228 and TYH2020245).

## References

Ackermann, H.W. (2009). Basic phage electron microscopy. Methods Mol Biol 501, 113-126.

Al-Hendy, A., Toivanen, P., and Skurnik, M. (1992). Lipopolysaccharide O side chain of *Yersinia enterocolitica* O:3 is an essential virulence factor in an orally infected murine model. Infect Immun 60, 870-875.

Coil, D., Jospin, G., and Darling, A.E. (2014). A5-miseq: an updated pipeline to assemble microbial genomes from Illumina MiSeq data. In arXiv:14015130 [q-bioQM].

Garneau, J.R., Depardieu, F., Fortier, L.C., Bikard, D., and Monot, M. (2017). PhageTerm: a tool for fast and accurate determination of phage termini and packaging mechanism using next-generation sequencing data. Scientific reports 7, 8292.

Hietala, V., Horsma-Heikkinen, J., Carron, A., Skurnik, M., and Kiljunen, S. (2019). The removal of endo- and enterotoxins from bacteriophage preparations. Front Microbiol 10.

Hyman, P. (2019). Phages for Phage Therapy: Isolation, characterization, and host range breadth. Pharmaceuticals (Basel) 12.

Kiljunen, S., Hakala, K., Pinta, E., Huttunen, S., Pluta, P., Gador, A., Lönnberg, H., and Skurnik, M. (2005a). Yersiniophage fR1-37 is a tailed bacteriophage having a 270 kb DNA genome with thymidine replaced by deoxyuridine. Microbiology 151, 4093-4102.

Kiljunen, S., Vilen, H., Pajunen, M., Savilahti, H., and Skurnik, M. (2005b). Nonessential genes of phage fYeO3-12 include genes involved in adaptation to growth on *Yersinia enterocolitica* serotype O:3. J Bacteriol 187, 1405-1414.

Leskinen, K., Pajunen, M.I., Vilanova, M.G.R., Kiljunen, S., Nelson, A., Smith, D., and Skurnik, M. (2020). YerA41, a *Yersinia ruckeri* bacteriophage: determination of a non-sequencable DNA bacteriophage genome via RNA-sequencing. Viruses, xx.

Manohar, P., and Ramesh, N. (2019). Improved lyophilization conditions for long-term storage of bacteriophages. Scientific reports 9, 15242.

Merabishvili, M., Vervaet, C., Pirnay, J.P., De Vos, D., Verbeken, G., Mast, J., Chanishvili, N., and Vaneechoutte, M. (2013). Stability of *Staphylococcus aureus* phage ISP after freeze-drying (lyophilization). PLoS One 8, e68797.

Oduor, J.M.O., Kadija, E., Nyachieo, A., Mureithi, M.W., and Skurnik, M. (2020). Bioprospecting *Staphylococcus* phages with therapeutic and bio-control potential. Viruses 12.

Pajunen, M., Kiljunen, S., and Skurnik, M. (2000). Bacteriophage fYeO3-12, specific for *Yersinia enterocolitica* serotype O:3, is related to coliphages T3 and T7. J Bacteriol 182, 5114-5120.

Pajunen, M.I., Elizondo, M.R., Skurnik, M., Kieleczawa, J., and Molineux, I.J. (2002). Complete nucleotide sequence and likely recombinatorial origin of bacteriophage T3. J Mol Biol 319, 1115-1132.

Pajunen, M.I., Kiljunen, S.J., Söderholm, M.-E.L., and Skurnik, M. (2001). Complete genomic sequence of the lytic bacteriophage fYeO3-12 of *Yersinia enterocolitica* serotype O:3. J Bacteriol 183, 1928-1937.

Philipson, C.W., Voegtly, L.J., Lueder, M.R., Long, K.A., Rice, G.K., Frey, K.G., Biswas, B., Cer, R.Z., Hamilton, T., and Bishop-Lilly, K.A. (2018). Characterizing phage genomes for therapeutic applications. Viruses 10.

Pulkkinen, E., Wicklund, A., Oduor, J.M.O., Skurnik, M., and Kiljunen, S. (2019). Characterization of vB_ApiM_fHyAci03, a novel lytic bacteriophage that infects clinical *Acinetobacter* strains. Arch Virol 164, 2197-2199.

Robinson, J.T., Thorvaldsdottir, H., Winckler, W., Guttman, M., Lander, E.S., Getz, G., and Mesirov, J.P. (2011). Integrative genomics viewer. Nat Biotechnol 29, 24-26.

Rutherford, K., Parkhill, J., Crook, J., Horsnell, T., Rice, P., Rajandream, M.A., and Barrell, B. (2000). Artemis: sequence visualization and annotation. Bioinformatics 16, 944-945.

Sacher, J.C., Zheng, J., and McCallin, S. (2019). Sourcing phages for compassionate use. Microbiol Aust 40, 24-27.

Salem, M., Pajunen, M.I., Jun, J.W., and Skurnik, M. (2020). Characterization and genomic analysis of fPS-2, fPS-65 and fPS-90 bacteriophages infecting *Yersinia pseudotuberculosis* serotype O:1a. Viruses, submitted.

Salem, M., and Skurnik, M. (2018). Genomic Characterization of Sixteen *Yersinia enterocolitica*-Infecting podoviruses of Pig Origin. Viruses 10.

Salem, M., Virtanen, S., Korkeala, H., and Skurnik, M. (2015). Isolation and characterization of *Yersinia*-specific bacteriophages from pig stools in Finland. J Appl Microbiol 118, 599-608.

Sambrook, J., and Russell, D.W. (2001). Molecular cloning: a laboratory manual, the third edition, 2nd edn (Cold Spring Harbor, N.Y.: Cold Spring Harbor Laboratory).

Skurnik, M., Hyytiäinen, H.J., Happonen, L.J., Kiljunen, S., Datta, N., Mattinen, L., Williamson, K., Kristo, P., Szeliga, M., Kalin-Mänttäri, L., et al. (2012). Characterization of the genome, proteome, and structure of yersiniophage fR1-37. J Virol 86, 12625-12642.

Skurnik, M., Kiljunen, S., and Pajunen, M. (2008). phage therapy. therapeutic microbiology: probiotics and related strategies, 373-389.

Skurnik, M., Pajunen, M., and Kiljunen, S. (2007). Biotechnological challenges of phage therapy. Biotechnol Lett 29, 995-1003.

Skurnik, M., and Strauch, E. (2006). Phage therapy: facts and fiction. Int J Med Microbiol 296, 5-14.

Skurnik, M., Venho, R., Bengoechea, J.-A., and Moriyón, I. (1999). The lipopolysaccharide outer core of *Yersinia enterocolitica* serotype O:3 is required for virulence and plays a role in outer membrane integrity. Mol Microbiol 31, 1443-1462.

Skurnik, M., Venho, R., Toivanen, P., and Al-Hendy, A. (1995). A novel locus of *Yersinia enterocolitica* serotype O:3 involved in lipopolysaccharide outer core biosynthesis. Mol Microbiol 17, 575-594.

Sybesma, W., Rohde, C., Bardy, P., Pirnay, J.P., Cooper, I., Caplin, J., Chanishvili, N., Coffey, A., De Vos, D., Scholz, A.H., et al. (2018). Silk Route to the Acceptance and Re-Implementation of Bacteriophage Therapy-Part II. Antibiotics-Basel 7.

Van Twest, R., and Kropinski, A.M. (2009). Bacteriophage enrichment from water and soil. Methods Mol Biol 501, 15-21.

Yerushalmy, O., Coppenhagen-Glazer, S., Nir-Paz, R., Tuomala, H., Skurnik, M., Kiljunen, S., and Hazan, R. (2019). Complete genome sequences of two *Klebsiella pneumoniae* phages isolated as part of an international effort. Microbiol Resour Announc 8.

Zhang, L., and Skurnik, M. (1994). Isolation of an R- M+ mutant of *Yersinia enterocolitica* serotype O:8 and its application in construction of rough mutants utilizing mini-Tn5 derivatives and lipopolysaccharide-specific phage. J Bacteriol 176, 1756-1760.

Zierdt, C.H. (1988). Stabilities of lyophilized *Staphylococcus aureus* typing bacteriophages. Appl Environ Microbiol 54, 2590.

## Chapter 19

# Commercialization of Phage Therapeutics: the Value of Intellectual Property and Patents

**Fabien Palazzoli[1], Michael Koeris[2] and Shawna McCallin[3]***

[1] Biotech IP analytics, IPStudies, Switzerland
[2] Keck Graduate Institute of Applied Life Sciences, Claremont, CA, USA
[3]Division of Infectious Diseases, Geneva University Hospitals and Faculty of Medicine, Geneva, Switzerland,

*Corresponding author: shawna.mccallin@gmail.com

DOI: https://doi.org/10.21775/9781913652517.19

**Abstract**

The time is ripe for a transition of bacteriophage (phage)-based products and services from a laboratory setting to commercial markets, yet there remain lingering doubts and questions as to if, how, and when this passage will occur. Exorbitant sunk costs to pay for the research and testing required to secure marketing authorizations for sales and distribution, and thus generate money, require high early-stage amounts of investment. The return on that initial investment, in turn, calls for high price tags and defence of intellectual property to ensure a certain market-share. Yet the market of antibacterials is conventionally low-cost, low-tech, and low-value, and phage therapies are furthermore not your standard pharmaceutical. Due to their historical discovery and use, many foundational aspects of phage R&D and applications fall in the public domain, therefore excluding patentability. Newer methods and techniques in areas in need of innovation for phage development, however, do offer a chance for intellectual property (IP) protection. While natural phages and phage combinations remain the most-heavily patented area, the volume of patent applications and their geographical origin show signs that there is indeed growing interest in protecting phage-based technologies, a key step before commercialization. The following chapter aims to provide the reader with a basic overview of different IP rights and agreements that are pertinent to the biotech industry and discuss their associated strengths and weaknesses. Current examples of product and business development in the areas of natural phages, genetically-modified phages, bioproduction and formulation, and therapeutic strategies will be highlighted. In the absence of evidence of efficacy from well-designed clinical trials, however, the value of patents and IP protection for phage products and their associated process is difficult to evaluate. The performance of phage therapy in the next few years will be decisive for sustainable and scalable commercialization.

## Introduction

The push for the development of phage into a commercialization product or treatment strategy is at an all-time high and is likely to continue for at least the near future. While phages were first discovered and used over a century ago, biotechnology has brought new innovation to the field, and with it, new opportunities for turning research into potential products and services. The increased need for phages due to antibiotic resistance and "one health" paradigms, as well as the growing interest in the human virome and its effect on the microbiome, has further pushed developmental priorities in diverse areas of human medicine, agri- and aquaculture, pastoral farming, and food safety. The ultimate evaluation of these applications will be their ability to obtain marketing authorizations for use, and thus the ability to generate sales and revenue. However, the testing required to obtain such authorizations requires significant financial investment, and one way to justify such investment is the ability to legally protect the ideas, processes, and things involved with the phage product or service.

According to the World Intellectual Property Organization (WIPO), Intellectual property (IP) refers to "creations of the mind, such as inventions; literary and artistic works; designs; and symbols, names, and images used in commerce" (WIPO). The main structures for IP are patents, copyrights, trademarks, and trade secrets, each with its own advantages and disadvantages that will be explained in detail further, with an emphasis on patents as the vehicle that is largely sought after for strengthening commercial development. The value of protecting IP for a business or entity ranges widely with the specific product, process, or invention, but the overarching doctrine is that possessing IP is highly valuable for inciting investment and ensuring eventual revenue. As more phage centric companies are forming, the question as to how IP plays a role in their business strategy, growth, and development is an important consideration for phage commercialization.

Commercialization of phage has encountered difficulties for IP protection in consideration that, firstly, much of phage knowledge and methods are in the public domain. Phages themselves, being natural entities by definition and easily isolatable from natural sources, are of limited protection individually. Second, few phage-composed products would have a fixed composition for an extended period of time due to changing epidemiological patterns, with unclear regulatory requirements for what that would mean for associated authorizations and/or patents. Even more importantly, phage-based products could be inherently inexpensive, which, while positive for prescribers and health costs, is viewed negatively in terms of justifying investment in IP protection and for generating investment to cover costs associated with R&D. Phage often finds itself in such "Catch-22" situations, caught between required costs and unforeseeable returns. Therefore, other aspects of commercial development, such as speed and expertise, may have more of an impact on phage development than traditional channels of IP protection. The true value of protecting IP is ultimately linked to the performance of the product or invention itself in a real-world setting.

This chapter aims to give an overview of how and what type of IP tools exist and may play a role in the commercialization of phage applications. After an introductory chapter on general IP concepts, the key areas for IP in relation to phage and phage applications will be outlined, within the context of therapeutic applications. A preliminary patent landscape analysis shows that there is a trend for increasing patent activity in the sector, showing that patenting may be becoming more important for commercial development, especially in some countries. This overview serves to provide a summary of current happenings in the phage field and give some indication of how commercialization may proceed.

## IP Basics

The domain of IP - and specifically how to protect inventions - is a field of its own and requires the support of specialists to ensure due diligence and defendable protection. Most academic institutions have available personnel or offices that deal specifically with technology transfer, and thus IP that result from academic scientists and students and their research. In order to facilitate further understanding of how IP plays a role in phage development, some basic terms and definitions will be stated below. Local or online seminars or tutorials covering these IP topics in greater depth, as well as websites from patent offices, such as WIPO, are helpful resources for gaining a deeper insight into IP (WIPO).

*Patents and related rights*

The creators of intellectual property and/or their associated institutions (e.g. universities, companies, foundations) have rights to that which they create that can be legally defined. IP is divided into two categories: Copyright comprising films, music, literary and artistic works (e.g. novels, poems, drawings and sculptures) and Industrial Property, including trademarks, industrial designs, and, most importantly, patents that will be the focus of this chapter.

The WIPO defines a patent as "an exclusive right granted for an invention, which is a product or a process that provides, in general, a new way of doing something, or offers a new technical solution to a problem. To get a patent, technical information about the invention must be disclosed to the public in a patent application". The WIPO specifies that the patent owner "has the exclusive right to prevent or stop others from commercially exploiting the patented invention" (WIPO). This protection is defined in both geographical space, with patent applications filed in designated countries or regions, and time, with generally a duration of 20 years after the filing date: If a patent expired or is abandoned before the expiry date, the covered invention becomes part of the public domain, meaning the said invention can now be commercially made, used, distributed or sold without the previous patent owner's consent. The Chapter I of the European Patent Convention specifies patentability requirements: patents shall be granted for any inventions that are new (no publication prior to filing such as an article, conference presentation or poster, etc), which involve an inventive step (non -obvious) and which are susceptible of industrial application (utility) (EPC, 2016). As novelty is a key patentability criterion, analyzing the state of the art, including articles, and also anterior patent documents, is particularly important to have an issued patent. Indeed, the filed patent application will be reviewed by a patent examiner from a patent office who will decide if the invention for which a protection is requested meets patentability requirements. Preliminary Patent searches can be accomplished using online search engines, such as espacenet (EPO) or patentscope (WIPO), or in assistance by professionals, with whom an "assisted patent search" can be conducted, using keywords, companies, or inventor names. The result of such patent searches results in a list of similar patents that should be thoroughly read to ensure that the overlap does not occur with the new application.

The patent itself is manifested in a written application that consists of bibliographic data (e.g. title, abstract, names, numbers, dates), claims defining the inventive subject matter, and a detailed description eventually including drawings or biological sequences. After writing the patent application with a patent attorney, the patent applicant will file it to a patent office which triggers the "priority date" and starts the processing of becoming a patent. The regional or national offices that will further the process, at which time patent applications could stay at the national level, or follow regional or international filing procedures, as structured by the Protection Cooperation Treaty (PCT; Figure 1), a ratified procedure for filing patents at the international level that has been accepted by over 150 countries since 1970 (WIPO). Filing an international patent application, commonly called a PCT application, has advantages for later in the process should the patent seek protection in multiple geographical designated states. Fees must be paid during all the life of the patent application and events after the patent is issued; costs for maintenance of the patent will occur periodically throughout the patent lifecycle (de Wilton, 2011).

Obtaining a patent is only the first step to ensuring that it brings about financial returns. The rights to a patent are valid for a specific amount of years in the countries where the patent has been granted and is in force (annual fees are required). During this time, the patent owners are responsible for protecting themselves against patent infringement i.e. persecution of patent violations by third-parties such as competitors. Litigation in these instances can be time-consuming and expensive. From the priority date to 20 years before expiring entails that obtaining marketing authorizations as soon as possible after patenting is desirable.

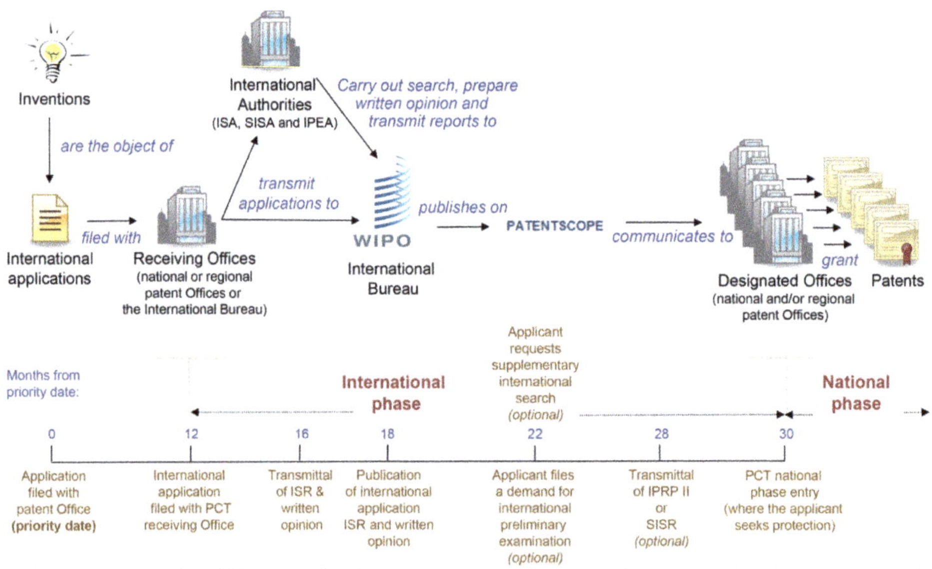


**Figure 1**. Overview of Patent Cooperation Treaty (PCT) and its associated procedure in terms of steps, entities, and timelines, as of October 2017. Source: WIPO at https://www.wipo.int/pct/en/faqs/faqs.html available for reproduction under a Creative Commons – Attribution (BY) 3.0 IGO – license.

A provisional patent application (PPA) is a strategic move to gain time before filing a complete patent application (Chitale et al., 2020). PPAs are mainly linked to the US-patent system, but are also used in Europe and in other states. Such a PPA does not grant or guarantee a future patent application, as it does not automatically turn into a patent at any point (the filings are separate). Rather an advantage of a PPA is to confer an earlier priority date if it were to come into competition with other patents filed after this date. The requirements, both in terms of information and financially, are lower than what is required for a patent application, and Its duration of validity is 12 months from the filing date, meaning that the actual patent application must be filed before a maximum of 12 months from the filing of the PPA. Besides being strategic, PPAs are a relevant strategy for inventors and their corresponding applicants who require additional time to perform further R&D of their invention. There is little consequence if the full patent application is not filed at the end of this 12-month period, and the contents of the PPA will not be made public unless they are followed by a future patent application.

For researchers, the question shouldn't be to publish or patent? The amount of time required for patenting can indeed create temporary conflicts for researchers who wish to quickly publish their research. As novelty is a patentability criterion, no information about the invention should be disclosed, whatever the format, before a patent application is filed. Like patents, the position of being first is also important for publishing: other researchers who are not pursuing patents may publish highly similar research, thus depriving the patent applicant the scientific prestige at the expense of possibly being awarded the patent. Once the patent application was filed, the invention covered in the patent application can be communicated.

*Trade secrets and know-how*

Trade secret is a type of IP that is linked directly to confidential information and those who have access to it, use it, or contribute to its conception. The classic example of an extremely commercially-valuable trade secret is the recipe for Coca-Cola, which remained secret for >100 years. Benefits of this type of protection are that there is no time limit to the duration of protection of trade secrets, but it does take a certain level of investment, and trust and creativity to maintain. This type of protection has many examples in foodstuffs, such as for KFC or Krispy Creme Doughnuts, where a key piece of information - the recipes - are known to few and kept under lock and key (often literally). Considering that expertise in the phage field remains relatively rare, know-how is a valuable attribute in a person, and trade secrets can offer some degree of IP protection.

This is interesting in the sense that, due to so much of phage processes being present in the public domain, trade secrets are an alternative or a complementary option for protecting information, especially in companies. For information to merit being considered a trade secret, it needs to have commercial value, and that information needs to be protected by having restricted access to specific persons or groups of persons and constructing policies and infrastructure to maintain its protection (Hull, 2019). Protection of trade secrets is difficult to uphold, as it does not protect against competitors developing similar, or identical, processes or information, but it is rather enforced at the individual level through non-disclosure, non-competition, and other such contracts and agreements. This type of information is more applicable to companies than research institutions, whose goal is to produce public knowledge and is liable to reproducibility.

Know-how and pure skill can be valuable assets. Anyone having worked with phages is familiar with how finicky phages can be and some researchers more than others are particularly skilled at performing more complex experiments. Animal models are a good example of know-how and expertise that are difficult to reproduce and possible points of collaboration between industry and academia OECH, (Galtier et al., 2017; Oechslin et al., 2016). The methods used for manufacturing, for example, such as tangential flow filtration and anion-exchange chromatography, are widely known and therefore not patentable. However, the right combination of specific parameters, steps, and materials that allows for good manufacturing output does not need to be disclosed. Having experience in this area is an asset to those that possess it by saving time and resources, which are important considerations for commercial endeavours.

*IP Transfer and licensing*

Those who create IP are not necessarily those who will want, or be able, to exploit it commercially. In an academic setting, licensing agreements for IP to another party that is interested in developing it further is the ideal situation. Licensing agreements take on a variety of forms, offering non-exclusive to exclusive rights, with payment at different time points through the development cycle (Ferguson and Kaudinya, 2014). It is not strictly necessary to pursue a license right away in order to achieve certain commercial outcomes, however, which is where the utility of options or evaluation agreements can be utilised. How does an option differ from a license? While an option is not a license, it is, at best, the right to negotiate in good faith for a license. Options also come in many flavours, such as exclusive and non-exclusive, and are further segmented by application space, indication, geography and other factors. While in that sense they are analogous to licenses, the principal differences between an option and a license are that it is both cheaper, and less specific, which makes it easier to negotiate because there are fewer terms and details to be worked through than for any license. Early on in development, options are a vehicle to save money and time, which are particularly scarce resources for young companies. Having an exclusive option is functionally similar to having an exclusive license for the purposes of a) evaluating the utility of said patent or b) preventing anyone else from licensing that patent. Companies intent on applying to grants as a commercial (start-up) entity, there is a certain utility in having optioned patents; similarly, if seeking outside funding from venture capital sources, having options is the next best thing to having patents. The time horizon for an option is necessarily restricted, or else it would deter the pursuit of an actual license. Observed option horizons fall into the 6 to 12 months range, with rare cases extending to 15 or 18 months. Less than six months is usually too short, while more than 12 months starts to be economically suboptimal for the licensing institution.

Evaluation agreements are another other major activity to pursue within the timeframe of an option agreement. This is usually connected to a material transfer agreement (MTA), to allow for the transfer of materials for validation purposes. For any IP or patent that has not explicitly emanated from work directly supervised or carried out by the prospective licensee, it is critically necessary to validate the utility of the patent in an appropriate scientific or clinical setting. For these purposes, an option would be appropriate, but if there is a low risk of licensure to another party or no immediate need to attract funding and build the foundation of a patent portfolio, an evaluation agreement is usually even cheaper than options. Within an evaluation agreement, the use of the material will be tightly defined, sometimes to specify the experiments or work that can be carried out, and define who has access to the material. Like options, they are also time-bound and usually carry with them stipulations wherein the recipient of material has to show that the material has been returned or destroyed, no copies or other duplications have occurred.

In summary, options are useful for both offensive purposes - attracting funding - as well as for defensive purposes - blocking other parties from licensing the IP. They are cheaper than licenses and faster to negotiate and execute, but they don't run for more than 12 months. Secondly, it is essential to validate the disclosed invention for utility within the commercial context, which can be achieved by establishing an option or an evaluation agreement.

### Other IP rights

Other ways of protecting IP is through copyrights, trademarks, and trade secrets. Creating copyrighted works, denoted by the symbol ©, is the protection conferred typically for artworks or other intellectual creations concerning the rights for distribution, copying, use, or expression of the work. This type of vehicle is not important for phage development but does apply to such works, such as figures in articles, books, or books themselves. The book, "The Perfect Predator" published about the experience of an emergency treatment for an *Acinetobacter baumannii* infection is protected by copyright (Stradthdee and Patterson, 2019). Trademarks are also not inherently important to phage development itself, although companies can seek the protection of elements that are used to support their brand, such as Adaptive Phage Therapeutics' (APT) PhageBank™ (APT, 2019). This type of protection can protect logos or slogans and are valid for unrestricted renewable periods of 10 years in applied countries.

## Inventions and innovations in phage therapy

Considering the widespread nature of phage research, different areas are more conducive to, or in need of, innovation and can confer IP protection than others. The basic process of developing phage for therapeutic applications starts with environmental isolation of natural phage, identity and activity characterization, manufacturing, and formulation until ultimately conducting efficacy studies to support claims of activity and safety (Table 1). The categories of patents that are currently held can be divided into separate themes, starting with the most frequent and common: natural phages.

## Natural phages

Natural phages and phage combinations remain the most frequently patented element of phage development, where the phages themselves compose the treatment and confer direct antibacterial activity. The ease and abundance of phages from natural sources is on one hand at benefit to having ample diversity and starting material that creates a low barrier to entry for development, and on the other, a barrier to IP protection thus weakening financial justifications for development. While theoretically, it may be possible to patent a single phage, the importance is that claims of the patent need to support its industrial application. Genetic sequences may be part of the patent application, although it is not possible to patent solely the genetic sequence of a phage: novel genetic sequences have been deemed a discovery, not an invention (Association for Molecular Pathology v. Myriad Genetics US Supreme Court case; (Cartwright-Smith, 2014)). The fact that phages can be sensitive to genetic drift over time and are distinguished by a (somewhat) arbitrary threshold of 95% sequence similarity weakens the defence of genetic identity for patent protection (Adriaenssens and Brister, 2017).

Phages may be given individually, but are more likely to be combined together into cocktails in order to cover enough strains of a particular bacterial species or for the treatment of polymicrobial infections. The products currently in or that have gone through clinical development contain four to 12 individual phages, while commercial products from Eastern Europe are more complex mixtures

(Jault et al., 2019; McCallin et al., 2013; McCallin et al., 2018; Petrovic Fabijan et al., 2020). Ten T4-like phages from an internal collection were combined for the treatment of acute pediatric *Escherchia coli* diarrhoea (Sarker et al., 2016). This phage collection containing ~50 T4-like phages was patented by Nestle Inc. prior to this trial (Patent US8021657B2). The clinical trial did not show an expected effect of phage treatment in diarrhoea patients, and its further development has not been pursued.

Another product that has been tested in clinical trials and that was covered by patent was BioPhage-PA from Biocontrol Ltd (CA2680108C). This six-phage cocktail against *Pseudomonas aeruginosa* was demonstrated to be effective for the treatment of chronic otitis in a phase I/II trial (Wright et al., 2009). Despite sufficient efficacy data to continue clinical development, a phase III study for this product has yet to be conducted ten years later. The reason for the lack of follow-through lies in financial viability: the market for chronic *P. aeruginosa* otitis did not justify the cost of a phase III trial. The biotech was soon bought-out by another phage biotech who was interested in developing a *Pseudomonas* phage product against more life-threatening conditions (sepsis, respiratory infections), and who subsequently merged with yet another SME (small to medium enterprise) to form Armata Pharmaceuticals.

The isolation and characterization of phages have become activities that are done frequently at academic and non-commercial institutions. Several large phage collections exist throughout the world, to name a few: Bacteriophage Collection of the Institute of Immunology and Experimental Therapy of the Polish Academy of Sciences in Wroclaw, PhageBank of the United States Department of Defense's Naval Medical Research Center, The George Eliava Institute of Bacteriophages, Microbiology and Virology (IBMV) in Tbilisi, Georgia, The Leibniz Institute German Collection of Microorganisms and Cell Cultures GmbH (DSMZ), Félix d'Hérelle Reference Center for bacterial viruses of the Université Laval, Bacteriophage Bank of Korea. One way to finance the development and maintenance of such collections is through either selling of individual phages (i.e. DSMZ, Laval) or through licensing to a commercial partner. This last strategy has been utilized through a variety of partnerships. APT holds an exclusive license to the phages collected by the US Navy for commercial development. Vesale Pharma has connections with Queen Astrids to allow access through a non-exclusive license to their phage collection. A third collaborative arrangement exists between Ferring Pharmaceuticals, Intralytix, and the Eliava Foundation. Considering the difficulties to assess the monetary value of a single phage, or any phage at early stages of development that are within the means of academic laboratories, these types of partnerships allow the commercialization value to grow as a function of results achieved and develop over time.

The protection of natural phages may be further complicated by the Nagoya Protocol, an international agreement to promote biodiversity conservation that stipulates ownership of genetic resources by the geographical location of origin. While some researchers argue that this should not apply to microbiological organisms, particularly phage, due to their abundance and diversity, the DSMZ that comprises a phage collection and sells phages for research purposes has upheld the agreement (Yurkov et al., 2019). Members of the DSMZ have reflected that their adoption of Nagoya compliance was expensive (estimated ~200,000 Euros) and has made strain deposition more complex, despite the legal clarity and increased quality control it confers. The potential legal challenge of phages obtained from “origin unknown” or from countries where proper permits were not previously obtained is unknown and may turn out to be more bark than bite in order to try to encourage compliance and goodwill.

Indeed, the abundance of phages in environmental sources makes it difficult to establish their origins without a doubt, as it does to protect others from isolating similar - or better - phages for industrial applications. Once a natural phage has shown potential, options to enhance or introduce new attributes may also be an option to increase the protectability.

### Genetic modifications and synthetic biology

A more recent - and more protectable - innovation in phages is the use of genetically- modified phages and their creation through synthetic biology. Not only can genetic modification help to overcome limitations of natural phages, such as narrow host range, it also makes the resulting recombinant entity patentable circumvents. Introducing mutations, removing sequences in part or in whole, or inserting transgenic material into the phage as genetic “payloads” can be used to increase the utility of phages significantly. A few examples include the increase of host range, the novel ability to remove biofilms, or increasing the antimicrobial activity through CRISPR or antimicrobial peptides (AMPs). These different genetic modifications, as well as the genetic engineering platforms used to create them, are patentable.

Genetic modification capability has developed greatly since homologous-recombination innovation of the early nineties, although it is still used today. There are now a number of different approaches to make genetic engineering easier, faster, and more precise (Bardy et al., 2016; Lemire et al., 2018). The ability to engineer phages opens a floodgate to the possible effects they can have and strengthens their patentability. Examples of the commercial value and interest in phage engineering are evidenced by the multiple phage companies that have included such enhancements and platforms in their pipelines. There currently exists a mix of companies in the market that develop natural phage only, while others are using a mix of both natural and/or modified phage, and a few are using only modified. Locus Bioscience has a mixed model where they combine natural phages with CRISPR Cas3-containing engineered phage (crPhages) that degrades bacterial DNA. They recently announced the first FDA-approved clinical trial with their crPhages for the treatment of urinary tract infections (Clinical Trial Identifier NCT04191148). This approach has generated investment >40 million USD in the company, which will surely help to overcome some of the financial bottlenecks experienced by some phage companies, and may lead to larger future instalments based on the achievement of milestones. Eligo Biosciences, an early-stage company that is based in Paris, is an example of both a modified-phage only approach and how academic research can be transitioned into commercial technology. The creation of Sequence-Specific AntiMicrobials (SSAMs) from the Massachusetts Institute of Technology (MIT) serves as the basis of their proprietary phage-encoded antimicrobial delivery systems (Bikard et al., 2014).

Governmental legislation and public push-back against the use of GMOs has long been thought to be a barrier to developing the aforementioned innovations. This is less of a barrier in the US and more of a concern for European markets. However, the world's first treatment of a patient with a phage that had been genetically-modified was published in 2019 (Dedrick et al., 2019). One of the three applied phages were modified in order to remove their lysogenic capacity through Bacteriophage Recombineering of Electroporated DNA (BRED) (Marinelli et al., 2008). The patient's condition drastically improved and, despite having received >32 weeks of intravenous phage, phage treatment resulted in no deleterious effects or safety concerns.

**Table 1.** Opportunities and challenges in different areas of invention and innovation for phage therapeutic development.

| Area | Opportunities | Challenges |
|---|---|---|
| **Natural phages** | Abundant raw material<br>Familiarity<br>Some clinical evidence documented in case reports | Difficult to protect<br>Easy to circumvent<br>Manual screening<br>Low early-stage predictive value<br>Nagoya protocol (permits) |
| **GMO phage & synthetic biology** | Greater coverage<br>Higher efficacy<br>Novel activity<br>Protectable<br>Versatility<br>Specificity<br>High level of expertise | Limited by biological factors<br>No clinical data (soon to change)<br>Technical difficulty<br>Environmental protection laws/Regulatory<br>Platform setup |
| **Manufacturing** | Faster production<br>Higher yields<br>Higher purity<br>Limited competence | Many known techniques<br>Value linked to product performance (downstream)<br>Expensive infrastructure |

| | | |
|---|---|---|
| **Formulation & delivery** | Better stability and shelf-life<br>New delivery strategies<br>Efficient delivery<br>Durability at the site of infection | Extra costs and processing<br>Additional partner or competence needed |
| **Therapeutic Indication** | More lucrative markets (i.e. microbiome)<br>Prolonged treatments (vs acute)<br>Orphan indications | Patient selection<br>Small patient populations<br>Unknown or indirect causation |

## Bioproduction and formulation

The historical paucity of phage treatments in development has led to a concomitant paucity in manufacturing capacity for clinical trials first, and certainly for approved products. The requirement of phage therapeutics - like all other human therapeutics - to be produced under GMP (Good Manufacturing Practices) conditions means that investment into cGMP facilities and processes has to be made, and that is what has happened in the last decade as the field started to be revitalized through investment. The majority of commercial developers of phage therapeutics have decided to in-source that process (APT, Armata, Pherecydes, Technophage, BiomX), yet there also exist a small number of contract research and manufacturing organizations (CRO/CMO). While there are several commercial entities that offer these services (Jafral, CleanCells), academic researchers also greatly contribute to the field (Malik et al., 2017; Mancuso et al., 2018; Merabishvili et al., 2019; Vinner et al., 2019). These contract organizations can usually provide some of the development services across the spectrum of process development, upstream (fermentation) and downstream (purification) drug substance manufacturing, quality control, drug product development, manufacture, fill-finish, and labelling capacity.

The first GMP-produced phage product was realized by a partnership between Pherecydes and the CMO CleanCells for the phage cocktails tested in

Phagoburn clinical trial (Jault et al., 2019). The trial incurred delays and setbacks due to the production process, which ultimately suffered severe stability issues that may have occurred in relation to the manufacturing process. The final concentration of the phage product used on patient wounds was 1,000-10,000-fold weaker than what was intended, which was cited as a likely cause of failure for the clinical trial. This unfortunate occurrence highlights the need to address the complete value chain of creating a properly formulated and stable product even at the early stages of clinical testing.

There is intellectual property that can be generated in all parts of the process, though it might not all be patentable, nor may it be desirable to be patented. It is more common to keep manufacturing developments as trade secrets. Experience with optimizing phage amplification, purification and formulation, is limited to a few competent individuals and companies and highly valuable to ensuring the feasibility of project budgets. Not all CROs/CMOs will work with phages due to the fact that they have not previously worked with these entities and there is a significant burden to update the accreditation information, which further restricts the universe of potential partners.

It is important to note that the challenge in the development of cGMP phage products is of a different nature than for classical microbial products or single entity biologics, such as monoclonal antibodies. Phages often are composited into a cocktail for efficacy, meaning a phage-based product shares some similarity to that most familiar of viral products - the vaccine. That means that manufacture of a certain number of phage drug substance components has to be undertaken individually, each requiring optimization, and then composited into the FDP. There is dialogue between industry and regulators to enable the inclusion / exclusion / change of components of a phage cocktail, currently up to and including licensure, and will likely evolve with the needs of the field. That means it might be foreseeable that changes be made to the manufacturing process throughout the clinical trial development. The recent approval by the FDA of the first clinical trial using APT's PhageBank is a good illustration of how appropriate and adaptable legislation can support commercial development for phage therapy; while APT has GMP capacity, it will use different phages or different phage combinations personalized to the patient's bacterial isolate(s).

Formulation for delivery is no more or less challenging than it is for other biomolecules and principally depends on the route of administration, with parenteral administration being the simplest, followed by topical and oral delivery in ascending order of difficulty. Most formulations are protectable as IP, again either in patent or trade secret format. The value of formulation is demonstrated through the significant financial losses due to stability issues of the Phagoburn phage cocktails (Jault et al., 2019). As another example of the need and value of formulation, one can look at GI delivery. The acid barrier of the gut needs to be crossed, and a certain release profile (where in the GI tract and how much) is desired. For that, phages need to be encapsulated or otherwise formulated such that oral administration can be achieved. Incorporation of phage into hydrogels for the delivery of phage for the treatment or prevention of articular or bone infections is another example of formulation strategies that may increase the efficacy of phage treatments (Akanda et al., 2018; Brown et al., 2018).

**Other (lysins, diagnostics, device-coatings)**

The use of phages is not limited to therapeutics, albeit that being the focus of this book chapter. Phages have been used as tools and have been patented as such for decades. It might be useful to the reader to think briefly through the use of phages as diagnostic elements for either clinical or non-clinical uses. Several commercial entities have used natural and engineered versions of phages to enable enhanced bacterial detection. More recent approaches focused on using engineered versions, but natural phages can be patentable within the context of a diagnostic device. The important point is that the phage component is the critical enabling piece, and even a natural phage becomes protectable because of the non-obviousness of the complete diagnostic assembly and workflow, i.e. the whole is greater than the sum of its parts. Other novel uses that have been successfully pursued are using phages as biocontrol agents on catheters or implants on the one hand, or use parts of phages for the same purpose, notably the class of molecules effecting cell lysis, holins and lysins. The latter type of entity is clearly patentable as it is usually produced recombinantly.

## Non-clinical phage applications

The development of phage for human therapeutic purposes is one of the strongest drivers for commercialization, although it also represents the highest barrier to entry due to the costs required for producing clinical-grade products and for conducting large-scale clinical research. There is an active development for phage products for other purposes, such as agri- and aqua- culture, livestock treatment, and food decontamination that is beyond exploring within the current chapter, and not included for the current analyses, but certainly merits attention (Holtappels et al., 2019). Moreover, non-clinical development can be a gateway market entry strategy. A number of phage companies focused on development in these areas before branching into more human-related markets, and several regulatory-approved products exist on the market. Intralytix, a US-based phage company has a variety of patented phage products for food safety or protection, with its Listshield™ product used for decontamination of *Listeria monocytogenes* on meats holds a generally-recognized-as-safe (GRAS) status from the FDA. Within the past few years, Intralytix has partnered with Ferring Pharmaceuticals to test a phage cocktail with activity against *E. coli* in Crohn's patients, for which enrolment is currently open for a phase I/II clinical trial in the US (ClinicalTrials.gov Identifier: NCT03808103). Micreos, a Dutch-based company, also created a phage product against *Listeria*, and now other pathogens, as a foundation for commercial development (their PhageGaurd brand). They have recently been developing an endo-lysin based product called GladSkin against *Staphylococcus aureus* for dermatology. The success these companies experienced with non-therapeutic product development helped diversify their product profile and gain entry to a market with lower barriers to generate revenue earlier than what could be expected for medicinal products.

## Preliminary data from the patent landscape

As described above, the first function of the patent system is protection. The second one is disclosure. In order to stimulate innovation and contribute to economic growth, patent applications and granted patents are published after a delay of 18 months after the filing, in whatever the language of countries in which they were filed and eventually granted. Thus, the details are available to the scientific community and to the public at large. Such technical and legal information can be searched and analyzed within a Freedom-to-Operate analysis (qualitative analysis, generally around dozens of patent families) or a Patent Landscape Analysis (quantitative analysis, around generally hundreds or thousands of patent families, depending on the topic to be studied) that are useful for understanding the commercial interest of an industry.

One indication of the advancement of phage in the direction of commercialization is patent activity in the field. A preliminary patent landscape was performed to analyze the number of patent filings relating to phage therapeutics by searching relevant keywords. Terms related to (bacterio)phage(s) combined to therapy, antibacterial, for defined pathogens (e.g. *Escherichia*, *Staphylococcus*), other synonyms or related keywords and dedicated classifications codes were used to search patent documents in the Orbit professional patent database (from the Questel company). Statistical data were then processed using database-specific analytical tools. A patent landscape provides a "snapshot of the patent situation of a specific technology, either within a given country or region, or globally. They can inform policy discussions, strategic research planning or technology transfer. They may also be used to analyze the validity of patents based on data about their legal status".

Most of the phage compositions are focused on prevention and/or treatment of bacterial infections in humans, although some of them also include the use of phages for animals (e.g. aquaculture), plants or in soil remediation. Historical patent families were based on preventing or treating bacterial infections, but more recent patent families now also cover methods of treating or preventing bacterial-associated cancers or microbiota diseases, methods for modulating inflammatory bowel diseases, and more chronic diseases. Compositions comprising phages include antibiotic compositions, feed additive compositions, disinfectants and inhibition of biofilm formation, detergents and compositions for microorganism detection. Some patents also protect isolated proteins derived from phages with antibacterial activity against specific bacteria such as depolymerases, lyases or endolysins.

Between 1995 and 2019, a total of 1640 patent families, meaning all the filings/ applications for "same invention" (e.g. patent applications filed for one invention in several countries), were retrieved (Figure 2A), covering the aforementioned topics. There is a clear increase in the filings per year since 2010; the true numbers for 2018 and 2019 are higher than those reported here due to the incompressible publication delay of 18 months.

The United States, China, and South Korea are by far the countries with the highest number of priority filings over this time period, with 562, 345, and 239 patent families, respectively (Figure 2B). European countries collectively hold 272 patent families, with additional ones originating from Russia (*n*=84), Georgia (*n*=8) and Israel (*n*=2). The United Kingdom is the leader of priority countries for Europe, holding 84 priority filings. That is not surprising, as the UK was home to the first modern clinical trial in humans using a *Pseudomonas* phage cocktail for the treatment of chronic otitis (Wright et al., 2009), showing that it was an early-adopter in phage therapy. Germany holds the second highest number of priority filings in Europe (*n*=44), followed by Poland (*n*=24) where phage therapy has been used for patients since the 1970s (Miedzybrodzki et al., 2012; Międzybrodzki et al., 2018; Weber-Dabrowska et al., 2000). China & US are in competition as for most biotech industries, such as CRISPR-based technologies. Priority filings in Russia were fewer than what was expected due to their long-standing experience in the field; the Russian Federation collectively holds 84 priority filings dispersed between multiple assignees. Indeed, the priority country is likely to reflect the geographical origin of the invention in most cases, with the exception of the United States, where it includes not only-US players, but also non-US players who choose to file in the US for strategic purposes (known general trend for all industries). It can also be noted that most of the Russian and Chinese priority filings were not extended in other countries with international patent applications (PCT), to be compared to the US or European priority filings (data not shown).

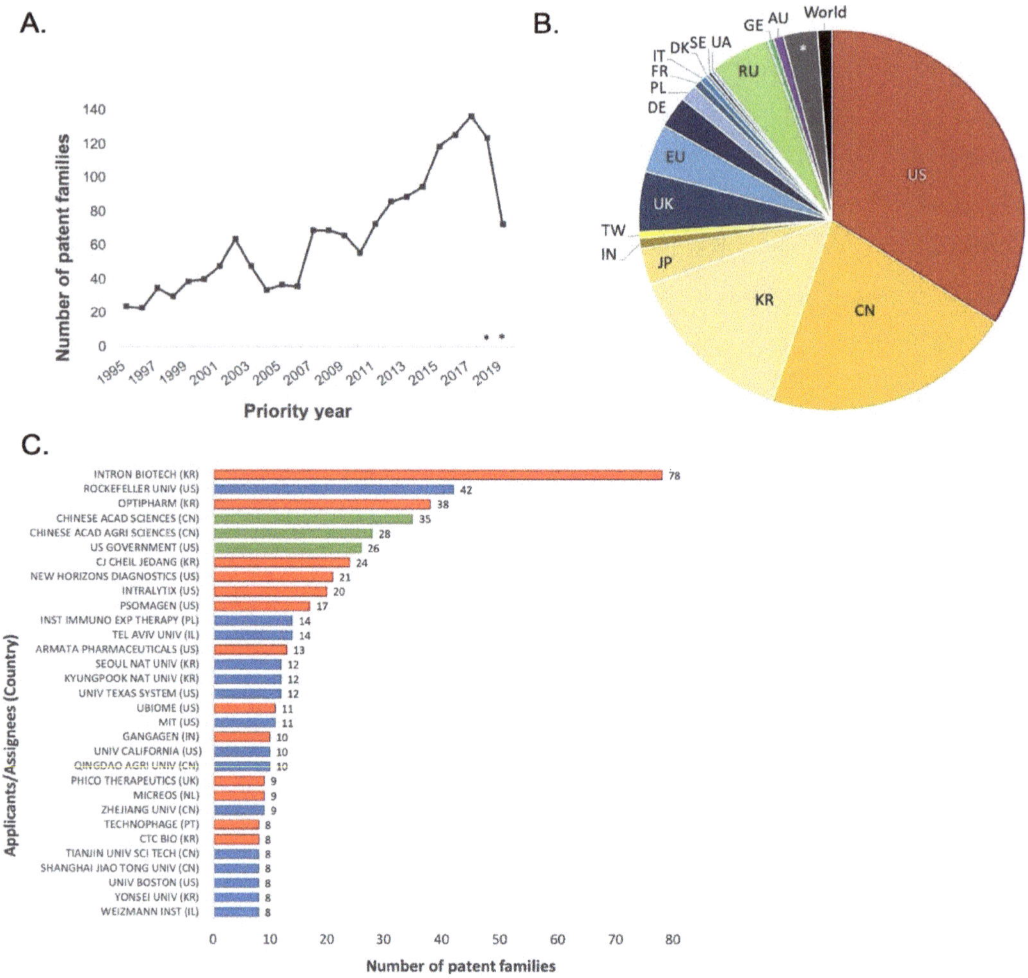


**Figure 2**. Results of a preliminary patent landscape for phage therapeutics **A.** Temporal distribution of patent families overtime (1995-2019) showing the number of priority filings filed per year over the past 25 year period. *Number of priority filings from 2018-2019 are incomplete due to the 18 month publishing delay. **B.** Pie chart distribution of priority filings per countries/regions. Patents classified as "EU" denote priority filings filed via the European system rather than a priority filing filed in a national country; "World" designates patents filed at the PTC. * included patenting families classified from 20 additional countries. **C.** Main patent assignees owning ≥ 8 patent families; Academic entities are shown in blue, commercial companies are shown in red, and governmental institutions are in green. (Country codes: US United States, CN China, KR Korea, JP Japan, IN India, TW Taiwan, UK United Kingdom, EU Europe, DE Germany, PL Poland, FR France, IT Italy, DK Denmark, SE Sweden, UA Ukraine, RU Russia, GE Georgia, AU Australia).

While patenting is typically associated with commercial entities, academia creates a great deal of innovation for the phage field (Figure 2C). Only 31 assignees, commercial (*n*=13), academic (*n*=15), or government-affiliated (*n*=3), possess at least eight patent families, representing 33% of total patent families (*n*=541) identified in this landscape. The distribution of total patent-family holdings between the major holders for academic and industrial entities is 34% (*n*=186) and 49% (*n*=266) of patents, respectively. The academic labs affiliated to the US Government (*n*=26) or those universities associated with the Chinese-governmental (*n*=63) account for 16% patent assignees in the landscape. The top five assignees of patent families are IntronBiotech, Rockefeller University, Optipharm, the Chinese Academy of Science, and the Chinese Academy of Agricultural Sciences. IntronBiotech and Optipharm are two Korean companies that show the highest level of patent activity related to phage development. Both IntronBiotech and Rockefeller have experience in developing phage lysins, which were also included in this landscape analysis. Most of the Korean and Chinese priority filings have been filed since the beginning of the 2000s with an acceleration since 2010 (data not shown).

**Importance to business model**

The stark reality of creating a biotech company is that it is expensive. A lot of initial investment is required in order to bring a product or service to the state where it can be sold in order to generate revenue. While governments may offer initial funding that is low-risk in the form of grants or no-interest loans to early-stage companies, larger projects often require venture or angel capital investment. A contradiction to this, however, was the funding source for the Phagoburn clinical trial that tested the use of a natural phage cocktail for the treatment of burn wound infections. Pherecydes (France) was awarded 3.8 million Euros as a grant by the European Commission to finance the manufacture and clinically evaluate their phage product between 2013-2017. Unfortunately, the trial was a failure in multiple aspects, and also highlights the reliance of companies on governmental, rather than investment capital, for funding. Countries have their own ecosystem of financing biotech and start-up companies that have an impact on how fast and how easily money can be brought into projects.

Where this initial capital comes from (and if it comes), is inherently linked to the potential return-on-investment (ROI) that must be generated from future sales of the developed product or service. The price tag of phage-related sales is difficult to estimate: while phages are inherently inexpensive in terms of raw materials, treatments would need to be expensive in order to attract investment for further development. At a recent conference, it was stated that investors want to see that treatments will cost tens of thousands of dollars - not less (Sacher and Zheng, 2020). Acute infections, which have been thus far the target of most clinical trials, have the disadvantage of being short-lived and sparse epidemiology: finding funding to do these trials has also proved difficult. This explains why companies that have focused on chronic or common afflictions would have greater volume for treatments and have likewise generated more interest from larger investors. Expansion into the area of the microbiome has attracted investment from more established pharmaceutical companies looking to branch into the field. Emphasizing the ability to combat antibiotic-resistant infections, while desperately needed, might not fetch the highest price tag, and explains why a number of academics and governments are active in addressing these needs.

A review of the current patent landscape of phage therapeutics provided here shows that commercial interest is increasing. The number of new patent filings has gone up since 2010 and is likely to continue to grow. This, in tandem with the investment and/or collaborations from larger pharma companies (Ferring, Janssen, Merk, Roviant Sciences) putting into phage SMEs (Locus Biosciences, BiomX, Armata Pharmaceuticals, Intralytix, IntronBiotech) is a healthy sign for commercial development. The approval of several clinical studies by the FDA and the more frequent approvals for experimental ("compassionate") case treatments also hints that regulatory partners will support this growth.

## Conclusions and perspectives

Phage-based products and phage therapy are in the midst of the commercialization process as evidenced by the growing number of phage companies, some of which are attracting considerable amounts of investment.

The future sustainability and the best business model to support this will require more time to reveal itself. The pursuit of IP protection is in no doubt a conventional and potentially lucrative cornerstone for creating a business. The trends in patenting phage-based innovations reflect this growing interest in developing phage into a business. In the phage field, we have outlined the various areas where patent activity is highest, and overall there is an increasing trend of patent filings for phage applications. Companies and countries with previous experience are being met with newcomers that are quickly increasing their patent profiles, predominantly from Asian countries. Despite any number of patents in relation to phage therapeutics, however, patents are no guarantee for the marketing authorizations that are required for phage therapy to be a viable commercial enterprise.

The current hype and excitement surrounding phage therapy at this time will only be sustained - and merited - if phages are shown to reliably and effectively treat bacterial infections. Marketing authorizations are granted for a specific pharmaceutical product by regulatory authorities based on the results of clinical trials. The strain-specific nature of phages makes natural phages difficult to design fixed-composition products and regulatory frameworks make it difficult (although far from impossible) to do personalized treatments. The approval to start several clinical trials in the US, especially one using a personalised approach and another using GMO-phage, shows that there is still confidence in and a need for phage therapy in modern medicine. The failure of recent clinical trials for the treatment of burn wound infections or for acute pediatric diarrhoea sheds light on where the priority should be for phage commercialization: obtaining clear signs of clinical efficacy.

**Conflict of interests:** Shawna McCallin is an independent consultant for clinical trials and co-founder of Phages4All, Fabien Palazzoli is Head of Biotech IP analytics, IPStudies, Switzerland, who provided the preliminary data from the patent landscape on phages comprised in the part 3. Patent Landscape, and Michael Koeris is the founder of several phage-enabled biotech start-ups including Sample6. None of the co-authors have any commercial agreements with one another.

## References

Adriaenssens, E., and Brister, J.R. (2017). How to name and classify your phage: an informal guide. Viruses *9*.

Akanda, Z.Z., Taha, M., and Abdelbary, H. (2018). Current review - the rise of bacteriophage as a unique therapeutic platform in treating peri-prosthetic joint infections. J Orthop Res *36*, 1051-1060.

APT (2019). The Science: PhageBank and HRQT (Adaptive Phage Therapeutics).

Bardy, P., Pantucek, R., Benesik, M., and Doskar, J. (2016). Genetically modified bacteriophages in applied microbiology. J Appl Microbiol *121*, 618-633.

Bikard, D., Euler, C.W., Jiang, W., Nussenzweig, P.M., Goldberg, G.W., Duportet, X., Fischetti, V.A., and Marraffini, L.A. (2014). Exploiting CRISPR-Cas nucleases to produce sequence-specific antimicrobials. Nat Biotechnol *32*, 1146-1150.

Brown, T.L., Petrovski, S., Chan, H.T., Angove, M.J., and Tucci, J. (2018). Semi-solid and solid dosage forms for the delivery of phage therapy to epithelia. Pharmaceuticals (Basel) *11*.

Cartwright-Smith, L. (2014). Patenting genes: what does Association for Molecular Pathology v. Myriad Genetics mean for genetic testing and research? Public Health Rep *129*, 289-292.

Chitale, S., Lawler, C., and Macfarlane, S. (2020). Understanding the basics of patenting. Nat Biotechnol *38*, 263-270.

de Wilton, A. (2011). Patent Value: A business perspective for technology startups. technology innovation management review, 6.

Dedrick, R.M., Guerrero-Bustamante, C.A., Garlena, R.A., Russell, D.A., Ford, K., Harris, K., Gilmour, K.C., Soothill, J., Jacobs-Sera, D., Schooley, R.T., *et al.* (2019). Engineered bacteriophages for treatment of a patient with a disseminated drug-resistant Mycobacterium abscessus. Nat Med *25*, 730-733.

EPC (2016). The European Patent Convention, E.P. Office, ed.

EPO. Espacenet Patent Search (European Patent Office).

Ferguson, S.M., and Kaudinya, U.S. (2014). Licensing the Technology: Biotechnology Commercialization Strategies Using University and Federal

Labs. In Biotechnology Entrepreneurship Starting, Managing, and Leading Biotech Companies (Academic Press), p. 21.

Galtier, M., De Sordi, L., Sivignon, A., de Vallee, A., Maura, D., Neut, C., Rahmouni, O., Wannerberger, K., Darfeuille-Michaud, A., Desreumaux, P., *et al.* (2017). Bacteriophages Targeting Adherent Invasive *Escherichia coli* strains as a Promising New Treatment for Crohn's Disease. J Crohns Colitis *11*, 840-847.

Holtappels, D., Lavigne, R., Huys, I., and Wagemans, J. (2019). Protection of phage applications in crop production: a patent landscape. Viruses *11*.

Hull, J. (2019). Protecting trade secrets: how organizations can meet the challenge of taking "reasonable steps". In WIP Magazine (WIPO).

Jault, P., Leclerc, T., Jennes, S., Pirnay, J.P., Que, Y.A., Resch, G., Rousseau, A.F., Ravat, F., Carsin, H., Le Floch, R., *et al.* (2019). Efficacy and tolerability of a cocktail of bacteriophages to treat burn wounds infected by *Pseudomonas aeruginosa* (PhagoBurn): a randomised, controlled, double-blind phase 1/2 trial. Lancet Infect Dis *19*, 35-45.

Lemire, S., Yehl, K.M., and Lu, T.K. (2018). Phage-Based Applications in Synthetic Biology. Annu Rev Virol *5*, 453-476.

Malik, D.J., Sokolov, I.J., Vinner, G.K., Mancuso, F., Cinquerrui, S., Vladisavljevic, G.T., Clokie, M.R.J., Garton, N.J., Stapley, A.G.F., and Kirpichnikova, A. (2017). Formulation, stabilisation and encapsulation of bacteriophage for phage therapy. Adv Colloid Interface Sci *249*, 100-133.

Mancuso, F., Shi, J., and Malik, D.J. (2018). High throughput manufacturing of bacteriophages using continuous stirred tank bioreactors connected in series to ensure optimum host bacteria physiology for phage production. Viruses *10*.

Marinelli, L.J., Piuri, M., Swigonova, Z., Balachandran, A., Oldfield, L.M., van Kessel, J.C., and Hatfull, G.F. (2008). BRED: a simple and powerful tool for constructing mutant and recombinant bacteriophage genomes. PLoS One *3*, e3957.

McCallin, S., Alam Sarker, S., Barretto, C., Sultana, S., Berger, B., Huq, S., Krause, L., Bibiloni, R., Schmitt, B., Reuteler, G., *et al.* (2013). Safety analysis of a Russian phage cocktail: from metagenomic analysis to oral application in healthy human subjects. Virology *443*, 187-196.

McCallin, S., Sarker, S.A., Sultana, S., Oechslin, F., and Brussow, H. (2018). Metagenome analysis of Russian and Georgian Pyophage cocktails and a placebo-controlled safety trial of single phage versus phage cocktail in healthy Staphylococcus aureus carriers. Environ Microbiol *20*, 3278-3293.

Merabishvili, M., Pirnay, J.P., Vogele, K., and Malik, D.J. (2019). Production of Phage Therapeutics

and Formulations: Innovative Approaches. In Phage Therapy: A Practical Approach, A. Gorski, ed. (Switzerland: Springer Nature).

Miedzybrodzki, R., Borysowski, J., Weber-Dabrowska, B., Fortuna, W., Letkiewicz, S., Szufnarowski, K., Pawelczyk, Z., Rogoz, P., Klak, M., Wojtasik, E., *et al.* (2012). Clinical aspects of phage therapy. Adv Virus Res *83*, 73-121.

Międzybrodzki, R., Hoyle, N., Zhvania, F., Lusiak-Szelachowska, M., Dabrowska, B.W., Lobocka, M., Borysowski, J., Alavidze, Z., Kutter, E., Gorski, A., *et al.* (2018). Current updates from the long-standing phage research centers in Georgia, Poland, and Russia. In Bacteriophages, Harper D., Abedon S., Burrowes B., and M. M., eds. (Springer, Cham).

Oechslin, F., Piccardi, P., Mancini, S., Gabard, J., Moreillon, P., Entenza, J.M., Resch, G., and Que, Y.A. (2016). Synergistic interaction between phage therapy and antibiotics clears *Pseudomonas aeruginosa* infection in endocarditis and reduces virulence. J Infect Dis.

Petrovic Fabijan, A., Khalid, A., Maddocks, S., Ho, J., Gilbey, T., Sandaradura, I., Lin, R.C., Ben Zakour, N., Venturini, C., Bowring, B., *et al.* (2020). Phage therapy for severe bacterial infections: a narrative review. Med J Aust *212*, 279-285.

Sacher, J.C., and Zheng, J. (2020). Phage Futures Congress 2020: Meeting Recap. Capsid & Tail *70*.

Sarker, S.A., Sultana, S., Reuteler, G., Moine, D., Descombes, P., Charton, F., Bourdin, G., McCallin, S., Ngom-Bru, C., Neville, T., *et al.* (2016). Oral phage therapy of acute bacterial diarrhea with two coliphage preparations: a randomized trial in children from Bangladesh. EBioMedicine *4*, 124-137.

Stradthdee, S., and Patterson, T. (2019). The Perfect Predator. A Scientist's Race to Save her Husband from a Deadly Superbug (Hachette Books).

Vinner, G.K., Richards, K., Leppanen, M., Sagona, A.P., and Malik, D.J. (2019). Microencapsulation of enteric bacteriophages in a ph-responsive solid oral dosage formulation using a scalable membrane emulsification process. Pharmaceutics *11*.

Weber-Dabrowska, B., Mulczyk, M., and Gorski, A. (2000). Bacteriophage therapy of bacterial infections: an update of our institute's experience. Arch Immunol Ther Exp (Warsz) *48*, 547-551.

WIPO. Homepage (World Intellectual Property Organization).

WIPO. PATENTSCOPE (WIPO IP Portal).

WIPO. Protecting your Inventions Abroad: Frequently Asked Questions About the Patent Cooperation Treaty (PCT) (WIPO).

Wright, A., Hawkins, C.H., Anggard, E.E., and Harper, D.R. (2009). A controlled clinical trial of a therapeutic bacteriophage preparation in chronic otitis due to antibiotic-resistant *Pseudomonas aeruginosa*; a preliminary report of efficacy. Clin Otolaryngol *34*, 349-357.

Yurkov, A., Püschner, H.M., and Hartman Scholz, A. (2019). DSMZ: the European Union's first Registered Collection under the Nagoya Protocol. Microbiology Australia *40*, 108-113.

www.ingramcontent.com/pod-product-compliance
Lightning Source LLC
LaVergne TN
LVHW070844210926
848482LV00016B/863